The Routledge Handbook of Landscape Ecology

The Handbook provides a supporting guide to key aspects and applications of landscape ecology to underpin its research and teaching. A wide range of contributions written by expert researchers in the field summarize the latest knowledge on landscape ecology theory and concepts, landscape processes, methods and tools, and emerging frontiers.

Landscape ecology is an interdisciplinary and holistic discipline, and this is reflected in the chapters contained in this Handbook. Authors from varying disciplinary backgrounds tackle key concepts such as landscape structure and function, scale and connectivity; landscape processes such as disturbance, flows, and fragmentation; methods such as remote sensing and mapping, fieldwork, pattern analysis, modelling, and participation and engagement in landscape planning; and emerging frontiers such as ecosystem services, landscape approaches to biodiversity conservation, and climate change. Each chapter provides a blend of the latest scientific understanding of its focal topics along with considerations and examples of their application from around the world.

An invaluable guide to the concepts, methods, and applications of landscape ecology, this book will be an important reference text for a wide range of students and academics in ecology, geography, biology, and interdisciplinary environmental studies.

Robert A. Francis is Reader in Ecology in the Department of Geography at King's College London. His research focuses on urban ecology, freshwater ecology, and nature and society interactions. He edited *A Handbook of Global Freshwater Invasive Species* and co-edited *Urban Landscape Ecology: Science Policy and Practice* (with James D.A. Millington and Michael A. Chadwick) and *The Routledge Handbook of Biosecurity and Invasive Species* (with Kezia Barker), all by Routledge.

James D.A. Millington is Senior Lecturer in the Department of Geography at King's College London. He is a geographer and landscape ecologist with expertise in using computational and statistical modelling tools to investigate spatial ecological and socio-economic processes and their interaction. He is the co-editor of *Urban Landscape Ecology: Science Policy and Practice* (with Robert A. Francis and Michael A. Chadwick).

George L.W. Perry is Professor in the School of Environment at the University of Auckland. His research is focused on the dynamics of forest ecosystems at spatial scales from the population to the landscape and at temporal scales from decades to millennia.

Emily S. Minor is Associate Professor of Biological Sciences at the University of Illinois at Chicago. Her research explores human alteration of the landscape and how this can affect ecological communities and processes at the landscape scale.

Contents

Contents

Figures

Figures

Figures

Tables

Contributors

Marc Antrop is Honorary Professor in the Department of Geography at the University of Ghent. He has been a leading researcher in landscape ecology for decades, with specialisms in landscapes sciences, remote sensing, geographic information systems, and planning. His research is particularly focused on a holistic approach to the landscape, combining landscape ecology, historical geography, landscape perception, and landscape visualization and design.

Víctor Arroyo-Rodríguez is a researcher at the Universidad Nacional Autónoma de México. He studies the impact of land-use change on tropical plants and animals across different spatial scales, particularly to inform biological conservation efforts.

Robert F. Baldwin is Endowed Chair and Professor of Conservation Biology in the Forestry and Environmental Conservation Department at Clemson University. His research interests include landscape ecology, ecosystem services, conservation biology, urbanization, and conservation education.

Justine A. Becker is a researcher in the Department of Zoology and Physiology at the University of Wyoming. Her work focuses on animal behavior, particularly in relation to movement ecology and how this is influenced by spatio-temporal heterogeneity.

Joseph R. Bennett is Associate Professor in the Department of Biology at Carleton University. His current research areas include conservation prioritization, threatened species, invasion ecology, environmental policy, and spatial statistics. He has a particular interest in practical questions regarding invasive species control and management to protect threatened species.

Richard E. Brazier is Professor of Earth Surface Processes in the Department of Geography at the University of Exeter. His research interests are in geomorphology and hydrology with an emphasis on soil erosion, sediment and nutrient mobilization and delivery, water quality, and landform evolution from hillslope to landscape scales.

James M.R. Brock is a Professional Teaching Fellow in the School of Biological Sciences at the University of Auckland. His research concentrates on how the composition and structure of vegetation communities are influenced by abiotic and anthropic factors, including plant invasions and wildfires.

Calum Brown is a researcher at the Karlsruhe Institute of Technology. He investigates the processes that cause change in land management and ecosystems, using statistical and computational methods.

Andrew Butler is Associate Professor in landscape architecture and planning at the Swedish University of Agricultural Sciences. He is interested in discourses on landscapes, landscape identity, participation in landscape issues, and landscape assessment.

C. Alina Cansler is a researcher at the School of Environmental and Forest Sciences at University of Washington who studies the effects of disturbance – particularly wildfires – on ecosystems in western North America. Her research interests are broad, and include plant community ecology, plant conservation biology, ecosystem biomass/fuel structure, management and reanalysis of large datasets, and to use of active remote sensing for research and management.

Eliana Cazetta is Full Professor at Santa Cruz State University. Her research interests are in the impacts of anthropogenic disturbances such as habitat loss, fragmentation, and defaunation on plant–animal interactions, especially seed dispersal by animals.

Jeremy S. Dertien is a researcher and PhD candidate in the Forestry and Environmental Conservation Department at Clemson University. His research interests include population and community ecology, conservation planning, quantitative ecology, and spatial statistics.

Lenore Fahrig is Chancellor's Professor in the Department of Biology and co-director of the Geomatics and Landscape Ecology Research Laboratory at Carleton University. Her research uses spatial simulation modelling and field studies to examine the effects of landscape structure on the abundance, distribution, and persistence of organisms.

Adam Ford is Canada Research Chair in Wildlife Restoration Ecology and Assistant Professor of Biology at the University of British Columbia. His research focuses on ecology, connectivity, and human–wildlife conflict.

Robert A. Francis is Reader in Ecology in the Department of Geography at King's College London. His research focuses on urban ecology, freshwater ecology, and nature and society interactions. He edited *A Handbook of Global Freshwater Invasive Species* and co-edited *Urban Landscape Ecology: Science Policy and Practice* (with James D.A. Millington and Michael A. Chadwick) and *The Routledge Handbook of Biosecurity and Invasive Species* (with Kezia Barker), all by Routledge.

Megan M. Friggens is a Research Ecologist at the U.S. Forest Service. She works on a variety of projects that model or assess ecosystem and species' response to climate change and other disturbances.

R. Daniel Hanks is a researcher in the Department of Forestry and Environmental Conservation at Clemson University. His research examines the spatio-temporal gradients associated with the response of fauna to stressors on the landscape/riverscape.

Sarah J. Hart is an Assistant Professor in the Department of Forest and Rangeland Stewardship at Colorado State University. Her research examines the causes and consequences of forest disturbances in the context of global change.

Brian J. Harvey is Assistant Professor in the School of Environmental and Forest Sciences at the University of Washington. His research examines how disturbances and climate interact to shape forest ecosystems, with a particular focus on fire and insect outbreaks over multiple spatial and temporal scales.

Aveliina Helm is Associate Professor in the Institute of Ecology and Earth Sciences at the University of Tartu. Her research focuses on the impact of landscape structure on biodiversity and provision of ecosystem services, on temporal delays in community dynamics in response to global changes, and on conservation and restoration of biodiversity and related ecosystem services.

G. Darrel Jenerette is Professor in the Department of Botany and Plant Sciences at the University of California Riverside. He researches the coupling between biodiversity, energy fluxes, and biogeochemical cycling embedded within ecological landscapes.

Jason P. Julian is Professor in the Department of Geography at Texas State University. He studies human–environment interactions across a variety of landscapes, focusing particularly on water resources, stream and wetlands mitigation, land use change, ecosystem services, and protected places.

Alisa R. Keyser is a Data Manager and Analyst at the Natural Resource Ecology Laboratory, Colorado State University. Her research has focused on forest ecosystem ecology, applied climatology, fire severity and forest recovery after fire.

Michael J. Koontz is a researcher at the University of Colorado at Boulder. He investigates how spatial patterns of vegetation structure and composition affect the resilience of forests to wildfire and bark beetle disturbance in order to guide effective management action.

Jill Lancaster is Principal Research Fellow at the University of Melbourne. Her research interests are in the biology of aquatic insects, their population and community ecologies, and specifically, how the persistence and coexistence of stream-dwelling insects are influenced by the fluvial landscape of stream channels.

Finnbar Lee is a researcher in the School of Environment at the University of Auckland. He is currently investigating how metacommunity structure and connectivity are influenced by network topologies in dendritic systems.

Rachel A. Loehman is Research Landscape Ecologist with the United States Geological Survey. Her research focuses on landscape and fire ecology, climate changes and impacts, and fire effects on cultural and natural resources in a variety of forest and tundra ecosystems.

Todd Lookingbill is Associate Professor in the Department of Geography and the Environment at the University of Richmond. He uses spatial landscape models to explore landscape connectivity and ecology, habitat modelling, and watershed assessment.

Mailys Lopes is a researcher at the Institute of Zoology, Zoological Society of London, and LaSTIG, University of Paris East. Her research focuses on developing remote sensing tools to support ecological and biodiversity-related studies.

Amanda E. Martin is a Research Scientist at Environment and Climate Change Canada. She conducts research to inform conservation decision making when resources for action are limited, including evaluation of the relative effects of different management options on wildlife and conservation prioritization.

Kimberly Meitzen is Associate Professor in the Department of Geography at Texas State University. Her interests include fluvial geomorphology, river and society interactions, environmental flow management, and conservation biogeography.

B.C. Meyer is a researcher and Private Docent in the Institute for Geography at the University of Leipzig. His research interests include soil science, agricultural economics, geography, and hydrology.

G. Mezosi is Professor in the Department of Physical Geography and Geoinformatics at the University of Szeged.

Jesse E.D. Miller is Lecturer in Biology at Stanford University. He conducts research on the effects of global change on ecological communities to inform conservation management decisions, particularly in relation to the impacts of altered fire regimes on biodiversity.

James D.A. Millington is Senior Lecturer in the Department of Geography at King's College London. He is a geographer and landscape ecologist with expertise in using computational and statistical modelling tools to investigate spatial ecological and socio-economic processes and their interaction. He is the co-editor of *Urban Landscape Ecology: Science Policy and Practice* (with Robert A. Francis and Michael A. Chadwick).

Emily S. Minor is Associate Professor of Biological Sciences at the University of Illinois at Chicago. Her research explores human alteration of the landscape and how this can affect ecological communities and processes at the landscape scale.

Matthew Mitchell is a Research Associate in the Faculty of Land and Food Systems at the University of British Columbia. His research investigates how human actions at multiple spatial scales affect biodiversity and ecosystem services.

Scott W. Mitchell is Co-director of the Geomatics and Landscape Ecology Laboratory at Carleton University. His research interests include a variety of spatial analysis techniques to assist with environmental decision making, particularly in relation to uncertainty in land cover classification, measurement and modelling of primary productivity, and characterization of landscape pattern.

George L.W. Perry is Professor in the School of Environment at the University of Auckland. His research is focused on the dynamics of forest ecosystems at spatial scales from the population to the landscape and at temporal scales from decades to millennia. He is the co-author (with David O'Sullivan) of *Spatial Simulation: Exploring Pattern and Process*, published by Wiley-Blackwell.

Nathalie Pettorelli is Senior Research Fellow at the Institute of Zoology, Zoological Society of London. She heads the Environmental Monitoring and Conservation Modelling team and conducts research on the impacts of global environmental change on biodiversity and ecosystem services, with a particular focus on climate change.

John H. Porter is Research Associate Professor in the Department of Environmental Sciences at the University of Virginia. He teaches courses on Geographical Information Systems and applies them to studies of the dynamic barrier islands of Virginia, USA.

Thomas Ranius is Professor in Conservation Biology at the Swedish University of Agricultural Sciences. He conducts applied research into effects on biodiversity and ecosystem services of forestry, nature conservation, and climate change; along with spatial ecological work on species of conservation interest.

Tarmo K. Remmel is Associate Professor in the Faculty of Environmental and Urban Change at York University. His research incorporates aspects of spatial pattern, accuracy and uncertainty, and boreal disturbance analyses.

Karin L. Riley is Research Ecologist at the U.S. Forest Service. Her current research focuses on better understanding the relationship between climate and wildfire, particularly in the context of climate change, and how spatial planning can inform fire and landscape management options.

Ingrid Sarlöv Herlin is Professor of Landscape Planning and Head of the Department of Landscape Architecture, Planning and Management at the Swedish University of Agricultural Sciences. Her research interests include sustainable foodscapes, urban agriculture, landscape character assessment, landscape identity and landscape changes, cultural heritage, and animals in the landscape.

Henrike Schulte to Bühne is a researcher at the Institute of Zoology, Zoological Society of London, and Imperial College London. Her research utilizes satellite remote sensing data to map human impacts and ecosystem responses, focusing in particular on disturbance dynamics and vegetation structure.

Rosemary L. Sherriff is Professor in the Department of Geography, Environment and Spatial Analysis at Humboldt State University. She directs the dendroecology lab and conducts research in the areas of biogeography, landscape ecology, forest disturbances (wildfire, insect outbreaks), dendroecology, and climate change.

Craig Eric Simpkins is a researcher at the University of Göttingen, with research interests in landscape connectivity and virtual ecology.

Jennifer E. Smith is a researcher at the Institute of Zoology, Zoological Society of London.

Erica A.H. Smithwick is Distinguished Professor of Geography at The Pennsylvania State University, College of Earth and Mineral Sciences and Associate Faculty in the Earth and Environmental Systems Institute. She is an Associate Director of the Institutes of Energy and Environment. Her laboratory group (Landscape Ecology at Penn State) is actively involved in understanding how a wide range of disturbances, especially fire, affect ecosystem function at landscape scales and the influence of these changes on socio-ecological resilience and sustainability.

Janet M. Wilmshurst is Associate Professor in the School of Environment at the University of Auckland and Principal Scientist at Manaaki Whenua-Landcare Research. Her palaeoecological research is focused on reconstructing past vegetation and plant-animal interactions and learning how they have been shaped by natural and anthropogenic disturbance.

Sarah V. Wyse is a researcher in the Bio-Protection Research Centre at Lincoln University. Her research interests include plant invasions, forest ecology, and fire ecology.

Jingjing Zhang is Lecturer in Marine Sciences at Auckland University of Technology. Her most recent research focuses on modelling animal movement and landscape connectivity.

Carly D. Ziter is Assistant Professor of Biology at Concordia University. Her lab conducts solutions-oriented science to enhance biodiversity conservation and ecosystem service provision in urban and urbanizing landscapes.

Preface

Way back in 2002, eminent landscape ecologists Jianguo Wu and Richard Hobbs published a position paper in the international journal *Landscape Ecology*, stating:

> Comprehensive and integrative university curricula ... in landscape ecology need to be established and strengthened. These curricula and training programs must emphasize the interdisciplinarity and holistic nature of landscape ecology, as well as the integration between science and applications.
>
> *(Wu and Hobbs 2002, p. 357)*

This sentiment, which is as valid today as it was then, served as the inspiration for this Handbook, in which we hope to provide a supporting guide to key aspects and applications of landscape ecology to underpin its research and teaching. The chapters in this volume speak to landscape ecology's holistic and interdisciplinary nature in the range of key concepts, processes, methods, and frontiers that are covered; and each provides a blend of the latest scientific understanding of its focal topics along with considerations and examples of their application.

Following an overview of landscape ecology and its development as a discipline by Francis and Antrop (Chapter 1), the book is organized in four sections. The opening section explores key theory and concepts, with Antrop providing an authoritative discussion of the seminal patch-corridor-matrix (PCM) model that was central to early ideas and concepts in landscape ecology (Chapter 2). This chapter is followed by a treatment of the concepts of scale and hierarchy, which are central to ecology and geography in general but especially important for landscape ecology, by Millington (Chapter 3). Chapter 4, by Fahrig et al., introduces us to the central concept of landscape connectivity, which has been instrumental in driving our understanding of how spatial linkages in landscape structure facilitate flows and other landscape processes, and which has increasingly translated into landscape management applications that attempt to maintain or enhance connectivity.

The following section contains chapters on landscape processes. The theme of spatial structure and connectivity is continued by Lee et al. (Chapter 5), who, in particular, address how it influences the movement of organisms. The loss of connectivity and habitat is addressed more generally by Martin et al. (Chapter 6), who consider the evolving meaning and interpretations of fragmentation in landscape ecology – from the breaking up (but not necessarily loss) of habitat through to patch size and isolation effects. Smithwick (Chapter 7) then extends this discussion by focusing on nutrient flows in the landscape, proposing a framework of 'nutrient geography' that can be used to understand these impacts and integrate their dynamics. Three chapters are then dedicated to the all-important processes of disturbance in the landscape. Harvey et al. (Chapter 8) present the disturbance regime concept, exploring repeated patterns of disturbance over space and time and how these link to concepts like resilience and scale. Top-down landscape disturbances and the ecological impacts they can have are examined by Loehman et al. (Chapter 9) through the lens of anthropogenic climate change and the interactive disturbances it can cause. Finally, Brock and Wyse (Chapter 10) tackle bottom-up landscape disturbances, looking at factors like fire and species invasions and how these may transform the landscape through incremental and sometimes interactive or synergistic processes.

The largest section of the book is on methods and tools for landscape ecology. We start with a chapter on fieldwork as a fundamental element of landscape ecology research, noting, in particular, advances in technology such as drones and other sensors, alongside the value of more traditional field methods (Chapter 11, by Miller et al.). This discussion is followed by exploration of another central approach in landscape ecology: remote sensing and mapping (Chapter 12, by Pettorelli et al.). The chapter examines both the theory of remote sensing of landscapes and the tools that are available to implement it. Chapter 13 (Porter) goes beyond satellite and aerial remote sensing to look at in situ sensor networks that allow investigation of ecological change and their increasing utility across landscapes. Chapter 14 (Perry et al.) examines the potential of paleoecological analysis to elucidate long-term landscape dynamics and how this understanding might be used to contextualize contemporary changes and contribute to landscape management. Remmel and Mitchell (Chapter 15) then discuss landscape pattern analysis, particularly the application and interpretation of landscape metrics. Another key method, quantitative modelling, is examined by Brown in Chapter 16, with a focus on some of the main techniques, particularly computer simulation. Butler and Sarlöv Herlin present a discussion of participation and engagement in landscape planning (Chapter 17) by examining how the public are included in landscape character assessment processes. The section concludes with a chapter on experimentation in landscape ecology (Chapter 18, by Jenerette) with multiple case studies and examples of creative and ambitious landscape-scale experiments.

The final section presents some selected frontiers in landscape ecology. These include how landscape ecology can contribute to biodiversity conservation (Chapter 19, by Baldwin et al.), which brings us back to some of the considerations in previous chapters around (for example) fragmentation and connectivity; ecosystem services in the landscape (Chapter 20, by Mitchell); and the particular landscape characteristics and processes of riverscapes (Chapter 21, by Lookingbill et al.). There then follows a discussion of landscape restoration, and why restoration needs to focus on landscapes, with a summary of the landscape ecological aspects that are relevant for successful restoration (Chapter 22, by Helm). The Handbook ends with a final chapter on landscapes and climate change, focusing on a range of European case studies that explore climate-related processes such as aridification, drought, and wind erosion (Chapter 23, by Meyer and Mezosi).

A large, sprawling discipline such as landscape ecology cannot be comprehensively detailed in 23 chapters, or even double that. Nevertheless, we hope to have presented some of the key aspects of the discipline to help guide educators and researchers. There are certainly topics that we would have liked to include, from concepts (e.g. gradient theory) through processes (e.g. landscape evolution), methods (e.g. cutting-edge techniques for measuring lightscapes and soundscapes), and some of the wide and exciting range of frontiers (such as urban landscapes, seascapes, biosecurity, and so on). Our intention was to incorporate some of these topics, but the compilation of the book took place during 2020, when the world was struggling to respond and adapt to the SARS-CoV-2 epidemic. Some authors who had been commissioned to provide chapters found themselves unable to work on them as a result, because many universities and research institutes had to rapidly switch to online or more flexible delivery of educational programs, and research projects were thrown into uncertainty or chaos. Many researchers had to juggle childcare and other family responsibilities alongside their academic duties. The compilation and editing process was also delayed for these reasons. As a result, we decided to proceed with the body of chapters incorporated here, with the potential to expand coverage in the event of a second edition down the line.

We are grateful to all the authors who were able to contribute to the Handbook, especially in such challenging times, as well as everyone who provided peer review for the chapters and those who offered constructive comments or support. We would particularly like to thank Robert Fletcher, Karen Holl, Matt McGlone, Brenden McNeil, Carol Miller, David O'Sullivan, Jonathan Porter, Derek Robinson, Daniel Schillereff, Thomas Smith, Emma Tebbs, David Theobald, Maria Trivino de la Cal, Peter Vogt, and Katherine Zeller. We are also grateful to everyone at Routledge who contributed to the book's commissioning and production, especially Andrew Mould and Egle Zigaite, who waited very patiently for it to arrive. We hope that this addition to the Routledge Handbooks series will prove interesting and useful to readers around the world.

Robert A Francis, London
James D.A. Millington, London
George L. W. Perry, Auckland
Emily S. Minor, Chicago
3 January 2021

Reference

Wu, J. and Hobbs, R. (2002) 'Key issues and research priorities in landscape ecology: An idiosyncratic synthesis', *Landscape Ecology*, vol 17, pp355–365.

Acknowledgments

The editors would like to thank the chapter authors for taking the time to contribute to this volume, as well as those colleagues who have supported us with advice, peer reviews of chapters, and other guidance during the book's journey to completion. At Routledge, we would like to thank Andrew Mould and Egle Zigaite for their tireless support throughout.

A brief history and overview of landscape ecology

Robert A. Francis and Marc Antrop

What is landscape ecology?

Landscape ecology is an interdisciplinary science that incorporates elements of ecology, geography, and social science to investigate how interactions between landscape form and process determine how landscapes function over space and time. Landscape ecology aims to contribute to transdisciplinary applied landscape research significant for policy. A summary of some of the different definitions and descriptions of landscape ecology given over the years is provided in Table 1.1. These show repeated emphasis on pattern and process (with reciprocity sometimes acknowledged) and interdisciplinarity, with later definitions in particular emphasizing social and cultural aspects of the field. Some ambiguity and fluidity of definition and interpretation do occur, of course: Kirchhoff et al. (2013) highlight that landscape ecology is 'an ambiguous term commonly used to refer to different research agendas in different disciplines', therefore resisting simple definitions, and consider it to be 'ecology guided by cultural meanings of lifeworldly landscapes' (p. 33).

Landscape ecology emerged gradually as ecologists began to study broader spatial areas beyond local ecotopes and became increasingly aware of the importance of distance and scale in shaping ecological patterns and processes. Landscape ecology's defining characteristic is that it examines such relationships at broad spatial scales, usually from several hectares to many kilometers. Landscape ecology borrowed many theories, concepts, and methods from spatial analysis in quantitative geography and geographical information systems (GIS). As a discipline, landscape ecology contributed largely to the operationalization and application of many of these, in particular by using landscape metrics (e.g. Turner and Gardner, 1990, 2015). Investigations over broad spatial scales can be notably difficult and complex because of the many different factors driving patterns and processes and the problems associated with obtaining data and conducting scientific experiments over large areas. However, explorations at these scales, which can be achieved using a landscape ecology framework, are important for effectively addressing many environmental issues, from biological conservation to resource management. As a result, landscape ecology has been, and will continue to be, a discipline with growing demand and application.

It is now widely acknowledged that a comprehensive understanding of landscapes and how they function is essential for addressing many environmental issues and developing effective and sustainable management techniques (e.g. Opdam et al., 2018), though achieving this is a

Table 1.1 Selected definitions and descriptions of 'Landscape Ecology' as a discipline, and the term 'Landscape' within the context of landscape ecology, over time

Definition or description	Reference
Landscape ecology is an aspect of geographical study which considers the landscape as a holistic entity, made up of different elements, all influencing each other.	Zonneveld (1972)
Landscape ecology is the synthetic intersection of many related disciplines that focus on the spatial and temporal pattern of the landscape.	Risser (1987)
Landscape ecology is the study of the reciprocal effects of spatial pattern on ecological processes.	Pickett and Cadenasso (2002)
Landscape is a complex of relationship systems, together forming a recognizable part of the earth's surface, and is formed and maintained by the mutual action of abiotic forces as well as human action.	Zonneveld (1995)
Landscape ecology is the study of how spatial pattern affects ecological process.	With (2002)
Landscape ecology offers a spatially explicit perspective on the relationships between ecological patterns and processes that can be applied across a range of scales.	Turner (2005)
Landscape ecology is concerned with the generation and dynamics of pattern in ecosystems, and the implications of pattern for population, community, and ecosystem-level processes.	Urban (2006)
The core business of landscape ecology is the interaction of landscape patterns and processes.	Opdam et al. (2009)
Landscape ecology is the study of relationships between spatial pattern and ecological process.	McKenzie et al. (2011)
Ecology guided by cultural meanings of lifeworldly landscapes.	Kirchhoff et al. (2013)
Landscape ecology is a transdisciplinary, problem-solving, human ecosystem science.	Naveh and Lieberman (2013)
A geographic area in which variables of interest are spatially heterogeneous. The boundary of a landscape may be delineated based on geographic, ecological, or administrative units (e.g., a watershed, an urban area, or a county) which are relevant to the research questions and objectives.	Wu (2013)
There are many appropriate ways to define landscape depending on the phenomenon under consideration. The important point is that a landscape is not necessarily defined by its size; rather, it is defined by an interacting mosaic of patches relevant to the phenomenon under consideration (at any scale). It is incumbent upon the investigator or manager to define landscape in an appropriate manner. The essential first step in any landscape-level research or management endeavour is to define the landscape, and this is of course prerequisite to quantifying landscape patterns.	McGarigal (2015)
Landscape ecology is an interdisciplinary field of research and practice that deals with the mutual association between the spatial configuration and ecological functioning of landscapes, exploring and describing processes involved in the differentiation of spaces within landscapes, and the ecological significance of the patterns which are generated by such processes.	Christensen et al. (2017)
Landscape ecology is an interdisciplinary field, drawing on theories and methods from across the physical, natural, and social sciences.	Frazier (2019)
Landscape ecology is predicated on the assumption that spatial patterns influence ecological processes, and those processes, in turn, feedback to form and alter landscape patterns.	Frazier et al. (2019)

complex and difficult task. Landscape ecology is an expanding global discipline that has so far remained somewhat fragmented due to both its interdisciplinary and transdisciplinary nature and geographical contrasts in approach to landscape research and management (Wu and Hobbs, 2002; Kirchhoff et al., 2013; Christensen et al., 2017).

A brief history of landscape ecology

Landscape ecology is a relatively new discipline, at least in its more formal conceptualization. Landscape ecological principles and concepts were used long before Troll (1939) introduced the name for the first time. For example, many of the writings of Alexander von Humboldt at the end of the nineteenth century are ecology and biogeography with the environment ('nature') being the hierarchical structured whole, for which Smuts introduced holism as the key concept in 1926 (Antrop and Van Eetvelde, 2017). The science of studying large spatial extents of the natural environment has existed in some form since the early biogeographical work of Sir Joseph Dalton Hooker and Alfred Russell Wallace in the nineteenth century.

The more formal investigation of the interactions between landscape structure and dynamics and the ecological processes that influence, and are influenced by, landscapes began to emerge in the 1970s, inspired partially by the spatial implications of MacArthur and Wilson's (1967) Equilibrium Theory of Island Biogeography. Indirectly, early ecologists such as Aldo Leopold also included an inherently spatial approach to the landscape in their works, in Leopold's case in the context of nature conservation and environmental ethics (see Silbernagel, 2003).

Troll (1939) introduced the term 'landscape ecology' (*Landschaftsoekologie*) within the context of studying vegetation patterns from aerial photographs, i.e. linked to the first bird's eye perspective on the landscape, which fundamentally changed the perception and observation of the environment. The Second World War and the shifting focus in the post-war scientific community halted the further development of holistic landscape ecology. It was only revitalized by the 'Working Group Landscape-Ecological Research' (Werkgroep Landschapsecologisch Onderzoek, WLO) in The Netherlands and consolidated with the first international conference in Veldhoven (Tjallingii and de Veer, 1982). This was a seminal event, drawing together a wide variety of landscape researchers from around the world (and during the Cold War period) and was instrumental in propagating the central ideas of landscape ecology to the United States (Antrop, 2005).

The earliest papers explicitly using the term 'landscape ecology' recorded in Web of Science (WoS) are from the early 1970s (e.g. Bauer, 1970), while the International Association of Landscape Ecology (IALE) was formed in 1982. The roots of the development of landscape ecology are in Europe, stemming from broad-scale spatial considerations of land evaluation, planning, and management (Zonneveld, 1972, 1995; Schreiber, 1990; Jongman, 2005). The 'formal' discipline had been emerging for some time during the 1970s–1980s but was established more firmly by the work of Forman and Godron (1986), who wrote the first major book dedicated to the subject and set out a framework for the subsequent and ongoing explosion of landscape-scale research (Wu and Hobbs, 2002).

Landscape ecology emerged in North America, and the United States in particular, only in the years 1972–1987 (Forman, 2015). Before American ecologists introduced landscape ecology as a new subdiscipline of ecology, the landscape was already studied by ecologists such as Aldo Leopold, in the perspective of wildlife management and environmental ethics, and by cultural geographers such as Carl Sauer (1925). The international conference of IALE in Münster in 1987 and the book by Forman and Godron (1986) triggered the rapid expansion of landscape ecology in North America, which is now the stronghold of landscape ecological research (Turner, 2005; Wu, 2017; Muderere et al., 2018).

Total Publications
4,269

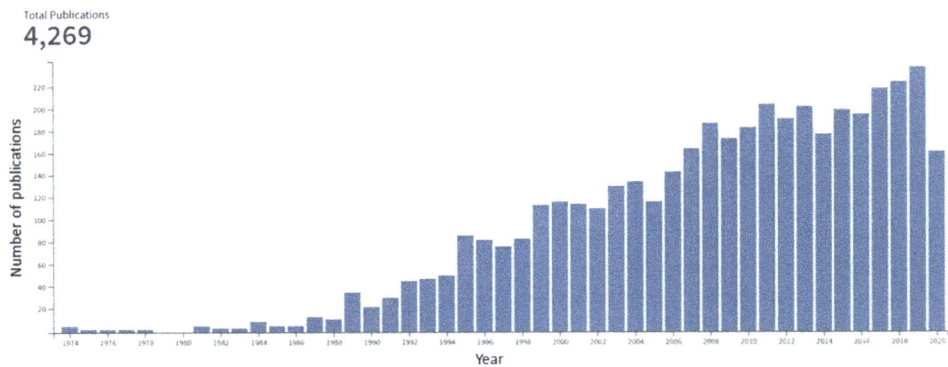

Figure 1.1 The number of papers published each year containing the term 'landscape ecology' as a topic, from ISI's Web of Science, ranging from 1974 to 2020 (search performed September 4, 2020, so the 2020 total is lower than the full year).

The number of papers published containing the term 'landscape ecology' in WoS from 1970 onwards is shown in Figure 1.1. This demonstrates sustained interest in the discipline by scientists and practitioners, particularly as a realization of the need to understand and manage systems at broad spatial scales has become apparent, and appropriate methods and technologies have developed to meet this need (e.g. remote sensing and GIS technologies).

Currently, landscape ecology is strongly interdisciplinary in both the concepts it uses to address broad-scale ecological investigations and the methods and techniques applied. It is supported by several dedicated scientific journals, including *Landscape Research*, *Landscape and Urban Planning*, and *Landscape Ecology*, and the IALE has 28 regional and national branches around the world. International conferences on landscape ecology take place every year, attracting hundreds of researchers (e.g. McIntyre et al., 2013; Young et al., 2020). A broad-scale approach to environmental problems is integral to many large international environmental programs, such as the EU Water Framework Directive (2000/60/EC), which necessitates management at the river basin scale; the Convention on Biological Diversity, which requires the conservation of ecosystems as one form of biological diversity; and the prevention, control, and management of species invasions, which occur over broad scales and are directly influenced by spatial patterns and processes within the landscape. These are just a few notable examples, and landscape ecology and its researchers and practitioners have important roles to play in addressing these and other issues.

The geography of landscape ecology as a discipline

Currently, two 'schools' of landscape ecology are dominant and interact with each other: the European school, grounded in the cultural and historical roots of the landscapes, and the North American school, based on systems theory and quantitative spatial analysis of patterns and processes, in particular in a rapidly urbanizing world. The diffusion of landscape ecology in the world follows either of these schools (Antrop and Van Eetvelde, 2019).

The development in Europe is also closely linked to its long history of human settlement and the use of natural resources. Facing issues of sustainability, degradation, and fragmentation of (and the need to reconnect) natural and semi-natural ecosystems and landscapes required changing stakeholders' views on landscape management, natural resource production, and environmental preservation (e.g. Van der Sluis et al., 2019). Consequently, the generalized development of landscape ecology in Europe has focused on the broad spatial arrangement and geography of discrete ecosystems and how these are structured and may change, interact, and ultimately be managed. An early example is the Kromme Rijn Project in The Netherlands (Tjallingii, 1974), wherein a 300 km² area of the river catchment (a tributary of the Rhine, or Rijn) was extensively mapped to delineate the spatial pattern of soils and vegetation in order to determine the diversity of such ecosystems and the potential impacts of development in the area. Interestingly, even in this early work, questions regarding the appropriateness of different ecological and biogeographical definitions (ecosystems, units, biocoenoses) arose; questions that still resonate today. This geographical work determined the importance of examining the natural world at broad spatial scales to understand not just how ecosystems are arranged but how abundant and diverse they are, and how equally broad-scale developments (e.g. road networks, land use change) may influence them.

In North America, landscape ecology developed, rather, within a 'systems science' framework, or the examination of how a system may function (including characteristics and rates of biomass production, energy fluxes, nutrient cycles, maintenance of biodiversity, and so on) depending upon the spatial configuration of its components and the spatial scale under examination (Turner, 2005). This more quantitative, statistical approach demonstrated that the historical tendency of ecological investigations to focus at fine spatial (and temporal) scales and on relatively discrete, homogeneous ecosystems was insufficient to address key environmental problems that operated at broad spatial scales, such as climate change, habitat and ecosystem fragmentation, and pollution. A milestone was the introduction of tools, such as Fragstats (McGarigal and Marks, 1995), that allowed calculation of landscape metrics for describing patterns and testing theories. Broad- and multi-scale investigations became more popular, particularly from 1990 onwards, and this has been responsible to a great extent for the expansion of the discipline. These investigations have been facilitated in North America by the fact that this region has extensive natural and semi-natural landscapes that remain less influenced by human development (at least directly), and so it has been possible to conduct detailed investigations of landscapes that do not vary dramatically at broad spatial scales (e.g. large areas of wilderness or grasslands; Mladenoff et al., 1993), which is much more difficult to achieve in Europe. The National Science Foundation's Long-term Ecological Research (LTER) Program has also been a boon in this regard, providing novel datasets from broad spatial and temporal scales that have been used to answer key ecological questions (Brazel, 2000; Kratz et al., 2003; Redman et al., 2004). Young et al. (2020) note that 'the European tradition of concern about character and the human role in shaping landscapes [is different from the] North American perspective of humans as a disturbance factor' (p. 18).

The more rigorous, quantitative approach to landscape ecology in North America has to some extent blended with the more descriptive geographical approach from Europe to form the core of the discipline as it stands today, though the marriage is not an easy one: many aspects of landscape form and function remain to be quantified (particularly in relation to the highly variable and heterogeneous landscape mosaics found around the world), and debates over the relevance and applicability of key concepts and methods still rage (Kirchhoff et al., 2013). As an example, a wooded landscape that has been fragmented so that only patches of woodland remain might be addressed, from a geographical perspective, by the creation of woodland corridors to reconnect the separate patches and thereby allow organisms that rely on the woodland to move more freely around the landscape. But ecologically, many important aspects need to be considered to determine whether this type of intervention is likely to be successful: the dispersal abilities of the organisms in question (including whether they can move across gaps between woodland without corridors, and if so, what the movement thresholds are), the spatial scales at which the organisms operate (will they recognize the corridors as movement vectors or consider them a low-quality habitat, for example?), the type of trees required, the width of the woodland corridors required to allow movement, the spatial and temporal corridor dynamics, the number of corridors required to allow movement and maintenance of the species population or metap-opulations throughout the landscape, and so on (see Chapter 5). Taking this further to include the social dimension, questions arise regarding what the landscape means to the regional community, whether people are willing to invest resources in supporting the species under consideration, and what the historical background of the landscape is, among other things. For example, some kinds of fragmented landscape are desirable to the populace, such as the traditional landscapes in Europe, which integrate cultural, historical, economic, and ecological values. Despite these many kinds of complexity, the principle of investigating and quantifying ecological patterns and pro-cesses at broad spatial scales has been adopted since the 1990s by researchers in many countries, mainly for biological resource management (such as biological conservation) (e.g. Burbidge, 1994; Hobbs, 1994), with some notable success, though problems of transferability remain.

The application of landscape ecology has perhaps the most potential in those countries that are currently going through socio-ecological transitions that are dramatically impacting their ecosystems on a broad scale and are likely to continue to do so. In contrast to the more devel-oped parts of the world, where societies have been living alongside extensive and intensive landscape change that has impacted ecological quality for centuries (e.g. Europe), landscape ecology has a substantial opportunity to have a mitigating influence on landscape and land use impacts in developing regions. Landscape ecological studies have increased across many Asian countries in recent years, particularly in relation to (for example) urban landscapes (Francis et al., 2016; Muderere et al., 2018), though African and South American studies are less common. In some countries (e.g. China), there has been a particular focus on the use of landscape metrics to characterize and quantify landscape pattern and change. In some cases, a landscape ecology perspective is allowing innovative contributions to be made to key aspects of development; for example, the planning and construction of more ecologically sound cities for rapidly expanding populations (e.g. Zhang and Wang, 2006; Beninde et al., 2015). For reviews of trends in landscape ecology in different regions, see Turner (2005) for North America, Fu and Lu (2006) for China, Young et al. (2020) for the United Kingdom, and Antrop (2007) for the discipline more generally.

The vast majority of landscape ecology research has a direct or indirect focus on human impacts on the systems under investigation, and this is in essence the applied value of landscape ecology. The asking of ecological questions is a worthy endeavor in itself, but it is in the need for application at the most appropriate scales to solve environmental problems that landscape ecology finds its greatest value.

Avoiding semantic confusion: clarifying some key terms and concepts within the discipline

> Define your terms … or we shall never understand one another
>
> *(Voltaire, 1764, p. 52)*

The lack of standardized, discrete terminology plagues the ecological sciences, as it does many others. Many concepts have roots in various disciplines, and the semantic meaning of the words may shift a lot. Landscape ecology is a way of seeing and thinking the landscape (Zonneveld, 1982). It aims for a holistic approach and as noted earlier, is interdisciplinary and even trans-disciplinary (Naveh, 2000, 2007). This section discusses and defines most of the commonly used terms in the landscape ecological literature, not to achieve a consensus of meaning but to acknowledge common interpretations and encourage further exploration. Table 1.2 briefly

Table 1.2 Sample definitions of some common terms used in landscape ecology literature

Terminology	Definition and description
Assemblage	Populations of different species coexisting within a defined spatial area (see 'community').
Biocoenosis	A spatially-defined ecological community, essentially the organisms associated with a particular biotope.
Biodiversity	The most formal and widely accepted definition comes from the Convention on Biological Diversity: The variability among living organisms from all sources, including, *inter alia*, terrestrial, marine and other aquatic ecosystems and the ecological complexes of which they are part: this includes diversity within species, between species and of ecosystems.
Biotope	An area of relatively uniform environmental and biological conditions. Usually defined by the particular biological assemblages that require such conditions and therefore characterize the biotope. Generally considered to exist at fine spatial scales.
Community	Utilized in two ways. First, it can be an expression of two or more populations of species coexisting within a defined area (i.e. a discrete spatial unit). Second, it can be the formal classification of particular species associations, such as a typical old-growth woodland community for a given biogeographical region. This latter definition does not, therefore, have a spatial dimension. In both cases, the definition usually includes species within similar taxa (so plant community or microbial community), but it does not have to be so restricted. To avoid confusion, and to acknowledge the non-typical and stochastic aggregations of species that are actually found in nature, especially where influenced by humans, the term 'assemblage' is sometimes used instead of the first definition.
Connectedness	Refers to the fact that two adjacent patches of the same type are spatially joined. Sometimes, the term 'structural connectivity' is also used.
Connectivity	Functional connectivity refers to the possible movement of an individual of a given species between patches, whether or not they are spatially connected. Structural and functional connectivity may occur concurrently, but this is not always the case. For example, patches of woodland linked by hedgerow corridors may be structurally connected, but this may not allow movement of a given organism between the patches (and so lacks functional connectivity). Likewise, a structurally disconnected landscape may still be functionally connected if organisms can cross gaps between patches.
Ecotope	May refer to the smallest ecologically distinct spatial unit within a mapping or classification system (and may therefore be synonymous with e.g. 'patch') or an area that exhibits a particular suite of environmental characteristics required to support a species or community (and therefore overlaps with 'biotope').
Edge	Refers to the border of a landscape unit (e.g. patch or corridor), i.e. the different properties along its perimeter.
Edge effect	Differences in ecological characteristics that exist within the edge of a landscape unit (e.g. patch or corridor). Usually refers to differences in biological diversity or community composition but can also be applied to abiotic factors such as changes in wind speed at woodland edges.

(Continued)

Table 1.2 (Continued) Sample definitions of some common terms used in landscape ecology literature

Terminology	Definition and description
Function	Usually expressed in two ways: first, as particular kinds of activity appropriate to the subject under question, incorporating the modes of action by which it fulfils its purpose; second, an expression of the various factors that create an activity or characteristic. To use an example: the *function* of an organism is to metabolize, develop, and reproduce, but the organism itself is a *function of* biological and environmental factors interacting (e.g. flows of material and energy, organization of components). The terms 'process' and 'function' are somewhat conflated. The first definition of function given here inherently includes processes (metabolization, development, and reproduction are all processes [activities], after all). The second definition (with origins in mathematics) considers function as a *result* of processes. In landscape ecology, spatial structure and function are the two poles of the fundamental landscape ecological paradigm: the study of spatial landscape patterns and their relationship to ecological processes and functioning. Spatial structure controls dynamics in the landscape, which simultaneously transforms the structure into a better-adapted new one, or as Forman and Godron (1986, p. 3) put it elegantly, 'An endless feedback loop: Past functioning has produced today's structure; Today's structure produces today's functioning; Today's functioning will produce future structure'. 'Landscape functions' refer to the capacity of a landscape to provide goods and services to society (see e.g. Willemen et al., 2008). These vary spatially and depend on land use, planning, and management.
Grain	The grain, or grain size, of an image refers to the finest spatial resolution that is achievable. In vector maps, it is defined as the Minimum Mapping Unit (MMU).
Guild	A group of species that may be taxonomically different but have similar resource requirements and consequently, share an ecological niche. An example would be birds that feed on insects in the foliage of oak woodlands, or granivorous (seed-eating) rodents, ants, and birds in a desert ecosystem.
Heterogeneity and homogeneity	Heterogeneous refers to something (e.g. a system) containing components that differ in form and/or function. The more varied and numerous the components, the more heterogeneous the system is. Landscapes and ecosystems are usually highly heterogeneous, for example, because they contain many different abiotic and biotic components that vary in their form and function (e.g. rocks, soils, dead wood, species, communities). Homogeneous refers to the opposite: something that consists of many identical components. Landscape heterogeneity is linked to concepts of entropy, order, and chaos in the landscape, and with coherence and autocorrelation (see text and Figures 1.2 and 1.3).
Holon	Holons are open systems that have a certain autonomy but are assembled in a hierarchy. A holon can be part of larger ones and can embed smaller ones. The whole forms a hierarchically organized or ordered structure. Each holon has its proper significant scale and context, which define the significance of variables for studying in a comprehensive way. The concept allows the holistic paradox to be overcome.

(Continued)

Table 1.2 (Continued) Sample definitions of some common terms used in landscape ecology literature

Terminology	Definition and description
Land(scape) element	A discrete (tangible, material) component of a landscape. Sometimes used in landscape classification and evaluation to describe a spatial land unit (similar to 'ecotope').
Land unit	Generic name of any kind of spatial landscape unit. In some land evaluation and classification systems, it has a specific scale definition. Zonneveld (1989, 2005) considers it as a fundamental (holistic) concept.
Network	An interconnected system that allows flows of material, species, or energy between components.
Percolation	The percolation theory relates the connectedness of the matrix to the area the patches occupy. It defines when a matrix inversion occurs (Gardner and O'Neill 1990).
Permeability	A measure of the capacity of an area to allow any kind of movement through the landscape. Often applied to the capacity for the boundary of an ecosystem to allow species or materials (e.g. water, soils) to pass through in one or both directions. Permeability will vary dependent upon individual species or type of material.
Population	The number of individuals of the same species occupying a defined area.
Process	Some form of action, usually resulting in something being moved or changed. Related to function depending upon use (see earlier). The relationship between landscape pattern and process is a fundamental aspect of landscape ecology, as discussed previously.
Sink	For a population, this is usually an area (usually a particular landscape element) where something is destroyed or removed from a system (at least in its current form). An example would be a patch of woodland with very low resource abundance or suitability for small mammal populations that results in excessive mortality and therefore removes those individuals from the surrounding landscape. For materials, a sink is usually an area of accumulation; for example, a river may act as a sink for sediments that arrive through erosive processes, or a pond may act as a sink for pollutants.
Source	An area (usually a particular landscape element) where an excess of something is produced, from whence it may disperse into the wider system. An example would be a patch of woodland with very high resource abundance that results in high small mammal populations, which then spread from the woodland into the surrounding landscape.

summarizes other terms that may often be used within the context of landscape ecology. For further examples of definitions, see Wu (2013).

Landscape

The word 'landscape' has multiple meanings in most languages. It refers not only to the tangible elements we perceive in our environment but also to its scenic manifestation and all the subjective, existential, and aesthetic experiences it evokes. The landscape is deeply rooted in culture and history. It reflects the character and the identity of a tract of land, which gains meaning as a territory for the community that lives there and shaped it. Hence, it is also a social construct, which is reflected in ancient administrative divisions and custom laws (Cosgrove, 2004). As a result, there are multiple interpretations of the landscape, and these will vary between disciplines; for discussions of these different interpretations, and a more specific history of the term, see (for example) Cosgrove and Daniels (1988), Cosgrove (2002, 2004), Olwig (2002, 2004), and Antrop and Van Eetvelde (2017). Given this, it is useful if disciplines, or indeed individual studies, can define as far as possible their interpretation of the term. Table 1.1 gives some definitions of 'landscape' from the landscape ecology literature.

Two conventional formal definitions of landscape currently exist: (1) that for cultural landscapes, defined by the UNESCO World Heritage Convention (2008, p. 14), which considers cultural landscapes to be the 'combined works of nature and of man' and explores various categories and subcategories; and (2) that of the European Landscape Convention by the Council of Europe (2008, p. 9): 'Landscape means an area, as perceived by people, whose character is the result of the action and interaction of natural and/or human factors'. As most countries are signatories to these conventions, they have had to adopt these definitions in policy; hence, many researchers and practitioners use these definitions so as to apply their knowledge in society.

In the context of landscape ecology, there are two main interpretations of the landscape: (1) a spatially delineated physical tract of land 'with its distinguishing characteristics and features' (see e.g. Oxford English Dictionary, 2020 and the European Landscape Convention); in this sense, the 'land' can be considered a spatial building block in the sense of land units (cf. Zonneveld, 1995); and (2) a territory meaningful for people living in it (the 'actors' or 'agents' that shape it) and significant for policymakers (for management, planning, and legalization, e.g. conservation, protection).

In some ways, this becomes a distinction between 'land' and 'landscape', with *land* referring mainly to a delineated tract of the terrestrial surface. In many cultures, land is a property owned by an individual or social group that possesses use rights over it, which determines its instrumental and economical value (Antrop van Eetvelde, 2017).

Ies Zonneveld emphasized the important difference between land and landscape, and in his book *Land Ecology* (Zonneveld, 1995), he explained why. As a pioneer in land classification and land evaluation, he chose *land* instead of *landscape* because of the ambiguity in the word 'landscape' due to its multiple meanings. The word 'land' fitted better when defining holistic land units by interpretation of aerial photographs. In a similar reasoning, Richard Forman spoke of *Land Mosaics – The Ecology of Landscapes and Regions* (Forman, 1996). Both show that landscape ecologists use the word 'landscape' in a more specific sense of spatial pattern or mosaic composed of interacting elements of different nature at a specific scale level. It is worthwhile to note that the definition of landscape in landscape ecology is not completely consistent with the widely accepted definition in the European Landscape Convention as noted earlier (Antrop and van Eetvelde, 2017). For the European Landscape Convention definition, the organism-centered perspective is only that of humans, whereas for the scientific discipline, the organism can be non-human.

It is also worth noting that the '-scape' suffix has recently been applied to a range of contexts in landscape or spatial ecology, further extending the fluidity of interpretations. In aquatic ecology, the term 'riverscape' is sometimes used (Haslam, 2008; see Chapter 21) to encapsulate not only the physical area (and its components) that contributes to a fluvial system but also the cultural and social ways in which the river and its surroundings are interpreted. Likewise, the terms 'oceanscape' and 'seascape' are sometimes used to describe spatial variation in ocean characteristics such as water temperature (Royer et al., 2004; Pittman et al., 2011). Other ecological studies relate '-scapes' to particular broad-scale physical features (e.g. wallscapes; Francis and Hoggart, 2012) or environmental characteristics (such as soundscapes; Pijanowski et al., 2011 and lightscapes; Bennie et al., 2014).

Scale and landscape level

Landscape ecology defines scale as a combination of two parameters: extent and grain. The extent of the landscape being studied should be ecologically meaningful given the scale at which the target phenomenon (organism, species) operates and should include the broader surroundings, i.e. the landscape context if it is likely to affect the processes being studied (Wiens, 1976). In the case of human actions, the extent may have to correspond to a project planning area or an administrative management authority, for example.

The grain refers to the smallest information unit that contains reliable information. This corresponds to the pixel size in raster remote sensing data and the size of the grid cell in digital raster maps. The analysis of landscape patterns often starts from maps or aerial imagery as data sources. The resolution of these documents defines the smallest area of the landscape with significant information. For analogue and digital vector maps, this is given by the minimum mapping unit (MMU). This is the size of the smallest areal feature that can be reliably mapped as a discrete entity at a given map scale.

Note here that the scale concept may cause a lot of confusion. Cartographers and geographers define scale differently. A cartographic scale is the ratio between a length on the map and its true length on the terrain. Consequently, a map scale of 1:20,000 is called larger than a map at scale 1:100,000, but as a printed document, it has a smaller extent than the 1 100,000 map printed on the same size. Also, a map at 1:20,000 scale shows more detail (i.e. fine grain) than one at a scale of 1:100,000. The confusion increases as, for example, planners speak of a large project, meaning one with a large extent and most likely represented on a small-scale map. To avoid confusion, landscape ecologists speak about coarse-grained and fine-grained landscapes, which are related to landscape heterogeneity and important for assessing ecological processes. A coarse-grained landscape containing fine-grained areas is considered optimal to provide for large-patch ecological benefits, multihabitat species, including humans, and a breadth of environmental resources and conditions (Forman, 1996; Cassar, 2019).

In the same way, 'broad-scale' is sometimes used (as in this chapter) to indicate an areal extent of several kilometers (a more intuitive landscape area), whereas fine-scale is smaller, e.g. considering individual landscape elements, though the terms are also somewhat subjective and open to interpretation. The usage of these terms typically relates to *extent* (horizontal distance) rather than *grain* (level of detail).

The grain should be balanced with the extent, considering the desired scale to study the landscape pattern and the computational efficiency. On the one hand, the grain should be kept as fine as possible to ensure that small and narrow, yet meaningful, features of the landscape are preserved in the model (Wiens, 1976). On the other hand, the grain should be increased in relation to the extent so that unnecessary 'noisy' detail is not confused with the important coarse-scale patterns over large spatial extents. This may be achieved by increasing the minimum mapping unit above the resolution set by the grain. Forman proposes the following thumb-rule to calculate the grain size of a landscape mosaic: 'the average diameter or area of all patches present' (Forman and Godron, 1986).

Often, the term *landscape scale* is used besides *landscape level*; these are fundamentally different concepts that are often confused. Landscape scale refers to the extent of the study or the size of the project in terms of the spatial area. Landscape level refers to the level of organization in a hierarchically organized ecological system (King, 2005). Similarly, one can speak of patch-level, class-level, and population-level (Fahrig, 2005). This distinction between scale and landscape is discussed further in Chapter 3.

Ecosystem and ecotopes

The term *ecosystem* refers to the functional properties of a system as a set of relations. This definition requires further specification as to how the system is spatio-temporally delineated, as the ecosystem concept itself is scale-independent. An ecosystem may be a small pond, a patch of woodland, or an entire landscape. Ecosystems are open systems (i.e. they can exchange materials or energy with other systems) and are typically embedded in each other in a holarchy (e.g. van Leeuwen, 1982; Jørgensen and Müller, 2000). Ecosystems are also complex (with multiple components interacting over different scales) and adaptive (with the capacity to change).

An ecotope is the (smallest) tangible spatial building block of the landscape. These are sometimes referred to as landscape 'components' or 'elements'. Examples include a patch of woodland, a park, a pond, or a hedgerow, though they do not have to include a biotic community (for example, a building may represent an ecotope). A 'landscape' may ultimately be considered an ecosystem composed of many ecotopes.

Habitat

The terms 'ecosystem' and 'habitat' are sometimes used almost as synonyms, and the two concepts are conflated to an extent. 'Habitat' may be defined in various ways, and different measures of what habitat is and how species use it have led to some confusion in ecology and difficulty in comparing and contrasting studies. Essentially, habitat may be defined as 'the resources and conditions present in an area that produce occupancy – including survival and reproduction – by a given organism' (Hall et al., 1997). Importantly, habitat may only be defined by reference to a particular organism; though 'woodland' may in general act as a habitat for many species, it is more accurate to say that it is an ecosystem, while the conditions found within the woodland (and perhaps elsewhere) form the habitat for a given species. For example, the habitat for barn owls (*Tyto alba*) in the United Kingdom will include the resources provided by many different ecosystems at different spatial scales, some of which are interchangeable: farmland and grassland for hunting; woodland, individual trees, and buildings for nesting; and so on (Taylor, 1994). It is also essential to note that habitat 'selection' (remember that this is not always a conscious or even semi-conscious process for many species) takes place over a hierarchy of scale, beginning at the broader scales of resource distribution (such as climatic conditions, soil and vegetation types, etc.) and becoming finer as species select for localized food or environmental resources (e.g. choice of tree for foraging/nesting, germination of seeds in soil patches with sufficient moisture or nutrients). Typically in landscape ecology, patches and corridors are defined as 'favorable habitat' for a given species, while 'unfavorable habitat' would be considered as the matrix (see Chapter 2). The frequently used term 'habitat fragmentation' more accurately refers to the fragmentation of discrete habitats or landscape elements (e.g. areas of woodland) rather than the resources available for a given species, though the two may occur simultaneously.

Mosaic

Forman and Godron's (1986) patch-corridor-matrix (PCM) mosaic model, discussed in more depth in Chapter 2, is perhaps the oldest conceptual model for interpreting landscape structure. A mosaic may simply be regarded as 'a pattern of adjacent and connecting landscape units'. Forman (1996) notes that the pattern includes patches, corridors, and a matrix, but not all of these elements need necessarily be present; a typical leopard-spot semi-arid landscape may be composed of the dominant matrix of sparse vegetation or bare soil, with distinct patches of denser vegetation, but no distinct corridors (Aguiar and Sala, 1999). Patches, corridors, and other aspects of the PCM model are covered in Chapter 2.

Landscape composition and configuration

The landscape consists of features ('elements') of different nature distributed in a given geographical space. The composition refers to the nature of these elements and properties, such as their number, size, and shape. Composition defines the landscape diversity. The configuration refers to the way these elements are arranged and distributed in patterns of space and form. The combination of both gives the complexity of the landscape, expressed by characteristics such as heterogeneity, coherence, and order, which can be measured by information-entropy.

Figure 1.2 summarizes some of the basic concepts related to the composition and configuration of the landscape. Sub-figures a', a'', a''', b, c, d, and c' show different configurations of a landscape composed of only a few patch types (A, B, and C). The central series with cases a to d shows the effects of spatial configuration. Landscapes a, b, and c have the same composition

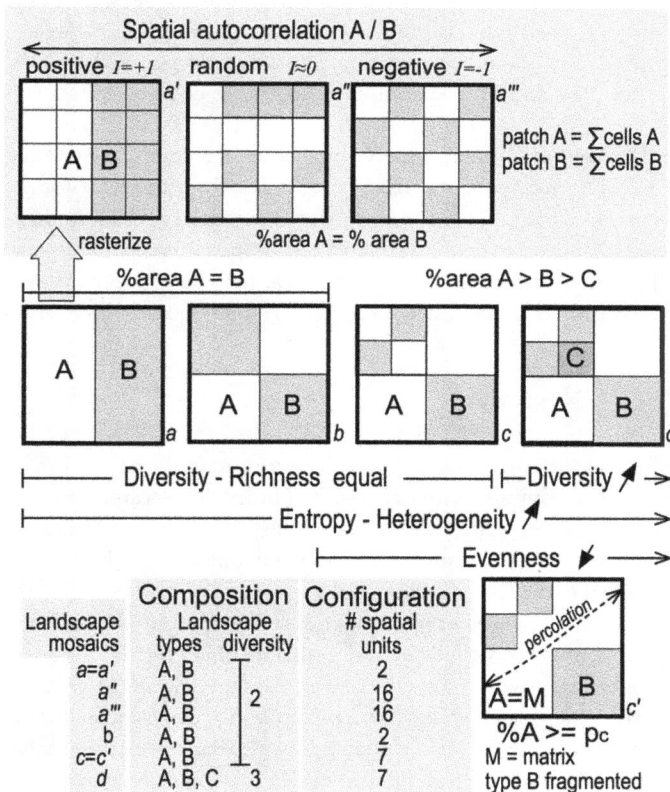

Figure 1.2 Summary of some basic concepts related to the composition and configuration of the landscape: see text for explanation.

LANDSCAPE PATTERNS

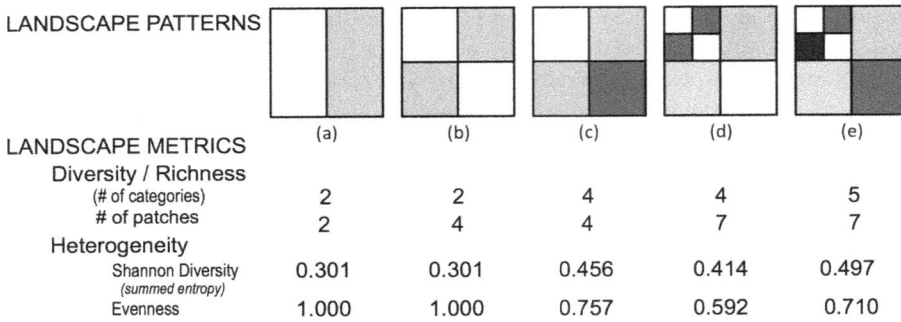

LANDSCAPE METRICS	(a)	(b)	(c)	(d)	(e)
Diversity / Richness					
(# of categories)	2	2	4	4	5
# of patches	2	4	4	7	7
Heterogeneity					
Shannon Diversity (summed entropy)	0.301	0.301	0.456	0.414	0.497
Evenness	1.000	1.000	0.757	0.592	0.710

Figure 1.3 Examples of landscape metrics of diversity and heterogeneity for patterns of different composition and configuration. After Antrop and Van Eetvelde (2017).

of two elements A and B; hence, they have the same landscape diversity or richness. In cases a and b, the types A and B have the same coverage within the landscape as a whole, but the configuration is more complex, consisting of more separated spatial units. The complexity increases with further fragmentation in cases c and d. Cases c and d have the same spatial configuration but not the same composition, as the richness (diversity) increases to 3. From case a to d, the heterogeneity increases, which is expressed by an increase in the entropy and a decrease of the evenness (Shannon's entropy and evenness) (see also Figure 1.3). As the proportional area of the patches of type A increases, the mosaic transforms into a patch-matrix model. When the proportional area of type A reaches the critical percolation threshold (case c'), type A becomes the matrix M.

Figure 1.3 shows a rasterized version of a mosaic where the patches are composed of adjacent grid cells with the same value. The areal proportions of types A and B are the same for the patterns a', a'', and a'''. The way units of the same type are spatially distributed defines the fragmentation, contagion, interspersion, and autocorrelation between the patch types. The spatial autocorrelation of the grid cell patches varies from maximal positive (Moran's I metric = +1) over random (most chaotic, I = 0) to maximally distributed, here in the quadratic tessellation (I = −1).

Borders and transitions

Borders separate patches from the matrix, and zones in a mosaic, from each other. They can be tangible (material) or intangible, such as administrative or property borders. They can be crisp or fuzzy, permeable or impermeable to different organisms or flows of materials, or form gradients and ecotones (Figure 1.4). When borders become more interdigitated, one can speak of fractal landscapes (Milne, 1990; Farina, 2006; Turner and Gardner, 2015).

An ecotone refers to a transitional area between two ecological communities (often referred to in the context of adjacent ecosystems, e.g. woodland and grassland) that contains a mix of species characteristic to each community as well as some that may exist only within the ecotone. A classic example is a riparian (riverside) zone, which may contain a mix of characteristically terrestrial and aquatic or semi-aquatic species as well as specifically riparian species. Ecotones are usually very biodiverse relative to the adjacent communities. They are characterized by an abrupt change, so that it is clear where the transition between the communities occurs. In contrast, an ecocline is a transitional zone between ecological communities that extends over a wide spatial area and is essentially a gradient of different species assemblages.

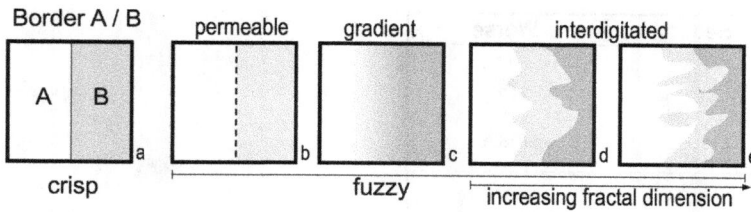

Figure 1.4 Border types between land units.

In these cases, the mosaic and PCM models may not be appropriate for describing the landscape. Instead, gradient models should be used, focusing on ecotones, permeability, eco-fields, flows, diffusion processes, and so on. For further discussion of ecotones, ecoclines, eco-fields, and gradients and their usage see McGarigal and Cushman (2002), Farina and Belgrano (2006), and Wu (2014).

Landscape coherence

Coherence refers to the degree of correspondence between features of different kinds in space or time (Mander and Murka, 2003; Mander et al., 2010). Coherence is one of the holistic meta-properties of the landscape. It is perceived and experienced as a degree of order ('fitting well together', planned, 'well-managed') or chaos ('neglected', 'derelict') when looking at the landscape. Phipps (1984) relates coherence to information-entropy, which allows measurement of the degree of order (or chaos). He makes the distinction between ecological order (based on vertical relations within the terrain) and topological order (based on horizontal relations). Van Mansvelt (1997) used 'ecological coherence' in the rural landscape in three ways: vertical (on site), horizontal (landscape-level), and cyclical (temporal) coherence. Opdam et al. (2003) developed a *landscape cohesion index* indicating the conservation potential of landscapes for biodiversity.

Essentially, coherence expresses spatial autocorrelation between landscape components (see the top section of Figure 1.2). Spatial autocorrelation is a multi-dimensional form of autocorrelation and can be measured by Moran's *I* statistic. Uuemaa et al. (2008) used autocorrelation and correlograms to assess landscape heterogeneity, and Mander et al. (2010) to characterize landscape patterns and assess fragmentation.

Shannon (1948) introduced information theory to quantify the information that is processed by compression, signal transmission, and data loss in communication. The key concept is entropy. The unit of information depends on the logarithmic base in the calculation of the entropy. Usually, information units are expressed in *bit* (binary logarithm) or *nat* (natural logarithm). Information theory has been applied in many domains. In geography and landscape research, it has been applied by Kilchenmann (1973) to quantify the information in thematic maps and calculate correlations between them to make regional classifications. Phipps (1981) used it in ecological landscape analysis. Shannon's information-entropy allows the quantification of holistic landscape concepts such as diversity, complexity, and heterogeneity. It has been used frequently in landscape metrics and is available as routines in the FRAGSTATS-package. Vranken et al. (2015) gave a review of the use of the concept in landscape ecology.

Fragmentation

The term *fragmentation* initially referred to 'habitat fragmentation' as a process during which a large habitat is transformed into smaller patches that are isolated from each other. Possible effects

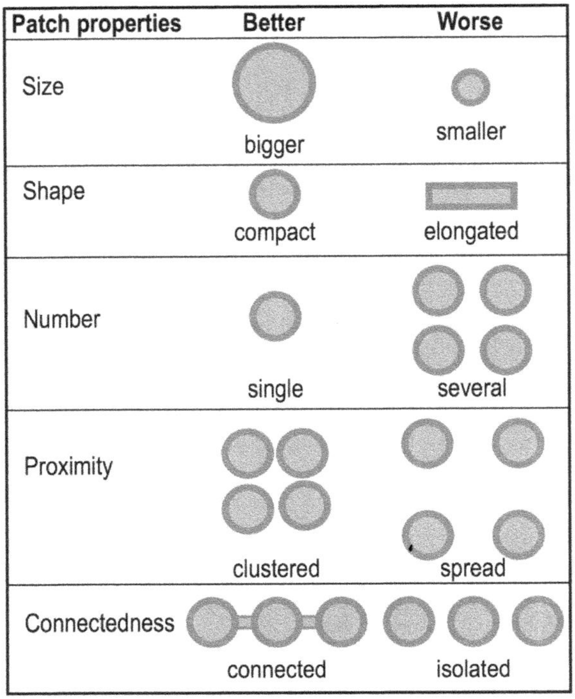

Patch properties	Better	Worse
Size	bigger	smaller
Shape	compact	elongated
Number	single	several
Proximity	clustered	spread
Connectedness	connected	isolated

Figure 1.5 Design principles for patch properties related to effects of landscape fragmentation. After Diamond (1975).

of this process are habitat loss, increase of the number of patches, decrease of patch size, decrease of the core/edge ratio, and increase of isolation. Fahrig (2003) conducted a literature review of the use of the concept. In general, a strong negative effect of habitat loss on biodiversity is accepted, but also, a positive or weaker negative effect of fragmentation per se on biodiversity is observed. To avoid confusion, Fahrig suggested that the term *fragmentation* should not be used for *habitat loss* but should be limited to changes in the spatial configuration. Although the process affects patches, it is a landscape-level characteristic. Design principles to mitigate fragmentation refer to the spatial properties of patches and to their configuration (Figure 1.5) (see Diamond, 1975; Fahrig, 2003; Opdam, 1991, 2006; Donaldson et al., 2017).

Some strengths and weaknesses of landscape ecology

With any sphere of investigation, establishing what it is best placed to do is an important step. A focus on the landscape scale is necessary for developing a more holistic understanding of ecological patterns and processes, and how human impacts and resultant environmental problems might be managed and mitigated. River management, ecosystem restoration, biological conservation, nature reserve design, water resource management, urban design and planning, sustainability planning, mitigation of climate change, and natural disaster management are all examples of practices that have been subject to calls for holistic and landscape-scale understanding (e.g. Cook, 1991; Wiens, 2002; Opdam and Wascher, 2004; Musacchio, 2009; Goddard et al., 2009).

Research at the landscape scale is most directly applicable to major environmental issues; within river ecology, for example, it is acknowledged that a broad-scale approach is a requirement for successful management, including rehabilitation and restoration efforts, because the river system is highly connected. Restoration at the finer spatial scale of individual reaches is unlikely to be successful in the long term because there is no consideration of the position of the reach in the wider hierarchy of structure and process (Francis, 2009). For example, restoring habitat in a given reach will not aid biodiversity if conditions further upstream mean that species are absent or cannot be dispersed to the restored downstream location.

The interdisciplinarity of landscape ecology is also a major strength of the discipline. It draws heavily on the spatial principles of geography and geographical techniques, as well as the quantitative approach of ecology and many aspects of the social sciences, in its approach to environmental issues. Although problems posed by this interdisciplinarity can be difficult to overcome in many cases, landscape ecology at least embraces these differing disciplines and recognizes that they are essential for the development of effective solutions to problems. It is also at the forefront of the development of novel methods of investigation across disciplines (e.g. remote sensing and GIS) as well as ecological metrics and modelling.

However, investigations at the landscape scale present scientific challenges. A staple of good ecological science is replication of experimental setup and empirical observation, so that false conclusions are not drawn and generalizations not made based on a single finding that may be flawed or may have occurred through chance (Hurlbert, 1984). At broad spatial scales, this can be very difficult to achieve – it is not easy to find or measure comparable landscapes, and replicating experiments (while avoiding false replication, or pseudoreplication [Hurlbert, 1984]) is almost impossible. All landscapes are complex, all are different with their own individual contexts, and so while some generalizations can be drawn across different landscapes (e.g. the principles of the PCM model), it is important to remember that no two landscapes are the same. This means that the transferability of findings between landscapes is not always as easy as we would like and serves to highlight the utility of modelling approaches for exploring variation in pattern and process across different landscapes (Chapter 16).

Establishing an appropriate or most suitable spatial scale to meet the objectives of an investigation can also be difficult, as there are so many patterns and processes operating at different spatial scales that key drivers can be difficult to identify. This will also depend on how the extent and grain of the landscape under investigation are delineated or determined. For example, in a semi-arid plain with a patchy vegetation configuration, it can be challenging to determine whether patches of vegetation exist because soil moisture allowed those patches to establish or whether plants managed to establish randomly and have then acted to trap and store moisture in their soil microclimate as a form of ecosystem engineering (e.g. Aguiar and Sala, 1999). The reality may lie somewhere in-between.

This complexity means that there are many aspects of form and function to be assessed, often with limited resources. There is a reliance on remote sensing and modelling at the landscape scale that may be compromised by a lack of field data or evaluation. If field data is collected, it is often a subset that is then extrapolated, meaning that very few processes, organisms, etc., are actually measured. Developing methods to address these issues is a key challenge for landscape ecology. Landscape ecology has been, and continues to be, a tremendously successful field, but it comes with some inherent challenges and limitations. The varying strengths and weaknesses of landscape ecology concepts, methods, and applications are discussed in various chapters throughout this book.

References

Aguiar, M.R. and Sala, O.E. (1999) 'Patch structure, dynamics and implications for the functioning of arid ecosystems', *Trends in Ecology and Evolution*, vol 14, no 7, pp273–277.

Antrop, M. (2005) 'From holistic landscape synthesis to transdisciplinary landscape management', in Tress, B., Tress, G., Fry, G. and Opdam, P. (eds) *From Landscape Research to Landscape Planning: Aspects of Integration, Education and Application*. Springer, Heidelberg.

Antrop, M. (2007) 'Reflecting upon 25 years of landscape ecology', *Landscape Ecology*, vol 22, pp1441–1443.

Antrop, M. and Van Eetvelde, V. (2017) *Landscape Perspectives: The Holistic Nature of Landscape*. Springer, Heidelberg, Landscape Series 23.

Antrop M. and Van Eetvelde V. (2019) 'Territory and/or scenery: concepts and prospects of Western landscape research', in Mueller, L. and Eulenstein, F. (eds) *Current Trends in Landscape Research. Innovations in Landscape Research*. Springer Nature, New York.

Bauer, G. (1970) 'Studies of landscape ecology of the natural protection area Entenfang near Wesseling', *Decheniana*, vol 123, no 1–2, pp165–198.

Beninde, J., Veith, M. and Hochkirch, A. (2015) 'Biodiversity in cities needs space: a meta-analysis of factors determining intra-urban biodiversity variation', *Ecology Letters*, vol 18, no 6, pp581–592.

Bennie, J. Davies, T.W., Inger, R. and Gaston, K.J. (2014) 'Mapping artificial lightscapes for ecological studies', *Methods in Ecology and Evolution*, vol 5, pp534–540.

Brazel, A., Selover, N., Vose, R. and Heisler, G. (2000) 'The tale of two climates: Baltimore and Phoenix urban LTER sites', *Climate Research*, vol 15, pp123–135.

Burbidge, A.A. (1994) 'Conservation biology in Australia: where should it be heading, will it be applied?', in C. Moritz and J. Kikkawa (eds) *Conservation biology in Australia and Oceania*, Surrey Beatty and Sons, Chipping Norton.

Cassar, L.F. (2019) 'Landscape and ecology: the need for a holistic approach to the conservation of habitats and biota', in Howard, P., Thompson, I., Waterton, E. and Atha, M. (eds.) *The Routledge Companion to Landscape Studies*. Routledge, London and New York.

Christensen, A.A., Brandt, J. and Svenningsen, S.R. (2017) 'Landscape ecology', in Richardson, D., Castree, N., Kwan, M-P., Kobayahsi A, Liu, W. and Marston, R.A. (eds) *The International Encyclopedia of Geography: People, the Earth, Environment, and Technology*. Wiley, Hoboken.

Cook, E.A. (1991) 'Urban landscape networks: an ecological planning framework', *Landscape Research*, vol 16, no 3, pp7–15.

Cosgrove, D. (2002) 'Landscape and the European sense of sight: Eyeing nature', in Anderson, K., Domosh, M., Pile, S. and Thrift N. (eds) *Handbook of Cultural Geography*. SAGE Publications, London.

Cosgrove, D. (2004) 'Landscape and Landschaft. Lecture at the "Spatial Turn in History" Symposium German Historical Institute', *GHI Bulletin*, vol 35, pp57–71

Cosgrove, D. and Daniels, S. (1988) *The Iconography of Landscape*. Cambridge University Press, Cambridge.

Council of Europe (2008) European Landscape Convention *and* Explanatory Report, Council of Europe, Document by the Secretary General established by the General Directorate of Education, Culture, Sport and Youth, and Environment: Florence.

Diamond, J.M. (1975) 'The island dilemma: lessons of modern biogeographic studies for the design of natural reserves', *Biological Conservation*, vol 7, pp129–146.

Donaldson, L., Wilson, R.J. and Maclean, I.M.D. (2017) 'Old concepts, new challenges: adapting landscape-scale conservation to the twenty-first century', *Biodiversity Conservation*, vol 26, pp527–552.

Fahrig, L. (2003) 'Effects of habitat fragmentation on biodiversity', *Annual Review of Ecology, Evolution, and Systematics*, vol 34, pp487–515.

Fahrig, L. (2005) 'When is a landscape perspective important?', in Wiens, J.A. and Moss, M.R. (eds.) *Issues and Perspectives in Landscape Ecology*. Cambridge University Press, Cambridge.

Farina, A. (2006) *Principles and Methods in Landscape Ecology*. Springer, Heidelberg.

Farina, A. and Belgrano, A. (2006) 'The eco-field hypothesis: toward a cognitive landscape', *Landscape Ecology*, vol 21, no 1, pp5–17.

Forman, R.T.T. (1990) 'The Beginnings of landscape Ecology in America', in Zonneveld, I.S. and Forman R.T.T. (eds) *Changing Landscapes: An Ecological Perspective*. Springer Verlag, New York.

Forman, R.T.T. (1996) *Land Mosaics: The Ecology of Landscapes and Regions*. Cambridge University Press, Cambridge.

Forman, R.T.T. (2015) 'Launching Landscape Ecology in America and Learning from Europe', in Barrett, G.W., Barrett, T.L. and Wu, J. (eds) *History of Landscape Ecology in the United States*. Springer, New York.

Forman, R.T.T. and Godron, M. (1986) *Landscape Ecology*. Wiley, New York.

Francis R.A. (2009) 'Perspectives on the potential for reconciliation ecology in urban riverscapes', *CAB Reviews: Perspectives in Agriculture, Veterinary Science, Nutrition and Natural Resources*, vol 4, 73.

Francis, R.A and Hoggart, S.P.G. (2012) 'The flora of urban river wallscapes', *River Research and Applications*, vol 28, pp1200–1216.

Francis, R.A., Millington, D.A. and Chadwick, M.A. (2016) 'Introduction: An overview of landscape ecology in cities', in Francis, R.A., Millington, D.A. and Chadwick, M.A. (eds) *Urban Landscape Ecology: Science, Policy and Practice*. Routledge, London.

Frazier, A.E. (2019) 'Emerging trajectories for spatial pattern analysis in landscape ecology', *Landscape Ecology*, vol 34, pp2073–2082

Frazier, A.E., Bryan, B.A., Buyantuev, A., Chen, L., Echeverria, C., Jia, P., Liu, L., Li, Q., Ouyang, Z., Wu, J., Xiang, W-N., Yang, J., Yang, L. and Zhao, S. (2019) 'Ecological civilization: perspectives from landscape ecology and landscape sustainability science', *Landscape Ecology*, vol 34, pp1–8

Fu, B. and Lu, Y. (2006) 'The progress and perspectives of landscape ecology in China', *Progress in Physical Geography*, vol 30, no 2, pp232–244.

Gardner and O'Neill (1990) 'Pattern, process and predictability: the use of neutral models for landscape analysis', in Turner, M.G., Gardner, R.H. (eds.), *Quantitative methods in landscape ecology*, New York: Springer-Verlag, pp289–307.

Goddard, M.A., Dougill, A.J. and Benton, T.G. (2009) 'Scaling up from gardens: biodiversity conservation in urban environments', *Trends in Ecology and Evolution*, vol 25, no 2, pp90–98.

Hall, L.S., Krausman, P.R. and Morrison, M.L. (1997) 'The habitat concept and a plea for standard terminology', *Wildlife Society Bulletin*, vol 25, no 1, pp173–182.

Haslam, S.M. (2008) *The Riverscape and the River*. Cambridge University Press, Cambridge.

Hobbs, R.J. (1994) 'Landscape ecology and conservation: moving from description to application', *Pacific Conservation Biology*, vol 1, no 3, pp170–176.

Hurlbert, S.H. (1984) 'Pseudoreplication and the design of ecological field experiments', *Ecological Monographs*, vol 54, no 2, pp187–211.

Jongman, R.G.H. (2005) 'Landscape ecology in land-use planning', in Wiens J. and Moss M. (eds) *Issues and Perspectives in Landscape Ecology*. Cambridge University Press, Cambridge.

Jørgensen, S.E and Müller F. (eds) (2000) *Handbook of Ecosystem Theories and Management*. Lewis Publishers, London.

Kilchenmann, A. (1973) 'Die Merkmalanayse für Nominaldaten: eine Methode zur Analyse von Qualitativen geographischen Daten', *Geoforum*, vol 15, pp33–45.

King, A.W. (2005) 'Hierarchy theory and the landscape…level? Or, Words do matter', in Wiens, J.A. and Moss, M.R. (eds.), *Issues and Perspectives in Landscape Ecology*. Cambridge University Press, Cambridge

Kirchhoff, T., Trepl, L. and Vicenzotti, V. (2013) 'What is landscape ecology? An analysis and evaluation of six different conceptions', *Landscape Research*, vol 38, no 1, pp33–51.

Kratz, T.K., Deegan, L.A., Harmon, M.E. and Lauenroth, W.K. (2003) 'Ecological variability in space and time: insights gained from the US LTER Program', *BioScience*, vol 53, no 1, pp57–67.

MacArthur, R.H. and Wilson, E.O. (1967) *The Theory of Island Biogeography*. Princeton University Press, Princeton.

Mander, Ü. and Murka, M. (2003) 'Landscape coherence: a new criterion for evaluating impacts of land use changes', in Mander, U. and Antrop, M. (eds) *Multifunctional Landscapes Vol. III: Continuity and Change*. WIT Press, Southampton.

Mander, Ü., Uuemaa, E., Roosaare, J., Aunap, R. and Antrop, M. (2010) 'Coherence and fragmentation of landscape patterns as characterized by correlograms: A case study of Estonia', *Landscape and Urban Planning*, vol 94, no 1, pp31–37.

McGarigal, K. (2015) 'FRAGSTATS Help. Department of Environmental Conservation University of Massachusetts, Amherst', https://www.umass.edu/landeco/research/fragstats/documents/fragstats.help.4.2.pdf, accessed 17 Dec 2020

McGarigal, K. and Cushman, S.A. (2002) *The Gradient Concept of Landscape Structure: Or, Why are There so Many Patches*. University of Massachusetts, Amherst.

McGarigal, K. and Marks B.J. (1995) *FRAGSTATS: Spatial Pattern Analysis Program for Quantifying Landscape Structure*, USDA For. Serv. Gen. Tech. Rep. PNW-351.

McIntyre, N.E., Iverson, L.R. and Turner, M.G. (2013) 'A 27-year perspective on landscape ecology from the US-IALE annual meeting', *Landscape Ecology*, vol 28, pp1845–1848.

McKenzie, D., Miller, C. and Falk, D.A. (2011) 'Toward a theory of landscape fire', in McKenzie, D., Miller, C. and Falk, D.A. (eds.), *The Landscape Ecology of Fire*. Springer, New York.

Milne, B.T. (1990) 'Lessons from applying fractal models to landscape patterns', in Turner M.G. and Gardner R.H. (eds) *Quantitative Methods in Landscape Ecology*. Springer, New York.

Mladenoff, D.J., White, M.A., Pastor, J. and Crow, T.R. (1993) 'Comparing spatial pattern in unaltered Old-Growth and disturbed forest landscapes', *Ecological Applications*, vol 3, no 2, pp294–306.

Muderere, T. Murwira A., and Tagwireyi, P. (2018) 'An analysis of trends in urban landscape ecology research in spatial ecological literature between 1986 and 2016', *Current Landscape Ecology Reports*, vol 3, pp43–56.

Musacchio, L.R. (2009) 'The ecology and culture of landscape sustainability: emerging knowledge and innovation in landscape research and practice', *Landscape Ecology*, vol 24, no 8, pp989–992.

Naveh, Z. (2000) 'What is holistic landscape ecology? A conceptual introduction', *Landscape and Urban Planning*, vol 50, no 1–3, pp7–26.

Naveh, Z. (2007) *Transdisciplinary Challenges in Landscape Ecology and Restoration Ecology: An Anthology*. Springer, Heidelberg.

Naveh, Z. and Lieberman, A.S. (2013) *Landscape Ecology: Theory and Application*, 2nd edition. Springer, New York

Olwig, K.R. (2002) *Landscape, Nature and the Body Politic: From Britain's Renaissance to America's New World*. University of Wisconsin Press, Madison.

Olwig, K.R. (2004) '"This is not a landscape": Circulating reference and land shaping', in Palang, H., Sooväli, H., Antrop, M. and Setten, S. (eds) *European Rural Landscapes: Persistence and Change in a Globalising Environment*. Kluwer Academic Publishers, Dordrecht.

Opdam, P. (1991) 'Metapopulation theory and habitat fragmentation: a review of holarctic breeding bird studies', *Landscape Ecology*, vol 5, no 2, pp93–106.

Opdam, P. (2006) 'Ecosystem networks: a spatial concept for integrative research and planning of landscapes', in Tress, B., Tress, G., Fry, G. and Opdam, P. (eds) *From Landscape Research to Landscape Planning: Aspects of Integration, Education and Application*. Wageningen UR Frontis Series, vol 12, pp51–65. http://library.wur.nl/frontis/landscape_research/04_opdam.pdf.

Opdam, P., Luque, S. and Jones, K.B. (2009) 'Changing landscapes to accommodate for climate change impacts: a call for landscape ecology', *Landscape Ecology*, vol 24, pp715–721

Opdam, P. and Wascher, D. (2004) 'Climate change meets habitat fragmentation: linking landscape and biogeographical scale levels in research and conservation', *Biological Conservation*, vol 117, no 3, pp285–297.

Opdam, P., Verboom, J. and Pouwels, R. (2003) 'Landscape cohesion: an index for the conservation potential of landscapes for biodiversity', *Landscape Ecology*, vol 18, no 2, pp113–126.

Opdam, P., Luque, S., Nassauer, J., Verburg, P.H. and Wu, J. (2018) 'How can landscape ecology contribute to sustainability science?', *Landscape Ecology*, vol 33, pp1–7.

Phipps, M. (1981) 'Information theory and landscape analysis', in Tjallingii, S. and de Veer, A.A. (eds) Perspectives in Landscape Ecology: proceedings of International Congress organized by Netherlands Society for Landscape Ecology, Veldhoven, the Netherlands, April 6–11. Pudoc, Wageningen.

Phipps, M. (1984) 'Rural landscape dynamics: the illustration of some concepts', in Brandt, J. and Agger, P. (eds), Methodology in Landscape Research and Planning. Proceedings of the 1st International Seminar of the International Association of Landscape Ecology, vol. I. Roskilde University Centre, Roskilde, October 15–19, 1984, pp47–54.

Pickett, S.T.A. and Cadenasso, M.L. (2002) 'The ecosystem as a multidimensional concept: meaning, model, and metaphor', *Ecosystems*, vol 5, no 1, pp1–10.

Pijanowski, B.C., Farina, A., Gage, S.H., Dumyahn, S.L. and Krause, B.L. (2011) 'What is soundscape ecology? An introduction and overview of an emerging new science', *Landscape Ecology*, vol 26, pp1213–1232.

Pittman, S.J., Kneib, R.T. and Simenstad, C.A. (2011) 'Practicing coastal seascape ecology', *Marine Ecology Progress Series*, vol 427, pp187–190.

Redman, C.L., Grove, J.M. and Kuby, L.H. (2004) 'Integrating social science into the Long-Term Ecological Research (LTER) network: social dimensions of ecological change and ecological dimensions of social change', *Ecosystems*, vol 7, no 2, pp161–171.

Risser, P.G. (1987) 'Landscape ecology: state of the art', in Turner, M.G. (ed.), *Landscape Heterogeneity and Disturbance*. Springer, New York.

Royer, F., Fromentin, J.M. and Gaspar, P. (2004) 'Association between bluefin tuna schools and oceanic features in the western Mediterranean', *Marine Ecology Progress Series*, vol 269, pp249–263.

Sauer, C.O. (1925) 'The morphology of landscape', in Wiens, J.H., Moss, M.R., Turner, M.G and Mladenoff, D.J. (eds) *Foundation Papers in Landscape Ecology*. Columbia University Press, New York.

Schreiber K.-F. (1990) 'The history of landscape ecology in Europe', in Zonneveld, I.S. and Forman, R.T.T. (eds) *Changing Landscapes: An Ecological Perspective*. Springer, New York.

Shannon, C.E. (1948) 'A mathematical theory of communication', *Bell System Technical Journal*, vol 27, no 3, pp379–423 & 623–656.

Silbernagel, J. (2003) 'Spatial theory in early conservation design: examples from Aldo Leopold's work', *Landscape Ecology*, vol 18, pp635–646.

Taylor, I. (1994) *Barn Owls: Predator-Prey Relationships and Conservation*. Cambridge University Press, Cambridge.

Tjallingii, S.P. (1974) 'Unity and diversity in landscape', *Landscape Planning*, 1: 7–34.

Troll, C. (1939) 'Luftbildplan and okologische bodenforschung', *Zeitschraft der Gesellschaft jur Erdkunde Zu*, 241–298.

Tjallingii, S.P. and de Veer, A.A. (eds) (1982) *Perspectives in Landscape Ecology*. Centre for Agricultural Publishing and Documentation, Wageningen.

Turner, M.G. (2005) 'Landscape ecology in North America: past, present and future', *Ecology*, vol 86, no 8, pp1967–1974.

Turner, M.G. and Gardner, R.H. (eds) (1990) *Quantitative Methods in Landscape Ecology*. Springer, New York.

Turner, M.G. and Gardner, R.H. (2015) *Landscape Ecology in Theory and Practice: Pattern and Process*. Springer, New York.

United Nations Educational, Scientific and Cultural Organisation (UNESCO) (2003) 'Operational guidelines for the implementation of the World Heritage Convention', whc.unesco.org/archive/op guide08-en.pdf#annex3, accessed 4 Dec 2020.

Uuemaa, E., Roosaare, J., Kanal, A. and Mander Ü. (2008) 'Spatial correlograms of soil cover as an indicator of landscape heterogeneity', *Ecological Indicators*, vol 8, no 6, pp783–794.

Wiens, J.A. (1976) 'Population responses to patchy environments', *Annual Review of Ecology and Systematics*, vol 7, pp81–120.

Wiens, J.A. (2002) 'Riverine landscapes: taking landscape ecology into the water', *Freshwater Biology*, vol 46, no 4, pp501–515.

Willemen, L., Verburg, P.H., Hein, L. and van Mensvoort, M.E.F. (2008) 'Spatial characterization of landscape functions', *Landscape and Urban Planning*, vol 88, pp34–43.

With, K.A. (2002) 'The landscape ecology of invasive spread', *Conservation Biology*, vol 16, no 5, pp1192–1203

Wu, J. (2017) 'Thirty years of *Landscape Ecology* (1987–2017): retrospects and prospects', *Landscape Ecology*, vol 32, pp2225–2239

Wu, J. (2013) 'Landscape Ecology', in Leemans, R. (ed) *Ecological Systems: Selected Entries from the Encyclopedia of Sustainability Science and Technology*, New York, Springer, pp179–200.

Wu, J. (2014) 'Urban ecology and sustainability: The state-of-the-science and future directions', *Landscape and Urban Planning*, vol 125, pp209–221.

Wu, J. and Hobbs, R. (2002) 'Key issues and research priorities in landscape ecology: an idiosyncratic synthesis', *Landscape Ecology*, vol 17, pp355–365.

Urban, D.L. (2006) 'Landscape Ecology', in El-Shaarawi, A.H. and Piegorsch, W.W. (eds), *Encyclopedia of Environmetrics*. John Wiley & Sons, New York.

Van der Sluis, T., Arts, B.J.M., Kok, K., Bogers, M., Gravsholt Busck, A., Sepp, K., Loupa Ramos, I., Pavlis, E., Geamana, N. and Crouzat, E. (2019) 'Drivers of European landscape change: stakeholders' perspectives through Fuzzy Cognitive Mapping', *Landscape Research*, vol 44, no 4, pp458–476.

van Leeuwen, C.G. (1982) 'From ecosystem to ecodevice', in Tjallingii, S.P. and de Veer, A.A. (eds), *Perspectives in Landscape Ecology*. Pudoc, Wageningen.

Van Mansvelt, J.D. (1997) 'An interdisciplinary approach to integrate a range of agro- landscape values as proposed by representatives of various disciplines', *Agriculture, Ecosystems and Environment*, vol 63, pp233–250.

Voltaire, F.M.A. (1764) 'Dictionnaire philosophique', trans. Anon (1824) *A Philosophical Dictionary: From the French*. Hunt and Hunt, London.

Vranken, I., Baudry, J., Aubinet, M., Visser, M. and Bogaert, J. (2015) 'A review on the use of entropy in landscape ecology: heterogeneity, unpredictability, scale dependence and their links with thermodynamics', *Landscape Ecology*, vol 30, pp51–65.

Young, C., Bellamy, C., Burton, V., Griffiths, G., Metzger, M.J., Neumann, J., Porter, J. and Millington, J.D.A. (2020) 'UK landscape ecology: trends and perspectives from the first 25 years of ialeUK', *Landscape Ecology*, vol 35, pp11–22.

Zhang, L. and Wang, H. (2006) 'Planning an ecological network of Xiamen Island (China) using landscape metrics and network analysis', *Landscape and Urban Planning*, vol 78, pp449–456.

Zonneveld, I.S. (1972) *Landevaluation and Land(scape) Scienc*e. ITC textbook VII-4, Enschede.

Zonneveld, I.S. (1989) 'The land unit: A fundamental concept in landscape ecology and its applications', *Landscape Ecology*, vol 3, no 2, pp67–89.

Zonneveld, I.S. (1995) *Land Ecology: An Introduction to Landscape Ecology as a Base for Land Evaluation, Land Management and Conservation*. SPB Academic Publishing, Amsterdam.

Zonneveld, I.S. (1982) 'Land(scape) ecology, a science or a state of mind', in Tjallingii, S.P. and de Veer, A.A. (eds.), *Perspectives in Landscape Ecology. Proceedings of the International Congress of the Netherlands Society for Landscape Ecology*, Centre for Agricultural Publishing and Documentation, Wageningen, pp9–15

Zonneveld, I.S. (2005) 'The land unit as a black box: a Pandora's box?', in Wiens, J.A. and Moss, M.R. (eds) *Issues in Landscape Ecology*. Cambridge University Press, Cambridge.09601200Information Classification: General00Information Classification: General

Part I
Theory and concepts in landscape ecology

2

Landscape mosaics and the patch-corridor-matrix model

Marc Antrop

Introduction

The landscape is a holistic, complex, and highly dynamic manifestation of the environment that people experience in very different ways. In order to understand the landscape, we reduce its complexity using models according to a certain perspective, a way of seeing (Antrop and Van Eetvelde 2017, Cosgrove 2002). Hence, different landscape representations are constructed mentally for any given place.

Different disciplines study the landscape, and each one uses specific approaches and models. The models used in landscape ecology were developed relatively recently. The reduction of holistic landscape complexity into mosaics and patches, corridors and matrix are two intertwined models that became popular among landscape ecologists in the 1980s, mainly because these offered a way for spatial analysis of new data that became available through remote sensing and grid maps.

First, this chapter briefly presents the models commonly used in landscape research in order to situate the landscape ecological models properly. Second, it discusses the essential concepts and properties of the mosaic model and the patch-corridor-matrix model in two separate sections. Third, the origins and underlying theories of the models used in landscape ecology are presented. The next section focusses on some analytical tools for landscape patterns and how the information technology (IT) development interacted with the methodological development of landscape ecology. Examples are spatial analysis using geographical information systems (GIS), geostatistics, and landscape metrics. Finally, an assessment is given of the strengths, restrictions, and weaknesses of both landscape models, and some challenges for the future are presented.

Conceptual models of space

Zonal entities or continuous fields?

The conceptual models of space consist of two fundamentally different approaches: discrete entities and continuous fields. Land use and land cover maps are typically conceived as zonal entities. Climate, relief, slope, and groundwater depth are typically continuous fields and represented on maps by isolines and as surfaces. Most often, these are constructed from discrete sampling and statistical interpolation.

The choice between an entity-based approach and a field-based approach defines the kind of study to go on, i.e. the primitives, data encoding and format, methods for analysis, visualization, and representation (Burrough and McDonnell 1998). In reality, the borders between landscape units are seldom crisp, patterns may be fuzzy, and transitions may form ecological gradients and ecotones (Farina 2006, Farina and Belgrano 2006).

Models commonly used in landscape research

Antrop and Van Eetvelde (2017) discuss five models that are commonly used in landscape studies. Table 2.1 summarizes and compares their properties. The primitives are the key concepts that make the model. This chapter focusses only on the third group, the mosaic and patch-corridor-matrix models, which are mainly applied in landscape ecology. They will be discussed here as two separate models, as they derive from different paradigms and theories.

Geometric properties define the different primitives in all these models. However, their semantic meaning differs between the models.

Essentially, the mosaic model and the patch-corridor-matrix model are typical *zonal models*, i.e. they consist of discrete, well-delineated landscape entities. These land units are defined by selecting themes, which are classified into categories, very often land cover types. McGarigal and Marks (1995) speak of *categorical landscapes*. However, in the description of the constituting primitives, gradients and fuzzy borders are recognized.

Overview

Model 1: elements, components, structures

This model is the oldest and was developed by geographers and cartographers at the beginning of the 20th century in order to use maps as a scientific tool for analyzing spatial patterns of geographical phenomena (Eckert 1907). These were codified into three primitives according to their spatial dimensions: elements, components, and structures. For each, specific methods for analysis were developed (Unwin 1981). In the study of landscape, the main issues were how to

Table 2.1 Comparison of models commonly used in landscape research

Model	Primitives	Spatial types	Observation perspective	Typical application domain
1	elements, components, structures	objects, zonal, surfaces, networks, tessellations	vertical, bird's eye, distant	geography, cartography
2	points, lines, polygons, surfaces	objects, zonal, surfaces, networks	vertical, bird's eye, distant	GIS, digital modelling
3	mosaic, patch, corridor, matrix	zonal, networks	vertical, bird's eye, distant	landscape ecology
4	mass, screen, space	zonal, surface, viewshed	horizontal, terrain observation	visual landscape analysis, perception, scenery
5	landmark, path, node, district, edge	objects, zonal	horizontal, terrain observation	mental mapping, planning

codify and present the broad diversity of landscape phenomena and how to delineate and map spatial units as regions (Granö 1929, Granö and Paasi 1997).

Elements are discrete phenomena with well-defined borders. Material elements are objects, such as a tree, a house, or a field. The edges or borders are clear and tangible. Elements can also be non-material and abstract, such as formally defined administrative units or property boundaries. The scale of the study defines how the elements can be described. At a fine scale, details of a house and its footprint can be mapped. On broad scales, houses are symbolized by dots or aggregated to build up areas represented as a polygon. On a broad scale, adjacent fields with similar land cover are aggregated in one polygon, while elements smaller than the pre-set Minimum Mappable Unit (MMU) will be omitted (Longley et al. 1999, Openshaw 1984).

Components are phenomena that vary continuously in space and only occasionally may present clear edges or borders. The terrain surface with continuously varying elevation is a typical example. Other components are variations of slopes, exposition, soils, groundwater, and microclimate. In order to describe and map components, they must be sampled and subdivided into well-defined classes or categories. The choice of the descriptive attributes, the operational formulation of variables, the sampling scheme, the number of classes, the class-limits, and the classification method are chosen according to the objectives of the analysis and ultimately define the outcome. The level or scale of measurement is critical here. For example, a slope can be described at a nominal scale using such categories as flat, convex, or concave, or at an ordinal scale using such classes as flat, gentle, moderate, and steep, but can also be expressed at quantitative scales using percentages or degrees.

Structures are the ways the elements or components are spatially and functionally linked and related. They have a meaning that transcends that of their component parts. In fact, structures are conceived mentally to reduce the complexity for a better understanding and to facilitate the use of the landscape. Defining structures is an essential step in landscape modelling. Different structures can be defined according to the goals or functions considered. Generally, four main types of structure are recognized: spatial structures, relational structures, temporal structures, and functional structures.

Spatial structures deal with the geometry, distribution, and topology between elements in geographical space. The geometrical dimensions define the size and shape of the selected features and the way they are encoded as points, lines, areas, or surfaces in modelling. Spatial distributions can be random, clustered, or regular. Topology defines accessibility and isolation of elements and connectivity.

Relational structures focus on the nature and strength of the interactions between the composing elements. In landscape studies, the distinction between vertical and horizontal relations is fundamental. The vertical relations between different phenomena located at the same place often express ecological dependency and coherence. The horizontal relations express fluxes of material, energy, or information through space and define a topological dependency.

Temporal structures describe the coherence and transformation between successive events or states. Examples are crop calendars and phenology, seasonal movements of people and cattle, and landscape change trajectories.

Functional structures focus on how specific processes work. Examples of how these can be practically applied are found in ecodevices (see later).

In the landscape ecological perspective and depending on the scale of the study, patches and corridors are objects defined by polygons with clear borders. Corridors can also be considered networks. The matrix is essentially seen as a component. Often, vertical relations define a patch, and processes in the landscape express horizontal relations.

Model 2: point, line, polygon, surface

In a digital GIS-environment, all real objects and landscape phenomena are geocoded in a format that is the most suitable for the purpose of the analysis at a given scale (Longley et al. 1999, Burrough and McDonnell 1998). Basically, two formats of encoding are used: vector and raster. The *vector format* uses four primitives: point (node), line (polyline, arc), polygon, and surface (volume). In the *raster format*, a grid overlay in an analogue landscape map is used to sample the landscape, or – in the case of remote sensing data – the pixels are used. The grid cell or pixel is the smallest spatial unit carrying information (the *grain*). Each grid cell is assigned a value or code corresponding to the thematic attribute. In raster format, any object, such as a patch, is an aggregate of connected, adjacent grid cells with the same value. Its shape, size, and edge are distorted by the size and shape of the grain (see also Figure 2.4).

The choice of the geocoding determines what further analyses are possible. For example, in a broad-scale regional analysis of settlement patterns, buildings can be coded as points, having X-, Y-, and Z-coordinates for their location and an ID that links to an attribute table. For a fine-scale analysis, the shape and height of the buildings can be important, and then, geocoding will use polygons instead. Points, lines, and polygons are used to represent objects and patterns. Landscape mosaics, as for example formed by land cover or territories, are represented by patterns of polygons. Surfaces are constructed from discrete observations, which are used to interpolate continuous phenomena, such as the terrain and landform. Surfaces are used in spatial and three-dimensional (3-D) modelling.

Maps in raster format have become very popular for spatial analysis and the calculation of spatial statistics and landscape metrics.

Model 3: mosaic and patch-corridor-matrix

Forman (1982) recognized three basic categories of landscape elements that configure the ecological mosaic of the landscape (Figure 2.1). A *patch* is an area in the landscape that is beneficial for the species under consideration, while the space that is not is referred to as *matrix*. The *matrix* is the background ecosystem or land use in the landscape mosaic and has the most dominant coverage and the highest connectedness. Linear elements are called *corridors* and are also considered beneficial for the species studied. Corridors can connect patches and form spatial ecological networks. While patches and corridors focus on the selected organisms or species, a mosaic can be seen as a multi-organism variant where all patches have different values for different species. Richness, heterogeneity, and connectedness characterize the mosaic and may be defined formally in different ways (Merriam 1984, Baudry and Merriam 1988, Gardner and O'Neill 1990, McGarigal 2015, Antrop and Van Eetvelde 2017).

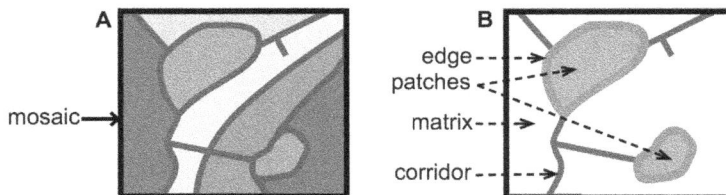

Figure 2.1   The mosaic and the patch-corridor-matrix model for landscape analysis.

Model 4: mass, screen, space

In this model, mass, screen, and space are the primitives. These are used in particular for describing the visual structure of the landscape scenery, which is useful for visual impact assessment, landscape character assessment, and landscape planning and design (De Veer and Burrough 1978). *Mass* consists of volumes that block the view, such as buildings and woods. *Screens* are tall linear elements that in some way obstruct the view, such as hedges or tree rows. A *space* is an open area without any elements obstructing the view for at least the distance of the critical viewing distance, which is approximately 1000–1500 m (Antrop and Van Eetvelde 2017). Generally, spaces are bordered by masses or screens, which can be either biotic or abiotic. Screens often have some 'transparency'. They may also form networks, which according to screen density and orientation, can result in a 'filtered' or 'closed' landscape scenery. In some cases, screens are the dominant elements characterizing the landscape type, e.g. a bocage landscape.

In the landscape ecological perspective, masses can be regarded as patches, screens as corridors, and spaces as the matrix.

Model 5: landmark, path, node, district, edge

The primitives are landmark, district, path, node, and edge and describe the mental image when we construct a mental map of a place. The importance and significance of these primitives may vary for different observers, such as inhabitants and tourists. The cognitive map of a place defines its legibility or *imageability* (Lynch 1960). Cognitive maps support urban and landscape design and spatial planning. For the landscape ecological analysis of spatial patterns, this model is less relevant.

The mosaic model

Paradigm

The observation that the landscape shows a heterogeneity that varies with scale was strikingly revealed by observations from high viewpoints and from the air. Aerial photography gave an important boost to the study of landscape patterns. It even caused Carl Troll to say 'aerial photography was in a large extent landscape ecology' (Troll 1939). Aerial photographs reveal landscape mosaics directly, and photointerpretation allows their mapping for land(scape) classification and evaluation. Using different scales and resolutions, a hierarchy of land units can be mapped and described. In a holistic approach, land units can serve as black boxes. Zonneveld (1995, 2005) used this term in the common meaning of any entity that has unknown internal functions or mechanisms but is useful when building hierarchical systems.

The most obvious forms of representing *spatial structures* are maps. These are two-dimensional structures with a spatial basis defined by the scale and a map projection (georeference). Without a spatial basis, they become schemes. Gradually, more 3-D and even four-dimensional representations have become common. Spatial structures in the landscape are often represented as patterns, networks, or configurations. Examples are field systems, settlement patterns, road networks, networks of hedgerows, and blue-green infrastructure. Many are tangible, but intangible structures also exist, such as territorial subdivisions and property boundaries.

The mosaic immediately reveals a whole series of properties of the landscape: the composition and configuration as well as diversity, heterogeneity, fragmentation, and spatial autocorrelation. These properties allow the inference of processes that determine functional flows and movements through the landscape and change the patterns.

Making the mosaic

Land classification aims to define and map homogeneous landscape units. This results in mosaics of zonal patterns. Zonneveld (1995) described two methods to do this: by subdivision and by aggregation. The *holistic method* often starts from aerial photographs that directly show landscape patterns (Figure 2.2). It is a fast and reliable method to create a hierarchy of spatial land units, which can be used as *black boxes* that will become gradually more defined by adding additional information (Zonneveld 2005). The *parametric method* aggregates information of different landscape components by overlaying thematic maps to form a composite of land units. No hierarchy is made, but each unit is defined by the composing themes. In GIS-overlaying, meaningless sliver-polygons are often formed, which have to be cleaned up (see also Figure 2.3 and the tools section later).

With both methods, landscape mosaics are created. When it comes to ecological pattern analysis of patches, corridors, and matrix, the parametric method allows the direct selection of specific themes. The holistic land units (black boxes) are more suitable to analyze meta-characteristics of the landscape such as diversity, heterogeneity, and connectivity.

Primitives in the mosaic model

In the mosaic model, the meaning of the zones derives directly from the data used for their identification and delineation on the map as *land units* (Zonneveld 1995). This meaning can be

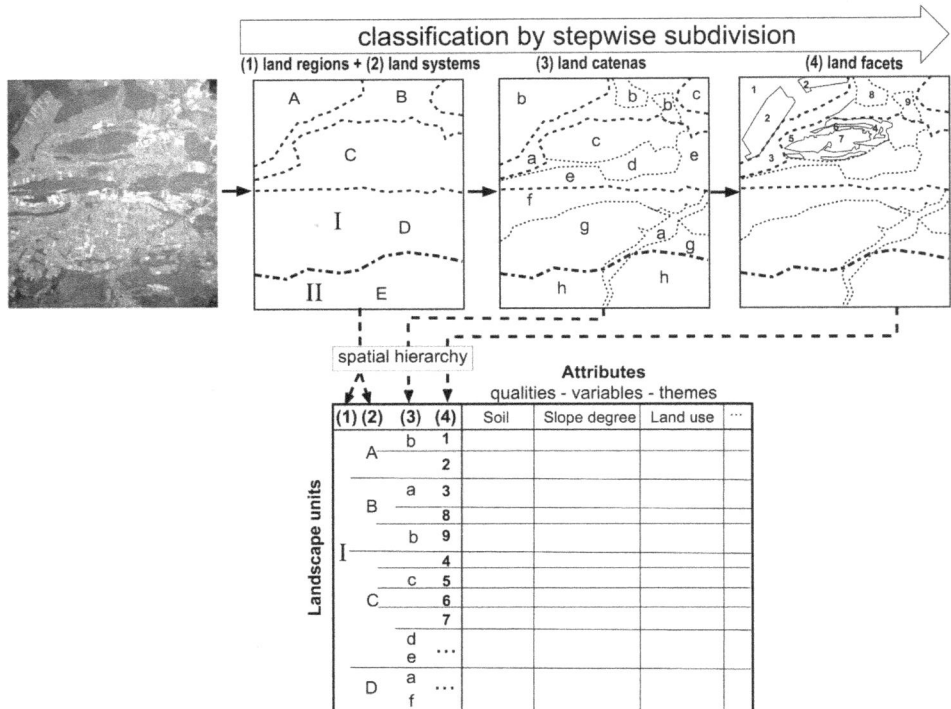

Figure 2.2 The holistic method of landscape classification by stepwise subdivision of land units. (Antrop and Van Eetvelde 2017.)

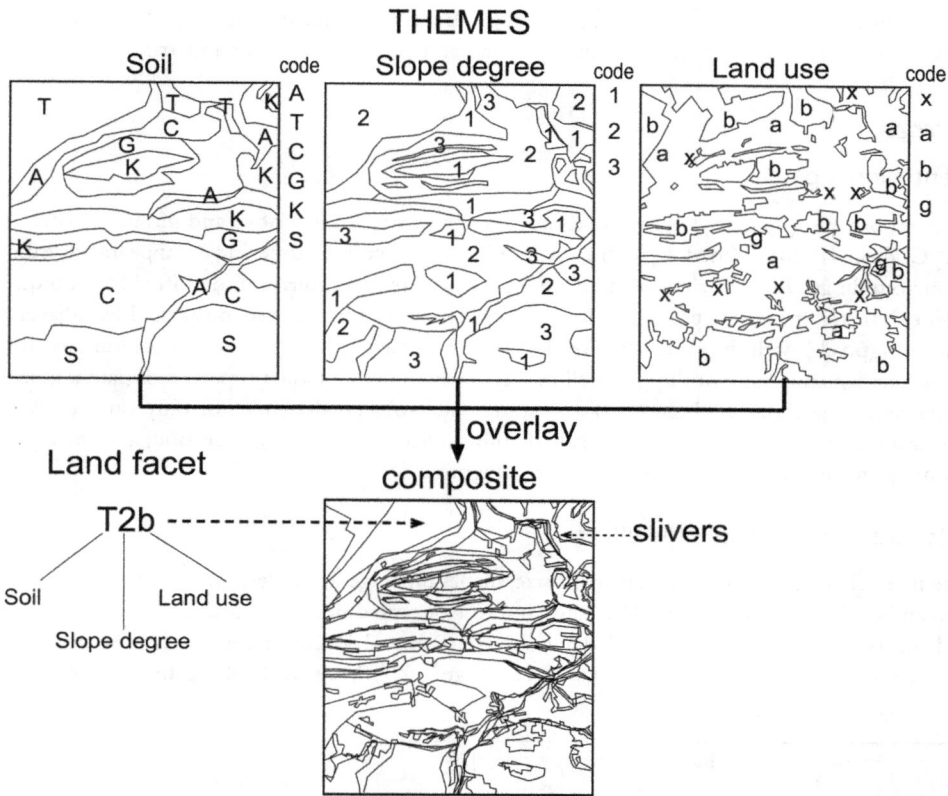

Figure 2.3 Parametric landscape classification by overlaying thematic maps to make a map composite. (Antrop and Van Eetvelde 2017.)

monothematic, such as land cover or soils, or multi-thematic when using composite or synthetic land units or landscape types. The land units are conceived as internally homogeneous and have crisp borders. In maps, they are represented as polygons and treated as choropleths (Antrop and Van Eetvelde 2017).

The patch-corridor-matrix model

Paradigm

The arrangement or structural pattern of patches, corridors, and a matrix that constitute a landscape is a major determinant of functional flows and movements through the landscape and of changes in its pattern and process over time (Cassar 2019, Forman 1995a).

Primitives in the patch-corridor-matrix model

The definition of the primitives demands *a priori* assessment of the landscape patterns according to the suitability as habitat for the species under consideration. The approach is essentially an organism-centered perspective, which will be discussed in detail further on. Essentially, the mosaic model is a multi-species variant of patches and corridors without a matrix.

Patches

Identifying a patch

With an organism-centered perspective, the concept of a patch is relative and intuitive. Forman and Godron (1986) defined a patch as 'a nonlinear surface area differing in appearance from its surroundings'. In vector maps, these are represented by polygons and identified by a unique ID-field. In raster maps, a patch is not one object but a pattern that is constituted by adjacent grid cells (pixels) with the same value or ID. When rasterizing analogue maps or gridding vector maps, the size and shape of the grid cell matter, as these define basic properties of patches: size, shape, and edge (border) (Figure 2.4). Creating patch-objects depends not only on the grain (resolution, pixel size), but also on the rule defining adjacency (the neighborhood) and the algorithm to create patterns (or image segments) (Figure 2.5).

Patch size, shape, and edge effects

The main characteristics of a patch are its size, shape, and edge. Both determine the species richness and ecological functioning. The patch interior or core differs from the patch edges. A large and compact patch has a large core/edge ratio size (Figure 2.6). Generally, the richness (number of species, diversity) increases with patch size (Figure 2.7). Very small, elongated patches with

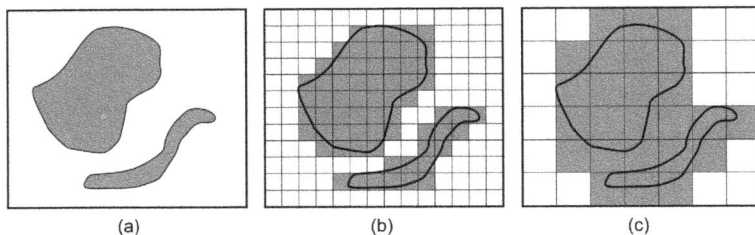

(a)   (b)   (c)

Figure 2.4   (a) Two patches in vector format; (b) these patches after rasterizing: size, shape, and edge (border) are distorted; (c) only one patch remains when the grain size is doubled.

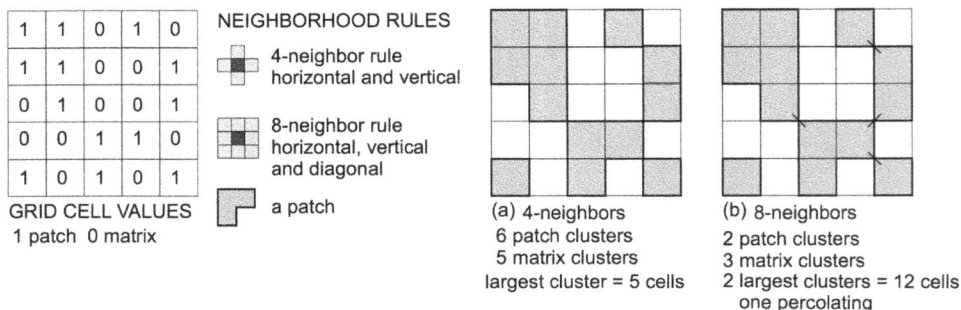

Figure 2.5   Identifying patches by clustering of grid cells according to neighborhood rules. (Turner and Gardner 2001.)

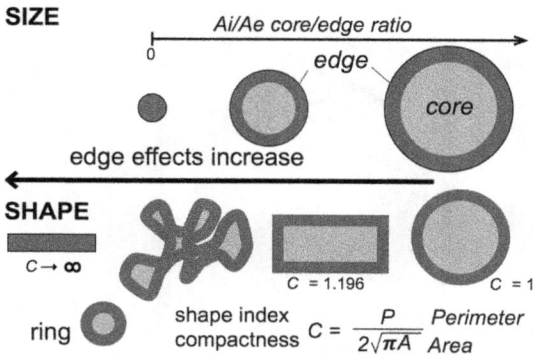

SIZE

Ai/Ae core/edge ratio

edge

core

edge effects increase

SHAPE

$C \to \infty$

$C = 1.196$

$C = 1$

ring

shape index
compactness $C = \dfrac{P}{2\sqrt{\pi A}}$ $\dfrac{Perimeter}{Area}$

Figure 2.6 Variables to describe patches: the ratio between the edge and core depends on the size (area) and shape. A shape index expresses the compactness of the patch in comparison to a circle. (Antrop and Van Eetvelde 2017.)

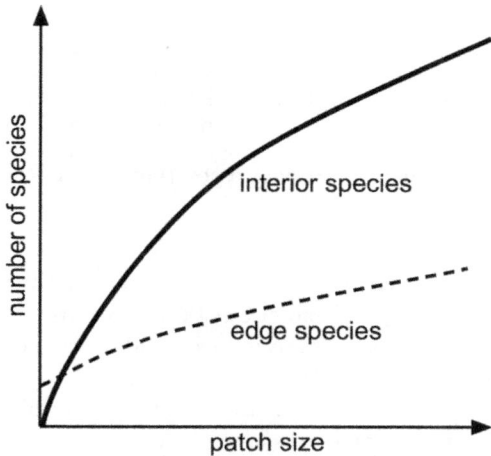

Figure 2.7 Generally, the richness of a patch increases with its size. Simultaneously, the proportion of core species increases faster than the edge species. (After Wiens 1976; Forman 1976; Forman and Godron 1986.)

very convoluted edges may even have no core. Edge effects increase when the size and compactness of the patch decrease. Edge effects also define the interaction between the patch as a spatial object and its surroundings. The border can be crisp or fuzzy in various ways. See Chapter 1 for a general discussion (see Figure 1.4).

An ecologically optimum patch shape usually has a large core with some curvilinear boundaries and narrow lobes (Forman 1995a). Patch properties proved to be important in landscape management and planning (see also Figure 1.5).

Corridors

Forman (1995a) defines a corridor as 'a strip of a particular type that differs from the adjacent land on both sides'. Based on the width of the corridor and the edge effects that go with it,

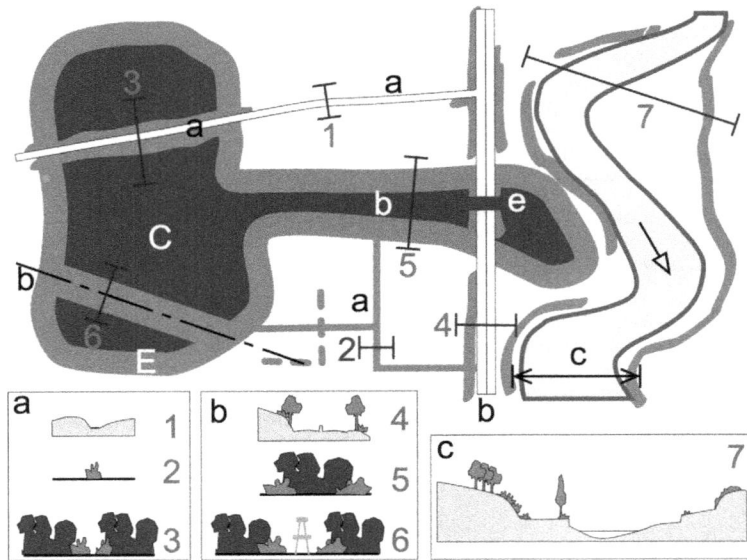

Figure 2.8 Different types of corridors: (a) linear, (b) strip corridors, (c) stream corridors. C: core, E edge, e: ecoduct. 1–7: cross-sections: 1, 3, 6 are corridors made by space bordered by vegetation, 2, 5 are corridors made by vegetation bordered by open space, 7 are stream corridors, often complexes of previous types. (Antrop and Van Eetvelde 2017.)

Forman and Godron (1981) made the distinction between *line* and *strip* corridors and added a special case, the stream corridor, where processes are oriented and directed by the water flow (Figure 2.8). Examples of corridors are hedgerows, embankments, rivers, roads, and powerlines. They may form complex, special ecosystems (Forman et al. 2003)

Burel and Baudry (2003) recognize the following corridor functions: a conduit, a barrier, a filter, and a habitat. As patches, corridors can also function as sources or sinks.

Corridors can form *ecological networks*. Networks can simultaneously connect and fragment the landscape. Policy and planning frequently recognize *blue-green infrastructure* as ecological networks to be managed (Brears 2018).

Matrix

The *matrix* is defined as the landscape feature that has the largest proportional area in the extent of the study area and that has a dissimilar structure or composition to the patches. The matrix is spatially dominant and is considered a background-patch. It has the highest connectedness, i.e. it occupies most of the space in a continuous way. See also Chapter 4.

Merriam (1984) introduced the concept of landscape connectivity, which he defined as the degree to which the landscape facilitates or impedes the movement of species among resource patches. Baudry and Merriam (1988) used the term *connectedness* to refer to spatial joining of distinct patches and used the term *connectivity* for the possible movement of organisms between patches, whether or not they are spatially connected. However, both concepts have been used often without clear distinction (Antrop and Van Eetvelde 2017).

Kindlmann and Burel (2008) gave a review of the connectivity measures. They distinguish two groups of definitions to describe connectivity. *Structural connectivity* is based on the spatial properties of the landscape structure regardless of any behavioral aspect of the organisms. Functional connectivity considers the behavioral responses of organisms to the characteristics of the patches and corridors and their spatial configuration and includes the matrix. Tichendorf and Fahrig (2000a, b, 2001) advised drawing a distinction between *landscape connectivity*, seen as a property of the entire landscape, and *patch connectivity* as a property of the patch, relevant in metapopulation ecology. See also Chapter 4.

The *percolation theory* (Gardner and O'Neill 1990) relates the connectedness of the matrix to the proportion of area occupied by patches. This proportion defines when a *matrix inversion* occurs, i.e. when the background-patch stops being the most dominant and most connected landscape type (Figure 2.9). This *percolation critical threshold* depends on the method by which adjacent cells are aggregated to define a patch (Figure 2.10). An example of a matrix inversion is the transformation of a rural landscape into an urbanized one, or a forest turning into a savannah.

Origins and theories of the landscape ecological models

An organism-centered perspective on the landscape

The organism-centered perspective on the landscape is the first fundamental paradigm of landscape ecology. John Wiens (1976) formulated it concisely as follows:

> In the real world, environments are patchy. Factors influencing the proximate physiological or behavioural state or the ultimate fitness of individuals exhibit discontinuities on many scales in time and space. The patterns of these continuities produce an environmental patchwork which exerts powerful influences on the distributions of organisms, their interactions, and their adaptations
>
> *(p. 81)*

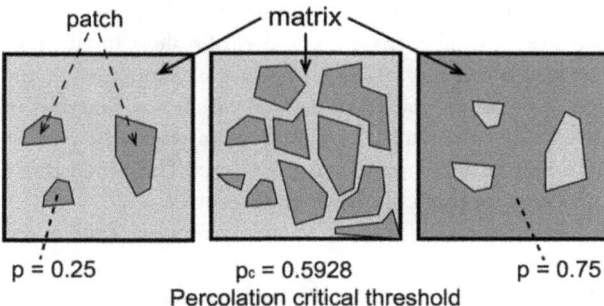

patch   matrix

p = 0.25 p_c = 0.5928   p = 0.75
Percolation critical threshold

Figure 2.9 Matrix inversion in a landscape conceived as patches in a matrix. The matrix is the landscape type with the largest coverage and connectedness. When the number of patches increases and their summed area reaches the proportion of 0.5928 (in the case of a square grid cell with a 4-neighbors clustering rule), then the aggregated patches get the largest connectedness and become the matrix. (Antrop and Van Eetvelde 2017 after Gardner and O'Neill 1990.)

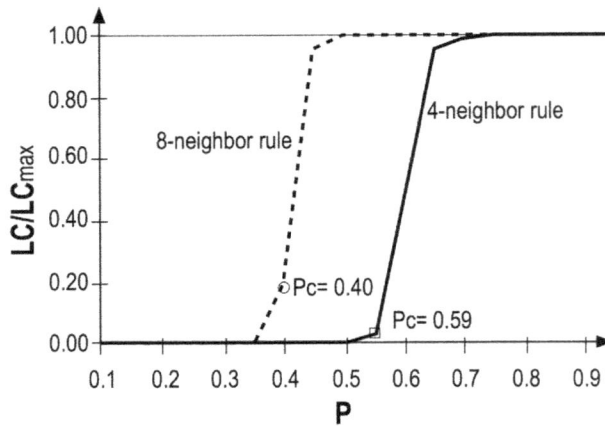

Figure 2.10 Critical percolation thresholds for a 4-neighbor and 8-neighbor clustering rule. LC/LCmax is the ratio between the cluster size to the maximal possible cluster size, and P is the summed real proportion of the clusters. Once the percolation threshold is reached, the cluster size rapidly increases to its maximal size. (After Turner and Gardner 2001.)

and

> the patch structure of an environment is that which is recognized by or relevant to the organisms under consideration. Patchiness is thus *organism-defined*, and must be considered in terms of the perceptions of the organism rather than those of the investigator.
>
> *(Wiens 1976, p. 83)*

Richard Forman defined three basic landscape elements that make up the structural components of the ecological mosaic:

> Landscapes vary from a few kilometers to several hundred kilometers in diameter yet may be considered to represent a single level of scale in space. Each level of scale is, in effect, an ecological mosaic composed of patches which vary in size, shape, origin, distinctness, number and configuration. Most mosaics contain narrow strips or lines which may function, in part, as corridors. Similarly, most mosaics contain a background type in which the patches and corridors are embedded, and which may be considered the matrix. Thus, I recognize three basic categories of landscape elements: patches, corridors and matrix.
>
> *(Forman 1982, p. 36)*

The organism-centered perspective in landscape ecology defines landscape as a tract of land containing a mosaic of different habitat patches and focusses on a particular 'target' habitat according to the species under study (Dunning et al. 1992, McGarigal 2015). As the habitat patches are defined relative to a particular organism's perception and scaling of the environment (Wiens 1976), the size of the area of interest, i.e. the landscape to be studied, depends on the spatial pattern that is meaningful to that particular organism.

An important consequence of the organism-centered perspective is that a full range of spatial scales needs to be considered when studying a region as a whole. A hierarchical multi-scale approach becomes necessary. A landscape is not defined by its size but rather, by an interacting mosaic of patches at any scale.

The interaction between landscape structure and functioning

Another essential paradigm links spatial patterns, or landscape structure, to the ongoing processes, which are considered as functions and more recently, as landscape services

Both the mosaic and the patch-corridor-matrix model are dynamic. The spatial structure carries and defines the ecological processes, while simultaneously, these processes transform and shape the spatial pattern (Figure 2.11). Therefore, a second fundamental paradigm of landscape ecology is:

> An endless feedback loop:
> Past functioning has produced today's structure;
> Today's structure produces today's functioning;
> Today's functioning will produce future structure.

(Forman and Godron 1986)

Forman and Godron recognized five disturbance mechanisms that generate five types of patches differing in dynamics and stability: (1) a spot disturbance patch, (2) a remnant patch, (3) an environmental resource patch, (4) an introduced patch, and (5) an ephemeral patch (Forman and Godron 1981, Forman 1979).

Island theory and metapopulation

The island theory in biogeography and the metapopulation theory inspired the development of the patch-corridor-matrix model. Based on biodiversity studies in islands of the Pacific Ocean

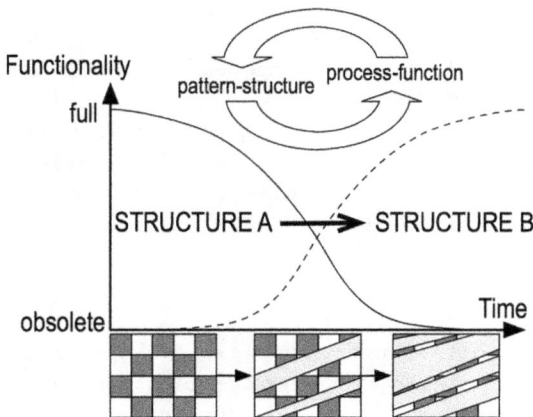

Figure 2.11 When the functionality of processes in a spatial structure decreases, the structure is gradually replaced by a new one, increasing the functionality.

by MacArthur and Wilson (1967), the island theory, in particular the Equilibrium Theory of Island Biogeography, relates species richness and the spatial distribution of organisms to distance, connectedness, and size of favorable places in an otherwise hostile environment. Patches are seen as suitable 'island' habitats for certain species and are scattered in a 'hostile' background environment, the matrix. The distance between them defines whether they can serve as 'stepping-stones' for the dispersal of organisms. Eventually, corridors may connect patches, facilitating this. The properties of the patches are related to diversity and the survival or extinction of species. Forman formulated some of these hypotheses as follows:

> For subpopulations on separate patches, the local extinction rate decreases with greater habitat quality or patch size, and recolonization increases with corridors, stepping stones, a suitable matrix habitat, or short inter-patch distance.
>
> *(Forman 1995b)*

The importance of distance

A first 'law' of geography states that 'everything is related to everything else, but near things are more related than distant things' (Tobler 1970). Essentially, it stresses the importance of distance for the effects of the surrounding 'things' on the place under consideration (Waters 2017). Likewise, the influence of a 'thing' on its environment decreases with distance. This holds for ecosystem processes and human behavior, and combined, it provides a general spatial flow principle in the landscape mosaic that is useful in planning and management (Forman 1995a). This 'law' also forms the basis of geostatistics, gravitation models, field theory, and models of dispersion and diffusion.

The map is not the landscape

Landscape ecological analysis mostly uses digital maps as representation of the landscape. All maps are human constructs, i.e. interpretations and representations of the real world from a particular perspective and on a predefined scale. They are derived from available data, which are most often imperfect because they are incomplete, outdated, or inaccurate. Map errors are a general concern in any quantitative landscape analysis, in particular when using categorical data, because of the many choices the investigator has to make during the classification, which adds uncertainty.

Besides the choices to be made in geocoding the objects and features to be represented, the landscape is often sliced into thematic layers. The choice and definition of the themes are essential for the analysis, as this determines the *thematic resolution*, which influences how finely the map classes resolve differences in the environment (McGarigal 2015).

Scale and heterogeneity

In all paradigms described here, the landscape is observed using *a vertical bird's eye perspective*. This means that the observer is somewhat disconnected from the landscape. The observation distance defines the scale by which the landscape is seen in a holistic way (Antrop and Van Eetvelde 2017). Zooming the scale in and out, which Forman and Godron (1986) called the *shuttle analysis,* allows the hierarchic structure of the landscape to be discovered. The *extent* viewed at one glance and the resolution or *grain* by which objects and details are seen define the scale. Extent and grain set the boundaries for describing the landscape. The scale is the primary factor defining the landscape heterogeneity and its information content. Thus, it also defines the meaning and significance of the elements and structures and the way they are represented on a map: for example, the scale defines whether or not zones in the mosaic are seen as heterogeneous and whether the borders between the land units are seen as crisp or fuzzy gradients.

Scale and heterogeneity are fundamental concepts, as 'landscape ecology emphasizes the interactions between spatial patterns and ecological processes, that is, the causes and consequences of spatial heterogeneity in a range of scales' (Turner et al. 2001). The distinction between composition and configuration at different scales is essential here. Both concepts have been clarified in detail in the opening chapter (see Figures 1.2 and 1.3).

Hierarchy theory and holons and ecodevices

Hierarchy theory is part of the general systems theory focusing upon the levels of organization and issues of scale in complex systems (Allen 2001). According to hierarchy theory, processes in an ecosystem are organized into discrete scales of interaction, which impose discrete spatial scales on the landscape; hence, the paradigm that the spatial hierarchy of landscape patterns allows an explanation of ecological processes.

Koestler (1970) conceived a hierarchy as a living tree composed of intermediary structures or sub-assemblies organized on a series of levels in ascending order of complexity. He called these a *holon*, characterized by a remarkable degree of autonomy or self-government. Each is a sub-assembly, which behaves towards its subordinated parts as a self-contained whole and towards its superior controls as a dependent part.

Naveh and Lieberman (2013) introduced the concept of holons in landscape ecology and see them as building blocks in the Total Human Ecosystem (THE). Holons are self-regulating subsystems that have a certain degree of freedom and autonomy. Hence, they control functions that sustain the holon in its environment. Each holon belongs to a holon on a superior level in the hierarchy and may contain embedded holons of a lower level. Essentially, each landscape and land unit is a complex open system with energy, materials, organisms, and information moving in and out.

Van Leeuwen (1982) and Van Wirdum (1982) called such an 'upkeeping or protective machinery' an *ecodevice*. To sustain a habitat, an ecodevice must be able to control the incoming and the outgoing flows. For this, it uses four basic functions: supply, disposal, resistance, and retention (Figure 2.12). The concept is used not only in ecology but also in environmental planning

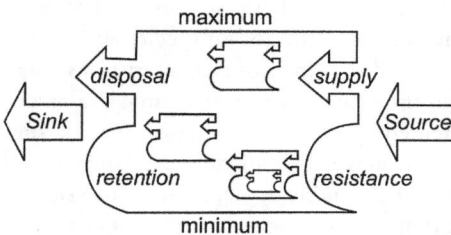

Figure 2.12 A holon as an ecodevice with its regulating functions and embedded ones at a lower hierarchical level.

(Tsjallingii 1996) and in characterizing cultural landscapes (Antrop and Van Eetvelde 2017). The most typical example of such an improved habitat-protecting device is the human house.

Disturbance: the cause of patterns, good and bad

For landscape ecologists, the cause of landscape patterns and heterogeneity was initially a matter of disturbance (Turner et al. 2001). Disturbance dynamics occur at all spatial and temporal scales and have natural and/or anthropogenic origins. They vary in frequency, intensity, and magnitude. They define landscape changes and ecological succession and make history. From the perspective of nature conservation, landscape protection, and environmental impact assessment, disturbances are often seen as disruptive and as threats to existing values (see Chapters 8, 9, and 10).

However, many of these ecological disturbances are seen differently by cultural and historical geographers and archaeologists. They see human actions and impacts as different adaptations to the environment to create sustainable lifeways (Buttimer 2001, Thomas 1956). This depends on environment capabilities, the cultural tradition, and available technology in the period of creation. In time, a broad diversity of cultural landscapes came into existence. The human impact can be 'good' and 'bad', but the assessment is essentially a matter of axiology (Antrop 2012).

The UNESCO World Heritage Convention (1995) defined 'cultural landscape' as the expression of the 'combined works of nature and man' and recognized it as a new category allowing landscapes to be listed for their 'outstanding universal value'.

Tools for analyzing mosaic and patch-corridor-matrix patterns

The geographical origins of landscape ecological models

Chorology refers to the approach in geography to making classifications of regions and landscapes. These were initially landscape mosaics at multiple scales in a hierarchical system. Later, these spatial units were applied for land classification and land evaluation, in particular for the purpose of mapping uncharted areas in developing countries (Christian and Stewart 1968). Many spatial models were developed in the 1960s for studying a broad variety of features that result from the 'circularly causal relationship between spatial structure and spatial processes' (Abler et al. 1971). Most of these models of spatial organization found applications in human and social themes, such as the influence of 'central places' on the surrounding land use and defining spheres of influence in a spatial hierarchy. Simultaneously, network analysis developed, focusing on the flow of fluxes and the diffusion of innovation through space.

In landscape ecology, spatial models were first introduced when ecologists 'upscaled' their perspective on the world to the landscape level (Wiens 1989). Early studies of vegetation patterns focused on relations between land qualities and the diffusion of species and habitat fragmentation (Phipps 1981). Landscape mosaics enabled the study of characteristics such as heterogeneity and fragmentation in landscapes and their changes in time. Later, the patch-corridor-matrix model allowed a more target-oriented approach, focusing on specific species or habitats for improving their subsistence in an increasingly disturbed urbanized and industrialized human environment. The purpose of the landscape ecological analysis was not only oriented towards an academic understanding of landscape as a system but rapidly also broadened to offer practical applications for management, conservation, planning, and policy.

How geo-information technology helped the development of landscape ecology and vice versa

Most geographical models on spatial organization from the 1960s did not deal with highly complex patterns such as the ones found in the natural world. The reasons were twofold: the lack of detailed, fine-grained maps of landscape features and the lack of computing power in the pre-computer age. Most digital map data were manually sampled from course analogue thematic maps or aerial photographs, and calculating spatial statistics was tedious (Kilchenmann 1973). The first Landsat space image was available in 1976, but image processing and recognition only became practical a decade later. The first raster-based GIS allowing spatial analysis, Tomlin's Map Analysis Package (MAP) and the concept of map algebra, dated from the mid-1980s (Tomlin 1994).

As landscape ecologists enlarged their perspective to the landscape scale only in the early 1980s, they had the 'late adopter's advantage'. At that time, the development of available digital data was rapidly expanding, as was the spatial information technology based on affordable personal computers, which offered many new opportunities that had not existed before. Landscape ecology took these opportunities to create its own specific methodology and approach based on theories and models developed earlier.

However, the transition was not sudden, as noted by John Wiens, editor-in-chief of the new journal *Landscape Ecology*, established in 1987. He made a survey of the papers published in the first five volumes of the journal (Wiens 1992) and found that three-fourths of the studies were conducted at spatial scales ranging from a few hectares to many square kilometers, i.e. the scale of human landscapes. Nearly half were concerned with landscape structure, and many focused on aspects of human land use or on the analysis of spatial patterns. He also found there only a small but well-defined emphasis on computer modelling and simulation. However, most studies were predominately descriptive or conceptual. He concluded that landscape ecology was not yet a quantitative discipline and not concerned with theory, experimenting, and hypothesis testing. He suggested that if landscape ecology were also to become an applied discipline, its empirical foundation should become quantitative rather than qualitative and its theoretical framework predictive rather than solely heuristic.

Three decades later, Wu (2015) updated the review to assess the state-of-the-art of landscape ecology again. Clearly, landscape ecology had matured at a fast pace. The research topics and had broadened and became more interdisciplinary. Multiple-scale and hierarchical approaches had become the norm. The analysis relied increasingly on remote sensing data and GIS and on quantitative spatial statistics.

Most innovative was the development of tools like landscape metrics allowing the quantitative description of landscape patterns and properties of land mosaics. They became used as indicators for heterogeneity, fragmentation, and change in the pattern–process interaction. The patch-corridor-matrix model was specifically designed to allow this novel approach.

The impact of geographical information systems

Modern landscape ecology developed simultaneously with the production of digital data, the creation of GIS for mapping, and geostatistical analysis. The properties of digital data set the requirements and possibilities for spatial analysis and visual representation of the land. This had a significant impact on the concept of landscape and on modelling in general. Basically, a GIS consists of a multiple layered database of different themes covering the same area. There are three modes of analysis. The horizontal mode analyzes the features in one data layer, looking for topological order. The vertical mode looks for relations between data layers at the same location to assess ecological order or coherence (Phipps 1984). If data layers refer to the same theme at different time moments, the vertical comparison becomes a multitemporal analysis. This is achieved by map overlaying, and tools for map algebra were developed early on. Exemplary are the 'Map Analysis Package' by Tomlin in 1988 (see Tomlin 1994), the 'Principles of Geographical Information Systems for Land Resources Assessment' by Burrough in 1986 (see Burrough and McDonnell 1998), and the application of information theory for the analysis of thematical maps and field sampling by Kilchenmann (1973) and Phipps (1981).

The early digital representation of the landscape was a raster image. Each pixel became a sample of the landscape. The pixel was the ultimate spatial information unit, determined by its size or resolution, and containing indirect information on the landscape properties, mainly defined spectral signatures of the land cover. Landscape units had to be constructed from the pixel patterns. Essentially, this was creating patches in categorical raster maps. Spatial analysis consisted in the description of geometrical properties of these patches (size, shape, density, distribution, connectedness, etc.).

When vector graphic displays became available in 1958, cartographic mapping rapidly shifted to scalable vector graphics offering new possibilities for spatial modelling. The first vector-GIS emerged in the mid-1970s. A milestone was Esri's ARC/INFO for microcomputers in 1982 and for PC in 1986.

Although landscape ecological modelling and analysis parallels the development of desktop computing and mapping, spatial analysis, and geostatistics, the specific tools for landscape ecology, such as *landscape metrics* and landscape indicators, emerged in the 1990s.

Landscape metrics and indicators

A breakthrough: the FRAGSTATS package

An important moment was when the software package FRAGSTATS (McGarigal and Marks 1995) became available. The package was designed for analyzing spatial patterns on categorical raster maps. Four levels were proposed for analyzing landscape patterns: cell-level, patch-level, class-level, and landscape-level. Each level results in a specific set of landscape metrics, many of which are correlated.

Cell-level metrics are used in categorical raster maps representing a continuous theme or geographical surface. They allow for description of the characteristics of the spatial context or neighborhood of the resolution cells, i.e. the grain, in relation to dispersal, movement, and percolation.

Patch-level metrics are calculated for individual patches of a specific type. They allow measurements of characteristics related to size, shape, and core/edge properties.

Class-level metrics aggregate all patches of a given type (class). This can be done by summing and averaging properties and studying their statistical and spatial distributions. Class metrics quantify the density and spatial configuration of each patch type. These metrics are considered significant to study habitat fragmentation, which leads to changes in functions and reduces biodiversity.

Landscape-level metrics integrate all patch types or classes over the full extent of the landscape studied. Essentially, they describe the characteristics of the landscape mosaic.

These focus on pattern properties such as composition and configuration and express meta-properties of the landscape such as landscape diversity, heterogeneity, fragmentation, contagion, resistance, and disturbance.

Other packages

More recently, new software packages for calculating landscape metrics have become available, such as the open-source R tool for raster maps (Hesselbarth et al. 2019). Also, packages focusing on specific tasks have become available, such as PolyFrag for calculating fragmentation in vector-based maps (MacLean and Congalton 2013) and ZonalMetrics for calculating landscape metrics in user-defined zones (Adamczyk and Tiede 2017). Many are used as plug-ins in the ArcInfo software (Rempel et al. 2012). The most important innovation is the shift from grid-based analytics, which originated with raster imagery from remote sensing, to vector-based analytics.

Using landscape metrics

Uuemaa et al. (2009) performed a review of the evolution and applications of landscape metrics between 1994 and 2008 based on international peer-reviewed articles. This showed that the term 'landscape metrics' was mainly used in analyses using the FRAGSTATS software or similar programs. The term 'landscape indices' was used in a broader sense and was named differently, although similar formulae were used (Figure 2.13). The term 'landscape indices' was more often used in the 1990s, while now, the term 'landscape metrics' is prevalent.

Landscape metrics/indicators are used in broad domains, including, in diminishing order of importance, biodiversity and habitat analysis, assessment of water quality, the evaluation of landscape pattern and change, urbanization processes, road network quantification, measurement of landscape aesthetics, and landscape management, planning, and monitoring

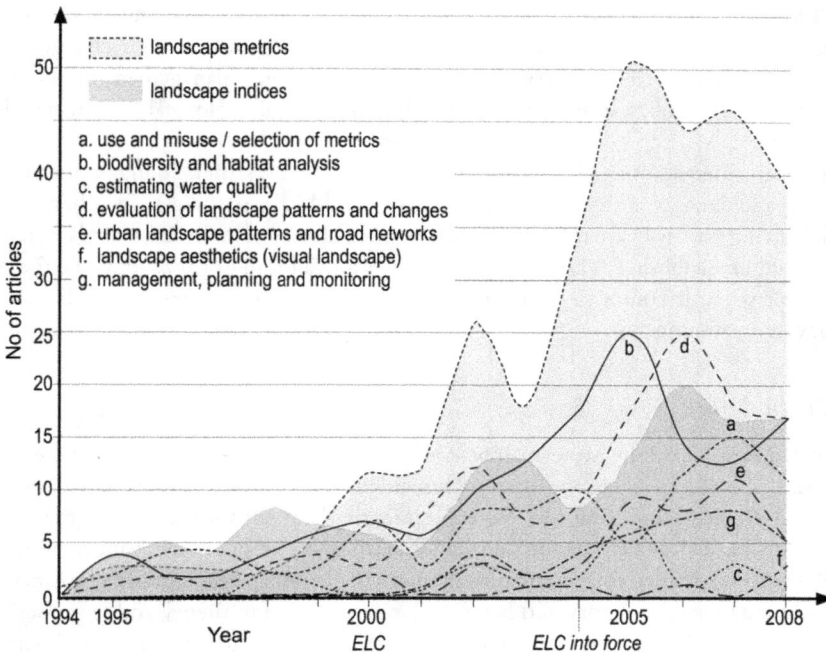

Figure 2.13 Use and applications of landscape metrics and indices according to peer-reviewed publications from 1994 to 2008. (After Uuemaa et al. 2009, 2013.)

(Uuemaa et al. 2013). Since the European Landscape Convention, landscape research has boomed in general, and so has the use of landscape metrics (Figure 2.13). Landscape metrics is one of the key topics in landscape ecological research in both academic and applied research. However, a decreasing trend can be noted since 2005 in the number of publications on this topic.

Most publications discussing landscape metrics/indices deal with biodiversity and habitat analysis, landscape assessment, and change. Social aspects, the visual landscape, and landscape perception are less frequent. After the hype of the introduction and success in relating pattern and processes, the focus shifted somewhat to more critical discussions about the use and misuse of landscape metrics and their real meaning and significance (Haines-Young and Chopping 1996, Li and Wu 2004, 2007). The high expectations were not realized in particular in the field of making policy-relevant indicators, analytically sound and allowing transparent interpretation (Parris 2003, Dramstad and Sogge 2003). The lack of absolute thresholds and difficulties in comparing cases over a variety of scales and geographical regions hampered their use. Many important aspects of the landscape are not quantifiable, and the existing metrics are not always viable and are difficult to interpret, causing uncertainty in decision making.

Conclusions

Strengths and weaknesses of the models

The premises for developing the patch-corridor-matrix model are soundly based in ecological theories. It is a coherent model specifically developed for landscape ecology and spatial pattern analysis, in particular for target-oriented research. Specific tools and landscape metrics have been developed for patch and corridor analysis. In landscape ecology, landscape mosaic models are used to study characteristics such as heterogeneity and fragmentation and their changes in time. Applications of patch-corridor-matrix are found in planning and management of nature conservation and landscape policy, in particular in the development of ecological networks and green-blue corridors.

Although landscape metrics opened new and interesting perspectives, their abundant use also showed limitations and dangers in their use. Causal relationships between metrics and changing patterns and the driving processes are difficult to establish. No absolute thresholds or universally applicable laws could be established. Their abstract character and a lot of uncertainty make their explanation difficult when it comes to formulating reliable and transparent indicators that are relevant in policy and planning.

Challenges for the future

Early studies in landscape ecology were mainly concerned with the description of patterns and processes. Also, the use of landscape metrics remained mainly descriptive. Wiens (1992) already emphasized the need for a shift towards experimenting and statistical testing of predictive hypotheses, allowing better explanation and understanding of the processes as mechanisms of change. O'Neill (2005) defined four theoretical domains the landscape ecological explanation should be based upon: hierarchy theory and landscape scale, percolation theory and hypothesis testing, spatial population theory, and economic geography with links to central place theory and location theory.

The success of landscape ecological models initiated new applications in other research domains. As such, landscape ecology became an 'umbrella' for interdisciplinary and transdisciplinary landscape research (Zonneveld 1995, Naveh 1999) and contributed to bridging natural and social sciences dealing with the landscape (Bastian 2001, Opdam et al. 2001, Wiens 2005, Farina 2009). Naveh (2007) advocated that landscape ecology should become more proactive to increase its social relevance and to face the challenges for the future.

Bibliography

Abler, R.F., Adams, J.S. and Gould, P. (1971) *Spatial Organization: The Geographer's View of the World*. Prentice-Hall, Englewood Cliffs, NJ

Adamczyk, J. and Tiede, D. (2017) 'ZonalMetrics: a Python toolbox for calculating landscape metrics in user defined zones', *Computers and Geosciences*, vol 99, S91–99.

Allen, T.F.H. (2001) 'A summary of the principles of hierarchy theory', https://web.archive.org/web/20011218001638/http://isss.org/hierarchy.htm, accessed 17 Dec 2020.

Antrop, M. (2012) 'Intrinsic values of landscapes', in T. Papayannis and P. Howard (eds) *Reclaiming the Greek Landscape*. Med-INA, Athens.

Antrop, M. and Van Eetvelde, V. (2017) *Landscape Perspectives: The Holistic Nature of Landscape*. Springer, Heidelberg.

Bastian, O. (2001) 'Landscape ecology: towards a unified discipline?', *Landscape Ecology*, vol 6, pp757–766

Baudry, J. and Merriam, H.G. (1988) 'Connectivity and connectedness: functional versus structural patterns in landscapes', in K. F. Schreiber (ed) 'Connectivity in Landscape Ecology', Proceedings of the 2nd International Association for Landscape Ecology. Münstersche Geographische Arbeiten, 29. Ferdinand Schöningh, Paderborn.

Brears, R.C. (2018) *Blue and Green Cities: The Role of Blue-Green Infrastructure in Managing Urban Water Resources*. Palgrave Macmillan, London.

Burel, F. and Baudry, J. (2003) *Landscape Ecology: Concepts, Methods and Applications*. Science Publishers, Inc., Enfield, New Hampshire.

Burrough P.A. and McDonnell R.A. (1998) *Principles of Geographical Information Systems*. Oxford University Press, Oxford.

Buttimer A. (ed) (2001) *Sustainable Landscapes and Lifeways: Scale and Appropriateness*. Cork University Press, Cork.

Cassar, L.F. (2019) 'Landscape and ecology: the need for a holistic approach to the conservation of habitats and biota', in P. Howard, I. Thompson, E. Waterton and M. Atha (eds) *The Routledge Companion to Landscape Studies*. Routledge, London and New York.

Christian, C.S. and Stewart, G.A. (1968) 'Methodology of integrated surveys', in Proceedings of the Conference on Aerial Surveys and Integrated Studies. UNESCO, Toulouse.

Cosgrove, D. (2002) 'Landscape and the European sense of sight: eyeing nature', in K. Anderson, M. Domosh, S. Pile and N. Thrift (eds) *Handbook of Cultural Geography*. SAGE Publications, London.

Cosgrove, D. (2003) 'Landscape: ecology and semiosis', in H. Palang and G. Fry (2003) *Landscape Interfaces. Cultural Heritage in Changing Landscapes*. Kluwer Academic Publishers, Dordrecht.

Cosgrove, D. (2004) 'Landscape and landschaft. Lecture at the "spatial turn in history" symposium german historical institute', *GHI Bulletin*, vol 35, pp57–71.

Council of Europe (2000) 'European landscape convention. Firenze, 20.10.2000', http://conventions.coe.int/Treaty/en/Treaties/Html/176.htm, accessed 17 Dec 2020.

De Veer A. and Burrough P. (1978) 'Physiographic landscape mapping in the Netherlands', *Landscape Planning*, vol 5, pp45–62.

Diamond, J.M. (1975) 'The island dilemma: lessons of modern biogeographic studies for the design of natural reserves', *Biological Conservation*, vol 7, pp129–146.

Donaldson, L., Wilson, R.J. and Maclean, I.M.D. (2017) 'Old concepts, new challenges: adapting landscape-scale conservation to the twenty-first century', *Biodiversity and Conservation*, vol 26, pp527–552.

Dramstad, W. and Sogge, C. (eds) (2003) 'Agricultural impacts on landscapes; developing indicators for policy analysis', in Proceedings from the NIJOS/OECD Expert Meeting on Agricultural Landscape Indicators in Oslo, Norway, October 7-9, 2002.

Dunning, J.B., Danielson, B.J. and Pulliam H.R. (1992) 'Ecological processes that affect populations in complex landscapes', *Oikos*, vol 65, no 1, pp169–175.

Eckert, M. (1907) 'Die Kartographie als Wissenschaft', *Zeitschrift der Gesellschaft für Erdkunde zu Berlin*, vol 8, pp539–55.

Fahrig, L. (2003) 'Effects of habitat fragmentation on biodiversity', *Annual Review of Ecology, Evolution, and Systematics*, vol 34, pp487–515.

Fahrig, L. (2005) 'When is a landscape perspective important?', in J.A. Wiens and M.R. Moss (eds) *Issues in Landscape Ecology*. Cambridge University Press, Cambridge.

Farina, A. (2006) *Principles and Methods in Landscape Ecology*. Springer, Dordrecht.

Farina, A. (2009) *Ecology, Cognition and Landscape: Linking Natural and Social Systems*, Springer Science & Business Media, Dordrecht.

Farina, A. and Belgrano, A. (2006) 'The eco-field hypothesis: toward a cognitive landscape', *Landscape Ecology*, vol 21, no 1, pp5–17.

Forman, R.T.T (1979) 'The Pine Barrens of New Jersey: an ecological mosaic', in R.T.T. Forman (ed) *Pine Barrens: Ecosystem and Landscape*. Academic Press, New York.

Forman, R.T.T. (1982) 'Integrating among landscape elements: a core of landscape ecology', in S.P. Tsjallingii and A.A. De Veer (eds), *Perspectives in Landscape Ecology, Proceedings of the International Congress of the Netherlands Society of Landscape Ecology*. PUDOC, Wageningen.

Forman, R.T.T. (1995a) *Land Mosaics: The Ecology of Landscapes and Regions*. Cambridge University Press, Cambridge.

Forman, R.T.T. (1995b) 'Some general principles of landscape and regional ecology', *Landscape Ecology*, vol 10, no 3, pp133–142.

Forman, R.T.T. and Godron, M. (1981) 'Patches and structural components for a landscape ecology', *BioScience*, vol 31, no 10, pp733–740.

Forman, R.T.T. and Godron, M. (1986) *Landscape Ecology*. Wiley, New York.

Forman, R.T.T., Galli, A.E. and Leck, C.F. (1976) 'Forest size and avian diversity in New Jersey woodlots with some landuse implications', *Oecologia*, vol 26, pp1–8.

Forman, R.T.T., *et al*. (ed) (2003) *Road Ecology: Science and Solutions*. Island Press, Washington.

Gardner, R.H. and O'Neill, R.V. (1990) 'Pattern, process and predictability: the use of neutral models for landscape analysis', in M.G. Turner and R.H. Gardner (eds) *Quantitative Methods in Landscape Ecology*. Springer, New York.

Granö, J.G. (1929) ,Reine Geographie. Eine methodologische Studie beleuchtet mit Beispielen aus Finnland und Estland. Helsinki, Helsingfors', *Acta Geographica*, vol 2, no 2, pp202.

Granö, O. and Paasi, A. (eds) (1997) 'Pure Geography', J.G. Granö 1929, translated by M.Hicks. The John Hopkins University Press, Baltimore and London.

Haines-Young, R. and Chopping, M. (1996) 'Quantifying landscape structure: a review of landscape indices and their application to forested landscapes', *Progress in Physical Geography*, vol 20, no 4, pp418–445.

Hesselbarth, M.H.K., Sciaini, M., With, K.A., Wiegand, K. and Nowosad, J. (2019) 'Landscape metrics: an opensource R tool to calculate landscape metrics', *Ecography*, vol 42, pp1648–1657.

Kemp, K. (ed) (2007) *Encyclopedia of Geographic Information Science*. SAGE Publications, London.

Kilchenmann, A. (1973) ,Die Merkmalanayse für Nominaldaten: eine Methode zur Analyse von Qualitativen geographischen Daten' *Geoforum*, vol 15, pp33–45.

Kindlmann, P. and Burel, F. (2008) 'Connectivity measures: a review', *Landscape Ecology*, vol 23, pp879–890.

King, A.W. (2005) 'Hierarchy theory and the landscape … level? Or, Words do matter', in J.A. Wiens and M.R. Moss (eds) *Issues in Landscape Ecology*. Cambridge University Press, Cambridge.

Koestler, A. (1970) 'Beyond atomism and holism: the concept of the holon', *Perspectives in Biology and Medicine*, vol 13, no 2, pp131–154.

Li, H. and Wu, J. (2004) 'Use and misuse of landscape metrics', *Landscape Ecology*, vol 19, pp389–399.

Li, H. and Wu, J. (2007) 'Landscape pattern analysis: key issues and challenges', in J. Wu and R.J. Hobbs (eds) *Key Topics in Landscape Ecology*. Cambridge University Press, Cambridge.

Longley, P.A., Goodchild, M.F., Maguire, D.J. and Rhind, D.W. (eds) (1999) *Geographical Information Systems: Principles, Techniques, Management and Applications*. Wiley, New York.

Lynch, K. (1960) *The Image of the City*. The MIT Press, Cambridge.

MacArthur, R.H. and Wilson, E.O. (1967) *The Theory of Island Biogeography*. Princeton University Press, Princeton, NJ.

Mander, Ü. and Murka, M. (2003) 'Landscape coherence: a new criterion for evaluating impacts of land use changes', in U. Mander and M. Antrop (eds) *Multifunctional Landscapes Vol. III: Continuity and Change*. WIT Press, Southampton.

Mander, Ü., Uuemaa, E., Roosaare, J., Aunap, R. and Antrop, M. (2010) 'Coherence and fragmentation of landscape patterns as characterized by correlograms: a case study of Estonia', *Landscape and Urban Planning*, vol 94, no 1, pp31–37.

McGarigal, K. (2015) 'FRAGSTATS Help. Department of environmental conservation University of Massachusetts, Amherst', https://www.umass.edu/landeco/research/fragstats/documents/fragstats.help .4.2.pdf, accessed 17 Dec 2020.

McGarigal, K. and Marks B.J. (1995) 'FRAGSTATS: spatial pattern analysis program for quantifying landscape structure', *USDA Forest Service General Technical Report*, PNW-351.

MacLean, M.G. and Congalton, R.G. (2013) 'PolyFrag: a vector-based program for computing landscape metrics', *GIScience & Remote Sensing*, vol 50, no 6, pp591–603.

Merriam, G. (1984) 'Connectivity: a fundamental ecological characteristic of landscape pattern', in J. Brandt and P. Agger (eds), *Proceedings of the 1st International Seminar on Methodology in Landscape Ecological Research and Planning*. Roskilde University, Roskilde.

Milne, B.T. (1990) 'Lessons from applying fractal models to landscape patterns', in M.G. Turner and R.H. Gardner (eds) *Quantitative Methods in Landscape Ecology*. Springer-Verlag, New York.

Naveh, Z. (1999) 'Toward a transdisciplinary landscape science', in J.A. Wiens and M.R. Moss (eds) *Issues in Landscape Ecology*. Cambridge University Press, Cambridge.

Naveh, Z. (2007) *Transdisciplinary Challenges in landscape ecology and restoration ecology: An Anthology*. Springer, Dordrecht.

Naveh, Z. and Lieberman, A.S. (2013) *Landscape Ecology: Theory and Application*, 2nd edition Springer, New York.

O'Neill, R.V. (2005) 'Theory in landscape ecology', in J.A. Wiens and M.R. Moss (eds), *PART II: Theory, Experiments, and Models in Landscape Ecology*. Cambridge University Press, Cambridge.

Opdam, P. (1991) 'Metapopulation theory and habitat fragmentation: a review of holarctic breeding bird studies', *Landscape Ecology*, vol 5, no 2, pp93–106.

Opdam, P. (2006) 'Ecosystem networks: a spatial concept for integrative research and planning of landscapes', in B. Tress, G. Tress, G. Fry and P. Opdam (eds) *From Landscape Research to Landscape Planning: Aspects of Integration, Education and Application*. Wageningen UR Frontis Series, vol. 12, 51–65, http://library.wur.n l/frontis/landscape_research/04_opdam.pdf.

Opdam, P., Foppen, R. and Vos, C. (2001) 'Bridging the gap between ecology and spatial planning in landscape ecology', *Landscape Ecology*, vol 16, no 8, pp767–779.

Opdam, P., Verboom, J. and Pouwels, R. (2003) 'Landscape cohesion: an index for the conservation potential of landscapes for biodiversity', *Landscape Ecology*, vol 18, no 2, pp113–126.

Openshaw, S. (1984) *The Modifiable Areal Unit Problem. Concepts and Techniques in Modern Geography*, vol. 38. GeoBooks, Norwich.

Openshaw, S. and Taylor, P. (1979) 'A million or so correlation coefficients: three experiments on the modifiable areal unit problem', in N. Wingley (ed.) *Statistical Methods in the Spatial Sciences*. Pion, London.

Parris, K. (2003) 'Agricultural landscape indicators in the context of the OECD work on Agri-10 environmental indicators', in W. Dramstad and C. Sogge (eds) Agricultural Impacts on Landscapes; Developing Indicators for Policy Analysis. Proceedings from the NIJOS/OECD Expert Meeting on Agricultural Landscape Indicators in Oslo, Norway, October 7-9, 2002, pp10–18.

Phipps, M. (1981) 'Information theory and landscape analysis', in S. Tjallingii and A.A. de Veer (eds) Perspectives in Landscape Ecology: Proceedings of International Congress organized by Netherlands Society for Landscape Ecology, Veldhoven, the Netherlands, April 6-11, 1981. Pudoc, Wageningen.

Phipps, M. (1984) 'Rural landscape dynamics: the illustration of some concepts', in J. Brandt and P. Agger (eds) Methodology in Landscape Research and Planning. Proceedings of the 1st International Seminar of the International Association of Landscape Ecology, vol. I. Roskilde University Centre, Roskilde, October 15–19, 1984, pp47–54.

Rempel, R.S., Kaukinen, D. and Carr, A.P. (2012) *Patch Analyst and Patch Grid*. Ontario Ministry of Natural Resources, Centre for Northern Forest Ecosystem Research, Thunder Bay.

Shannon, C.E. (1948) 'A mathematical theory of communication', *Bell System Technical Journal*, vol 27, no 3, pp379–423 & 623–656.

Thomas, W.L. (ed) (1956) *Man's Role in the Changing Face of the Earth*. The University of Chicago Press, Chicago.

Tischendorf, L. and Fahrig, L. (2000a) 'How should we measure landscape connectivity?', *Landscape Ecology*, vol 15, pp633–641.

Tischendorf, L. and Fahrig, L. (2000b) 'On the usage and measurement of landscape connectivity', *Oikos*, vol 90, pp7–19.

Tischendorf, L. and Fahrig, L. (2001) 'On the use of connectivity measures in spatial ecology: a reply', *Oikos*, vol 95, pp152–155.

Tobler, W. (1970) 'A computer movie simulating urban growth in the Detroit region', *Economic Geography*, vol 46(Supplement), pp234–240.

Tomlin, C.D. (1994) 'Map algebra: one perspective', *Landscape and Urban Planning*, vol 30, pp3–12.

Troll, C. (1939) *Luftbildforschung und Landeskundige Forschung. Erdkundliches Wissen.* Schriftenreihe für Forschung und Praxis, Heft 12. F. Steiner Verlag, Wiesbaden.

Tsjallingii, S. (1996) *Ecological Conditions. Strategies and Structures in Environmental Planning.* IBN Scientific Contributions, vol. 2. IBN-DLO, Wageningen.

Turner, M.G. and Gardner, R.H. (2001) *Landscape Ecology in Theory and Practice, Pattern and Process.* Springer, New York.

Turner S.J., O'Neill R.V., Gardner R.H. and Milne B.T. (1991) 'Effects of changing spatial scale on the analysis of landscape pattern', *Landscape Ecology*, vol 3, pp153–162.

Unwin D. (1981) *Introductory Spatial Analysis.* Methuen, London.

Uuemaa, E., Roosaare, J., Kanal, A. and Mander, Ü. (2008) 'Spatial correlograms of soil cover as an indicator of landscape heterogeneity', *Ecological Indicators*, vol 8, no 6, pp783–794.

Uuemaa, E., Antrop, M., Jüri, R., Riho, M. and Mander, Ü. (2009) 'Landscape metrics and indices: an overview of their use in landscape research', *Living Reviews in Landscape Research*, vol 3, no 1, pp1–28.

Uuemaa, E., Mander, Ü. and Marja, R. (2013) 'Trends in the use of landscape spatial metrics as landscape indicators: a review', *Ecological Indicators*, vol 28, pp100–106.

Van Leeuwen, C.G. (1982) 'From ecosystem to ecodevice', in S.P. Tsjalingii and A.A. de Veer, (eds) Perspectives in Landscape Ecology: Contributions to Research, Planning and Management of Our Environment, Proceedings of the International Congress Organized by the Netherlands Society for Landscape Ecology, Veldhoven, the Netherlands, April 6-11, 1981, pp29–34.

Van Mansvelt, J.D. (1997) 'An interdisciplinary approach to integrate a range of agro-landscape values as proposed by representatives of various disciplines', *Agriculture, Ecosystems and Environment*, vol 63, pp233–250.

Van Wirdum, G. (1982) 'Design for a land ecological survey of nature protection', in S.P. Tsjalingii and A.A. de Veer (eds) Perspectives in Landscape Ecology: Contributions to Research, Planning and Management of Our Environment, Proceedings of the International Congress organized by the Netherlands Society for Landscape Ecology, Veldhoven, the Netherlands, April 6-11, 1981, pp245–252.

Vranken, I., Baudry, J., Aubinet, M., Visser, M., and Bogaert, J. (2014) 'A review on the use of entropy in landscape ecology: heterogeneity, unpredictability, scale dependence and their links with thermodynamics', *Landscape Ecology*, vol 30, pp51–65.

Waters, N. (2017) 'Tobler's first law of geography', in D. Richardson, N Castree, M.F. Goodchild, A. Kobayashi, W. Liu and R.A. Marston (eds) *The International Encyclopedia of Geography.* Wiley, Hoboken.

Wiens, J.A. (1976) 'Population responses to patchy environments', *Annual Review of Ecology and Systematics*, vol 7, pp81–120.

Wiens, J.A. (1989) 'Spatial scaling in ecology', *Functional Ecology*, vol 3, no 4, pp385–397.

Wiens, J.A. (1992) 'What is landscape ecology really?', *Landscape Ecology*, vol 7, pp149–150.

Wiens, J.A. (2005) 'Toward a unified landscape ecology', in J.A. Wiens and M.R. Moss (eds) *Issues in Landscape Ecology.* Cambridge University Press, Cambridge.

Wu, J. (2015) 'Key concepts and research topics in landscape ecology revisited: 30 years after the Allerton Park workshop', *Landscape Ecology*, vol 28, pp1–11.

Zonneveld, I.S. (1982) 'Land(scape) ecology, a science or a state of mind', in S.P. Tsjallingii and A.A. de Veer (eds) Perspectives in Landscape Ecology. Proceedings of the Intern. Congress of the Netherlands Society for Landscape Ecology, Centre for Agricultural Publishing and Documentation, Wageningen, pp9–15.

Zonneveld, I.S. (1995) *Land Ecology: An Introduction to Landscape Ecology as a Base for Land Evaluation. Land Management and Conservation.* SPB Academic Publishing, Amsterdam.

Zonneveld, I.S. (2005) 'The land unit as a black box: a Pandora's box?', in J.A. Wiens and M.R. Moss (eds) *Issues in Landscape Ecology.* Cambridge University Press, Cambridge.

3

Scale and hierarchy in landscape ecology

James D.A. Millington

Introduction

What is the landscape of a beetle? How does it differ from the landscape of a white-tailed deer? When is vegetation in a landscape considered fragmented? How do we identify and understand feedbacks between vegetation, physical substrate, and ecological disturbance? Underlying these questions are issues of *scale*, whether we are considering how the beetle and deer perceive their landscapes, how ecologists observe the beetle and deer to investigate how they interact with the world around them, or how ecologists use data collected at one scale to understand processes operating at another. Although scale refers to the physical dimensions of the world, organisms (including humans) perceive and respond to patterns and processes differently. Consequently, it is vital that when describing or investigating landscapes and the environment, we think carefully about our scales of observation and the range of scales that are of particular relevance to the organisms and processes being examined (Wiens 1989; Levin 1992). The emphasis throughout this chapter is on spatial dimensions because of the close links between spatial ecology and landscape ecology, but considerations of scale are equally important in the temporal dimension, and the two are often related and interact to shape ecological systems (Wolkovich et al. 2014; Ryo et al. 2019). The chapter starts with some definitions, first considering components of scale and then examining the differences between 'scale' and hierarchical 'level'. Issues of scale selection and scale dependence are then discussed before the challenges and opportunities of scaling and multi-scale analysis are outlined.

What is 'scale'?

Components of scale

A primary distinction to be made when thinking about scale is the conceptual difference between the *intrinsic* scales and the *imposed* scales of an object, pattern, or process (Wu and Li 2006). The difference enables us to consider the importance of observation and measurement; intrinsic scales might be thought of as independent of observation or measurement, whereas imposed scales depend on – and vary with – outside observers, measurement devices, and study objectives. Intrinsic scales are thus inherent in a phenomenon, characterizing their behavior and dynamics (e.g. typical spatial extent or event frequency). We use scale in the plural here, as multiple scales are often important for a phenomenon of interest independently of any observation of it. In contrast to intrinsic scales, imposed scales are due to the observer rather than the phenomenon being observed itself. However, for all organisms other than humans, these imposed scales are determined in some way by the intrinsic scale of the observer as discussed below. For example, the phenomena a beetle can sense through its antennae on the ground are very different from those that an eagle soaring in the sky can observe with its eyes. However, humans have developed ways to observe phenomena at a variety of imposed scales and independently of our own intrinsic (human) scale. For example, microscopes allow humans to see things that are far smaller than we would otherwise be able to see given our intrinsic human scale (i.e. the properties of our eyes and visual cognition), and aerial photography allows us to view the world in ways that we would not otherwise be able to (contributing to the establishment of landscape ecology itself; see Chapters 1 and 12). Thus, the terms *observational* scale, *experimental* scale, *analysis* scale, and *policy* scale are usually used directly in association with human investigation and can be independent of the intrinsic scale of the phenomenon being studied. All are closely related, and it could be argued that observational scale – the scale at which measurements or samples are made – provides the basis for all the other imposed scale types (experiments, analysis, and policy all relying in some way on observation).

Imposed (e.g. observational) scales have a variety of components, including *cartographic scale*, *grain*, and *extent* (Wu and Li 2006; Wu 2007). The *cartographic scale* component is possibly the most widely understood in general terms and refers to the ratio of the size of an object in the empirical world to the size of the object as represented on the map. In cartographic terms, 'small scale' means many units in the real world for every unit on the map, and vice versa for large scale. Thus, a map with scale 1:1,000,000 (1 cm on the map represents 1,000,000 cm – or 10 km – on the Earth) is smaller scale than a map with scale 1:1000 (1 cm on the map represents 10 m on the Earth). Correspondingly, this nomenclature means that for the same size of map (e.g. printed on paper 1 m square), 'small-scale' maps provide less detail than 'large-scale' maps, but the area they represent is greater. These cartographic scale terms refer to the detail *of the map*; the map is a representation of data about the landscape with the cartographic scale determining what can be represented on it. In contrast, for landscape ecologists, scale terms are usually used to refer directly to *the data themselves* meaning that an ecologist's 'small-scale' data would be plotted on a 'large-scale' cartographic map.

In landscape ecology, the scale components of *grain* and *extent* are more important than the scale elements of a map. Grain refers to the resolution of data, which in turn, is the smallest unit (whether spatial or temporal) within which measured values are assumed to be homogeneous (i.e. constant). For example, even though rainfall may vary during an hour, if the grain of a time series is 1 hour, then each measurement will represent an entire hour (e.g. total rainfall during an hour). Extent refers to the total measurement area (space) or duration (time) of a data set. Just as the cartographic scale determines what is represented on a map (in terms of detail and area), the grain and extent of data influence what is observed and what information can be gained. As demonstrated in Figure 3.1, landscape ecologists generally use 'fine scale' to refer to data that are collected at fine (or, high) resolution (i.e. small grain, often over a small extent) and 'broad scale' to refer to data that are collected over a large extent (often with large, or 'coarse', grain). The grain and extent of data are often co-dependent (small grain with small extent, coarse grain with large extent) because of logistical constraints of measurement and the information content of the data. Generally, fine-resolution data require more effort or technical resource to collect, thereby limiting the extent over which they can feasibly be collected. Furthermore, as fine-grain data

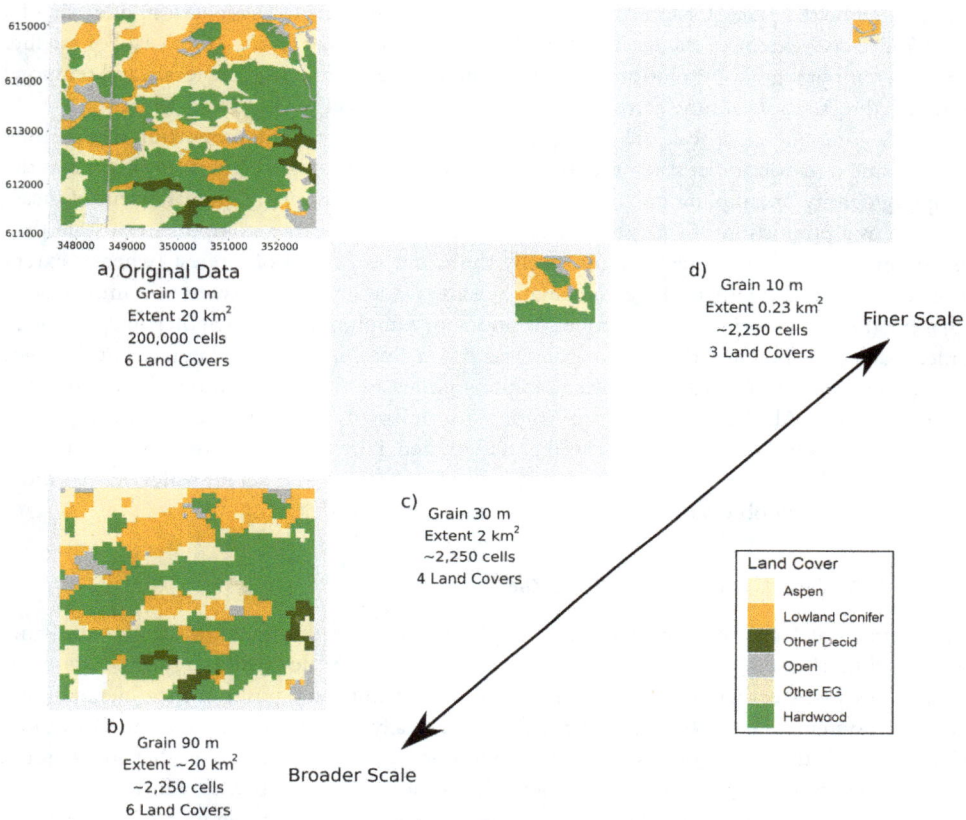

Figure 3.1 Varying scales of data for a forest study area. The original data in (a) from Millington et al. (2010) have been aggregated to a coarser grain but with identical extent to create data in (b). The number of map cells is maintained in (c) and (d), which have a progressively finer scale (i.e. smaller grain and smaller extent). Note how the number of land covers decreases with finer scale; this is by chance in this example, but in general, the number of discrete classes is expected to decrease. (After Wiens 1989.)

accumulate over large extents, they can contain huge amounts of information, which in turn produce challenges for data storage, transmission, and analysis. As technical and analytical limits are pushed back, these logistical constraints are being diminished, but challenges do still remain.

The grain and extent of data depend on the measurement tools being used. For example, remotely sensed satellite image data are widely used in landscape ecology because they can be collected automatically, globally, and regularly, providing a unique perspective on biophysical characteristics at the Earth's surface that we would not otherwise have access to (Fassnacht et al. 2006; Kennedy et al. 2014; Chapter 12). Images from satellite remote sensing are raster data; that is, they are a grid of pixels each with a value indicating some property at the Earth's surface (e.g. reflectance of radiation at a given wavelength). Raw satellite image data may be 'classified' so that each pixel is assigned to a single land cover (or other categorical variables) based on characteristics of the reflected radiation (Giri 2012), or pixels may be assigned numerical measures of characteristics such as elevation, canopy height, or productivity, for example, in the area represented by the pixel (Chapter 12). For raster data, grain refers to the length of one side of a pixel and spatial extent to the total area of all pixels in the image (e.g. Figure 3.1). For satellite imagery, temporal grain (resolution) is the time interval between image retrievals, and temporal extent is the duration of time across which a series of images cover from start to finish.

Other components of scale are *coverage* and *spacing*. Coverage is the proportion of the study area or time duration under study that is actually measured or sampled and is also known as the sampling density or sampling intensity. Coverage is a relative measure (e.g. areal extent of measurement as a proportion of total study area) and is a more subjective component of scale than extent, given that an investigator must specify the wider study area of interest (whereas extent is an absolute measurement of area or time). Spacing is the distance between two immediately adjacent (i.e. neighboring) measurement locations or sampling points, whether in space or in time. Spacing is known as the sampling interval or lag. Spacing is similar to grain but does not require that unmeasured values between sampling points are the same as those identified at the sampling points. All the components of imposed scale briefly defined here lead to important constraints on what patterns are observed and identified. However, as discussed below, intrinsic and theoretical considerations of what patterns can be observed at different scales may be more important than any observational or technological constraints that exist.

Hierarchical 'level' is distinct from 'scale'

The terms 'scale' and 'level' may seem to imply similar concepts, but they are not the same thing, and their distinction must be recognized (King 2005). 'Scale' (including the components of scale discussed earlier) is distinct from 'level', which implies a position in a hierarchically organized system (e.g. O'Neill et al. 1986). A hierarchically organized system is one that can be divided into distinct sub-systems that interact with one another to compose other sub-systems. A sub-system is at the same level in the hierarchy as those sub-systems with which it functionally interacts but in a lower level than the sub-system within which it is embedded. Conversely, sub-systems have a higher position (level) in the hierarchy than those embedded within them (e.g. see Figure 3.2). Sub-systems at the same hierarchical level interact at faster rates with one another than with sub-systems in other hierarchical levels, potentially with orders-of-magnitude differences in rates between levels (Simon 1962; Wu 2013). To demonstrate the possible evolutionary roots for why ecological systems are hierarchically organized like this – and how using this organization can be useful for understanding system properties – Simon (1962, p. 470) presented the parable of two watchmakers. There is not room here to re-present the parable (and see Wu 1999 for a further exposition), but the key point is that the watchmaker taking a hierarchical approach to construct her watches (constructing sub-units from individual components that are then coupled with other sub-units to compose the final product) has a greater chance of completing a watch before an interruption can disrupt the process than the other watchmaker not taking a hierarchical approach (i.e. who tries coupling all components individually to create the final product without creating sub-units). The implication is that the development of biologically complex organisms through a hierarchical evolutionary process (e.g. cells composing organs composing humans, etc.) had a greater chance of succeeding than a non-hierarchical one.

When referring to hierarchically organized systems, embedded means 'faster than' and/or 'constrained by', such that the functioning of a lower-level sub-system may be faster than and/or constrained by a higher-level sub-system. However, for the more general concept of 'a hierarchy', embedded can mean 'smaller than' and/or 'contained within', such that a lower-level sub-system may be physically smaller and physically contained within a higher-level sub-system. Note that the key difference between these ideas is that in hierarchically organized systems, sub-systems

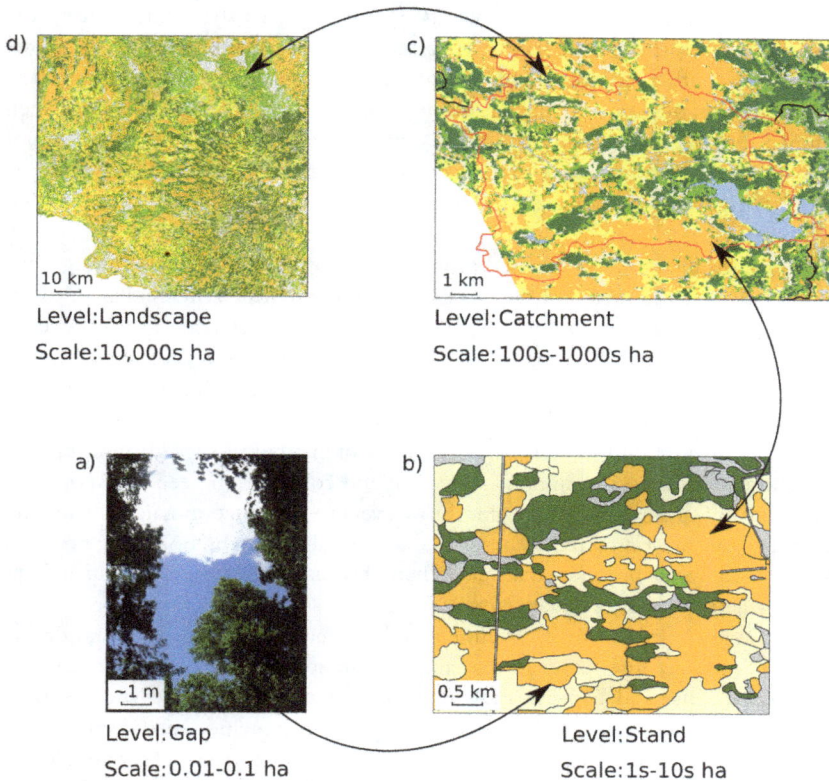

Figure 3.2   Four levels in a (nested) forest hierarchy. At the lowest level, *Gaps* (a) in the forest canopy are defined by the influence of large trees. Multiple gaps will be found in *Stands* (b) at the next level, defined by species composition and age structure. Many stands will be found within *Catchments* (c) at the next level, defined by local drainage basins. In turn, multiple catchments may be found at the *landscape* level (d), which may be defined by changes in land use and disturbance regime. (After Urban et al. 1987.)

at the same level *interact functionally* to compose the next higher sub-system (see Holland 1998 for how this relates to the concept of 'emergence' in complex systems). Furthermore, where sub-systems are physically contained within one another, the hierarchy is said to be 'nested' (e.g. see Wu 2013). Most of the hierarchically organized systems considered in landscape ecology are 'nested' in this way and thus have both sets of attributes. That is, sub-systems lower in the hierarchy are both functionally 'faster than' and 'constrained by' and physically 'smaller than' and 'contained within' sub-systems higher in the hierarchy. It is the combination of both sets of attributes, due to the nested character of physical environments (MacArthur 1972), that can lead to confusion about the meaning and usage of the terms 'level' and 'scale' in landscape ecology.

An archetypal example of a nested, hierarchically organized system in landscape ecology, concerning forest landscapes, is described by Urban et al. (1987). In particular, they outline a hierarchy to describe species-composition dynamics in deciduous forests of the eastern United States (Figure 3.2). The hierarchy ranges from forest gaps at the lowest level, through forest 'stands' or seral stages and river catchments, to the forest landscape itself. In these forests, when large, old trees die, they fall to leave a gap in the canopy (characteristically 0.01–0.1 ha in extent).

This gap subsequently provides smaller, subordinate trees that had been struggling beneath the larger tree (or even seeds waiting in the soil) with greater resources (e.g. light), which allows them to thrive. These trees then compete with one another until one or a few of them manage to dominate the others, thereby winning the competition for the gap. Interactions between trees mark the functional and spatial boundary of the gap – those surrounding trees that are not influenced by the change in resources (due to the fallen tree) are not 'in' the gap. Those trees are, however, in the next level of organization up the hierarchy established by Urban et al. (1987) – the forest 'stand' or seral stage.

This next level in the hierarchy is composed of mosaics of gaps and their surrounding trees, which interact (for example via seed and nutrient exchange) and which are similar in terms of their species composition, density, and size-class distribution (characteristically 1–10s ha in extent). Foresters use the term 'stand' to define these areas and often focus their management plans at this level, for example by aiming to mimic the natural processes of gap creation (by harvesting individual large trees) and subsequent tree regeneration in the stand. Where forest management is minimal or non-existent, and processes of ecological succession dominate, natural processes and conditions – including gap creation and competition, seed dispersal, local soil types, topography, and disturbances (e.g. windthrow events) – produce mosaics of gaps and surrounding trees that can be considered to be in the same seral stage. The boundary between these areas is determined by the processes that make them internally similar (i.e. human activity, local soil, topography and local disturbances, etc.).

Urban et al. (1987) propose that river catchments (characteristically 100–1000 ha) define the next higher level, both because stands in the same catchment will share a similar resource base (water availability, soils) and because they will interact more with one another than with those in other catchments. Interactions between stands might include seed dispersal and nutrient flux, and the boundary between different catchments is delineated functionally by the watershed defining them. Finally, at the highest level in this hierarchy, the landscape is defined as a mosaic of interacting river catchments (with a spatial extent of 10,000s ha). The boundary of the landscape is indicated by areas with non-interacting catchments and different physiographic characteristics, disturbance regimes, and/or human activity and land use conditions.

This example helps explain why the terms 'level' and 'scale' are often confused and (incorrectly) used interchangeably in (landscape) ecology. As in this example, hierarchies in landscapes (and biological systems in general; e.g. Berry and Kindlmann 2008) are usually nested – with components of the different levels contained, physically, within one another – meaning that it is inevitable that components in a lower level of the hierarchy also have a smaller spatial extent than those in higher levels (see King 1997). What is not inevitable is that functional interactions exist at each of these levels. For example, as King (2005) highlights, if interactions at the level below the landscape (in the preceding example, the catchments) do not produce 'emergent, holistic, aggregate' properties that would not otherwise exist, there is no 'landscape level' – there is simply a landscape (the definition of which is organism dependent). Of course, the 'landscape scale' might still be referred to in this case, but 'scale' refers only to the physical extent of the area that encompasses the mosaic of catchments. Consequently, the terms 'level' and 'scale' can be differentiated; 'level' is a position in a hierarchy which, if enumerated, is unitless and simply reflects the rank ordering relative to other levels in the hierarchy, whereas 'scale' refers to the spatial and/or temporal dimensions of an entity or event and in contrast to 'level', can be enumerated as a quantity with units of space (e.g. m^2) or time (e.g. s). Furthermore, if entities in a hierarchy do not depend on the functional interactions of entities at lower levels, 'level' should not be used as a descriptor (e.g. we should simply refer to the 'landscape' and not the 'landscape level'). And finally, we should only refer to the 'landscape scale' if we are willing to define its spatial extent and the characteristic duration of temporal processes.

One of the primary benefits of taking a hierarchical perspective – and being concerned with distinguishing different levels – is that it helps to simplify complex systems for analysis and understanding (Simon 1962). For example, a hierarchical perspective facilitates the use of a 'triadic approach' in which a focal level (often designated 'level 0') is selected, and then adjacent levels are considered: the embedded (faster/smaller, 'level −1') for understanding dynamics influencing the focal level, and embedding (slower/larger, 'level 1') for understanding constraints on processes within the focal level (O'Neill 1989). In the hierarchical forest landscape structure outlined here, for instance, treating stands as level 0 facilitates understanding about not only gap dynamics (level −1) to investigate stand processes and patterns, but also how those processes and patterns are constrained by the hydrological balance and position of the stand within the catchment (level 1). A hierarchical approach may also be useful when developing disaggregated, 'bottom-up' models that represent fine-scale elements to investigate ecosystem dynamics (e.g. individual- and agent-based models). The 'pattern-oriented modelling' suggested by Grimm et al. (2005) argues that such a model is useful if the processes specified at one level are able to produce patterns observed at other levels of organization. For example, continuing to use the hierarchical forest landscape as a case, a model of gap dynamics should be able to reproduce the age structure of a stand when simulated over time to provide confidence that mechanisms are appropriately represented (e.g. Rademacher et al. 2004). General criticisms of the hierarchy perspective have been argued, including a lack of clear definitions and principles (e.g. Wilby 1994), and in reality, multiple hierarchies often exist and overlap one another. But the successes of applications across multiple disciplines indicates that a hierarchical perspective provides analytic tractability in the types of complex systems that landscape ecology is concerned with, and its continuing use is expected (e.g. Wu 2013).

Scale dependence and selection

There is no single 'correct' scale

Possibly the most important scale concept in ecology is the understanding that there is no single 'correct' scale at which ecological phenomena should be studied (Levin 1992, 2000). This is because patterns and heterogeneity (see Chapter 15) are scale dependent, varying with the scale of observation (Turner 1989). Although humans have developed techniques and technologies to observe and measure the physical world at different grains and extents (e.g. from microscopes to satellite imagery), other organisms' perceptions of the physical world are dictated by their intrinsic (characteristic) scale. The difference between intrinsic and imposed scales (as discussed above) has been described by Mac Nally (2005) as the difference between *organism-centric* scale problems and *probing* scale problems. Organism-centric scale problems arise from the influence of organisms' intrinsic scale on their perception and response to the physical world, while probing scale problems arise from the different possible scales at which humans can observe and measure that world (using tools that are independent of our own intrinsic scale). Interactions between probing and organism-centric scale problems are inherent in how humans understand the relationships of organisms (including ourselves!) to pattern and heterogeneity in the physical world.

Organism-centric scale problems can thus dictate how an organism's 'landscape' should be defined. Heterogeneity and pattern in a beetle's landscape may be created by clumps of grass (e.g. tens of centimeters in diameter) in a matrix of unvegetated ground, as the difference between these two land covers can influence their movements greatly Wiens and Milne (1989). But for a larger organism, the white-tailed deer, for example, this same area will be perceived differently. Patterns and heterogeneity at a broader scale, for example between areas of grassland, shrubland, and forest (e.g. hundreds of meters in diameter), will be more important for the deer than for the beetle. It is patterns at this broader scale that influence deer movement and other behaviors (e.g. winter sheltering; Millington et al. 2010), much more so than fine-grained differences (such as between clumps of grass and a bare ground matrix) that influence the beetle. From the perspective of the habitat-forming vegetation in these examples, it is the scale of fragmentation that is key to defining landscape. Analogous to grain, 'dispersion' is a measure of the spatial arrangement of fragments of vegetation that lies on a continuum from 'geographical' (broad scale with greater isolation of fragments) to 'structural' (finer scale with less isolation; Lord and Norton 1990). Detecting critical scales in fragmented landscapes is important for understanding how organisms with differing dispersal behavior perceive the connectivity of a landscape (Keitt et al. 1997).

With regard to probing scale, a raster, categorical land-cover data set with a larger pixel size (i.e. larger grain) will likely represent fewer distinct patches of habitat (e.g. clumps of grass for the beetle or forest stands for the deer) across a given area of the Earth's surface than a raster data set with smaller pixel size. This is because habitat patches with area smaller than an individual pixel may be dominated by other land surface categories (in terms of area), and so that pixel will not be classified as habitat (assuming categories of land cover are either viable habitat or not, as in the patch-mosaic model of landscape ecology; see Chapter 2). The larger pixels get, the more often this will happen, and so the fewer pixels will be classified as habitat (as shown, for example, in Figure 3.1). If habitat patches are evenly distributed spatially across the area, the extent of the image will also influence both the number of habitat types and the number of individual patches observed; an image with smaller extent will observe fewer habitat patches, as more will lie outside the image boundary. Wiens (1989) formalized this pattern by saying that as grain increases (i.e. pixel size increases) for a constant extent, the proportion of spatial heterogeneity in the observed landscape contained within a sample (i.e. pixel) increases and is lost to the observer (Figure 3.3). For a raster land-cover map, this means that larger pixels are more likely

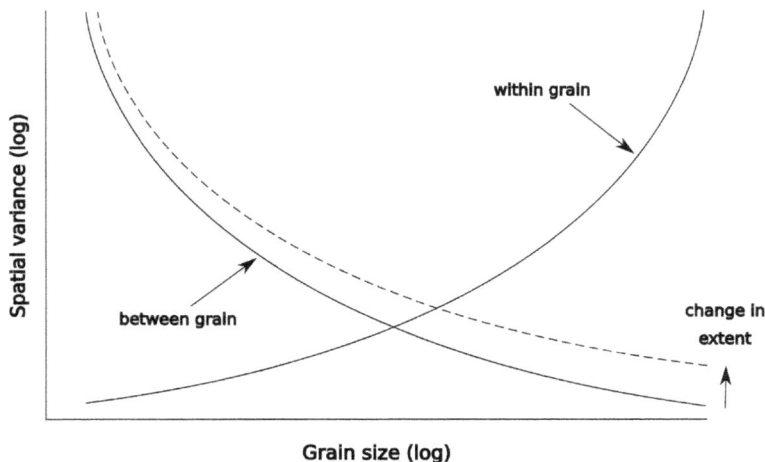

Figure 3.3 The effect of changing grain and extent on spatial variance. As grain increases, more spatial variance is included within samples (e.g. pixels), and vice versa. A lower effect, increasing extent, may increase the number of classes (hence, spatial variance). (After Wiens 1989.)

to contain multiple land-cover types, which we are unable to see (because they are classified as a single type). Conversely, between-sample heterogeneity (i.e. pixel variance) decreases as grain increases for a constant extent, because spatial heterogeneity in the landscape becomes 'averaged out' (i.e. fewer land-cover types are represented in the final map). Finally, as extent increases (i.e. if we look at a larger landscape), between-sample variance will increase, but within-sample variance will likely remain unchanged. That is, more land-cover types will be present in the final map, but each pixel will likely contain just as many different land-cover types (but we are unable to see these, as in the final map, each pixel is assigned to only a single type). These issues contribute, in part, to land-cover classification errors in remotely sensed satellite imagery (Moody and Woodcock 1995; Foody 2002) and have become known as the Modifiable Areal Unit Problem (MAUP; Gehlke and Biehl 1934; Openshaw 1984; Wu 2004; Bissonette 2017). The MAUP recognizes that statistical analysis of data for aggregated units of areas (e.g. 'pixels') can be affected not only by grain size but also by the different ways units of a given grain size might be configured (e.g. triangular vs. square vs. hexagonal pixels).

Beyond influencing patterns in raster data, scale dependence means that the imposed scale of observation can influence our understanding of relationships that are due to the intrinsic scale of organisms or patterns being observed. For example, the relationship between the body size of an animal species and its geographic range (i.e. the spatial extent over which that species is found) is generally expected to be positive (as one increases, so does the other), but some studies have reported a negative or no relationship (Gaston and Blackburn 1996). Examining multiple studies, Gaston and Blackburn (1996) showed how studies that examined comprehensive (i.e. full) coverage of species' geographic ranges did indeed usually find positive body size–geographic range relationships, and that it was mainly those with only partial coverage that found negative or variable relationships. Similarly, Palmer and White (1994) examined the importance of scale dependence on the relationship between the number of species found in a region and the absolute area of that region. Palmer and White found that both grain and extent of observation were important in identifying species diversity hotspots (hotspots increase as a function of both grain and extent) and that diversity hotspots are observable over a wide range of grains.

Thus, when observing, measuring, monitoring, or experimenting to investigate questions in landscape ecology, we must remember not only that is there no single correct scale to adopt (because of organism-centric scale problems) but also that the scale we use may artifactually influence results (because of probing scale problems). So, how do we establish what scale or scales we should use to collect and analyze data? Ultimately, the scale or scales we work at will be dictated by the motivations of our study's research questions and practical limitations of data and other factors. Considerations include (Meentmeyer 1989; Bissonette 2017) the size and speed (i.e. rate) of phenomena (e.g. organisms) and processes being studied, the amount of observed variance or heterogeneity (e.g. in land covers) in the landscape, the size of spatial units within a given organizational level (e.g. gap, patch, etc.), scales of existing data and data collection sources (e.g. remotely sensing), and technological constraints on data collection (e.g. data logger capacity). Generalizing this still further, approaches for selecting a scale or set of scales to work at might be thought of as 'process oriented', 'pattern oriented', and 'data oriented'.

Process-oriented scale selection

The process-oriented approach to scale selection starts by considering the intrinsic scale of the process or organism of interest. This approach includes considering, for example, the rate at which a process characteristically operates (e.g. a landscape fire burning over days) or the characteristic size of an organism (e.g. a 30 m tall pine tree). In turn, characteristic rates and sizes can help to establish the scales of patterns or environmental heterogeneity that are likely to influence the function of the process or organism. For example, Mac Nally (2005) suggests an organism-centric approach that uses the somatic size and lifetime of an organism to estimate a characteristic measure of scale, λ:

$$\lambda = O(\alpha\delta)$$

where α is the characteristic length of the organism, δ is the characteristic lifetime of the organism, and O means 'on the order of'. In turn, the total distance moved by the organism over its entire lifetime, stated as a multiple of λ, provides a measure that can be understood as the 'experience' (E) of the organism. For example, for the endangered Australian swift parrot (*Lathamus discolor*), Mac Nally (2005) estimates $\lambda \sim 5$ m/yr, as the birds are about 0.25 m in length and live for around 20 years. The swift parrot migrates from Tasmania to central Victoria for the winter each year, moving around 1,000 km in the 6 months it is on the mainland, giving an estimate for E of 2×10^6 λ on the mainland. The perception of the organism can then be related to measures of spatial landscape variation (e.g. distance between patches of potential habitat) and temporal resource fluctuation (e.g. quality of those habitat patches for breeding), also scaled in units of λ and denoted L. For instance, the swift parrot forages on flowering eucalyptus, which Mac Nally (2005) estimates to vary at spatial extents of tens of kilometers over 3–6 months, giving an estimate of L of $1–2 \times 10^3$ λ. It is important to understand that L is a measure of *variation* in the landscape and that E and L are measures of space *and* time. These measures can be used to identify theoretical scales of landscape that the organism can perceive and respond to. An organism should be readily able to perceive and respond to the landscape variation when E ~ L (the scales are concordant) and when E > L (although not as well). However, this theory predicts that when L > E, the organism will find it difficult to perceive and will be unable to respond to the landscape variation (and landscape resources at this scale will be 'unreachable'), and when E >> L and L >> E, the organism will be unable to perceive the landscape variation. Thus, the quoted values of E and L mean that the swift parrot is likely able to perceive and respond to the flowering eucalyptus but not to resources with smaller extent but with longer availability. For human investigators and managers, it also implies that measurements and management might be best targeted in the concordant zone where E ~ L.

Pattern-oriented scale selection

A good understanding of the processes (and/or organisms) to be studied is needed for process-oriented approaches to scale selection, as the approach described by Mac Nally (2005) illustrates. Thus, the process-oriented approach is less appropriate when previous understanding and data about the process or organism of interest are unavailable or when processes are initially inferred from pattern. A pattern-oriented approach seeks to identify the scales at which the spatial (or temporal) variability of an attribute or phenomenon is maximized (Meentmeyer 1989), based on the relationships shown in Figure 3.3. For example, if a spatial pattern in vegetation cover is believed to be regularly spaced, blocking techniques or an analysis of spatial lags can be performed (Turner et al. 1991; Dale 1999). This approach compares within- and between-block variance for blocks of varying size (uniformly located across a landscape) to identify the grain at which variance is maximized. When pattern is irregular, moving window analyses are more appropriate (Turner et al. 1991). These work in a similar fashion to blocking but compare variance between halves of a block (window) that is moved across the landscape; variance is maximized where patch edges are found, potentially allowing identification of characteristic patch sizes. Similarly, pattern-oriented approaches are appropriate when processes operating within a particular level of a hierarchy are of concern. For example, if forest gap processes (e.g. tree regeneration) are of interest, the characteristic size of the gap will govern the choice of scale (primarily grain, but also possibly extent to ensure that a sufficient number of gaps is included in data collection and analysis). Note that these methods are primarily appropriate for univariate measures of landscape pattern, and other approaches, such as multi-scale pattern analysis (Jombart et al. 2009) or wavelet analysis (e.g. Carl et al. 2016) are needed in multivariate cases.

Beyond identifying appropriate grain, other approaches are useful for identifying appropriate landscape extents or distances between study units in a landscape (Holland and Yang 2016). For example, Amici et al. (2015) examined the importance of landscape extent empirically for forest plant species richness by examining regression models and variance partitioning for landscapes of different sizes. Autocorrelation methods are frequently useful to identify the lag distance beyond which independence can be assumed, which is often important for appropriate statistical sampling (e.g. Millington et al. 2007). Similarly, Pasher et al. (2013) considered a fixed landscape size (100 ha), but locations of these landscapes in a broader region were selected to maximize heterogeneity (in the sample of landscapes) and such that the distance between landscapes meant they were not spatially autocorrelated. Regardless of the specific approaches employed, some preliminary analysis of pattern will often be useful to help guide the scale or scales of study most appropriate for the research questions in hand.

Data-oriented scale selection

Available existing data and technological constraints often determine the scales of observation and analysis. For example, the spatial and temporal grain and extent of satellite image data (whether raw or classified) has a strong influence on what can be observed when using such data. There is often a trade-off in spatial and temporal scales for satellite remote sensing imagery, as sensors that have high temporal resolution have low spatial resolution (and vice versa). Satellites in geostationary orbit (such as weather satellites) can retrieve images every 15 minutes but are a relatively long way from Earth and so have a low spatial resolution (1 km). Such coarse spatial resolution can make these data inappropriate for studying many organisms or processes. In contrast, polar-orbiting satellites such as Landsat are much closer to Earth and so have a higher spatial resolution (e.g. 30 m), but can only retrieve images for a location on Earth's surface when they pass overhead every 16 days or so. These characteristics again influence what is observed. For example, a large wildfire that burns across a landscape for a short time (e.g. a few days) may be detected multiple times by a geostationary satellite but not at all by a polar-orbiting satellite. Conversely, a small wildfire may go undetected by a geostationary satellite (due to its low spatial resolution) but be detected by a polar-orbiting satellite (with higher spatial resolution). However, as discussed further in Chapter 12, advances in remote sensing technologies that mean such restrictions are now diminishing, for example with the introduction of the Sentinel satellites (10 m spatial resolution with a maximum of 5 days between overpasses) and methods to detect processes acting at sub-pixel scale (e.g. for fire detection; Wooster 2012). Advances in object-based image analysis are also developing useful tools to minimize problems of artificial patch boundaries due to fixed data resolution (Karl and Maurer 2010; Lechner and Rhodes 2016).

Trade-offs may also exist in data collection between spatial and temporal grain and extent due to the power and data constraints of the technology used. Devices used to collect data remotely must be powered in some way (often by an electric battery) and be able to transmit data to a receiving station or store data for later physical retrieval. A smaller grain for any fixed extent will increase the coverage (sampling density) of collected data, decrease the spacing (sample interval) of the data, and result in an overall increase in the amount of data that must be transmitted or stored. Furthermore, if power is used each time a data collection is made, more frequent collection will require more energy. Thus, data collected at a smaller grain (i.e. higher resolution), whether spatial or temporal, often demands increased power and data transmission/ storage resources. For example, Webb et al. (2010) used global positioning system (GPS) collars to track the movement of white-tailed deer in Oklahoma, United States, but faced a trade-off between relatively high-frequency data collection (to understand the potential effects of moon phase and short-term weather patterns on movement patterns) against data storage capacity and battery duration of the GPS device (high frequency of data collection consuming the resources more quickly than lower frequency). In some circumstances, such technical restrictions may be a problem (depending on the questions we wish to address), but as with satellite sensing advances in terrestrial sensor networks (e.g. see Chapter 13), power storage/production and other technologies means that such restrictions are diminishing.

Pattern and process across scales

Scaling and extrapolation

The challenges of scale dependence have contributed to landscape ecologists' interest in *scaling* (e.g. Wiens 1989; Ludwig et al. 2000; Denny and Benedetti-Cecchi 2012). Indeed, some have argued that understanding scaling is a foundational challenge for improving scientific understanding; "the problem of relating phenomena across scales is the central problem in biology and in all of science" (Levin 1992, p. 1961). This definition of scaling – the identification of relationships between phenomena at different scales – is closely related to the term *extrapolation*; the adjustment of a value measured at one scale to match what the value would be if measured at another scale. Scaling from fine scale to broad scale, known as 'scaling-up', is important for understanding how regional patterns and processes are produced from fine-scale, local data collection (Turner et al. 1989). Scaling-up is challenged by the expense and logistics of collecting fine-resolution (detailed) data collection across large extents (but see improvements in sensor networks). Local spatial variation can be controlled and accounted for through appropriate sampling and experimental designs, but this becomes increasingly difficult across broader extents. The converse of scaling-up, 'downscaling', is important for understanding how the context of broad-scale patterns constrains local processes and for interpreting the outputs of broad-scale ecosystem and physical (e.g. climate) models at local scales. For example, the position of a tree of a given species in a forest stand (whether near the center or the edge of the stand) may influence its chances of survival (e.g. because of differences in resource availability or browse pressure at the edge of a stand compared with the center). Downscaling is challenging because broad-scale relationships become less coherent (i.e. identifiable) at local scales as fine-grained variation and historical contingencies become more important. These scaling challenges lie at the heart of understanding ecological patterns and processes in landscapes; scaling-up the effects of local processes often does not neatly account for observed landscape patterns, while scaling-down often fails to account for contingencies that affect local patterns and processes (Cushman et al. 2010b).

Because our understanding of patterns and process is so closely related to our data about them, the terms 'scaling-up' and 'downscaling' are often used synonymously for the process of adjusting values between scales. The identification of consistent relationships between values at different scales (i.e. scaling relationships) is desirable to evaluate possible problems in analysis due to the MAUP (e.g. when analyzing wildlife movements; Bissonette 2017), to enable estimation of fine-scale patterns from broad-scale data (e.g. remote sensing; Riiters 2005; Anderson 2018), and for integrating disparate data sets (e.g. at different resolutions; Atkinson 2013). Others have suggested that scaling relationships may be useful as conceptual models in the processes of assessing the robustness of understanding by, for example, comparing the performance of a model against data at multiple scales to identify where, when, and at what scales relationships hold or collapse (Miller et al. 2004). In landscape ecology, more effort has been devoted to developing methods to extrapolate values from fine to broad scales (scaling-up) than vice versa. Turner et al. (2001) attribute this emphasis to the immense data requirements needed to verify the predicted results of scaling relationships from data at larger grain (across large extents) to finer grain in multiple local locations (in contrast, for example, to the downscaling of general circulation model outputs using measurements from a large number of automated weather stations or stream-flow gauges; Wilby and Wigley 1997). Three general methods to extrapolate by aggregating values about patterns and processes from finer to broader scales have been considered (King 1991; Rastetter et al. 1992): linear 'lumping', the statistical expectation operator, and the calibration approach. 'Lumping' is the most straightforward approach, in which estimates of values at the larger grain are simply averaged over the finer-grain data falling within the aggregate area (e.g. total photosynthetic activity for an entire tree calculated from the mean value of a sample of individual leaves). The statistical expectation operator extends 'lumping' by considering the probability density of fine-grain data and using this with weighting to calculate expected values at the larger grain (e.g. accounting for the distribution of photosynthetic activity across a sample of leaves to calculate a weighted sum for total tree photosynthetic activity). The calibration approach uses regression with weighted sums of fine-grain data with data (for other variables; e.g. leaf size in the previous examples) from the larger grain to estimate scaling relationships between the scales. Cushman et al. (2010b) list several difficulties associated with these traditional approaches, many associated with the criticism that they are tied to a hierarchical view (as in above) of scaling in which large-grain properties can be readily aggregated from finer-grain data. For example, they argue that aggregating approaches assume that the finer-grain data and processes being extrapolated can be treated as random variables, that none of these methods can account for disequilibrial dynamics, and that these methods result in a loss of information about pattern and processes. More recently, Hamil et al. (2016) have shown how mixed-effect regression models can be used in a calibration-type approach to better account for the varying effects of spatial heterogeneity at different scales.

Given their criticisms of the scaling challenges that emerge from the patch-mosaic model of landscape structure (which assumes that discrete patches of land or habitat are categorically defined), Cushman et al. (2010b) argued for a different approach they call the 'gradient perspective'. This approach does not require a hierarchical perspective for examining landscapes across different scales and instead, emphasizes the depiction and analysis of continuously varying ecological phenomena (McGarigal and Cushman 2005; Cushman 2010a). The gradient perspective considers direct measurements of variables and processes, meaning that redefinition (of entities, variables, or units of observation) is not needed when moving between two different scales; Cushman et al. (2010b, p. 60) argue that this enables better consideration of non-linear and multivariate interactions, thereby 'greatly simplifying the task of robustly linking patterns with processes across scale[s]'. One area where the consequences of these differences in perspective for scaling are readily apparent is in a comparison of scaling categorical pattern metrics (e.g. Wu et al. 2002; Wu 2004) against the more recent scaling of gradient surface models (e.g. Frazier 2016; Frazier and Kedron 2017). As discussed in detail in Chapter 15, landscape pattern metrics enable the identification, quantification, and interpretation of spatial heterogeneity and its change. In their classic studies of categorical pattern metrics, Wu et al. (2002) and Wu (2004) demonstrated a range of responses of pattern metrics with changing map grain and extent: no clear scaling response; simple scaling functions; and 'stepped' or 'staircase' behavior. In contrast, taking a gradient perspective, Frazier (2016) found that polynomial functions were appropriate for extrapolating in many cases and argued that (continuous) surface metrics might be better suited than (categorical) pattern metrics for scaling from broad to finer scales. However, Kedron et al. (2018) highlight that finding appropriate analogues between pattern and surface metrics may not be possible in many cases (meaning that both are needed), and once again, we are reminded that the appropriate method or perspective will depend not only on the questions and organisms in hand but also on the scale or scales at which they might most appropriately be studied.

The continuing need for multi-scale research

If process or pattern are not considered when selecting the scale at which to observe or collect data, scale is essentially being selected arbitrarily. Unfortunately, this arbitrariness seems too frequent in landscape ecology studies. For example, in a review of wildlife-habitat research from 1993 to 2007, Wheatley and Johnson (2009) found that 70% of observational studies used an arbitrarily chosen scale with no apparent biological rationale. In their review of papers on habitat selection between 2009 and 2014, McGarigal et al. (2016) reported that only 5% explicitly examined multiple scales and levels. And in a review of 348 studies published between 2004 and 2014, Estes et al. (2018) found that most studies did not even clearly report the scale they were working at. Despite clear recognition of the importance of scale by ecologists since the late 1980s, it appears that comprehensive consideration of scale remains an outstanding challenge. Furthermore, continuing improvements in our understanding of scale dependence, observational/MAUP issues, and scaling challenges mean that intentionally identifying a *single* scale of study may prove inadequate. For example, Martin (2018) argued that we cannot assume that species respond to their landscape context at a single scale and that different biological responses may need to be examined at different scales. Others have argued that because the pressing challenges of the Anthropocene are caused by human influence on physical systems at multiple scales, improved understanding of ecological scaling is imperative (Scholes 2017). Improvements in technology mean that multi-method approaches to multi-scale research should prove fruitful in this regard (Bissonette 2017), and multi-hypothesis and multi-model approaches (e.g. Miguet et al. 2016) may facilitate thoughtful and systematic understanding of scale effects.

Multi-scale research will be particularly important if multiple processes or organisms occur in a landscape (which seems inevitable), as each will likely have its own set of intrinsic scales, which will be further compounded in studies that involve human–environment interactions. Data for human phenomena are often not available at the same spatial and temporal resolution, with human data frequently aggregated to political or other administrative units that have no, or limited, correspondence to the scale of physical data or processes (e.g. Millington et al. 2007). Such situations result in a scale mismatch between the scale of social organization and the scale of ecological processes, with potentially detrimental impacts on ecological and social function and resilience (Cumming et al. 2006). Furthermore, notwithstanding the problem of whether social data aggregated to political and administrative units suffer from problems associated with the MAUP, different people interpret and understand the world based on different perspectives of what constitutes scale, including those different from the definitions of scale offered here (what Manson 2008 termed the 'epistemological scale continuum'). For example, to reflect the representative nature of social structures, sociologists might add 'representational' and 'organizational' to the spatial and temporal dimensions of scale (Cumming et al. 2006). If social and natural sciences are to become integrated to ensure sustainability and resilience, there is still work to be done in improving our understanding of how different disciplines perceive space and scale (Vogt et al. 2002; Higgins 2012), and the ecological perspective presented here should be seen as only one of many.

Summary

1. There is no single 'correct' scale in landscape ecology.
2. The notion of 'scale' is not the same as 'level of organization', although in landscape ecology, the two are often closely related.
3. Patterns in time and space are scale dependent.
4. Scaling is the identification of relationships between phenomena at different scales.
5. There are different ways to establish the appropriate scale to work at depending on process, pattern, data, and perspectives.
6. Multi-scale analyses will increasingly be needed in future to understand landscapes in the Anthropocene.

References

Amici, V., Rocchini, D., Filibeck, G., Bacaro, G., Santi, E., Geri, F., Landi, S., Scoppola, A. and Chiarucci, A. (2015) 'Landscape structure effects on forest plant diversity at local scale: exploring the role of spatial extent', *Ecological Complexity*, vol 21, pp44–52.

Anderson, C.B. (2018) 'Biodiversity monitoring, earth observations and the ecology of scale', *Ecology Letters*, vol 21, no 10, pp1572–1585.

Atkinson, P.M. (2013) 'Downscaling in remote sensing', *International Journal of Applied Earth Observation and Geoinformation*, vol 22, pp106–114.

Berryman, A.A. and Kindlmann, P. (2008) *Population Systems: A General Introduction*. Springer Netherlands, Heidelberg.

Bissonette, J.A. (2017) 'Avoiding the scale sampling problem: a consilient solution', *The Journal of Wildlife Management*, vol 81, no 2, pp192–205.

Carl, G., Doktor, D., Schweiger, O. and Kuehn, I. (2016) 'Assessing relative variable importance across different spatial scales: a two-dimensional wavelet analysis', *Journal of Biogeography*, vol 43, no 12, pp2502–2512.

Cumming, G., Cumming, D.H. and Redman, C. (2006) 'Scale mismatches in social-ecological systems: causes, consequences, and solutions', *Ecology and Society*, vol 11, no 1, art 14.

Cushman, S.A., Gutzweiler, K., Evans, J.S. and McGarigal, K. (2010a) 'The gradient paradigm: a conceptual and analytical framework for landscape ecology', in Cushman, S.A and Huettmann, F. (eds) *Spatial Complexity, Informatics, and Wildlife Conservation*. Springer, Tokyo.

Cushman, S.A., Littell, J. and McGarigal, K. (2010b) 'The problem of ecological scaling in spatially complex, nonequilibrium ecological systems', in Cushman, S.A and Huettmann, F. (eds) *Spatial Complexity, Informatics, and Wildlife Conservation*. Springer, Tokyo.

Dale M.R.T. (1999) *Spatial Pattern Analysis in Plant Ecology*. Cambridge University Press, Cambridge.

Denny, M. and Benedetti-Cecchi, L. (2012) 'Scaling up in ecology: mechanistic approaches', *Annual Review of Ecology, Evolution, and Systematics*, vol 43, pp1–22.

Estes L., Elsen P.R., Treuer T., Ahmed L., Caylor, K., Chang, J. Choi, J.J. and Elis, E.C. (2018) 'The spatial and temporal domains of modern ecology', *Nature Ecology & Evolution*, vol 2, pp819–826.

Fassnacht, K.S., Cohen, W.B., and Spies, T.A. (2006) 'Key issues in making and using satellite-based maps in ecology: a primer', *Forest Ecology and Management*, vol 222, no 1, pp167–181.

Foody, G.M. (2002) 'Status of land cover classification accuracy assessment', *Remote Sensing of Environment*, vol 80, no 1, pp185–201.

Frazier, A.E. (2016) 'Surface metrics: scaling relationships and downscaling behavior', *Landscape Ecology*, vol 31, no 2, pp351–363.

Frazier, A.E. and Kedron, P. (2017) 'Landscape metrics: past progress and future directions', *Current Landscape Ecology Reports*, vol 2, no 3, pp63–72.

Gaston, K.J. and Blackburn, T.M. (1996) 'Range size-body size relationships: evidence of scale dependence', *Oikos*, vol 75, no 3, pp479–485.

Gehlke, C.E. and Biehl, K. (1934) 'Certain effects of grouping upon the size of the correlation coefficient in census tract material', *Journal of the American Statistical Association*, vol 29, no 185A, pp169–170.

Giri, C.P. (2012) *Remote Sensing of Land Use and Land Cover: Principles and Applications*. CRC Press, London.

Grimm, V., Revilla, E., Berger, U., Jeltsch, F., Mooij, W.M., Railsback, S.F., Thulke, H.H., Weiner, J., Wiegand, T. and DeAngelis, D.L. (2005) 'Pattern-oriented modeling of agent-based complex systems: lessons from ecology', *Science*, vol 310, no 5750, pp987–991.

Hamil, K.A.D., Iannone III, B.V., Huang, W.K., Fei, S. and Zhang, H. (2016) 'Cross-scale contradictions in ecological relationships', *Landscape Ecology*, vol 31, no 1, pp7–18.

Higgins, S., Mahon, M. and McDonagh, J. (2012) 'Interdisciplinary interpretations and applications of the concept of scale in landscape research', *Journal of Environmental Management*, vol 113, pp137–145.

Holland, J.D. and Yang, S. (2016) 'Multi-scale studies and the ecological neighborhood', *Current Landscape Ecology Reports*, vol 1, no 4, pp135–145.

Holland, J.H. (1998) *Emergence: From Chaos to Order*. Oxford University Press, Oxford.

Jombart, T., Dray, S. and Dufour, A.B. (2009) 'Finding essential scales of spatial variation in ecological data: a multivariate approach', *Ecography*, vol 32, no 1, pp161–168.

Karl, J.W. and Maurer, B.A. (2010) 'Spatial dependence of predictions from image segmentation: a variogram-based method to determine appropriate scales for producing land-management information', *Ecological Informatics*, vol 5, no 3, pp194–202.

Kedron, P.J., Frazier, A.E., Ovando-Montejo, G.A. and Wang, J. (2018) 'Surface metrics for landscape ecology: a comparison of landscape models across ecoregions and scales', *Landscape Ecology*, vol 33, no 9, pp1489–1504.

Keitt, T.H., D.L. Urban, and B.T. Milne (1997) 'Detecting critical scales in fragmented landscapes', *Conservation Ecology*, vol 1, no 1, p4 [online] Available at: https://www.ecologyandsociety.org/vol1/iss1/art4/

Kennedy, R.E., Andréfouët, S., Cohen, W.B., Gómez, C., Griffiths, P., Hais, M., Healey, S.P., Helmer, E.H., Hostert, P., Lyons, M.B. and Meigs, G.W. (2014) 'Bringing an ecological view of change to Landsat-based remote sensing', *Frontiers in Ecology and the Environment*, vol 12, no 6, pp339–346.

King, A.W. (1991) 'Translating models across scales in the landscape', in Turner, M.G. and Gardner R.H. (eds) *Quantitative Methods in Landscape Ecology*. Springer, New York.

King, A.W. (1997) 'Hierarchy theory: a guide to system structure for wildlife biologists', in Bissonette, J.A. (ed) *Wildlife and Landscape Ecology: Effects of Pattern and Scale*. Springer, New York.

King, A.W. (2005) 'Hierarchy theory and the landscape … level? or, Words do matter', in Wiens, J.A. and Moss, M.R. (eds) *Issues and Perspectives in Landscape Ecology*. Cambridge University Press, Cambridge.

Lechner, A.M. and Rhodes, J.R. (2016) 'Recent progress on spatial and thematic resolution in landscape ecology', *Current Landscape Ecology Reports*, vol 1, no 2, pp98–105.

Levin, S.A. (1992) 'The problem of pattern and scale in ecology: the Robert H. MacArthur award lecture', *Ecology*, vol 73, no 6, pp1943–1967.

Lord, J.M. and Norton, D.A. (1990) 'Scale and the spatial concept of fragmentation', *Conservation Biology*, vol 4, no 2, pp197–202.

Ludwig, J.A., Wiens, J.A. and Tongway, D.J. (2000) 'A scaling rule for landscape patches and how it applies to conserving soil resources in savannas', *Ecosystems*, vol 3, pp84–97.

Mac Nally, R. (2005) 'Scale and an organism-centric focus for studying interspecific interactions in landscapes', in Wiens, J.A. and Moss, M.R. (eds) *Issues and Perspectives in Landscape Ecology*. Cambridge University Press, Cambridge.

MacArthur, R.H. (1972) *Geographical Ecology: Patterns in the Distribution of Species*. Princeton University Press, Princeton.

Manson, S.M. (2008) 'Does scale exist? An epistemological scale continuum for complex human–environment systems', *Geoforum*, vol 39, no 2, pp776–788.

Martin, A.E. (2018) 'The spatial scale of a species' response to the landscape context depends on which biological response you measure', *Current Landscape Ecology Reports*, vol 3, no 1, pp23–33.

McGarigal, K. and Cushman, S.A. (2005) 'The gradient concept of landscape structure', in Wiens J and Moss M (eds) *Issues and Perspectives in Landscape Ecology*. Cambridge University Press, Cambridge.

McGarigal, K., Wan, H.Y., Zeller, K.A., Timm, B.C. and Cushman, S.A. (2016) 'Multi-scale habitat selection modeling: a review and outlook', *Landscape Ecology*, vol 31, no 6, pp1161–1175.

Meentemeyer, V. (1989) 'Geographical perspectives of space, time, and scale', *Landscape Ecology*, vol 3, no 3–4, pp163–173.

Miguet, P., Jackson, H.B., Jackson, N.D., Martin, A.E. and Fahrig, L. (2016) 'What determines the spatial extent of landscape effects on species?', *Landscape Ecology*, vol 31, no 6, pp1177–1194.

Miller, J.R., Turner, M.G., Smithwick, E.A., Dent, C.L. and Stanley, E.H. (2004) 'Spatial extrapolation: the science of predicting ecological patterns and processes', *BioScience*, vol 54, no 4, pp310–320.

Millington, J.D., Perry, G.L. and Romero-Calcerrada, R. (2007) 'Regression techniques for examining land use/cover change: a case study of a Mediterranean landscape', *Ecosystems*, vol 10, no 4, pp562–578.

Millington, J.D., Walters, M.B., Matonis, M.S. and Liu, J. (2010) 'Effects of local and regional landscape characteristics on wildlife distribution across managed forests', *Forest Ecology and Management*, vol 259, no 6, pp1102–1110.

Moody, A. and Woodcock, C.E. (1995) 'The influence of scale and the spatial characteristics of landscapes on land-cover mapping using remote sensing', *Landscape Ecology*, vol 10, no 6, pp363–379.

Openshaw, S. (1984) 'The modifiable areal unit problem', *CATMOG* (38).

O'Neill, R.V. (1989) 'Perspectives in hierarchy and scale', in Roughgarden, J., May, R.M. and Levin, S.A. (eds) *Perspectives in Ecological Theory*. Princeton University Press, Princeton.

O'Neill, R.V., DeAngelis, D.L., Waide, J.B. and Allen, T.F.H. (1986) *A Hierarchical Concept of the Ecosystem*. Princeton University Press, Princeton.

Palmer, M.W. and White, P.S. (1994) 'Scale dependence and the species-area relationship', *The American Naturalist*, vol 144, no 5, pp717–740.

Pasher, J., Mitchell, S.W., King, D.J., Fahrig, L., Smith, A.C. and Lindsay, K.E. (2013) 'Optimizing landscape selection for estimating relative effects of landscape variables on ecological responses', *Landscape Ecology*, vol 28, no 3, pp371–383.

Rademacher C., Neuert C., Grundmann V., Wissel C. and Grimm V. (2004) 'Reconstructing spatiotemporal dynamics of Central European natural beech forests: the rule-based forest model BEFORE', *Forest Ecology and Management*, vol 194, pp249–268.

Rastetter, E.B., King, A.W., Cosby, B.J., Hornberger, G.M., O'Neill, R.V. and Hobbie, J.E. (1992) 'Aggregating fine-scale ecological knowledge to model coarser-scale attributes of ecosystems', *Ecological Applications*, vol 2, no 1, pp55–70.

Riitters, K.H. (2005) 'Downscaling indicators of forest habitat structure from national assessments', *Ecological Indicators*, vol 5, no 4, pp273–279.

Ryo M., Aguilar-Trigueros C.A., Pinek L., Muller L.A.H. and Rillig M.C. (2019) 'Basic principles of temporal dynamics', *Trends in Ecology and Evolution*, vol 34, pp723–733.

Scholes, R.J. (2017) 'Taking the mumbo out of the jumbo: progress towards a robust basis for ecological scaling', *Ecosystems*, vol 20, no 1, pp4–13.

Simon, H.A. (1962) 'The Architecture of Complexity', *Proceedings of the American Philosophical Society*, vol 106, no 6, pp467–482.

Turner, M.G., O'Neill, R.V., Gardner, R.H. and Milne, B.T. (1989) 'Effects of changing spatial scale on the analysis of landscape pattern', *Landscape Ecology*, vol 3, no 3, pp153–162.

Turner, M.G., Gardner, R.H., O'neill, R.V. and O'Neill, R.V. (2001) *Landscape Ecology in Theory and Practice*. Springer, New York.

Turner S.J., O'Neill, R.V., Conley, W., Conley, M.R. and Humphries, H.C. (1991) 'Pattern and scale: statistics for landscape ecology', in Turner, M.G. and Gardner R.H (eds) *Quantitative Methods in Landscape Ecology*. Springer, New York.

Urban, D.L., O'Neill, R.V. and Shugart Jr., H.H. (1987) 'Landscape ecology: a hierarchical perspective can help scientists understand spatial patterns', *BioScience*, vol 37, no 2, pp119–127.

Vogt, K.A., Grove, M., Asbjornsen, H., Maxwell, K.B., Vogt, D.J., Sigurdardottir, R. and Dove, M. (2002) 'Linking ecological and social scales for natural resource management', in Liu, J. and Taylor, W. (eds) *Integrating Landscape Ecology into Natural Resource Management*. Cambridge University Press, Cambridge.

Webb, S.L., Gee, K.L., Strickland, B.K., Demarais, S. and DeYoung, R.W. (2010) 'Measuring fine-scale white-tailed deer movements and environmental influences using GPS collars', *International Journal of Ecology*, vol 2010, art 459610.

Wheatley, M. and Johnson, C. (2009) 'Factors limiting our understanding of ecological scale', *Ecological Complexity*, vol 6, no 2, pp150–159.

Wiens, J.A. (1989) 'Spatial scaling in ecology', *Functional Ecology*, vol 3, no 4, pp385–397.

Wiens, J.A. and Milne, B.T. (1989) 'Scaling of 'landscapes' in landscape ecology, or, landscape ecology from a beetle's perspective', *Landscape Ecology*, vol 3, no 2, pp87–96.

Wilby, J. (1994) 'A critique of hierarchy theory', *Systems Practice*, vol 7, no (6), pp653–670.

Wilby, R.L. and Wigley, T.M.L. (1997) 'Downscaling general circulation model output: a review of methods and limitations', *Progress in Physical Geography*, vol 21, no 4, pp530–548.

Wolkovich, E.M., B.I. Cook, K.K. McLauchlan, and Davies, T.J. (2014) 'Temporal ecology in the Anthropocene', *Ecology Letters*, vol 17, no 11, pp1365–1379.

Wooster, M.J., Xu, W. and Nightingale, T. (2012) 'Sentinel-3 SLSTR active fire detection and FRP product: pre-launch algorithm development and performance evaluation using MODIS and ASTER datasets', *Remote Sensing of Environment*, vol 120, pp236–254.

Wu, J. (1999) 'Hierarchy and scaling: extrapolating information along a scaling ladder', *Canadian Journal of Remote Sensing*, vol 25, no 4, pp367–380.

Wu, J. (2004) 'Effects of changing scale on landscape pattern analysis: scaling relations', *Landscape Ecology*, vol 19, no 2, pp125–138.

Wu, J. (2007) 'Scale and scaling: a cross-disciplinary perspective', in Wu, J. and Hobbs, R.J. (eds) *Key Topics in Landscape Ecology*. Cambridge University Press, Cambridge.

Wu, J. (2013) 'Hierarchy theory: an overview', in Rozzi, R., Pickett, S.T.A., Palmer, C., Armesto, J.J. and Callicott, J.B. (eds) *Linking Ecology and Ethics for a Changing World*. Springer Netherlands, Dordrecht.

Wu, J. and Li, H. (2006) 'Concepts of scale and scaling', in Wu, J., Jones, K.B., Li, H. and Loucks, O.L. (eds) *Scaling and Uncertainty Analysis in Ecology: Methods and Applications*. Springer Nature, Heidelberg.

Wu, J., Shen, W., Sun, W. and Tueller, P.T. (2002) 'Empirical patterns of the effects of changing scale on landscape metrics', *Landscape Ecology*, vol 17, no 8, pp761–782.

4

Landscape connectivity

*Lenore Fahrig, Víctor Arroyo-Rodríguez, Eliana Cazetta,
Adam Ford, Jill Lancaster, and Thomas Ranius*

Introduction: movement success and population persistence

The concept of landscape connectivity arose from three interconnected ideas that took root during the 1970s and 1980s. First, populations of many species are distributed across disjunct patches of habitat (den Boer 1968). Second, the persistence of such populations depends not only on reproduction and survival but also on movement of individuals between the patches of habitat (Levins 1969). And third, there are features of landscapes – both natural and human-made – that can help or hinder movements between the patches (Merriam 1984). A landscape with high connectivity is one that facilitates movement of organisms among habitat patches (Taylor et al. 1993). Movement success is thus central to the concept of landscape connectivity.

Organisms move for different reasons: to search for food, mates, or nesting sites, to avoid danger from predators or bad weather, to disperse to new sites, and to make annual migrations. Not all organisms make all of these movement types; for example, plants move only for dispersal. Movements vary from passive (via wind or currents) to active locomotion, and different life stages of a species can move for different reasons and in different ways. For example, in many aquatic insects, immature stages swim, walk, or move via water currents, whereas mature stages fly. Here, we use the term 'movement' to refer to any type of movement, for any motivation, by any type of organism, including its gametes or propagules.

Irrespective of the proximate reason for moving, the ultimate reason is to increase individual fitness by increasing reproduction. Not all individual movements are 'successful', i.e. not all movements increase fitness. Some even lead to the death of the moving individual, e.g. when passive dispersers such as seeds land in non-habitat or when the moving individual is taken by a predator. Movements not leading to death can also be unsuccessful if, for example, the energy used during movement is not outweighed by energy acquired during foraging movements. In other words, movements entail both benefits and costs. Movement is successful when an individual who moves reaps the benefit of that movement, such that the benefit is higher than the cost, and the individual's reproductive success is increased. As such, movement success is a per-individual concept.

Although movement success is defined at the level of the individual, it has implications for population persistence (Fahrig 2007). Natural selection works to increase movement success of individuals, and an increase in average movement success across individuals in a population will lead to an increase in population abundance and distribution, and thus, population persistence

(Bowne and Bowers 2004). Species vary in the importance of movement success to population persistence relative to other population processes. For example, for species that depend on ephemeral habitats, such as dung beetles, movement success is a driving factor in population dynamics and population persistence. For species inhabiting more stable habitats, local factors affecting reproduction and mortality may play a larger role in population dynamics than movement success. Therefore, individual movement success and population persistence are linked, but they are not equivalent.

Movement success and landscape structure

Given the link between movement success and population persistence, and given the ongoing human alteration of landscapes, it is important to know whether there are any general relationships between landscape structure (e.g. forest cover or fragmentation, landscape heterogeneity, edge density, etc.) and movement success. The most obvious landscape alteration is the ongoing conversion of natural covers to human land-uses, such as agriculture and development. The amount of habitat is trivially related to the total number of moving individuals of a given species that is associated with that habitat. More habitat leads to a larger population, which increases the sheer number of individuals moving.

However, opportunities to increase natural habitat as a way of increasing the number of movements (i.e. by increasing the population size and thus, the number of emigrants) are increasingly limited by the growing demand for land area to feed and house the human population. This leads to the question of whether we can increase population persistence by increasing individual movement success by landscape management, such as through the judicious addition of small bits of habitat (e.g. corridors, stepping stones), or by changing other aspects of landscape structure, such as matrix quality or landscape heterogeneity. As noted by Baguette et al. (2013), 'The *best* way to curb such extinctions would be to increase the carrying capacity … by increasing either the habitat area or the habitat quality. … An *alternative* (or complementary) strategy would be to increase the exchange of individuals among local populations, to reduce their functional isolation' (our italics).

Much of the landscape connectivity literature implies, explicitly or implicitly, that movement success is tightly linked to habitat configuration, on the assumption that movement is highly constrained by habitat, i.e. organisms are reluctant to leave habitat and venture into the matrix (non-habitat), and that organisms will cross only small distances between habitat patches. This assumption has led to the dominant emphasis in connectivity application on habitat corridors, stepping stones, and between-patch distances.

The assumption that movements are constrained by habitat may be valid for organisms that are physiologically restricted to a certain habitat type, such as many aquatic species that simply cannot survive out of water and so cannot cross the aquatic–terrestrial boundary. However, both theory and empirical work suggest that this model of highly constrained movement is much less general than is often assumed (see later). Theory indicates that a species' movement attributes can evolve such that the risk associated with moving through the matrix is reduced, thus increasing the benefit/cost ratio of such movements. In particular, theory consistently predicts the evolution of straighter, faster movement trajectories through more risky cover types, reducing the time spent there and thereby, increasing movement success (Bartoń et al. 2009, Travis et al. 2012, Martin and Fahrig 2015). If movement cost is reduced by the evolution of straighter, faster movement trajectories in the matrix than in preferred habitat, this increases the benefit/cost ratio of making these movements through the matrix (Martin and Fahrig 2015).

Many empirical studies have challenged the notion that habitat boundaries strongly limit movements (reviews in Harrison 1991, Bowne and Bowers 2004). Animals make frequent movements into and through the matrix, including not only dispersal movements but also seasonal movements and even daily movements (e.g. butterflies – Baguette et al. 2000, Schultz et al. 2012, birds – Fraser and Stutchbury 2004, turtles – Roe et al. 2009, and forest primates – Arroyo-Rodríguez et al. 2017, Galán-Acedo et al. 2019). In a global synthesis, Tucker et al. (2018) showed that movement distances of mammals are actually longer in resource-poor environments. Similarly, wind-dispersed plants move farther across openings than in forests (Nathan et al. 2002), and depending on pollinator behavior, gametes of forest plants can travel faster across open areas than within forests (Kam et al. 2010). Even many habitat specialists can readily cross the habitat boundary into the matrix during regular movements. For example, tracking data on the fisher, a forest-specialist mammal, show that home ranges can contain multiple forest patches with regularly used movement routes between them through cropland, developed open spaces such as golf courses and cemeteries, and pastures (LaPoint et al. 2013). The same is true for some forest small mammals; Bowman and Fahrig (2002) found that chipmunks, a forest specialist, crossed at least 600 m over open spaces between forest patches to return to their home territories following translocation. As predicted by theory, empirically documented movements through the matrix are typically straight and fast in comparison to movements through habitat (Schultz et al. 2012, LaPoint et al. 2013), reducing the amount of time spent in the matrix. Note that the idea that the edges of habitat patches do not represent movement boundaries is consistent with a view of the landscape as a gradient of use or movement rather than discrete patches. This more nuanced view of the landscape has not yet been widely incorporated into the concepts and application of landscape connectivity (see next section), which generally divide the landscape into habitat and matrix.

Connectivity concepts

There are over 3,000 papers on habitat connectivity. Searches conducted on 14 January 2019, using the search terms 'landscape connectivity' OR 'habitat connectivity' OR 'patch connectivity', returned 3,580 items on Google Scholar and 3,212 items on Web of Knowledge. The initial concept of landscape connectivity (Merriam 1984) was developed to capture the notions discussed earlier: depending on the movement attributes of a species, some landscape patterns will have high benefit/cost ratios of between-patch movements, i.e. high movement success, and therefore, will have high population abundance and population persistence. Such landscapes are said to have high connectivity (Figure 4.1a). However, practical difficulties in estimating connectivity according to this initial concept have led to a divergence in the ways that researchers measure and even conceptualize connectivity (Figure 4.1b–d). Each of these concepts has its uses and limitations. Here, we describe the three main connectivity concepts, in chronological order, and we label them according to their originators: *Merriam connectivity*, *Noss connectivity*, and *Hanski connectivity*. We focus mainly on describing the concepts as they were defined by their originators and as they are generally applied today. We also include some discussion of ongoing modifications to these original concepts.

Merriam connectivity: emphasis on movement success

Merriam (1984) first introduced the concept of landscape connectivity as the interaction between movement attributes and landscape structure that influences between-patch movements and ultimately, population persistence. Merriam specified that connectivity is 'defined not just by environmental features but also by species behaviour', i.e. 'functional connectivity'

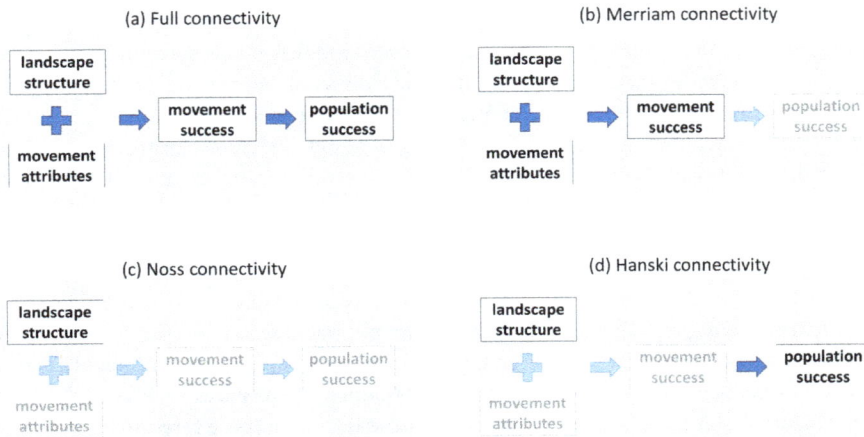

Figure 4.1   Illustration of the full concept of connectivity (a) and three major variants on this concept (b–d). Boxes in dark text represent components that are explicitly included, while boxes in light text represent components that are either ignored or implicitly assumed. We label the three main connectivity concepts for their originators, Gray Merriam, Reed Noss, and Ilkka Hanski, but note that a variety of alterations have been proposed by later authors.

sensu Tischendorf and Fahrig (2000). This initial concept explicitly linked movement success to population size and population persistence (Fahrig et al. 1983, Fahrig and Merriam 1985) (Figure 4.1a). A subtle shift in Merriam connectivity occurred when Merriam and co-authors simplified the definition to 'the degree to which a landscape facilitates or impedes movement of organisms among resource patches' (Taylor et al. 1993). Here, the emphasis became limited to movement success, while the link between movement success and population success became implicit (Figure 4.1b).

Empirical estimates of Merriam connectivity focus on estimating individual movement success in response to landscape structure. There are two main approaches to doing this. In the first approach (Figure 4.2a; e.g. Goodwin and Fahrig 2002b, Cline and Hunter 2016, Fletcher et al. 2019), one begins by making detailed field studies to estimate individual movement attributes – boundary-crossing tendency, movement speed and step lengths, movement tortuosity, and mortality during movement – in all possible cover types and boundary types (e.g. Ricketts 2001, Goodwin and Fahrig 2002a, Mueller et al. 2014, Tucker et al. 2018). These movement attributes are then scaled up to infer between-patch movement success in the landscape for a variety of different landscapes that vary systematically in their structures. The results are then summarized into relationships between movement success and metrics of landscape structure (e.g. landscape heterogeneity, edge density, percentages of different cover types, etc.).

In the second approach to estimating Merriam connectivity (Figure 4.2b), between-patch movement success is estimated directly at the landscape scale rather than by scaling up from individual movement attributes. Movement success is measured in many different landscapes, varying widely in landscape structure, and the results are again summarized into relationships between movement success and metrics of landscape structure. In principle, this approach could be followed by conducting mark-recapture experiments in all resource patches in all the landscapes. While in many situations, this would require a sampling effort far greater than would be possible, there are some situations where it may be possible. These include cases where the

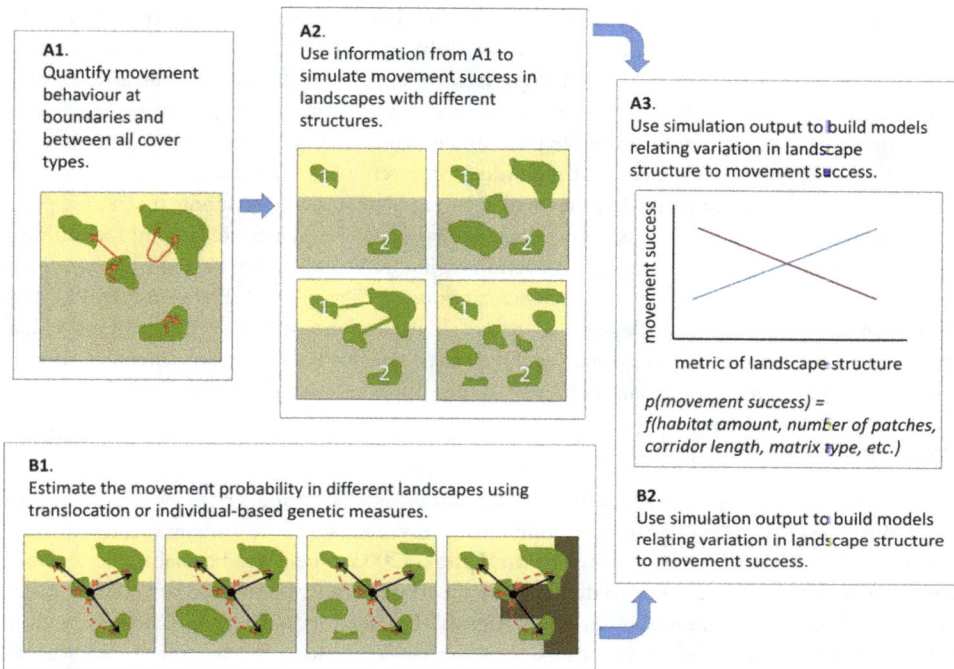

Figure 4.2 Illustration of two methods (A and B) for estimating landscape connectivity as originally defined (see Figure 4.1a).

provenance of very many individuals can be determined over very large areas, as can sometimes occur for genetic tagging (Lamb et al. 2019), parentage analysis of seeds (Aufrett et al. 2017), or stable isotope labels (Herrmann et al. 2016). For example, Flores-Manzanero et al. [2019) related genetic distance to landscape structure variables in order to estimate the determinants of connectivity in a species of small mammals. It may also be possible to estimate movement success across multiple landscapes for large, collared mammals, where huge quantities of movement data can be collected and analyzed over very large regions. Finally, it may be possible in situations where homing animals can be translocated within multiple landscapes with different structures (shown in Figure 4.2b; e.g. Bender and Fahrig 2005, Geoffroy et al. 2019).

We also note that estimating Merriam connectivity may be more feasible for largely one-dimensional systems. In rivers and streams, for example, movements can be in primarily up- and down-stream directions, especially for species that move long distances along channels, such as some fish (salmon, eels, lamprey) and snails (Schneider and Lyons 1993). Estimating movement success in such one-dimensional systems is more straightforward than in two-dimensional systems, because in one-dimensional systems, there is only one possible movement axis, i.e. up and down stream or along the coast. This simplifies the task of identifying landscape elements that help or hinder movement, in comparison to two-dimensional systems, where movements can go in any direction, making the role of landscape features in movement success much more difficult to estimate. Many features can influence movement success along a river, e.g. natural features such as riffles, pools, and waterfalls, and anthropogenic features such as dams, weirs, and road culverts. These features typically span the entire channel width, so the structure of these landscapes can be depicted as a linear sequence of different patch types that may allow or hinder movement. Movement success of aquatic organisms has been measured directly over short river lengths (Erman 1986, Hayashi and Nakane 1989, Bubb et al. 2004) and across individual structures or patch types (Belford and Gould 1989, Jackson et al. 1999, Lancaster et al. 2011, Thiem et al. 2013, Brooks et al. 2017). Such direct measures may be suitable for estimating movement success in larger landscapes with multiple patches, i.e. whole rivers or watersheds (analogous to Figure 4.2a). The movement success of individuals in rivers with different landscape structures can be estimated also from the densities and life stages of individuals along migration paths (Tamario et al. 2019). Alternatively, movement success could be measured directly for multiple linear landscapes using mark-recapture, genetic tagging, radio telemetry, or radio transponders attached to large-bodied aquatic organisms (analogous to Figure 4.2b). For example, PIT tags (Passive Integrated Transponder devices) are routinely used to monitor movements of individual fish in rivers and often over long distances. Thus, river systems may be more amenable to estimating Merriam connectivity than many two-dimensional terrestrial systems.

Limitations of Merriam connectivity

While measuring Merriam connectivity is sometimes possible, it is often, perhaps usually, impossible due to the challenges of estimating individual movement success for many species in many cover types. This problem has been noted before. Brooks (2003) stated: 'Traditional studies of landscape connectivity have attempted to discern individual behavioral responses to landscape features, but this methodology is intractable for many species'. Tischendorf and Fahrig (2000) were even more blunt: 'The effort required to estimate connectivity empirically likely exceeds any feasible project'. In addition to the difficulty of empirically measuring movement attributes or movement success, Merriam connectivity requires modelling to extrapolate these empirical results to a landscape scale (rightmost panel in Figure 4.2; e.g. Hauenstein et al. 2019). Such extrapolation involves assumptions and errors (Fahrig et al. 2019). Even if it is possible to estimate Merriam connectivity for a few species or in a few environments, deriving cross-species generalities about the relationships between landscape structure and Merriam connectivity is likely close to impossible.

Noss connectivity: emphasis on habitat and corridors

Not long after Merriam (1984) introduced the concept of connectivity, Noss (1987) suggested a different definition: 'the extent to which patches are connected to one another by similar habitat or corridors' (Figure 4.1c). Noss connectivity is entirely about landscape structure; no information on animal movement success is necessary here, because animals are assumed to move only or mainly through their preferred habitat. While this may be an appropriate assumption for some species, it appears to be inappropriate for many species, with significant practical consequences (see 'Limitations of Noss connectivity' later). However, due to its ease of measurement, Noss connectivity is by far the most commonly used concept of connectivity in applied situations. In its simplest applications, it only requires land cover maps. In species-specific applications, it also requires information about species habitat associations and limited assumptions about the species' ability to cross gaps in its habitat. Using this information, connectivity tools (see later) are used to prioritize habitat for preservation. Because Noss connectivity assumes that movement is constrained by habitat, large-scale maps of Noss connectivity generally look a lot like habitat maps (e.g. Pelletier et al. 2017).

Others have noted that Noss connectivity is fundamentally different from Merriam's original concept. Baudry and Merriam (1988) called Noss connectivity 'connectedness', noting that it 'refers to structural links between elements of a landscape', and differentiating it from Merriam connectivity, which '[is] a parameter of landscape function', i.e. individual movement success. To avoid confusion with Merriam connectivity, Tischendorf and Fahrig (2000) called Noss connectivity 'structural connectivity' as opposed to Merriam connectivity, which they called 'functional connectivity', again referring to the lack of explicit estimation of individual movement success in Noss connectivity. Finally, Fischer and Lindenmayer (2007) referred to Noss connectivity as 'landscape connectivity' to highlight that it does not consider individual movement success.

It is important to note that the use of Noss connectivity in single-species applications typically involves implicit assumptions about animal movement, i.e. that animals move only or mainly through the habitat and that they can only cross small gaps in habitat. For this reason, it is often assumed that when estimated for a single species, Noss connectivity is a measure of functional connectivity (i.e. Merriam connectivity). This assumption is evident in a review of connectivity applications (Ayram et al. 2016), which concluded that most such studies are about functional connectivity, because these studies refer to particular species. However, these studies are in fact using Noss connectivity, because most of them identify only the distribution of the species habitat and do not include or measure movement or movement success in response to landscape features.

Noss connectivity includes a strong emphasis on corridors. The review by Ayram et al. (2016) found that '[n]early half of the articles explicitly raised the issue of identifying or proposing potential corridors'. In fact, for many researchers and practitioners, connectivity has become essentially synonymous with habitat corridors and therefore, with Noss connectivity. Noss connectivity is not only the most commonly used connectivity concept in applied situations; it is also very prevalent in the scientific literature, on a par with Merriam connectivity. A citations search in Web of Science combining 'Merriam' as author and 'connectivity' as topic resulted in 8 articles, which have been cited 2,084 times, while a search of Noss as author and 'connectivity' as topic resulted in 15 articles, which have been cited 2,224 times.

Limitations of Noss connectivity

The biggest limitations of Noss connectivity are its inherent assumptions that (i) movements are largely limited to habitat and (ii) when individuals do move through the matrix, their success is lower than when they move through habitat. It is these assumptions that allow estimation of Noss connectivity without measured movement attributes or movement success except for estimation of gap-crossing distances (e.g. Desrochers and Hannon 1997). While these two assumptions are likely valid in some situations, evidence to date suggests that this may not be the norm. Given that Noss connectivity is the primary concept in applied situations, this calls into question the validity of many, if not most, applications of connectivity in conservation planning.

The first assumption inherent in Noss connectivity is that animal movement paths are constrained by habitat; in other words, animals turn back towards habitat when they encounter habitat boundaries. There are several situations in which this appears to be a reasonable assumption. These typically involve a prey species or a microclimate-sensitive species, where the cover type on the other side of the habitat boundary is perceived to lack protection from predators or exposure. For example, Rittenhouse and Semlitsch (2006) found no radio-tagged individuals of a forest-specialist salamander more than a few meters into the adjacent grassland. Several studies have shown that small mammals turn back at boundaries between habitat and roads (Ford and Fahrig 2008, Rico et al. 2007), and Sieving et al. (1996) demonstrated through playbacks that several understory forest birds do not move from temperate rainforests into open matrix. In addition, as mentioned earlier, obligate aquatic species always turn back at the water–land boundary.

However, this behavior, where animals avoid moving from habitat into matrix, may be the exception rather than the rule. Movements of forest-dwelling small mammals are not nearly as constrained by wooded cover (Bowman and Fahrig 2002) as had been assumed by Fahrig and Merriam (1985). The same is true for many large mammals, especially large predators. Based on their studies of individual movement of the fisher, a forest-habitat specialist, LaPoint et al. (2013) advised against using habitat associations to identify corridors, as fishers appear to actually prefer moving through open areas. Vanbianchi et al. (2018) came to a similar conclusion for lynx, another forest specialist, stating that 'maintaining connectivity will require preserving habitats and linkages that would previously have been deemed unsuitable for lynx'. Similarly, Scharf et al. (2018) studied space use by four mammalian predators – black bears (*Ursus americanus*), bobcats (*Lynx rufus*), coyotes (*Canis latrans*), and wolves (*Canis lupus*) – and concluded: 'We could not find a direct correspondence between corridors chosen and used by wildlife on the one hand, and *a priori* habitat suitability measurements on the other hand … We suggest future studies to rely more on movement data to directly identify wildlife corridors based on the observed behavior of the animals'. Even the arboreal kinkajou, a tropical forest-specialist mammal, readily moves through farms and pastures during natal and adult dispersal movements: 'farms and pastures did not pose higher resistance to dispersal movements than forests' (Keeley et al. 2017). In other words, for these animals, Noss connectivity does not coincide with animal movements. Worryingly, some large-scale connectivity projects focus on large predators, i.e. species for which Noss connectivity is likely to fail (e.g. 'Yukon to Yellowstone' – grizzly bear; 'Algonquin to Adirondacks' – wolf).

The assumption of habitat-constrained movement is likely inappropriate for other species as well. A global review of primate data indicates movements through various kinds of human-modified covers (Galán-Acedo et al. 2019). Roe et al. (2009) studied turtle movements between wetlands and found that they were completely unrelated to Noss connectivity, stating:

> Neither network nor relative connectivity was related to any physical landscape attribute commonly used as a surrogate for actual connectivity,' and 'information [on movement] can potentially yield more important insight on connectivity than measures of landscape structural features alone.

Similarly, for an endangered butterfly, Schultz et al. (2012) found no congruence between habitat distribution and movement, stating: 'The implicit assumption in most other studies is that *a priori* designation of habitat based on physical structure or resources alone will adequately characterize movement.' Noss connectivity may even be inappropriate as an indicator of movement for some plants. Aavik et al. (2014) found no relationship between gene flow and Noss connectivity in a grassland plant. Thus, the assumption of habitat-constrained movement greatly limits the applicability of Noss connectivity. As Cushman et al. (2013) state:

> suitability for occupancy and suitability for dispersal may not be driven by the same factors. … Few studies have formally evaluated the performance of habitat suitability models as surrogates for landscape resistance, but those that have, generally have found them to perform poorly. This highlights the importance of not assuming that habitat relationships optimally reflect the landscape features governing population connectivity.

In general, Noss connectivity is expected to correspond to movement success whenever movement is clearly constrained by habitat. For example, as mentioned earlier, for aquatic species in linear systems, it is often reasonable to assume that a continuous flow of water along river channels is sufficient to ensure movement. An exception is aquatic species with terrestrial stages that may have very little movement during the aquatic stage (Jackson et al. 1999), and so connectivity may depend on movement of terrestrial stages, which may or may not be constrained by the distribution of terrestrial habitat. Noss connectivity may also correspond to movement success for plant species whose seeds will only germinate in their preferred habitat. For example, Cushman et al. (2014) found a strong relationship between river corridor networks and movement success, as indicated by gene flow, for a species of cottonwood found only in riparian zones. However, for terrestrial animal species, the assumption of habitat-constrained movement is often not met.

Some authors have recognized that the assumption of habitat-constrained movement limits the applicability of Noss connectivity and have made alterations to deal with this. Anadón et al. (2018) studied the separate effects of habitat suitability and connectivity on the spread of ungulates in Spain. They simultaneously estimated a least cost path model along with habitat suitability, thus effectively controlling for habitat availability when estimating the effect of connectivity. McClure et al. (2016) used data from global positioning system (GPS) collared animals to estimate habitat suitability specifically for movement. They then showed that circuit theory and least cost models based on movement-defined habitat suitability (rather than species presence data) performed well in predicting the remainder of the movement data. Similarly, Zeller et al. (2018) showed that resistance maps using species presence data do not correspond to actual movement but that this can be overcome by building resistance maps using actual movement data.

The second inherent assumption in Noss connectivity is that when animals do move through the matrix, their movements are less successful than when they move through the habitat. This is likely true in some situations, e.g. when the matrix cover is particularly deadly, such as a high-traffic road. However, many studies have found that a higher risk in the matrix can be compensated by faster and straighter movements through the matrix, reducing the time spent there for a given distance travelled or equivalently, increasing the distance travelled for a given time spent. This relationship between movement risk per time and movement speed will determine whether movement success is higher through habitat or through matrix and likely explains why some species actually prefer to move through matrix (reviewed earlier). Many species have greater movement speeds and travel longer distances in matrix than in habitat. Nowicki et al. (2014) found a tenfold difference for butterflies. For dragonflies, Chin and Taylor (2009) found that '[l]ong distance movements were more likely, and short-distance movements were less likely, when there were larger amounts of cut matrix between peatlands'. Goodwin and Fahrig (2002b) found that between-patch movements of a beetle were lower in landscapes containing more habitat, because movements were slower in habitat than in matrix. Hass et al. (2018) found that pollinators moved farther along non-vegetated field boundaries than vegetated ones. And Tucker et al. (2019) found that birds move farther in homogeneous than heterogeneous landscapes, likely due to a sparser distribution of resources.

The widespread emphasis on habitat corridors (sensu Noss 1987) in conservation planning is particularly worrying if habitat distribution is not a good indicator of animal movement preferences or success. This concern is not new. Franklin (1993) stated: 'While we intuitively expect that corridors are important, their effectiveness has not been proven and there is almost certainly a large proportion of the species for which corridors are not likely to be very useful.' Despite the proliferation of connectivity studies since then, the general conclusion has not changed. Ayram et al. (2016) concluded that 'the empirical validation of corridors is scarce because of the difficulty in obtaining field data with regard to dispersal.' There are counter-examples (e.g. Haddad and Tewksbury 2005), but the accumulating evidence appears to support the notion that habitat corridors often do not coincide with species movement paths, again calling into question the widespread application of Noss connectivity, at least in its original form. In addition, in discussing corridors, it is important to note that the literature has two distinct meanings of the word. Noss's original definition of 'corridor' was a physical linear strip of habitat connecting patches of habitat. As data have accumulated on actual movement paths of individual organisms, the term 'corridor' has also been used to describe movement routes that are frequently used through the landscape. These movement routes or corridors are often not through the preferred habitat of the species, so this use of the term 'corridor' is very different from Noss's meaning.

Despite the lack of evidence supporting the general application of Noss connectivity, this does not negate the important role of small bits of habitat, including linear strips ('corridors') and small patches ('stepping stones'), for population persistence in human-modified landscapes. For example, this has been shown for reptiles and amphibians (Mendenhall et al. 2014), for plants (Bennett and Arcese 2013, Horskins et al. 2006), and for species richness in general (Wintle et al. 2019). This is because these bits of habitat increase the total amount of habitat in the landscape and therefore, the potential population size of the species in that landscape. This benefit accrues even if these bits of habitat do not increase individual movement success. Protection of all bits of habitat is important for population persistence, irrespective of whether they play a special role in facilitating connectivity (Fahrig 2017).

Connectivity tools

Connectivity tools, as they are currently applied, generally measure Noss connectivity. In their review of applied studies, Ayram et al. (2016) found 23 methods used to measure connectivity. The most common were least cost path analysis, graph theory, and habitat availability from habitat suitability models. They found a striking increase in connectivity studies since 2008, which they linked to software for circuit theory, graph theory, and conservation prioritization. Almost all studies estimated connectivity either directly by habitat suitability or by using habitat suitability as input into connectivity algorithms. Such tools are easy to apply because they only require land cover data; no species movement data are needed except for limited assumptions about species gap-crossing distances.

Least cost path approaches started to be applied in connectivity work after algorithms became available in geographical information system (GIS) software (e.g. Meegan and Maehr 2002). Least cost paths are based on habitat suitability, with the assumption that less suitable habitat is more costly for movement and therefore, movements follow more suitable habitat. An important limitation to this approach is that there is no option to include cases where movement success through unsuitable habitat is high due to faster and straighter movements. As discussed earlier, the assumed equivalency between movement success and habitat suitability is not valid for many species, and so this calls into question the usefulness of least cost path analyses in estimating movement success. The least cost path approach is also problematic because it inherently assumes that the animal has complete knowledge of the landscape between its current location and where it 'wants' to go and that it uses that information to estimate and move along the least costly path. This assumption may be reasonable in some situations, but it is very unlikely to be true for natal dispersal movements, where the animal is moving through novel areas of the landscape. Finally, the parameterization of least cost path models can be problematic; most often, it is based on either expert opinion or habitat suitability modelling, again on the assumption that habitat constrains movement.

Graphs (Fahrig et al. 1983, Lefkovitch and Fahrig 1985, Urban and Keitt 2001) and circuits (McRae et al. 2008) also generally measure Noss connectivity, as applications usually assume that movement between nodes or points is more likely when they are linked by habitat or when they are closer together. For example, the 'equivalent connected area index' is built on the 'probability of connectivity', which 'is defined as the probability that two points randomly placed within the landscape fall into habitat areas that are reachable from each other' (Saura et al. 2011, Tarabon et al. 2019). As such, it is similar to the 'effective mesh size' of Jaeger (2000). In principle, graphs and circuits could be used to estimate Merriam connectivity if data on boundary responses and individual movement rates through different cover types were incorporated into the graph or circuit structure, but researchers almost always make the same assumptions as are made in least cost path applications, i.e. distance-based and habitat suitability–based movement success. Moilanen (2011) argues that there is currently an over-use or uncritical use of graph-theoretic methods in landscape planning and that their ease of use is resulting in over-confidence in the relevance of the results to conservation. Such over-confidence is particularly problematic because there is a wide array of different metrics for defining connectivity from graphs with different inherent assumptions about movement behavior. These can lead to very different estimates of connectivity (e.g. d'Acampora et al. 2018).

Hanski connectivity: emphasis on immigration

Hanski connectivity emerged from metapopulation theory. In 1994, Hanski introduced a measure, S_i, which is a user-parameterized negative function of the distances from patch i to other patches and a positive function of the sizes of those other patches. Hanski did not give S_i a name in 1994, but later, Moilanen and Hanski (1998) referred to S_i as 'isolation (actually connectivity but we use isolation to confer to common practice)', and then in 1999, Hanski began explicitly referring to S_i as a metric of connectivity. Originally, S_i was used to predict patch colonization-extinction dynamics, but its use has broadened to predict species occurrence patterns and patch-level species richness. The spatial scale of the Hanski connectivity function is obtained by fitting the connectivity model with occurrence data (e.g. Ranius et al. 2010). In Hanski's original function, connectivity was assumed to decrease with between-patch distance according to a negative exponential function. Other functions are possible, and the best function for a given species is assumed to be related to its dispersal biology.

Hanski connectivity has important similarities to and differences from the previous connectivity concepts. Unlike the previous concepts, Hanski connectivity is an attribute of a given focal patch, not a whole landscape; however, a whole landscape measure of Hanski connectivity can be obtained by averaging patch connectivity values across the patches in a landscape. The original Hanski connectivity concept is similar to Noss connectivity and different from Merriam connectivity in that it avoids directly measuring individual movement attributes or individual movement success, and it ignores the effect of matrix quality and spatial pattern on movement success (Howell et al. 2018). However, it is different from Noss connectivity in two important ways. First, it does not consider the particular habitat configurations that dominate the measurement and application of Noss connectivity, namely, corridors and stepping stones. Second, while Noss connectivity simply assumes a relationship between habitat pattern and movement, in the application of Hanski connectivity, this relationship is validated by modelling the relationship between S_i and patterns of patch occupancy or colonization-extinction events (Figure 4.1d). If there is a strong relationship between S_i and patch occupancy, one infers that movement is an important, potentially limiting factor for population success. However, it is important to note that in Hanski's metapopulation models, connectivity affects patch immigration rate rather than individual movement success. This distinction is important because immigration rate is affected not only by the movement success of individuals but also by the number of individuals that potentially can move, which is affected by population size (Figure 4.3). Note that Calabrese and Fagan's (2004) 'potential connectivity' is conceptually equivalent to Hanski connectivity.

Hanski's emphasis on distances between the focal patch and other patches rather than on aspects of the matrix that might influence movement success may result from his work with butterflies, where this assumption seems to hold. For example, Moilanen and Hanski (1998) concluded that incorporating matrix type into their Hanski connectivity metric did not improve prediction of patch occupancy by the Glanville fritillary butterfly. Ouin et al. (2008) came to the same conclusion for the meadow brown butterfly, noting that forest cover is the only matrix element that impedes their movement, which is otherwise unaffected by the matrix. The same seems to be true for several other butterflies (Fahrig and Paloheimo 1988, Leidner and Haddad 2010). In regions where forest cover is relatively low, this would lead to generally unconstrained movement across the landscape for butterflies, such that inter-patch movement success is only a function of inter-patch distance, though this would not be the case when the matrix is entirely forest, as in Haddad and Tewksbury (2005).

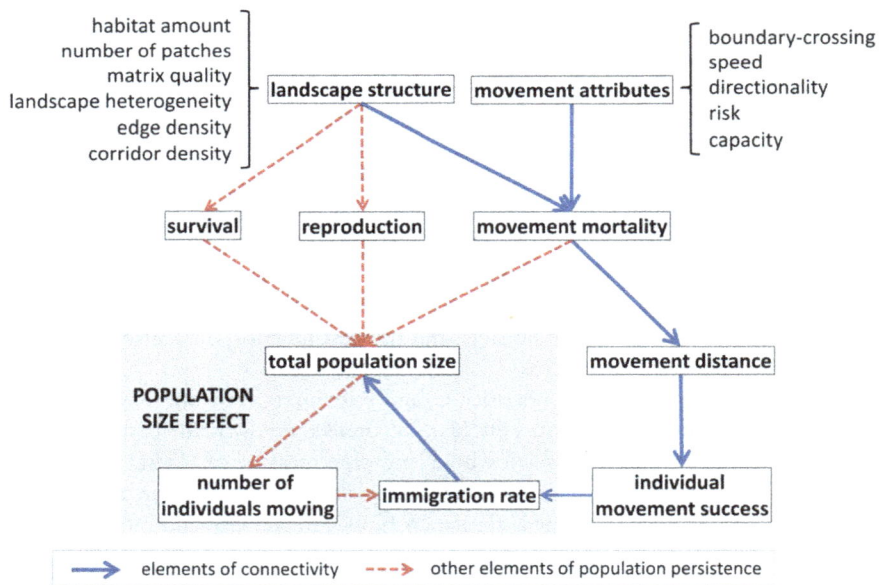

Figure 4.3 The effect of landscape structure on population size (and therefore on population persistence) is only partly determined by connectivity (blue arrows). Landscape structure also affects population size through its effects on reproduction and survival (dashed red arrows). Hanski connectivity estimates patch immigration rate by linking patch isolation to population occurrence or colonization. However, immigration rate is affected not only by movement success but also by population size, which determines the number of emigrants and therefore, the number of potential immigrants to the patch – the population size effect. Therefore, Hanski connectivity combines effects of landscape structure on movement, reproduction, and survival.

78

Limitations of Hanski connectivity

Hanski intended his measure of connectivity as an estimate of patch immigration rate, not as a measure of movement success. This creates challenges when trying to understand the role of movement success using Hanski connectivity, because Hanski connectivity inherently confounds movement success with population parameters, i.e. reproduction and survival. This is because patch immigration rate is determined not only by individual movement success but also by population size, which is itself influenced by reproductive success and survival, and reproductive success and survival are also affected by landscape structure (Figure 4.3).

In fact, it is possible to find a strong relationship between Hanski connectivity and population success (occurrence/colonization) even in situations where there is no actual relationship between movement success and Hanski connectivity. To see this, note that measures of Hanski connectivity are strongly correlated with (even redundant with) the amount of habitat surrounding the patch (Moilanen and Nieminen 2002, Bender et al. 2003, Tischendorf et al. 2003, Prugh 2009, Fahrig 2013). This is because the closer other patches are to patch i and the larger those other patches are (i.e. the higher the Hanski connectivity), the more habitat there is near to patch i (see figure 7 in Fahrig 2003). This means that Hanski connectivity is strongly related to the total number of potential immigrants to patch i (more surrounding habitat = higher population = more potential immigrants). The more potential immigrants there are, the more likely they are to land on patch i by chance, irrespective of individual movement success. Thus, Hanski connectivity confounds movement success with a population size effect (Figure 4.3). To avoid this confounding, Duflot et al. (2018) included both amount of habitat in the landscape and Hanski connectivity in models predicting the abundance and richness of grassland species. They found no effect of connectivity once habitat amount was accounted for. As Hanski connectivity predicts the immigration rate to a patch and not individual movement success, some authors do not consider Hanski connectivity to be a measure of landscape connectivity, e.g. Tischendorf and Fahrig (2000): 'Any measure of connectivity must be based on movement of an organism through a landscape. ... demographic indicators, such as species abundance and distribution, while potentially related to connectivity, are not measures of connectivity.'

A second limitation of Hanski connectivity, at least in its original formulation, is that it ignores the effect of matrix pattern and focuses only on inter-patch distances and sizes of patches. While often appropriate for butterflies, this is likely not appropriate for most species. Gilbert and Levine (2013) included a term for matrix quality in a Hanski connectivity model, but the change in quality was applied to the whole matrix uniformly; they did not include spatial variation in the matrix that might influence movement paths. Bender and Fahrig (2005) showed that spatial variation in matrix pattern can greatly obscure or even eliminate the relationship between individual movement success and connectivity measures that are based on patch size and isolation. Note, however, that more recent versions of Hanski connectivity do allow spatial variation in matrix quality. For example, Howell et al. (2018) incorporated least cost path modelling into the estimates of colonization probability in a metapopulation model, and Kärvemo et al. (2016) included variation in matrix quality using multiple Hanski connectivity estimates based on distances not only to habitat but also to matrix covers.

Aquatic systems: cross-ecosystem contrasts

As an observation (not a criticism), we note that movement and connectivity within marine and freshwater systems seldom feature in the landscape ecology literature, and vice versa. Nevertheless, contrasting systems in the same theoretical framework can sometimes yield fruitful or unexpected insights. For example, logistical difficulties in one system may be absent in others, and indeed, earlier, we suggest that measuring Merriam connectivity may be relatively straightforward in largely one-dimensional systems such as rivers.

In fact, there is a rich literature from marine systems that aligns closely with the original concept of landscape connectivity (Figure 4.1a). This research focuses on organisms with complex life cycles in which juveniles and adults occupy quite different and often distant habitat patches in naturally fragmented seascapes. Such organisms include barnacles and mussels, with pelagic larvae that live in the open ocean and sessile adults that inhabit rocky shores, and reef fish that also have pelagic larvae while adults are sedentary around reefs. Thus, larvae inhabit and disperse through the matrix that connects patches of sessile adults. Although some self-recruitment does occur (Jones et al. 1999), larval movement from outside the local area is required to sustain many populations; the notion that movement success influences demographic rates of the local population(s) is implicit in these systems (Cowen and Sponaugle 2009) and often assessed directly (Carson et al. 2011).

Although it is virtually impossible to follow movements of individual larvae, geochemical signatures (Levin 2006, Thorrold et al. 2007) can determine the probability that larvae from a particular source population will disperse to different patches of adult habitat within a landscape. Such data can be used to construct a 'connectivity matrix,' which describes the probability of movement between source and settlement nodes (Cowen et al. 2006), analogous to the approach suggested in Figure 4.2 b1. It is conceivable that metrics of movement success can be derived from such matrices. However, describing the physical structure of these marine landscapes is unlikely to be straightforward and is difficult to capture by a simple metric (Figure 4.2 a3). Larval dispersal is not determined simply by the distance between adult patches (cf. Hanski's butterflies), because dispersal distance and direction are strongly contingent upon water currents (Connolly et al. 2001), which are themselves influenced by complex interactions between coastal topography, wind, tidal forces, surface waves, and water stratification. Although models can describe the ocean currents that create transport corridors and influence the dispersal of marine larvae (Werner et al. 2007), deriving a metric of landscape structure or ranking different landscapes with respect to structural complexity has challenges. However, heterogeneity, anisotropy, and advective transport (e.g. wind- and current-assisted movement) are common to other ecosystems, and this is an active area of research (Vandermeer and Carvajal 2001, Shima et al. 2010).

SAMC: a hybrid approach

Recently, Fletcher et al. (2019) proposed a new approach for estimating landscape connectivity. Like circuit theory, their Markov chain–based method ('SAMC') is based on resistance maps. However, the SAMC approach incorporates two important additional elements. First, it decomposes landscape resistance into its two components – resistance due to behavioral avoidance of a cover type, and resistance due to mortality of individuals that do enter the cover type. This is particularly important because land cover types with high resistance to movement via behavioral avoidance can in some cases also have low mortality risk (Fletcher et al. 2019). This means that there can be a trade-off between dispersal mortality and avoidance of a given cover type when estimating connectivity. The second important element included in SAMC is that it can incorporate the initial distribution of the population, thus partly accounting for the effect of population size on the number of emigrants and therefore, on immigration.

Conceptually, SAMC is a hybrid approach. Its underlying structure is related to the circuit theory approach to estimating Noss connectivity, as it is based on resistance maps. However, by separating behavioral resistance from mortality resistance, SAMC becomes much more similar to Merriam connectivity. If different cover types can have low behavioral resistance but high mortality resistance, or vice versa, and if this affects the inferred connectivity, the application of SAMC requires detailed species-specific information about animal movement behavior and movement risk across all cover types in the landscape, just as is required for Merriam connectivity. By incorporating the initial distribution of individuals into the estimated movement probabilities, SAMC also becomes related to Hanski connectivity, because it estimates immigration rate, at least over the short term, as in Hanski connectivity. Note, however, that SAMC is not a full demographic model, as it does not include reproduction, and so it only partly includes the population size effect that is implicit in Hanski connectivity (Figure 4.3). Thus, at least in principle, SAMC is an amalgam of aspects of the three connectivity concepts. However, in practice, SAMC relies on the collection of detailed, species-specific information on population distribution and on movement and mortality in all cover types in a landscape. This may make it as data-hungry as a realistic individual-based simulation model of animal movement (e.g. Hauenstein et al. 2019, Trapp et al. 2019). It thus remains to be seen whether the introduction of SAMC will encourage collection of those data or whether in practice, most applications of connectivity will continue to use the basic Noss connectivity concept, where movement success is assumed to be determined by habitat distribution.

Implications for conservation and future research

Given the large differences among the three major concepts of connectivity, and their various alterations and combinations, it is confusing to both the practitioner and the researcher that they all have the same label. Our main conclusion is that practitioners and researchers should be aware that 'connectivity' is not a single thing but rather, a wide range of concepts that differ in important ways. It is important to be cognizant of which concept (or modified concept) one is using and its aims, assumptions, and appropriate uses. Although Taylor et al. (2006) argued that we should return to the original definition of connectivity, i.e. Merriam connectivity, given that all three concepts are in wide and varied use under the same name, we suggest that this is no longer a realistic goal.

Given the limitations surrounding all three concepts of landscape connectivity, there are clearly many areas needing further research. Noss connectivity and Hanski connectivity are the simplest to apply, because movement success is assumed to be constrained by habitat in the former, and immigration rate is assumed to be determined by patch isolation in the latter. In principle, both of these can be measured using only knowledge of habitat and species distribution and possibly an estimate of the species' gap-crossing ability. In contrast, Merriam connectivity requires information about how the species responds to all cover types in the landscape or its movement success in different landscapes. Research is needed to identify the situations in which the different concepts of connectivity are meaningful and can be applied.

In the applied world, Noss connectivity is ruling the day with the application of least cost path analysis, graph theory, and circuit theory. What are the implications of this for conservation? Conservation policies that increase Noss connectivity by applying habitat-based connectivity metrics for land planning will increase habitat conservation because the dominant assumption in Noss connectivity is that habitat = connectivity. This is usually a good outcome for species conservation, because habitat loss is probably the main threat to species conservation.

However, if the assumptions inherent in Noss connectivity do not apply in a particular situation, this can lead to less-than-optimal and constrained land planning decisions for conservation.

The use of Noss connectivity constrains the prioritization and selection of habitats for preservation/restoration to those that have been identified as important in a connectivity analysis. This limits options and increases cost. For example, Beier and Noss (1998) argued that evidence of animal movements in corridors is the same as evidence that corridors are important for conservation. This is a leap in logic that effectively undervalues the protection of habitat that is not in a habitat corridor (Keeley et al. 2017). The benefit of corridors for conservation needs to be demonstrated and not simply assumed. Second, Noss connectivity is used as a reason to limit conservation to contiguous habitat and to offer little or no protection to small patches of habitat even when they are numerous. This prioritization of large, contiguous habitat areas is not supported by evidence (Fahrig 2017, Wintle et al. 2019). Third, the dominant focus on habitat means that there is too little attention placed on matrix quality and pattern. This is problematic for conservation because many species appear to move preferentially through the matrix (see examples earlier).

The assumed strong linkage between movement and population success that underlies conservation decisions based on connectivity is also risky for conservation in some situations. For example, wildlife overpasses and underpasses have become common approaches to mitigate the impacts of roads on wildlife populations, based on the assumption that the movement barrier effect of roads is critical. However, if the mortality effect of roads on population persistence is larger than the barrier effect, then the emphasis should be on measures that keep animals off roads rather than measures to facilitate movement across them (Teixeira et al. 2020). In general, research is needed to identify the relative importance of barriers to movement versus other landscape impacts for population persistence. For example, fencing likely has large effects on connectivity for many large animals (Jakes et al. 2018), but there is little research that would allow us to understand the situations in which this has, versus does not have, a net negative impact on population persistence. Current connectivity research ignores the fact that the effect of the landscape on population success is not only, or even mostly, determined by movement success (Figure 4.3), because landscape structure influences other processes, e.g. species interactions, that also affect reproduction and mortality.

On the other hand, a benefit of Noss connectivity and Hanski connectivity over Merriam connectivity is that they can be measured with reference only to habitat distribution and without reference to a particular species. Connectivity metrics are assumed to apply to all species associated with a particular habitat type. This is very helpful in the context of landscape management. It avoids the dilemma in applying Merriam connectivity, namely, that connectivity is highly species-specific while land management must be cross-species. To get around this problem, Cushman et al. (2013) suggest selecting a set of umbrella species, doing a Merriam connectivity estimate for each one (Figure 4.1), and combining the results into a cross-species connectivity map. However, research is needed to determine whether this approach can work, as even closely related species can have very different responses to landscape structure (Henry et al. 2019). In addition, it may be difficult to simultaneously maximize connectivity for desirable species without increasing the spread of undesirable or exotic species and without increasing biotic homogenization (Olden and Rooney 2006).

All this begs the question of whether connectivity is an effective paradigm for land management aimed at species conservation. Is there a more practical approach that does not require species-level information about movement success? For example, Gagné et al. (2015) suggest an alternative approach that does not explicitly consider connectivity but rather, involves a prioritized sequence of decisions to maximize conservation of natural habitats, maximize the diversity of natural habitats, minimize impacts on fresh water systems, and minimize the marginal impacts of new human activities. In the meantime, in situations where connectivity metrics are used to make land management decisions, we recommend that researchers and practitioners be aware of the particular concept of connectivity within which they are working (Figure 4.1) and especially of its assumptions and limitations. Before a particular approach is applied, evidence should be provided demonstrating that the selected approach is actually appropriate in the particular situation.

Acknowledgments

We are grateful for comments from the GLEL Friday discussion group, including Joe Bennett, Andrea Clouston, Daniel Cook, Joan Freeman, Josie Hughes, Jochen Jaeger, Sahebeh Karimi, Hsien-Yung Lin, Iman Momeni, Amanda Martin, Jamie McLaren, Peter Morrison, Anna Tran Nguyen, Dave Omond, Karine Pigeon, Richard Pither, Lutz Tischendorf, Elise Urness, and Jaimie Vincent.

References

Aavik, T., Holderegger, R. and Bolliger, J. (2014) 'The structural and functional connectivity of the grassland plant *Lychnis flos-cuculi*', *Heredity*, vol 112, pp471–478.

Anadón, J.D., Pérez-García, J.M., Pérez, I., Royo, J. and Sánchez-Zapata, J.A. (2018) 'Disentangling the effects of habitat, connectivity and interspecific competition in the range expansion of exotic and native ungulates', *Landscape Ecology*, vol 33, pp597–608.

Arroyo-Rodríguez, V., Pérez-Elissetche, G.K., Ordóñez-Gómez, J.D., González-Zamora, A., Chaves, O.M., Sánchez-López, S., Chapman, C.A., Morales-Hernández, K., Pablo-Rodríguez, M. and Ramos-Fernández, G. (2017) 'Spider monkeys in human-modified landscapes: the importance of the matrix', *Tropical Conservation Science*, vol 10, pp1–30.

Aufrett, A.G., Rico, Y., Bullock, J.M., Hooftman, D.A.P., Pakeman, R.J., Soons, M.B., Suárez-Esteban, A., Traveset, A., Wagner, H.H. and Cousins, S.A.O. (2017) 'Plant functional connectivity – integrating landscape structure and effective dispersal', *Journal of Ecology*, vol 105, pp1648–1656.

Ayram, C.A.C., Mendoza, M.E., Etter, A. and Salicrup, D.R.P. (2016) 'Habitat connectivity in biodiversity conservation: a review of recent studies and applications', *Progress in Physical Geography*, vol 40, pp7–37.

Baguette, M., Petit, S. and Quéva, F. (2000) 'Population spatial structure and migration of three butterfly species within the same habitat network: consequences for conservation', *Journal of Applied Ecology*, vol 37, pp100–108.

Baguette, M., Blanchet, S., Legrand, D., Stevens, V.M. and Turlure, C. (2013) 'Individual dispersal, landscape connectivity and ecological networks', *Biological Reviews*, vol 88, pp310–326.

Baudry, J. and Merriam, H.G. (1988) 'Connectivity and connectedness: functional versus structural patterns in landscapes', in Schreiber, K.F. (ed) Connectivity in Landscape Ecology. Proceedings of the 2nd seminar of the International Association for Landscape Ecology, Munster.

Bartoń, K.A., Phillips, B.L., Morales, J.M. and Travis, M.J. (2009) 'The evolution of an "intelligent" dispersal strategy: biased, correlated random walks in patchy landscapes', *Oikos*, vol 118, pp309–319.

Beier, P. and Noss, R.F. (1998) 'Do habitat corridors provide connectivity?', *Conservation Biology*, vol 12, pp1241–1252.

Belford, D.A. and Gould, W.R. (1989) 'An evaluation of trout passage through six highway culverts in Montana', *North American Journal of Fisheries Management*, vol 9, pp437–445.

Bender, D.J. and Fahrig, L. (2005) 'Matrix spatial structure obscures the relationship between inter-patch movement and patch size and isolation', *Ecology*, vol 86, pp1023–1033.

Bender, D.J., Tischendorf, L. and Fahrig, L. (2003) 'Using patch isolation metrics to predict animal movement in binary landscapes', *Landscape Ecology*, vol 18, pp17–39.

Bennett, J.R. and Arcese, P. (2013) 'Human influence and classical biogeographic predictors of rare species occurrence', *Conservation Biology*, vol 27, pp417–421.

Bowman, J. and Fahrig, L. (2002) 'Gap crossing by chipmunks: an experimental test of landscape connectivity', *Canadian Journal of Zoology*, vol 80, pp1556–1561.

Bowne, D.R. and Bowers, M.A. (2004) 'Interpatch movements in spatially structured populations: a literature review', *Landscape Ecology*, vol 19, pp1–20.

Brooks, A.J., Wolfenden, B., Downes, B.J. and Lancaster, J. (2017) 'Do pools impede drift dispersal by stream insects?', *Freshwater Biology*, vol 62, pp1578–1586.

Brooks, C.P. (2003) 'A scalar analysis of landscape connectivity', *Oikos*, vol 102, pp433–439.

Bubb, D.H., Thom, T.J. and Lucas, M.C. (2004) 'Movement and dispersal of the invasive signal crayfish *Pacifastacus leniusculus* in upland rivers', *Freshwater Biology*, vol 49, pp357–368.

Calabrese, J.M. and Fagan, W.F. (2004) 'A comparison-shopper's guide to connectivity metrics', *Frontiers in Ecology and Environment*, vol 2, pp529–536.

Carson, H.S., Cook, G.S., López-Duarte, P.C. and Levin, L.A. (2011) 'Evaluating the importance of demographic connectivity in a marine metapopulation', *Ecology*, vol 92, pp1972–1984.

Chin, K.S. and Taylor, P.D. (2009) 'Interactive effects of distance and matrix on the movements of a peatland dragonfly', *Ecography*, vol 32, pp715–722.

Cline, B.B. and Hunter, M.L. (2016) 'Movement in the matrix: substrates and distance-to-forest edge affect postmetamorphic movements of a forest amphibian', *Ecosphere*, vol 7, art e01202.

Connolly, S.R., Menge, B.A. and Roughgarden, J. (2001) 'A latitudinal gradient in recruitment of intertidal invertebrates in the northeast Pacific Ocean', *Ecology*, vol 82, pp1799–1813.

Cowen, R.K. and Sponaugle, S. (2009) 'Larval dispersal and marine population connectivity', *Annual Review of Marine Science*, vol 1, pp443–466.

Cowen, R.K., Paris, C.B. and Srinivasan, A. (2006) 'Scaling of connectivity in marine populations', *Science*, vol 311, pp522–527.

Cushman, S.A., McRae, B., Adriaensen, F., Beier, P., Shirley, M. and Zeller, K. (2013) 'Biological corridors and connectivity', in Macdonald, D.W. and Willis, K.J. (eds) *Key Topics in Conservation Biology 2*, First edition. Wiley, Hoboken.

Cushman, S.A., Max, T., Meneses, N., Evans, L.M., Ferrier, S., Honchak, B., Whitham, T.G. and Allan, G.J. (2014) 'Landscape genetic connectivity in a riparian foundation tree is jointly driven by climatic gradients and river networks', *Ecological Applications*, vol 24, pp1000–1014.

d'Acampora, B.H.A., Higueras, E. and Román, E. (2018) 'Combining different metrics to measure the ecological connectivity of two mangrove landscapes in the Municipality of Florianópolis, Southern Brazil', *Ecological Modelling*, vol 384, pp103–110.

den Boer, P.J. (1968) 'Spreading of risk and stabilization of animal numbers', *Acta Biotheoretica*, vol 18, pp165–192.

Desrochers, A. and Hannon, S.J. (1997) 'Gap crossing by forest songbirds during the post-fledging period', *Conservation Biology*, vol 11 pp1204–1210.

Duflot, R., Daniel, H., Aviron, S., Alignier, A., Beaujouan, V., Burel, F., Cochard, A., Ernoult, A., Pain, G. and Pithon, J.A. (2018) 'Adjacent woodlands rather than habitat connectivity influence grassland plant, carabid and bird assemblages in farmland landscapes', *Biodiversity Conservation*, vol 27, pp1925–1942.

Erman, N.A. (1986) 'Movements of self-marked caddisfly larvae, *Chyranda centralis* (Trichoptera: Limnephilidae) in a Sierran spring stream', *Freshwater Biology*, vol 16, pp455–464.

Fahrig, L. (2003) 'Effects of habitat fragmentation on biodiversity', *Annual Reviews of Ecology, Evolution and Systematics*, vol 34, pp487–515.

Fahrig, L. (2007) 'Non-optimal animal movement in human-altered landscapes', *Functional Ecology*, vol 21, pp1003–1015.

Fahrig, L. (2013) 'Rethinking patch size and isolation effects: the habitat amount hypothesis', *Journal of Biogeography*, vol 40, pp1649–1663.

Fahrig, L. (2017) 'Ecological responses to habitat fragmentation per se', *Annual Reviews of Ecology, Evolution and Systematics*, vol 48, pp1–23.

Fahrig, L. and Merriam, H.G. (1985) 'Habitat patch connectivity and population survival'. *Ecology*, vol 66, pp1762–1768.

Fahrig, L. and Paloheimo, J.E. (1988) 'Effect of spatial arrangement of habitat patches on local population size', *Ecology*, vol 69, pp468–475.

Fahrig, L., Lefkovitch, L.P. and Merriam, H.G. (1983). 'Population stability in a patchy environment', in Lauenroth, W.K., Skogerboe, G.V. and Flug, M. (eds) *Analysis of Ecological Systems: State-of-the-art in Ecological Modelling*. Elsevier, New York.

Fahrig, L., Arroyo-Rodríguez, V., Bennett, J.R., Boucher-Lalonde, V., Cazetta, E., Currie, D.J., Eigenbrod, F., Ford, A.T., Harrison, S.P., Jaeger, J.A.G., Koper, N., Martin, A.E., Martin, J-L., Metzger, J.P., Morrison, P., Rhodes, J.R., Saunders, D.A., Simberloff, D., Smith, A.C., Tischendorf, L., Vellend, M. and Watling, J.I. (2019) 'Is habitat fragmentation bad for biodiversity?', *Biological Conservation*, vol 230, pp179–186.

Fischer, J. and Lindenmayer, D.B. (2007) 'Landscape modification and habitat fragmentation: a synthesis', *Global Ecology and Biogeography*, vol 16, pp265–280.

Fletcher, R.J., Sefair, J.A., Wang, C., Poli, C.L., Smith, T.A.H., Bruna, E.M., Holt, R.D., Barfield, M., Marx, A.J. and Acevedo, M.A. (2019) 'Towards a unified framework for connectivity that disentangles movement and mortality in space and time', *Ecology Letters*, vol 22, pp1680–1689.

Flores-Manzanero, A., Luna-Bárcenas, M.A., Dyer, R.J. and Vázquez-Domínguez, E. (2019) 'Functional connectivity and home range inferred at a microgeographic landscape genetics scale in a desert-dwelling rodent', *Ecology and Evolution*, vol 9, pp437–453.

Ford, A.T. and Fahrig, L. (2008) 'Movement patterns of eastern chipmunks (*Tamias strictus*) near roads', *Journal of Mammalogy*, vol 89, pp895–903.

Franklin, J.F. (1993) 'Preserving biodiversity: species, ecosystems, or landscapes?', *Ecological Applications*, vol 3, pp202–205.

Fraser, G.S. and Stutchbury, B.J.M. (2004) 'Area-sensitive forest birds move extensively among forest patches', *Biological Conservation*, vol 118, pp377–387.

Gagné, S.A., Eigenbrod, F., Bert, D., Cunnington, G.M., Olson, L.T., Smith, A.C. and Fahrig, L. (2015) 'A simple landscape design framework for biodiversity conservation', *Landscape and Urban Planning*, vol 136, pp13–27.

Galán-Acedo, C., Arroyo-Rodríguez, V., Andresen, E., Arregoitia, L.V., Vega, E., Peres, C.A. and Ewers, R.M. (2019) 'The conservation value of human-modified landscapes for the world's primates', *Nature Communications*, vol 10, art 152.

Geoffroy, C., Fiola, M-L., Bélisle, M. and Villard, M-A. (2019) 'Functional connectivity in forest birds: evidence for species specificity and anisotropy', *Landscape Ecology*, vol 34, pp1363–1377.

Gilbert, B. and Levine, J.M. (2013) 'Plant invasions and extinction debts', *PNAS*, vol 110, pp1744–1749.

Goodwin, B.J. and Fahrig, L. (2002a) 'Effect of landscape structure on the movement behaviour of a specialized goldenrod beetle, *Trirhabda borealis*', *Canadian Journal of Zoology*, vol 80, pp24–35.

Goodwin, B.J. and Fahrig, L. (2002b) 'How does landscape structure influence landscape connectivity?', *Oikos*, vol 99, pp552–570.

Haddad, N.M. and Tewksbury, J.J. (2005) 'Low-quality habitat corridors as movement conduits for two butterfly species', *Ecological Applications*, vol 15, pp250–257.

Hanski, I. (1994) 'A practical model of metapopulation dynamics', *Journal of Animal Ecology*, vol 63, pp151–162.

Hanski, I. (1999) 'Habitat connectivity, habitat continuity, and metapopulations in dynamic landscapes', *Oikos*, vol 87, pp209–219.

Harrison, S. (1991) 'Local extinction in a metapopulation context: an empirical evaluation', *Biological Journal of the Linnean Society*, vol 42, pp73–88.

Hass, A.L., Kormann, U., Tscharntke, T., Clough, Y., Baillod, A.B., Sirami, C., Fahrig, L., Martin, J-L., Baudry, J., Bertrand, C., Bosch, J., Brotons, L., Burel, F., Georges, R., Giralt, D., Marcos-García, M.A., Ricarte, A., Siriwardena, G. and Batáry, P. (2018) 'Landscape configurational heterogeneity by small-scale agriculture, not crop diversity, maintains pollinators and plant reproduction in Western Europe', *Proceedings of the Royal Society of London B*, vol 285, art 20172242.

Hauenstein, S., Fattebert, J., Grüebler, M.U., Naef-Daenzer, B., Pe'er, G. and Hartig, F. (2019) 'Calibrating an individual-based movement model to predict functional connectivity for little owls', *Ecological Applications*, vol 29, art e01873.

Hayashi, F. and Nakane, M. (1989) 'Radio-tracking and activity monitoring of the dobsonfly larva, *Protohermes grandis* (Megaloptera: Corydalidae)', *Oecologia*, vol 78, pp468–472.

Henry, E., Brammer-Robbins, E., Aschehoug, E. and Haddad, N. (2019) 'Do substitute species help or hinder endangered species management?', *Biological Conservation*, vol 232, pp127–130.

Herrmann, J.D., Carlo, T.A., Brudvig, L.A., Damschen, E.I., Haddad, N.M., Levey, D.J., Orrock, J.L. and Tewksbury, J.J. (2016) 'Connectivity from a different perspective: comparing seeds dispersal kernels in connected versus unfragmented landscape', *Ecology*, vol 97, pp1274–1282.

Horskins, K., Mather, P.B. and Wilson, J.C. (2006) 'Corridors and connectivity: when use and function do not equate', *Landscape Ecology*, vol 21, pp641–655.

Howell, P.E., Muths, E., Hossack, B.R., Sigafus, B.H. and Chandler, R.B. (2018) 'Increasing connectivity between metapopulation ecology and landscape ecology', *Ecology*, vol 99, pp1119–1128.

Jackson, J.K., Mcelravy, E.P. and Resh, V.H. (1999) 'Long-term movements of self-marked caddisfly larvae (Trichoptera: Sericostomatidae) in a California coastal mountain stream', *Freshwater Biology*, vol 42, pp525–536.

Jaeger, J.A.G. (2000) 'Landscape division, splitting index, and effective mesh size: new measures of landscape fragmentation', *Landscape Ecology*, vol 15, pp115–130.

Jakes, A.F., Jones, P.F., Paige, L.C., Seidler, R.G., Juijser, M.P. (2018) 'A fence runs through it: a call for greater attention to the influence of fences on wildlife and ecosystems', *Biological Conservation*, vol 227, pp310–318.

Jones, G., Milicich, M., Emslie, M. and Lunow, C. (1999) 'Self-recruitment in a coral reef fish population', *Nature*, vol 402, art 802.

Kamm, U., Gugerli, F., Rotach, P., Edwards, P. and Holderegger, R. (2010) 'Open areas in a landscape enhance pollen-mediate gene flow of a tree species: evidence from northern Switzerland', *Landscape Ecology*, vol 25, pp903–911.

Kärvemo, S., Johansson, V., Schroeder, M. and Ranius, T. (2016) 'Local colonization-extinction dynamics of a tree-killing bark beetle during a large-scale outbreak', *Ecosphere*, vol 7, art e01257.

Keeley, A.T.H., Beier, P., Keeley, B.W. and Fagan, M.E. (2017) 'Habitat suitability is a poor proxy for landscape connectivity during dispersal and mating movements', *Landscape and Urban Planning*, vol 161, pp90–102.

Lamb, C.T., Ford, A.T., Proctor, M.F., Royle, J.A., Mowat, G. and Boutin, S. (2019) 'Genetic tagging in the Anthropocene: scaling ecology from alleles to ecosystems', *Ecological Applications*, vol 29, art e01876.

Lancaster, J., Downes, B.J. and Arnold, A. (2011) 'Lasting effects of maternal behaviour on the distribution of a dispersive stream insect', *Journal of Animal Ecology*, vol 80, pp1061–1069.

LaPoint, S., Gallery, P., Wikelski, M. and Kays, R. (2013) 'Animal behavior, cost-based corridor models, and real corridors', *Landscape Ecology*, vol 28, pp1615–1630.

Lefkovitch, L.P. and Fahrig, L. (1985) 'Spatial characteristics of habitat patches and population survival', *Ecological Modelling*, vol 30, pp297–308.

Leidner, A.K. and Haddad, N.M. (2010) 'Natural, not urban, barriers define population structure for a coastal endemic butterfly', *Conservation Genetics*, vol 11, pp2311–2320.

Levin, L.A. (2006) 'Recent progress in understanding larval dispersal: new directions and digressions', *Integrative and Comparative Biology*, vol 46, pp282–297.

Levins, R. (1969) 'Some demographic and genetic consequences of environmental heterogeneity for biological control', *Bulletin of the Entomological Society of America*, vol 15, pp237–240.

Martin, A.E. and Fahrig, L. (2015) 'Matrix quality and disturbance frequency drive evolution of species behaviour at habitat boundaries', *Ecology and Evolution*, vol 5, pp5792–5800.

McClure, M.L., Hansen, A.J. and Inman, R.M. (2016) 'Connecting models to movements: testing connectivity model predictions against empirical migration and dispersal data', *Landscape Ecology*, vol 31, pp1419–1432.

McRae, B.H., Dickson, B.G., Keitt, T.H. and Shah, V.B. (2008). 'Using circuit theory to model connectivity in ecology and conservation', *Ecology*, vol 10, pp2712–2724.

Meegan, R.P. and Maehr, D.S. (2002) 'Landscape conservation and regional planning for the Florida panther', *Southeastern Naturalist*, vol 1, pp217–232.

Mendenhall, C.D., Frishkoff, L.O., Santos-Barrera, G., Pacheco, J., Mesfun, E., Quijano, F.M., Ehrlich, P.R., Ceballos, G., Daily, G.C. and Pringle, R.M. (2014) 'Countryside biogeography of Neotropical reptiles and amphibians', *Ecology*, vol 95, pp856–870.

Merriam, G. (1984) 'Connectivity: a fundamental ecological characteristic of landscape pattern', in Brandt, J. and Agger, P.A. (eds) Proceedings of the First International Seminar on Methodology in Landscape Ecological Research and Planning. Rosskilde University Centre, Rosskilde.

Moilanen, A. (2011) 'On the limitations of graph-theoretic connectivity in spatial ecology and conservation', *Journal of Applied Ecology*, vol 48, pp1543–1547.

Moilanen, A. and Hanski, I. (1998) 'Metapopulation dynamics: effects of habitat quality and landscape structure', *Ecology*, vol 79, pp2503–2515.

Moilanen, A. and Nieminen, M. (2002) 'Simple connectivity measures in spatial ecology', *Ecology*, vol 83, pp1131–1145.

Mueller, T., Lenz, J., Caprano, T., Fiedler, W. and Böhning-Gaese, K. (2014) 'Large frugivorous birds facilitate connectivity of fragmented landscapes', *Journal of Applied Ecology*, vol 51, pp684–692.

Nathan, R., Horn, H.S., Chave, J. and Levin, S.A. (2002) '5 mechanistic models for tree seed dispersal by wind in dense forest and open landscapes', in Levey, D.J., Silva, W.R. and Galetti, M. (eds) *Seed Dispersal and Frugivery: Ecology, Evolution and Conservation*. CABI Publishing, Wallingford.

Noss, R.F. (1987) 'From plant communities to landscapes in conservation inventories: a look at The Nature Conservancy (USA)', *Biological Conservation*, vol 41, pp11–37.

Nowicki, P., Vrabec, V., Binzenhöfer, B., Feil, J., Zakšek, B., Hovestadt, T. and Settele, J. (2014) 'Butterfly dispersal in inhospitable matrix: rare, risky, but long-distance', *Landscape Ecology*, vol 29, pp401–412.

Olden, J.D. and Rooney, T.P. (2006) 'On defining and quantifying biotic homogenization', *Global Ecology and Biogeography*, vol 15, pp113–120.

Ouin, A., Martin, M. and Burel, F. (2008) 'Agricultural landscape connectivity for the meadow brown butterfly (*Maniola jurtina*)', *Agriculture, Ecosystems and Environment*, vol 124, pp193–199.

Pelletier, D., Lapointe, M-E., Wulder, M.A., White, J.C. and Cardille, J.A. (2017) 'Forest connectivity regions of Canada using circuit theory and image analysis', *PLOS ONE*, vol 12, art e0169428.

Prugh, LR. (2009) 'An evaluation of patch connectivity measures', *Ecological Applications*, vol 19, pp1300–1310.

Ranius, T., Johansson, V. and Fahrig, L. (2010) 'A comparison of patch connectivity measures using data on invertebrates in hollow oaks', *Ecography*, vol 33, pp1–8.

Ricketts, T.H. (2001) 'The matrix matters: effective isolation in fragmented landscapes', *American Naturalist*, vol 158, pp87–99.

Rico, A., Kindlmann, P. and Sedlácek, F. (2007) 'Barrier effects of roads on movements of small mammals', *Folia Zoologica*, vol 56, pp1–12.

Rittenhouse, T.A.G. and Semlitsch, R.D. (2006) 'Grasslands as movement barriers for a forest-associated salamander: migration behavior of adult and juvenile salamanders at a distinct habitat edge', *Biological Conservation*, vol 131, pp14–22.

Roe, J.H., Brinton, A.C. and Georges, A. (2009) 'Temporal and spatial variation in landscape connectivity for a freshwater turtle in a temporally dynamic wetland system', *Ecological Applications*, vol 19, pp1288–1299.

Saura, S., Estreguil, C., Mouton, C. and Rodríguez-Freire, M. (2011) 'Network analysis to assess landscape connectivity trends: application to European forests (1990–2000)', *Ecological Indicators*, vol 11, pp407–416.

Scharf, A.K., Belant, J.L., Beyer, D.E., Wikelski, M. and Safi, K. (2018) 'Habitat suitability does not capture the essence of animal-defined corridors', *Movement Ecology*, vol 6, art 18.

Schneider, D.W. and Lyons, J. (1993) 'Dynamics of upstream migration in two species of tropical freshwater snails', *Journal of the North American Benthological Society*, vol 12, pp3–16.

Schultz, C.B., Franco, A.M.A. and Crone, E.E. (2012) 'Response of butterflies to structural and resource boundaries', *Journal of Animal Ecology*, vol 81, pp724–734.

Shima, J.S., Noonburg, E.G. and Phillips, N.E. (2010) 'Life history and matrix heterogeneity interact to shape metapopulation connectivity in spatially structured environments', *Ecology*, vol 91, pp1215–1224.

Sieving, K.E., Willson, M.F. and De Santo, T.L. (1996) 'Habitat barriers to movement of understory birds in fragmented south-temperate rainforest', *The Auk*, vol 113, pp944–949.

Tamario, C., Calles, O., Watz, J., Nilsson, P.A. and Degerman, E. (2019) 'Coastal river connectivity and the distribution of ascending juvenile European eel (*Anguilla anguilla* L.): Implications for conservation strategies regarding fish-passage solutions', *Aquatic Conservation: Marine and Freshwater Ecosystems*, vol 29, art 3064.

Tarabon, S., Bergès, L., Dutoit, T. and Isselin-Nondedeu, F. (2019) 'Maximizing habitat connectivity in the mitigation hierarchy. A case study on three terrestrial mammals in an urban environment', *Journal of Environmental Management*, vol 243, pp340–349.

Taylor, P.D., Fahrig, L., Henein, K. and Merriam, G. (1993) 'Connectivity is a vital element of landscape structure', *Oikos*, vol 68, pp571–573.

Taylor, P.D., Fahrig, L. and With, K. (2006) 'Landscape connectivity: a return to basics', in Crooks, K.R. and Sanjayan, M. (eds) *Connectivity Conservation*. Cambridge University Press, Cambridge.

Teixeira, F.Z., Rytwinski, T. and Fahrig, L. (2020) 'Inference in road ecology research: what we know versus what we think we know', *Biology Letters*, vol 16, 20200140.

Thiem, J., Binder, T., Dumont, P., Hatin, D., Hatry, C., Katopodis, C., Stamplecoskie, K. and Cooke, S.J. (2013) 'Multispecies fish passage behaviour in a vertical slot fishway on the Richelieu River, Quebec, Canada', *River Research and Applications*, vol 29, pp582–592.

Thorrold, S.R., Zacherl, D.C. and Levin, L.A. (2007) 'Population connectivity and larval dispersal: using geochemical signatures in calcified structures', *Oceanography*, vol 20, pp80–89.

Tischendorf, L. and Fahrig, L. (2000) 'On the usage and measurement of landscape connectivity', *Oikos*, vol 90, pp7–19.

Tischendorf, L., Bender, D.J. and Fahrig, L. (2003) 'Evaluation of patch isolation metrics in mosaic landscapes for specialist versus generalist dispersers', *Landscape Ecology*, vol 18, pp41–50.

Trapp, S.E., Day, C.C., Flaherty, E.A., Zollner, P.A. and Smith, W.P. (2019) 'Modeling impacts of landscape connectivity on dispersal movements of northern flying squirrels (*Glaucomys sabrinus griseifrons*)', *Ecological Modelling*, vol 394, pp44–52.

Travis, J.M.J., Mustin, K., Bartoń, K.A., Benton, T.G., Clobert, J., Delgado, M.M., Dytham, C., Hovestadt, T., Palmer, S.C.F., Van Dyck, H. and Bonte, D. (2012) 'Modelling dispersal: an eco-evolutionary framework incorporating emigration, movement, settlement behavior and the multiple costs involved', *Methods in Ecology and Evolution*, vol 3, pp628–641.

Tucker, M.A., Böhning-Gaese, K., Fagan, W.F., et al. (2018) 'Moving in the Anthropocene: global reductions in terrestrial mammalian movements', *Science*, vol 359, pp466–469.

Tucker, M.A., Alexandrou, O., Bierregaard, R.O., *et al.* (2019) 'Large birds travel farther in homogeneous environments', *Global Ecology and Biogeography*, vol 28, art 12875.

Urban, D. and Keitt, T. (2001) 'Landscape connectivity: a graph-theoretic perspective', *Ecology*, vol 82, pp1205–1218.

Vanbianchi, C., Gaines, W.L., Murphy, M.A. and Hodges, K.E. (2018) 'Navigating fragmented landscapes: Canada lynx brave poor quality habitats while traveling', *Ecology and Evolution*, vol 8, pp11293–11308.

Vandermeer, J. and Carvajal, R. (2001) 'Metapopulation dynamics and the quality of the matrix', *American Naturalist*, vol 158, pp211–220.

Werner, F.E., Cowen, R.K. and Paris, C.B. (2007) 'Coupled biological and physical models: present capabilities and necessary developments for future studies of population connectivity', *Oceanography*, vol 20, pp54–69.

Wintle, B.A., Kujala, H., Whitehead, A., Cameron, A., Veloz, S., Kukkala, A., Moilanen, A., Gordon, A., Lentini, P.E., Cadenhead, N.C.R. and Bekessy, S.A. (2019) 'Global synthesis of conservation studies reveals the importance of small habitat patches for biodiversity', *PNAS*, vol 116, pp909–914.

Zeller, K.A., Jennings, M.K., Vickers, T.W., Ernest, H.B., Cushman, S.A. and Boyce, W.M. (2018) 'Are all data types and connectivity models created equal? Validating common connectivity approaches with dispersal data', *Diversity and Distributions*, vol 24, pp868–879.

Part II

Landscape processes

5

Spatially structured ecosystems, connectivity, and movement

*Finnbar Lee, Jingjing Zhang, Craig Eric Simpkins,
Justine A. Becker, and George L.W. Perry*

Introduction

Central to landscape ecology is the view that spatial heterogeneity is ubiquitous, occurs at multiple scales, and reciprocally interacts with ecological processes. This spatial structure influences the demographic processes that determine population and community structure and function. As a result, understanding these dynamics has become ever more pressing as stressors such as habitat loss and fragmentation increase in intensity. Many populations are spatially structured (Figure 5.1), which reflects the distribution of the resources that limit species and the emergent variation in demographic rates (e.g. reproduction, growth, and mortality). Spatial structuring of populations can take many forms, however; at one extreme lies populations of the same species in the same landscape that are entirely isolated from each other, and at the other are networks of species with local populations that frequently exchange individuals. As Kritzer and Sale (2004) point out, these local populations are linked by dispersal, and hence, the spatio-temporal scales at which organisms perceive and interact with the landscape are crucial. During an organism's lifetime, it may undertake a range of movement processes, including foraging for resources, seasonal migration, and natal dispersal, all of which link populations and the landscapes they inhabit. At this point, it is worth noting that much of the work on spatially structured populations has focused on animals. However, many of the dynamics we review in this chapter are relevant to plant populations and communities (Freckleton and Watkinson 2002), even if their scaling, especially in time, is different.

In this chapter, we will describe metapopulation (networks of local populations linked by dispersal) ecology and its extension to communities and ecosystems, emphasizing how models of these systems link local (patch) and regional (landscape) dynamics. We will then consider the concept of connectivity and the methods that landscape ecologists have used to quantify it. Finally, we will review 'movement ecology' and consider how understanding the movement of individual organisms can help us to explain spatial dynamics at the landscape-level. Having described the concepts that underpin spatially structured ecological systems, we will conclude by considering the implications of this research for conservation and restoration activities. Although the body of research we review is focused on the emergence and consequences of spatial structure on population and community processes, much of it has not considered

Isolated populations **Metapopulation** **Patchy populations**

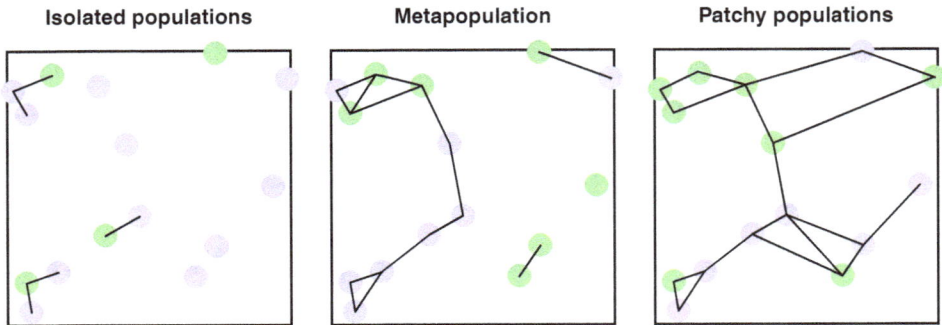

Figure 5.1 Different types of spatially structured populations along a gradient of dispersal rate from isolated populations with very infrequent immigration/emigration to patchy populations where movement between habitat patches is frequent; green patches are occupied by the species of interest and purple are unoccupied. The lines connecting habitats show dispersal with the grey-scale showing frequency.

landscape heterogeneity explicitly (e.g. by considering the landscape as habitat vs. non-habitat) and has developed largely independently of landscape ecology.

Metapopulations, metacommunities, and metaecosystems

A recurring image in ecology and evolution is that the physical landscape provides a stage upon which the eco-evolutionary drama performed by species plays out (Hutchinson 1967; Hand et al. 2015). Landscape ecology's concern with this drama has focused on how patterns on the stage influence the performance of the actors. However, much of the conceptual work we discuss in this chapter has looked more generally at the consequences of considering organisms in their spatial context. Classical ecological theory tended to ignore space, either because it led to analytical difficulties or because it meant that equilibrial solutions to models could not be resolved (Perry 2002). Many of the topics we review in this chapter have arisen more recently following the recognition that 'space matters' and most ecological systems are not equilibrial (Tilman and Kareiva 1997).

A word on the equilibrium theory of biogeography

To a lesser or greater extent, McArthur and Wilson's (1967) equilibrium theory of island biogeography (ETIB) underpins many of the concepts that we discuss in this chapter. To that end, we will succinctly summarize its key features. In its simplest form, the ETIB is concerned with the question of how many species an island will hold at equilibrium (in this case, where immigration and extinction rates are the same). McArthur and Wilson (1967) argue that island isolation will drive the rate of immigration, and island area will determine the rate of extinction; hence, small and isolated islands will hold few species, whereas large, non-isolated ones will be relatively species rich. If we imagine an archipelago of islands of different sizes and different distances from a mainland source pool, then we would expect the number of species on those islands to be a function of the rate of immigration and the rate of extinction. Species richness will equilibrate when the rate of immigration and extinction balance, even if the identity of the species themselves continues to change (Figure 5.2). The ETIB model is important for (at least!) three reasons. First, it links regional and local dynamics, although it emphasizes the

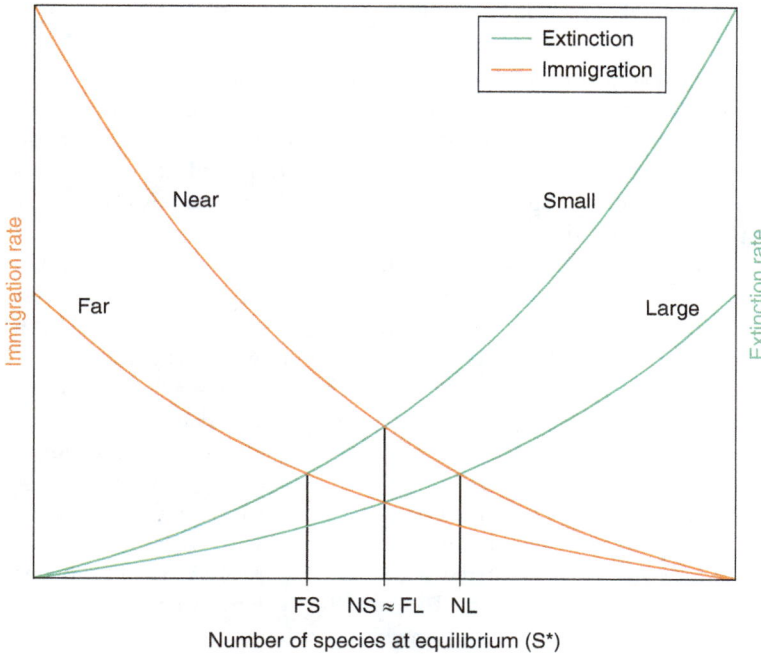

Figure 5.2   McArthur and Wilson's (1967) ETIB predicts equilibrium species richness (S*) on islands as a function of immigration (I) and extinction (E) rates on islands of different size and isolation. Large islands that are near (NL) the mainland will hold more species than near-small (NS) or far-large (FL) islands, which will be similar (but not necessarily identical), and far-small (FS) islands will hold fewer still.

former. Second, it provides an expectation of how species richness will change with island area (remembering that an island could also be a habitat patch in a sea of some other land-use type). This expectation is related to the much-studied relationship between habitat area (A) and species richness (S), where $S = cA^z$; that is, the species–area relationship (Lomolino 2000). Finally, the ETIB demonstrates the importance of connectivity (isolation) and dispersal in determining ecosystem composition. As we will discuss, this emphasis on connectivity has been recurrent in the conservation biology and landscape ecology literature since McArthur and Wilson's seminal work.

Metapopulations and source-sink dynamics

A metapopulation is 'a "population" of unstable local populations, inhabiting discrete habitat patches' (Hanski 1998, p. 41) whose dynamics are driven by local extinction and recolonization of patches. The earliest metapopulation model was described by Levins (1969) in the context of understanding the control of insect pest populations in heterogeneous environments. Levins's model is spatially implicit and represents the dynamics (in terms of occupancy) of multiple local populations that experience recurrent extinction and recolonization:

$$\frac{dp}{dt} = cp(1 - p) - mp \tag{5.1}$$

where p is the proportion of patches occupied, m is the local extinction rate, and c is the colonization rate.

In this system, patch occupancy will stabilize at $p^\star = 1 - m/c$, and so the dynamics are determined by the balance between the rate at which patches suffer local extinction and the rate at which they are recolonized (such that if $m > c$, the metapopulation suffers extinction). Occupied patches could be seen either as short-lived semelparous individuals with dispersing offspring or as local populations. Irrespective of this, the model is spatially *implicit* because although there is a spatial element, the populations are not explicitly located in a spatial context (the proportion of patches occupied is of interest, irrespective of their identity). As Warren et al. (2020) point out, whereas the ETIB is concerned with the question of 'How many species occupy an island?', Levins asks 'How many islands does a species occupy?' The ETIB describes a specific form of spatially structured population in which dispersal is from a single large patch (the core or mainland) to a group of smaller ones (the satellites or islands). Interestingly, Hanski (1998) considers metapopulation theory as intermediate between theoretical and landscape ecology on a gradient of approaches to understanding spatial ecological dynamics.

Levins's model can be expanded to consider multiple species and to be spatially explicit. The spatially explicit form is potentially more relevant to landscape ecology in terms of understanding pattern formation, but the former is of interest regarding how space per se can influence species coexistence, so we describe it here briefly. Tilman (1994) extended Levins's model to multiple species and showed that assuming strong competition–colonization trade-offs (i.e. the best competitor is the weakest disperser), many species can coexist in a spatially structured environment; this is the 'spatial competition hypothesis of diversity' (Tilman 1994, p. 7). Although this analytic approach is quite different from those most frequently used in landscape ecology, it reminds us of the fundamental ways in which recognizing space can influence our understanding and interpretation of ecological dynamics. A fundamental assumption of the spatially implicit form of the Levins model is that dispersing organisms can arrive anywhere. In reality, however, dispersal tends to be local; for example, the seed dispersal kernel (the probability density of the distance seeds travel from their parents) is typically leptokurtic (Levin et al. 2003). Lehman and Tilman (1997) explored a stochastic formulation of the Levins model on a discrete grid with the extension that the distance that dispersers travel from their origin can be constrained. This form of the model yields two important results: (i) the realized equilibrium patch occupancy under local dispersal is lower than the theoretical expectation, as propagules disperse to occupied sites and are lost, and (ii) spatial clustering of occupied cells emerges from local dispersal – showing that ecological patterns can arise in the absence of underlying abiotic heterogeneity (Figure 5.3).

Many other spatially structured population models have been developed, a particularly influential one being the source-sink dynamics model described by Pulliam (1988). In source-sink models, local populations inhabit patches of varying quality. In some of these patches (the 'sources'), the quality will be sufficiently high to support population growth ($\lambda > 1$), but in others (the 'sinks'), it will be poor, and so the population will decline ($\lambda < 1$). It is this inter-patch variation in habitat quality that distinguishes source-sink dynamics from metapopulations as described by Levins (Dias 1996). Over time, the population in the source patches will exceed the carrying capacity, and so individuals will have to leave them; those individuals may disperse to sink habitats and so buffer the local population within the sink habitats against local extinction. A critical ecological implication of source-sink dynamics is that populations can persist in sub-optimal habitat; hence, inferring quality from patch occupancy and local population size is problematic (see also van Horne 1983).

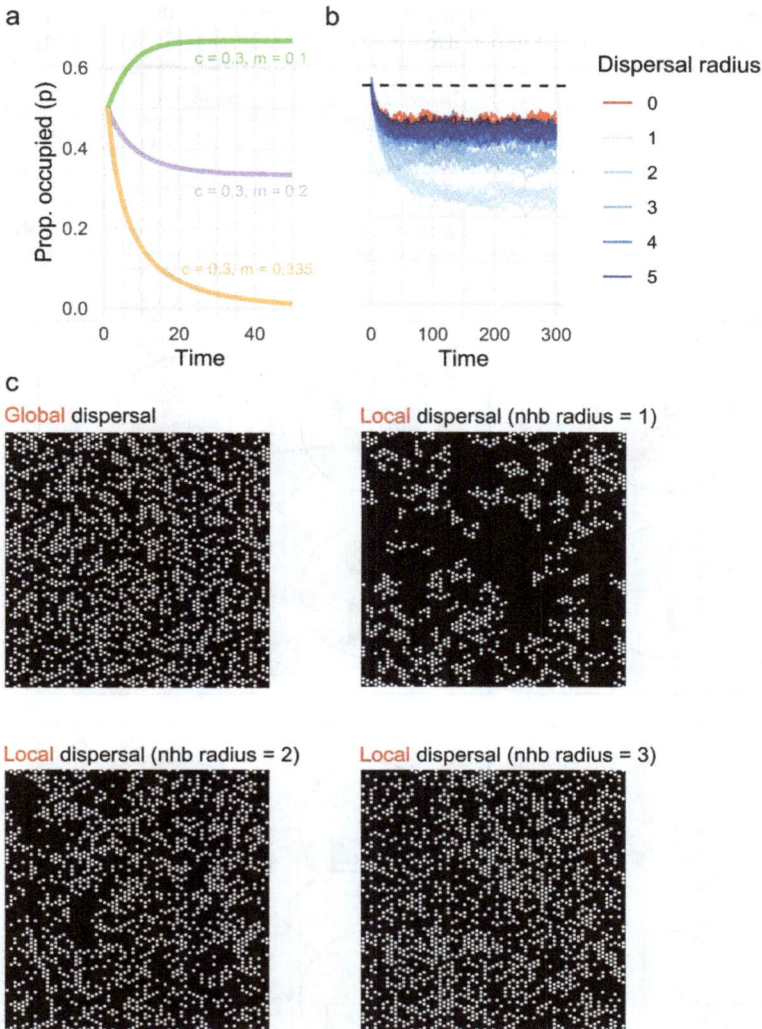

Figure 5.3 (a) Temporal dynamics of the Levins metapopulation model with different c/m ratios, (b) the effects of limited dispersal on metapopulation dynamics ($c = 0.3$, $m = 0.15$), and (c) maps showing spatial pattern of occupancy (white) under dispersal limitation.

Metacommunity theory

Metacommunity theory is an effort to link community dynamics at regional and local scales. Thompson et al. (2020, p. 3) define metacommunities as 'sets of local habitats that are connected by dispersal, and that species within each local habitat interact with each other and respond to local environmental conditions'. The key elements in a metacommunity are the local populations of species, the habitats they inhabit, and the dispersal that connects them. Leibold et al. (2004) and Leibold and Miller (2004) describe four archetypal metacommunity mechanistic processes that vary in the extent of habitat heterogeneity and strength of dispersal (Table 5.1 and

Table 5.1 There are four broad types of metacommunity dynamics depending on the level of patch homogeneity and the importance of dispersal in the system (see Leibold and Miller 2004)

		Dispersal can affect local abundance?	
		No	*Yes*
Patches homogeneous?	No	Species sorting	Mass effect
	Yes	Patch dynamics	Neutral models

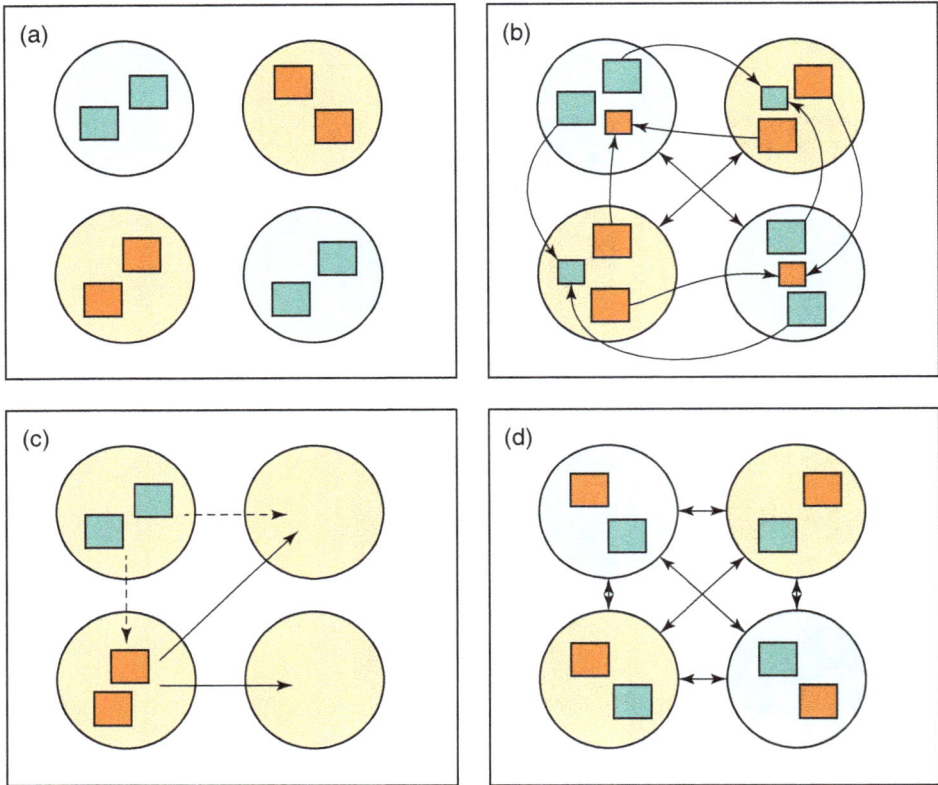

Figure 5.4 Representation of the four metacommunity archetypes: (a) species sorting, (b) mass effects, (c) patch dynamics, and (d) neutral dynamics. Circles represent habitat patches of two types (colors), and squares represent populations of two species (colors). Larger squares are locally competitively superior, and solid lines are stronger dispersers than dashed lines. (Adapted from Leibold et al. 2004.)

Figure 5.4). These archetypes are species sorting, patch dynamics, mass effect (source-sink), and neutral models. The *species sorting* perspective describes variation in species composition in the metacommunity via individual species' responses to environmental heterogeneity, where different species are favored in patches with different local environmental conditions. This perspective builds on theories that community change over environmental gradients will result in a strong correlation between local species composition and the environment (Whittaker 1962; Tilman 1982). *Patch dynamics* models assume that species diversity among multiple, identical patches that undergo stochastic and deterministic extinction events is maintained via trade-offs. Typically, a trade-off in colonization and competitive ability is considered, where less competitive species can colonize patches faster than they are driven extinct by superior competitors in other patches (Leibold et al. 2004). *Mass effects* occur where dispersal and environmental heterogeneity interactively determine community composition. As in the species sorting model, patches have varying environmental conditions favoring different species. This heterogeneity is coupled with high levels of dispersal among habitat patches and results in immigration supplementing local birth rates and enhancing densities in patches where species are unfavored. The high immigration rates facilitate source-sink dynamics allowing populations to be maintained in environments that are usually outside the species' environmental range (Shmida and Wilson 1985). Finally, the *neutral perspective* describes a framework where species are identical in their dispersal, competitive, and fitness abilities. Local and regional community composition are determined by stochastic demographic processes and dispersal limitation (Hubbell 2001).

The four archetypes represent seemingly discrete mechanisms rather than a series of continuous processes; thus, there has been a need for a general framework describing the continuous nature of metacommunity processes (Logue et al. 2011; Thompson et al. 2020). The framework proposed by Thompson et al. (2020) is an effort to draw together the four endpoints described in Table 5.1. They argue that metacommunity dynamics are a function of density-independent responses to environmental conditions (species sorting by niche) and density-dependent responses to biotic conditions (species interactions) and dispersal (immigration and emigration). This framework can be expressed as follows (Thompson et al. 2020):

$$N_{ix}(t+1) = N_{ix}(t) \frac{r_{ix}(t)}{1 + \sum_{j=1}^{S} \alpha_{ij} N_{jx}(t)} + I_{ix}(t) - E_{ix}(t) \tag{5.2}$$

$N_{ix}(t+1)$ is the population size at time $t+1$, the second term describes each species' (i) response to abiotic (numerator [r_{ix}]) and biotic (denominator) conditions (i.e. density-dependent interactions, α_{ij}) and the third term immigration [I] and emigration [E] in each patch x at time t. By adjusting the terms in Equation 5.2, a continuous gradient of metacommunity structuring processes can be represented rather than just four discrete endpoints. Analysis of this model led Thompson et al. (2020, p. 1324) to a series of hypotheses about how interactions between heterogeneity, competition, and dispersal generate landscape-level patterns in composition:

- α-Diversity (local richness) increases with dispersal and niche breadth but declines as interspecific competition intensifies;
- Spatial β-diversity (species turnover in space or time) peaks under low dispersal (irrespective of niche breadth and interspecific competition) but declines under high dispersal rates;
- Temporal β-diversity is highest under intermediate dispersal but declines under high dispersal rates;
- γ-Diversity (landscape richness) is eroded under high dispersal rates, with the rate of loss controlled by the strength of competitive interactions (slower when weak).

Metaecosystem theory

Many ecosystems benefit from allochthonous materials (so-called 'spatial subsidies'); for example, many island ecosystems receive nutrients from marine ecosystems in the form of feces, regurgitated food waste, carcasses, and so forth (Polis et al. 1997). The recognition that these subsidies can exert powerful controls on ecosystem structure and function is central to the development of metaecosystem theory (Loreau et al. 2003). For example, Montago et al. (2019) estimate that allochthonous inputs increase abundance in recipient ecosystems by on average c. 45%; these increases are driven by direct and indirect effects, with the latter varying latitudinally. Animal movement underpins these flows of matter and nutrients (Schmitz et al. 2018). Different types of movement occur on different spatial and temporal scales (from daily foraging to seasonal migration to natal dispersal), and these will have different effects on resource fluxes (Gounand et al. 2018). For example, daily foraging between different environments may effectively couple them into a single ecosystem (again, this will be affected by scale), whereas movements such as natal dispersal will lead to resource flows between ecosystems that rarely exchange matter.

Putting the landscape into meta* ecology?

So, where is the landscape in metapopulation, metacommunity, and metaecosystem theory? As noted by several authors, landscape ecology and meta* theory have developed somewhat in isolation from, or at least parallel to, each other (Biswas and Wagner 2012; Almeida-Gomes et al. 2020). Although Levins's (1969) original formulation did not explicitly consider the landscape, the implications of landscape heterogeneity for metapopulation dynamics have been explored (With 2004; Fahrig 2007a) but remain much less well known for metacommunity and metaecosystem theory.

Thompson et al. (2020) describe the view of the landscape as a network of inhabitable patches embedded in a hostile matrix as 'canonical' in metacommunity theory, but they also comment on the need to go beyond this perspective. Landscape heterogeneity mediates species sorting and competitive influences, and the success of dispersal between patches, and hence, influences all three of the processes that determine metacommunity dynamics. However, incorporating landscape heterogeneity in these theories, although necessary, is not easy. On the one hand, as we will describe more later, measuring habitat quality and rates of dispersal between patches is challenging even in an age of rapid advances in animal-borne telemetry (Jacobson and Peres-Neto 2010) and remote earth observation (Giezendanner et al. 2020). However, perhaps more fundamental are questions about the landscape models adopted when studying spatially structured ecosystems (Talley 2007) and how this decision affects inferences about metacommunity dynamics (Biswas and Wagner 2012). Biswas and Wagner (2012) argued that conceiving the landscape in terms of 'contrast' – the difference in habitat quality between adjacent patches – may be more useful than distance or resistance when considering the four metacommunity archetypes (Table 5.1) described by Leibold et al. (2004). Ryberg and Fitzgerald (2016) demonstrated that landscape contrast could help to explain structure in metacommunities of reptiles. However, they also note that a challenge in moving beyond the binary view of the landscape is that it poses difficult questions about the perceptual scales at which different organisms operate (see also Chapter 3).

Studies in freshwater environments also suggest that considering landscape connectivity from a more biological perspective can provide insights into metacommunity organization (Almeida-Gomes et al. 2020). For example, Kärnä et al. (2015) assessed whether overland, watercourse, or environmental distance better predicted community turnover. They reported that environmental distance slightly outperforms the other distances but also that selecting an appropriate distance requires a good understanding of the dispersal and connectivity biology in the systems they consider. Similar questions remain about how we can understand the flow of energy and matter between ecosystems more generally (Hillman et al. 2018; Erős and Lowe 2019), and addressing these will be crucial if metaecosystem ecology is to integrate community and landscape ecology approaches as envisaged by Loreau et al. (2003). Experimental studies show that landscape structure influences metaecosystem flows (Harvey et al. 2020), and recent tools such as stoichiometric distribution models (Leroux et al. 2017) provide a route to quantifying these effects over broad spatial extents. Understanding how landscape ecologists have conceptualized the deceptively complex property of connectivity and sought to quantify it is the subject of the next section of this chapter.

Connectivity

The movement of organisms links the local entities that make up a metapopulation, metacommunity, or metaecosystem. For movement between locations to be successful, the locations must be 'connected'; hence, there is an emphasis in contemporary conservation and restoration on reinstating and maintaining connectivity between landscape elements (Littlefield et al. 2019). However, precisely what constitutes connectivity and how it can be measured are contested and remain areas of active interest for spatial and landscape ecology (Calabrese and Fagan 2004).

What is connectivity?

Connectivity is central to spatially structured ecosystem dynamics because it is assumed to be related to dispersal success, and dispersal provides the critical link between elements in spatial ecological systems. Taylor et al. (1993, p. 571) define landscape connectivity as 'the degree to which the landscape facilitates or impedes movement among resource patches', and they distinguish between the processes of structural and functional connectivity. Structural connectivity describes the arrangement of the landscape elements, whereas functional connectivity is the degree to which organisms move between them. The distinction between structural and functional connectivity is fundamental. Structural connectivity focuses on the spatial relationships between landscape elements; it is not determined by a species habitat preferences or movement capability (Tischendorf and Fahrig 2000; Taylor et al. 2006; Correa Ayram et al. 2016). Functional connectivity integrates landscape structure with the perceptual, behavioral, and dispersal characteristics of a particular species and is, therefore, species *and* landscape specific (Adriaensen et al. 2003; Taylor et al. 2006). A related but occasionally conflated term, habitat connectivity, specifically refers to the degree of functional connectivity between patches of optimal habitat for an individual species.

An important determinant of how ecologists perceive connectivity (and isolation) is how they represent the landscape, whether terrestrial, aquatic, or marine. A classic representation has been to divide the landscape into discrete habitat patches that can each support a population (e.g. above a minimum size) embedded in some hostile (uninhabitable) matrix; this binary view is adopted by the theory of island biogeography and classical metapopulation theory, for example. As we describe later, this model is often further abstracted such that the patches/nodes are points (e.g. patch centers) rather than polygons. In these cases, functional connectivity may be defined based on the distance between two patches and a dispersal kernel (the probability of successful dispersal as a function of distance), which could be a step function or a continuous curve. In some instances (e.g. amphibians occupying ponds; Brooks 2004), this binary model may adequately represent the landscape. However, even if we hold to a matrix versus patch model, the matrix is unlikely to be uniformly uninhabitable – variations in matrix quality may still facilitate movement through it, and patches that are too small to support a population may still act as 'stepping stones for dispersal'. Relaxing the binary view of the landscape leads to a more realistic model of its spatial structure but opens a further series of questions about how to quantify its connectivity, especially in terms of scale.

How do we measure connectivity?

Although connectivity and its corollary, isolation, occupy central places in modern conservation theory and practice, quantifying connectivity has proven to be technically and conceptually challenging (Calabrese and Fagan 2004; Taylor et al. 2006). If we restrict ourselves to thinking about the dispersal of organisms from a patch in search of new breeding habitat (e.g. natal dispersal), then what do we need to measure to quantify connectivity? Ideally, we would have information for many organisms about the location they started from and high-resolution information about their movement trajectory and the location they 'finished' in (or their point of mortality if that occurred). Depending on the question of interest, the endpoint of movement could represent a range of alternatives from the short-term movement to access food resources to finding and establishing a territory and successfully breeding. From this information, we may be able to infer the elements in the landscape that organisms prefer to move through and how their behavior changes in them, drawing on behavioral landscape ecology (Lima and Zollner 1996) and statistical tools such as statistical state-space models (Jonsen et al. 2013).

While high-resolution biotelemetry has made tracking animal movement at high resolution possible (Kays et al. 2015), it is still logistically challenging to track many animals over long periods at high spatial resolution. Furthermore, we are often interested in predicting the consequences of changes in structural connectivity before they occur. Many disciplines (e.g. geography, archaeology, ecology, physics, and chemistry) are interested in how matter and energy flow through heterogeneous media (including landscapes), and hence, many methods have been developed to quantify connectivity. As described earlier, because dispersal is a fundamental component of metapopulation, metacommunity, and metaecosystem theory, adequately describing connectivity is vital for their empirical evaluation.

From a measurement (data) perspective, Calabrese and Fagan (2004) distinguish between three types of connectivity: structural, potential, and actual. Structural connectivity refers to the configuration and composition of the landscape (as per Taylor et al. 1993), and potential connectivity integrates structural connectivity with dispersal ability (similar to Taylor et al.'s functional connectivity). Potential connectivity measures often rely on surfaces that describe the difficulty (resistance) for organisms to move through them. Functional and potential connectivity can be measured without access to movement data and so can provide indirect estimates of connectivity. Actual connectivity draws on data describing individual movement between locations (habitats, patches) and so provides a direct measure of connectivity. To illustrate how these methods have been developed and implemented, we will consider five approaches: landscape metrics (structural); graph-theoretic approaches, including those based on least-cost paths and circuit-based methods (potential); and inference from movement and genetic information (actual).

Estimating connectivity using landscape metrics (structural)

There is a long history of landscape ecologists designing metrics to quantify various aspects of landscape structure, decomposed into composition (what is there?) and configuration (where is it?). These metrics describe the structural connectivity of a landscape in terms of, for example, the amount of different habitat types, the average distance between patches, and the average patch size (Gustafson 1998). It is hoped that quantifying the composition and configuration of the landscape can allow us to infer the ease with which an organism might move through it. However, because it does not directly incorporate biological information (e.g. movement capacity, perceptual scale), it is very much an indirect approach. Schumaker (1996) used a dispersal model to test whether landscape metrics could predict the success of virtual dispersers. Of the nine metrics evaluated, only three provided more information than patch area alone (and all three included information about patch area and perimeter), but only one of those – patch cohesion – was robust to small shifts (uncertainties) in patch perimeter. Schumaker (1996) concludes that it is crucial to think carefully about the biology of the organism of interest and the details of specific landscapes when computing geometric measures of structural connectivity. As ever with such metrics, the critical challenge is mechanistically linking ecological process and pattern.

Estimating connectivity using graph-theoretic approaches (potential)

A key component of most methods that estimate potential connectivity is a resistance surface (Zeller et al. 2012). A resistance surface is a grid-based representation of a landscape that characterizes the difficulty for an individual of some species of interest to traverse a grid cell (Etherington 2012, 2016). The value assigned to each cell is based on its landscape features or elements and represents a range of species-specific factors that influence movement, such as behavioral aversion. The first step in creating a resistance surface is obtaining data for the study landscape. Landscape data must cover a sufficiently large extent to answer the question of the study without adding artificial edge effects (Spear et al. 2010) while not being so large as to obscure key details and become computationally unwieldy. The grain of these data should also match that perceived by the target species, with Simpkins et al. (2017) showing that low-resolution landscape data substantially alter the accuracy of connectivity estimates. Landscape grids are then grouped into thematically similar categories, with the aim being to capture all the key landscape variables guiding species movements. Excluding important variables at this stage can decrease the accuracy of estimates (Simpkins et al. 2018), although as much of the landscape data used in connectivity assessments is not created solely for these studies, it is often not possible to include all the variables desired (Sawyer et al. 2011; Zeller et al. 2012).

The value for each cell in the resistance surface is further informed using empirical movement data (Etherington et al. 2014). Species movements may be captured in one of five broad categories (Zeller et al. 2012, 2018):

1. Expert opinion, which while not being genuinely empirical, is widely used, although it can introduce uncertainties that are difficult to quantify (Krueger et al. 2012);
2. Detection data, comprising single point locations of unknown individuals;
3. Relocation data, which comprises a set of multiple sequential locations of the same individual but at too low a frequency for a movement track to be inferred;
4. Pathway data, similar to relocation data but collected at a higher temporal frequency, allowing a movement track to be reliably created, e.g. geographical positioning system (GPS) tracking data; and
5. Genetic data, consisting of genetic samples from multiple individuals and locations used to calculate the genetic distances between populations.

The patch dynamics view of the landscape (i.e. binary division into habitat and matrix) is frequently adopted in spatial and landscape ecology. This view provides an obvious link to graph-theoretic and network approaches to quantifying landscape connectivity (Figure 5.5). In a spatial graph (or network), nodes are joined by links (edges) with the nodes having an exact location in space and potentially other attributes such as habitat quality (Dale and Fortin 2010; O'Sullivan 2014). The links between patches can be binary (i.e. connected or not) or weighted (e.g. a value

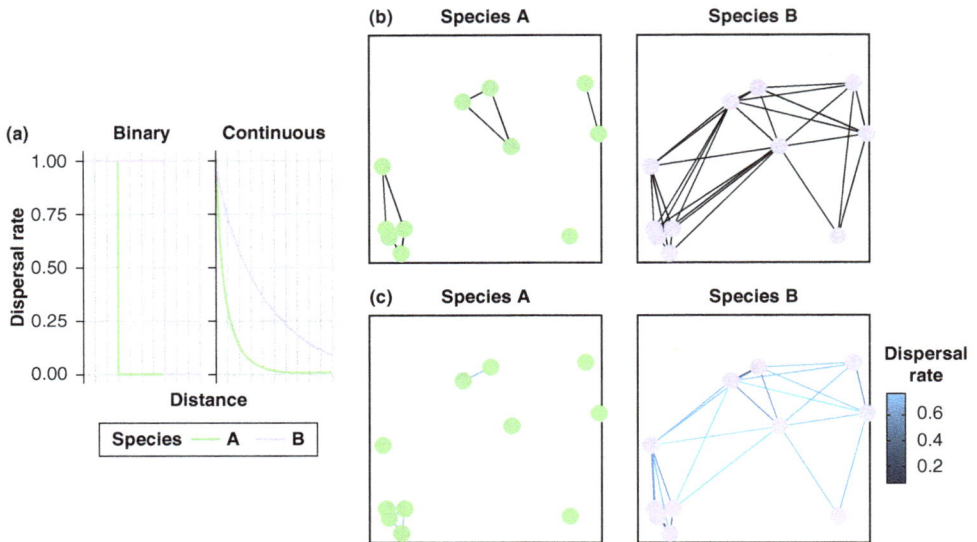

Figure 5.5 Representing functional connectivity as a spatial graph. (a) Binary and continuous dispersal kernels for two species (binary and continuous), (b) and (c) patches (nodes) and connections between them (edges) based on the binary and continuous dispersal kernels, respectively. In (c), the strength of the link is represented by the color of the edge (edges with strength less than 0.15 not shown for clarity).

between 0 and 1 representing the strength of connection). Landscape ecologists have drawn on this graph-theoretic model to depict ecological connectivity (Urban and Keitt 2001; Urban et al. 2009); resistance surfaces provide a way to estimate the presence and strength of the links in a network.

1. Graph theory via least-cost paths

Least-cost paths (LCPs) are among the most widely used methods for quantifying landscape connectivity (the most frequent in the 162 studies assessed by Correa Ayram et al. 2016). Initially conceptualized in the 1950s and 1960s, this method became much more practical with the rise of easily accessible computer power and the implementation of these methods in open-source libraries. The idea behind LCPs is that in heterogeneous landscapes, different elements will have higher or lower costs associated with their traversal, and organisms try to minimize these costs (e.g., energy use, risk of mortality) when moving through a landscape. For example, these elements could be related to topography (slope angle), habitat (wetlands vs. grasslands), or impermeable barriers (e.g. rivers, roads). The LCP is a trajectory that minimizes the travel cost between two points (there may be multiple equal LCPs between two points, and in a homogeneous landscape, the LCP will be the straight-line path). Thus, an important assumption in this approach is that the disperser has perfect knowledge of the entire landscape.

We can estimate the connectivity between two points in the landscape based on the accumulated cost of the LCPs between them (Figure 5.6). Likewise, we can construct an accumulated cost surface depicting the cost from a single point to all other locations in the landscape. This accumulated cost surface can be used to calculate the catchment (reachable area) around some fixed point, assuming a maximum allowable cost. In some cases, the costs of traversal may not be

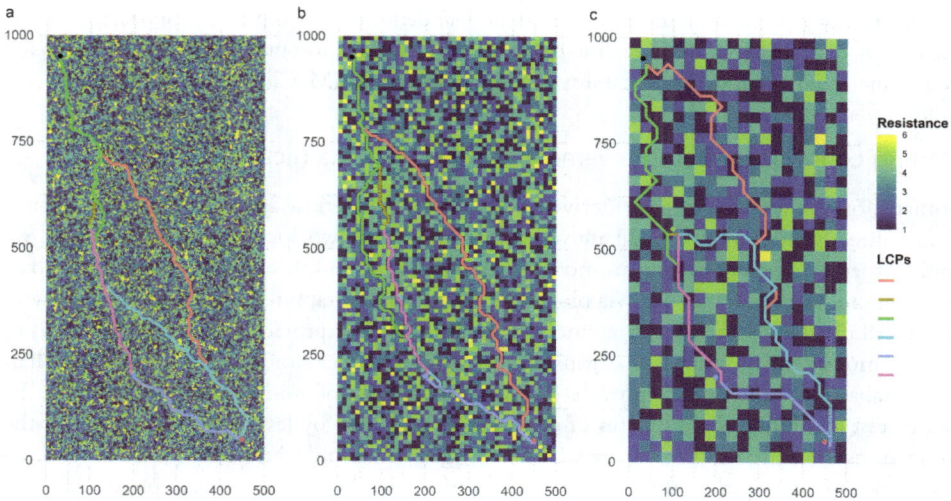

Figure 5.6 Least-cost paths between four points in a landscape generated using the modified random clusters approach at three different grains: (a) original, (b) aggregated by a factor of 10, and (c) aggregated by a factor of 25. The lines show LCPs between a suite of locations in the landscape and the background coloring the patch-level resistance (blue [low] to yellow [high]). Note the changes in LCP with spatial grain (coarsening from a to c).

anisotropic (e.g. the costs of upstream vs. downstream movement in streams or with vs. against ocean currents; Treml et al. 2008), and so the estimation of the LCP may need to consider the direction of travel. While (landscape) ecologists do not seem to have frequently considered this potential isotropy, other disciplines have (Alberti 2019), and this is something that ecologists could usefully consider.

2. Graph theory via circuit-based approaches

A complementary approach to quantifying landscape connectivity for genes and organisms is grounded in circuit theory (McRae and Beier 2007; McRae et al. 2008). In this model of connectivity, the landscape is imagined as a network of nodes connected by resistors. A key difference between circuit-based approaches and LCP is that the former considers all possible routes simultaneously and so does not assume that a disperser has perfect knowledge of the landscape (McRae et al. 2008). The two key predictions of the circuit-based approach are the resistance distance and the current. The resistance distance describes the isolation between pairs of locations, but unlike LCPs, it incorporates multiple pathways. Two locations connected by many different pathways with low resistance will have a low resistance distance and vice versa; if there is just a single pathway connecting locations, then the resistance distance will be the same as the LCP distance. The current shows the concentration of flow through different parts of the landscape and so highlights landscape elements important for movement (and landscape management), such as corridors and bottlenecks (Roever et al. 2013; Pelletier et al. 2014). An interesting application of the circuit-based approach is provided by Deith and Brodie (2020), who sought to estimate habitat accessibility to bushmeat hunters in Malaysian Borneo. Using a model selection framework, Deith and Brodie compared the predictions of different models of accessibility (e.g. based on the distance to specific features, resistance) with an extensive database of camera images of the hunters (nearly 1600 independent images from a pool of nearly 20,000). Their study demonstrated that a circuit-based approach tended to outperform the other models even if fine-grained landscape variability remained important (McClure et al. 2016)

Inferring connectivity from movement and genetic data (actual)

Estimates of connectivity can be derived from graph-theoretical approaches in the absence of data directly describing animal movement. However, if two locations are connected, we would expect to observe organisms moving between them, and this should be evident in data describing movement tracks (e.g. via biotelemetry or camera trapping) or genetic connectivity. As we will discuss further, recent technological advances have provided stunning new ways to 'see' the movement of individual organisms. Biotelemetry data alongside information such as large databases of camera traps provide direct measurements of animal movement and can be used to test and refine the estimates of connectivity provided by less direct approaches. If the connections identified indirectly are plausible, organisms should be observed moving along them more frequently than through other parts of the landscape! As an example of this type of approach, Stewart et al. (2019) compared the movement of GPS-tracked fisher (*Pekania pennanti*) with the predictions of three models of functional connectivity: use of structurally similar habitat (e.g. habitat corridors), movement on least-cost patches, and use of stepping stone patches to traverse the matrix. Predictions of the types of movement behavior (e.g. movement speed and correlation in direction between movements) expected under the three hypotheses were developed and confronted with the biotelemetry data. Based on a model selection approach, Stewart et al. found that for six of the ten fishers, the model based on the use of structurally similar habitat was the best supported given the data, and it was the second best supported model for two more individuals. The best supported model for the other four was the LCP; no individuals' movement were supported by the stepping stone model. Thus, this study provides an example of how biotelemetry data can be used to test hypotheses developed from different conceptual models of landscape connectivity. These data are invaluable in making informed decisions about which model of landscape connectivity is most appropriate from an organism perspective on a gradient from random walkers (circuit theory) to perfectly informed dispersers (LCP).

Wright (1943) described an isolation by distance (IBD) model in which spatially separated (whether by barriers to movement or poor dispersal ability) local populations will become genetically separated from each other. Geographic distance alone may not capture connectivity as well as resistance-based approaches, leading McRae (2006) to propose the isolation by resistance (IBR) model. Even if geographic distance alone does not predict connectivity, the IBD has a valuable role as a null model. Thus, the IBD and IBR models suggest that if we can measure the genetic distance between populations, we can infer their connectivity; this endeavor is central to landscape genetics (Storfer et al. 2007; Waits et al. 2015). A typical approach might be to build distance matrices describing geographic, environmental, and genetic distances between populations and evaluate these using some form of Mantel test to evaluate correlations between the matrices.

Landscape genetic approaches have been used to look at barriers and corridors for individual movement (including those in the past) and also to understand spatially structured population dynamics. For example, Epps et al. (2013) used microsatellite markers to evaluate past and current connectivity in isolated populations of the African elephant (*Loxodonta africana*) in Tanzania. They demonstrated declines in connectivity and that rates of flow were predictable from topography and patterns of human settlement. Based on Mantel tests, Epps et al. rejected an IBD model but found strong support for an IBR model (based on slope) and concluded that contemporary patterns of connectivity in the elephant populations represented both past and current gene flow. The time-scale over which genetic patterns change is an important one in the application of these approaches. Genetic structure may represent genetic exchange in the distant past and is unlikely to respond instantaneously to landscape changes (e.g. imposition or removal of barriers) (Landguth et al. 2010; Epps and Keyghobadi 2015).

A second application of landscape genetic data has been to evaluate and refine the predictions of connectivity made using graph-theoretic methods (Storfer et al. 2007). If the predictions of LCPs (for example) are valid, we would expect to see correlations between genetic distance and resistance distance. This type of evaluation is especially valuable given the uncertainties in identifying the variables to include in the estimation of resistance surfaces and their relative weights. For example, Etherington et al. (2014) used genetic distances between isolated populations of the brush-tailed possum (*Trichosorus vulpecula*) to select between different model structures and parameterizations for least-cost path models. In a similar application, Wang et al. (2009) integrated genetic information into LCP models to develop a better understanding of the barriers to dispersal between habitats of the California tiger salamander (*Ambystoma californiense*) and identified surprising new dispersal opportunities.

Making connectivity dynamic and forward-looking

The methods for estimating connectivity that we have described usually assume a static map from which to estimate resistance. However, connectivity is dynamic across multiple spatial and temporal scales, driven by short-term processes, such as meteorological conditions, alongside much longer-term dynamics, such as climate shifts and land-use change (Zeigler and Fagan 2014). As landscape structure changes, so too will its connectivity. However, a single snapshot of landscape structure underpins many evaluations of connectivity, and it is uncertain how useful they are as a landscape changes. Using a virtual ecology approach, Simpkins and Perry (2017) demonstrated that under increasing temporal variability, cost-surfaces estimated using snapshots of landscape structure will become increasingly error-laden. In a more empirically grounded approach, Albert et al. (2017) used graph-theoretic methods for landscapes with different scenarios of land-use change to estimate the priority habitat to protect for the maintenance of connectivity for a suite of vertebrate species (developing effective multi-species approaches is another significant challenge for connectivity quantification).

A second issue central to future connectivity is how landscapes will respond to climate change. The term 'climate connectivity' is used to describe the ability of organisms to move through the landscape as they track changing climates (McGuire et al. 2016). As climates change, organisms will be forced to shift ranges, and there has been concern as to whether (and which) taxa will be able to 'keep up' (Corlett and Westcott 2013; Burrows et al. 2014). Even for taxa that are not dispersal-limited, habitat fragmentation may reduce climate connectivity. Such interactions between climate and connectivity are not restricted to terrestrial ecosystems; for example, Herrera-R et al. (2020) demonstrate how in the Amazon, fish seeking cooler headwater environments may have their movement impeded by dams and culverts. Climate velocity is a measure that helps to quantify climate connectivity – it is the speed and direction (vector) an organism would need to follow to track climate change (Loarie et al. 2009). Loarie et al. estimated that across the globe, climate velocities range from less than 0.1 to more than 1.25 km. yr^{-1}, varying with ecosystem type and topography. However, habitat loss and fragmentation pose further problems that potentially interact with climate change (i.e. habitat change is unlikely to be random with respect to climate trends). Given the extent of habitat loss and fragmentation, and projections of future climate change, climate corridors are likely to be essential components of efforts to protect and reinstate landscape-level connectivity (McGuire et al. 2016).

Movement ecology

The movement of individual organisms is at the heart of meta★ ecology and connectivity. Animal movement influences many ecological functions, including seed dispersal (Nield et al. 2020), disease spread (Scherer et al. 2020), and biogeochemical fluxes (Schmitz et al. 2018). Nathan et al. (2008) outlined a framework to unite the disparate strands of research on organismal movement and its ecological consequences. This body of research is relevant to landscape ecology, as it provides crucial insights into the processes (dispersal, movement) that link spatially structured ecological entities and how these interact with landscape structure. The framework described by Nathan et al. (2008) seeks to integrate hierarchically structured (individual movements → movement phases → lifetime movement track) movement trajectories with the environmental context in which these activities are embedded. They argue that movement paths emerge from the integration of a range of external (environment) and individual (e.g. motion capacity, navigation capacity) factors and that the resulting trajectories will, in turn, feedback to those drivers (Figure 5.7). External (environmental) factors will influence different facets of movement (Doherty and Driscoll 2018; Riotte-Lambert and Matthiopoulos 2020); for example, habitat quality may mediate immigration and emigration, but the physical environment may mediate transfer (movement between patches). These external factors may not be just the physical conditions in the landscape but may include the presence of other species, as in the 'landscape of fear' (Gaynor et al. 2019). The central and reciprocal role of the landscape in determining

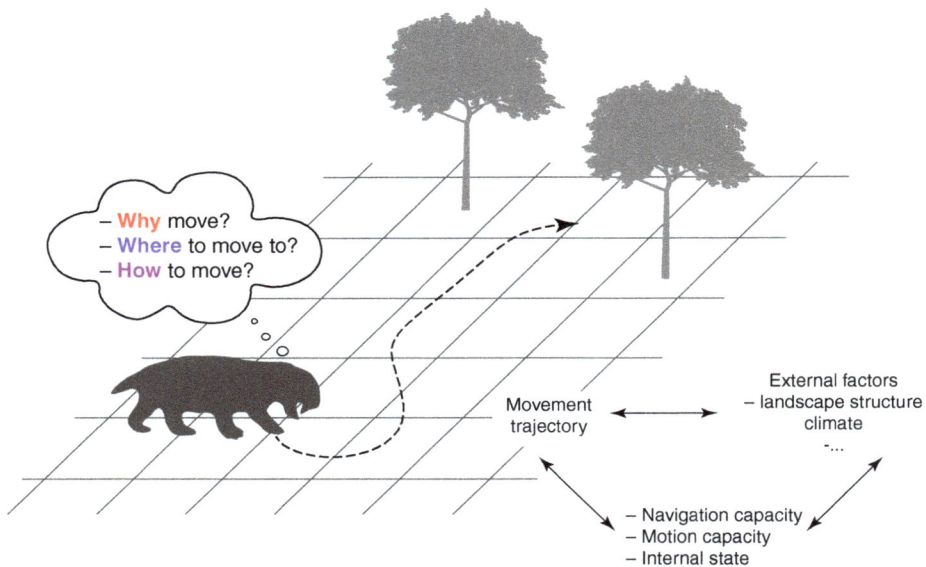

Figure 5.7 The movement ecology framework (Nathan et al. 2008) considers movement trajectories to emerge from interactions and feedbacks between an organism's internal state, external factors, and movement capacity. (Tree image from phylopics; LeonardoG (photography) and T. Michael Keesey (vectorization) under CC By-SA 3.0 license.)

individual movement means that there should be a natural partnership between movement and landscape ecology (Doherty et al. 2019).

The movement ecology framework has benefited from rapid advances in technology that enable ecologists to track the movement of individual animals at unprecedented resolutions in time and space (Kays et al. 2015), even if how it can most effectively be used to address fundamental and applied ecological questions is sometimes unclear (McGowan et al. 2017; Williams et al. 2020). Animal-borne telemetry has provided unprecedented detail about the activities of a broadening range of species in terrestrial, marine, and freshwater environments, often to infer the causes and consequences of individual movement (Cooke et al. 2013; Kays et al. 2015; Harcourt et al. 2019). A key focus of research in this area has been unpacking the direct and indirect consequences of terrestrial habitat loss and fragmentation on animal movement. In a synthesis of GPS telemetry data (more than 800 individuals across 57 species), Tucker et al. (2018) showed that the movement of terrestrial mammals was up to 50% reduced (distance travelled) in areas of intense human activity. This reduction arises from individual behavioral changes and shifts in species distributions in areas where the human footprint is high. Animals respond to specific anthropogenic effects such as the imposition of barriers to movement, but human activity can also reduce the need for animals to move when they are in areas of concentrated resources (e.g. agricultural areas). Fahrig (2007b) proposed that the response of individual species to habitat change will depend on the context they evolved in (e.g. continuous vs. patchy habitat) and in particular, the sensitivity of species to boundaries between habitat types (e.g. patch vs. mosaic). This framework predicts that species that evolved in continuous landscapes will suffer more from fragmentation than those that evolved in patchy landscapes (Table 5.2), and that changes in movement will ultimately affect population dynamics. Such changes are likely to flow-on to the ecological functions that emerge from animal movement. For example, Nield et al. (2020) show, using simulation, that seed dispersal distances may show threshold responses to habitat loss partly because of the response of mobile organisms (in their case, the large ratite, the emu) to habitat boundaries.

Conservation applications

Up to this point, we have primarily reviewed the literature around spatially structured populations, connectivity, and movement from a conceptual perspective. However, these concepts also

Table 5.2 Predicted Movement and Population Response to Habitat Change (Fahrig 2007b)

| Matrix quality | Landscape type evolved in | |
	Patchy	Continuous
Low	Evolved: Low movement rate with strong boundary response Response: decreased patch occupancy with possible increased risk of local extinction	Evolved: Flexible movement behavior with low boundary response
High	Evolved: High movement rate with low boundary response Response: increased dispersal mortality and increased risk of local extinction	Response: increased dispersal mortality and increased risk of local extinction

Evolved shows the movement behavior likely to be shown by species in those landscape types, and response is how those species may respond to habitat loss and fragmentation. Boundary response indicates how sensitive an organism is to boundaries between contrasting landscape elements.

underpin much of modern conservation biology. For example, the ETIB and metapopulation theory have been used to understand the effects of habitat loss and fragmentation on ecological populations and communities (Haila 2002). Likewise, insights from efforts to quantify landscape (terrestrial, freshwater, and marine) connectivity and understand how it influences fluxes of matter and energy movement are central to efforts to optimize the design of reserves and landscape-level management activities (Littlefield et al. 2019). In this section, we will first look at how the theories presented elsewhere in the chapter are relevant to conservation and landscape management and then illustrate this through a series of case studies.

Translating theoretical concepts to landscape management

The two main threats to ecosystems – habitat loss and fragmentation, and invasive species – are inherently spatial. The fundamental message from the ETIB, metapopulation, metacommunity, and metaecosystem theory is that the spatial context for ecological processes at multiple levels is an essential determinant of ecological dynamics. As a result, many conservation and landscape management initiatives seek to maintain large areas of intact habitat and to preserve or reinstate connectivity. However, translating the conceptual understanding provided by ecological theory into concrete/tactical management activities has proven challenging. As an example, the intuitive sense of the argument that maintaining connectivity is crucial has led to the development of landscape-level interventions such as corridors, which can take many forms (Figure 5.8).

Figure 5.8 Three habitat corridors: (a) 'classical' forest corridor between isolated patches (Brazil), (b) wildlife overpass (Singapore), and (c) fish passage (River Otter, United Kingdom). (Images (a) and (b) come from Wlimedia Commons users Ipe-institutodepesquisasecologicas and Benjamin P.Y-H. Lee (University of Kent); both CC BY-SA.)

However, there has been a long-running debate about the pros and cons of the use of corridors. On one side, they may promote the flow of individuals and genes between isolated populations and provide some habitat, but on the other, they may promote the flow of deleterious genes and pathogens, facilitate the spread of invasive species, and even *inhibit* the movement of some species. Meta-analyses of corridor effectiveness suggest that they are useful conservation tools, but the lack of long-term studies makes this evaluation only partial (Gilbert-Norton et al. 2010; Haddad et al. 2014). This debate highlights the challenges in translating spatial ecological theory to specific landscapes and species, and how the difficulties in conducting landscape-level experiments make it difficult to evaluate such interventions (Steinberg and Kareiva 1997).

We conclude this chapter by describing a series of case studies that highlight the application of landscape and spatial ecological theory in conservation management and the challenges associated with translating ecological theory into practice.

Source-sink metapopulation dynamics of an amphidromous fish

Many species of freshwater fish native to oceanic islands exhibit amphidromous life cycles (McDowall 2003). Amphidromy is a form of diadromy in which individuals travel between rivers and the ocean to complete their life-cycle. Many amphidromous fish disperse widely when returning to rivers from the sea (as opposed to natal homing, which is a strategy common among salmonids). This dispersal behavior means that species exist as metapopulations where individual rivers function as 'patches' linked by dispersal to sea. Amphidromy is beneficial for species inhabiting isolated islands, as it allows colonization of new islands as they emerge from hot spots along plate boundaries, as has happened in Hawaii, and allows recolonization after disturbance (McDowall 2007), but it also makes species particularly vulnerable to anthropogenic stressors, invasive species, and loss of connectivity. One such example is the extinction of the upokororo or New Zealand grayling (*Prototroctes oxyrhynchus*), the only known historic freshwater fish extinction to have occurred in New Zealand. In 1860, the species was widespread and abundant, but it went extinct sometime during the early–mid-20th century. Its extinction coincided with the introduction of trout and a period of rapid land transformation following European colonization, which resulted in predation (trout) and habitat degradation (land clearing), two of the primary cited causes of the upokororo's extinction (Lee and Perry 2019). These drivers alone fail, however, to explain how the species went extinct from uninvaded, pristine, isolated rivers. Lee and Perry (2019) showed that accounting for the species' amphidromous dispersal behavior (i.e. considering metapopulation dynamics) explains how the species went extinct from isolated, pristine rivers. The presence of trout and habitat degradation in some rivers resulted in these rivers becoming sinks. Fish in the pristine rivers dispersed out to sea, and a fraction returned to pristine rivers, but others ended up in sink rivers and were predated on by trout or were unable to reproduce due to degraded habitat. As the proportion of sink rivers in the landscape increased, fewer and fewer fish returning from the sea would end up in pristine rivers, resulting in fewer individuals being produced each subsequent generation and ultimately, resulting in extinction. This simple case study shows that by accounting for metapopulation dynamics, the presumed causes of extinction (predation and habitat degradation) in some habitats could have resulted in the extinction of upokororo from all habitats.

Maintenance of landscape connectivity in the context of long-distance migrations and individual variability

Long-distance animal migrations are critical to the persistence and resiliency of both populations and ecosystems (Fryxell and Sinclair 1988). Many large herbivores migrate between seasonal home ranges, in turn influencing the transfer of subsidies and the degree of connectivity between spatially distant landscapes. The recent declines of several migratory populations have prompted strong calls for the conservation of these long-distance movements (Dobson et al. 2010; Berger and Cain 2014). However, large mammals are characterized by high mobility, which presents significant challenges for conservation planning and management; their vast ranges often encompass several habitat types, different jurisdictions or ownerships, and multiple roads and fences.

Conserving the entirety of many ungulate ranges is often infeasible; researchers instead often draw on landscape ecology theory to optimize the maintenance of spatial connectivity for these populations (Bolger et al. 2007). Recent advances in high-resolution GPS tracking and remote sensing technology have enabled ecologists to develop fine-scale maps of key ungulate migration routes (Middleton et al. 2020). These maps combine both habitat-based and actual measures (from animal relocation data) of connectivity and movement corridors that can then be directly targeted in planning processes. In the Greater Yellowstone Ecosystem, mapping of specific migration corridors such as the 'Path of the Pronghorn' and the 'Red Desert to Hoback' mule deer migration has facilitated scientific outreach and initiated the engagement from diverse groups of stakeholders – private landowners, federal and state agencies, and industry – in the conservation of these populations and their movements (Middleton et al. 2020).

While high-resolution tracking of individual animals has revolutionized the study and conservation of animal migration, it has also revealed that these individuals can vary extensively in their migratory behavior. Many ungulate populations have now been characterized as 'partially migratory', which describes the tendency of some individuals in a population to migrate while others remain resident in a single range throughout the year (Berg et al. 2019). These individual differences in movement tendencies complicate the designation of migratory corridors and highlight some of the shortcomings of the spatial ecology framework in real-world applications. A single corridor or conservation strategy is unlikely to suffice for populations with high inter-individual variability in behavior. Understanding and incorporating behavioral differences among individuals into theoretical models of landscape and habitat connectivity will be necessary for effective conservation management.

An integrated network approach for biodiversity conservation in agroecosystems

Farmland is an archetypal representative of a landscape fragmented by human modification (Fahrig et al. 2011) and potentially a novel ecosystem worthy of conservation for its own sake (Krauss et al. 2010). Agriculture landscapes commonly contain cropped fields and natural and semi-natural habitat, and the ecological interactions between species and flows between communities produce the emergent structure and dynamics of the agroecosystem (Bohan et al. 2013). Fostering native biodiversity is a key goal in the sustainable management of agroecosystems (Keinath et al. 2017; Kremen and Merenlender 2018). Enhancing functional connectivity promotes ecosystem functioning and can lead to more efficient resource use and improve ecosystem stability (Martin et al. 2019; Montoya et al. 2019).

Agriculture landscapes contain ecological networks of interacting species that are linked to one another via movements in a spatial network (Hagen et al. 2012; Martin et al. 2019). Therefore, patterns of animal–environmental interactions among species are scale dependent and are governed by species-specific resource requirements, such as shelter, nesting, and food, and their movement abilities (Nathan et al. 2008; Keinath et al. 2017). Conservation plans based on estimates of connectivity for just a single species may be ineffective as they fail to consider the potential adverse effects it may have on the other species in the ecosystem.

Recently, there has been an upsurge in using integrative network analysis to prioritize connectivity restoration in agroecosystem management (Tixier et al. 2013). Instead of focusing on a single scale and life-stage (habitat use, migration, or dispersal), multiple indicator species at multiple spatial scales are being considered during the network modelling process. In these models, indicator groups are chosen to reflect particular protection goals, such as agricultural sustainability (Bailey and Muths 2019), ecological processes or functions (Grass et al. 2019), or cultural service provisioning (Assandri et al. 2018). Clauzel and Godet (2020) applied a multi-scale and multi-species approach to identify key habitats for connected functional networks of amphibians in an agroecosystem near Paris, France. First, they modelled the ecological network of each species and identified areas that could improve multi-species connectivity. Their approach provides 'no net loss' biodiversity measures and is, therefore, a promising guide for conservation actions. Another example of a multi-species approach is provided by Hunter-Ayad and Hassall (2020), who applied hybrid pattern-process models to six key taxa in wetland ecosystems in the United Kingdom. Habitat suitability models were developed first, which then provided the spatial context for individual-based models that predicted metapopulation dynamics under different post-management conditions. These examples demonstrate that quantifying how landscape configurations shape species-specific movement and dispersal patterns is a crucial first step in identifying the factors that most strongly influence functional connectivity and can inform strategies for biodiversity conservation in agroecosystems (Adriaensen et al. 2003; Duflot et al. 2017).

Summary

While spatial heterogeneity is a central and organizing theme in landscape ecology, other sub-disciplines of ecology have also paid considerable attention to 'space'. Some of this development has occurred in isolation from (or at least parallel to) landscape ecology, while some has been more deeply intertwined. A recurrent theme in this chapter has been the central importance of spatial context to population, community, and ecosystem dynamics, and a rich body of theory has been developed to address this issue. However, fundamental questions remain about how best to integrate the 'landscape' into some of this theory and the extent to which classic models of the landscape grounded in island biogeography (e.g. patch vs. matrix views) are adequate. Addressing how best to integrate the complementary perspectives on space and spatial heterogeneity taken by different sub-disciplines of ecology is likely to be a fruitful area for future research.

References

Adriaensen, F., Chardon, J.P., De Blust, G., Swinnen, E., Villalba, S., Gulinck, H. and Matthysen, E. (2003) 'The application of 'least-cost' modelling as a functional landscape model', *Landscape and Urban Planning*, vol 64, pp233–247.

Albert, C.H., Rayfield, B., Dumitru, M. and Gonzalez, A. (2017) 'Applying network theory to prioritise multi-species habitat networks that are robust to climate and land-use change', *Conservation Biology*, vol 31, pp1383–1396.

Alberti G. (2019) 'movecost: An R package for calculating accumulated slope-dependent anisotropic cost-surfaces and least-cost paths', *SoftwareX* vol 10, p100331.

Almeida-Gomes, M., Valente-Neto, F., Pacheco, E.O., Ganci, C.C., Leibold, M.A., Melo, A.S. and Provete, D.B. (2020) 'How does the landscape affect metacommunity structure? A quantitative review for lentic environments', *Current Landscape Ecology Reports*, vol 5, pp68–75.

Assandri, G., Bogliani, G., Pedrini, P. and Brambilla, M. (2018) 'Beautiful agricultural landscapes promote cultural ecosystem services and biodiversity conservation', *Agriculture, Ecosystems and Environment*, vol 256, pp200–210.

Bailey, L.L. and Muths, E. (2019) 'Integrating amphibian movement studies across scales better informs conservation decisions', *Biological Conservation*, vol 236, pp261–268.

Berg, J.E., Hebblewhite, M., St. Clair, C.C. and Merrill, E.H. (2019) 'Prevalence and mechanisms of partial migration in ungulates', *Frontiers in Ecology and Evolution*, vol 7, art 325.

Berger, J. and Cain, S.L. (2014) 'Moving beyond science to protect a mammalian migration corridor', *Conservation Biology*, vol 28, pp1142–1150.

Biswas, S.R. and Wagner, H.H. (2012) 'Landscape contrast: a solution to hidden assumptions in the metacommunity concept?', *Landscape Ecology*, vol 27, pp621–631.

Bohan, D.A., Raybould, A., Mulder, C., Woodward, G., Tamaddoni-Nezhad, A., Bluthgen, N., Pocock, M.J.O., Muggleton, S., Evans, D.M., Astegiano, J., Massol, F., Loeuille, N., Petit, S. and Macfadyen, S. (2013). 'Chapter one: Networking agroecology: integrating the diversity of agroecosystem interactions', in Woodward, G. and Bohan, D.A. (eds) *Advances in Ecological Research, Ecological Networks in an Agricultural World*. Academic Press, Cambridge, MA.

Bolger, D.T., Newmark, W.D., Morrison, T.A. and Doak, D.F. (2007) 'The need for integrative approaches to understand and conserve migratory ungulates', *Ecology Letters*, vol 11, pp63–77.

Brooks, R.T. (2004) 'Weather-related effects on woodland vernal pool hydrology and hydroperiod', *Wetlands*, vol 24, pp104–114.

Burrows, M.T., Schoeman, D.S., Buckley, L.B., Moore, P., Poloczanska, E.S., Brander, K.M., *et al.* (2014) 'The pace of shifting climate in marine and terrestrial ecosystems', *Science*, vol 334, pp652–655.

Calabrese, J.M. and Fagan, W.F. (2004) 'A comparison-shopper's guide to connectivity metrics', *Frontiers in Ecology and the Environment*, vol 2, pp529–536.

Clauzel, C. and Godet, C. (2020) 'Combining spatial modeling tools and biological data for improved multi-species assessment in restoration areas', *Biological Conservation*, vol 250, art 108713.

Cooke, S.J., Midwood, J.D., Thiem, J.D., Klimley, P., Lucas, M.C., Thorstad, E.B., Eiler, J., Holbrook, C. and Ebner, B.C. (2013) 'Tracking animals in freshwater with electronic tags: past, present and future', *Animal Biotelemetry*, vol 1, art 5.

Corlett, R.T. and Westcott, D.A. (2013) 'Will plant movements keep up with climate change?', *Trends in Ecology and Evolution*, vol 28, pp482–488.

Correa Ayram, C.A., Mendoza, M.E., Etter, A. and Salicrup, D.R.P. (2016) 'Habitat connectivity in biodiversity conservation: a review of recent studies and applications', *Progress in Physical Geography*, vol 40, pp7–37.

Dale, M.R.T. and Fortin, M.-J. (2010) 'From graphs to spatial graphs', *Annual Review of Ecology, Evolution, and Systematics*, vol 41, pp21–38.

Deith, M.C.M. and Brodie, J.F. (2020) 'Predicting defaunation: accurately mapping bushmeat hunting pressure over large areas', *Proceedings of the Royal Society B: Biological Sciences*, vol 287, art 20192677.

Dias, P.C. (1996) 'Sources and sinks in population biology', *Trends in Ecology and Evolution*, vol 11, pp326–330.

Dobson, A.P., Borner, M., Sinclair, A.R.E., Hudson, P.J., Anderson, T.M., Bigurube, G., et al. (2010) 'Road will ruin Serengeti', *Nature*, vol 467, pp272–273.

Doherty, T.S. and Driscoll, D.A. (2018) 'Coupling movement and landscape ecology for animal conservation in production landscapes', *Proceedings of the Royal Society B: Biological Sciences*, vol 285, art 20172272.

Doherty, T.S., Fist, C.N. and Driscoll, D.A. (2019) 'Animal movement varies with resource availability, landscape configuration and body size: a conceptual model and empirical example', *Landscape Ecology*, vol 34, pp603–614.

Duflot, R., Ernoult, A., Aviron, S., Fahrig, L. and Burel, F. (2017) 'Relative effects of landscape composition and configuration on multi-habitat gamma diversity in agricultural landscapes', *Agriculture, Ecosystems and Environment*, vol 241, pp62–69.

Epps, C.W. and Keyghobadi, N. (2015) 'Landscape genetics in a changing world: disentangling historical and contemporary influences and inferring change', *Molecular Ecology*, vol 24, pp6021–6040.

Epps, C.W., Wasser, S.K., Keim, J.L., Mutayoba, B.M. and Brashares, J.S. (2013) 'Quantifying past and present connectivity illuminates a rapidly changing landscape for the African elephant', *Molecular Ecology*, vol 22, pp1574–1588.

Erős, T. and Lowe, W.H. (2019) 'The landscape ecology of rivers: from patch-based to spatial network analyses', *Current Landscape Ecology Reports*, vol 4, pp103–112.

Etherington, T.R. (2012) 'Mapping organism spread potential by integrating dispersal and transportation processes using graph theory and catchment areas', *International Journal of Geographical Information Science*, vol 26, pp541–556.

Etherington, T.R. (2016) 'Least-cost modelling and landscape ecology: concepts, applications, and opportunities', *Current Landscape Ecology Reports*, vol 1, pp40–53.

Etherington, T.R., Perry, G.L.W., Cowan, P.E. and Clout, M.N. (2014) 'Quantifying the direct transfer costs of common brushtail possum dispersal using least-cost modelling: a combined cost-surface and accumulated-cost dispersal kernel approach', *PLoS ONE*, vol 9, art e88293.

Fahrig, L. (2007a) 'Landscape heterogeneity and metapopulation dynamics', in Wu, J. and Hobbs, R.J. (eds) *Key Topics in Landscape Ecology*. Cambridge University Press, Cambridge.

Fahrig, L. (2007b) 'Non-optimal animal movement in human-altered landscapes', *Functional Ecology*, vol 21, pp1003–1015.

Fahrig, L., Baudry, J., Brotons, L., Burel, F.G., Crist, T.O., Fuller, R.J., Sirami, C., Siriwardena, G.M. and Martin, J-L. (2011) 'Functional landscape heterogeneity and animal biodiversity in agricultural landscapes', *Ecology Letters*, vol 14, pp101–112.

Freckleton, R.P. and Watkinson, A.R. (2002) 'Large-scale spatial dynamics of plants: metapopulations, regional ensembles and patchy populations', *Journal of Ecology*, vol 90, pp419–434.

Fryxell, J.M. and Sinclair, A.R.E. (1988) 'Causes and consequences of migration by large herbivores', *Trends in Ecology and Evolution*, vol 3, pp237–241.

Gaynor, K.M., Brown, J.S., Middleton, A.D., Power, M.E. and Brashares, J.S. (2019) 'Landscapes of fear: spatial patterns of risk perception and response', *Trends in Ecology and Evolution*, vol 34, pp355–368.

Giezendanner, J., Pasetto, D., Perez-Saez, J., Cerrato, C., Viterbi, R., Terzago, S., Palazzi, E. and Rinaldo, A. (2020) 'Earth and field observations underpin metapopulation dynamics in complex landscapes: near-term study on carabids', *Proceedings of the National Academy of Sciences*, vol 117, pp12877–12884.

Gilbert-Norton, L., Wilson, R., Stevens, J.R. and Beard, K.H. (2010) 'A meta-analytic review of corridor effectiveness: corridor meta-analysis', *Conservation Biology*, vol 24, pp660–668.

Gounand, I., Harvey, E., Little, C.J. and Altermatt, F. (2018) 'Meta-ecosystems 2.0: rooting the theory into the field', *Trends in Ecology and Evolution*, vol 33, pp36–46.

Grass, I., Loos, J., Baensch, S., Batáry, P., Librán-Embid, F., Ficiciyan, A., Klaus, F., Riechers, M., Rosa, J., Tiede, J., Udy, K., Westphal, C., Wurz, A. and Tscharntke, T. (2019) 'Land-sharing/-sparing connectivity landscapes for ecosystem services and biodiversity conservation', *People and Nature*, vol 1, pp262–272.

Gustafson, E.J. (1998) 'Quantifying landscape spatial pattern: what is the state of the art?', *Ecosystems*, vol 1, pp143–156.

Haddad, N.M., Brudvig, L.A., Damschen, E.I., Evans, D.M., Johnson, B.L., Levey, D.J., et al. (2014) 'Potential negative ecological effects of corridors', *Conservation Biology*, vol 28, pp1178–1187.

Hagen, M., Kissling, W.D., Rasmussen, C., De Aguiar, M.A., Brown, L.E., Carstensen, D.W., et al. (2012). 'Biodiversity, species interactions and ecological networks in a fragmented world', *Advances in Ecological Research*, vol 46, pp 89–210.

Haila, Y. (2002) 'A conceptual genealogy of fragmentation research: from island biogeography to landscape ecology', *Ecological Applications*, vol 12, pp321–334.

Hand, B.K., Lowe, W.H., Kovach, R.P., Muhlfeld, C.C. and Luikart, G. (2015) 'Landscape community genomics: understanding eco-evolutionary processes in complex environments', *Trends in Ecology and Evolution*, vol 30, pp161–168.

Hanski, I. (1998) 'Metapopulation dynamics', *Nature*, vol 396, pp41–49.

Harcourt, R., Sequeira, A.M.M., Zhang, X., Roquet, F., Komatsu, K., Heupel, M., et al. (2019) 'Animal-borne telemetry: an integral component of the ocean observing toolkit', *Frontiers in Marine Science*, art 326.

Harvey, E., Gounand, I., Fronhofer, E.A. and Altermatt, F. (2020) 'Metaecosystem dynamics drive community composition in experimental, multi-layered spatial networks', *Oikos*, vol 129, pp402–412.

Herrera-R, G.A., Oberdorff, T., Anderson, E.P., Brosse, S., Carvajal-Vallejos, F.M., Frederico, R.G., Hidalgo, M., Jézéquel, C., Maldonado, M., Maldonado-Ocampo, J.A., Ortega, H., Radinger, J., Torrente-Vilara, G., Zuanon, J. and Tedesco, P.A. (2020) 'The combined effects of climate change and river fragmentation on the distribution of Andean Amazon fishes', *Global Change Biology*, vol 26, pp 5509–5523.

Hillman, J.R., Lundquist, C.J. and Thrush, S.F. (2018) 'The challenges associated with connectivity in ecosystem processes', *Frontiers in Marine Science*, vol 5, art 364.

van Horne, B. (1983) 'Density as a misleading indicator of habitat quality', *Journal of Wildlife Management*, vol 47, pp893–901.

Hubbell, S.P. (2001) *The Unified Neutral Theory of Biodiversity and Biogeography*. Princeton University Press, Princeton.

Hunter-Ayad, J. and Hassall, C. (2020). 'An empirical, cross-taxon evaluation of landscape-scale connectivity', *Biodiversity and Conservation*, vol 29, pp1339–1359.

Hutchinson, G.E. (1967) *The Ecological Theater and the Evolutionary Play*. Yale University Press, New Haven, London.

Jacobson, B. and Peres-Neto, P.R. (2010) 'Quantifying and disentangling dispersal in metacommunities: how close have we come? How far is there to go?', *Landscape Ecology*, vol 25, pp495–507.

Jonsen, I.D., Basson, M., Bestley, S., Bravington, M.V., Patterson, T.A., Pedersen, M.W., Thomson, R., Thygesen, U.H. and Wotherspoon, S.J. (2013) 'State-space models for bio-loggers: a methodological road map', *Deep Sea Research Part II: Topical Studies in Oceanography*, vol 88–89, pp34–46.

Kärnä, O.-M., Grönroos, M., Antikainen, H., Hjort, J., Ilmonen, J., Paasivirta, L. and Heino, J. (2015) 'Inferring the effects of potential dispersal routes on the metacommunity structure of stream insects: as the crow flies, as the fish swims or as the fox runs?', *Journal of Animal Ecology*, vol 84, pp1342–1353.

Kays, R., Crofoot, M.C., Jetz, W. and Wikelski, M. (2015) 'Terrestrial animal tracking as an eye on life and planet', *Science*, vol 348, art aaa2478.

Keinath, D.A., Doak, D.F., Hodges, K.E., Prugh, L.R., Fagan, W., Sekercioglu, C.H., Buchart, S.H.M. and Kauffman, M. (2017) 'A global analysis of traits predicting species sensitivity to habitat fragmentation', *Global Ecology and Biogeography*, vol 26, pp115–127.

Krauss, J., Bommarco, R., Guardiola, M., Heikkinen, R.K., Helm, A., Kuussaari, M., Lindborg, R., Öckinger, E., Pärtel, M., Pino, J., Pöyry, J., Raatikainen, K.M., Sang, A., Stefanescu, C., Teder, T., Zobel, M. and Steffan-Dewenter, I. (2010) 'Habitat fragmentation causes immediate and time-delayed biodiversity loss at different trophic levels', *Ecology Letters*, vol 13, pp597–605.

Kremen, C. and Merenlender, A.M. (2018) 'Landscapes that work for biodiversity and people', *Science*, vol 362, art eaau6020.

Kritzer, J.P. and Sale, P.F. (2004) 'Metapopulation ecology in the sea: from Levins' model to marine ecology and fisheries science', *Fish Fisheries*, vol 5, pp131–140.

Krueger, T., Page, T., Hubacek, K., Smith, L. and Hiscock, K. (2012) 'The role of expert opinion in environmental modelling', *Environmental Modelling and Software*, vol 36, pp4–18.

Landguth, E.L., Cushman, S.A., Schwartz, M.K., McKelvey, K.S., Murphy, M. and Luikart, G. (2010) 'Quantifying the lag time to detect barriers in landscape genetics', *Molecular Ecology*, vol 19, pp4179–4191.

Lee, F. and Perry, G.L. (2019) 'Assessing the role of off-take and source-sink dynamics in the extinction of the amphidromous New Zealand grayling (*Prototroctes oxyrhynchus*)', *Freshwater Biology*, vol 64, pp1747–1754.

Lehman, C.L. and Tilman, D.A. (1997) 'Competition in spatial habitats', in *Spatial Ecology: The Role of Space in Population Dynamics and Interspecific Interactions*. Princeton University Press, Princeton, NJ, pp185–204.

Leibold, M.A. and Miller, T.E. (2004) 'From metapopulations to metacommunities', in Hanski, I. and Gaggiotti, O.E. (eds) *Ecology, Genetics and Evolution of Metapopulations*. Academic Press, Cambridge, MA.

Leibold, M.A., Holyoak, M., Mouquet, N., Amarasekare, P., Chase, J.M., Hoopes, M.F., Holt, R.D., Shurin, J.B., Law, R., Tilman, D., Loreau, M. and Gonzalez, A. (2004) 'The metacommunity concept: a framework for multi-scale community ecology', *Ecology Letters*, vol 7, pp601–613.

Leroux, S.J., Wal, E.V., Wiersma, Y.F., Charron, L., Ebel, J.D., Ellis, N.M., Hart, C., Kissler, E., Saunders, P.W., Moudrá, L., Tanner, A.L. and Yalcin, S. (2017) 'Stoichiometric distribution models: ecological stoichiometry at the landscape extent', *Ecology Letters*, vol 20, pp1495–1506.

Levins, R. (1969) 'Some demographic and genetic consequences of environmental heterogeneity for biological control', *Bulletin of the Entomological Society of America*, vol 15, pp237–240.

Levin, S.A., Muller-Landau, H.C., Nathan, R., and Chave, J. (2003) 'The ecology and evolution of seed dispersal: a theoretical perspective', *Annual Review of Ecology, Evolution, and Systematics*, vol 34, no. 1, pp575–604.

Lima, S.L. and Zollner, P.A. (1996) 'Towards a behavioral ecology of ecological landscapes', *Trends in Ecology and Evolution*, vol 11, pp131–136.

Littlefield, C.E., Krosby, M., Michalak, J.L. and Lawler, J.J. (2019) 'Connectivity for species on the move: supporting climate-driven range shifts', *Frontiers in Ecology and the Environment*, vol 17, pp270–278.

Loarie, S.R., Duffy, P.B., Hamilton, H., Asner, G.P., Field, C.B. and Ackerly, D.D. (2009) 'The velocity of climate change', *Nature*, vol 462, pp1052–1055.

Logue, J.B., Mouquet, N., Peter, H., Hillebrand, H. and Group, M.W. (2011) 'Empirical approaches to metacommunities: a review and comparison with theory', *Trends in Ecology and Evolution*, vol 26, pp482–491.

Lomolino, M.V. (2000) 'Ecology's most general, yet protean pattern: the species-area relationship', *Journal of Biogeography*, vol 27, pp17–26.

Loreau, M., Mouquet, N. and Holt, R.D. (2003) 'Meta-ecosystems: a theoretical framework for a spatial ecosystem ecology', *Ecology Letters*, vol 6, pp673–679.

MacArthur, R.H. and Wilson, E.O. (1967) *The Theory of Island Biogeography*. Princeton University Press, Princeton, NJ.

Martin, E.A., Dainese, M., Clough, Y., Báldi, A., Bommarco, R., Gagic, V., et al. (2019) 'The interplay of landscape composition and configuration: new pathways to manage functional biodiversity and agroecosystem services across Europe', *Ecology Letters*, vol 22, pp1083–1094.

McClure, M.L., Hansen, A.J. and Inman, R.M. (2016) 'Connecting models to movements: testing connectivity model predictions against empirical migration and dispersal data', *Landscape Ecology*, vol 31, pp1419–1432.

McDowall, R.M. (2003) 'Hawaiian biogeography and the islands' freshwater fish fauna', *Journal of Biogeography*, vol 30, pp703–710.

McDowall, R.M. (2007) 'On amphidromy, a distinct form of diadromy in aquatic organisms', *Fish and Fisheries*, vol 8, pp1–13.

McGowan, J., Beger, M., Lewison, R.L., Harcourt, R., Campbell, H., Priest, M., Dwyer, R.G., Lin, H-Y., Lentini, P., Dudgeon, C., McMahon, C., Watts, M. and Possingham, H.P. (2017) 'Integrating research using animal-borne telemetry with the needs of conservation management', *Journal of Applied Ecology*, vol 54, pp423–429.

McGuire, J.L., Lawler, J.J., McRae, B.H., Nuñez, T.A. and Theobald, D.M. (2016) 'Achieving climate connectivity in a fragmented landscape', *Proceedings of the National Academy of Sciences*, vol 113, pp7195–7200.

McRae, B.H. (2006) 'Isolation by resistance', *Evolution*, vol 60, pp1551–1561.

McRae, B.H. and Beier, P. (2007) 'Circuit theory predicts gene flow in plant and animal populations', *Proceedings of the National Academy of Sciences*, vol 104, pp19885–19890.

McRae, B.H., Dickson, B.G., Keitt, T.H. and Shah, V.B. (2008) 'Using circuit theory to model connectivity in ecology, evolution, and conservation', *Ecology*, vol 89, pp2712–2724.

Middleton, A.D., Sawyer, H., Merkle, J.A., Kauffman, M.J., Cole, E.K., Dewey, S.R., Gude, J.A., Gustine, D.D., McWhirter, D.E., Proffitt, K.M. and White, P.J. (2020) 'Conserving transboundary wildlife migrations: recent insights from the Greater Yellowstone Ecosystem', *Frontiers in Ecology and the Environment*, vol 18, pp83–91.

Montoya, D., Haegeman, B., Gaba, S., De Mazancourt, C., Bretagnolle, V. and Loreau, M (2019) 'Trade-offs in the provisioning and stability of ecosystem services in agroecosystems', *Ecological Applications*, vol 29, art e01853.

Nathan, R., Getz, W.M., Revilla, E., Holyoak, M., Kadmon, R., Saltz, D. and Smouse, P.E. (2008) 'A movement ecology paradigm for unifying organismal movement research', *Proceedings of the National Academy of Sciences*, vol 105, pp19052–19059.

Nield, A.P., Nathan, R., Enright, N.J., Ladd, P.G. and Perry, G.L.W. (2020) 'The spatial complexity of seed movement: animal-generated seed dispersal patterns in fragmented landscapes revealed by animal movement models', *Journal of Ecology*, vol 108, pp687–701.

O'Sullivan, D. (2014) 'Spatial network analysis', in Fischer, M.M. and Nijkamp, P. (eds) *Handbook of Regional Science*. Springer, Berlin.

Pelletier, D., Clark, M., Anderson, M.G., Rayfield, B., Wulder, M.A. and Cardille, J.A. (2014) 'Applying circuit theory for corridor expansion and management at regional scales: tiling, pinch points, and omnidirectional connectivity', *PLoS ONE*, vol 9, art e84135.

Perry, G.L.W. (2002) 'Landscapes, space and equilibrium: shifting viewpoints', *Progress in Physical Geography*, vol 26, pp339–359.

Polis, G.A., Anderson, W.B. and Holt, R.D. (1997) 'Toward an integration of landscape and food web ecology: the dynamics of spatially subsidised food webs', *Annual Review of Ecology and Systematics*, vol 28, pp289–316.

Pulliam, H.R. (1988) 'Sources, sinks and population regulation', *American Naturalist*, vol 132, pp652–661.

Riotte-Lambert, L. and Matthiopoulos, J. (2020) 'Environmental predictability as a cause and consequence of animal movement', *Trends in Ecology and Evolution*, vol 35, pp163–174.

Roever, C.L., van Aarde, R.J. and Leggett, K. (2013) 'Functional connectivity within conservation networks: delineating corridors for African elephants', *Biological Conservation*, vol 157, pp128–135.

Ryberg, W.A. and Fitzgerald, L.A. (2016) 'Landscape composition, not connectivity, determines metacommunity structure across multiple scales', *Ecography*, vol 39, pp932–941.

Sawyer, S.C., Epps, C.W. and Brashares, J.S. (2011) 'Placing linkages among fragmented habitats: do least-cost models reflect how animals use landscapes?: least-cost modelling for habitat linkage design', *Journal of Applied Ecology*, vol 48, pp668–678.

Scherer, C., Radchuk, V., Franz, M., Thulke, H., Lange, M., Grimm, V. and Kramer-Schadt, S. (2020) 'Moving infections: individual movement decisions drive disease persistence in spatially structured landscapes', *Oikos*, vol 129, pp651–667.

Schmitz, O.J., Wilmers, C.C., Leroux, S.J., Doughty, C.E., Atwood, T.B., Galetti, M., Davies, A.B. and Goetz, S.J. (2018) 'Animals and the zoogeochemistry of the carbon cycle', *Science*, vol 362, art eaar3213.

Schumaker, N.H. (1996) 'Using landscape indices to predict habitat connectivity', *Ecology*, vol 77, pp1210–1225.

Shmida, A.V.I. and Wilson, M.V. (1985) 'Biological determinants of species diversity', *Journal of Biogeography*, vol 12, pp1–20.

Simpkins, C.E. and Perry, G.L.W. (2017) 'Understanding the impacts of temporal variability on estimates of landscape connectivity', *Ecological Indicators*, vol 83, pp243–248.

Simpkins, C.E., Dennis, T.E., Etherington, T.R. and Perry, G.L.W. (2017) 'Effects of uncertain cost-surface specification on landscape connectivity measures', *Ecological Informatics*, vol 38, pp1–11.

Simpkins, C.E., Dennis, T.E., Etherington, T.R. and Perry, G.L.W. (2018) 'Assessing the performance of common landscape connectivity metrics using a virtual ecologist approach', *Ecological Modelling*, vol 367, pp13–23.

Spear, S.F., Balkenhol, N., Fortin, M.-J., Mcrae, B.H. and Scribner, K. (2010) 'Use of resistance surfaces for landscape genetic studies: considerations for parameterisation and analysis', *Molecular Ecology*, vol 19, pp3576–3591.

Steinberg, E.K. and Kareiva, P.M. (1997) 'Challenges and opportunities for empirical evolution of "spatial theory"', in Tilman, D. and Kareiva, P. (eds) *Spatial Ecology: The Role of Space in Population Dynamics and Interspecific Interactions*. Princeton University Press, Princeton.

Stewart, F.E.C., Darlington, S., Volpe, J.P., McAdie, M. and Fisher, J.T. (2019) 'Corridors best facilitate functional connectivity across a protected area network', *Scientific Reports*, art 10852.

Storfer, A., Murphy, M.A., Evans, J.S., Goldberg, C.S., Robinson, S., Spear, S.F., Dezzani, R., Delmelle, E., Vierling, L. and Waits, L.P. (2007) 'Putting the 'landscape' in landscape genetics', *Heredity*, vol 98, pp128–142.

Talley, T.S. (2007) 'Which spatial heterogeneity framework? Consequences for conclusions about patchy population distributions', *Ecology*, vol 88, pp1476–1489.

Taylor, P.D., Fahrig, L., Henin, K. and Merriam, G. (1993) 'Connectivity is a vital element of landscape structure', *Oikos*, vol 68, pp571–573.

Taylor, P.D., Fahrig, L. and With, K.A. (2006) 'Landscape connectivity: a return to the basics', in Crooks, K.R. and Sanjayan, M. (eds) *Connectivity Conservation*. Cambridge University Press, Cambridge.

Thompson, P.L., Guzman, L.M., De Meester, L., Horváth, Z., Ptacnik, R., Vanschoenwinkel, B., Viana, D.S. and Chase, J.M. (2020) 'A process-based metacommunity framework linking local and regional scale community ecology', *Ecology Letters*, vol 23, pp1314–1329.

Tilman, D. (1982) *Resource Competition and Community Structure*. Princeton University Press, Princeton.

Tilman, D.A. (1994) 'Competition and biodiversity in spatially structured habitats', *Ecology*, vol 75, pp2–16.

Tischendorf, L. and Fahrig, L. (2000) 'On the usage and measurement of landscape connectivity', *Oikos*, vol 90, pp7–19.

Tixier, P., Peyrard, N., Aubertot, J.-N., Gaba, S., Radoszycki, J., Caron-Lormier, G., Vinatier, F., Mollot, G. and Sabbadin, R. (2013) 'Modelling interaction networks for enhanced ecosystem services in agroecosystems', *Advances in Ecological Research*, vol 49, pp437–480.

Treml, E.A., Halpin, P.N., Urban, D.L. and Pratson, L.F. (2008) 'Modeling population connectivity by ocean currents, a graph-theoretic approach for marine conservation', *Landscape Ecology*, vol 23, pp19–36.

Tucker, M.A., Böhning-Gaese, K., Fagan, W.F., Fryxell, J.M., Van Moorter, B., Alberts, S.C., et al. (2018) 'Moving in the Anthropocene: global reductions in terrestrial mammalian movements', *Science*, vol 359, pp466–469.

Urban, D. and Keitt, T. (2001) 'Landscape connectivity: a graph-theoretic perspective', *Ecology*, vol 82, pp1205–1218.

Urban, D.L., Minor, E.S., Treml, E.A. and Schick, R.S. (2009) 'Graph models of habitat mosaics', *Ecology Letters*, vol 12, pp260–273.

Waits, L.P., Cushman, S.A. and Spear, S.F. (2015) 'Applications of landscape genetics to connectivity research in terrestrial animals', in Balkenhol, N., Cushman, S.A., Storfer, A.T. and Waits, L.P (eds) *Landscape Genetics*. Wiley, Chichester.

Wang, I.J., Savage, W.K. and Bradley Shaffer, H. (2009) 'Landscape genetics and least-cost path analysis reveal unexpected dispersal routes in the California tiger salamander (*Ambystoma californiense*). *Molecular Ecology*, vol 18, pp1365–1374.

Warren, B.H., Ricklefs, R.E., Thébaud, C., Gravel, D. and Mouquet, N. (2020) 'How consideration of islands has inspired mainstream ecology: links between the theory of island biogeography and some other key theories', in *Encyclopedia of the World's Biomes*. Elsevier, Amsterdam, pp. 57–6C.

Whittaker, R.H. (1962) 'Classification of natural communities', *Botanical Review*, vol 28, pp1–239.

Williams, H.J., Taylor, L.A., Benhamou, S., Bijleveld, A.I., Clay, T.A., Grissac, S., et al. (2020) 'Optimising the use of biologgers for movement ecology research', *Journal of Animal Ecology*, vol 89, pp186–206.

With, K.A. (2004) 'Metapopulation dynamics: perspectives from landscape ecology', in Hanski, I. and Gaggiotti, O.E. (eds) *Ecology, Genetics and Evolution of Metapopulations*. Academic Press, Cambridge, MA.

Wright, S. (1943) 'Isolation by distance', *Genetics*, vol 28, pp114–138.

Zeigler, S.L. and Fagan, W.F. (2014) 'Transient windows for connectivity in a changing world', *Movement Ecology*, vol 2, art 1

Zeller, K.A., McGarigal, K. and Whiteley, A.R. (2012) 'Estimating landscape resistance to movement: a review', *Landscape Ecology*, vol 27, pp777–797.

Zeller, K.A., Jennings, M.K., Vickers, T.W., Ernest, H.B., Cushman, S.A. and Boyce, W.M. (2018) 'Are all data types and connectivity models created equal? Validating common connectivity approaches with dispersal data', *Diversity and Distributions*, vol 24, pp868–879.

6

Habitat fragmentation

Amanda E. Martin, Joseph R. Bennett, and Lenore Fahrig

What is habitat fragmentation?

The answer to this question is not simple. Although landscape ecologists frequently study habitat fragmentation (Wu, 2013), what the term actually means—and how it should be measured—is widely debated (Fahrig, 2003, 2019; Lindenmayer and Fischer, 2007; Didham et al., 2012; Fletcher et al., 2018; Fahrig et al., 2019). Habitat fragmentation means different things to different people, and this has a big effect on the conclusions different people draw about the effect of habitat fragmentation on wildlife. This has led some to suggest that the term has become a panchreston, a term used in so many ways that it has become almost meaningless (Bunnell, 1999; Lindenmayer and Fischer, 2007). In this section, we introduce the three major conceptualizations of habitat fragmentation and explain why they lead to highly diverging conclusions about the effects of habitat fragmentation on wildlife. For each, we describe how effects of habitat fragmentation on wildlife are evaluated, as the differences in study design clearly illustrate the differences in concepts. Please see Box 6.1 for definitions of key terms used in this section.

Box 6.1 Definitions of key terms used to describe the different conceptualizations of habitat fragmentation

Habitat: the area containing the features (e.g. shelter, food) needed for a given wildlife species to persist. In most studies, habitat is represented by one or several land cover types (e.g. forest or wetland). All other land cover types are thus non-habitat, often referred to as 'matrix' in landscape ecology.

Landscape: the area surrounding the sampling site(s) where the ecological response is measured. The size of this area is usually assumed to be related to the scale at which the species perceives and interacts with the landscape.

Habitat loss: reduction in the area of habitat within a landscape.

Habitat subdivision: the breaking apart of a given amount of habitat into more, smaller patches (or fragments).

Habitat degradation: reduction in the quality of habitat (or some portion of habitat), reducing provisioning of the features needed for a given wildlife species to persist within that habitat.

> **Reduction in matrix quality:** changes in the non-habitat portion of the landscape that increase the risk of mortality or sub-lethal effects to individuals moving outside of habitat, e.g. during dispersal, and/or contribute to degradation of adjacent areas of habitat.

Habitat fragmentation as the breaking apart of habitat: fragmentation per se

The idea of habitat fragmentation was first introduced by Curtis in 1956 and again, independently, by Moore in 1962. Curtis (1956) described the change in pattern of wooded area in Cadiz Township, Wisconsin, United States from 1831 to 1950. He noted that 'only 0.8 per cent of the land under forest in 1831 is still in what might be called a seminatural state, and this tiny portion is broken up into even more minute *fragments*, widely scattered throughout the area' (p. 137; our italics). Similarly, Moore (1962) described the change in pattern of heathland in Dorset, England, writing that '[s]ince 1811 the total area of heath in "Dorset" has been reduced by about 66%. It has been broken up into over 100 *fragments*' (p. 379; our italics).

There are two important aspects to Curtis's and Moore's concept of habitat fragmentation. First, they described habitat fragmentation as a landscape-scale phenomenon. They conceptualized habitat fragmentation as the breaking apart of habitat and measured it as the number of fragments—i.e. habitat patches—remaining in a landscape. Curtis (1956) also included two other measures of fragmentation still commonly used today (Fahrig, 2017): average size of all woodlots in the landscape and total length of edge between all woodlots and other land cover types in the landscape. Thus, in these seminal papers, habitat fragmentation was measured at the landscape scale as the number of patches, mean patch size, and total edge length in the landscape.

Second, Curtis and Moore described habitat fragmentation as a change in pattern caused by, but different from, habitat loss. For example, Moore (1962) explicitly separated habitat loss from fragmentation, referring to these as reduction and fragmentation, respectively. He noted that '[t]he *reduction* of heath continued at approximately the same rate, to 45 000 acres (18 000 ha) in 1934; the process of *fragmentation* progressed much faster' (p. 374; our italics). Differences between rates of habitat loss and rates of fragmentation can occur because habitat loss does not necessarily result in more fragments. If habitat removal subdivides habitat patches, then there is both habitat loss and fragmentation (Figure 6.1a). However, if habitat is removed by making large patches smaller, then there is habitat loss but no change in fragmentation (Figure 6.1b). If entire habitat patches are removed, then habitat loss is accompanied by a decrease in fragmentation because the resulting landscape has fewer patches (Figure 6.1c).

In summary, habitat fragmentation was originally conceptualized as the breaking apart (or subdivision) of habitat into multiple patches in a landscape. Habitat fragmentation may increase, remain constant, or decrease with habitat loss: it depends on how habitat is removed (Figure 6.1). Hereafter, we refer to this conceptualization as habitat fragmentation per se (sensu Haila and Hanski, 1984; Fahrig, 2003; Table 6.1).

Estimating effects of habitat fragmentation per se

The challenge in estimating effects of habitat fragmentation per se is to isolate the effects of fragmentation from the effects of habitat loss. There are four main methods for doing this. First,

(a) Habitat loss, increasing fragmentation

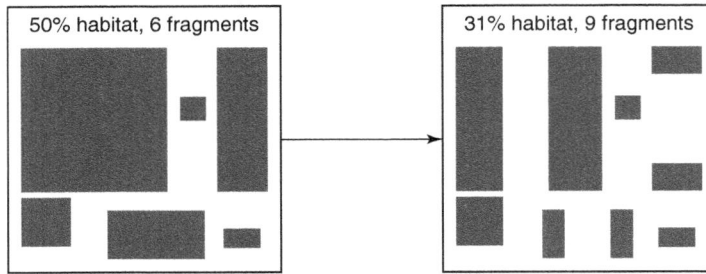

(b) Habitat loss, no fragmentation

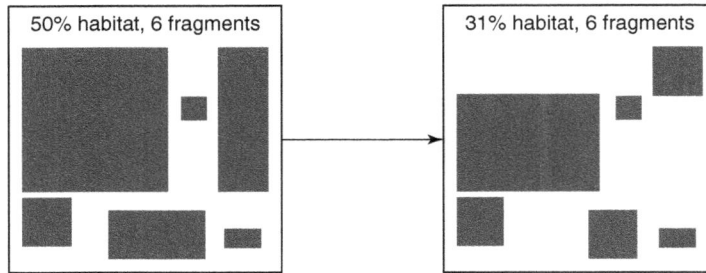

(c) Habitat loss, decreasing fragmentation

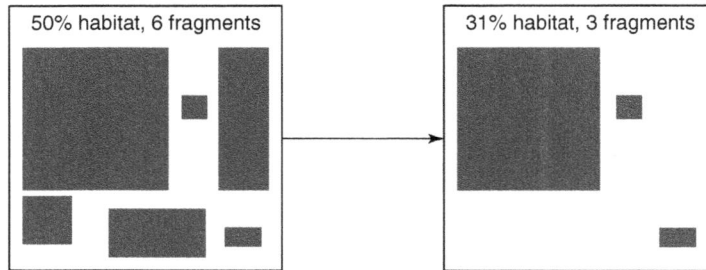

Figure 6.1   Removal of a given amount of habitat can result in (a) a more fragmented distribution of remnant habitat, if habitat removal results in subdivision of habitat patches, (b) no change in fragmentation, if habitat is removed by making patches smaller, or (c) a less fragmented distribution of remnant habitat, if entire habitat patches are removed. Dark grey = habitat, white = non-habitat.

Table 6.1 Summary of the features, as defined in Box 6.1, included in each of the three major conceptualizations of habitat fragmentation

Feature	Habitat fragmentation per se	Habitat fragmentation as loss and subdivision	Habitat fragmentation as human land use intensity
Habitat loss		×	×
Habitat subdivision	×	×	×
Habitat degradation			×
Reduction in matrix quality			×

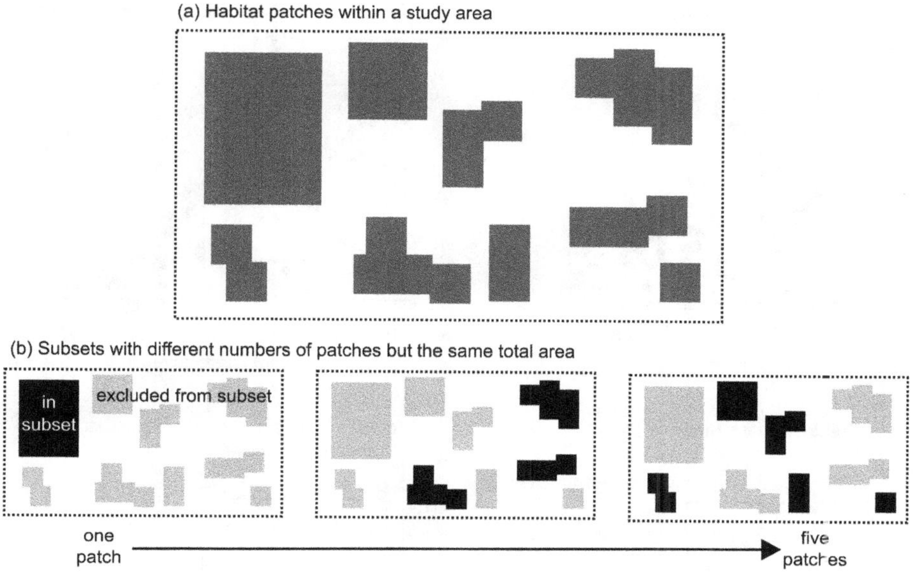

Figure 6.2 Depiction of a SLOSS (single large or several small) study design, which requires (a) a set of different-sized habitat patches and a list of the species found in each patch. Dark grey = habitat, white = non-habitat. (b) Subsets of patches are created, each with a different number of patches but the same total area of habitat, and the number of species in each subset is tallied. Black = patch included in subset, light grey = patch excluded from subset.

when species richness is the ecological response of interest, a SLOSS (single large or several small) study can be used. This method requires a list of species found in each of a number of patches that vary in size (Figure 6.2a). Then, subsets of patches are created, each with a different number of patches but the same total area of habitat, and the number of species is compared across subsets (Figure 6.2b). A second approach is to create replicate experimental landscapes that vary independently in fragmentation per se and habitat amount (e.g. With and Pavuk, 2011). Third, one can use a stratified sampling design to reduce correlations between fragmentation per se and habitat amount across a set of sample landscapes by selecting landscapes with (a) low fragmentation per se and low habitat amount, (b) low fragmentation per se and high habitat amount, (c) high fragmentation per se and low habitat amount, and (d) high fragmentation per se and high habitat amount (Figure 6.3; e.g. Ethier and Fahrig, 2011). Fourth, statistical approaches can be used to disentangle effects of fragmentation per se from effects of habitat amount, for example, by using standardized partial regression coefficients (Smith et al., 2009).

Habitat fragmentation as habitat loss and subdivision

Although habitat fragmentation was initially conceptualized as a particular change in pattern that may or may not accompany habitat loss, the definition of habitat fragmentation changed subtly but importantly over the ensuing decades. This is seen in Wilcove et al. (1986) who defined habitat fragmentation as the process through which 'a large expanse of habitat is transformed into a number of smaller patches of smaller total area, isolated from each other by a matrix of habitat unlike the original' (p. 237). They go on to explicitly specify that habitat fragmentation

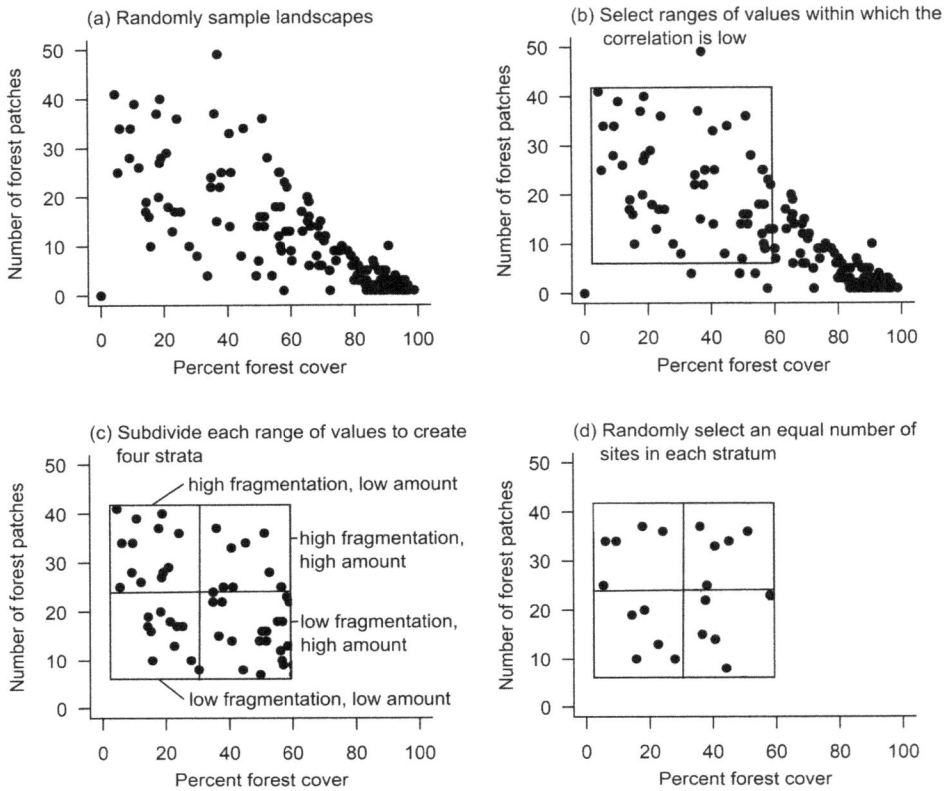

Figure 6.3   Example site selection for a mensurative study designed to minimize correlations between habitat fragmentation per se (e.g. number of forest patches within a landscape) and habitat amount (e.g. the percentage of landscape covered by forest) by (a) randomly sampling landscapes within the study region; (b) plotting the relationship between habitat fragmentation per se and amount, and selecting ranges of values within which the correlation between variables is low; (c) subdividing each range of values to create four strata (low fragmentation–low amount, low fragmentation–high amount, etc.); and (d) randomly selecting an equal number of sampling sites from each stratum.

'has two components, both of which cause extinctions: (1) reduction in total habitat area … and (2) redistribution of the remaining area into disjunct fragments' (p. 238). Thus, the definition of fragmentation changed to encompass habitat loss *and* subdivision (Table 6.1).

This definition of habitat fragmentation was largely predicated on the fact that in most situations, habitat subdivision occurs through habitat loss, i.e. habitat is subdivided by removal of habitat from the landscape. And—although removal of a given amount of habitat can result in an increase, no change, or a decrease in the number of habitat fragments (Figure 6.1)—human-caused habitat loss tends to drive habitat fragmentation along a typical trajectory. In particular, landscapes with less habitat tend to have smaller habitat patches, while both the number of habitat patches and the total edge length (or edge density) tend to peak at intermediate amounts of habitat (Fahrig, 2003; Figure 6.4).

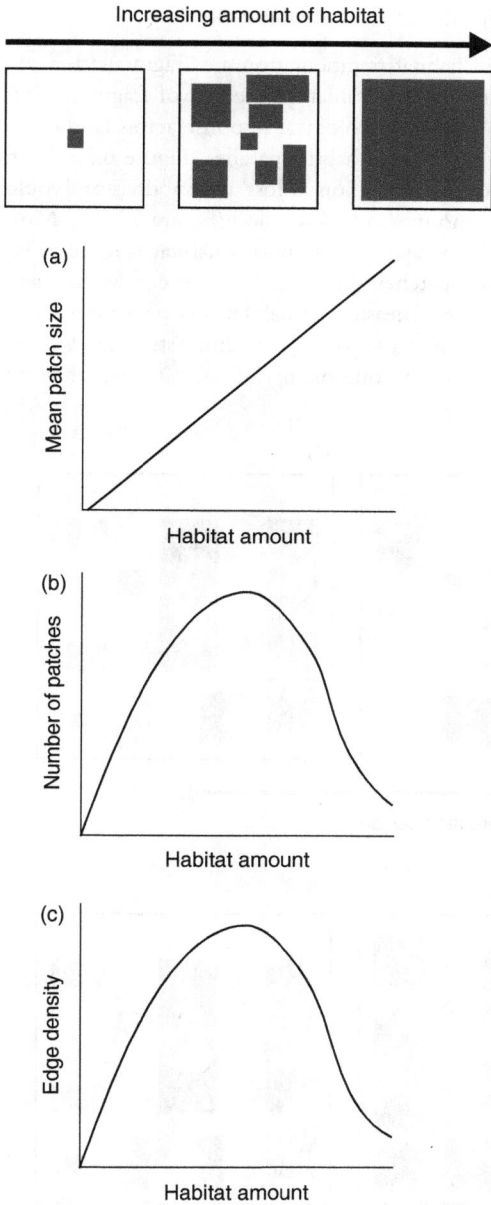

Figure 6.4 Typical relationships between habitat amount and three measures of habitat subdivision: (a) mean patch size (patch size averaged over all patches in the landscape), (b) number of patches, and (c) edge density (length of habitat–non-habitat edge / landscape size; see also Fahrig, 2003). Dark grey = habitat, white = non-habitat.

Amanda E. Martin, Joseph R. Bennett, and Lenore Fahrig

Estimating effects of habitat fragmentation as loss and subdivision

The distinction between the initial definition of habitat fragmentation, i.e. fragmentation per se, and this modified one can be seen in Figure 6.5. To evaluate the effect of fragmentation per se on an ecological response, we would compare that ecological response across landscapes with the same amount of habitat but different numbers of habitat patches (Figure 6.5a–c). In contrast, studies using the modified definition of fragmentation as loss and subdivision would compare landscapes that differed in their habitat amounts and subdivision (Figure 6.5d–f). Most commonly, such studies compare a 'reference' landscape of continuous habitat (Figure 6.5d) with a 'fragmented' landscape containing habitat patches (Figure 6.5e or f; e.g. Mahan and Yahner, 1999). Others evaluate the effect of either a measure of habitat loss or a measure of subdivision (not both) on an ecological response across a set of experimental/study landscapes, under the assumption that their measure indexes both components of habitat fragmentation

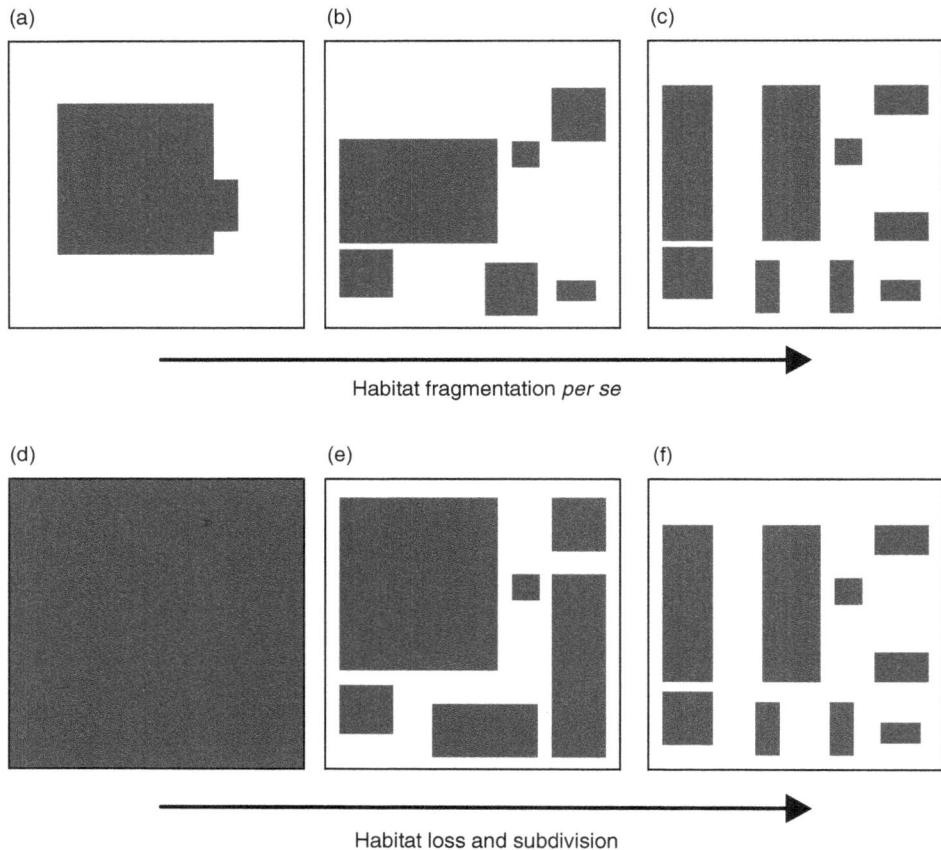

Figure 6.5 Structure of landscapes compared when evaluating the effects of habitat fragmentation per se (a–c) versus fragmentation as loss and subdivision (d–f) on an ecological response. To evaluate the effect of fragmentation per se on an ecological response, we would compare that ecological response across landscapes with the same amount of habitat but different numbers of habitat patches. In contrast, studies defining fragmentation as loss and subdivision would compare landscapes that differed in their habitat amounts and subdivision. Dark grey = habitat, white = non-habitat.

(e.g. Öckinger et al., 2009). Alternatively, studies may create a metric that combines habitat loss and subdivision. This can be done by measuring multiple landscape metrics (habitat amount, number of patches, edge density, etc.), combining them using a method such as principal components analysis, and measuring the effect of the new combined variable on the ecological responses (e.g. Nour et al., 1999). Finally, studies may evaluate the effects of habitat fragmentation as loss and subdivision using a time series of data of both the changing landscape pattern and the changing ecological response (Bohannon and Blinnikov, 2019).

Because studies evaluating the effects of habitat fragmentation as loss and subdivision on an ecological response do not model their separate effects, one cannot estimate the relative effects of habitat loss and subdivision. In other words, if there is a negative effect of habitat loss and subdivision on an ecological response, then one cannot tell whether it is due to habitat loss or habitat subdivision or both.

Evidence to date suggests that the effects of habitat loss and subdivision are driven mainly by habitat loss. Wildlife species occurrence and abundance are typically lower in landscapes with less habitat (e.g. Betts et al., 2007; Zuckerberg and Porter, 2010; Collins and Fahrig, 2017; Melo et al., 2017). This makes sense because, all else being equal, a landscape containing less habitat can support fewer individuals. Population declines can lead to extirpation or extinction when enough habitat is lost. Theory suggests that there is a threshold habitat amount below which a population cannot sustain itself and goes extinct (Fahrig, 2001; Table 1 in Swift and Hannon, 2010). Similarly, empirical studies have shown that wildlife occurrence and abundance can rapidly drop to zero below some threshold habitat amount (e.g. Homan et al., 2004; Betts et al., 2007; Zuckerberg and Porter, 2010). Nevertheless, it is not possible to determine this conclusively from studies that do not disentangle habitat loss from subdivision.

Habitat fragmentation as human land use intensity

The third common conceptualization of habitat fragmentation is as an umbrella term that encompasses not only habitat loss and subdivision but also other changes in the landscape caused by human land use, e.g. roads, intensive agriculture, urbanization, and degradation of remnant patches (Table 6.1).

Estimating effects of habitat fragmentation as human land use intensity

Studies of fragmentation as land use intensity typically study the changes in an ecological response along a gradient of human land use intensity, such as a rural to urban gradient (e.g. Gibbs, 1998; Bergerot et al., 2010). Similarly to studies of fragmentation as loss and subdivision, studies of fragmentation as land use intensity do not estimate the relative effects of habitat loss and subdivision, or the relative effects of other landscape changes such as habitat degradation or matrix quality decline, as all of these are combined into a single gradient. Thus, if there is a negative effect of the habitat fragmentation as land use intensity on an ecological response, then one can only conclude that one (or several) of these aspects negatively affects that response.

Are there benefits to disentangling habitat subdivision from habitat loss, habitat degradation, and declines in matrix quality?

As discussed earlier, only studies conceptualizing habitat fragmentation as habitat fragmentation per se can distinguish between the effect of habitat subdivision and the effects of habitat loss, habitat degradation, and declines in matrix quality on an ecological response. This begs the question: is there any benefit to disentangling habitat subdivision from these other processes?

Amanda E. Martin, Joseph R. Bennett, and Lenore Fahrig

We argue that there is, although we recognize that this viewpoint is not universal. This is made clear through comparison of studies that do and do not separate effects of habitat subdivision from habitat loss.

Studies that do not estimate the separate effects of habitat loss and subdivision on wildlife have likely contributed to the perception that both loss and subdivision have strong, negative effects on wildlife (e.g. see opening sentences of Buchmann et al., 2013; Wilson et al., 2016). However, results of studies of habitat fragmentation per se do not support this perception. In 2003, Fahrig published a review evaluating the evidence for effects of fragmentation per se. She found that the majority of effects of fragmentation per se were non-significant and weaker than those of habitat loss. She found only 12 studies with significant effects of fragmentation per se on ecological responses when controlling for effects of habitat amount (Fahrig, 2003). Of the significant responses to fragmentation per se in these studies, 68% (21/31) were 'positive' (in the statistical sense), i.e. they found higher species richness, abundance, movement success, etc. in landscapes with more fragmented habitat than in landscapes with less fragmented habitat. These findings were supported 14 years—and 106 additional studies—later. Fahrig (2017) found that 76% of 381 significant responses to habitat fragmentation per se were positive. More positive than negative responses to fragmentation per se were found in all studied taxa (plants, microorganisms, invertebrates, fish/herptiles, birds, mammals), and most effects of fragmentation per se on threatened species (6/7 significant responses) and declining species (43/72 significant responses) were positive. Thus disentangling habitat loss from habitat subdivision can help expand our understanding of wildlife responses to landscape change, because it allows detection of different directions and magnitudes of effects of each aspect of landscape change.

A question of scale: what do patch size and isolation studies tell us about the effects of habitat fragmentation?

One commonality among habitat fragmentation per se, habitat fragmentation as loss and subdivision, and habitat fragmentation as land use intensity is that fragmentation is conceptualized as a landscape-scale phenomenon (see Box 6.1, Table 6.1). However, many studies aiming to evaluate the effects of habitat fragmentation on wildlife measure habitat fragmentation at the scale of an individual patch, for example as the size or isolation of the sampled patch (Fardila et al., 2017).

This shift from landscape-scale concept to patch-scale measurement can be traced back to the theory of island biogeography (MacArthur and Wilson, 1967) and Levins's conceptual extrapolation of that theory from islands to habitat patches within a non-habitat 'matrix' (Levins, 1970). MacArthur and Wilson proposed that the number of species on an island would depend on its size because, all else being equal, the extinction rate would be lower on a large than on a small island. Thus, a larger island should have more species than a smaller island. They also predicted that the colonization rate should be higher on a less isolated island, resulting in more species on an island that is closer to the mainland. Levins (1970) extrapolated this theory to habitat patches, proposing that habitat patches surrounded by an inhospitable matrix (e.g. agricultural fields, urban development) would function in the same way as islands surrounded by water.

However, MacArthur, Wilson, and Levins were not the ones to extrapolate from individual habitat patches to groups of patches or landscapes: Diamond did this when developing his principles for the design of nature reserves (Diamond, 1975, p. 144):

> Given a certain total area available for reserves in a homogenous habitat, the reserve should generally be divided into as few disjunctive pieces as possible (principle B), for essentially

the reasons underlying principle A. Many species that would have a good chance of surviving in a single large reserve would have their survival chances reduced if the same area were apportioned among several smaller reserves.

Principle A states that a single large habitat patch is better than a single small patch. Principle B states that it is better to have a single large patch than several smaller ones of the same total area, i.e. that there is a negative effect of habitat fragmentation on wildlife, independent of the amount of habitat, at the landscape scale. Thus, Diamond justified principle B based on extrapolation of patch size effects on wildlife to effects of fragmentation per se. This series of extrapolations from islands to habitat patches and from habitat patches to landscapes led to the assumption that patch-scale studies provide evidence of the effects of landscape-scale habitat fragmentation per se (e.g. Haddad et al., 2015; Fletcher et al., 2018).

The resulting studies of the relationships between patch size/isolation and ecological responses have provided important information on wildlife responses at the patch scale. However, extrapolation of these effects to landscape scales is problematic. Empirical patch-scale studies of the effects of patch size/isolation on wildlife typically come to different conclusions than empirical landscape-scale studies of effects of fragmentation per se on wildlife. For example, long-term experimental studies have demonstrated strong effects of patch size (positive) and/or isolation (negative) on a number of ecological responses, including species richness, population persistence, and abundance (Haddad et al., 2015). Observational studies have similarly found that patch size and/or isolation can have strong effects on wildlife (e.g. Öckinger et al., 2009; Keinath et al., 2017). Extrapolation of these results to the landscape scale leads to a conclusion that fragmentation per se negatively affects wildlife (e.g. Haddad et al., 2015; Fletcher et al., 2018). In contrast, as discussed earlier, landscape-scale studies generally find weak or positive effects of fragmentation per se on ecological responses (Fahrig, 2003, 2017). Therefore, it appears that the extrapolation from patch-scale studies to landscape-scale inferences about habitat fragmentation per se is not valid.

One likely explanation for the conflicting conclusions between patch-scale studies and landscape-scale studies of habitat fragmentation per se is that most, if not all, purely patch-scale studies (i.e. ones that do not also include measurements at the landscape scale) confound the effect of habitat fragmentation per se on wildlife with the effect of habitat loss. This is because both patch size and isolation tend to be correlated with the amount of habitat in the surrounding landscape. A patch tends to be larger in an area with more habitat. A patch also tends to be more isolated—i.e. farther from other patches in the landscape—when there is less habitat in that landscape. As discussed earlier, habitat loss has large effects on a variety of ecological responses. Thus, patch-scale empirical studies may very well be indexing the negative effects of habitat loss on wildlife rather than effects of fragmentation per se.

However, evidence to date suggests that the correlation between patch size/isolation and habitat amount is not the sole explanation for conflicting conclusions from studies at the patch versus landscape scale. This evidence comes from studies that evaluate effects of patch size or isolation while controlling for the amount of habitat in the surrounding landscape (through study design and/or statistical methods). Some such studies have found no effect of patch size/ isolation when controlling for habitat amount (e.g. Melo et al., 2017; Merckx et al., 2019), which is consistent with the idea that patch size/isolation is indexing the effect of habitat loss (Fahrig, 2013). However, others have found significant effects of patch size (positive) or isolation (negative) on wildlife independent of effects of habitat amount (e.g. Evju and Sverdrup-Thygeson, 2016; Lindgren and Cousins, 2017; Seibold et al., 2017; Torrenta and Villard, 2017). Nevertheless, such results do not validate extrapolation of patch size/isolation effects to negative

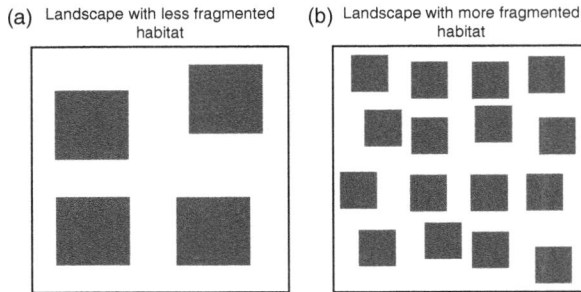

Figure 6.6   Landscapes with less fragmented habitat, i.e. fewer and larger patches (a), tend to have, on average, longer distances among patches than landscapes with more fragmented habitat (b). Note that the total amount of habitat is identical in (a) and (b). Dark grey = habitat, white = non-habitat.

effects of fragmentation per se at the landscape scale. This extrapolation relies on the assumption that relationships (and mechanisms underlying those relationships) observed at the patch scale are the same as those at the landscape scale. Conflicting conclusions from patch-scale studies that control for habitat amount and landscape-scale studies of fragmentation per se do not support this assumption.

Differences in the inferred effects of habitat fragmentation per se on wildlife from studies of patch isolation (when controlling for habitat amount) versus landscape-scale measures of fragmentation per se may be explained by differences in the perceived relationship between the distance among patches and habitat fragmentation per se at the two spatial scales. At the patch scale, greater patch isolation, i.e. a larger distance between the sampled patch and its neighbors, is interpreted as evidence of greater fragmentation per se. However, for a given habitat amount and landscape size, distances between patches are generally *shorter*, not longer, in landscapes with more habitat fragments (Figure 6.6). Thus, studies showing fewer species, lower abundances, etc. in more isolated patches than in less isolated patches may translate to more species, higher abundances, etc. in landscapes with more fragmented habitat than in landscapes with less fragmented habitat. This is a clear example of how inconsistency in definition and measurement can hinder our shared understanding of the effects of habitat fragmentation, with shorter inter-patch distances equating to less fragmentation per se to some people and more fragmentation per se to others. However, this cannot explain the conflicting conclusions from studies of patch size (when controlling for habitat amount) and fragmentation per se at the landscape scale.

The conflicting conclusions from patch-scale and landscape-scale studies could occur because the mechanisms that drive effects at one scale are different from those driving effects at the other scale (Fahrig et al., 2019). Mechanisms at the patch scale (e.g. negative edge effects) may drive positive effects of patch size on wildlife (Figure 6.7a). It is possible that mechanisms functioning at the scale of individual patches could also affect responses to habitat fragmentation per se at the landscape scale. However, this may not translate across scales to a negative effect of fragmentation per se if other mechanisms involving interactions of individuals over multiple habitat patches (e.g. reduced competition) drive positive effects of fragmentation per se that outweigh the patch-scale effects (Figure 6.7b). Various mechanisms have been proposed to explain both positive and negative relationships between fragmentation per se and ecological responses (Fahrig, 2017; Fahrig et al., 2019). These are shown in Figure 6.7 and discussed in the following sections.

Mechanisms driving effects of habitat fragmentation per se on wildlife

There are two primary patch-scale mechanisms that should drive a negative effect of habitat fragmentation per se on wildlife at the landscape scale. First, a negative effect of fragmentation per se could be driven by the effect of habitat patch size on local population persistence. If individuals avoid moving between habitat patches, then they will have access to more habitat within a single patch than the equivalent area of habitat subdivided into multiple patches. This could translate into small populations in small patches, which can have a high probability of local extinction (Matthies et al., 2004). If a patch becomes small enough, and if individuals cannot use multiple patches as part of their territories, the patch may not be able to support even one territory. Declines in population persistence with declining patch size may translate into negative effects of fragmentation per se on wildlife because more fragmented landscapes have smaller patches, albeit more of them.

The second patch-scale mechanism that could produce negative effects of habitat fragmentation per se on wildlife is negative edge effects, i.e. the phenomenon whereby some ecological response (e.g. nest predation rates, abundance, species richness) increases from the edge to the interior of a habitat patch (e.g. Watson et al., 2004). In this case, the quality of habitat within a patch varies, with lower-quality habitat near the edge than in the interior of a patch. This could translate into a negative effect of fragmentation per se because, for a given habitat amount, a landscape with more fragmented habitat has more edge than a landscape with less fragmented habitat (see Figure 6.5a–c). Thus, a landscape with more fragmented habitat will have less high-quality habitat. Negative edge effects may occur, for example, because temperature, moisture, wind speed, etc. change from habitat edge to interior (Davies-Colley et al., 2000). There can also be increased rates of predation or parasitism near habitat edges (e.g. Sosa and Lopez de Casenave, 2017) because individuals near edges are at risk not only from predators/parasites that have the same habitat but also from those that use the adjacent matrix as habitat and hunt along edges.

However, not all edge effects are negative and, thus, edge effects could result in positive effects of habitat fragmentation per se on wildlife species that show positive responses to habitat edges (e.g. Stone et al., 2018). As earlier, the existence of a positive edge effect implies that habitat quality varies from the patch edge (high quality) to patch interior (low quality). These species should benefit from increasing fragmentation per se because landscapes with more fragmented habitat have more edge and thus, more high-quality habitat. Positive responses to edges between habitat and matrix may occur, for example, because food resources are more abundant in edges than in the patch interior (Harding and Gomez, 2006; Macreadie et al., 2010).

The majority of mechanisms involving interactions of individuals over multiple habitat patches (i.e. landscape-scale mechanisms) should drive positive relationships between habitat fragmentation per se and ecological responses (Figure 6.7b). For example, fragmentation per se could have a positive effect on wildlife because there are more opportunities for risk-spreading (also called bet-hedging) in space in landscapes with more habitat patches.

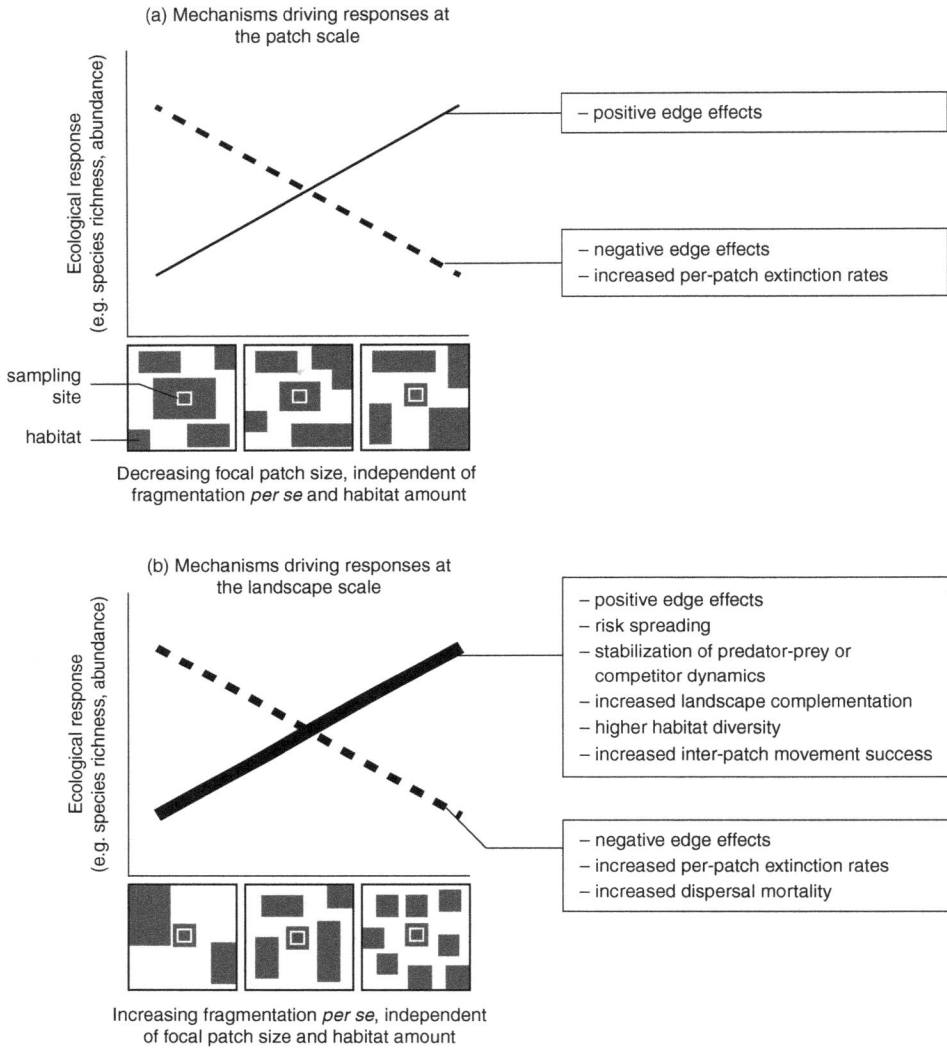

(a) Mechanisms driving responses at the patch scale

Ecological response (e.g. species richness, abundance)

– positive edge effects

– negative edge effects
– increased per-patch extinction rates

sampling site

habitat

Decreasing focal patch size, independent of fragmentation *per se* and habitat amount

(b) Mechanisms driving responses at the landscape scale

Ecological response (e.g. species richness, abundance)

– positive edge effects
– risk spreading
– stabilization of predator-prey or competitor dynamics
– increased landscape complementation
– higher habitat diversity
– increased inter-patch movement success

– negative edge effects
– increased per-patch extinction rates
– increased dispersal mortality

Increasing fragmentation *per se*, independent of focal patch size and habitat amount

Figure 6.7   Relationships between ecological responses and measures of (a) patch size (independent of fragmentation per se and habitat amount) and (b) habitat fragmentation per se (e.g. number of patches, independent of focal patch size and habitat amount) predicted for each proposed mechanism. Line thickness is proportional to the number of mechanisms that should produce a given relationship. Mechanisms functioning at the scale of individual patches could also affect responses to habitat fragmentation per se at the landscape scale. However, other mechanisms function only at the landscape scale, i.e. they involve interactions of individuals over multiple habitat patches. Positive effects of patch size would not translate across scales to negative effects of fragmentation per se if other mechanisms involving interactions of individuals over multiple habitat patches typically drive positive effects of habitat fragmentation per se that outweigh the patch-scale effects. Note that we show the ecological response being measured within the same-sized area within each focal patch, which avoids confounding effects of patch size/fragmentation per se on the ecological response with effects of sampling effort. Dark grey = habitat, white = non-habitat.

This is supported by theoretical work showing that having more habitat patches can reduce the risk of an entire population going extinct because the risk of simultaneous extinction of all local populations is lower when there are more patches (den Boer, 1968; Bascompte et al., 2002).

If habitat patches function as temporary refugia from predators, parasites, or competitors, then having a larger number of (albeit smaller) patches should benefit the wildlife community. Huffaker (1958) was the first to demonstrate this, using an experimental system comprised of oranges as 'habitat', the six-spotted mite (*Eotetranychus sexmaculatus*), which feeds on oranges, and a predatory mite (*Typhlodromus occidentalis*), which feeds on the six-spotted mite. Huffaker showed that, for a given total surface area of orange, persistence of this predator–prey system was more likely when there were more, smaller orange patches. This finding was replicated in later theoretical and empirical studies (Ellner et al., 2001; Karsai and Kampis, 2011). This most likely occurs in systems where the dispersal ability of a prey species is greater than the dispersal ability of its predator, allowing prey to colonize a habitat patch after predator extirpation there, increase in numbers, and then disperse to other patches before the predator reaches the patch. Similarly, fragmentation per se may allow inferior competitor(s) to coexist with superior competitor(s) on a landscape if the inferior competitor is better able to move among patches and exploit those that are temporarily free of the superior competitor(s) when there are more patches on the landscape.

Habitat fragmentation *per se* could have positive effects on wildlife species at the landscape scale if those species require multiple land cover types as habitat (e.g. species that breed in ponds but overwinter in forests). This is because patches of different habitat types tend to be more intermixed and closer together in a more fragmented landscape. Thus, fragmentation per se can increase accessibility of different habitat types, i.e. landscape complementation (Dunning et al., 1992).

Habitat fragmentation per se could also have positive relationships with measures of species diversity (e.g. Shannon diversity of species, species richness, genetic diversity) because there are generally more diverse environmental conditions/habitat types in landscapes with greater fragmentation per se. The general idea here is that environmental conditions vary in space, and a wider range of these conditions are covered when habitat is more fragmented than when habitat is less fragmented (Figure 6.8).

Finally, habitat fragmentation per se may have a positive relationship with an ecological response because, for a given habitat amount, distances among habitat patches are typically shorter in a more fragmented landscape (Figure 6.6). This should translate into increased inter-patch movement success if an individual's probability of mortality increases with the distance moved (e.g. Smith and Batzli, 2006; Johnson et al., 2009). Shorter distances between patches may also translate into increased inter-patch movement success because individuals of some species are more likely to cross gaps between habitat patches when those distances are shorter than when they are longer (Lees and Peres, 2009; Robertson and Radford, 2009). Theory suggests that wildlife species are more likely to persist in a landscape undergoing habitat loss and fragmentation if they have higher rates of successful inter-patch dispersal (Martin and Fahrig, 2016). Thus, for some wildlife species, the benefits of successful inter-patch dispersal could translate into positive effects of fragmentation per se.

In contrast, increased habitat fragmentation per se could result in increased dispersal mortality in some cases, driving negative effects of fragmentation per se on wildlife. This is because, all else being equal, the probability of a disperser encountering a habitat boundary and entering the matrix is higher when habitat patches are smaller (Fahrig, 2007). This should increase per capita

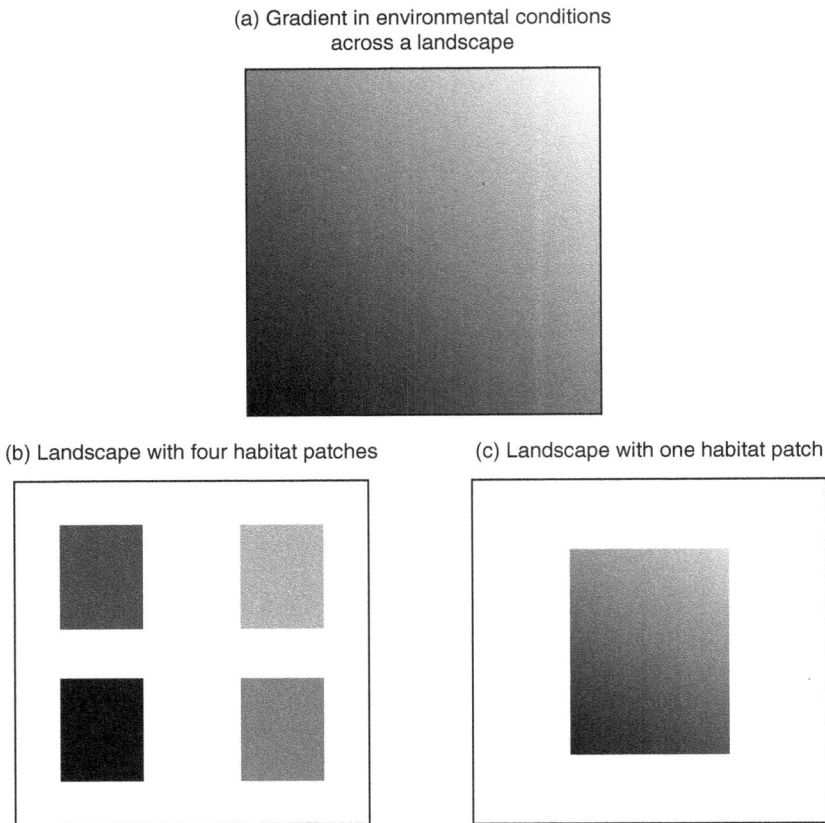

Figure 6.8 Illustration to show that when (a) environmental conditions vary in space (depicted here as a gradient from black to light grey), a wider range of these conditions is covered when a given habitat amount is subdivided into more patches (b) than when it is subdivided into fewer patches (c). Thus, habitat is more heterogeneous in a landscape with more fragmented habitat. White = non-habitat.

mortality rates because the probability of dispersal mortality is generally higher in matrix than in habitat. This is supported by simulation modeling showing that wildlife species are less likely to persist in a landscape undergoing habitat loss and fragmentation per se if they have higher per capita rates of dispersal through matrix (Martin and Fahrig, 2016).

Evaluating the effects of patch size versus habitat fragmentation per se on wildlife

Conclusions about the effects of habitat fragmentation per se on wildlife from patch-scale and landscape-scale studies can be contradictory, even when controlling for correlations between patch size and habitat amount. Thus, to date, evidence does not support extrapolation of patch-scale studies to landscape-scale inferences about effects of fragmentation per se on wildlife. Instead, further study is needed to understand the independent effects of patch size versus fragmentation per se on wildlife.

Studies estimating the independent effects of patch size versus habitat fragmentation per se on wildlife must be designed in a way that avoids confounding the effect of patch size on an ecological response with the effect of either fragmentation per se or habitat amount. This is because patch size can be correlated with fragmentation per se (e.g. the number of patches) in the surrounding landscape (Figure 6.9a), as well as habitat amount (Figure 6.9b). One could disentangle effects of patch size, fragmentation per se, and habitat amount through experimental design, creating replicate landscapes that vary independently in their focal patch size (where the response is measured), landscape-scale fragmentation per se (e.g. number of patches, including the sampled patch), and habitat amount (including the sampled patch; Figure 6.10). Sampling effort should be kept constant across all landscapes (e.g. by sampling the same-sized area in each focal patch) to avoid confounding an effect of patch size with an effect of sampling effort on the response. Mensurative experiments are another option, with a stratified sampling design to reduce correlations between patch size, number of patches, and habitat amount across a set of sample landscapes. Statistical approaches (alone, or combined with an experimental or mensurative study design) can also be used to disentangle effects of patch size, number of patches, and habitat amount (e.g. Thornton et al., 2011). There may be no effect of patch size on an ecological response when controlling for habitat amount in such a study (Fahrig, 2013; Melo et al., 2017; Merckx et al., 2019). However, given that two-thirds of patch-scale mechanisms should drive positive effects of patch size independently of habitat amount and fragmentation per se (Figure 6.7a), when there is an effect of patch size, we expect it to generally be positive. When there is an effect of fragmentation per se on an ecological response, it is predicted to generally be positive, because two-thirds of landscape-scale mechanisms should drive positive effects of fragmentation per se (Figure 6.7b).

A study design that disentangles the effects of patch size and landscape-scale measures of habitat fragmentation per se on wildlife also allows direct comparison of their relative effects. This could have important implications both for our understanding of these phenomena and for conservation. For example, a study may find that effects of landscape-scale measures of fragmentation per se on a species are weaker than those of focal patch size. This would suggest that fragmentation per se is less important for that species than patch size effects and that, from a management perspective, patch-based management is likely to be more effective than landscape-based management.

We highlight three considerations that are key to proper evaluation of the independent effects of patch size and landscape-scale measures of habitat fragmentation per se. First, selection of an appropriate landscape size is crucial, because the effects of landscape-scale measures of fragmentation per se and habitat amount may be underestimated if measured at a suboptimal scale (e.g. Ethier and Fahrig, 2011; Smith et al., 2011). There are a number of hypotheses for what should determine the appropriate landscape size; however, empirical support for these hypotheses is generally weak (Jackson and Fahrig, 2015; Miguet et al., 2016). Instead, one can empirically estimate the landscape size via a multi-scale analysis (see Figure 6.11). Second, patch size and the measure of fragmentation per se should be derived from the same data. This avoids the potential case where there is more accurate definition of habitat at one scale than the other. Third, studies should strive to use the best possible data to index habitat. In many cases, researchers do not have detailed data on the locations of the resources (e.g. shelter, food) that comprise a species' actual habitat; instead, habitat is represented by one or several land cover types. In these cases, researchers should ensure, to the greatest degree possible, that the land cover type reasonably approximates the species' habitat. For example, if habitat is indexed by forest, then the species should be known to use all types of forests as habitat but not other land cover types (e.g. wetland, urban). Preferable to this approach would be one that uses empirical data to delineate habitat.

Amanda E. Martin, Joseph R. Bennett, and Lenore Fahrig

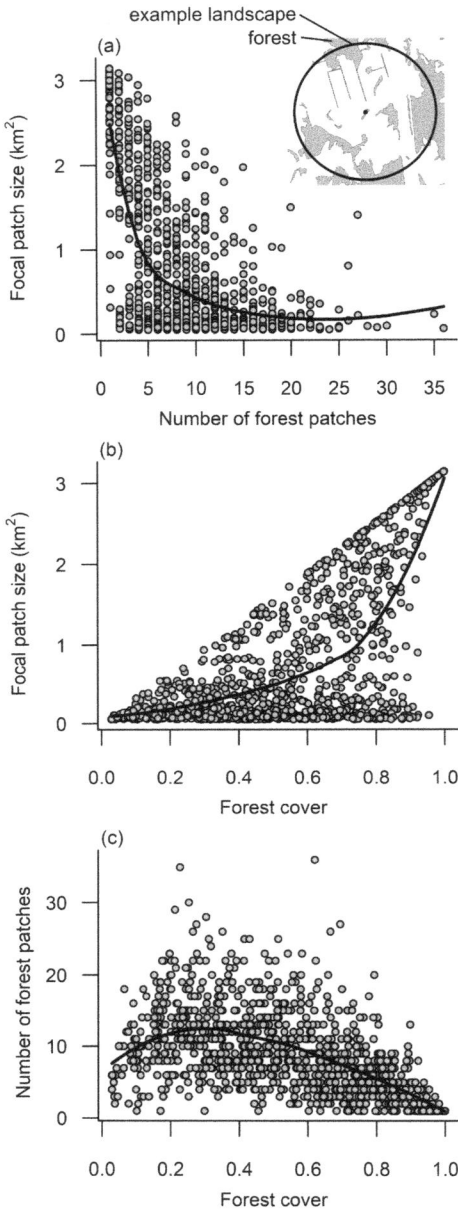

Figure 6.9 Example of the relationship between (a) focal patch size and habitat fragmentation per se, measured as the number of forest patches, (b) patch size and habitat amount, measured as the proportion of the landscape in forest, and (c) fragmentation per se and habitat amount. Empirical relationships were derived from a sample of 1000 circular landscapes (1-km radius) in southeastern Ontario, Canada. Each point represents a landscape. Contains information licensed under the Open Government Licence—Ontario (https://geohub.lio.gov.on.ca/datasets/wooded-area, accessed 23 November 2019).

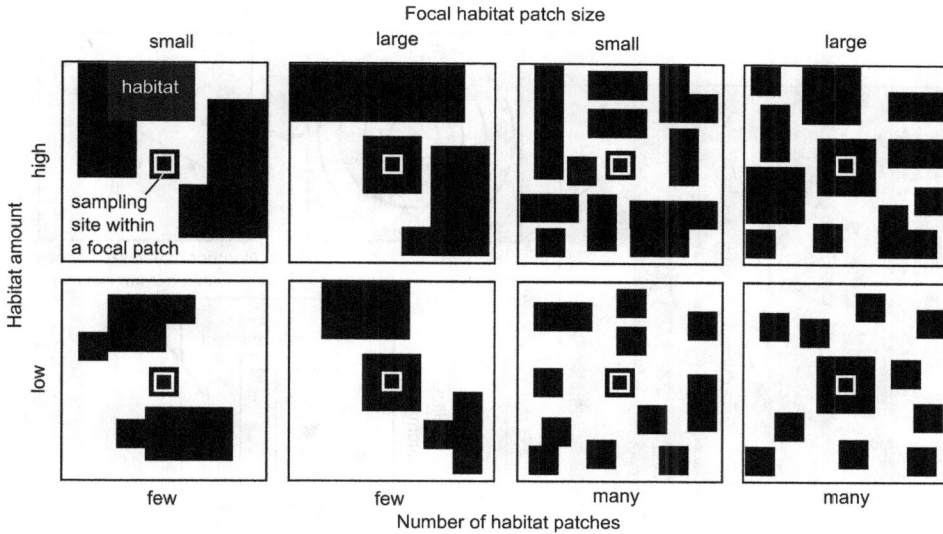

Figure 6.10 Example of the strata that can be used to design replicate landscapes in an experimental study or to select sites for a mensurative study (see also Figure 6.3), when the goal is to minimize correlations between the focal patch size, a landscape-scale measure of habitat fragmentation per se (e.g. number of forest patches within a landscape), and habitat amount (e.g. the percentage of the landscape covered by forest). Here, the eight strata are defined by subdividing the range of values for each of the three variables (small patch size–few patches–low amount, large patch size–few patches–low amount, etc.). Sampling effort is kept constant across all landscapes by sampling the same-sized area in each focal patch. Black = habitat, white = non-habitat.

Conclusions

In this chapter, we introduced the concept of habitat fragmentation and how its original definition—as the breaking apart of habitat, a phenomenon caused by, but distinct from, habitat loss—evolved over time to encompass other phenomena related to anthropogenic impacts on the landscape and species. We then highlight the reason why consideration of habitat subdivision independent of habitat loss, i.e. habitat fragmentation per se, could enhance our understanding of the processes and spatial patterns underlying anthropogenic (or natural disturbance) impacts on the landscape and wildlife. We followed this with a discussion of how fragmentation per se, a landscape-scale phenomenon, became synonymous with effects of patch size and patch isolation on wildlife. We also showed that evidence to date does not support extrapolation of the observed effects of patch size/isolation on wildlife to negative effects of habitat fragmentation per se on wildlife. Instead, studies at the landscape scale typically show that while habitat loss has strong, negative effects on wildlife, fragmentation per se tends to have weaker effects that are more likely to be positive than negative (Fahrig, 2003, 2017). We concluded with a discussion of the possible reasons for these conflicting conclusions from patch-scale and landscape-scale studies, and we provide recommendations for future research to increase our understanding of the effects of patch size versus landscape-scale fragmentation per se on wildlife.

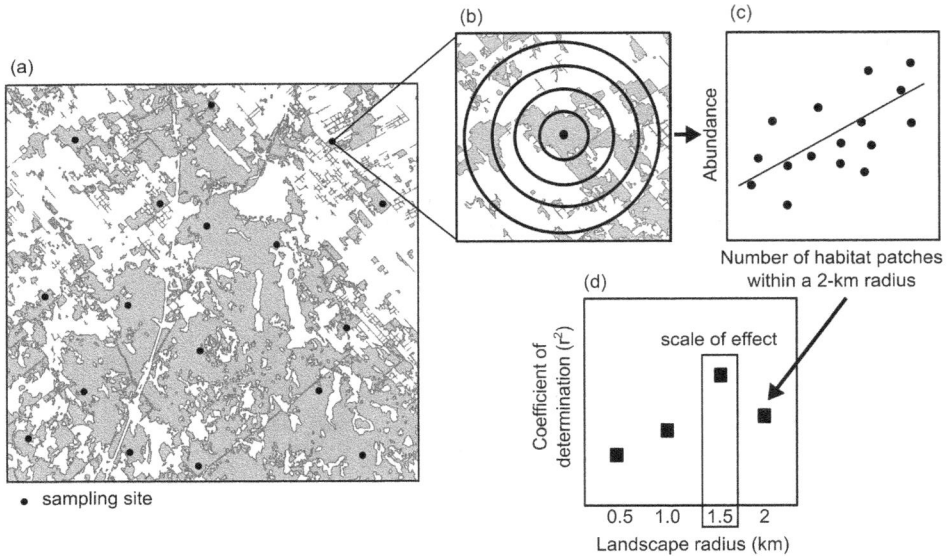

Figure 6.11 Illustration of the design of a study to empirically estimate the appropriate landscape size. This is done by (a) measuring the ecological response (e.g. abundance) at sampling sites within habitat; (b) measuring the landscape-scale variable (e.g. number of habitat patches) within multiple landscape extents, represented by different-sized areas centered on each sampling site; (c) estimating the strength of relationship between the landscape variable and the ecological response measured at the central sampling site at each landscape extent; and (d) comparing the strength of relationship across extents and then selecting the extent where the relationship is strongest. Grey = habitat, white = non-habitat. *Source*: Martin (2018; modified with permission).

Although some suggest that the term *habitat fragmentation* is a panchreston, a term used in so many ways that it has become almost meaningless (Bunnell, 1999; Lindenmayer and Fischer, 2007), the implications of studies of the effects of habitat fragmentation on wildlife for conservation and land management are not meaningless. First and foremost, they suggest that our focus should remain on reducing habitat loss and on habitat restoration. We should not expect that manipulating the spatial pattern of habitat loss—to either increase or decrease fragmentation per se—will mitigate the negative effects of habitat loss on wildlife. However, when habitat loss is inevitable, management actions can be designed to increase or decrease fragmentation per se. Whether management seeks to increase or decrease fragmentation per se will depend on the management objective. If management is species-based (e.g., planning for an at-risk species), then managers should use information on the species-specific response to fragmentation per se, because some species respond positively to fragmentation per se, while others respond negatively. If the objective is to increase or maintain biodiversity through landscape-scale management, the evidence to date provides no biological justification for an assumption that a landscape with many, small patches has less value than a landscape with the same habitat amount but few, large patches. On the contrary, more wildlife species appear to respond positively than negatively to fragmentation per se (Fahrig, 2017).

Additionally, conflicting conclusions regarding the effects of habitat fragmentation per se on wildlife drawn from patch-scale versus landscape-scale studies highlight the importance of matching the scale of management action to the scale of ecological information used to inform that action. If land is being managed at the scale of individual patches, then it should be informed by studies of the effects of patch size/isolation on wildlife. Landscape-scale management should be informed by landscape-scale studies of the effects of habitat amount and fragmentation per se on wildlife.

In summary, protecting and restoring habitat for at-risk wildlife species is paramount. Just because habitat patches within a landscape are small does not mean that they have small conservation value. Instead, many species may benefit from having more, smaller patches in a landscape than an equivalent area in a few, large patches.

References

Bascompte, J., Possingham, H. and Roughgarden, J. (2002) 'Patchy populations in stochastic environments: critical number of patches for persistence', *American Naturalist*, vol 159, no 2, pp128–137.

Bergerot, B., Julliard, R. and Baguette, M. (2010) 'Metacommunity dynamics: decline of functional relationship along a habitat fragmentation gradient', *PLoS ONE*, vol 5 no 6, art e11294.

Betts, M.G., Forbes, G.J. and Diamond, A.W. (2007) 'Thresholds in songbird occurrence in relation to landscape structure', *Conservation Biology*, vol 21, no 4, pp1046–1058.

Bohannon, R. and Blinnikov, M. (2019) 'Habitat fragmentation and breeding bird populations in western North Dakota after the introduction of hydraulic fracturing', *Annals of the American Association of Geographers*, vol 109 no 5, pp1471–1492.

Buchmann, C.M., Schurr, F.M., Nathan, R. and Jeltsch, F. (2013) 'Habitat loss and fragmentation affecting mammal and bird communities: the role of interspecific competition and individual space use', *Ecological Informatics*, vol 14, pp90–98.

Bunnell, F.L. (1999) 'Let's kill a panchreston: giving fragmentation meaning', in J.A. Rochelle, L.A. Lehman and J. Wisniewski (eds) *Forest Fragmentation. Wildlife and Management Implications*. Brill, Leiden, Boston and Koln.

Collins, S.J. and Fahrig, L. (2017) 'Responses of anurans to composition and configuration of agricultural landscapes', *Agriculture, Ecosystems and Environment*, vol 239, no 2, pp399–409.

Curtis, J.T. (1956) 'The modification of mid-latitude grasslands and forests by man', in J.A. Wiens, M.R. Moss, M.G. Turner and D.J. Mladenoff (eds) *Foundation Papers in Landscape Ecology* (2007). Columbia University Press, New York.

Davies-Colley, R.J., Payne, G.W. and van Elswijk, M. (2000) 'Microclimate gradients across a forest edge', *New Zealand Journal of Ecology*, vol 24, no 2, pp111–121.

den Boer, P.J. (1968) 'Spreading of risk and stabilization of animal numbers', *Acta Biotheoretica*, vol 18, no 1–4, pp165–194.

Diamond, J.M. (1975) 'The island dilemma: lessons of modern biogeographic studies for the design of natural reserves', *Biological Conservation*, vol 7, no 2, pp129–146.

Didham, R.K., Kapos, V. and Ewers, R.M. (2012) 'Rethinking the conceptual foundations of habitat fragmentation research', *Oikos*, vol 121, no 2, pp161–170.

Dunning, J.B., Danielson, B.J. and Pulliam, H.R. (1992) 'Ecological processes that affect populations in complex landscapes', *Oikos*, vol 65, no 1, pp169–175.

Ellner, S.P., McCauley, E., Kendall, B.E., Briggs, C.J., Hosseini, P.R., Wood, S.N., Janssen, A., Sabelis, M.W., Turchin, P., Nisbet, R.M. and Murdoch, W.W. (2001) 'Habitat structure and population persistence in an experimental community', *Nature*, vol 412, pp538–543.

Ethier, K. and Fahrig, L. (2011) 'Positive effects of forest fragmentation, independent of forest amount, on bat abundance in eastern Ontario, Canada', *Landscape Ecology*, vol 26, no 6, pp865–876.

Evju, M. and Sverdrup-Thygeson, A. (2016) 'Spatial configuration matters: a test of the habitat amount hypothesis for plants in calcareous grasslands', *Landscape Ecology*, vol 31, no 9, pp1891–1902.

Fahrig, L. (2001) 'How much habitat is enough?', *Biological Conservation*, vol 100, no 1, pp65–74.

Fahrig, L. (2003) 'Effects of habitat fragmentation on biodiversity', *Annual Review of Ecology, Evolution, and Systematics*, vol 34, pp487–515.

Fahrig, L. (2007) 'Estimating minimum habitat for population persistence', in D.B. Lindenmayer and R.J. Hobbs (eds) *Managing and Designing Landscapes for Conservation: Moving from Perspectives to Principles*. Blackwell Publishing Ltd., Oxford.

Fahrig, L. (2013) 'Rethinking patch size and isolation effects: the habitat amount hypothesis', *Journal of Biogeography*, vol 40, no 9, pp1649–1663.

Fahrig, L. (2017) 'Ecological responses to habitat fragmentation per se', *Annual Review of Ecology, Evolution, and Systematics*, vol 48, pp1–23.

Fahrig, L. (2019) 'Habitat fragmentation: a long and tangled tale', *Global Ecology and Biogeography*, vol 28, no 1, pp33–41.

Fahrig, L., Arroyo-Rodríguez, V., Bennett, J.R., Boucher-Lalonde, V., Cazetta, E., Currie, D.J., Eigenbrod, F., Ford, A.T., Harrison, S.P., Jaeger, J.A.G., Koper, N., Martin, A. E., Martin, J.-L., Metzger, J.P., Morrison, P., Rhodes, J.R., Saunders, D.A., Simberloff, D. Smith, A.C., Tischendorf, L., Vellend, M. and Watling, J.I. (2019) 'Is habitat fragmentation bad for biodiversity?', *Biological Conservation*, vol 230, no 2, pp179–186.

Fardila, D., Kelly, L.T., Moore, J.L. and McCarthy, M.A. (2017) 'A systematic review reveals changes in where and how we have studied habitat loss and fragmentation over 20 years', *Biological Conservation*, vol 212, part A, pp130–138.

Fletcher, Jr., R.J., Didham, R.K., Banks-Leite, C., Barlow, J., Ewers, R.M., Rosindell, J., Holt, R.D., Gonzalez, A., Pardini, R., Damschen, E.I., Melo, F.P.L., Ries, L., Prevedello, J.A., Tscharntke, T., Laurance, W.F., Lovejoy, T. and Haddad, N.M. (2018) 'Is habitat fragmentation good for biodiversity?', *Biological Conservation*, vol 226, October, pp9–15.

Gibbs, J. P. (1998) 'Distribution of woodland amphibians along a forest fragmentation gradient', *Landscape Ecology*, vol 13, no 4, pp263–268.

Haddad, N.M., Brudvig, L.A., Clobert, J., Davies, K.F., Gonzalez, A., Holt, R.D., Lovejoy, T.E., Sexton, J.O., Austin, M.P., Collins, C.D., Cook, W.M., Damschen, E.I., Ewers, R.M., Foster, B.L., Jenkins, C.N., King, A.J., Laurance, W.F., Levey, D.J., Margules, C.R., Melbourne, B.A., Nicholls, A.O., Orrock, J.L., Song, D.-X. and Townshend, J.R. (2015) 'Habitat fragmentation and its lasting impact on Earth's ecosystems', *Science Advances*, vol 1, no 2, art e1500052.

Haila, Y. and Hanski, I.K. (1984) 'Methodology for studying the effect of habitat fragmentation on land birds', *Annales Zoologici Fennici*, vol 21, no 3, pp393–397.

Harding, E.K. and Gomez, S. (2006) 'Positive edge effects for arboreal marsupials: an assessment of potential mechanisms', *Wildlife Research*, vol 33, no 2, pp121–129.

Homan, R.N., Windmiller, B.S. and Reed, J.M. (2004) 'Critical thresholds associated with habitat loss for two vernal pool-breeding amphibians', *Ecological Applications*, vol 14, no 5, pp1547–1553.

Huffaker, C.B. (1958) 'Experimental studies on predation: dispersion factors and predator-prey oscillations', *Hilgardia*, vol 27, no 14, pp795–835.

Jackson, H.B. and Fahrig, L. (2015) 'Are ecologists conducting research at the optimal scale?', *Global Ecology and Biogeography*, vol 24, no 1, pp52–63.

Johnson, C.A., Fryxell, J.M., Thompson, I.D. and Baker, J.A. (2009) 'Mortality risk increases with natal dispersal distance in American martens', *Proceedings of the Royal Society B: Biological Sciences*, vol 276, no 1671, pp3361–3367.

Karsai, I. and Kampis, G. (2011) 'Connected fragmented habitats facilitate stable coexistence dynamics', *Ecological Modelling*, vol 222, no 3, pp447–455.

Keinath, D.A., Doak, D.F., Hodges, K.E., Prugh, L.R., Fagan, W., Sekercioglu, C.H., Buchart, S.H.M. and Kauffman, M. (2017) 'A global analysis of traits predicting species sensitivity to habitat fragmentation', *Global Ecology and Biogeography*, vol 26, no 1, pp115–127.

Lees, A.C. and Peres, C.A. (2009) 'Gap-crossing movements predict species occupancy in Amazonian forest fragments', *Oikos*, vol 118, no 2, pp280–290.

Levins, R. (1970) 'Extinction', *Lecture Notes in Mathematics*, vol 2, pp75–107.

Lindenmayer, D.B. and Fischer, J. (2007) 'Tackling the habitat fragmentation panchreston', *Trends in Ecology and Evolution*, vol 22, no 3, pp127–132.

Lindgren, J.P. and Cousins, S.A.O. (2017) 'Island biogeography theory outweighs habitat amount hypothesis in predicting plant species richness in small grassland remnants', *Landscape Ecology*, vol 32, no 9, pp1895–1906.

MacArthur, R.H. and Wilson, E.O. (1967) *The Theory of Island Biogeography*. Princeton University Press: Princeton.

Macreadie, P.I., Hindell, J.S., Keough, M.J., Jenkins, G.P. and Connolly, R.M. (2010) 'Resource distribution influences positive edge effects in a seagrass fish', *Ecology*, vol 91, no 7, pp2013–2021.

Mahan, C.G. and Yahner, R.H. (1999) 'Effects of forest fragmentation on behaviour patterns in the eastern chipmunk (*Tamias striatus*)', *Canadian Journal of Zoology*, vol 77, no 12, pp1991–1997.

Martin, A.E. (2018) 'The spatial scale of a species' response to the landscape context depends on which biological response you measure', *Current Landscape Ecology Reports*, vol 3, no 1, pp23–33.

Martin, A.E. and Fahrig, L. (2016) 'Reconciling contradictory relationships between mobility and extinction risk in human-altered landscapes', *Functional Ecology*, vol 30, no 9, pp1558–1567.

Matthies, D., Bräuer, I., Maibom, W. and Tscharntke, T. (2004) 'Population size and the risk of local extinction: empirical evidence from rare plants', *Oikos*, vol 105, no 3, pp481–488.

Melo, G.L., Sponchiado, J., Cáceres, N.C. and Fahrig, L. (2017) 'Testing the habitat amount hypothesis for South American small mammals', *Biological Conservation*, vol 209, no 5, pp304–314.

Merckx, T., Dantas de Miranda, M. and Pereira, H.M. (2019) 'Habitat amount, not patch size and isolation, drives species richness of macro-moth communities in countryside landscapes', *Journal of Biogeography*, vol 46, no 5, pp956–967.

Miguet, P., Jackson, H.B., Jackson, N.D., Martin, A.E. and Fahrig, L. (2016) 'What determines the spatial extent of landscape effects on species?', *Landscape Ecology*, vol 31, no 6, pp1177–1194.

Moore, N.W. (1962) 'The heaths of Dorset and their conservation', *Journal of Ecology* vol 50, no 2, pp369–391.

Nour, N. Van Damme, R., Matthysen, E. and Dhondt, A.A. (1999) 'Forest birds in forest fragments: are fragmentation effects independent of season?', *Bird Study*, vol 46, no 3, pp279–288.

Öckinger, E., Franzén, M., Rundlöf, M. and Smith, H.G. (2009) 'Mobility-dependent effects on species richness in fragmented landscapes', *Basic and Applied Ecology*, vol 10, no 6, pp573–578.

Robertson, O.J. and Radford, J.Q. (2009) 'Gap-crossing decisions of forest birds in a fragmented landscape', *Austral Ecology*, vol 34, no 4, pp435–446.

Seibold, S., Bässler, C., Brandl, R., Fahrig, L., Förster, B., Heurich, M., Hothorn, T., Scheipl, F., Thorn, S. and Müller, J. (2017) 'An experimental test of the habitat-amount hypothesis for saproxylic beetles in a forested region', *Ecology*, vol 98, no 6, pp1613–1622.

Smith, A.C., Koper, N., Francis, C.M. and Fahrig, L. (2009) 'Confronting collinearity: comparing methods for disentangling the effects of habitat loss and fragmentation', *Landscape Ecology*, vol 24, no 10, pp1271–1285.

Smith, A.C., Fahrig, L. and Francis, C.M. (2011) 'Landscape size affects the relative importance of habitat amount, habitat fragmentation, and matrix quality on forest birds', *Ecography*, vol 34, no 1, pp103–113.

Smith, J.E. and Batzli, G.O. (2006) 'Dispersal and mortality of prairie voles (*Microtus ochrogaster*) in fragmented landscapes: a field experiment', *Oikos*, vol 112, no 1, pp209–217.

Sosa, R.A. and Lopez de Casenave, J. (2017) 'Edge effect on bird nest predation in the fragmented caldén (*Prosopis caldenia*) forest of central Argentina: an experimental analysis', *Ecological Research*, vol 32, no 2, pp129–134.

Stone, M.J., Catterall, C.P. and Stork, N.E. (2018) 'Edge effects and beta diversity in ground and canopy beetle communities of fragmented subtropical forest', *PLoS ONE*, vol 13, no 3, art e0193369.

Swift, T.L. and Hannon, S.J. (2010) 'Critical thresholds associated with habitat loss: a review of the concepts, evidence, and applications', *Biological Reviews*, vol 85, no 1, pp35–53.

Thornton, D.H., Branch, L.C. and Sunquist, M.E. (2011) 'The relative influence of habitat loss and fragmentation: do tropical mammals meet the temperate paradigm? *Ecological Applications*, no 21, vol 6, pp2324–2333.

Torrenta, R. and Villard, M.A. (2017) 'A test of the habitat amount hypothesis as an explanation for the species richness of forest bird assemblages', *Journal of Biogeography*, vol 44, no 8, pp1791–1801.

Watson, J.E.M., Whittaker, R.J. and Dawson, T.P. (2004) 'Habitat structure and proximity to forest edge affect the abundance and distribution of forest-dependent birds in tropical coastal forests of southeastern Madagascar', *Biological Conservation*, vol 120, no 3, pp311–327.

Wilcove, D.S., McLellan, C.H. and Dobson, A.P. (1986) 'Habitat Fragmentation in the temperate zone', in M. E. Soulé (ed) *Conservation Biology: The Science of Scarcity and Diversity*. Sinauer Associates Inc., Sunderland.

Wilson, M.C., Chen, X.-Y., Corlett, R.T., Didham, R.K., Ding, P., Holt, R.D., Holyoak, M., Hu, G., Hughes, A.C., Jiang, L., Laurance, W.F., Liu, J., Pimm, S.L., Robinson, S.K., Russo, S.E., Si, X., Wilcove, D.S., Wu, J. and Yu, M. (2016) 'Habitat fragmentation and biodiversity conservation: key findings and future challenges', *Landscape Ecology*, vol 31, no 2, pp219–227.

With, K.A. and Pavuk, D.M. (2011) 'Habitat area trumps fragmentation effects on arthropods in an experimental landscape system', *Landscape Ecology*, vol 26, no 7, pp1035–1048.

Wu, J. (2013) 'Key concepts and research topics in landscape ecology revisited: 30 years after the Allerton Park workshop', *Landscape Ecology*, vol 28, no 1, pp1–11.

Zuckerberg, B. and Porter, W.F. (2010) 'Thresholds in the long-term responses of breeding birds to forest cover and fragmentation', *Biological Conservation*, vol 143, no 4, pp952–962.

7

Nutrient flows in the landscape

Erica A.H. Smithwick

Introduction

Lateral flows underlie a broad set of ecological processes, including species movement and dispersal, aquatic flows, and genetic exchange (Allan et al. 1997, Manel et al. 2003, Reiners and Driese 2004). Lateral flows of biogeochemical matter, including nutrients, water, and air, regulate the exchange of organic matter among ecosystems, govern nutrient exchange between terrestrial and aquatic systems, and can determine the movement and location of plants and mobile organisms (Augustine and Frank 2001, Naiman et al. 1994). At landscape scales, feedbacks among biogeochemical and species' flows and their interaction are influenced by heterogeneity in the environment (Chapin et al. 2012, Turner and Gardner 2015). However, predicting exactly how spatial patterns influence nutrient flows at landscape scales, and thresholds of these interactions, remains a research frontier in landscape ecology (Turner and Gardner 2015). This is particularly true in the Anthropocene, in which the magnitude and rates of nutrient flows are accelerating.

Broadly, nutrient cycling refers to a set of linked biochemical processes that occur in ecosystems. These nutrient transformations ultimately supply living tissues with the elements required for organism structure and function. For terrestrial ecosystems, critical elements for development and growth include carbon (C), hydrogen (H), and oxygen (O), which form the basis of energy production (i.e., carbohydrates); but other elements are also required for chemical reactions within organisms, including nitrogen (N), phosphorus (P), potassium (K), magnesium (Mg), sodium (Na), iron (Fe), and sulfur (S), among others. These latter elements are considered *critical nutrients*; i.e., when these elements are in limited supply, growth and other critical life functions (e.g., reproduction, defense) are constrained. The exchange of elements between the atmosphere, lithosphere, rhizosphere, and biosphere comprises the broad study of terrestrial biogeochemical cycling, reviewed in depth elsewhere (see Chapin et al. 2012 and Schlesinger and Bernhardt 2013 for especially comprehensive reviews on the topic).

In landscape ecology, where the focus is on pattern–process interactions at multiple scales (Turner 1989), biogeochemical cycling (a *process*) is understood to be a product of spatial arrangement (a *pattern*) from micro (10^{-6}) to mega (10^6) scales, encompassing from nutrient cycling at the grain of microbial organisms to how these processes are expressed across broad landscape extents. Nutrient flows, as defined here, refer to the transport of nutrient elements vertically or laterally across space. Vertical flows include, for example, the flow of N (e.g., fixation or denitrification) or C (e.g., photosynthesis or respiration) between the terrestrial biosphere and the atmosphere and are often studied by ecosystem ecologists (Figure 7.1a). Of particular interest to landscape ecologists, however, are *lateral* flows (Figure 7.1b). These flows include processes in which elements move horizontally across space, as a result of dissolved aquatic transport (e.g., leaching), physical transport (e.g., erosion), or aeolian transport (e.g., exchange of dust or smoke across regions). Some elements, such as N, are more likely to be moved through aquatic transport and others, like P, through physical transport. Recent attention to lateral transport processes in ecosystem studies (e.g., Lovett et al. 2005, Chapin et al. 2012) highlight their significance for understanding elemental net balance of whole ecosystems at landscape to regional scales and signal an opportunity for landscape and ecosystem ecologists to further engage.

Understanding the importance of lateral flows for nutrient cycling emerged from early watershed studies begun in the 1950s and 1960s, notably in the Hubbard Brook Ecosystem Study in the White Mountains of New Hampshire, United States (Likens et al. 1970, Whittaker et al. 1974). This study was seminal in exploring whole-watershed export of nutrients from experimental manipulation (removal) of vegetation at broad extents. This study and others like it demonstrated the importance of plant demand for the regulation of nutrients available for export. Specifically, nutrient losses from the watershed were considered to be in excess of plant and microbial demand. The balance of nutrient uptake by plants, or immobilization by microbial organisms, is not constant through vegetation development. As with young, disturbed stands in Hubbard Brook, older forest stands are also considered 'leaky', as demand for nutrients to sustain growth declines in late succession, and nutrients are leached from the system (Vitousek 1977, Sollins et al. 1980, Hedin et al. 1995). Nitrogen saturation, in which N inputs from atmospheric deposition exceed demands for N by plants and microbes, can also lead to N loss and contribute to associated leaching of critical nutrients (e.g., Ca) and elevated toxicity of other elements (e.g., aluminum [Al]) (Aber et al. 1998). Nutrient leaching may be rarer in other forested systems. Turner et al. (2007) found that lodgepole pine stands did not experience nutrient loss following stand-replacing wildfire, likely due to strong microbial immobilization; further, these forests can hold on to large N stocks through early forest post-fire development (Turner et al. 2019).

Occurring over the course of many decades, these studies have deepened our appreciation for the integrated nature of nutrient cycling dynamics across space and across scales. The regulation of nutrient balances within terrestrial ecosystems is now seen as a function of a complex set of regulating processes involving the atmosphere, biosphere, hydrosphere, and lithosphere.

Figure 7.1 Schematic figure describing nutrient fluxes on heterogeneous landscapes: (a) the occurrence of a process at a point (cell or patch), represented by columns (where the magnitude of the rate or flux is represented by the height of the column), and (b) lateral flows, the actual flow paths of nutrients across points (cell or patch), represented by arrows. Source: From Turner and Gardner (2015).

Inputs and outputs of nutrients to and from ecosystems are now recognized to be a function of local conditions (vegetation type or successional stage, or geology) as well as local context, the composition and arrangement of the surrounding environment that can influence nutrient flows across space. Interpreting the implications and causes of lateral flows resulting from these dynamics remains a ripe area of inquiry for landscape ecologists.

How are nutrient geographies changing in the Anthropocene?

Increasing magnitudes of nutrient inputs and outputs

Humans have dramatically altered the amounts of nutrients that are entering and leaving ecosystems (Figure 7.2a). For example, mean annual N atmospheric deposition has increased sevenfold since pre-industrialization (Galloway 2004, Gilliam et al. 2019). Many terrestrial forested systems are assumed to be N-limited, and growth enhancement by N deposition has been observed (Magill 2004). On the other hand, in other places, N saturation can exceed nutrient demand, and is associated with acid deposition, leading to tree mortality (Schulze 1989, McNulty et al. 2005, Wallace et al. 2007) and changes in plant biodiversity (Clark and Tilman 2008, Clark et al. 2013). Despite recent declines in atmospheric N deposition in the United States, the long-term impact on forested ecosystems remains unclear due to hysteresis in forest recovery (Gilliam et al. 2019).

In addition to atmospheric deposition, point- and non-point-source pollution has contributed to large flows of N and P into estuaries such as the Chesapeake Bay, the Gulf of Mexico, and the Balkan sea, resulting in algal blooms and dead zones (Diaz and Rosenberg 2008). Agricultural production accounts for the vast majority of N entering the biosphere (Smil 1999) but is largely dependent on fertilizer inputs. As global demand for agricultural production continues to grow, even larger flows of nutrients into coastal areas are likely.

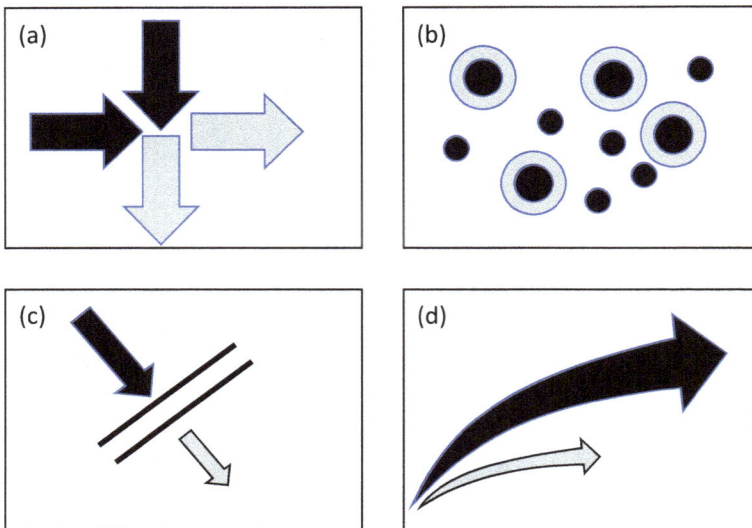

Figure 7.2 Schematics of changes of nutrient flows in the Anthropocene: (a) increasing magnitudes of nutrient inputs (black) and outputs (gray), (b) increasing variability of nutrient distribution across space (e.g., hot spots), (c) increasing mediation of flows due to changes in landscape structure such as fragmentation or infrastructure, and (d) increasing flow lengths due to teleconnections.

Increasing spatial and temporal variability of nutrient distribution

In addition to changes in the inputs and outputs, the spatial and temporal variability of nutrient availability is increasing in the Anthropocene (Figure 7.2b). Hot spots and hot moments (HSHM; or, cool spots and cool moments) has had a notable impact in nutrient cycling studies by characterizing this variability and quantifying its impact on whole-ecosystem nutrient cycling. For example, hotspots of denitrification at micro-scales have been shown to be influential on ecosystem nutrient budgets (Groffman et al. 2009). Hotspots of methane emissions from inland ponds and lakes are known to significantly impact carbon cycling (Cole et al. 2007) and may have a larger impact than previously thought (Davidson et al. 2018). Tree islands in the Florida Everglades are known to be maintained by high levels of phosphorus that can be 6–100 times greater than levels found in surrounding environments (Wetzel et al. 2005). Nutrient cycling in savanna systems has quantifiable spatial patchiness (Smithwick et al. 2016). As these studies demonstrate, the HSHM concept has been fruitful for stimulating research on spatial patterns of nutrient cycling. Bernhardt et al. (2017) recently called for a more inclusive framing of HSHM as 'control points', a broadened definition that would acknowledge the fundamental linkage between spatial and temporal patterns of nutrient dynamics while recognizing expanded gradients of nutrient patterns outside specific hot spot locations.

Importantly, human land use patterns exacerbate natural variability in nutrient processing. For example, elevated hot spots of livestock and oil/gas emissions can now be detected from satellite imagery (Buchwitz et al. 2017). Du et al. (2015) also documented significant declines in acidity (H^+, pH, SO_4^{-2}, NO_3^-) with increasing distance from urban centers in China (Figure 7.3). Thus, the geospatial patterns of nutrient modification are evident across a variety of landscapes, from urban to agricultural, altering patterns of nutrient flows.

Increasing mediation of flows

Lateral nutrient flows are increasingly governed by landscape structure, especially as modified by humans (Figure 7.2c). Two factors, habitat fragmentation and infrastructural barriers, are

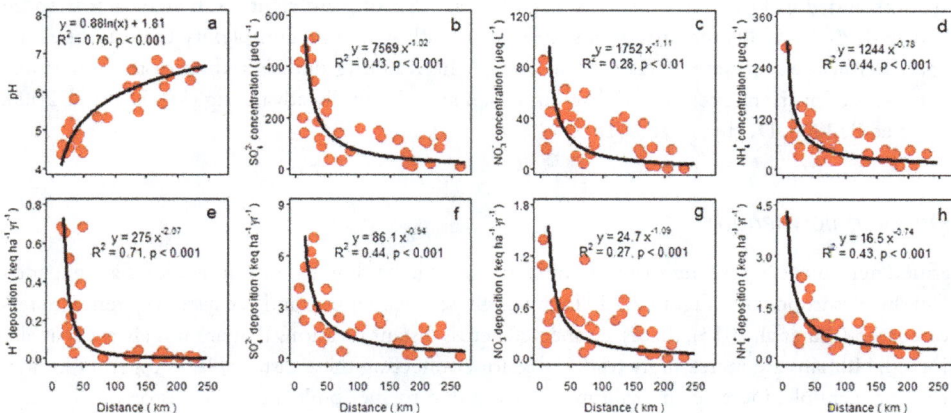

Figure 7.3 Changes in (a) pH, concentrations (μeq L^{-1}) of (b) sulfate (SO_4^{2-}), (c) nitrate (NO_3^-), and (d) ammonium (NH_4^+) in bulk precipitation and bulk deposition (keq ha^{-1} yr^{-1}) of (e) hydrogen ion (H^+), (f) sulfate, (g), and (h) ammonium with a closer distance (km) to the center of the nearest large cities. Source: From Du *et al.* (2015).

particularly responsible for impeding or altering the flow of nutrients across space. Fragmentation, the breakup of habitat into smaller and more isolated patches (see Chapter 6), is associated with the development of edge environments at patch boundaries. It is estimated that over 70% of the world's remaining forest is within 1 km of an edge (Haddad et al. 2015). The edge is characterized by a zone of biophysical characteristics differentiable from the interior environments of the two conjoining patches. As a result, differences in nutrient stocks, cycling rates, and outputs are expected in edge environments (Weathers et al. 2001, Schroder and Fleig 2017). Relative amounts of N or P limitation have been shown to vary across narrow ecotonal gradients (Feller et al. 2003).

Yet, precisely because of these nutrient accumulation properties, humans artificially impose edge environments on landscapes to achieve particular benefits, i.e., as riparian buffers, to impede and filter nutrient exchange between terrestrial and aquatic ecosystems. The importance of riparian buffers for nutrient removals has been recognized by ecologists since the seminal paper of Peterjohn and Correll (1984), an idea subsequently broadened in the concept of 'the ecology of interfaces' by Naiman and Decamps (1997), which situated the understanding of riparian zones in watershed-level planning. Newer efforts have focused on understanding how landscape context, including land use, physiognomy, and arrangement, influence watershed nutrient export (King et al. 2005, Baker et al. 2006); how spatial patterning within riparian zones can influence hotspots of nutrient mineralization or immobilization (Moon et al. 2019); the role of riparian zones in supporting a broader set of ecosystem services (Zheng et al. 2016); and cross-disciplinary perspectives on multiscalar hydrologic connectivity (Covino 2017). At the heart of these studies is the recognition that nutrient flows on landscapes are a product of the spatial pattern and arrangement of patches on the landscape as well as barriers and facilitators to flow across patch boundaries.

Infrastructure is an increasingly common component of humanized landscapes and seascapes, resulting in increasing structural complexity for nutrient flows at landscape scales. Artificial structures can alter the flow of nutrients in both land and sea (Bishop et al. 2017). When impermeable, as in the case of water containment structures or dams, they can act as barriers to nutrient or organism movement. Even when permeable, or avoidable, they can alter flow paths, the timing of flows, or the chemical nature of the exchange. When designed to act as conduits for flow, such as stormwater drainage systems, they purposefully act to direct the movement of runoff and dissolved material transport, thus accelerating nutrient exchanges across space (Hale et al. 2015). Urban environments are replete with structural complexity that modifies the biogeochemical environment (Kaye et al. 2006). The resulting understanding of the functional complexity of nutrient cycling and flows in urban areas is rapidly advancing (Decker et al. 2000, Pickett et al. 2008, Decina et al. 2018).

Expanding flow lengths

Defined here as the 'operating space' of nutrient cycling, the flow lengths of nutrients are expanding in the Anthropocene (Figure 7.2d). Increasingly, socioeconomic and ecosystem systems are teleconnected (Liu et al. 2013), that is, connected across distant locations that are not always spatially adjoined. Humans are increasingly responsible for connecting these distant places together in new ways. For example, increases in wildfire intensity due to fuel build-up from historical fire suppression, antecedent disturbances, and increasingly conducive climate conditions have led to the emergence of mega-fires in the western United States and internationally (Stephens et al. 2014). These mega-fires emit particular matter, aerosols, and associated chemicals into the planetary boundary layer and the troposphere (Kahn et al. 2008). Horizontal transport of these plumes can affect air quality and human health in distant regions (Reisen et al. 2015). Spatial predictions of these impacts (Lassman et al. 2017) can occur through integration of *in situ* and remotely sensed observations and simulation modeling but remain challenging due to complex interactions among plume height, circulation patterns, downdraft changes in smoke chemistry, and surface patterns in vegetation or elevation (for example: Likens et al. 1979, Swart et al. 2014, Jaffe et al. 1999).

Another example of the expanding flow length of nutrient exchange is the transport of goods and food through global trade (Galloway et al. 2008; Figure 7.4). Global agricultural market dynamics promote the spatial exchange of fertilizer, grain, and meat at a global scale, substantively

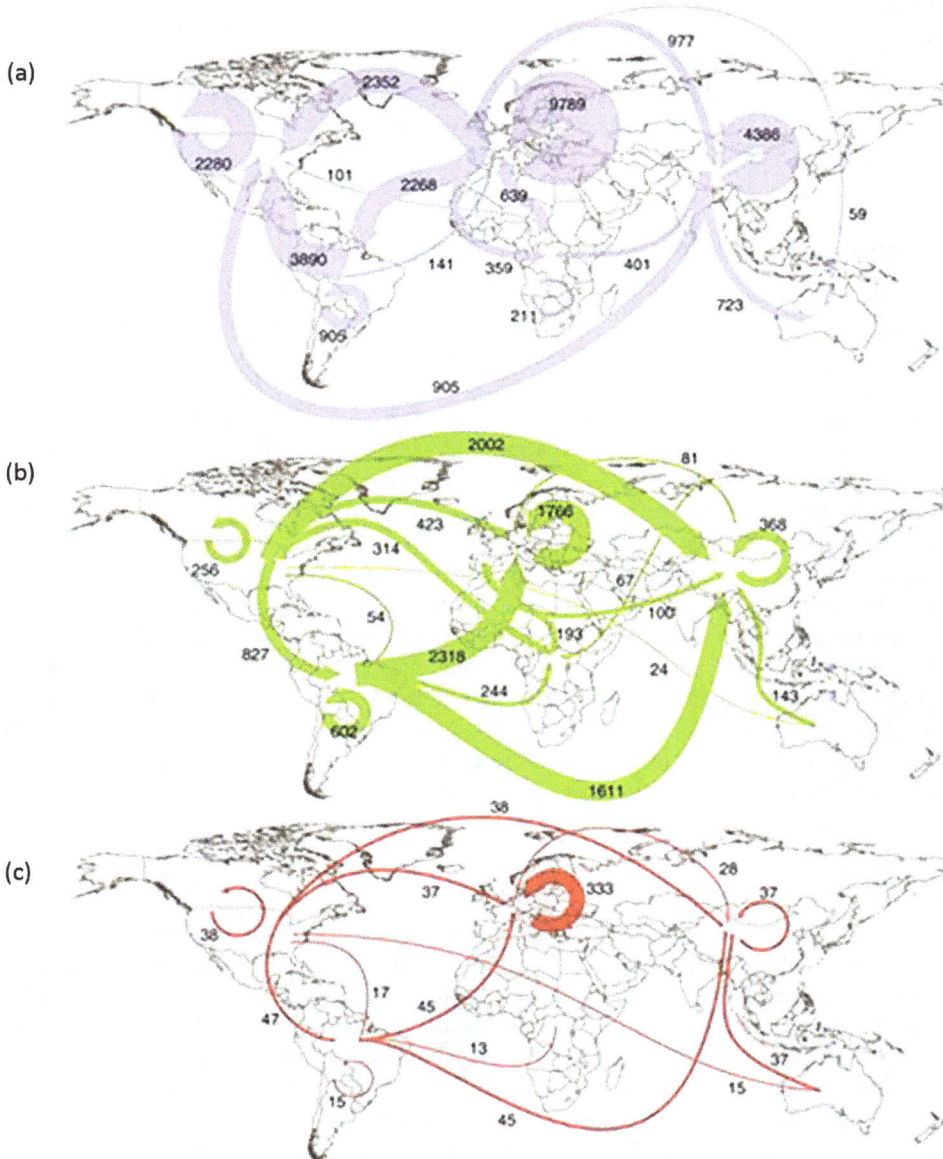

Figure 7.4 Nitrogen contained in internationally traded (a) fertilizer, (b) grain, and (c) meat. Source: From Galloway *et al.* (2008).

influencing the flows of elements like nitrogen and phosphorus. Regional agricultural diversity and farm size determine the availability of nutrient and food production, including the availability of micronutrients and protein to sustain human health (Herrero et al. 2017). Export of food and fertilizer, from nutrient-rich to nutrient-poor regions, is increasingly necessary to meet growing human demand (Hedberg 2019). Such long-recognized nutrient 'mining' (Stoorvogel et al. 1993) can lead to unsustainable input–output soil nutrient budgets in regions wherein extraction and trade of crops is greater than inputs. In addition to the flow of major staple crops such as meat and grains, the production of cash crops can also influence nutrient dynamics. For example, coffee production is known to influence a number of ecosystem services, including nutrient cycling; and as the spatial pattern of production areas shifts globally, as well as the way in which it is produced (i.e., shade-grown or not), the resultant nutrient flows and stocks will shift concomitantly (Jha et al. 2014). Nutrient flows via trade are, of course, coupled to the flow of embodied carbon in global trade, which has received substantial scrutiny in international climate policy (Sato et al. 2013).

Consequences for ecosystem services

As a result of the impacts described, ecosystem services provided by robust nutrient cycling processes are increasingly threatened in the Anthropocene. These services could include regulating services, such as oyster reefs or mussels in providing nutrient pollution control (DePiper et al. 2017, Vaughn 2018), or provisioning services such as agricultural productivity across land use gradients (Ma et al. 2019). Both excessive and inadequate nutrient inputs characterize global agricultural nutrient imbalances (Vitousek et al. 2009). Meeting the demands for food and fuel for a growing and developing population is a profound challenge in the Anthropocene (Rockstrom et al. 2017) and one that will continue to require large nutrient inputs or recycling efforts (Powers et al. 2019).

Given that landscapes provide a set of linked ecosystem services, i.e., ecosystem bundles (Raudsepp-Hearne et al. 2010, Bennett et al. 2016), it is important to recognize that nutrient flow regulation is integrally connected to a wide set of services at landscape scales. Alteration of the agents of nutrient cycling, the microbial community, deserves specific attention. Wetland disturbance has been found to modify the spatial pattern of microbial communities responsible for nutrient cycling (Moon et al. 2019). Thus, gradients in landscape development may alter the capacity of wetlands to regulate hydrologic flow or nutrient export, and this capacity may be determined by changes in microbial communities that occur at sub-meter scales.

Nutrient geography

Here, building on the recent understanding of nutrient flows across landscapes within ecosystem and landscape ecology, I explore how the conceptual framing of a *nutrient geography* perspective can help us understand changes in nutrient landscapes in the Anthropocene. Nutrient geography is defined here as a framework to explore the reciprocal interactions among patterns and processes governing nutrient flows. It involves two central components. First, there is an explicit attention to *scale*, the spatial or temporal dimension of an object (i.e., nutrient stock), or process (e.g., mineralization). Second, there is attention to the role of *agents* (e.g., organisms or actors that can influence change within a system) in creating or governing resultant patterns and processes. These two themes, scale and agent, can bridge key terms and emphases of various disciplines, including landscape ecology (Turner 1989, Wu and Hobbs 2002), ecosystem ecology (Chapin et al. 2012),

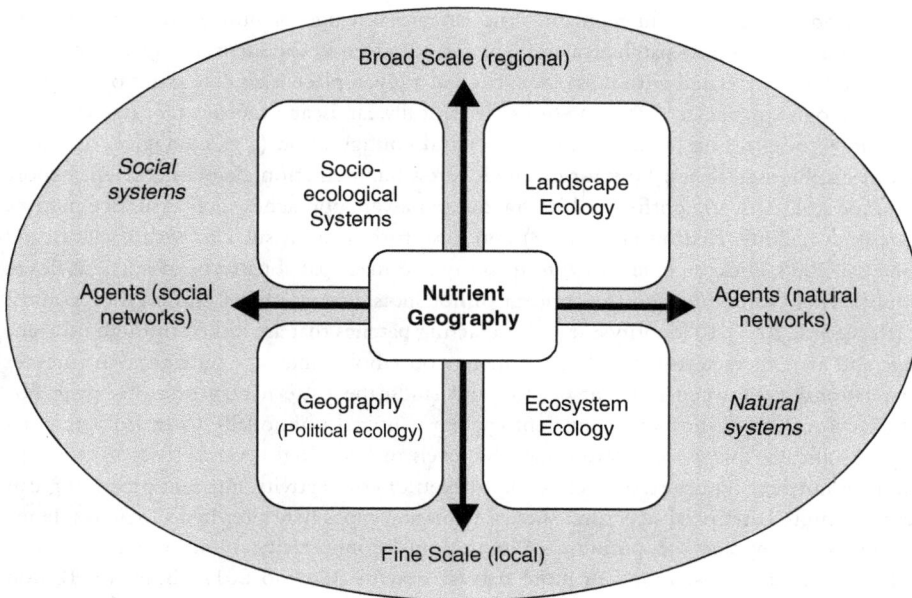

Figure 7.5 Generalized overview of the place of *Nutrient Geography* within other disciplinary domains (socio-ecological systems, geography, landscape ecology, and ecosystem ecology), and themes (scale and agents).

socio-ecological systems science (Walker et al. 2004, Folke et al. 2010), and geography (Adger 2000, Cote and Nightingale 2012, Brown 2014) (Figure 7.5). However, interdisciplinary approaches, especially across the social and ecological sciences, that address nutrient cycling and nutrient flows remain rare. By uniting perspectives from landscape and ecosystem ecology, geography, and socio-ecological systems, a focus on nutrient geography offers a new lens for examining the mechanisms and outcomes of altered nutrient flows in the Anthropocene. It allows, for example, an examination of how key actors within ecological *and* social systems collectively influence the movement of nutrients across space. It also promotes the distillation of the key frontiers for research and identifies opportunities for convergent science. In the following, I highlight some of the unique perspectives gained by focusing on scale and agents to understand nutrient flows.

Scale and scaling

Failing to account for lateral flows or spatial variability of nutrient dynamics may lead to ecological 'surprises' at landscape scales (Ludwig et al. 2000). For example, at fine levels, not properly accounting for lateral and horizontal flows of nutrients within the rhizosphere may lead to erroneous assumptions of whole-plant nutrient limitation or stress (Smithwick et al. 2013). At broader levels, geomorphic or hydrological gradients may differentially buffer or exacerbate ecological resilience. Ignoring variable nutrient dynamics across patch boundaries or ecotones may lead to ecological surprises at higher levels of organization if the edge dynamics are non-additive, i.e., non-linear or asymptotic (Smithwick et al. 2003, Cumming 2011).

As a result of these challenges, an expanded view of nutrient cycling on landscapes should move beyond classification of systems as simply 'open' or 'closed' and rather, explicitly acknowledge the role of landscape context in regulating the lateral exchange of nutrients to any specific location. Characterizing how patch arrangement and structure in the surrounding landscape (at multiple scales) are correlated with nutrient cycling at a given place is an easy point of entry for landscape ecologists to tackle these questions. Specifically, landscape metrics that describe and quantify patch composition (e.g., patch diversity) and configuration (e.g., contagion or fractal dimension) can be used to test how, for example, forest fragmentation changes nutrient export at a landscape scale. Less straightforward is how to use newer approaches such as surface metrics (McGarigal et al. 2009, Kedron et al. 2018) and flow path analysis such as with Circuitscape (McRae et al. 2008, Dickson et al. 2019) to quantify and map spatial patterns of nutrient flows. These approaches would broaden the concept of hot spots/hot moments, and even of 'control points' (Bernhardt 2017), to identify a set of interacting patches that are linked through nutrient exchange and are expected to have a large influence on whole-landscape nutrient dynamics.

The relationship between nutrient cycling and landscape pattern may be non-linear. For example, Covino (2017) shows that nutrient cycling is linked unimodally to hydrologic connectivity. Specifically, this work showed that low levels of hydrologic connectivity can lead to limitations in nutrient transport, whereas at high levels of connectivity, nutrient processing can be reactivity-limited. Artificial structures such as dams and pipes may alter hydrologic residence times differently, providing a continuum of longitudinal connectivity in anthropogenic landscapes that differs from that found in more natural systems (Covino 2017). Similarly, Bishop et al. (2017) found that dispersal of marine invertebrate larvae is a non-linear function of larval duration as well as physical barriers to connectivity, which has implications for understanding trophic influences on nutrient flows.

When scaling these non-linear dynamics to landscape or regional scales, it may be difficult to make robust assessments. For example, the relative influence of N deposition will be different in different regions depending on the initial soil nutrient capital, which itself is influenced by differences in the geophysical template and other environmental gradients (e.g., climate) and disturbance history (McNeil et al. 2012). Moreover, individual tree species have different functional traits (e.g., mycorrhizal associations, rooting depths, leaf mass, shade tolerance) that will result in differential responses in growth or survivorship (McNeil et al. 2007, McNeil et al. 2012, Pardo et al. 2011, Smithwick et al. 2013). Horn et al. (2018) explored how the functional traits (e.g., mycorrhizal association) of tree species affect growth and survivorship responses to N and S deposition in the conterminous United States. While many species had monotonically increasing or decreasing responses, a small set of species displayed threshold behaviors in response to increasing N and S deposition. Understanding the interactions between these species-specific physiological responses with spatial patterns of the geophysical template, which could act to buffer or accentuate the response, remains an important area of research. For example, trees growing on Ca-rich substrates are likely to buffer against Al toxicity resulting from Ca leaching from excess nitrate levels in the soil. Further, physical or chemical stress within tree roots may impede tree nutrient acquisition, although the mechanisms remain understudied (Smithwick et al. 2013). Such understanding could guide knowledge about the role of forests in regulating nutrient flows across landscapes, with important implications for controlling nutrient pollution.

Advances in modeling that link nutrient and hydrologic flows at landscape scales (Gbondo-Tugbawa et al. 2001, Shimizu et al. 2018, Yuan et al. 2018) is encouraging and suggests increasing potential for understanding the degree to which nutrient flows are governed by landscape characteristics and for incorporating non-linear relationships. Yet, spatially explicit landscape models remain computationally demanding (Keane et al. 2015), and data to parameterize and corroborate models are often insufficient at watershed scales, especially when both hydrologic and nutrient data are required.

Agents

Both human and non-human organisms mediate nutrient flows. Incorporating these agents and their behaviors into an understanding of nutrient cycling remains a research frontier in landscape ecology. Nutrient geography offers a lens through which both the social and natural sciences can be viewed as contributing to this understanding. This perspective is especially important in the Anthropocene, when nutrient flows are heavily modified by humans directly and strongly influenced by human modification of landscape patterns and other organisms.

Networks are common in non-human ecological systems within and across a broad set of animal species. Interactions within and across these networks (e.g., trophic interactions) directly or indirectly influence how nutrients move across space and time. As a result, exploring how metapopulation or metacommunity interactions are associated with nutrient exchange across landscapes would help elucidate how organism interactions across space govern nutrient flow. As another example, pollinator communities are influenced by landscape condition (Kennedy et al. 2013). It follows, then, that the internal behavior of pollinator networks likely has complex interactions with plant networks at landscape scales. The degree to which nutrient transfers are mediated by organismal networks and trophic interactions remains in a frontier in landscape ecology research.

For large mammals, we have gleaned many insights about the significant interactions between large mammal communities and ecosystem function. In intact systems, large mammals significantly modify the nutrient environment in a number of ways, e.g., through consumption, deposition of urine and feces, modification of vegetation type and productivity, modification of disturbance regimes such as fire, and direct effects of trampling and hoof action on the soil surface (Hobbs 1996, Schmitz 2008, Christenson et al. 2013). Recent work suggests that these effects are, at least in savanna ecosystems, approximately of the same order of magnitude as other drivers of nutrient exchange, including N deposition, fire, and N fixation (Veldhuis et al. 2018). In addition to overall effects, large mammals are known to influence the spatial pattern of nutrient cycling processes (Augustine and Frank 2001, Augustine et al. 2003). Moreover, mammal traits, including associated feeding preferences, can drive landscape-level outcomes of mammal–ecosystem interactions (Young et al. 2020).

No longer can we assume that these non-human dynamics operate in the absence of human influence. Of the 5,674 extant mammal species, approximately 25% are currently threatened or endangered (IUCN 2017), suggesting impending and significant changes in ecosystem function (Figure 7.6). Large mammals, such as carnivores and herbivores greater than 100 kg, are of particular concern (Bar-On et al. 2018) due to widespread habitat loss and poaching. For example, at least 60% of herbivores are threatened with extinction (Ripple et al. 2015). In natural systems, an immediate effect of increased herbivore mortality by predators would be enhanced nutrient hotspots at places of carrion, which could then feed back to influence vegetation and subsequent herbivore presence (Figure 7.7). Carcass ecology, that is, the study of carrion and its associated ecological impacts on ecosystem processes and trophic cascades (Olea et al. 2019), has recently gained attention for its role in influencing nutrient cycling. For example, whale falls are known to support a vast community of scavengers and provide important inputs into ocean nutrient cycling (Smith and Baco 2003). More generally, the shift in predator–prey dynamics can have cascading effects on nutrient cycling (Schmitz et al. 2010). Over longer timescales, the removal of ecosystem engineers such as large mammals could have vastly more impactful changes on vegetation state changes and associated ecosystem processes (Coggan et al. 2018).

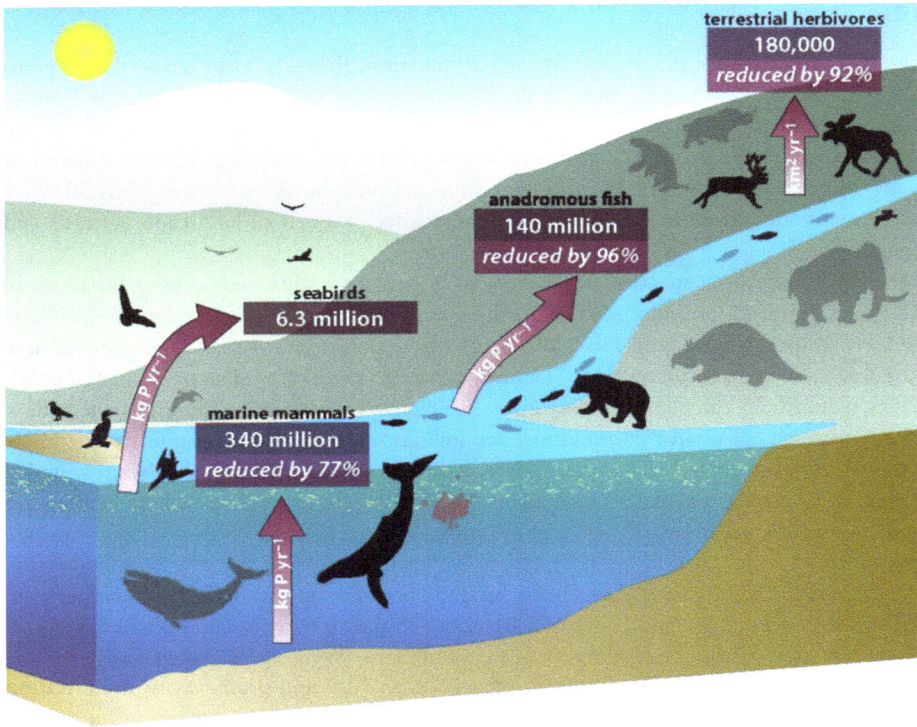

Figure 7.6 Changing lateral nutrient distribution capacity by terrestrial animals due to megafaunal extinctions. Source: From Doughty *et al.* (2016).

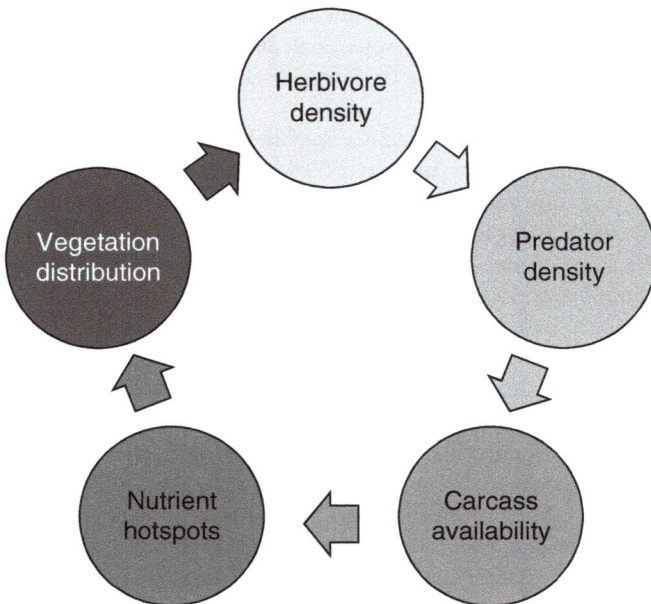

Figure 7.7 Interactions between predator–prey dynamics, carcass availability, the spatial distribution of nutrients, and vegetation.

Among the larger transfers of nutrients by animals are now known to be those that occur at aquatic–terrestrial boundaries, and changes in the magnitude of these nutrient flows could have large impacts on ecosystem function. For example, salmon are known to transport substantial amounts of nutrients from the marine environment into riparian systems; subsequent harvesting by predators, e.g., bears, then moves these nutrients to land. These transfers provide a conveyor belt of nutrient transport across ocean-riparian-terrestrial ecosystems (Naiman et al. 2002, Helfield and Naiman 2006) and significantly impact tree growth in the riparian zone (Quinn et al. 2018). Declines in salmon stocks due to pollution or overharvesting may have the consequence of slowing or eliminating this conveyor belt, fundamentally changing the productivity of these regions (Schindler et al. 2003). In the absence of nutrients from salmon, Hurteau et al. (2016) found greater phylogenetic clustering in riparian plants, suggesting that greater salmon density may influence plant diversity as well as productivity.

It is not only the removal of species but also the introduction of species that can alter nutrient cycling flows (Ehrenfeld 2003). For example, the introduction of invasive rats onto an ocean island was shown to decrease seabird abundance, leading to a negative influence on coral reef production (Graham et al. 2018; Figure 7.8). In another example, Hall et al. (2018) found a compound effect of invasion of an invasive tree (*Falcataria moluccana*) and the Caribbean tree frog (*Eleutherodactylus coqui*) into native Hawaiian rainforests. While it has been known that this non-native tree species increases N availability due to the fact that it is an N-fixing species, this study showed that the frog accelerated nitrous oxide (N_2O + NO) fluxes by enhancing decomposition processes in the high-N environments created by the invasive tree.

Outside human influences on ecological networks, human networks themselves are complex and multifaceted, and changes in these networks can have large implications for nutrient flows. The teleconnections associated with global agricultural markets, as described earlier, are an obvious example of how human economic networks can regulate nutrient flows at a global scale. Similarly, human decision-making in urban areas determines the spatial arrangement of key nutrients regulating infrastructure as well as the overall level of nutrient inputs. However, it is unknown how information exchange through human social networks governs decision-making

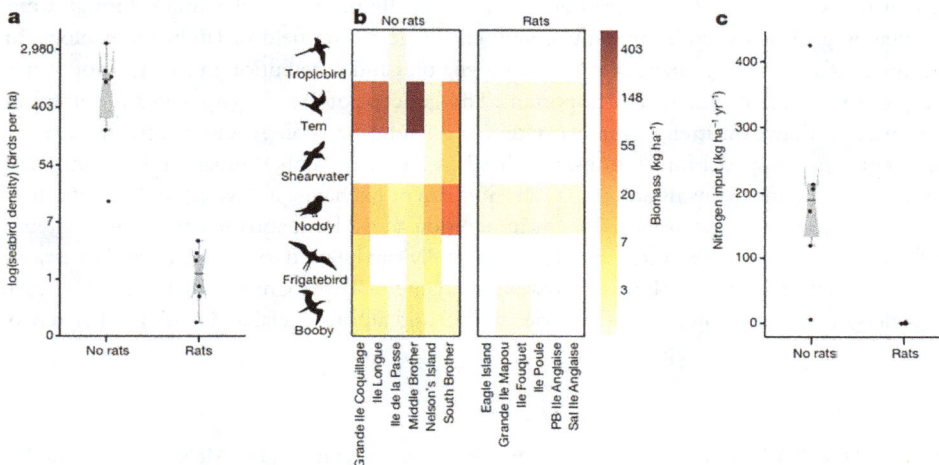

Figure 7.8 Seabird densities, biomass, and nitrogen input to islands with and without rats in the Chagos Archipelago. Source: From Graham *et al.* (2008).

about nutrient management, or how human movement along transport networks determines nutrient availability across space and time.

Frontiers in nutrient geography

The examples provided highlight emerging areas of inquiry that could be enhanced by convergence among ecologists, landscape ecologists, biogeochemists, geographers, and other disciplines and could help expand our understanding of the causes and consequences of nutrient flows on landscapes. They specifically highlight the role of scale and multiscalar processes, as well as the role of human and non-human agents within systems, in unraveling the mechanisms governing nutrient flows. Looking forward, broad questions to be asked using this perspective include: 'When do black box approaches to nutrient cycling fail?', 'Under what conditions do changes in landscape structure cause threshold behaviors in nutrient flows?', and 'How do networks (human and non-human) influence nutrient exchanges across space and time?' Extending these overarching research questions to specific research goals is a frontier in landscape ecology and will necessarily require convergent research approaches across many disciplines. Some example research questions may include:

(1)   Can metrics be used to identify the equivalent of 'pinch points' on the landscape that govern nutrient exchange, as is commonly done in animal studies?
(2)   How does human decision-making about the built environment influence nutrient flows in terrestrial and marine systems?
(3)   What characteristics of invaded systems or invasive agents are likely to cause the greatest changes in nutrient flows?
(4)   What are the consequences of large mammal declines for nutrient flows at local, landscape, and global scales?
(5)   How are ecosystem services of heterogeneous landscapes, such as nutrient retention, influenced by feedbacks among global change drivers and tree species composition?

Conclusion

Nutrient flows within landscapes are outcomes of spatially interactive dynamics through time. Understanding these dynamics presents a new challenge to the field of landscape ecology. In the context of accelerating changes in nutrient dynamics and distribution in the Anthropocene, the topic is particularly timely and important. The concept of *nutrient geographies* is helpful in this effort, as it draws on strengths in ecosystem and landscape ecology and geography, directly linking concepts of spatial interactions, thresholds, scale, and agents (human and non-human) to understand the mechanisms and outcomes of nutrient exchange across space. As in much of landscape ecology, the concepts are ripe for integration across both aquatic and marine systems. As a frontier in convergent research, nutrient geography can be used to address critical questions about elemental flows within the Earth system, fundamentally recognizing that they are spatial, scalar, and governed by complex interactions within and among social and ecological systems.

References

Aber, J., McDowell, W., Nadelhoffer, K., Magill, A., Berntson, G., Kamakea, M., McNulty, S., Currie, W., Rustad, L. and Fernandez, I. (1998) 'Nitrogen saturation in temperate forest ecosystems: hypotheses revisited', *BioScience*, vol 48, no 11, pp921–934.

Adger, W.N. (2000) 'Social and ecological resilience: are they related?', *Progress in Human Geography*, vol 24, no 3, pp347–364.

Allan, D., Erickson, D. and Fay, J. (1997) 'The influence of catchment land use on stream integrity across multiple spatial scales', *Freshwater Biology* vol 37, pp149–161.

Augustine, D.J. and Frank, D.A. (2001) 'Effects of migratory grazers in spatial heterogeneity of soil nitrogen properties in a grassland ecosystem', *Ecology*, vol 82, pp3149–3162.

Augustine, D.J., McNaughton, S.J. and Frank, D.A. (2003) 'Feedbacks between soil nutrients and large herbivores in a managed savanna ecosystem', *Ecological Applications*, vol 13, pp1325–1337.

Baker, M.E., Weller, D.E. and Jordan, T.E. (2006) 'Improved methods for quantifying potential nutrient interception by riparian buffers', *Landscape Ecology*, vol 21, pp1327–1345.

Bar-On, Y.M., Phillips, R. and Milo, R. (2018) 'The biomass distribution on Earth', *Proceedings of the National Academy of Sciences of the USA*, vol 115, pp6506–6511.

Bennett, E.M., Solan, M., Biggs, R., McPhearson, T., Norström, A.V., Olsson, P., Pereira, L., Peterson, G.D., Raudsepp-Hearne, C., Biermann, F., Carpenter, S.R., Ellis, E.C., Hichert, T., Galaz, V., Lahsen, M., Milkoreit, M., López, B.M., Nicholas, K.A., Preiser, R., Vince, G., Vervoort, J.M. and Xu, J. (2016) 'Bright spots: seeds of a good Anthropocene', *Frontiers in Ecology and the Environment*, vol 14, pp441–448.

Bernhardt, E.S., Blaszczak, J.R., Ficken, C.D., Fork, M.L., Kaiser, K.E. and Seybold, E.C. (2017) 'Control points in ecosystems: moving beyond the hot spot hot moment concept', *Ecosystems*, vol 20, pp665–682.

Bishop, M.J., Mayer-Pinto, M., Airoldi, L., Firth, L.B., Morris, R.L., Loke, L.H.L., Hawkins, S.J., Naylor, L.A., Coleman, R.A., Chee, S.Y. and Dafforn, K.A. (2017) 'Effects of ocean sprawl on ecological connectivity: impacts and solutions', *Journal of Experimental Marine Biology and Ecology*, vol 492, pp7–30.

Brown, K. (2014) 'Global environmental change I: a social turn for resilience?' *Progress in Human Geography*, vol 38, no 1, pp107–117.

Buchwitz, M., Schneising, O., Reuter, M., Heymann, J., Krautwurst, S., Bovensmann, H., Burrows, J.P., Boesch, H., Parker, R.J., Somkuti, P., Detmers, R.G., Hasekamp, O.P., Aben, I., Butz, A., Frankenberg, C. and Turner, A.J. (2017) 'Satellite-derived methane hotspot emission estimates using a fast data-driven method', *Atmospheric Chemistry and Physics*, vol 17, pp5751–5774.

Chapin, F.S., III, Matson, P.A. and Vitousek, P.M. (2012) *Principles of Terrestrial Ecosystem Ecology*. Springer, New York.

Christenson, L.M., Mitchell, M.J., Groffman, P.M. and Lovett, G.M. (2013) 'Cascading effects of climate change on forest ecosystems; biogeochemical links between trees and moose in the Northeast USA', *Ecosystems*, vol 17, pp442–457.

Clark, C.M. and Tilman, D. (2008) 'Loss of plant species after chronic low-level nitrogen deposition to prairie grasslands', *Nature*, vol 451, pp712–715.

Clark, C.M., Morefield, P.E., Gilliam, F.S. and Pardo, L.H. (2013) 'Estimated losses of plant biodiversity in the United States from historical N deposition (1985–2010)', *Ecology*, vol 94, pp1441–1448.

Coggan, N.V., Hayward, M.W. and Gibb, H. (2018) 'A global database and "state of the field" review of research into ecosystem engineering by land animals', *Journal of Animal Ecology*, vol 87, pp974–994.

Cole, J.J., Prairie, Y.T., Caraco, N.F., McDowell, W.H., Tranvik, L.J. Striegl, R.G., Duarte, C.M., Kortelainen, P., Downing, J.A., Middelburg, J.J. and Melack, J. (2007) 'Plumbing the global carbon cycle: integrating inland waters into the terrestrial carbon budget', *Ecosystems*, vol 10, pp172–185.

Cote, M. and Nightingale, A.J. (2012) 'Resilience thinking meets social theory: situating social change in socio-ecological systems (SES) research', *Progress in Human Geography*, vol 36, no 4, pp475–489.

Covino, T. (2017) 'Hydrologic connectivity as a framework for understanding biogeochemical flux through watersheds and along fluvial networks', *Geomorphology*, vol 277, pp133–144.

Cumming, G.S. (2011) 'Spatial resilience: integrating landscape ecology, resilience, and sustainability', *Landscape Ecology*, vol 26, pp899–909.

Davidson, T.A., Audet, J., Jeppesen, E., Landkildehus, F., Lauridsen, T.L., Søndergaard, M and Syväranta, J. (2018) 'Synergy between nutrients and warming enhances methane ebullition from experimental lakes', *Nature Climate Change*, vol 8, pp156–160.

Decina, S.M., Templer, P.H. and Hutyra, L.R. (2018) 'Atmospheric inputs of nitrogen, carbon, and phosphorus across an urban area: unaccounted fluxes and canopy influences', *Earth's Future*, vol 6, pp134–148.

Decker, E.H., Elliott, S., Smith, F.A., Blake, D.R. and Rowland, F.S. (2000) 'Energy and material flow through the urban ecosystem', *Annual Review of Energy and the Environment*, vol 25, pp685–740.

DePiper, G.S., Lipton, D.W. and Lipcius, R.N. (2017) 'Valuing ecosystem services: oysters denitrification, and nutrient trading programs', *Marine Resource Economics*, vol 32, pp1–20.

Diaz, R.J. and Rosenberg, R. (2008) 'Spreading dead zones and consequences for marine ecosystems', *Science*, vol 321, pp926–929.

Dickson, B.G., Albano, C.M., Anantharaman, R., Beier, P., Fargione, J., Graves, T.A., Gray, M.E., Hall, K.R., Lawler, J.J., Leonard, P.B., Littlefield, C.E., McClure, M.L., Novembre, J., Schloss, C.A., Schumaker, N.H., Shah, V.B. and Theobald, D.M. (2019) 'Circuit-theory applications to connectivity science and conservation', *Conservation Biology*, vol 33, pp239–249.

Doughty, C.E., Roman, J., Faurby, S., Wolf, A., Haque, A., Bakker, E.S., Malhi, Y., Dunning, J.B. Jr. and Svenning, J.C. (2016) 'Global nutrient transport in a world of giants', *Proceedings of the National Academy of Sciences*, vol 113, no 4, pp868–873.

Du, E., de Vries, W., Liu, X., Fang, J., Galloway, J.N. and Jiang, Y. (2015) 'Spatial boundary of urban 'acid islands' in southern China', *Scientific Reports*, vol 5, art 12625.

Ehrenfeld, J.G. (2003) 'Effects of exotic plant invasions on soil nutrient cycling processes', *Ecosystems*, vol 6, no 6, pp503–523.

Feller, I.C., McKee, K.L., Whigham, D.F. and O'Neill, J.P. (2003) 'Nitrogen vs. phosphorus limitation across an ecotonal gradient in a mangrove forest', *Biogeochemistry*, vol 62, pp145–175.

Folke, C., Carpenter, S.R., Walker, B., Scheffer, M., Chapin, T. and Rockström, J. (2010) 'Resilience thinking: integrating resilience, adaptability and transformability', *Ecology and Society*, vol 15, no 4, pp20.

Galloway, J.N. (2004) 'Nitrogen cycles: past, present, and future, *Biogeochemistry*, vol 70, pp153–226.

Galloway, J.N., Townsend, A.R., Erisman, J.W., Bekunda, M., Cai, Z.C., Freney, J.R., Martinelli, L.A., Seitzinger, S.P. and Sutton, M.A. (2008) 'Transformation of the nitrogen cycle: recent trends, questions, and potential solutions', *Science*, vol 320, pp889–892.

Gbondo-Tugbawa, S.S., Driscoll, C.T., Aber, J.D. and Likens, G.E. (2001) 'Evaluation of an integrated biogeochemical model (PnET-BGC) at a northern hardwood forest ecosystem', *Water Resources Research*, vol 37, pp1057–1070.

Gilliam, F.S., Burns, D.A., Driscoll, C.T., Frey, S.D., Lovett, G.M. and Watmough, S.A. (2019) 'Decreased atmospheric nitrogen deposition in eastern North America: predicted responses of forest ecosystems', *Environmental Pollution*, vol 244, pp60–74.

Graham, N.A.J., Wilson, S.K., Carr, P., Hoey, A.S., Jennings, S. and MacNeil, M.A. (2018) 'Seabirds enhance coral reef productivity and functioning in the absence of invasive rats', *Nature*, vol 559, pp250–253.

Groffman, P.M., Butterbach-Bahl, K., Fulweiler, R.W., Gold, A.J., Morse, J.L., Stander, E.K., Tague, C., Tonitto, C. and Vidon, P. (2009) 'Challenges to incorporating spatially and temporally explicit phenomena (hotspots and hot moments) in denitrification models', *Biogeochemistry*, vol 93, pp49–77.

Haddad, N.M., Brudvig, L.A., Clobert, J., Davies, K.F., Gonzalez, A., Holt, R.D., Lovejoy, T.E., Sexton, J.O., Austin, M.P., Collins, C.D., Cook, W.M., Damschen, E.I., Ewers, R.M., Foster, B.L., Jenkins, C.N., King, A.J., Laurance, W.F., Levey, D.J., Margules, C.R., Melbourne, B.A., Nicholls, A.O., Orrock, J.L., Song, D.X. and Townshend, J.R. (2015) 'Habitat fragmentation and its lasting impact on Earth's ecosystems', *Science Advances*, vol 1, no 2, art e1500052.

Hale, R.L., Turnbull, L., Earl, S.R., Childers, D.L. and Grimm, N.B. (2015) 'Stormwater infrastructure controls runoff and dissolved material export from arid urban watersheds', *Ecosystems*, vol 18, pp62–75.

Hall, S.J., Huber, D.P. and Hughes, R.F. (2018) 'Invasion of Hawaiian rainforests by an introduced amphibian predator and N_2-fixing tree increases soil N_2O emissions', *Ecosphere*, vol 9, no 9, art e02416.

Hedberg, R.C. (2019) 'Coming out of the foodshed: phosphorus cycles and the many scales of local food', *Annals of the American Association of Geographers*, vol 110, no 3, pp1–21.

Hedin, L.O., Armesto, J.J. and Johnson, A.H. (1995) 'Patterns of nutrient loss from unpolluted, old-growth temperate forests: evaluation of biogeochemical theory', *Ecology*, vol 76, no 2, pp493–509.

Helfield, J.M. and Naiman, R.J. (2006) 'Keystone interactions: Salmon and bear in riparian forests of alaska', *Ecosystems*, vol 9, pp167–180.

Herrero, M., Thornton, P.K., Power, B., Bogard, J.R., Remans, R., Fritz, S., Gerber, J.S., Nelson, G., See, L., Waha, K., Watson, R.A., West, P.C., Samberg, L.H., van de Steeg, J., Stephenson, E., van Wijk, M. and Havlík, P. (2017) 'Farming and the geography of nutrient production for human use: a transdisciplinary analysis', *The Lancet Planetary Health*, vol 1, art e33-e42.

Hobbs, N.T. (1996) 'Modification of ecosystems by ungulates', *Journal of Wildlife Management*, vol 60, pp695–713.

Horn, K.J., Thomas, R.Q., Clark, C.M., Pardo, L.H., Fenn, M.E., Lawrence, G.B., Perakis, S.S., Smithwick, E.A.H., Baldwin, D., Braun, S., Nordin, A., Perry, C.H., Phelan, J.N., Schaberg, P.G., St Clair, S.B.,

Warby, R. and Watmough, S. (2018) 'Growth and survival relationships of 71 tree species with nitrogen and sulfur deposition across the conterminous U.S.', *PLoS ONE*, vol 13, no 10, art e0205296.

Hurteau, L.A., Mooers, A.O., Reynolds, J.D. and Hocking, M.D. (2016) 'Salmon nutrients are associated with the phylogenetic dispersion of riparian flowering-plant assemblages', *Ecology*, vol 97, pp450–460.

International Union for Conservation of Nature (IUCN) (2017) *International Union for Conservation of Nature Annual Report 2017*. IUCN, Gland, Switzerland.

Jaffe, D., Anderson, T., Covert, D., Kotchenruther, R., Trost, B., Danielson, J., Simpson, W., Berstsen, T., Karlsdottir, S., Blake, D., Harris, J., Carmichael, G. and Uno, I. (1999) 'Transport of asian air pollution to North America', *Geophysical Research Letters*, vol 26, no 6, pp711–714.

Jha, S., Bacon, C.M., Philpott, S.M., Mendez, V.E., Laderach, P. and Rice, R.A. (2014) 'Shade coffee: update on a disappearing refuge for biodiversity', *Bioscience*, vol 64, pp416–428.

Kahn, R.A., Chen, Y., Nelson, D.L., Leung, F.Y., Li, Q.B., Diner, D.J. and Logan, J.A. (2008) 'Wildfire smoke injection heights: two perspectives from space', *Geophysical Research Letters*, vol 35, no 4. art L04809.

Kaye, J.P., Groffman, P.M., Grimm, N.B., Baker, L.A. and Pouyat, R.V. (2006) 'A distinct urban biogeochemistry?', *Trends in Ecology & Evolution*, vol 21, pp192–199.

Keane, R.E., McKenzie, D., Falk, D.A., Smithwick, E.A.H., Miller, C. and Kellogg, L.K.B. (2015) 'Representing climate, disturbance, and vegetation interactions in landscape models', *Ecological Modelling*, vol 309, pp33–47.

Kedron, P.J., Frazier, A.E., Ovando-Montejo, G.A. and Wang, J. (2018) 'Surface metrics for landscape ecology: a comparison of landscape models across ecoregions and scales', *Landscape Ecology*, vol 33, pp1489–1504.

Kennedy, C.M., Lonsdorf, E., Neel, M.C., Williams, N.M., Ricketts, T.H., Winfree, R., *et al.* (2013) 'A global quantitative synthesis of local and landscape effects on wild bee pollinators in agroecosystems', *Ecology Letters*, vol 16, no 5, pp584–599.

King, R.S., Baker, M.E., Whigham, D.F., Weller, D.E., Jordan, T.E., Kazyak, P.F. and Hurd, M.K. (2005) 'Spatial considerations for linking watershed land cover to ecological indicators in streams', *Ecological Applications*, vol 15, pp137–153.

Lassman, W., Ford, B., Gan, R.W., Pfister, G., Magzamen, S., Fischer, E.V. and Pierce, J.R. (2017) 'Spatial and temporal estimates of population exposure to wildfire smoke during the Washington state 2012 wildfire season using blended model, satellite, and in situ data' *Geohealth*, vol 1, pp106–121.

Likens, G.E., Bormann, F.H., Johnson, N.M., Fisher, D.W. and Pierce, R.S. (1970) 'Effects of forest cutting and herbicide treatment on nutrient budgets in Hubbard Brook Watershed-Ecosystem', *Ecological Monographs*, vol 40, no 1, pp43–51.

Likens, G.E., Wright, R.F., Galloway, J.N. and Butler T.J. (1979) 'Acid Rain', *Scientific American*, vol 241, no 4, pp43–51.

Liu, J.Q., Hull, V., Batistella, M., DeFries, R., Dietz, T., Fu, F., Liu, J.Q., Hull, V., Batistella, M, DeFries R., Dietz T., Fu, F., Hertel, T.W., Izaurralde, R.C., Lambin, E.F., Li, S., Martinelli, L.A., McConnell, W., Moran, E.F., Naylor, R., Ouyang, Z., Polenske, K.R., Reenberg, A., de Miranda Rocha, G., Simmons, C.S., Verburg, P.H., Vitousek, P.M., Zhang, F. and Zhu, C. (2013) 'Framing sustainability in a telecoupled world', *Ecology and Society*, vol 18, no 2, art 26.

Lovett, G.M., Jones, C.G., Turner, M.G. and Weathers, K.C. (eds) (2005) *Ecosystem Function in Heterogeneous Landscapes*. Springer, New York.

Ludwig, J.A., Wiens, J.A. and Tongway, D.J. (2000) 'A scaling rule for landscape patches and how it applies to conserving soil resources in savannas', *Ecosystems*, vol 3, pp84–97.

Ma, L.W., Bicking, S. and Muller, F. (2019) 'Mapping and comparing ecosystem service indicators of global climate regulation in Schleswig-Holstein, Northern Germany', *Science of The Total Environment*, vol 648, pp1582–1597.

Manel, S., Schwartz, M.K., Luikart, G. and Taberlet, P. (2003) 'Landscape genetics: combining landscape ecology and population genetics', *Trends in Ecology and Evolution*, vol 18, no 4, pp189–197.

Magill, A.H. (2004) 'Ecosystem response to 15 years of chronic nitrogen additions at the Harvard Forest LTER, Massachusetts, USA', *Forest Ecology and Management*, vol 196, pp7–28.

McGarigal, K., Tagil, S. and Cushman, S.A. (2009) 'Surface metrics: an alternative to patch metrics for the quantification of landscape structure', *Landscape Ecology*, vol 24, pp433–450.

McNeil, B.E., Read, J.M. and Driscoll, C.T. (2007) 'Foliar nitrogen responses to elevated atmospherica nitrogen deposition in nine temperate forest canopy species', *Environmetnal Science and Technology*, 41, pp5191–5197.

McNeil, B.E., Read, J.M. and Driscoll, C.T. (2012) 'Foliar nitrogen responses to the environmental gradient matrix of the Adirondack Park, New York', *Annals of the Association of American Geographers*, vol 102, pp1–16.

McNulty, S.G., J. Boggs, J.D. Aber, L. Rustad, and A. Magill. (2005) 'Red spruce ecosystem level changes following 14 years of chronic N fertilization', *Forest Ecology and Management*, 219, pp279–291.

McRae, B.H., Dickson, B.G., Keitt, T.H. and Shah, V.B. (2008) 'Using circuit theory to model connectivity in ecology, evolution, and conservation', *Ecology*, vol 89, pp2712–2724.

Moon, J.B., Wardrop, D.H., Smithwick, E.A.H. and Naithani, K.J. (2019) 'Fine-scale spatial homogenization of microbial habitats: a multivariate index of headwater wetland complex condition', *Ecological Applications*, vol 29, art e01816.

Naiman, R.J. and Decamps, H. (1997) 'The ecology of interfaces: Riparian zones', *Annual Review of Ecology and Systematics*, vol 28, pp621–658.

Naiman, R.J., Pinay, G., Johnston, C.A. and Pastor, J. (1994) 'Beaver influences on the long-term biogeochemical characteristics of boreal forest drainage networks', *Ecology*, vol 75, pp905–921.

Naiman, R.J., Bilby, R.E., Schindler, D.E. and Helfield, J.M. (2002) 'Pacific salmon, nutrients, and the dynamics of freshwater and riparian ecosystems', *Ecosystems*, vol 5, pp399–417.

Olea P.P., Mateo-Tomás P. and Sánchez-Zapata J.A. (2019) 'Introduction to the topic of carrion ecology and management', in Olea P., Mateo-Tomás P. and Sánchez-Zapata J. (eds) *Carrion Ecology and Management*. Wildlife Research Monographs, vol 2. Springer, Cham.

Pardo, L.H., Fenn, M.E., Goodale, C.L., Geiser, L.H., Driscoll, C.T., Allen, E.B., Baron, J.S., Bobbink, R., Bowman, W.D., Clark, C.M., Emmett, B., Gilliam, F.S., Greaver, T.L., Hall, S.J., Lilleskov, E.A., Liu, L.L., Lynch, J.A., Nadelhoffer, K.J., Perakis, S.S., Robin-Abbott, M.J., Stoddard, J.L., Weathers, K.C. and Dennis, R.L. (2011) 'Effects of nitrogen deposition and empirical nitrogen critical loads for ecoregions of the United States', *Ecological Applications*, vol 21, pp3049–3082.

Peterjohn, W.T. and Correll, D.L. (1984) 'Nutrient dynamics in an agricultural watershed: observations on the role of a riparian forest', *Ecology*, vol 65, pp1466–1475.

Pickett, S.T.A., Cadenasso, M.L., Grove, J.M., Groffman, P.M., Band, L.E., Boone, C.G., Burch, W.R., Grimmond, C.S.B., Hom, J., Jenkins, J.C., Law, N.L. Nilon, C.H. Pouyat, R.V. Szlavecz, K., Warren, P.S. and Wilson, M.A. (2008) 'Beyond urban legends: an emerging framework of urban ecology, as illustrated by the Baltimore Ecosystem Study', *Bioscience*, vol 58, pp139–150.

Powers, S.M., Chowdhury, R.B., MacDonald, G.K., Metson, G.S., Beusen, A.H.W., Bouwman, A.F., Hampton, S.E., Mayer, B.K., McCrackin, M.L. and Vaccari, D.A. (2019, 'Global opportunities to increase agricultural independence through phosphorus recycling', *Earths Future*, vol 7, pp370–383.

Quinn, T.P., Helfield, J.M., Austin, C.S., Hovel, R.A. and Bunn, A.G. (2018) 'A multidecade experiment shows that fertilization by salmon carcasses enhanced tree growth in the riparian zone', *Ecology*, vol 99, pp2433–2441.

Raudsepp-Hearne, C., Peterson, G.D. and Bennett, E.M. (2010) 'Ecosystem service bundles for analyzing tradeoffs in diverse landscapes', *Proceedings of the National Academy of Sciences of the USA*, vol 107, pp5242–5247.

Reiners, W.A. and Driese, K.L. (2004) *Transport Processes in Nature: Propagation of Ecological Influences through Environmental Space*. Cambridge University Press, Cambridge.

Reisen, F., Duran, S.M., Flannigan, M., Elliott, C. and Rideout, K. (2015) 'Wildfire smoke and public health risk', *International Journal of Wildland Fire*, vol 24, pp1029–1044.

Ripple, W.J., Newsome, T.M., Wolf, C., Dirzo, R., Everatt, K.T., Galetti, M., Hayward, M.W., Kerley, G.I.H., Levi, T., Lindsey, P.A., Macdonald, D.W., Malhi, Y., Painter, L.E., Sandom, C.J., Terborgh, J. and Van Valkenburgh, B. (2015) 'Collapse of the world's largest herbivores', *Science Advances*, vol 1, no 4, art e1400103.

Rockstrom, J., Williams, J., Daily, G., Noble, A., Matthews, N., Gordon, L., Wetterstrand, H., DeClerck, F., Shah, M., Steduto, P., de Fraiture, C., Hatibu, N., Unver, O., Bird, J., Sibanda, L. and Smith, J. (2017) 'Sustainable intensification of agriculture for human prosperity and global sustainability', *Ambio*, vol 46, pp4–17.

Sato, M. (2013) 'Embodied carbon in trade: a survey of the empirical literature', *Journal of Economic Surveys*, vol 28, no 5, pp831–861.

Schindler, D.E., Scheuerell, M.D., Moore, J.W., Gende, S.M., Francis, T.B. and Palen, W.J. (2003) 'Pacific salmon and the ecology of coastal ecosystems', *Frontiers in Ecology and the Environment*, vol 1, pp31–37.

Schlesinger, W.H. and Bernhardt, E. (2013) *Biogeochemistry: An Analysis of Global Change*, 3rd Ed. Elsevier, Amsterdam.

Schmitz, O.J. (2008) 'Herbivory from individuals to ecosystems', *Annual Review of Ecology Evolution and Systematics*, vol 39, pp133–152.

Schmitz, O.J., Hawlena, D. and Trussell, G.C. (2010) 'Predator control of ecosystem nutrient dynamics', *Ecology Letters*, vol 13, pp1199–1209.

Schroder, T. and Fleig, F.D. (2017) 'Spatial patterns and edge effects on soil organic matter and nutrients in a forest fragment of southern Brazil', *Soil Research*, vol 55, pp649–656.

Schulze, E.-D. (1989) 'Air pollution and forest decline in a Spruce (*Picea abies*) forest', *Science*, vol 244, pp776–783.

Shimizu, M., Wentz, E.A., Merson, J., Kellogg, D.Q. and Gold, A.J. (2018) 'Modeling anthropogenic nitrogen flow for the Niantic River catchment in coastal New England', *Landscape Ecology*, vol 33, pp1385–1398.

Smil, V. (1999) 'Nitrogen in crop production: an account of global flows', *Global Biogeochemical Cycles*, vol 13, pp647–662.

Smith, C.R. and Baco, A.R. (2003) 'Ecology of whale falls at the deep-sea floor', *Oceanography and Marine Biology*, vol 41, pp311–354.

Smithwick, E.A.H., Harmon, M.E. and Domingo, J.B. (2003) 'Modeling multiscale effects of light limitations and edge-induced mortality on carbon stores in forest landscapes', *Landscape Ecology*, vol 18, pp701–721.

Smithwick, E.A.H., Eissensat, D.M., Lovett, G.M., Bowden, R.D., Rustad, L.E. and Driscoll, C.T. (2013) 'Root stress and nitrogen deposition: consequences and research priorities', *New Phytologist*, vol 197, pp1697–1708.

Smithwick, E.A., Baldwin, D.C. and Naithani, K.J. (2016) 'Grassland productivity in response to nutrient additions and herbivory is scale-dependent', *PeerJournal*, vol 4, art e2745.

Sollins, P., Grier, C.C., McCorison, F.M., Cromack Jr, K., Fogel, R. and Fredriksen, R.L. (1980) 'The internal element cycles of an old-growth Douglas-fir ecosystem in western Oregon', *Ecological Monographs*, vol 50, no 3, pp261–285.

Stephens, S.L., Burrows, N., Buyantuyev, A., Gray, R.W., Keane, R.E., Kubian, R., Liu, S., Seijo, F., Shu, L., Tolhurst, K.G. and van Wagtendonk, J.W. (2014) 'Temperate and boreal forest mega-fires: characteristics and challenges', *Frontiers in Ecology and the Environment*, vol 12, pp115–122.

Stoorvogel, J.J., Smaling, E.M.A. and Janssen, B.H. (1993) 'Calculating soil nutrient balances in Africa at difference scales.1. Supra-national scale', *Fertilizer Research*, vol 35, pp227–235.

Swart, P.K., Oehlert, A.M., Mackenzie, G.J., Eberlie, G.P. and Reigmer, J.J.G. (2014) 'The fertilization of the Bahamas by Saharan dust; A trigger for carbonate precipitation', *Geology*, vol 42, no 8, pp671–674.

Turner, M.G. (1989) 'Landscape ecology: the effect of pattern on process', *Annual Review of Ecology and Systematics*, vol 20, pp171–197.

Turner, M.G. and Gardner R.H. (2015) *Landscape Ecology in Theory and Practice: Pattern and Process*, 2nd Ed. Springer, Amsterdam.

Turner, M.G., Smithwick, E.A., Metzger, K.L., Tinker, D.B. and Romme, W.H. (2007) 'Inorganic nitrogen availability after severe stand-replacing fire in the Greater Yellowstone ecosystem', *Proceedings of the National Academy of Sciences*, vol 104, no 12, pp4782–4789.

Turner, M.G., Whitby, T.G. and Romme, W.H. (2019) 'Feast not famine: nitrogen pools recover rapidly in 25-yr-old postfire lodgepole pine', *Ecology*, vol 100, no 3, art e02626.

Vaughn, C.C. (2018) 'Ecosystem services provided by freshwater mussels', *Hydrobiologia*, vol 810, pp15–27.

Veldhuis, M.P., Gommers, M.I., Olff, H. and Berg, M.P. (2018) 'Spatial redistribution of nutrients by large herbivores and dung beetles in a savanna ecosystem', *Journal of Ecology*, vol 106, pp422–433.

Vitousek, P.M. (1977) 'The regulation of element concentrations in mountain streams in the northeastern United States', *Ecological Monographs*, vol 47, no 1, pp65–87.

Vitousek, P.M., Naylor, R., Crews, T., David, M.B., Drinkwater, L.E., Holland, E., Johnes, P.J., Katzenberger, J., Martinelli, L.A., Matson, P.A., Nziguheba, G., Ojima, D., Palm, C.A., Robertson, G.P., Sanchez, P.A., Townsend, A.R. and Zhang, F.S. (2009) 'Nutrient imbalances in agricultural development', *Science*, vol 324, no 5934, pp1519–1520.

Walker, B.H., Holling, C.S., Carpenter, S.R. and Kinzig, A. (2004) 'Resilience, adaptability and transformability in social–ecological systems', *Ecology and Society*, vol 9, no 2, art 5.

Wallace, Z., Lovett, G., Hart, J. and Machona, B. (2007) 'Effects of nitrogen saturation on tree growth and death in a mixed-oak forest', *Forest Ecology and Management*, vol 243, pp210–218.

Weathers, K.C., Cadenasso, M.L. and Pickett, S.T.A. (2001) 'Forest edges as nutrient and pollutant concentrators: potential synergisms between fragmentation, forest canopies and the atmosphere', *Conservation Biology*, vol 15, pp1506–1514.

Wetzel, P.R., van der Valk, A.G., Newman, S., Gawlik, D.E., Gann, T.T., Coronado-Molina, C.A., Childers, D.L. and Sklar, F.H. (2005) 'Maintaining tree islands in the Florida Everglades: nutrient redistribution is the key', *Frontiers in Ecology and the Environment*, vol 3, pp370–376.

Whittaker, R.H., Bormann, F.H., Likens, G.E. and Siccama, T.G. (1974) 'The Hubbard Brook ecosystem study: forest biomass and production', *Ecological Monographs*, vol 44, no 2, pp233–254.

Wu, J. and Hobbs, R. (2002) 'Key issues and research priorities in landscape ecology: an idiosyncratic synthesis', *Landscape ecology*, vol 17, no 4, pp355–365.

Young C, Fritz, H., Smithwick, E. and Venter, J. (2020) 'The landscape-scale drivers of herbivore assemblage distribution on the central basalt plains of Kruger National Park', *Journal of Tropical Ecology*, vol 36, no 1, pp13–28.

Yuan, Y.P., Wang, R.Y., Cooter, E., Ran, L.M., Daggupati, P., Yang, D.M., Srinivasan, R. and Jalowska, A. (2018) 'Integrating multimedia models to assess nitrogen losses from the Mississippi River basin to the Gulf of Mexico', *Biogeosciences*, vol 15, pp7059–7076.

Zheng, H., Li, Y.F., Robinson, B.E., Liu, G., Ma, D.C., Wang, F.C., Lu, F. Ouyang, Z.Y. and Daily, G.C. (2016) 'Using ecosystem service trade-offs to inform water conservation policies and management practices', *Frontiers in Ecology and the Environment*, vol 14, pp527–532.

8

The disturbance regime concept

Brian J. Harvey, Sarah J. Hart, and C. Alina Cansler

Introduction to disturbance in landscape ecology

A focal area of exploration using the lens of landscape ecology is the concept of *disturbance* as 'any relatively discrete event in time that disrupts ecosystem, community, or population structure and changes resources, substrate availability, or the physical environment' (White and Pickett 1985, 7). As natural and human-caused disturbances are key drivers of ecosystem dynamics, they have provided a ripe arena for ecological analyses of landscape change. Further, understanding and characterizing repeated patterns of disturbance along multiple dimensions (i.e., disturbance 'regimes'; Sousa 1984, Agee 1993, Turner 2010) is an area where landscape ecological theory has provided, and continues to provide, key insight.

Just as landscape ecology examines reciprocal interactions between spatial pattern and ecological process, the field of landscape ecology has an enduring series of reciprocal interactions with the field of disturbance ecology. The importance of the interconnectedness of landscape ecology with disturbance comes down to a fundamental characteristic of disturbances: disturbances and their repeated occurrence in ecosystems are inherently spatial processes that respond to and create spatial heterogeneity (Turner 2010). Insights about the causes and consequences of natural and anthropogenic disturbances have arisen from the application of landscape ecological theory and concepts, which have, in turn, been tested and refined on many large disturbances. A prime example of these interactions between fields can be found in a special issue published during the first year (1998) of the journal *Ecosystems* focused on '*Large Infrequent Disturbances*'. Using examples and lessons learned from disturbances that ranged from the 1980 eruption of Mt. St. Helens to the 1988 Yellowstone Fires to the 1993 floods in the Mississippi River, this key set of papers synthesized insights to date and laid out an ambitious research agenda and set of hypotheses about general landscape patterns of disturbance, ecosystem recovery, and management (Dale et al. 1998, Foster et al. 1998, Paine et al. 1998, Turner and Dale 1998, Turner et al. 1998, Romme et al. 1998).

Ecological inquiry into disturbance and subsequent ecological change across space and time has roots dating back to before the 20th century. Foundational work by (Cowles 1899) demonstrated the utility of 'space for time' substitution (i.e., chronosequence) studies to look at dynamic vegetation patterns through time in sand dunes near the Great Lakes—though disturbance was not explicitly mentioned (for a contemporary critical review of chronosequence methods, see Johnson and Miyanishi 2008). Early conceptualizations often posited disturbances as exogenous forces: ecosystems were in relative *equilibrium*, and disturbance events disrupted that equilibrium (Clements 1916, 1936). Studies considering disturbance were generally focused not on disturbance per se but on succession—the dynamics of ecosystem response to disturbance, which many ecologists viewed as a predictable sequence of changes that would eventually return the ecosystem to the pre-disturbance equilibrium (Clements 1936). With time came an increasing recognition of the vital role that disturbance has in maintaining ecosystems and their spatial mosaics of patches comprising different areas at different successional stages since the last disturbance (Wieslander 1935, Watt 1947)—that is, disturbances are fundamental to most natural ecosystems (Sousa 1984, Pickett and White 1985, Attiwill 1994). Along with the shift to a broader acceptance of disturbances as inherent in ecosystems came an intense focus on the processes governing ecological succession following disturbance, such as competition, facilitation, and chance events (or contingencies) (Gleason 1926, Egler 1954, Drury and Nisbet 1973, Connell and Slatyer 1977, Grime 1977). With progressively more study came the recognition that equilibrium is rare, and disturbance and *non-equilibrium* dynamics are the norm in most ecosystems (Drury and Nisbet 1973, Perry 2002). Moreover, equilibrium is dependent on the scale at which one examines ecological dynamics (Sprugel 1991, Turner et al. 1993, Jentsch and White 2019), and ecosystem dynamics often follow varied successional pathways rather than a single deterministic pathway (Cattelino et al. 1979, Fastie 1995, Kipfmueller and Kupfer 2005, Johnstone and Chapin 2006, Harvey and Holzman 2014).

Disturbance regime as a core concept

With the growing recognition of the ecological role of disturbance came focused attention on understanding the *disturbance regime*—the repeated pattern of disturbance over space and time, typically characterized by spatial, temporal, and mechanistic (e.g., intensity or magnitude) attributes (Sousa 1984, Agee 1993, Turner 2010). A disturbance event usually occurs over a discrete time period—minutes to months (though in some cases, droughts or insect outbreaks can last years). However, the disturbance regime is more appropriately characterized over broad time and space dimensions, allowing a fuller picture of the disturbance rotation (the time required to disturb an area equal to the landscape of interest) to unfold. Disturbance regime attributes can be, and often have been, characterized based on measures of central tendency (e.g., mean or median rotations). However, since rare and large disturbance events are often of disproportional importance for structuring ecosystems, the distributions (e.g., range, variance, or extreme values) of regime attributes across time and space are often of greater utility than measures of central tendency (Katz et al. 2005).

In a useful synthesis, Peters et al. (2011) separate disturbance events into three components that can help facilitate comparisons among mechanisms across systems. First, the *environmental drivers* of a disturbance event (climate, physical, biotic, and anthropogenic processes) each have their own characteristics (e.g., magnitude, duration, timing) and can be compared across systems. Second, the state of a system when disturbed (e.g., susceptibility, connectivity) is important to how the disturbance will unfold. Third, the mechanism of disturbance (e.g., combustion in fire, abrasion from wind, defoliation from organisms) may vary across disturbance types and systems. The interactions of drivers, system properties, and mechanisms produce a disturbance event, which then results in *legacies* (Figure 8.1)—the organic materials and organically derived patterns that persist through a disturbance (Franklin et al. 2000). Legacies can be of material (e.g., physical structures) or informational (e.g., genetic information) nature (Johnstone et al. 2016) and form the '*ecological memory*' that are the ingredients for post-disturbance ecosystem trajectories (Peters et al. 2011). Acceptance of repeated disturbances (i.e., regimes) as a part of system dynamics also coincided with concepts of *resilience*, or the capacity of a system to experience disturbance and maintain its essential structure and function (Holling 1973). If

Figure 8.1   Mature trees that survive disturbances are important biological legacies, providing seed sources for new tree establishment, increasing structural diversity of the forest over the course of post-disturbance succession, and providing wildlife habitat. (Photo from the Frank Church – River of No Return Wilderness, Idaho, United States, 2018, by C.A. Cansler.)

the adaptive traits present in an ecosystem are aligned with the disturbance regime, a system is more likely to be in a 'safe operating space' whereby disturbance and recovery cycles are part of the inherent system dynamics that promote resilience (Johnstone et al. 2016). Alternatively, if the disturbance-recovery cycle, changing environmental context, or exogenous factors (e.g., invasive species) move an ecosystem out of a state (e.g., forest), shifting among *alternative stable states* (e.g., repeated long-term cycles between forest and non-forest conditions) may be possible (Beisner et al. 2003).

The desire to merge knowledge regarding natural disturbance regimes with landscape ecology became a more important goal of research as ecosystem management rose to the forefront of natural resource management. Ecosystem management aims to maintain viable native species populations, representative amounts of native ecosystem types, and ecological processes—including disturbances—while also allowing long-term sustainable human use of ecosystem goods and services (Grumbine 1994). The implementation of ecosystem management benefits from a recognition of the dynamic nature of ecosystems, which emerges as a result of natural and anthropogenic disturbances as well as variation in climate (Morgan et al. 1994). The most common approach to this was to use reference conditions prior to major land-use changes due to European colonization and industrialization as a benchmark. These efforts are not without serious challenges, however, as ecosystems are in constant flux, and defining the temporal and spatial scale of reference conditions is non-trivial (Sprugel 1991). Often described as 'natural variability' (Landres et al. 1999), the '*natural range of variability*', or the 'historical range of variability' (Keane et al. 2009), these reference conditions strive to include more than a single static target, typically incorporate the full variation and range of conditions (e.g., communities, successional stages, disturbances) into a landscape, and often include spatial attributes of ecosystem components (Keane et al. 2002). There are two important ties to natural disturbance regimes. First, descriptions of the natural range of variability often incorporated the frequency or rotation of different disturbances and various severities of a single disturbance (e.g., 'surface fire' and 'stand-replacing fire') on a landscape. Second, because direct evidence—such as early aerial photography or stand age maps—of historical landscape composition is often lacking, simulation models are often used to identify the natural range of variability of different community types or successional stages on a landscape (see Chapter 14 for links between paleo ecology and landscape ecology). For simulation models to be parameterized, quantitative descriptions of the frequency and severity (etc.) of different disturbance types are needed. This, in turn, requires the collection of observational data on disturbance regime attributes.

Some of the important work on landscape patterns of disturbance proposed the concept of a '*shifting-mosaic steady state*' (Bormann and Likens 1979), which suggests that even though components of the landscape may change successional state and species composition due to disturbance, the landscape itself (at a sufficient spatial extent) is at equilibrium, as new patches created by disturbance are balanced by successional development in old patches. Fine-scale (small spatial grain at small spatial extents) spatial and temporal variation could produce broad-scale (small spatial grain over broad spatial extents) consistency in landscape composition. This conceptualization was revised to recognize that non-equilibrium landscapes can be created by large disturbances that impact most of the landscape (Sprugel 1991, Turner et al. 1993). Thus, not only was equilibrium scale-dependent (i.e., typically evident at larger scales but not at smaller—a concept sometimes referred to as meta-stability), but empirical studies (Romme and Despain 1989) and modeling studies (Turner et al. 1993) showed that landscape composition can fluctuate widely over time. Non-stationary climate, which causes non-stationarity in both species composition and disturbance regime attributes, further undermines the theoretical conceptualization of equilibrium (Jackson 2006, Blonder et al. 2018). Climate change also presents practical problems for managers trying to use the 'natural range of variability', since climate change will cause continual changes in decadal-scale characteristics of many disturbance regimes (Bebi et al. 2001, Allen et al. 2010, Bentz et al. 2010) and century-scale shifts in dominant species and locations of biomes, which will feed back on disturbance dynamics. Plans for adapting ecosystems and management practices to expected conditions under climate changes have begun to identify the 'future range of variability' (Thompson et al. 2009, Duncan et al. 2010) or 'future climate-analogue reference conditions' (Churchill et al. 2013). These methods can range from identifying regional-scale references that help managers anticipate future conditions in a different region (Keane et al. 2018) to stand-scale reference sites that provide pattern and compositional targets for restoration silviculture practices (Churchill et al. 2013).

A plethora of geospatial data and accompanying field measurements prior to, during, and after disturbances has catalyzed a renaissance of research examining the disturbances themselves and their various components. These data have allowed much greater investigation of what are often termed '*top-down*' and '*bottom-up*' *controls* on disturbance regimes. It is important to note that these terms have a different meaning in landscape ecology and disturbance ecology than in ecological studies of predator–prey relationships (Power 1992, Estes and Duggins 1995). In relation to disturbance, 'top-down' refers to broad-scale climatic controls on disturbance regimes, such as century-to-decadal-scale climate controlling the biogeography of plant species, and inter-annual variation in climate, which drives inter-annual variability in the extent and severity of disturbances (McKenzie and Kennedy 2011). In contrast, 'bottom-up' controls include biophysical drivers such as topography and soil moisture holding capacity as well as biotic factors such as plant species composition, or plant trait composition, and vegetation structure.

The top-down control of climate is an important driver of the distribution of many abiotic disturbance processes. Snow avalanche disturbances only occur where there is sufficient snowpack, wind-throw from tropical cyclones is much more likely in coastal areas in the Intertropical Convergence Zone, and fires are not an important disturbance process in barren alpine and desert environments where there is insufficient fuel to support fire spread. Inter-annual variability in weather also provides a strong control on many disturbance processes. Multi-year drought is a strong driver of widespread tree die-off events, during which the proximate cause of mortality is usually bark beetles (Allen et al. 2010)—the populations of which are also strongly influenced by climatic variation (Bentz et al. 2010) (Figure 8.2). Climatic variation also strongly influences fire frequency and extent, primarily by influencing fuel aridity and fire season length (Jolly et al. 2015) but also via fuel accumulation and the number of ignitions. In arid, fuel-limited systems, wet years lead to increased growth of grass and forbs, and the increase in fine fuels supports increased fire occurrences and extent (Littell et al. 2018). In contrast, flammability-limited systems experience strong increases in area burned in response to increased aridity (Littell et al. 2018).

The perceived importance of 'bottom-up' controls on disturbance regimes has increased with the increased availability of geospatial data that can be used to quantify the local biophysical

Figure 8.2 The 2006 Tatoosh Buttes fire (background slopes with little live tree cover) was a high-severity wind-driven fire that created extremely large homogeneous patches of almost complete tree mortality. A subsequent bark beetle outbreak (trees with red needles in the foreground) caused tree mortality outside the fire perimeter where host trees were available post-fire. These are two examples of top-down disturbances modified by bottom-up controls. Pasayten Wilderness, Washington, United States. (Photo from 2008 by C.A. Cansler.)

Figure 8.3 A complex mosaic of patches of rock, herbaceous vegetation, shrublands, and forest stands with differing structures can be seen in this photo of the Illilouette Creek Basin in Yosemite National Park, California, United States. A recently burned area in the foreground has biological legacies in the form of dead standing trees (snags) and mature trees. In the background, denser forests are found in ravines, and non-forest and open forest stands are predominantly on convex topography. These topographic controls are expressed after 40 years in which wildfires have been allowed to burn for resource benefit by the National Park Service. (Photo from 2012 by C.A. Cansler.)

setting and disturbance impacts. Early work on wind and fire disturbances (Heinselman 1973, Bormann and Likens 1979, Sprugel and Bormann 1981) examined the mosaic of disturbances across a landscape with little consideration of the local frequency or severity of disturbance. The 'shifting-mosaic-steady state' framework implied an axiomatic assumption that disturbances were stochastic. Since then, research has revealed strong topographical controls on wind-throw (Harcombe et al. 2004), avalanche impacts (Bebi et al. 2001), fire frequency (Kellogg et al. 2008), high-severity fire (Dillon et al. 2011), and fire refugia (Meddens et al. 2018). There is an increasing acknowledgment that disturbance impacts can be both predictable and stochastic (Meddens et al. 2016), depending on the scale being analyzed, the disturbance type, and the relative local strengths of top-down and bottom-up controls, and the relative strength of different controls may be non-stationary in time (Newman et al. 2019). For contagious disturbances (e.g., fire, pathogen spread) in particular, high variation in local topography and vegetation weakens the strength of the relationship between top-down controls and disturbance extent and severity (Turner and Romme 1994, McKenzie and Kennedy 2011, Cansler and McKenzie 2014, Harvey et al. 2016) (Figure 8.3).

Current themes and issues for research on disturbance regimes

While much has been learned about how disturbances are fundamental to ecosystems across the world, current themes in the field center around two key questions: How do multiple disturbances interact? and how are disturbance regimes changing? Collectively, these two questions feed into inquiries of how disturbances and disturbance-recovery cycles in ecosystems promote resilience in single states or shifts between alternative stable states.

Increasing disturbance activity in many regions of the world over the last few decades (Dale et al. 2001) has led to a drastic increase in the recognition that disturbances do not occur in isolation but instead, are always overlapping in space with varying time intervals between individual events. Further, such interactions produce lasting effects on patterns of ecosystem composition and structure (Peterson 2002). While accepting the paradigm of the disturbance regime concept (see earlier) means that individual disturbance events are constantly overlapping prior such events, until recently, there was less mechanistic understanding of how these interactions unfolded and how they affected ecosystems. Since the late 1990s and early 2000s, there has been a proliferation of research examining disturbance interactions and the development of multiple frameworks for conceptualizing how these interactions operate.

Several terms introduced since the late 1990s collectively provide a helpful framework for testing different ways that disturbances can interact. First, two disturbances may be *linked*, in that the occurrence of one disturbance influences the likelihood, size, or severity of a subsequent disturbance (Simard et al. 2011). In such settings, the focal response variable is the occurrence and/or characteristics of any subsequent disturbance. For example, recent increases in insect outbreak and fire disturbance across North America since the late 1990s led to researchers asking whether insect outbreaks influence the likelihood of subsequent fire occurrence (Meigs et al. 2015), area burned (Hart et al. 2015a), fire behavior (Simard et al. 2011, Schoennagel et al. 2012), or fire severity (e.g., ecological effects) (Harvey et al. 2014, Meigs et al. 2016). Linked disturbances either facilitate or impede the incidence of subsequent disturbances or either amplify or reduce the intensity of a subsequent disturbance (Kane et al. 2017).

Whether or not two disturbances are linked, sequential overlap of two disturbances (of the same or different disturbance type) may produce effects that are *synergistic*, or greater than the sum of the two disturbances individually; such an interaction has been termed *compound disturbances* (Paine et al. 1998). In such settings, the focal response variable is the ecosystem state, or response to the combinations of disturbances, rather than the second disturbance per se. Examples in North America have focused on how forest ecosystems can be impacted by compound effects of fire disturbance that follows wind blow-downs (Buma and Wessman 2011), previous fires (Turner et al. 2019), and insect outbreaks (Harvey et al. 2013). Finally, interacting disturbances can produce *cascading effects* wherein interactions among disturbances can increase the climate sensitivity of an ecosystem (Buma 2015, Seidl and Rammer 2017). Key factors that govern disturbance interactions and their outcomes are the time since the first disturbance (and therefore, the interval between the two disturbances when a second occurs), the severity/magnitude of the first disturbance, the order of the disturbance sequence (i.e., which disturbance comes first), and whether the first disturbance amplifies or attenuates the subsequent disturbance (Burton et al. 2020).

In addition to a focus on interactions among disturbances, much research in the first decades of the 21st century has focused on how disturbance regimes are changing (Turner 2010, Johnstone et al. 2016). A warming climate, changing land-use patterns, and human-aided transport of plants and animals into novel regions (i.e., invasive species), as well as interactions among these factors, have been identified as key drivers of disturbance regime change. Changing disturbance regimes will have profound implications for ecosystem services (Seidl et al. 2016), yet many impacts are currently poorly understood.

Many disturbances are sensitive to climate, and warming temperatures—as a key driver in releasing previous constraints on disturbance activity—are profoundly affecting many aspects of disturbance regimes. A warming climate is lengthening the 'season' for many disturbance regimes worldwide. For example, global climate warming increased the length of the fire season (i.e., the period of time during each year when weather is conducive to fire spread) by nearly

20% over the period from 1979 to 2013 (Jolly et al. 2015), which has been accompanied by a corresponding increase in the frequency of large fires in many regions (Westerling 2016). Biotic disturbances that are also driven by climate (e.g., bark beetle outbreaks) have increased in severity and extent since the late 1990s (Raffa et al. 2008). Continued warming is expected to drive continued changes in both abiotic and biotic disturbance regimes (Bentz et al. 2010, Westerling et al. 2011, Seidl et al. 2017), yet for many systems, feedbacks among climate, vegetation, and disturbance may modify the direct effects of warming on disturbance regimes (Hart et al. 2015b). Characterizing the spatial, or landscape, patterns of disturbances and how they are changing as the climate warms has also become a focus—yet this remains an area of current exploration. For example, satellite-derived burn severity atlases have been used to track changes in spatial heterogeneity in fire patterns over time and space (Cansler and McKenzie 2014, Harvey et al. 2016, Reilly et al. 2017, Collins et al. 2017). Finally, synergies among changing components of regimes have been documented in some systems. For example, increasing fire frequency in some conifer forests is leading to anomalously short intervals between severe fires (severe fires are inherent in system dynamics when occurring over long fire-free intervals), which is, in turn, driving increases in fire severity (e.g., fire-caused vegetation mortality and woody biomass combustion) (Turner et al. 2019).

The growing human population on the planet and the accompanying increasing mobility of people around the globe has driven land-use changes and introductions of novel species assemblages—ultimately affecting the ways that disturbances operate and how they impact ecosystems (Gaertner et al. 2014). For example, in many areas where widespread and frequent disturbances such as prairie fires historically occurred, broad-scale agricultural development and fragmentation of native vegetation has led to a cessation of disturbance and corresponding losses in native biodiversity (Leach and Givnish 1996). In such cases, land-use and land-cover change have dampened the occurrence and severity of disturbance. Alternatively, many historically fire-prone forested regions have been affected by nearly a century of fire suppression or exclusion, which has increased the likelihood of uncharacteristically large and severe fires through fuel accumulation (Hessburg et al. 2015). In these cases, land-use and land-cover change have amplified the potential for disturbance while also increasing human exposure to disturbance where settlements have expanded into previously fire-frequent landscapes (Moritz et al. 2014, Schoennagel et al. 2017). In many areas, the introduction of non-native vegetation that is more flammable than native vegetation has led to novel fire regimes and ecosystem dynamics. For example, the arrival of humans (providing an ignition source) and more recent exotic plantation forests and invasive species (providing highly flammable vegetation) in the southern hemisphere promotes the spread of large fires in systems that historically had infrequent or no fire, with native species lacking adaptations to frequent and severe fire (Perry et al. 2012, Kitzberger et al. 2016, McWethy et al. 2018). Finally, introduced pathogens (which are a novel disturbance) can alter subsequent fire disturbances, with implications for changing disturbance regimes (Metz et al. 2013, Simler et al. 2018). Continued land-use change and spread of non-native species are likely to play a major role in further alteration of disturbance regimes and landscape patterns worldwide.

Disturbance, resilience, and scale

With recognition of the inherent role of disturbance as a controlling ecological process came the conceptualization of ecological resilience, the capacity of an ecosystem to retain essentially the same structures and functions following disturbance (Holling 1973). Resilience is conferred by an ecosystem's *information legacies*, the *disturbance-adaptive traits* of the system, and *material legacies*, the individuals, propagules, and other remnants that persist through disturbance (Johnstone et al. 2016). Information legacies emerge as a result of long-term selective pressures and thus, broad spatial scales, rather than a single disturbance event. In contrast, material legacies are determined by characteristics of the disturbance event and the state of the system when disturbed and so may vary across finer spatial and temporal scales (Johnstone et al. 2016). Thus, resilience often emerges from processes at one scale interacting with processes at a different scale, often with complex nonlinear dynamics (i.e., cross-scale interactions; Peters et al. 2004, Reyer et al. 2015).

Changes in disturbance regimes that alter legacies may compromise resilience (Johnstone et al. 2016). A well-studied example is found in conifer forests of boreal North America, where stand-replacing wildfire is a key disturbance (Payette 1992). A central mechanism promoting resilience of conifer forest to wildfire is the production of serotinous cones in pine (*Pinus* spp.) and black spruce (*Picea mariana*), which allows rapid post-fire establishment of conifer trees (Greene et al. 1999). Resilience may be compromised when wildfires burn too intensely, leading to combustion of seed or soil organic layers. Exposure of mineral soils and local absence of conifer seed confer a competitive advantage on deciduous broadleaf species (Johnstone and Chapin 2006, Greene et al. 2007). When deciduous trees become established, they initiate changes in plant–soil feedbacks that alter fuels and lead to the establishment of a low-severity fire regime (Johnstone et al. 2010). Thus, predicting when and where disturbance may initiate shifts to alternative states requires a multi-scale understanding of the mechanisms that promote resilience.

In light of the potential for erosion of resilience to drive dramatic shifts in ecosystem state that are often hard to reverse, resilience has also become a central theme in environmental sustainability. In this context, resilience is often viewed as a desirable attribute of an ecosystem because it implies a predictable, although varying, supply of expected ecosystem services (Angeler and Allen 2016). This interpretation contrasts with Holling's (1973) original definition, which describes resilience as an emergent property of a system. In a recent effort to differentiate and link these two views, Higuera et al. (2019) suggest that socioecological systems would be best described in terms of the probability of a state change (i.e., *value-free resilience*) and the social acceptability of a state change (i.e., *value-laden resilience*). This separation is helpful, because ecological resilience and social values are not always in alignment (Higuera et al. 2019). For instance, cheatgrass (*Bromus tecotrum*) invasion of rangelands of the Great Basin led to the establishment of an ecosystem that exhibits high value-free resilience to wildfire, which is conferred by the life-history traits of cheatgrass (i.e., early-season growth and abundant seed production; D'Antonio and Vitousek 1992). However, because cheatgrass-dominated grasslands are characterized by high-frequency wildfire (Balch et al. 2013) and lower forage quality, these attributes are not highly valued by society (Brunson and Tanaka 2011). Further, ecosystems may be unlikely to change states, but the rate of return to the pre-disturbance state may be socially unacceptable. For example, subalpine forests of the Southern Rocky Mountains in the United States generally show high value-free resilience to infrequent high-severity wildfire (Minckley et al. 2012), but the recovery is often slow (Kipfmueller and Kupfer 2005, Rodman et al. 2019). Thus, when societal values emphasize forested conditions, silvicultural treatment may be used to expediate forest regeneration and stand development despite high ecological resilience (DeRose and Long 2014). Linking value-free resilience with the social acceptability of disturbance-driven change is critical to operationalizing the resilience framework in socio-ecological systems.

Looking ahead to the future: challenges and opportunities

Over recent decades, we have learned a lot about how environmental drivers, initial ecosystem conditions, and mechanisms underlying disturbance effects influence disturbance regimes. This research has revealed that many disturbances regimes are characterized by complex cross-scale interactions (Peters et al. 2004). For example, in their review of the drivers of bark beetle outbreaks, Raffa et al. (2008) describe how outbreaks occur in response to interactions among fine-scale processes, such as tree entry by an individual beetle, and broad-scale processes, like regional patterns in host abundance and susceptibility. Spatially extensive outbreaks occur when thresholds are surpassed and new positive feedbacks are established. Such nonlinear and cross-scale interactions represent a key challenge for disturbance ecology because they often result in emergent behavior that cannot be predicted from the individual processes or observations from single events (Peters et al. 2004). For instance, empirical models of bark beetle–induced tree mortality are often of little predictive capacity because they were developed under a limited set of conditions (Bentz et al. 1993, Hart et al. 2014). Combining long-term field data and spatially extensive remote sensing datasets can facilitate the prediction of ecological thresholds. However, the integration of such data remains a major challenge in understanding the current and future effects of disturbances (Lindenmayer et al. 2010, Weed et al. 2013, Peters et al. 2018).

Contemporary changes in climate, land-use practices, human population size, and biotic communities, and the resulting changes in disturbance regimes, have altered landscape spatial patterns in ecosystems around the world, with important consequences for the provisioning of ecosystem services (Turner 2010, Turner et al. 2013, Seidl et al. 2016). For example, the Great Barrier Reef, which is home to more than 1,700 aquatic animal species, contributes $6.4 billion to Australia's economy annually and is culturally important to many of Australia's aboriginal people (O'Mahoney et al. 2017). Across the Great Barrier Reef, coral cover has declined by more than half since 1985 due to cyclonic disturbance, population outbreaks of predatory crown-of-thorns starfish (*Acanthaster planci*), and temperature-driven coral bleaching events (De'ath et al. 2012). Concurrently, coral recovery rates have declined by more than 80% due to decreased water quality, warming, and the compound effects of frequent cyclones (Ortiz et al. 2018). Yet, both die-off and recovery rates vary in space and time across the Great Barrier Reef due to complex interactions among biological processes, environmental conditions, and geographic setting (De'ath et al. 2012, Ortiz et al. 2018). A central challenge for ecology is to predict how future global change and the ensuing changes in disturbance regimes will affect ecosystems around the world. There are many challenges to forecasting future disturbance regimes, including (1) uncertainty in future climate projections, (2) the potential for new drivers to be introduced (e.g., invasive species), (3) feedbacks between disturbances and climate, and (4) non-stationarity in environmental drivers.

As we head into the third decade of the 21st century, landscape ecology will continue to be an important lens through which to examine and understand disturbances. While much is still to be learned about the fundamental nature of many ecological disturbances, regimes themselves are rapidly changing as the climate warms and the human imprint on the biosphere expands. Questions and answers about how disturbances operate, how they are changing, and how we can manage social ecological systems are topics that are inherently spatial and that occur across a wide range of spatial and temporal scales. Landscape ecology is primed to play a critical role in bringing together ecological insights from the field with massive streams of remotely-sensed big data to aid in understanding, tracking, and predicting disturbance regimes of the Earth's future.

References

Agee, J.K. (1993) *Fire Ecology of Pacific Northwest Forests*, 2nd ed. Island Press, Seattle.

Allen, C.D., Macalady, A.K., Chenchouni, H., Bachelet, D., McDowell, N., Vennetier, M., Kitzberger, T., Rigling, A., Breshears, D.D. Hogg, E.H. (Ted), Gonzalez, P., Fensham, R., Zhang, Z., Castro, J., Demidova, N., Lim, J.-H., Allard, G., Running, S.W., Semerci, A. and Cobb, N. (2010) 'A global overview of drought and heat-induced tree mortality reveals emerging climate change risks for forests', *Forest Ecology and Management*, vol 259, pp660–684.

Angeler, D.G. and Allen C.R. (2016) 'Quantifying resilience', *Journal of Applied Ecology*, vol 53, pp617–624.

Attiwill, P.M. (1994) 'The disturbance of forest ecosystems: the ecological basis for conservative management', *Forest Ecology and Management*, vol 63, pp247–300.

Balch, J.K., Bradley, B.A., D'Antonio, C.M. and Gómez-Dans, J. (2013) 'Introduced annual grass increases regional fire activity across the arid western USA (1980–2009)', *Global Change Biology*, vol 19, pp173–183.

Bebi, P., Kienast, F. and Schönenberger, W. (2001) 'Assessing structures in mountain forests as a basis for investigating the forests' dynamics and protective function', *Forest Ecology and Management*, vol 145, pp3–14.

Beisner, B., Haydon, D. and Cuddington, K. (2003) 'Alternative stable states in ecology', *Frontiers in Ecology and the Environment*, vol 1, pp376–382.

Bentz, B.J., Amman, G.D. and Logan, J.A. (1993) 'A critical assessment of risk classification systems for the mountain pine beetle', *Forest Ecology and Management*, vol 61, pp349–366.

Bentz, B.J., Régnière, J., Fettig, C.J., Hansen, E.M., Hayes, J.L., Hicke, J.A., Kelsey, R.G., Negrón, J.F. and Seybold, S.J. (2010) 'Climate change and bark beetles of the Western United States and Canada: direct and indirect effects', *BioScience*, vol 60, pp602–613.

Blonder, B., Enquist, B.J., Graae, B.J., Kattge, J., Maitner, B.S., Morueta-Holme, N., Ordoñez, A., Šímová, I., Singarayer, J., Svenning, J.-C., Valdes, P.J. and Violle, C. (2018) 'Late quaternary climate legacies in contemporary plant functional composition', *Global Change Biology*, vol 24, pp4827–4840.

Bormann, F.H. and Likens, G.E. (1979) *Pattern and Process in a Forested Ecosystem: Disturbance, Development and the Steady State Based on the Hubbard Brook Ecosystem Study*. Springer, New York.

Brunson, M.W. and Tanaka, J. (2011) 'Economic and social impacts of wildfires and invasive plants in American deserts: lessons from the Great Basin', *Rangeland Ecology & Management*, vol 64, pp463–470.

Buma, B. (2015) 'Disturbance interactions: characterization, prediction, and the potential for cascading effects', *Ecosphere*, vol 6, art 70.

Buma, B. and Wessman, C.A. (2011) 'Disturbance interactions can impact resilience mechanisms of forests', *Ecosphere*, vol 2, art 64.

Burton, P. J., Jentsch, A. and Walker, L. R. (2020) 'The Ecology of Disturbance Interactions', *BioScience*, vol 70, pp854–870.

Cansler, C.A. and McKenzie, D. (2014) 'Climate, fire size, and biophysical setting control fire severity and spatial pattern in the northern Cascade Range, USA', *Ecological Applications*, vol 24, pp1037–1056.

Cattelino, P.J., Noble, I.R., Slatyer, R.O. and Kessell, S.R. (1979) 'Predicting the multiple pathways of plant succession', *Environmental Management*, vol 3, pp41–50.

Churchill, D.J., Larson, A.J., Dahlgreen, M.C., Franklin, J.F., Hessburg, P.F. and Lutz, J.A. (2013) 'Restoring forest resilience: from reference spatial patterns to silvicultural prescriptions and monitoring', *Forest Ecology and Management*, vol 291, pp442–457.

Clements, F.E. (1916) *Plant Succession: An Analysis of the Development of Vegetation*. Carnegie Institution of Washington, Washington.

Clements, F.E. (1936) 'Nature and structure of the climax', *Journal of Ecology*, vol 24, no 1, pp252–284.

Collins, B.M., Stevens, J.T., Miller, J.D., Stephens, S.L., Brown, P.M. and North, M.P. (2017) 'Alternative characterization of forest fire regimes: incorporating spatial patterns', *Landscape Ecology*, vol 32, pp1543–1552.

Connell, J.H. and Slatyer. R.O. (1977) 'Mechanisms of succession in natural communities and their role in community stability and organization', *American Naturalist*, vol 111, pp1119–1144.

Cowles, H.C. (1899) 'The ecological relations of the vegetation on the sand dunes of Lake Michigan. Part I: geographical relations of the dune floras', *Botanical Gazette*, vol 27, pp95–117.

D'Antonio, C.M., and Vitousek, P.M. (1992) 'Biological invasions by exotic grasses, the grass/fire cycle, and global change', *Annual Review of Ecology and Systematics* vol 23, pp63–87.

Dale, V.H., Lugo, A.E., MacMahon, J.A. and Pickett, S.T.A. (1998) 'Ecosystem management in the context of large, infrequent disturbances', *Ecosystems*, vol 1, pp546–557.

Dale, V.H., Joyce, L.A., McNulty, S., Neilson, R.P., Ayres, M.P., Flannigan, M.D., Hanson, P.J. Irland, L.C., Lugo, A.E., Peterson, C.J., Simberloff, D., Swanson, F.J., Stocks, B.J. and Wotton, B.M. (2001) 'Climate change and forest disturbances: climate change can affect forests by altering the frequency, intensity, duration, and timing of fire, drought, introduced species, insect and pathogen outbreaks, hurricanes, windstorms, ice storms, or landslides', *BioScience*, vol 51, pp723–734.

De'ath, G., Fabricius, K.E., Sweatman, H. and Puotinen, M. (2012) 'The 27-year decline of coral cover on the Great Barrier Reef and its causes', *Proceedings of the National Academy of Sciences*, vol 109, pp17995–17999.

DeRose, R.J. and Long, J.N. (2014) 'Resistance and resilience: a conceptual framework for silviculture', *Forest Science*, vol 60, pp1205–1212.

Dillon, G.K., Holden, Z.A., Morgan, P., Crimmins, M.A., Heyerdahl, E.K. and Luce, C.H. (2011) 'Both topography and climate affected forest and woodland burn severity in two regions of the western US, 1984 to 2006', *Ecosphere*, vol 2, art 130.

Drury, W.H. and Nisbet, I.C. (1973) 'Succession', *Journal of the Arnold Arboretum*, vol 54, pp331–368.

Duncan, S.L., McComb, B.C. and Johnson, K.N. (2010) 'Integrating ecological and social ranges of variability in conservation of biodiversity: past, present, and future', *Ecology and Society*, vol 15, no 1, art 5.

Egler, F.E. (1954) 'Vegetation science concepts I. Initial floristic composition, a factor in old-field vegetation development with 2 figs', *Vegetatio*, vol 4, pp412–417.

Estes, J.A., and Duggins, D.O. (1995) 'Sea otters and kelp forests in Alaska: generality and variation in a community ecological paradigm', *Ecological Monographs*, vol 65, pp75–100.

Fastie, C.L. (1995) 'Causes and ecosystem consequences of multiple pathways of primary succession at Glacier Bay, Alaska', *Ecology*, vol 76, pp1899–1916.

Foster, D.R., Knight, D.H. and Franklin, J.F. (1998) 'Landscape patterns and legacies resulting from large, infrequent forest disturbances', *Ecosystems*, vol 1, pp497–510.

Franklin, J.F., Lindenmayer, D., MacMahon, J.A., McKee, A., Magnuson, J., Perry, D.A., Waide, R. and Foster, D. (2000) 'Threads of continuity', *Conservation in Practice*, vol 1, pp8–17.

Gaertner, M., Biggs, R., Beest, M.T., Hui, C., Molofsky, J. and Richardson, D.M. (2014) 'Invasive plants as drivers of regime shifts: identifying high-priority invaders that alter feedback relationships', *Diversity and Distributions*, vol 20, pp733–744.

Gleason, H.A. (1926) 'The individualistic concept of the plant association', *Bulletin of the Torrey Botanical Club*, vol 53, no 1, pp7–26.

Greene, D.F., Zasada, J.C., Sirois, L., Kneeshaw, D., Morin, H., Charron, I. and Simard, M.-J. (1999) 'A review of the regeneration dynamics of North American boreal forest tree species', *Canadian Journal of Forest Research*, vol 29, pp824–839.

Greene, D.F., Macdonald, S.E., Haeussler, S., Domenicano, S., Noël, J., Jayen, K., Charron, I., Gauthier, S., Hunt, S., Gielau, E.T., Bergeron, Y. and Swift, L. (2007) 'The reduction of organic-layer depth by wildfire in the North American boreal forest and its effect on tree recruitment by seed', *Canadian Journal of Forest Research*, vol 37, pp1012–1023.

Grime, J.P. (1977) 'Evidence for the existence of three primary strategies in plants and its relevance to ecological and evolutionary theory', *American Naturalist*, vol 111, pp1169–1194.

Grumbine, R.E. (1994) 'What is ecosystem management?', *Conservation Biology*, vol 8, pp27–38.

Harcombe, P.A., Greene, S.E., Kramer, M.G., Acker, S.A., Spies, T.A. and Valentine, T. (2004) 'The influence of fire and windthrow dynamics on a coastal spruce–hemlock forest in Oregon, USA, based on aerial photographs spanning 40 years', *Forest Ecology and Management*, vol 194, pp71–82.

Hart, S.J., Veblen, T.T. and Kulakowski, D. (2014) 'Do tree and stand-level attributes determine susceptibility of spruce-fir forests to spruce beetle outbreaks in the early 21st century?', *Forest Ecology and Management*, vol 318, pp44–53.

Hart, S.J., Schoennagel, T., Veblen, T.T. and Chapman, T.B. (2015a) 'Area burned in the western United States is unaffected by recent mountain pine beetle outbreaks', *Proceedings of the National Academy of Sciences*, vol 112, pp4375–4380.

Hart, S.J., Veblen, T.T., Mietkiewicz, N. and Kulakowski, D. (2015b) 'Negative feedbacks on bark beetle outbreaks: widespread and severe spruce beetle infestation restricts subsequent infestation', *PLoS ONE*, vol 10, art e0127975.

Harvey, B.J. and Holzman, B.A. (2014) 'Divergent successional pathways of stand development following fire in a California closed-cone pine forest', *Journal of Vegetation Science*, vol 25, pp88–99.

Harvey, B.J., Donato, D.C., Romme, W.H. and Turner, M.G. (2013) 'Influence of recent bark beetle outbreak on fire severity and postfire tree regeneration in montane Douglas-fir forests', *Ecology*, vol 94, pp2475–2486.

Harvey, B.J., Donato, D.C. and Turner, M.G. (2014) 'Recent mountain pine beetle outbreaks, wildfire severity, and postfire tree regeneration in the US Northern Rockies', *Proceedings of the National Academy of Sciences*, vol 111, pp15120–15125.

Harvey, B.J., Donato, D.C. and Turner, M.G. (2016) 'Drivers and trends in landscape patterns of stand-replacing fire in forests of the US Northern Rocky Mountains (1984–2010)', *Landscape Ecology*, vol 31, pp2367–2383.

Heinselman, M.L. (1973) 'Fire in the virgin forests of the boundary waters canoe area, Minnesota', *Quaternary Research*, vol 3, pp329–382.

Hessburg, P.F., Churchill, D.J., Larson, A.J., Haugo, R.D., Miller, C., Spies, T.A., North, M.P., Povak, N.A., Belote, R.T., Singleton, P.H., Gaines, W.L., Keane, R.E., Aplet, G.H., Stephens, S.L., Morgan, P., Bisson, P.A., Rieman, B.E., Salter, R.B. and Reeves, G.H. (2015) 'Restoring fire-prone Inland Pacific landscapes: seven core principles', *Landscape Ecology*, vol 30, pp1805–1835.

Higuera, P.E., Metcalf, A.L., Miller, C., Buma, B., McWethy, D.B., Metcalf, E.C., Ratajczak, Z., Nelson, C.R., Chaffin, B.C., Stedman, R.C., McCaffrey, S., Schoennagel, T., Harvey, B.J., Hood, S.M., Schultz, C.A., Black, A.E., Campbell, D., Haggerty, J.H., Keane, R.E., Krawchuk, M.A., Kulig, J.C., Rafferty, R. and Virapongse, A. (2019) 'Integrating subjective and objective dimensions of resilience in fire-prone landscapes', *BioScience*, vol 69, pp379–388.

Holling, C.S. (1973) 'Resilience and stability of ecological systems', *Annual Review of Ecology and Systematics*, vol 4, pp1–23.

Jackson, S.T. (2006) 'Vegetation, environment, and time: the origination and termination of ecosystems', *Journal of Vegetation Science*, vol 17, pp549–557.

Jentsch, A. and White, P. (2019) 'A theory of pulse dynamics and disturbance in ecology', *Ecology*, vol 100, no 7, art e02734.

Johnson, E.A. and Miyanishi, K. (2008) 'Testing the assumptions of chronosequences in succession', *Ecology Letters*, vol 11, pp419–431.

Johnstone, J.F. and Chapin, F.S. (2006) 'Effects of soil burn severity on post-fire tree recruitment in boreal forest', *Ecosystems*, vol 9, pp14–31.

Johnstone, J.F., Hollingsworth, T.N., Chapin, F.S. and Mack, M.C. (2010) 'Changes in fire regime break the legacy lock on successional trajectories in Alaskan boreal forest', *Global Change Biology*, vol 16, pp1281–1295.

Johnstone, J.F., Allen, C.D., Franklin, J.F., Frelich, L.E., Harvey, B.J., Higuera, P.E., Mack, M.C., Meentemeyer, R.K., Metz, M.R., Perry, G.L., Schoennagel, T. and Turner, M.G. (2016) 'Changing disturbance regimes, ecological memory, and forest resilience', *Frontiers in Ecology and the Environment*, vol 14, pp369–378.

Jolly, W.M., Cochrane, M.A., Freeborn, P.H., Holden, Z.A., Brown, T.J., Williamson, G.J. and Bowman, D.M. (2015) 'Climate-induced variations in global wildfire danger from 1979 to 2013', *Nature Communications*, vol 6, art 7537.

Kane, J.M., Varner, J.M., Metz, M.R. and van Mantgem, P.J. (2017) 'Characterizing interactions between fire and other disturbances and their impacts on tree mortality in western U.S. Forests', *Forest Ecology and Management*, vol 405, pp188–199.

Katz, R.W., Brush, G.S. and Parlange, M.B. (2005) 'Statistics of extremes: modeling ecological disturbances', *Ecology*, vol 86, pp1124–1134.

Keane, R.E., Parsons, R.A. and Hessburg, P.F. (2002) 'Estimating historical range and variation of landscape patch dynamics: limitations of the simulation approach', *Ecological Modelling*, vol 151, pp29–49.

Keane, R.E., Hessburg, P.F., Landres, P.B. and Swanson, F.J. (2009) 'The use of historical range and variability (HRV) in landscape management', *Forest Ecology and Management*, vol 258, pp1025–1037.

Keane, R.E., Mahalovich, M.F., Bollenbacher, B.L., Manning, M.E., Loehman, R.A., Jain, T.B., Holsinger, L.M., Larson, A.J. and Webster, M.M. (2018) 'Effects of climate change on forest vegetation in the Northern Rockies Region', in J.E. Halofsky, D.L. Peterson, S.K. Dante-Wood, L. Hoang, J.J. Ho and L.A. Joyce (eds) *Climate Change Vulnerability and Adaptation in the Northern Rocky Mountains [Part 1]*. Gen. Tech. Rep. RMRS-GTR-374. US Department of Agriculture, Forest Service, Rocky Mountain Research Station, Fort Collins, CO.

Kellogg, L.-K.B., McKenzie, D., Peterson, D.L. and Hessl, A.E. (2008) 'Spatial models for inferring topographic controls on historical low-severity fire in the eastern Cascade Range of Washington, USA', *Landscape Ecology*, vol 23, pp227–240.

Kipfmueller, K.F. and Kupfer, J.A. (2005) 'Complexity of successional pathways in subalpine forests of the Selway-Bitterroot Wilderness Area', *Annals of the Association of American Geographers*, vol 95, pp495–510.

Kitzberger, T., Perry, G.L.W., Paritsis, J., Gowda, J.H., Tepley, A.J., Holz, A. and Veblen, T.T. (2016) 'Fire–vegetation feedbacks and alternative states: common mechanisms of temperate forest vulnerability to fire in southern South America and New Zealand', *New Zealand Journal of Botany*, vol 54, pp247–272.

Landres, P.B., Morgan, P. and Swanson, F.J. (1999) 'Overview of the use of natural variability concepts in managing ecological systems', *Ecological Applications*, vol 9, no 4, pp1179–1188.

Leach, M.K. and Givnish, T.J. (1996) 'Ecological determinants of species loss in remnant prairies', *Science*, vol 273, pp1555–1558.

Lindenmayer, D.B., Likens, G.E., Krebs, C.J. and Hobbs, R.J. (2010) 'Improved probability of detection of ecological "surprises"', *Proceedings of the National Academy of Sciences*, vol 107, pp21957–21962.

Littell, J.S., McKenzie, D., Wan, H.Y. and Cushman, S.A. (2018) 'Climate change and future wildfire in the Western United States: an ecological approach to nonstationarity', *Earth's Future*, vol 6, pp1097–1111.

McKenzie, D. and Kennedy, M.C. (2011) 'Scaling laws and complexity in fire regimes', in D. McKenzie, C. Miller and D. Falk (eds) *The Landscape Ecology of Fire. Ecological Studies (Analysis and Synthesis)*. Springer, Dordrecht.

McWethy, D.B., Pauchard, A., García, R.A., Holz, A., González, M.E., Veblen, T.T., Stahl, J. and Currey, B. (2018) 'Landscape drivers of recent fire activity (2001–2017) in south-central Chile', *PLoS ONE*, vol 13, art e0201195.

Meddens, A.J.H., Kolden, C.A. and Lutz, J.A. (2016) 'Detecting unburned areas within wildfire perimeters using Landsat and ancillary data across the northwestern United States', *Remote Sensing of Environment*, vol 186, pp275–285.

Meddens, A.J.H., Kolden, C.A., Lutz, J.A., Smith, A.M.S., Cansler, C.A., Abatzoglou, J.T., Meigs, G.W., Downing, W.M. and Krawchuk, M.A. (2018) 'Fire refugia: what are they, and why do they matter for global change?' *BioScience*, vol 68, no 12, pp944–954.

Meigs, G.W., Campbell, J.L., Zald, H.S., Bailey, J.D., Shaw, D.C. and Kennedy, R.E. (2015) 'Does wildfire likelihood increase following insect outbreaks in conifer forests?', *Ecosphere*, vol 6, pp1–24.

Meigs, G.W., Zald, H.S.J., Campbell, J.L., Keeton, W.S. and Kennedy, R.E. (2016) 'Do insect outbreaks reduce the severity of subsequent forest fires?', *Environmental Research Letters*, vol 11, art 045008.

Metz, M.R., Varner, J.M., Frangioso, K.M., Meentemeyer, R.K. and Rizzo, D.M. (2013) 'Unexpected redwood mortality from synergies between wildfire and an emerging infectious disease', *Ecology*, vol 94, pp2152–2159.

Minckley, T., Shriver, R.K. and Shuman, B. (2012) 'Resilience and regime change in a southern Rocky Mountain ecosystem during the past 1700 years', *Ecological Monographs*, vol 82, pp49–68.

Morgan, P., Aplet, G.H., Haufler, J.B., Humphries, H.C., Moore, M.M. and Wilson, W.D. (1994) 'Historical Range of Variability: a useful tool for evaluating ecosystem change', *Journal of Sustainable Forestry*, vol 2, pp87–111.

Moritz, M.A., Batllori, E., Bradstock, R.A., Gill, A.M., Handmer, J., Hessburg, P.F., Leonard, J., McCaffrey, S., Odion, D.C., Schoennagel, T. and Syphard, A.D. (2014) 'Learning to coexist with wildfire', *Nature*, vol 515, pp58–66.

Newman, E.A., Kennedy, M.C., Falk, D.A. and McKenzie, D. (2019) 'Scaling and complexity in landscape ecology', *Frontiers in Ecology and Evolution*, vol 7, art 293.

O'Mahoney, J., Simes, R., Redhill, D., Heaton, K., Atkinson, C., Hayward, E. and Nguyen, M. (2017) 'At what price? The economic, social and icon value of the Great Barrier Reef', http://146.116.27.35/jspui/bitstream/11017/3205/1/deloitte-au-economics-great-barrier-reef-230617.pdf, accessed 15 Dec 2020. Deloitte Access Economics.

Ortiz, J.-C., Wolff, N.H., Anthony, K.R.N., Devlin, M., Lewis, S. and Mumby. P.J. (2018) 'Impaired recovery of the Great Barrier Reef under cumulative stress', *Science Advances*, vol 4, art eaar6127.

Paine, R.T., Tegner, M.J. and Johnson, E.A. (1998) 'Compounded perturbations yield ecological surprises', *Ecosystems*, vol 1, pp535–545.

Payette, S. (1992) 'Fire as a controlling process in the North American boreal forest', in H.H. Shugart, R. Leemans, and G.B. Bonan (eds) *A Systems Analysis of the Global Boreal Forest*. Cambridge University Press, Cambridge.

Perry, G.L.W. (2002) 'Landscapes, space and equilibrium: shifting viewpoints', *Progress in Physical Geography*, vol 26, pp339–359.

Perry, G.L.W., Wilmshurst, J.M., McGlone, M.S., McWethy, D.B. and Whitlock, C. (2012) 'Explaining fire-driven landscape transformation during the initial burning period of New Zealand's prehistory', *Global Change Biology*, vol 18, pp1609–1621.

Peters, D.P.C., Pielke, R.A., Bestelmeyer, B.T., Allen, C.D., Munson-McGee, S. and Havstad, K.M. (2004) 'Cross-scale interactions, nonlinearities, and forecasting catastrophic events', *Proceedings of the National Academy of Sciences of the USA*, vol 101, pp15130–15135.

Peters, D.P.C., Lugo, A.E., Chapin, F.S., Pickett, S.T.A., Duniway, M., Rocha, A.V., Swanson, F.J., Laney, C. and Jones, J. (2011) 'Cross-system comparisons elucidate disturbance complexities and generalities', *Ecosphere*, vol 2, art 81.

Peters, D.P.C., Burruss, N.D., Rodriguez, L.L.,. McVey, D.S, Elias, E.H., Pelzel-McCluskey, A.M., Derner, J.D., Schrader, T.S., Yao, J. and Pauszek, S.J. (2018) 'An integrated view of complex landscapes: a big data-model integration approach to transdisciplinary science', *BioScience*, vol 68, pp653–669.

Peterson, G.D. (2002) 'Contagious disturbance, ecological memory, and the emergence of landscape pattern', *Ecosystems*, vol 5, pp329–338.

Pickett, S.T.A. and White, P.S. (1985) *The Ecology of Natural Disturbance and Patch Dynamics*. Academic Press, San Diego.

Power, M.E. (1992) 'Top-down and bottom-up forces in food webs: do plants have primacy', *Ecology*, vol 73, pp733–746.

Raffa, K.F., Aukema, B.H., Bentz, B.J., Carroll, A.L., Hicke, J.A., Turner, M.G. and Romme, W.H. (2008) 'Cross-scale drivers of natural disturbances prone to anthropogenic amplification: the dynamics of bark beetle eruptions', *BioScience*, vol 58, pp501–517.

Reilly, M.J., Dunn, C.J., Meigs, G.W., Spies, T.A., Kennedy, R.E., Bailey, J.D. and Briggs, K. (2017) 'Contemporary patterns of fire extent and severity in forests of the Pacific Northwest, USA (1985–2010)', *Ecosphere*, vol 8, art e01695.

Reyer, C.P.O., Brouwers, N., Rammig, A., Brook, B.W., Epila, J., Grant, R.F., Holmgren, M., Langerwisch, F., Leuzinger, S., Lucht, W., Medlyn, B., Pfeifer, M., Steinkamp, J., Vanderwel, M.C., Verbeeck, H. and Villela, D.M. (2015) 'Forest resilience and tipping points at different spatio-temporal scales: approaches and challenges', *Journal of Ecology*, vol 103, pp5–15.

Rodman, K.C., Veblen, T.T., Saraceni, S. and Chapman, T.B. (2019) 'Wildfire activity and land use drove 20th-century changes in forest cover in the Colorado front range', *Ecosphere*, vol 10, art e02594.

Romme, W.H. and Despain, D.G. (1989) 'Historical perspective on the yellowstone fires of 1988', *BioScience*, vol 39, pp695–699.

Romme, W.H., Everham, E.H., Frelich, L.E., Moritz, M.A. and Sparks, R.E. (1998) 'Are large, infrequent disturbances qualitatively different from small, frequent disturbances?', *Ecosystems*, vol 1, pp524–534.

Schoennagel, T., Veblen, T.T., Negron, J.F. and Smith, J.M. (2012) 'Effects of mountain pine beetle on fuels and expected fire behavior in lodgepole pine forests, Colorado, USA', *PLoS ONE*, vol 7, art e30002.

Schoennagel, T., Balch, J.K., Brenkert-Smith, H., Dennison, P.E., Harvey, B.J., Krawchuk, M.A., Mietkiewicz, N., Morgan, P., Moritz, M.A., Rasker, R., Turner, M.G. and Whitlock, C. (2017) 'Adapt to more wildfire in western North American forests as climate changes', *Proceedings of the National Academy of Sciences*, vol 114, no 18, pp4582–4590.

Seidl, R. and Rammer, W. (2017) 'Climate change amplifies the interactions between wind and bark beetle disturbances in forest landscapes', *Landscape Ecology*, vol 32, pp1485–1498.

Seidl, R., Spies, T.A., Peterson, D.L., Stephens, S.L. and Hicke, J.A. (2016) 'Review: searching for resilience: addressing the impacts of changing disturbance regimes on forest ecosystem services', *Journal of Applied Ecology*, vol 53, pp120–129.

Seidl, R., Thom, D., Kautz, M., Martin-Benito, D., Peltoniemi, M., Vacchiano, G., Wild, J., Ascoli, D., Petr, M., Honkaniemi, J., Lexer, M.J., Trotsiuk, V., Mairota, P., Svoboda, M., Fabrika, M., Nagel, T.A. and Reyer, C.P.O. (2017) 'Forest disturbances under climate change', *Nature Climate Change*, vol 7, pp395–402.

Simard, M., Romme, W.H., Griffin, J.M. and Turner, M.G. (2011) 'Do mountain pine beetle outbreaks change the probability of active crown fire in lodgepole pine forests?', *Ecological Monographs*, vol 81, pp3–24.

Simler, A.B., Metz, M.R., Frangioso, K.M., Meentemeyer, R.K. and Rizzo, D.M. (2018) 'Novel disturbance interactions between fire and an emerging disease impact survival and growth of resprouting trees', *Ecology*, vol 99, pp2217–2229.

Sousa, W.P. (1984) 'The role of disturbance in natural communities', *Annual Review of Ecology and Systematics*, vol 1, pp353–391.

Sprugel, D.G. (1991) 'Disturbance, equilibrium, and environmental variability: what is 'Natural' vegetation in a changing environment?', *Biological Conservation*, vol 58, pp1–18.

Sprugel, D.G. and Bormann, F.H. (1981) 'Natural disturbance and the steady state in high-altitude balsam fir forests', *Science*, vol 211, no 4480, pp390–393.

Thompson, J.R., Duncan, S.L. and Johnson, K.N. (2009) 'Is there potential for the historical range of variability to guide conservation given the social range of variability?', *Ecology and Society*, vol 14, no 1, art 18.

Turner, M.G. (2010) 'Disturbance and landscape dynamics in a changing world', *Ecology*, vol 91, pp2833–2849.

Turner, M.G. and Dale, V.H. (1998) 'Comparing large, infrequent disturbances: what have we learned?', *Ecosystems*, vol 1, pp493–496.

Turner, M.G. and Romme, W.H. (1994) 'Landscape dynamics in crown fire ecosystems', *Landscape Ecology*, vol 9, pp59–77.

Turner, M.G., Romme, W.H., Gardner, R.H., O'Neill, R.V. and Kratz, T.K. (1993) 'A revised concept of landscape equilibrium: disturbance and stability on scaled landscapes', *Landscape Ecology*, vol 8, pp213–227.

Turner, M.G., Baker, W.L., Peterson, C.J. and Peet, R.K. (1998) 'Factors influencing succession: lessons from large, infrequent natural disturbances', *Ecosystems*, vol 1, pp511–523.

Turner, M.G., Donato, D.C. and Romme, W.H. (2013) 'Consequences of spatial heterogeneity for ecosystem services in changing forest landscapes: priorities for future research', *Landscape Ecology*, vol 28, pp1081–1097.

Turner, M.G., Braziunas, K.H., Hansen, W.D. and Harvey, B.J. (2019) 'Short-interval severe fire erodes the resilience of subalpine lodgepole pine forests', *Proceedings of the National Academy of Sciences*, vol 116, no 23, pp11319–11328.

Watt, A.S. (1947) 'Pattern and process in the plant community', *Journal of Ecology*, vol 35, no 1/2, pp1–22.

Weed, A.S., Ayres, M.P. and Hicke, J.A. (2013) 'Consequences of climate change for biotic disturbances in North American forests', *Ecological Monographs*, vol 83, pp441–470.

Westerling, A.L. (2016) 'Increasing western US forest wildfire activity: sensitivity to changes in the timing of spring', *Philosophical Transactions of the Royal Society B: Biological Sciences*, vol 371, no 1696, art 20150178.

Westerling, A.L., Turner, M.G., Smithwick, E.A.H., Romme, W.H. and Ryan, M.G. (2011) 'Continued warming could transform greater yellowstone fire regimes by mid-21st century', *Proceedings of the National Academy of Sciences*, vol 108, pp13165–13170.

White, P.S. and Pickett, S.T.A. (1985) 'Natural disturbance and patch dynamics: an introduction', in S.T.A Pickett and P.S. White (eds) *The Ecology of Natural Disturbance and Patch Dynamics*. Academic Press, San Diego.

Wieslander, A.E. (1935) 'A vegetation type map of California', *Madroño*, vol 3, pp140–144.

9

Impacts of climate changes and amplified natural disturbance on global ecosystems

Rachel A. Loehman, Megan M. Friggens, Rosemary L. Sherriff, Alisa R. Keyser, and Karin L. Riley

Background

Temperature, precipitation, and solar radiation interact to determine biotic composition, distribution, and productivity at continental scales (Churkina and Running 1998, Seddon et al. 2016). Water availability is estimated to be the dominant control on vegetation growth over 40% of the global vegetated surface, whereas temperature limits growth over 33% and radiation over 27% of the global vegetated surface (Nemani et al. 2003). Climatic drivers of Earth's biomes vary geographically: water limitations (e.g., water use deficits) are the primary control on plant growth in arid and semi-arid climates, but where water is available (e.g., high elevation forests, high northern latitudes), temperature (e.g., growing season) is the more important limiting factor (Churkina and Running 1998).

Climate-driven constraints on plant and animal species distributions, phenology, and range dynamics have been extensively documented (Bumpus 1899, Grinnell 1917, Parmesan 2006, Puppi 2007). However, the role of natural climate variability in shaping environments has been largely overwhelmed by contemporary anthropogenic drivers of global change (e.g., atmospheric concentrations of carbon dioxide and other greenhouse gases) and associated changes in global climate that are unprecedented over millennial time scales (Cook et al. 2016). Anthropogenic climate changes and climate change impacts are a leading and widespread cause of the emergence of novel species assemblages, tree species range shifts and local extinctions, reduced biodiversity, loss of ecosystem resilience, and altered disturbance regimes of a magnitude not previously observed (Root et al. 2003, Chen et al. 2011, Abatzoglou and Williams 2016, Franklin et al. 2016, Stevens-Rumann et al. 2018). Thus, anthropogenic climate changes can be viewed as climate 'disturbances' that cause drastic changes in ecosystem structure and function (Perera et al. 2015) (Figure 9.1).

Natural disturbances, acting in tandem with and often exacerbated by climate changes, can trigger abrupt and persistent changes in ecosystems (Ratajczak et al. 2018) (Figure 9.1). Disturbances can be discrete, short-duration events (i.e., 'pulse' disturbances such as wildfires or hurricanes) or can exert persistent, cumulative stresses on an ecosystem (i.e., 'press' disturbances

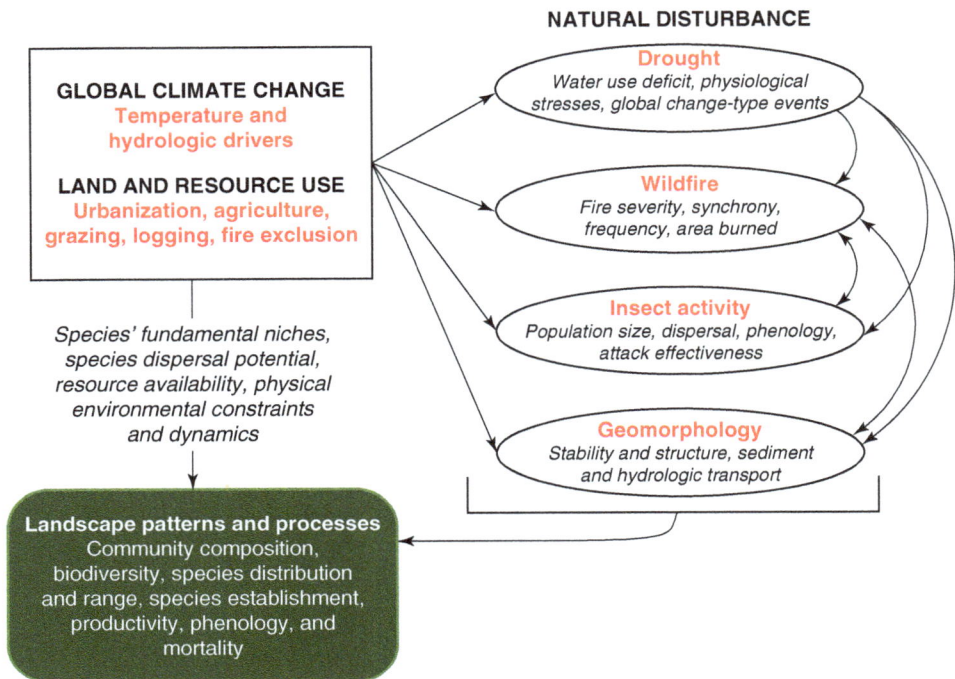

Figure 9.1 Climate (temperature and hydrological drivers) and human land and resource use shape landscape patterns and processes at broad scales (e.g., distributions of plant and animal species, biological diversity, and species life history). At finer spatial scales, natural disturbances such as wildfires, geomorphologic events, and insect activity respond to and interact with climate drivers to create and maintain biological diversity and landscape heterogeneity and initiate ecosystem renewal or reorganization; however, anthropogenic climate changes, unprecedented over millennial time scales, are the cause of substantial and often unanticipated landscape change(s). These climate perturbations influence the timing, frequency, extent, magnitude, and impacts of other disturbances and amplify disturbance interactions, often resulting in profound shifts in landscape patterns and processes.

such as increasing ocean or land surface temperature or decreased river flows) (Bender et al. 1984, Perera et al. 2015). Natural disturbances are a key, inherent element of the ecological environment and serve to maintain biological diversity and landscape heterogeneity and initiate ecosystem renewal or reorganization (Dornelas 2010, Turner 2010). However, the severity, frequency, and extent of many pulse-type disturbances have increased substantially in recent decades, with profound impacts on ecosystem functioning and resilience (Seidl et al. 2017). Large impacts to ecological systems can occur from a single pulse-type disturbance (e.g., a summer heatwave), from several disturbances acting independently (e.g., grazing and wildfire), or from the interactions of multiple, linked disturbances (e.g., increase in wildfire risk that occurs in forest stands with high levels of insect-caused tree mortality) (Hicke et al. 2012, Loehman et al. 2017, Harris et al. 2018) (Figure 9.2). Pulse-type disturbances that are superimposed on the underlying press of climate change amplify disturbance impacts on the distributions of populations and species (Harris et al. 2018).

Figure 9.2   Climate changes and other disturbances can cause highly visible, rapidly occurring, and/or persistent alteration of landscapes. (a) Uncharacteristically frequent fires favor increases in shrub cover, Wrangell-St. Elias National Park, Alaska. (b) Spruce bark beetle–caused mortality shifts forests from conifer-dominated to deciduous-dominated composition, Alaska Peninsula, Alaska. (c) Warming temperatures cause retrogressive thaw slumps along the Nigu River, Alaska. (d) Debris flow following wildfire transports sediments, changes hydrologic pathways, and alters landcover, Valles Caldera National Preserve, New Mexico. (From Rachel Loehman, U.S. Geological Survey (a), Amy Miller, National Park Service (b), Carson Baughman, U.S. Geological Survey (c), and Ana Steffen, National Park Service (d).)

Temperature and hydrologic disturbance

Each of the last three decades has been successively warmer at the Earth's surface than any preceding decade since 1850, and the planet has warmed by about 0.89 °C over the period from 1901 to 2012 (globally averaged land and ocean surface linear warming trend) (IPCC 2014), including higher minimum and maximum daytime and nighttime temperatures across all seasons (Alexander et al. 2006). The Intergovernmental Panel on Climate Change (IPCC) has stated unequivocally that human influence has been the dominant cause of the observed warming since the mid-20th century (Stocker et al. 2014), triggered by greenhouse gas concentrations that are the highest in history; further, observed changes in climate are unprecedented over decades to millennia, a consensus position shared by an overwhelming majority of climate scientists (Cook et al. 2016). Warming of the climate system has intensified precipitation patterns—dry (arid) systems are becoming drier, and wet (mesic) systems are becoming wetter (Dore 2005). This pattern is especially evident in the Mediterranean climate zone (drier) and in high latitudes (wetter) (Trenberth 2011, Knapp et al. 2015). Regional variability in precipitation has also increased, such that extreme wet years are distinguished by multiple extreme daily events and extreme dry years by increased time between precipitation events (Knapp et al. 2015). Earlier spring snowmelt and expansion of the rain/snow transition zone have resulted in increased runoff and decreased summer soil moisture in mountain ecosystems (Knowles et al. 2006, Kapnick and Hall 2012).

Some of the most pronounced disturbance-driven landscape changes are occurring at the boundaries between biomes and in high-latitude regions (Chapin et al. 1995, Serreze et al. 2000, Hinzman et al. 2005, IPCC 2014, National Climate Assessment [NCA] 2018). Visible, climate-driven changes in biomes include shifts in the latitude and elevation of the boreal forest–tundra ecotone in Alaska (Beck et al. 2011, Miller et al. 2017) and Siberia (Shiyatov et al. 2005); an elevational shift of the subalpine forest–tundra boundary in the Sierra Nevada of California (Millar et al. 2004); replacement of high-mountain grassland communities by formerly low-elevation shrubs in the Spanish Central Range in the Iberian Peninsula (Sanz-Elorza et al. 2003); and upward migration of vascular plants in the Rhaetian Alps of northern Italy (Parolo and Rossi 2008). High latitudes are warming twice as fast as the global average (0.6 °C per decade over the last 30 years; IPCC 2014). In continental high-latitude environments, warming temperatures have caused widespread degradation of persistently frozen ground (permafrost) and concomitant increased riverbank erosion, wetland dynamism, and changes in plant community structure (Rowland et al. 2010); shrub expansion and increased shrub growth in tundra areas (Sturm et al. 2001, Tape et al. 2006, Elmendorf et al. 2012); and increased landscape flammability (Higuera et al. 2011, Barrett et al. 2016). In coastal regions, land–sea ecotones are affected by land-based changes in patterns of precipitation and temperature and permafrost degradation, as well as oceanic influences such as altered ocean currents, rising sea levels, increasing sea surface temperatures, pronounced declines in sea ice extent, and incidence of storm-driven tides and saltwater intrusion (Jorgenson and Ely 2001, Post et al. 2009, Vermaire et al. 2013, Terenzi et al. 2014, Lantz et al. 2015, Passeri et al. 2015). The combined effects of inundation, salinization, and sedimentation from storm-driven tides, often reaching far inland, dramatically affect plant communities and alter ecosystem primary productivity. Changes in hydrologic regimes, such as decreased winter snowpack and increasing spring/summer drought, also significantly impact plant communities (Wipf et al. 2006, Hupp et al. 2013).

Earth's forests—including alpine, boreal, temperate, and tropical biomes—are highly sensitive to the impacts of global change (Malhi et al. 2004, Allen et al. 2015, Gauthier et al. 2015, Millar et al. 2015). Globally, major patterns of tree mortality and reduced forest productivity are associated with water limitations and warming temperatures (Allen et al. 2010, 2015, Zhao and Running 2010). Many studies show that warming alone is sufficient to cause changes in tree growth (Williams et al. 2010), biodiversity (Currie 2001), plant and animal species richness (Hansen et al. 2001, Garfin and Lenart 2007), and plant species distributions (Notaro et al. 2012), particularly in arid habitats. For tree species, increased temperatures reduce cold hardiness, influence the timing and synchronicity of budburst, and affect fruit and seed yields (Guak et al. 1998, Aber et al. 2001). Drought in combination with warming temperatures, sometimes called 'global-change-type drought,' can radically alter ecosystems via increased physiological stresses (Allen et al. 2015). Atmospheric warming and plant evapotranspiration increase in tandem, which amplifies water stress during drought, influences hydrological dynamics that drive water availability and increases ecosystem vulnerability to pests, pathogens, and severe wildfire (Breshears et al. 2005, Williams et al. 2013, Allen et al. 2015).

Mountain ecosystems, and particularly their treeline dynamics, have been identified as potential sentinels of climate changes for the past several decades, including observations of tree establishment at the leading (upper) edges of alpine biomes and extinctions at the trailing (lower) margins (Malanson et al. 2019). A global meta-analysis of treeline responses to climate warming found that treelines advanced in 52% of sites (particularly growth-limited sites with strong winter warming) and receded in 1% of sites (Harsch et al. 2009). European Alps forests have expanded and moved upward in elevation since the mid-20th century (Leonelli et al. 2011, Carlson et al. 2014). In Spain, stands of cold temperate beech (*Fagus sylvatica*) have shifted upward in elevation since the 1940s and were replaced at lower elevations with holm oak (*Quercus ilex*) forests (Peñuelas and Boada 2003). Treelines have also advanced in high-altitude forests in Ethiopia, the Himalaya of central Nepal, and the Changbai Mountains in China (Gaire et al. 2014, Jacob et al. 2015, Du et al. 2018). The dynamics of treelines are complex and influenced by fine-scale modulators such as soil physical and chemical properties, species-specific traits, disturbances, and human land use activities (Holtmeier and Broll 2005, Leonelli et al. 2010, HilleRisLambers et al. 2013). Thus, treeline expansion is not solely a factor of warming—movement can be influenced by fine-scale precipitation patterns and topographic controls on moisture availability (Elliott 2012, Elliott and Cowell 2015). For example, in the Putorana Mountains in northern Siberia, the upslope expansion of Siberian larch (*Larix sibirica* Ledeb.) is coincident with increasing winter precipitation (Kirdyanov et al. 2012). In the Sierra Nevada of California, United States, Millar et al. (2004) found that subalpine conifer growth and expansion were tightly coupled to both temperature and decadal climate variability, which modulated the precipitation regime, and snowfield and meadow tree invasions were correlated with minimum temperature and multi-year variability in available moisture, which modulated species responses to increasing temperature. In drier mountain ecosystems in the western Great Basin, United States, moisture availability and temperature control the treeline expansion of limber pine (*Pinus flexilis*) and bristlecone pine (*Pinus longaeva* D.K. Bailey) (Millar et al. 2015).

Non-forest plant species' ranges are also largely determined by climate; for example, in cold-limited biomes, warming temperatures may result in poleward and upward movement of species where geographic barriers (e.g., oceans, urban areas), competitive interactions, disturbances, or edaphic conditions do not limit dispersal (Walther et al. 2002, McKenney et al. 2007). In arid environments, rising temperatures are linked to increased evapotranspiration and the likelihood of water balance deficits, limiting plant growth and favoring drought-tolerant species (Raymond et al. 2014). Shrublands in arid environments can expand in response to increased annual and winter temperatures where warming is coincident with increased winter precipitation (Morgan et al. 2007, Notaro et al. 2012), whereas increased summer precipitation has been observed to favor forbs and grasses (Munson et al. 2013).

Extreme events

Extreme events are pulse-type climate disturbances that historically have been non-anthropogenic in origin but are becoming more frequent and severe due to anthropogenic climate changes. The occurrence and frequency of extreme, high-impact weather events (such as heatwaves, severe droughts, storms, hurricanes, and tornadoes) alter the short- and long-term stability of ecosystem structure, composition, and function (Frank et al. 2015, National Climate Assessment [NCA] 2018). Extreme events are important mechanisms of landscape change on most continents because they influence the movement of materials and the flow of energy through the environment (Peters et al. 2011, National Climate Assessment [NCA] 2018). Ecosystem impacts of extreme events can be long lasting or irreversible, especially when stabilizing ecological feedbacks such as predation, competition, and facilitation are altered or when repeated extreme events occur (Harris et al. 2018).

Globally, heatwaves (days- to weeks-long periods of abnormally hot weather) have generally become more frequent in recent decades, and regional heatwaves are expected as temperatures continue to warm (Perkins-Kirkpatrick and Gibson 2017). An increased frequency of extreme heat stress events has been reported in various parts of the world (Wang et al. 2008); in the southwestern United States and portions of Europe, long-term tree-ring data suggest that drought conditions during the late 20th and early 21st centuries represent the driest conditions for hundreds of years (Büntgen et al. 2010, Woodhouse et al. 2010). Temperature and water stress associated with heatwaves can influence plant growth across elevational gradients; for example, a summer heatwave in the Swiss Alps in 2003 increased plant water stress and decreased growth at lower elevations but enhanced growth at higher elevations (Jolly et al. 2005). The effects of extreme heating can be time-dependent: when arctic willow (*Salix arctica*), a small creeping shrub, was experimentally exposed to two consecutive 10-day heatwaves, growth initially increased, but subsequent growth was negatively impacted, possibly because of reduced cold weather tolerance (Marchand et al. 2005, 2006).

Hurricane intensity, frequency, and duration have increased since the early 1980s (IPCC 2014, National Climate Assessment [NCA] 2018). Hurricanes impact the landscape through the actions of rain and wind, and alter the physical structure, hydrology, and nutrient cycling of ecosystems. Short-term physical impacts include defoliation, branch breakage, blowdown, and altered microtopography; hydrologic impacts include an influx of water and nutrients and increases in runoff due to decreased canopy interception (Greening et al. 2006, Lugo 2008). Storm surges associated with hurricane events result in a pulse of marine salts, but most nutrient changes are from biomass moving from live to dead pools (McDowell 2011, Dahal et al. 2014). North Atlantic hurricanes that occurred from 1900 to 2011 resulted in a live biomass loss of 18.2 TgC/yr in forest ecosystems; the largest single-event loss was 59.5 TgC in 1969, but most single events resulted in ≤ 5 TgC loss (Dahal et al. 2014). Longer-term impacts can include rapid or slowed tree mortality, changes in successional direction, and increased species turnover and age class diversity, all of which impact long-term hydrologic and nutrient cycle recovery (Plotkin et al. 2013, Dahal et al. 2014, Tanner et al. 2014). Long-term studies of hurricane impacts indicate that even with extensive loss of live biomass, forest productivity can recover relatively quickly after storm disturbance when some live trees remain. For example, in Jamaica, the growth rates of trees that survived a hurricane increased over a 21 year period, and within 11 years the growth of undamaged trees was equal to the growth of undamaged trees (Tanner et al. 2014). In a simulated hurricane disturbance in New England (United States) where 80% of the canopy was damaged, stand productivity rebounded within 6 years, but basal area lagged behind control stands after two decades (Plotkin et al. 2013). Hurricane impacts on ecosystems depend on storm strength/intensity and recovery time between storms (Lugo 2008).

Geomorphic disturbance

Geomorphic disturbances can be pulse- or press-type disturbances, depending on the time scale (short-term or long-term) of both the disturbance and its effects (Glasby and Underwood 1996). They affect the shape and character of the Earth's surface, including its topography, surface, and near-surface features via movement of sediments, nutrients, and water; alteration of substrates, microenvironments, and microtopography; changes in resource availability; biomass losses; and mediation of other disturbances such as wildfires and disease spread (Parker and Bendix 1996, Virtanen et al. 2010). Some geomorphic disturbances (e.g., earthquakes, volcanoes) are independent of climate-related factors, whereas others, such as debris flows, floods and storm erosion, and permafrost thaw, can be strongly influenced by climate perturbations (Gariano and Guzzetti 2016). For example, shifting storm tracks that deliver uncharacteristic amounts of rainfall can saturate soils, leading to greater soil mass and increased potential for landslides, or increased overland flow and movement of soil and rock particles downslope (Crozier 2010, Allen 2014). Warming temperatures and changes in storm frequency and magnitude can destabilize soils and slopes via desertification or changes in vegetation (Benda and Dunne 1997, Cannon and DeGraff 2009), and glacial retreat can expose new sediments that are vulnerable to landslides (Huggel et al. 2012).

Permafrost—natural rock or soil material that remains below 0 °C for at least two consecutive years—is highly susceptible to warming-induced thaw (Serreze et al. 2000, Camill 2005, Schuur et al. 2008). Permafrost is estimated to cover 25% of the land surface in the Arctic and boreal regions of the northern hemisphere and also occurs in mountainous regions of the southern hemisphere (Zhang et al. 1999). Permafrost is a fundamental attribute of high-latitude landscapes, where it strongly influences landscape hydrology, ground surface patterns and stability, vegetation distribution, and land surface flammability (Rowland et al. 2010). As permafrost thaws, the overlying ground surface collapses, forming thermokarst features, including mounds, pits, troughs, and depressions that may or may not be filled with water. Permafrost thaw can result in partial or total conversion of vegetation communities; altered microtopography, hydrology, and ecosystem function; major release of stored carbon and methane (greenhouse gases); and feedbacks to further thaw and ground surface instability (Serreze et al. 2000, Rowland et al. 2010, Mack et al. 2011, Schuur et al. 2015). For example, warming temperatures and resulting loss of permafrost structures have caused shifts from shrub-dominated to wet graminoid–dominated, and from tree- and shrub-dominated to herb and sedge-dominated, vegetation communities in subarctic regions of Sweden (Christensen et al. 2004) and continental Alaska (Jorgenson et al. 2001), respectively. Where wildfires occur in permafrost, they can trigger cascading landscape changes that intensify for years after a fire event; combustion of insulating surface organic layers that overtop permafrost increases soil heat flux, decreases surface albedo, can accelerate further permafrost thaw, and may lead to changes in vegetation cover and community composition (Yoshikawa et al. 2002, Brown et al. 2015b, Helbig et al. 2016, Liu et al. 2019).

Disturbance interactions

The significance of interacting disturbances as agents of profound ecological change is globally evident. Highly visible, rapidly occurring, and persistent changes in landscape composition and structure can occur in systems in which both biological and physical elements are simultaneously or serially perturbed; e.g., wildfires of uncharacteristic severity that occur in conjunction with climate warming may catalyze major shifts in forest composition (Johnstone et al. 2010b, Brown et al. 2015a, Loehman et al. 2018, Stevens-Rumann et al. 2018). Here, we may see nonlinear feedbacks that produce novel landscape responses (Lauenroth et al. 1993, Temperli et al. 2013, Buma 2015, Coop et al. 2016) and profound shifts in successional pathways, species composition, and landscape carbon (Goetz et al. 2007, Johnstone et al. 2010a, b, Brown and Johnstone 2012, Thom et al. 2017). Natural disturbances whose spatial pattern, occurrence, or impacts have been substantially amplified by interactions with changing climate and other disturbances include wildfire, windthrow, debris flows, and insect outbreaks.

Climate changes and wildfire

Earth's ecosystems are highly influenced by fire disturbance (Bond et al. 2005, Krawchuk et al. 2009). Globally, anthropogenic climate change has emerged as a significant driver of increased fire, independent of natural climate variability (Abatzoglou et al. 2018). Observed 20th- and 21st-century anthropogenic climate changes of warming temperatures and earlier onset of snowmelt result in drier live and dead fuels, making large portions of the landscape flammable for longer periods of time (Westerling et al. 2006, Miller et al. 2011). Prolonged droughts are associated with regional fire years (Westerling and Swetnam 2003, Heyerdahl et al. 2008, Littell et al. 2016) and increased large fire activity and burned area (Abatzoglou and Kolden 2013, Riley et al. 2013). There is also evidence of increased fire activity associated with climate-driven changes in lightning activity; for example, in Tasmania, the proportion of lightning strikes that occurred in dry conditions increased ignition efficiency (Styger et al. 2018), and in Yosemite National Park, California, United States, decreased spring snowpack exponentially increased the number of lightning-ignited fires (Lutz et al. 2009).

Climate impacts on fire activity have already increased the area burned in Arctic tundra (Higuera et al. 2008, Chipman et al. 2015) and closed-canopy forests around the world (Andela et al. 2017) and are projected to continue with 21st-century climate change (Moritz et al. 2012, Flannigan et al. 2013, Littell et al. 2018). Climate changes, fire, and vegetation communities can interact in a complex of synergistic and antagonistic effects that amplify the negative impacts of natural disturbance on ecosystems (Figure 9.1). For example, in ecosystems with sufficient fuels to carry fire, warmer, drier climates are expected to increase fuel aridity, flammability, and fire activity (Gergel et al. 2017, McKenzie and Littell 2017). The resulting larger extent of wildfire area burned and higher severity of wildfires can increase the proportion of the landscape in the early stages of post-fire recovery (Falk et al. 2019). If the post-fire bioclimatic environment is unfavorable for seedling establishment (e.g., with severe drought), forests may transition to alternative states such as shrub- or grasslands (Guiterman et al. 2018, Davis et al. 2019). Thus, interactions of altered climate and changing fire regimes impact vegetation regeneration, community structure and composition, and the amount, type, and flammability of fuels (Abatzoglou and Williams 2016, Coop et al. 2016).

Windthrow and wildfire

Tree blowdown from storm events (windthrow) results in a significant change in forest fuel quantity and continuity. However, the interaction of wind disturbance followed by fire can be both amplifying and buffering. The risk of fire occurrence and severity can be increased after wind disturbance through microclimate drying and warming in gaps, increased fuel loading, and reduction in stem diameter, as larger trees are more likely to fall (Liu et al. 2008, Plotkin et al. 2013, Cannon et al. 2017). These severe fire events can delay regeneration and recovery through removal of adults, mortality of serotinous cones, and change to overstory composition (Liu et al. 2008, Buma and Wessman 2011, Tanner et al. 2014, Cannon et al. 2017). However, wind damage can also buffer against fire impacts through reduction in quantity of fine fuels or changes in fuel continuity that disrupt overstory fire spread. For example, seedlings growing on tip-up mounds can sometimes be protected from fire effects (Cannon et al. 2017, 2019). Fire in wind-damaged plots can result in increased basal sprouting, which can accelerate recovery and buffer damage from fire (Plotkin et al. 2013, Tanner et al. 2014, Cannon 2015,). The balance of amplifying and buffering interactions between wind and fire depends on forest type and the intensity and size of each disturbance (Kane et al. 2017, Cannon et al. 2019).

Wildfire and debris flows

The frequency of mudslides and debris flows increases following moderate- and high-severity fire because surface vegetation has been consumed, exposing sediments to transport by overland flow generated by rainfall and reducing the strength of the soil as the roots of dead plants decompose (Wondzell and King 2003, Hyde et al. 2007, Gartner et al. 2008, Cannon et al. 2010, Parise and Cannon 2012). Landslides increase landscape complexity in unburned areas by creating areas denuded of vegetation, which are then colonized by early seral plant communities. In burned areas, they contribute to the complexity of stream channels by transporting boulders and the trunks of dead trees into channels, where they may detrimentally impact fish populations in the short term but over the long term provide coarse woody debris for fish habitat (Lamberti et al. 1991, Hoffman and Gabet 2007).

Climate changes and insect-caused mortality

Insects are sensitive to short-term temperature fluctuations and longer-term warming trends (Paaijmans et al. 2013, Thackeray et al. 2016). Temperature governs insect survival, development, emergence, and dispersal, as well as the number of generations per year (Bentz et al. 2010). Since the late 1970s, irruptive outbreaks of forest insects have affected millions of hectares of trees in North America and Europe, with transformative consequences for ecological systems (Bentz et al. 2010, Seidl et al. 2014). The scale and severity of many recent outbreaks are widely believed to be unprecedented (Morris et al. 2017). For example, white spruce forests on the Kenai Peninsula in Alaska have undergone intensive attacks by spruce beetle (*Dendroctonus rufipennis*), resulting in widespread tree mortality across more than 2.4 million ha of forest—an expanse far exceeding outbreaks from any other time in the past 250 years (Berg et al. 2006, Sherriff et al. 2011). The resulting landscape changes can strongly impact forests and disrupt ecosystem services (Muller and Job 2009). Time since disturbance, type of disturbance (including fire, logging, land use change, wind, and/or drought; e.g., Bebi et al. 2003, Buma and Wessman 2011, 2012, Cannon et al. 2010, 2019, Morris et al. 2015), ecosystem structure and composition (Wermelinger 2004), and climate extremes (i.e., Berg et al. 2006, Sherriff et al. 2011) are important in predicting susceptibility to insect infestations and resulting landscape changes. Ongoing climate warming increasingly facilitates the potential for novel host and disturbance interactions at high latitudes and upper elevation forests (e.g., Bentz et al. 2010).

Anthropogenic impacts

Human impacts on the environment have triggered local to global-scale species range shifts, changes in phenology, and species extinctions, caused a general decline in diversity, and triggered functional shifts as sets of species with particular traits are replaced by other sets with different traits (Loreau et al. 2001, Thuiller et al. 2008). These impacts occur as the result of land use and resource consumption in addition to anthropogenic increases in greenhouse gas concentrations (Turner et al. 1991) (Figure 9.1). As summarized in Alberti et al. (2003) and elsewhere, humans use approximately 40% of global net primary production (Vitousek et al. 1997); globally, estimates of deforestation and other land use changes account for 53–58% of the difference between current and potential global vegetation biomass (Erb et al. 2018); some of the world's largest rivers (e.g., the Colorado, Yellow, and Nile Rivers) show a complete or nearly complete loss of perennial discharge to the ocean (Vörösmarty et al. 2004); at least half of the world's forests have disappeared as a result of human activity since AD 1700 (Harrison and Pearce 2000); and human activities have altered fire characteristics to the point where a new and unique 'human pyrome' (global expression of a fire regime) has been identified (Archibald et al. 2013).

Human activities can add complexity to landscape dynamics by altering relationships between climate and disturbance (Krawchuk et al. 2009, Bürgi et al. 2017). For example, plant water content in tropical biomes can be affected by agricultural activities (increased water content) and large-scale deforestation (decreased water content) (Liu et al. 2013). Temperate and tropical forests have been historically impacted by widespread low-intensity shifting cultivation and more recently impacted by intensive permanent agriculture, industrial logging, anthropogenic fire, and land fragmentation (Lewis et al. 2015). Likewise, urban expansion has substantial impacts on habitats, biogeochemistry, hydrology, land cover, and surface energy balances (d'Amour et al. 2017). Land use changes can feed back to climate and weather processes by affecting local albedo, available energy, and latent heat fluxes that ultimately alter local water and nutrient cycles and vegetation response to disturbance (Pielke et al. 2002, Brovkin et al. 2013); for example, on the North American Great Plains, natural grasslands have been converted to agricultural farmlands, resulting in changes to local temperatures and the near-surface hydrological cycle (Mahmood and Hubbard 2002, Mahmood et al. 2004).

In the absence of human influences, global wildfire activity is dependent primarily on climate, which acts as a top-down control on vegetation types and abundance and dictates the frequency of extreme weather episodes conducive to fire (Marlon et al. 2013). Over the past several thousand years, humans have played a fundamental role in shaping fire regimes worldwide, at times decoupling the fire–climate signal (Archibald et al. 2013, Scott et al. 2016). Recent studies of traditional burning practices from Australia (Bowman et al. 2011), New Zealand (McWethy et al. 2010), Central America (Avnery et al. 2011), Europe (Vannière et al. 2008), Asia (Mu et al. 2016), and North America (Swetnam et al. 2016) indicate that anthropogenic burning during the Holocene may have had significant impacts on biodiversity, landscape heterogeneity, and forest structure and composition. Humans can influence fire regimes via a variety of mechanisms; first, by altering the amount and distribution of ignition sources (Bowman et al. 2011). For example, excluding prescribed burning, human-ignited fires accounted for 44% of the area burned by wildfires in the contiguous United States from 1992 to 2012 (Balch et al. 2017). Conversely, despite recent increases in global temperature, the area burned has been declining globally as a consequence primarily of reduced burned area in the savannas of Africa, a decline likely driven by lifestyle changes of the human inhabitants (Andela et al. 2017). Second, humans alter vegetation structure, composition, and fuel chemical, physical, and spatial properties through land use (Pausas and Fernandez-Munoz 2012, Chergui et al. 2018), fragmentation of native vegetation (Syphard et al. 2007, Mann et al. 2016, Parisien et al. 2016), timber harvest (Naficy et al. 2010), and introduction of flammable invasive species (Paritsis et al. 2018). Third, active fire suppression has reduced the number and size of fires in some systems (North et al. 2015), contributing to changes in vegetation structure, composition, and fuel properties (Schoennagel et al. 2004) and often resulting in the occurrence of large fires that are difficult and expensive to contain and can include uncharacteristically large proportions of stand-replacing, high-severity burned area (Stephens et al. 2013, Calkin et al. 2015, North et al. 2015, Reilly et al. 2017).

Challenges and uncertainties

Climate changes are widely recognized as the largest threat to biodiversity, species survival, and ecosystem integrity across most biomes (Hulme 2005, Thuiller et al. 2008, Maclean and Wilson 2011), challenging historical interpretations, foundational assumptions, and attribution of ecological and evolutionary change. The rapid rate of contemporary climate change is exceeding the range of natural climate variability and accelerating the rate at which habitats are degraded and species are lost (Hannah et al. 2002, Overpeck et al. 2003, Thomas et al. 2004, Allen et al. 2015). From local to global scales, these changes occur in the context of anthropogenic ecosystem disruptions that include landscape fragmentation and urbanization, pollution, grazing, deforestation, nonnative species invasions, and a new and unique 'human pyrome' or global expression of a fire regime (Millar et al. 2007, Archibald et al. 2013, Seidl et al. 2014, Blackhall et al. 2015). There is a high likelihood that future environments will be very different from current or past environments, but we cannot be certain about the specifics of change (Millar et al. 2007). The extent to which changes in climate will lead to changes in plant and animal communities depends upon the rate, magnitude, and direction of changes in climate drivers. For example, in boreal forests, increasing temperature and decreasing precipitation are expected to reduce the productivity of spruce-fir forests in more southerly latitudes but increase productivity in latitudes north of 59 degrees (D'Orangeville et al. 2016). Importantly, our understanding of climate impacts on ecosystems is still growing and subject to change.

Predictions of climate and disturbance impacts on ecosystems rely on valid climate model projections, sufficiently complex ecosystem and vegetation models, and an understanding of the complexities and intricate processes of natural systems (Littell et al. 2011). Emerging environments that fall outside known historical ranges hinder our ability to develop prescient views of future conditions and change (Currie 2001, McKenney et al. 2007, Williams et al. 2013). Although continental-scale climate changes (e.g., increasing mean temperature) can be modeled with a high degree of accuracy and consistency, climate predictions at regional to local scales (scales that are relevant for land management) are more uncertain (Xie et al. 2015). This is particularly true for precipitation change, which is highly variable spatially in sign and amplitude. Despite variation in model performance, however, current global climate models produce a robust and consistent set of projections indicating that global climate change over the 21st century will probably exceed that observed over the past century, even for scenarios in which global greenhouse gas emissions are reduced by about 90% in 2100 compared with the present (Knutti and Sedláček 2013).

Fundamental research gaps challenge our ability to anticipate the future of Earth's ecosystems. First, we highlight the need to identify critical thresholds of climate or disturbance that induce rapid and persistent transformations of ecological systems (e.g., loss of resilience) (Holling 1973). In many cases, ecological attributes show minimal change until a critical environmental threshold is reached (Qian et al. 2003). For example, in climate-sensitive and fire-prone ecosystems, water limitations and warming temperatures can cause widespread tree mortality and trigger wildfires that are uncharacteristically severe or frequent, capable of abruptly reorganizing vegetation and fuel patterns and setting the stage for future, novel fire regimes (Turner 2010, Allen et al. 2015). Threshold detection is an important aspect of ecological risk assessment and environmental management intended to prevent severe social, economic, and environmental impacts that occur when biophysical thresholds are crossed (Kelly et al. 2015).

Second, research is needed on climate impacts to range-edge or ecotonal systems. Changes in habitat are likely to be most dramatic near ecotones (Allen and Breshears 1998, Kupfer et al. 2005, Joyce et al. 2009) and biome transition zones (e.g., Beck et al. 2011, Miller et al. 2017). Populations that live at the boundary of their current range are generally exposed to higher severity and frequency of extreme climate events than populations in their core ranges (Rehm et al. 2015). Range-edge populations may be better able to adapt to extreme climate and could play an important role in species persistence under future climate change (Rehm et al. 2015). However, range-edge populations also tend to be the least abundant and more at risk of extirpation due to anthropogenic activities (Zardi et al. 2015).

Third, prediction of future conditions is dependent on models' abilities to represent multi-scale interactions, feedbacks, and nonlinear effects in order to realistically represent complex and emergent ecosystem dynamics. For example, drier conditions are likely to reduce plant productivity, but increased CO_2 concentrations can support increased growth, water efficiency, and resistance to diseases (Aber et al. 2001, Sturrock et al. 2011, Notaro et al. 2012). Insect activity is expected to be influenced by changes in temperature—for example, spruce bark beetle development is likely to accelerate with projected increases in temperature, translating to more rapid reproduction and intensity of outbreaks (Lange et al. 2006, Bentz et al. 2010, Jönsson and Bärring 2011). The ranges of insects may also shift; for example, documented expansion of mountain pine beetle has already occurred in North America (Carroll et al. 2003, Raffa et al. 2008). Tree species that may already be more vulnerable due to drought stress could be more heavily impacted by insects, leading to increased tree mortality (Jactel et al. 2012, Anderegg et al. 2015, Trumbore et al. 2015). These complexities challenge our abilities to model ecosystem dynamics and predict future conditions.

At the center of future modeling research is a need for ongoing empirical studies that provide comprehensive calibration data and parameters that reflect emerging environments. Models developed using empirical data representative of historical conditions become less robust under climate change, because species dynamics—for example, seedling establishment rates after wildfire—are different in novel, non-equilibrium environments (Scheller 2018). Temporally deep, spatially explicit databases created from extensive field measurements are needed to quantify input parameters, describe initial conditions, and provide a reference for model testing and validation, especially as models are ported across large geographic areas and to new ecosystems (Cary et al. 2006). In total, expanding these frontiers of knowledge increases our ability to anticipate future shifts in landscape patterns and process and develop ecologically informed strategies for environmental management and conservation.

Disclaimer

Acknowledgments

We thank multiple reviewers (George Perry, School of Environment, University of Auckland; Carol Miller, U.S. Forest Service; Carson Baughman, U.S. Geological Survey), whose comments helped improve and clarify the manuscript.

References

Abatzoglou, J.T. and Kolden C.A. (2013) 'Relationships between climate and macroscale area burned in the western United States', *International Journal of Wildland Fire*, vol 22, pp1003–20.

Abatzoglou, J.T. and Williams, A.P. (2016) 'Impact of anthropogenic climate change on wildfire across western US forests', *Proceedings of the National Academy of Sciences*, vol 113, no 42, pp11770–11775.

Abatzoglou, J.T., Williams, A.P., Boschetti, L., Zubkova, M. and Kolden, C.A. (2018) 'Global patterns of interannual climate–fire relationships', *Global Change Biology*, vol 24, no 11, pp5164–5175.

Aber, J., Neilson, R.P., McNulty, S., Lenihan, J.M., Bachelet, D. and Drapek, R.J. (2001) 'Forest processes and global environmental change: Predicting the effects of individual and multiple stressors: We review the effects of several rapidly changing environmental drivers on ecosystem function, discuss interactions among them, and summarize predicted changes in productivity, carbon storage, and water balance', *BioScience*, vol 51, no 9, pp735–751.

Alberti, M., Marzluff, J.M., Shulenberger, E., Bradley, G., Ryan, C. and Zumbrunnen, C. (2003) 'Integrating humans into ecology: opportunities and challenges for studying urban ecosystems', BioScience, vol 53, no. 12, pp. 1169–1179.

Alexander, L.V., Zhang, X., Peterson, T.C., Caesar, J., Gleason, B., Klein Tank, A.M.G., Haylock, M., Collins, D., Trewin, B., Rahimzadeh, F. and Tagipour, A. (2006) 'Global observed changes in daily climate extremes of temperature and precipitation', Journal of Geophysical Research: Atmospheres, vol 111, art D05109.

Allen, C. (2014) 'Anthropogenic Earth-change: We are on a slippery slope, breaking new ground and it's our fault: A multi-disciplinary review and new unified earth-system hypothesis', Journal of Earth Science and Engineering, vol 4, no 1, pp1–53.

Allen, C.D. and Breshears, D.D. (1998) 'Drought-induced shift of a forest-woodland ecotone: Rapid landscape response to climate variation', Proceedings of the National Academy of Sciences, vol 95, pp14839–14842.

Allen, C.D., Macalady, A.K., Chenchouni, H., Bachelet, D., McDowell, N., Vennetier, M. Kitzberger, T., Rigling, A., Breshears, D.D., Hogg, E.T. and Gonzalez, P. (2010) 'A global overview of drought and heat-induced tree mortality reveals emerging climate change risks for forests', Forest Ecology and Management, vol 259, no 4, pp660–684.

Allen, C.D., Breshears, D.D. and McDowell, N.G. (2015) 'On underestimation of global vulnerability to tree mortality and forest die-off from hotter drought in the Anthropocene', Ecosphere, vol 6, pp1–55.

Andela, N., Morton, D.C., Giglio, L., Chen, Y., Van Der Werf, G.R., Kasibhatla, P.S. et al. (2017) 'A human-driven decline in global burned area', Science, vol 356, pp1356–1362.

Anderegg, W.R.L., Hicke, J.A., Fisher, R.A., Allen, C.D., Aukema, J., Bentz, B.J. et al. (2015) 'Tree mortality from drought, insects, and their interactions in a changing climate', New Phytologist, vol 208, pp674–683.

Archibald, S., Lehmann, C.E., Gómez-Dans, J.L. and Bradstock, R.A. (2013) 'Defining pyromes and global syndromes of fire regimes', Proceedings of the National Academy of Sciences, vol 110, no 16, pp6442–6447.

Avnery, S., Dull, R.A. and Keitt, T.H. (2011) 'Human versus climatic influences on late-Holocene fire regimes in southwestern Nicaragua', Holocene, vol 21, no 4, pp699–706.

Balch, J.K., Bradley, B.A., Abatzoglou, J.T., Nagy, R.C., Fusco, E.J. and Mahood, A.L. (2017) 'Human-started wildfires expand the fire niche across the United States', Proceedings of the National Academy of Sciences, vol 114, no 11, pp2946–2951.

Barrett, K., Loboda, T., McGuire, A.D., Genet, H., Hoy, E. and Kasischke, E. (2016) 'Static and dynamic controls on fire activity at moderate spatial and temporal scales in the Alaskan boreal forest', Ecosphere, vol 7, no 11, art e01572.

Bebi, P., Kulakowski, D. and Veblen, T.T. (2003) 'Interactions between fire and spruce beetles in a subalpine Rocky Mountain forest landscape', Ecology, vol 84, pp362–371.

Beck, P.S., Juday, G.P., Alix, C., Barber, V.A., Winslow, S.E., Sousa, E.E., Heiser, P., Herriges, J.D. and Goetz, S.J. (2011) 'Changes in forest productivity across Alaska consistent with biome shift', Ecology Letters, vol 14, no 4, pp373–379.

Benda, L. and Dunne, T. (1997) 'Stochastic forcing of sediment routing and storage in channel networks', Water Resources Research, vol 33, no 12, pp2865–2880.

Bender, E.A., Case, T.J. and Gilpin, M.E. (1984) 'Perturbation experiments in community ecology: Theory and practice', Ecology, vol 65, no 1, pp1–13.

Bentz, B.J., Regniere, J., Fettig, C.J., Hansen, E.M., Hayes, J.L., Hicke, J.A. et al. (2010) 'Climate change and bark beetles of the western United States and Canada: Direct and indirect effects', Bioscience, vol 60, pp602–13.

Berg, E.E., Henry, J.D., Fastie, C.L., DeVolder, A.D. and Matsuoka, S.M. (2006) 'Spruce beetle outbreaks on the Kenai Peninsula, Alaska, and Kluane National Park and Reserve, Yukon Territory: Relationship to summer temperatures and regional differences in disturbance regimes', Forest Ecology and Management, vol 227, pp219–232.

Blackhall, M., Veblen, T.T. and Raffaele, E. (2015) 'Recent fire and cattle herbivory enhance plant-level fuel flammability in shrublands', Journal of Vegetation Science, vol 26, no 1, pp123–133.

Bond, W.J., Woodward, F.I. and Midgley, G.F. (2005) 'The global distribution of ecosystems in a world without fire', New Phytologist, vol 165, pp525–537.

Bowman, D.M.J.S., Balch, J., Artaxo, P., Bond, W.J., Cochrane, M.A., D'Antonio, C.M., DeFries, R., Johnston, F.H., Keeley, J.E., Krawchuk, M.A., Kull, C.A., Mack, M., Moritz, M.A., Pyne, S., Roos, C.I., Scott, A.C., Sodhi, N.S. and Swetnam, T.W. (2011) 'The human dimension of fire regimes on Earth', Journal of Biogeography, vol 38, pp2223–2236.

Breshears, D.D., Cobb, N.S., Rich, P.M., Price, K.P., Allen, C.D. Balice, R.G., Romme, W.H., Kastens, J.H., Floyd, M.L., Belnap, J., Anderson, J.J., Myers, O.B. and Meyer, C.W. (2005). 'Regional vegetation die-off in response to global-change-type drought', *Proceedings of the National Academy of Sciences*, vol 102, pp15144–15148.

Brovkin, V., Boysen, L., Arora, V.K., Boisier, J.P., Cadule, P., Chini, L., Claussen, M., Friedlingstein, P., Gayler, V., Van Den Hurk, B.J.J.M. and Hurtt, G.C. (2013) 'Effect of anthropogenic land use and land-cover changes on climate and land carbon storage in CMIP5 projections for the twenty-first century', *Journal of Climate*, vol 26, no 18, pp6859–6881.

Brown, C.D. and Johnstone, J.F. (2012) 'Once burned, twice shy: Repeat fires reduce seed availability and alter substrate constraints on *Picea mariana* regeneration', *Forest Ecology and Management*, vol 266, pp34–41.

Brown, C.D., Liu, J., Yan, G. and Johnstone, J.F. (2015a) 'Disentangling legacy effects from environmental filters of postfire assembly of boreal tree assemblages', *Ecology*, vol 96, no 11, pp3023–3032.

Brown, D., Jorgenson, M.T., Douglas, T.A., Romanovsky, V.E., Kielland, K., Hiemstra, C., Euskirchen, E.S. and Ruess, R.W. (2015b) 'Interactive effects of wildfire and climate on permafrost degradation in Alaskan lowland forests', *Journal of Geophysical Research: Biogeosciences*, vol 120, no 8, pp1619–1637.

Buma, B. (2015) 'Disturbance interactions: Characterization, prediction, and the potential for cascading effects', *Ecosphere*, vol 6, no 4, pp1–15.

Buma, B. and Wessman, C.A. (2011) 'Disturbance interactions can impact resilience mechanisms of forests', *Ecosphere*, vol 2, no 5, pp1–3.

Buma, B. and Wessman, C.A. (2012) 'Differential species responses to compounded perturbations and implications for landscape heterogeneity and resilience', *Forest Ecology and Management*, vol 266, pp25–33.

Bumpus, H.C. (1899) 'The elimination of the unfit as illustrated by the introduced sparrow, *Passer domesicus*', in *Biological Lectures Delivered at the Marine Biological Laboratory of Wood's Hole*, 1896–97, pp209–26. Ginn & Co, Boston.

Büntgen, U., Trouet, V., Frank, D., Leuschner, H.H., Friedrichs, D., Luterbacher, J. and Esper, J. (2010) 'Tree-ring indicators of German summer drought over the last millennium', *Quaternary Science Reviews*, vol 29, no 7–8, pp1005–1016.

Bürgi, M., Östlund, L. and Mladenoff, D.J. (2017) 'Legacy effects of human land use: Ecosystems as time-lagged systems', *Ecosystems*, vol 20, no 1, pp94–103.

Calkin, D.E., Thompson, M.P. and Finney, M.A. (2015) 'Negative consequences of positive feedbacks in US wildfire management', *Forest Ecosystems*, vol 2, no 1, art 9.

Camill, P. (2005) 'Permafrost thaw accelerates in boreal peatlands during late-20th century climate warming', *Climatic Change*, vol 68, no 1–2, pp135–152.

Cannon, J.B. (2015) *Crash and Burn: Forest Tornado Damage, its Landscape Pattern, and its Interaction with Fire.* PhD diss., University of Georgia.

Cannon, J.B., Peterson, C.J., O'Brien, J.J. and Brewer, J.S. (2017) 'A review and classification of interactions between forest disturbance from wind and fire', *Forest Ecology and Management*, vol 406, pp381–390.

Cannon, J.B., Henderson, S.K., Bailey, M.H. and Peterson, C.J. (2019) 'Interactions between wind and fire disturbance in forests: Competing and buffering effects', *Forest Ecology and Management*, vol 436, pp117–128.

Cannon, S.H. and DeGraff, J. (2009) 'The increasing wildfire and post-fire debris-flow threat in western USA, and implications for consequences of climate change', in K. Sassa and P. Canuti (eds) *Landslides–Disaster Risk Reduction*. Springer, Heidelberg.

Cannon, S.H., Gartner, J.E., Rupert, M.G., Michael, J.A., Rea, A.H. and Parrett, C. (2010) 'Predicting the probability and volume of post-wildfire debris flows in the intermountain western United States', *GSA Bulletin*, vol 122, pp127–144.

Carlson, B.Z., Georges, D., Rabatel, A., Randin, C.F., Renaud, J., Delestrade, A., Zimmermann, N.E., Choler, P. and Thuiller, W. (2014) 'Accounting for tree line shift, glacier retreat and primary succession in mountain plant distribution models', *Diversity and Distributions*, vol 20, pp1379–1391.

Carroll, A.L., Taylor, S.W., Régnière, J. and Safranyik, L. (2003) 'Effect of climate change on range expansion by the mountain pine beetle in British Columbia', in T.L. Shore, J.E. Brooks, J.E. Stone (eds) Mountain Pine Beetle Symposium: Challenges and Solutions, Oct 30–31, Kelowna, BC. Victoria, BC: Natural Resources Canada, Information Report BC-X-399, pp223–32.

Cary, G.J., Keane, R.E., Gardner, R.H., Lavorel, S., Flannigan, M.D., Davies, I.D., Li, C., Lenihan, J.M., Rupp, T.S. and Mouillot, F. (2006) 'Comparison of the sensitivity of landscape-fire-succession models to variation in terrain, fuel pattern, climate and weather', *Landscape Ecology*, vol 21, no 1, pp121–137.

Chapin, F.S., III, G.R.Shaver, A.E. Giblin, K.G. Nadelhoffer, and J.A. Laundre. (1995) 'Responses of arctic tundra to experimental and observed changes in climate', *Ecology*, vol 76, pp694–711.

Chen, I.C., Hill, J.K., Ohlemüller, R., Roy, D.B. and Thomas, C.D. (2011) 'Rapid range shifts of species associated with high levels of climate warming', *Science*, vol 333, no 6045, pp1024–1026.

Chergui, B., Fahd, S., Santos, X., Pausas, J.G. (2018) 'Socioeconomic factors drive fire-regime variability in the mediterranean basin', *Ecosystems*, vol 21, pp619–628.

Chipman, M.L., Hudspith, V., Higuera, P.E., Duffy, P.A., Kelly, R., Oswald, W.W. and Hu, F.S. (2015) 'Spatiotemporal patterns of tundra fires: Late-quaternary charcoal records from Alaska', *Biogeosciences*, vol 12, pp4017–4027.

Christensen, T.R., Johansson, T., Åkerman, H.J., Mastepanov, M., Malmer, N., Friborg, T., Crill, P. and Svensson, B.H. (2004) 'Thawing sub-arctic permafrost: Effects on vegetation and methane emissions', *Geophysical Research Letters*, vol 31, no 4, art L04501.

Churkina, G. and Running, S.W. (1998) 'Contrasting climatic controls on the estimated productivity of global terrestrial biomes', *Ecosystems*, vol 1, pp206–215.

Cook, J., Oreskes, N., Doran, P.T., Anderegg, W.R., Verheggen, B., Maibach, E.W., Carlton, J.S., Lewandowsky, S., Skuce, A.G., Green, S.A. and Nuccitelli, D. (2016) 'Consensus on consensus: A synthesis of consensus estimates on human-caused global warming', *Environmental Research Letters*, vol 11, no 4, art 048002.

Coop, J.D., Parks, S.A., McClernan, S.R. and Holsinger, L.M. (2016) 'Influences of prior wildfires on vegetation response to subsequent fire in a reburned southwestern landscape', *Ecological Applications*, vol 26, no 2, pp346–354.

Crozier, M.J. (2010) 'Deciphering the effect of climate change on landslide activity: A review', *Geomorphology*, vol 124, pp260–267.

Currie, D.J. (2001) 'Projected effects of climate change on patterns of vertebrate and tree species richness in the conterminous United States', *Ecosystems*, vol 4, pp216–225.

Dahal, D., Liu, S. and Oeding, J. (2014) 'The carbon cycle and hurricanes in the United States between 1900 and 2011', *Scientific Reports*, vol 4, art 5197.

d'Amour, C.B., Reitsma, F., Baiocchi, G., Barthel, S., Guneralp, B., Erb, K., Haberl, H., Creutzig, F. and Seto, K.C. (2017) 'Future urban land expansion and implications for global croplands', *Proceedings of the National Academy of Sciences*, vol 114, no 34, pp8939–8944.

Davis, K.T., Dobrowski, S.Z., Higuera, P.E., Holden, Z.A., Veblen, T.T., Rother, M.T., Parks, S.A., Sala, A. and Maneta, M.P. (2019) 'Wildfires and climate change push low-elevation forests across a critical climate threshold for tree regeneration', *Proceedings of the National Academy of Sciences*, vol 116, no 13, pp6193–6198.

D'Orangeville, L., Duchesne, L., Houle, D., Kneeshaw, D., Côté, B. and Pederson, N. (2016) 'Northeastern North America as a potential refugium for boreal forests in a warming climate', *Science*, vol 352, no 6292, pp1452–1455.

Dore, M.H. (2005) 'Climate change and changes in global precipitation patterns: What do we know?', *Environment International*, vol 31, no 8, pp1167–1181.

Dornelas, M. (2010) 'Disturbance and change in biodiversity', *Philosophical Transactions of the Royal Society B: Biological Sciences*, vol 365, no 1558, pp3719–3727.

Du, H., Liu, J., Li, M.-H., Büntgen, U., Yang, Y., Wang, L., Wu, Z. and He, H.S. (2018) 'Warming-induced upward migration of the alpine treeline in the Changbai Mountains, northeast China', *Global Change Biology*, vol 24, pp1256–1266.

Elliott, G.P. (2012) 'Extrinsic regime shifts drive abrupt changes in regeneration dynamics at upper treeline in the Rocky Mountains, USA', *Ecology*, vol 93, no 7, pp1614–1625.

Elliott, G.P. and Cowell, C.M. (2015) 'Slope aspect mediates fine-scale tree establishment patterns at upper treeline during wet and dry periods of the 20th century', *Arctic, Antarctic, and Alpine Research*, vol 47, no 4, pp681–692.

Elmendorf, S.C., Henry, G.H., Hollister, R.D., Björk, R.G., Boulanger-Lapointe, N., Cooper, E.J., Cornelissen, J.H., Day, T.A., Dorrepaal, E., Elumeeva, T.G. and Gill, M. (2012) 'Plot-scale evidence of tundra vegetation change and links to recent summer warming', *Nature Climate Change*, vol 2, no 6, pp453–457.

Erb, K.H., Kastner, T., Plutzar, C., Bais, A.L.S., Carvalhais, N., Fetzel, T., Gingrich, S., Haberl, H., Lauk, C., Niedertscheider, M. and Pongratz, J., 2018. Unexpectedly large impact of forest management and grazing on global vegetation biomass. *Nature*, 553(7686), p.73.

Falk, D.A., Watts, A.C. and Thode A.E. (2019) 'Scaling ecological resilience', *Frontiers in Ecology and Evolution*, vol 7, art 275.

Flannigan, M., Cantin, A.S., de Groot, W.J., Wotton, M., Newbery, A. and Gowman, L.M. (2013) 'Global wildland fire season severity in the 21st century', *Forest Ecology and Management*, vol 294, pp54–61.

Frank, D., Reichstein, M., Bahn, M., Thonicke, K., Frank, D., Mahecha, M.D., Smith, P., Van der Velde, M., Vicca, S., Babst, F. and Beer, C. (2015) 'Effects of climate extremes on the terrestrial carbon cycle: Concepts, processes and potential future impacts', *Global Change Biology*, vol 21, no 8, pp2861–2880.

Franklin, J., Serra-Diaz, J.M., Syphard, A.D. and Regan, H.M. (2016) 'Global change and terrestrial plant community dynamics', *Proceedings of the National Academy of Sciences*, vol 113, no 14, pp3725–3734.

Gaire, N.P., Koirala, M., Bhuju, D.R. and Borgaonkar, H.P. (2014) 'Treeline dynamics with climate change at the central Nepal Himalaya', *Climate of the Past*, vol 10, no 4, pp1277–1290.

Garfin, G. and Lenart, M. (2007) 'Climate change effects on Southwest water resources', *Southwest Hydrology*, vol 6, no 16–17, art 34.

Gariano, S.L. and Guzzetti, F. (2016) 'Landslides in a changing climate', *Earth-Science Reviews*, vol 162, pp227–252.

Gartner, J.H., Cannon, S.H., Santi, P.M. and Dewolfe, V.G. (2008) 'Empirical models to predict the volumes of debris flows generated by recently burned basins in the Western U.S.', *Geomorphology*, vol 96, pp339–354.

Gauthier, S., Bernier, P., Kuuluvainen, T., Shvidenko, A.Z. and Schepaschenko, D.G. (2015) 'Boreal forest health and global change', *Science*, vol 349, no 6250, pp819–822.

Gergel, D.R., Nijssen, B., Abatzoglou, J.T., Lettenmaier, D.P. and Stumbaugh, M.R. (2017) 'Effects of climate change on snowpack and fire potential in the western USA', *Climatic Change*, vol 141, no 2, pp287–299.

Glasby, T.M. and Underwood, A.J. (1996) 'Sampling to differentiate between pulse and press perturbations', *Environmental Monitoring and Assessment*, vol 42, no 3, pp241–252.

Goetz, S., Mack, M., Gurney, K., Randerson, J. and Houghton, R. (2007) 'Ecosystem responses to recent climate change and fire disturbance at northern high latitudes: Observations and model results contrasting northern Eurasia and North America', Environmental Research Letters, vol 2, no 4, art 045031.

Greening, H., Doering, P. and Corbett, C. (2006) 'Hurricane impacts on coastal ecosystems', *Estuaries and Coasts*, vol 29, no 6, pp877–879.

Grinnell, J. (1917) 'Field tests of theories concerning distributional control', *American Naturalist*, vol 51, pp115–128.

Guak S., Olsyzk D.M. Fuchigani L.H. and Tingey D.T. (1998) 'Effects of elevated CO_2 and temperature on cold hardiness and spring bud burst and growth in Douglas-fir (*Pseudotsuga menziesii*)', *Tree Physiology*, vol 18, pp.671–679.

Guiterman, C.H., Margolis, E.Q., Allen, C.D., Falk, D.A. and Swetnam, T.W. (2018) 'Long-term persistence and fire resilience of oak shrubfields in dry conifer forests of northern New Mexico', *Ecosystems*, vol 21, no 5, pp943–959.

Hannah, L., Midgley, G.F., Lovejoy, T., Bond, W.J., Bush, M.L.J.C., Lovett, J.C., Scott, D. and Woodward, F.I. (2002) 'Conservation of biodiversity in a changing climate', *Conservation Biology*, vol 16, no 1, pp264–268.

Hansen, A.J., Neilson, R.P., Dale, V.H. Flather, C.H., Iverson, L.R., Currie, D.J., Shafer, S., Cook, R. and Bartlein, P.J. (2001) 'Global change in forests: Responses of species, communities, and biomes interactions between climate change and land use are projected to cause large shifts in biodiversity', *BioScience*, vol 51, pp765–779.

Harris, R.M., Beaumont, L.J., Vance, T.R., Tozer, C.R., Remenyi, T.A., Perkins-Kirkpatrick, S.E., Mitchell, P.J., Nicotra, A.B., McGregor, S., Andrew, N.R., Letnic, M., Kearney, M.R., Wernberg, T., Hutley, L.B., Chambers, L.E., Fletcher, M.-S., Keatley, M.R., Woodward, C.A., Williamson, G., Duke, N.C. and Bowman, D.M.J.S. (2018) 'Biological responses to the press and pulse of climate trends and extreme events', *Nature Climate Change*, vol 8, no 7, pp579–587.

Harrison, P. and Pearce, F. (2000) *AAAS Atlas of Population & Environment*. University of California Press, Berkeley.

Harsch, M.A., Hulme, P.E., McGlone, M.S. and Duncan, R.P. (2009) 'Are treelines advancing? A global meta-analysis of treeline response to climate warming', *Ecology Letters*, vol 12, pp1040–1049.

Helbig, M., Pappas, C. and Sonnentag, O. (2016) 'Permafrost thaw and wildfire: Equally important drivers of boreal tree cover changes in the Taiga Plains, Canada', *Geophysical Research Letters*, vol 43, no 4, pp1598–1606.

Heyerdahl, E.K., Morgan, P. and Riser, J.P. (2008) 'Multi-season climate synchronized historical fires in dry forests (1650–1900), northern Rockies, USA', *Ecology*, vol 89, no 3, pp705–716.

Hicke, J.A., Johnson, M.C., Hayes, J.L. and Preisler, H.K. (2012) 'Effects of bark beetle-caused tree mortality on wildfire', *Forest Ecology and Management*, vol 271, pp81–90.

Higuera, P.E., Brubaker, L.B., Anderson, P.M., Brown, T.A., Kennedy, A.T. and Hu, F.S. (2008) 'Frequent fires in ancient shrub tundra: Implications of paleorecords for Arctic environmental change', *PLoS ONE*, vol 3, no 3, art e0001744.

Higuera, P.E., Chipman, M.L., Barnes, J.L., Urban, M.A. and Hu, F.S. (2011) 'Variability of tundra fire regimes in Arctic Alaska: Millennial-scale patterns and ecological implications', *Ecological Applications*, vol 21, no 8, pp3211–3226.

HilleRisLambers, J., Harsch, M.A., Ettinger, A.K., Ford, K.R. and Theobald, E.J. (2013) 'How will biotic interactions influence climate change–induced range shifts?', *Annals of the New York Academy of Sciences*, vol 1297, no 1, pp112–125.

Hinzman, L.D., Bettez, N.D., Bolton, W.R., Chapin, F.S., Dyurgerov, M.B., Fastie, C.L., Griffith, B., Hollister, R.D., Hope, A., Huntington, H.P. and Jensen, A.M. (2005) 'Evidence and implications of recent climate change in northern Alaska and other arctic regions', *Climatic Change*, vol 72, no 3, pp251–298.

Hoffman, D.F. and Gabet, E.J. (2007) 'Effects of sediment pulses on channel morphology in a gravel-bed river', *Geological Society of America Bulletin*, vol 119, no 1–2, pp116–125.

Holling, C.S. (1973) 'Resilience and stability of ecological systems', *Annual Review of Ecology and Systematics*, vol 4, no 1, pp1–23.

Holtmeier, F.K. and Broll, G. (2005) 'Sensitivity and response of northern hemisphere altitudinal and polar treelines to environmental change at landscape and local scales', *Global Ecology and Biogeography*, vol 14, no 5, pp395–410.

Huggel, C., Clague, J.J. and Korup, O. (2012) 'Is climate change responsible for changing landslide activity in high mountains?', *Earth Surface Processes and Landforms*, vol 37, pp77–91.

Hulme, P.E. (2005) 'Adapting to climate change: Is there scope for ecological management in the face of a global threat?', *Journal of Applied Ecology*, vol 42, no 5, pp784–794.

Hupp, J.W., Safine, D.E. and Nielson, R.M. (2013) 'Response of cackling geese to spatial and temporal variation in production of crowberries on the Alaska Peninsula', *Polar Biology*, vol 36, no 9, pp1243–1255.

Hyde, K., Woods, S.W. and Donahue, J. (2007) 'Predicting gully rejuvenation after wildfire using remotely sensed burn severity data', *Geomorphology*, vol 86, pp496–511.

Intergovernmental Panel on Climate Change (2014) *Climate Change 2014: Synthesis Report. Contribution of Working Groups I, II and III to the Fifth Assessment Report of the Intergovernmental Panel on Climate Change* [Core Writing Team, R.K. Pachauri and L.A. Meyer (eds.)]. IPCC, Geneva, Switzerland, 151 pp.

Jacob, M., Frankl, A., Beeckman, H., Mesfin, G., Hendrickx, M., Guyassa, E. and Nyssen, J (2015) 'North Ethiopian afro-alpine tree line dynamics and forest-cover change since the early 20th century', *Land Degradation & Development*, vol 26, no 7, pp654–664.

Jactel, H., Petit, J., Desprez-Loustau, M.L., Delzon, S., Piou, D., Battisti, A. and Koricheva, J. (2012) 'Drought effects on damage by forest insects and pathogens: A meta-analysis', *Global Change Biology*, vol 18, pp267–76.

Johnstone, J.F., Chapin, F.S., Hollingsworth, T.N., Mack, M.C., Romanovsky, V. and Turetsky, M. (2010a) 'Fire, climate change, and forest resilience in interior Alaska', *Canadian Journal of Forest Research*, vol 40, no 7, pp1302–1312.

Johnstone, J.F., Hollingsworth, T.N., Chapin III, F.S. and Mack, M.C. (2010b) 'Changes in fire regime break the legacy lock on successional trajectories in Alaskan boreal forest', *Global Change Biology*, vol 16, no 4, pp1281–1295.

Jolly, W.M., Dobbertin, M., Zimmermann, N.E. and Reichstein, M. (2005) 'Divergent vegetation growth responses to the 2003 heat wave in the Swiss Alps', *Geophysical Research Letters*, vol 32, art L18409.

Jönsson, A.M. and Bärring, L. (2011) 'Future climate impact on spruce bark beetle life cycle in relation to uncertainties in regional climate model data ensembles', *Tellus A: Dynamic Meteorology and Oceanography*, vol 63, no 1, pp158–173.

Jorgenson, M.T. and Ely, C.R. (2001) 'Topography and flooding of coastal ecosystems on the Yukon-Kuskokwim Delta, Alaska: Implications for sea level rise', *Journal of Coastal Research*, vol 17, pp124–136.

Jorgenson, M.T., Racine, C.H., Walters, J.C. and Osterkamp, T.E. (2001) 'Permafrost degradation and ecological changes associated with a warming climate in central Alaska', *Climatic Change*, vol 48, no 4, pp551–579.

Joyce, L.A., Blate, G.M., McNulty, S.G., Millar, C.I., Moser, S., Neilson, R.P. and Peterson, D.L. (2009) 'Managing for multiple resources under climate change: National forests', *Environmental Management*, vol 44, no 6, art 1022.

Kane, J.M., Varner, J.M., Metz, M.R. and van Mantgem, P.J. (2017) 'Characterizing interactions between fire and other disturbances and their impacts on tree mortality in western U.S. Forests', *Forest Ecology and Management*, vol 405, pp188–199.

Kapnick, S. and Hall, A. (2012) 'Causes of recent changes in western North American snowpack', *Climate Dynamics*, vol 38, no 9–10, pp1885–1899.

Kelly, R.P., Erickson, A.L., Mease, L.A., Battista, W., Kittinger, J.N. and Fujita, R. (2015) 'Embracing thresholds for better environmental management', *Philosophical Transactions of the Royal Society B: Biological Sciences*, vol 370, no 1659, art 20130276.

Kirdyanov, A.V., Hagedorn, F., Knorre, A.A., Fedotova, E.V., Vaganov, E.A., Naurzbaev, M.M., Moiseev, P.A. and Rigling, A. (2012) '20th century tree-line advance and vegetation changes along an altitudinal transect in the Putorana Mountains, northern Siberia', *Boreas*, vol 41, pp56–67.

Knapp, A.K., Hoover, D.L., Wilcox, K.R., Avolio, M.L., Koerner, S.E., La Pierre, K.J., Loik, M.E., Luo, Y., Sala, O.E. and Smith, M.D. (2015) 'Characterizing differences in precipitation regimes of extreme wet and dry years: Implications for climate change experiments', *Global Change Biology*, vol 21, pp2624–2633.

Knowles, N., Dettinger, M.D. and Cayan, D.R. (2006) 'Trends in snowfall versus rainfall in the western United States', *Journal of Climate*, vol 19, pp4545–4559.

Knutti, R. and Sedláček, J. (2013) 'Robustness and uncertainties in the new CMIP5 climate model projections', *Nature Climate Change*, vol 3, no 4, pp369–373.

Krawchuk, M.A., Moritz, M.A., Parisien, M.A., Van Dorn, J. and Hayhoe, K. (2009) 'Global pyrogeography: The current and future distribution of wildfire', *PLoS One*, vol 4, art e5102.

Kupfer, J.A., Balmat, J. and Smith, J.L. (2005) 'Shifts in the potential distribution of Sky Island plant communities in response to climate change', in G.J. Gottfried, B.S. Gebow, L.G. Eskew and C.B. Edminster (eds). Connecting Mountain Islands and Desert Seas: Biodiversity and Management of the Madrean Archipelago II. Proc. RMRS-P-36. Fort Collins, CO: US Department of Agriculture, Forest Service, Rocky Mountain Research Station, pp.485–490.

Lamberti, G.A., Gregory, S.V., Ashkenas, L.R., Wildman, R.C. and Moore, K.M.S. (1991) 'Stream ecosystem recovery following a catastrophic debris flow', *Canadian Journal of Fisheries and Aquatic Sciences*, vol 48, no 2, pp196–208.

Lange, H., Økland, B. and Krokene, P. (2006) 'Thresholds in the life cycle of the spruce bark beetle under climate change', *Interjournal for Complex Systems*, vol 1648, pp1–10.

Lantz, T.C., Kokelj, S.V. and Fraser, R.H. (2015) 'Ecological recovery in an Arctic delta following widespread saline incursion', *Ecological Applications*, vol 25, pp172–185.

Lauenroth, W.K., Urban, D.L., Coffin, D.P., Parton, W.J., Shugart, H.H., Kirchner, T.B. and Smith, T.M. (1993) 'Modeling vegetation structure-ecosystem process interactions across sites and ecosystems', *Ecological Modelling*, vol 67, pp49–80.

Leonelli, G., Pelfini, M. and Morra di Cella, U. (2011) 'Climate warming and the recent treeline shift in the European Alps: The role of geomorphological factors in high-altitude sites', *AMBIO*, vol 40, pp264–273.

Lewis, S.L., Edwards, D.P. and Galbraith, D. (2015) 'Increasing human dominance of tropical forests', *Science*, vol 349, no 6250, pp827–832.

Littell, J.S. (2018) 'Drought and fire in the western USA: is climate attribution enough?' *Current Climate Change Reports*, vol 4, no. 4, pp. 396–406.

Littell, J.S., McKenzie, D., Kerns, B.K., Cushman, S. and Shaw, C.G. (2011) 'Managing uncertainty in climate-driven ecological models to inform adaptation to climate change', *Ecosphere*, vol 2, no 9, pp1–19.

Littell, J.S., Peterson, D.L., Riley, K.L., Liu, Y. and Luce, C.H. (2016) 'A review of the relationships between drought and forest fire in the United States', *Global Change Biology*, vol 22, no 7, pp2353–2369.

Liu, K., Lu, H. and Shen, C. (2008) 'A 1200-year proxy record of hurricanes and fires from the Gulf of Mexico coast: Testing the hypothesis of hurricane–fire interactions', *Quaternary Research*, vol 69, no 1, pp29–41.

Liu, Y.Y., van Dijk, A.I., McCabe, M.F., Evans, J.P. and de Jeu, R.A. (2013) 'Global vegetation biomass change (1988–2008) and attribution to environmental and human drivers', *Global Ecology and Biogeography*, vol 22, no 6, pp692–705.

Liu, Z., Ballantyne, A.P. and Cooper, L.A. (2019) 'Biophysical feedback of global forest fires on surface temperature', *Nature Communications*, vol 10, art 214.

Loehman, R.A., Keane, R.E., Holsinger, L.M. and Wu, Z. (2017) 'Interactions of landscape disturbances and climate change dictate ecological pattern and process: Spatial modeling of wildfire, insect, and disease dynamics under future climates', *Landscape Ecology*, vol 32, no 7, pp1447–1459.

Loehman, R., Flatley, W., Holsinger, L. and Thode, A. (2018) 'Can land management buffer impacts of climate changes and altered fire regimes on ecosystems of the southwestern United States?', *Forests*, vol 9, no 4, art 192.

Loreau, M., Naeem, S., Inchausti, P., Bengtsson, J., Grime, J.P., Hector, A., Hooper, D.U., Huston, M.A., Raffaelli, D., Schmid, B. and Tilman, D. (2001) 'Biodiversity and ecosystem functioning: Current knowledge and future challenges', *Science*, vol 294, no 5543, pp804–808.

Lugo, AE. (2008) 'Visible and invisible effects of hurricanes on forest ecosystems: An international review', *Austral Ecology*, vol 33, pp368–398.

Lutz, J.A., Van Wagtendonk, J.W., Thode, A.E., Miller, J.D. and Franklin, J.F. (2009) 'Climate, lightning ignitions, and fire severity in Yosemite National Park, California, USA', *International Journal of Wildland Fire*, vol 18, no 7, pp765–774.

Mack, M.C., Bret-Harte, M.S., Hollingsworth, T.N., Jandt, R.R., Schuur, E.A., Shaver, G.R. and Verbyla, D.L. (2011) 'Carbon loss from an unprecedented Arctic tundra wildfire', *Nature*, vol 475, no 7357, p489–492.

Maclean, I.M. and Wilson, R.J. (2011) 'Recent ecological responses to climate change support predictions of high extinction risk', *Proceedings of the National Academy of Sciences*, vol 108, no 30, pp12337–12342.

Mahmood, R. and Hubbard, K.G. (2002) 'Anthropogenic land-use change in the North American tall grass-short grass transition and modification of near-surface hydrologic cycle', *Climate Research*, vol 21, no 1, pp83–90.

Mahmood, R., Hubbard, K.G. and Carlson, C. (2004) 'Modification of growing-season surface temperature records in the northern great plains due to land-use transformation: Verification of modelling results and implication for global climate change', *International Journal of Climatology*, vol 24, no 3, pp311–327.

Malanson, G.P., Resler, L.M., Butler, D.R. and Fagre, D.B. (2019) 'Mountain plant communities: Uncertain sentinels?', *Progress in Physical Geography: Earth and Environment*, vol 41, no 4, pp521–543.

Malhi, Y. and Phillips, O.L. (2004) 'Tropical forests and global atmospheric change: A synthesis', *Philosophical Transactions of the Royal Society of London. Series B: Biological Sciences*, vol 359, no 1443, pp549–555.

Mann, M.L., Batllori, E., Moritz, M.A., Waller, E.K., Berck, P., Flint, A.L., Flint, L.E. and Dolfi, E. (2016) 'Incorporating anthropogenic influences into fire probability models: Effects of human activity and climate change on fire activity in California', *PLoS One*, vol 11, no 4, art e0153589.

Marchand, F.L., Mertens, S., Kockelbergh, F., Beyens, L. and Nijs, I. (2005) 'Performance of high Arctic tundra plants improved during but deteriorated after exposure to a simulated extreme temperature event', *Global Change Biology*, vol 11, no 12, pp2078–2089.

Marchand, F.L., Kockelbergh, F., Van De Vijver, B., Beyens, L. and Nijs, I. (2006) 'Are heat and cold resistance of arctic species affected by successive extreme temperature events?', *New Phytologist*, vol 170, no 2, pp291–300.

Marlon, J.R., Bartlein, P.J., Daniau, A.L., Harrison, S.P., Maezumi, S.Y., Power, M.J., Tinner, W. and Vanniére, B. (2013) 'Global biomass burning: A synthesis and review of Holocene paleofire records and their controls', *Quaternary Science Reviews*, vol 65, pp5–25.

McDowell, W.H. (2011). 'Impacts of hurricanes on forest hydrology and biogeochemistry', in *Forest Hydrology and Biogeochemistry* (pp. 643–657). Springer, Dordrecht.

McKenney, D.W., Pedlar, J.H., Lawrence, K., Campbell, K. and Hutchinson, M.F. (2007) 'Potential impacts of climate change on the distribution of North American trees', *BioScience*, vol 57, no 11, pp939–948.

McKenzie, D. and Littell, J.S. (2017) 'Climate change and the eco-hydrology of fire: Will area burned increase in a warming western USA?', *Ecological Applications*, vol 27, no 1, pp26–36.

McWethy, D.B., Whitlock, C., Wilmshurst, J.M., McGlone, M.S., Fromont, M., Li, X., Dieffenbacher-Krall, A., Hobbs, W.O., Fritz, S.C. and Cook, E.R. (2010) 'Rapid landscape transformation in South Island, New Zealand following initial Polynesian settlement', *Proceedings of the National Academy of Sciences USA*, vol 107, pp21343–21348.

Millar, C.I., Westfall, R.D., Delany, D.L., King, J.C. and Graumlich, L.J. (2004) 'Response of subalpine conifers in the Sierra Nevada, California, USA, to 20th-century warming and decadal climate variability', *Arctic, Antarctic, and Alpine Research*, vol 36, pp181–200.

Millar, C.I., Stephenson, N.L. and Stephens, S.L. (2007) 'Climate change and forests of the future: Managing in the face of uncertainty', *Ecological Applications*, vol 17, no 8, pp2145–2151.

Millar, C.I., Westfall, R.D., Delany, D.L., Flint, A.L. and Flint, L.E. (2015) 'Recruitment patterns and growth of high-elevation pines in response to climatic variability (1883–2013), in the western Great Basin, USA', *Canadian Journal of Forest Research*, vol 45, pp1299–1312.

Miller, A.E., Wilson, T.L., Sherriff, R.L. and Walton, J. (2017) 'Warming drives a front of white spruce recruitment near western treeline, Alaska', *Global Change Biology*, vol 23, pp5509–5522.

Miller, C., Abatzoglou, J., Brown, T. and Syphard, A.D. (2011) 'Wilderness fire management in a changing environment', in D. McKenzie, C. Miller and D. Falk (eds) *The Landscape Ecology of Fire*. Springer, Dordrecht.

Morgan, J.A., Milchunas, D.G., LeCain, D.R., West, M. and Mosier, A.R. (2007) 'Carbon dioxide enrichment alters plant community structure and accelerates shrub growth in the shortgrass steppe', *Proceedings of the National Academy of Sciences*, vol 104, pp14724–14729.

Moritz, M.A., Parisien, M.A., Batllori, E., Krawchuk, M.A., Van Dorn, J., Ganz, D.J. and Hayhoe, K. (2012) 'Climate change and disruptions to global fire activity', *Ecosphere*, vol 3, no 6, pp1–22.

Morris, J.L., DeRose, R.J. and Brunelle, A.R. (2015) 'Long-term landscape changes in a subalpine spruce-fir forest in central Utah, USA', *Forest Ecosystems*, vol 2, pp1–12.

Morris J., Cottrell, S., Fettig, C., Hansen, W., Sherriff, R., Carter, V., Clear, J., Clement, J., DeRose, R., Higuera, P., Mattor, K., Seddon, A., Seppa, H., Stednick, J. and Seybold, S. (2017) 'Managing bark beetle impacts on ecosystems and society: Priority questions to motivate future research', *Journal of Applied Ecology*, vol 54, pp750–760.

Mu, Y., Qin, X., Zhang, L. and Xu, B. (2016) 'Holocene climate change evidence from high-resolution loess/paleosol records and the linkage to fire–climate change–human activities in the Horqin dunefield in northern China', *Journal of Asian Earth Sciences*, vol 121, pp1–8.

Muller, M. and Job, H. (2009) 'Managing natural disturbance in protected areas: Tourists' attitude towards the bark beetle in a German national park', *Biological Conservation*, vol 142, pp375–383.

Munson, S.M., Muldavin, E.H., Belnap, J., Peters, D.P., Anderson, J.P., Reiser, M.H., Gallo, K., Melgoza-Castillo, A., Herrick, J.E. and Christiansen, T.A. (2013) 'Regional signatures of plant response to drought and elevated temperature across a desert ecosystem', *Ecology*, vol 94, no 9, pp2030–2041.

Naficy, C., Sala, A., Keeling, E.G., Graham, J. and DeLuca, T.H. (2010) 'Interactive effects of historical logging and fire exclusion on ponderosa pine forest structure in the northern Rockies', *Ecological Applications*, vol 20, no 7, pp1851–1864.

National Climate Assessment (NCA) (2018) 'National climate assessment', https://nca2014.globalchange.gov/, accessed 14 Dec 2020.

Nemani, R.R., Keeling, C.D., Hashimoto, H., Jolly, W.M., Piper, S.C., Tucker C.J., Myneni, R.B. and Running, S.W. (2003) 'Climate-driven increases in global terrestrial net primary production from 1982 to 1999', *Science*, vol 300, no 5625, pp1560–1563.

North, M.P., Stephens, S.L., Collins, B.M., Agee, J.K., Aplet, G., Franklin, J.F. and Fule, P.Z. (2015) 'Reform forest fire management', *Science*, vol 349, no 6254, pp1280–1281.

Notaro, M., Mauss, A., and Williams, J.W. (2012) 'Projected vegetation changes for the American Southwest: Combined dynamic modeling and bioclimatic-envelope approach', *Ecological Applications*, vol 22, pp1365–1388.

Overpeck, J., Whitlock, C. and Huntley, B. (2003) 'Terrestrial biosphere dynamics in the climate system: Past and future', in K.D. Alverson, T.F. Pedersen and R.S. Bradley (eds) *Paleoclimate, Global Change and the Future. Global Change: The IGBP Series*. Springer, Heidelberg.

Paaijmans, K.P., Heinig, R.L., Seliga, R.A., Blanford, J.I., Blanford, S., Murdock, C.C. and Thomas, M.B. (2013) 'Temperature variation makes ectotherms more sensitive to climate change', *Global Change Biology*, vol 19, no 8, pp2373–2380.

Parise, M. and Cannon, S.H. (2012) 'Wildfire impacts on the processes that generate debris flows in burned watersheds', *Natural Hazards*, vol 61, pp217–227.

Parisien, M.A., Miller, C., Parks, S.A., DeLancey, E.R., Robinne, F.N. and Flannigan, M.D. (2016) 'The spatially varying influence of humans on fire probability in North America', *Environmental Research Letters*, vol 11, no 7, art 075005.

Paritsis, J., Landesmann, J.B., Kitzberger, T., Tiribelli, F., Sasal, Y., Quintero, C., Dimarco, R.D., Barrios-García, M.N., Iglesias, A.L., Diez, J.P., Sarasola, M. and Nuñez, M.A. (2018) 'Pine plantations and invasion alter fuel structure and potential fire behavior in a Patagonian forest-steppe ecotone', *Forests*, vol 9, no 3, art 117.

Parker, K.C. and Bendix, J. (1996) 'Landscape-scale geomorphic influences on vegetation patterns in four environments', *Physical Geography*, vol 17, no 2, pp113–141.

Parmesan, C. (2006) 'Ecological and evolutionary responses to recent climate change', *Annual Review of Ecology, Evolution, and Systematics*, vol 37, pp637–669.

Parolo, G. and Rossi, G. (2008) 'Upward migration of vascular plants following a climate warming trend in the Alps', *Basic and Applied Ecology*, vol 9, no 2, pp100–107.

Passeri, D.L., Hagen, S.C., Medeiros, S.C., Bilskie, M.V., Alizad, K. and Wang, D. (2015) 'The dynamic effects of sea level rise on low-gradient coastal landscapes: A review', *Earth's Future*, vol 3, pp159–181.

Pausas, J.G. and Fernandez-Munoz, S. (2012) 'Fire regime changes in the Western Mediterranean Basin: From fuel-limited to drought-driven fire regime', *Climatic Change*, vol 110, pp215–226.

Peñuelas, J. and Boada, M. (2003) 'A global change-induced biome shift in the Montseny mountains (NE Spain)', *Global Change Biology*, vol 9, pp131–140.

Perera, A.H., Sturtevant, B.R. and Buse, L.J. (2015) 'Simulation modeling of forest landscape disturbances: An overview', in A.H. Perera, B.R. Sturtevant and L.J. Buse (eds) *Simulation Modeling Forest Landscape Disturbances*. Springer, Geneva, Switzerland.

Perkins-Kirkpatrick, S.E. and Gibson, P.B. (2017) 'Changes in regional heatwave characteristics as a function of increasing global temperature', *Scientific Reports*, vol 7, art 12256.

Peters, D.P.C., Lugo, A.E., Chapin, F.S. III, Pickett, S.T.A, Duniway, M., Rocha, A.V., Swanson, F.J., Laney, C. and Jones, J. (2011) 'Cross-system comparisons elucidate disturbance complexities and generalities', *Ecosphere*, vol 2, no 7, pp1–26.

Pielke Sr, R.A., Marland, G., Betts, R.A., Chase, T.N., Eastman, J.L., Niles, J.O., Niyogi, D.D.S. and Running, S.W. (2002) 'The influence of land-use change and landscape dynamics on the climate system: Relevance to climate-change policy beyond the radiative effect of greenhouse gases', *Philosophical Transactions of the Royal Society of London. Series A: Mathematical, Physical and Engineering Sciences*, vol 360, no 1797, pp1705–1719.

Plotkin, A.B., Foster, D., Carlson, J. and Magill, A. (2013) 'Survivors, not invaders, control forest development following simulated hurricane', *Ecology*, vol 94, no 2, pp414–423.

Post, E., Forchhammer, M.C., Bret-Harte, M.S., Callaghan, T.V., Christensen, T.R., Elberling, B., Fox, A.D., Gilg, O., Hik, D.S., Høye, T.T. and Ims, R.A. (2009) 'Ecological dynamics across the Arctic associated with recent climate change', *Science*, vol 325, no 5946, pp1355–1358.

Puppi, G. (2007) 'Origin and development of phenology as a science'. *Italian Journal of Agrometeorology*, vol 12, pp24–29.

Qian, S.S., King, R.S. and Richardson, C.J. (2003) 'Two statistical methods for the detection of environmental thresholds', *Ecological Modelling*, vol 166, no 1–2, pp87–97.

Raffa, K.F., Aukema, B.H., Bentz, B.J., Carroll, A.L., Hicke, J.A., Turner, M.G. and Romme, W.H. (2008) 'Cross-scale drivers of natural disturbances prone to anthropogenic amplification: The dynamics of bark beetle eruptions', *Bioscience*, vol 58, pp501–517.

Ratajczak, Z., Carpenter, S.R., Ives, A.R., Kucharik, C.J., Ramiadantsoa, T., Stegner, M.A., Williams, J.W., Zhang, J. and Turner, M.G. (2018) 'Abrupt change in ecological systems: Inference and diagnosis', *Trends in Ecology and Evolution*, vol 33, no 7, pp513–526.

Raymond, C.L., Peterson, D. and Rochefort, R.M. (2014) 'Climate change vulnerability and adaptation in the North Cascades region, Washington', *General Technical Report PNW-GTR-892*. US Department of Agriculture, Forest Service, Pacific Northwest Research Station, Portland, OR, 279, p892.

Rehm, E.M., Olivas, P., Stroud, J. and Feeley, K.J. (2015) 'Losing your edge: Climate change and the conservation value of range-edge populations', *Ecology and Evolution*, vol 5, no 19, pp4315–4326.

Reilly, M.J., Dunn, C.J., Meigs, G.W., Spies, T.A., Kennedy, R.E., Bailey, J.D. and Briggs, K. (2017) 'Contemporary patterns of fire extent and severity in forests of the Pacific Northwest, USA (1985–2010)', *Ecosphere*, vol 8, no 3, art e01695.

Riley, K.L., Abatzoglou, J.T., Grenfell, I.C., Klene, A.E. and Heinsch, F.A. (2013) 'The relationship of large fire occurrence with drought and fire danger indices in the western USA, 1984–2008: The role of temporal scale', *International Journal of Wildland Fire*, vol 22, pp894–909.

Root, T.L., Price, J.T., Hall, K.R., Schneider, S.H., Rosenzweig, C. and Pounds, J.A. (2003) 'Fingerprints of global warming on wild animals and plants', *Nature*, vol 421, pp57–60.

Rowland, J.C., Jones, C.E., Altmann, G., Bryan, R., Crosby, B.T., Hinzman, L.D., Kane, D.L., Lawrence, D.M., Mancino, A., Marsh, P. and McNamara, J.P. (2010) 'Arctic landscapes in transition: Responses to thawing permafrost', *Eos, Transactions American Geophysical Union*, vol 91, no 26, pp229–230.

Sanz-Elorza, M., Dana, E.D., González, A. and Sobrino, E. (2003) 'Changes in the high-mountain vegetation of the central Iberian Peninsula as a probable sign of global warming', *Annals of Botany*, vol 92, no 2, pp273–280.

Scheller, R.M. (2018) 'The challenges of forest modeling given climate change', *Landscape Ecology*, vol 33, no 9, pp1481–1488.

Schoennagel, T., Veblen, T.T. and Romme, W.H. (2004) 'The interaction of fire, fuels, and climate across Rocky Mountain forests', *BioScience*, vol 54, no 7, pp661–676.

Schuur, E.A., Bockheim, J., Canadell, J.G., Euskirchen, E., Field, C.B., Goryachkin, S.V., Hagemann, S., Kuhry, P., Lafleur, P.M., Lee, H. and Mazhitova, G. (2008) 'Vulnerability of permafrost carbon to climate change: Implications for the global carbon cycle', *BioScience*, vol 58, no 8, pp701–714.

Schuur, E.A., McGuire, A.D., Schädel, C., Grosse, G., Harden, J.W., Hayes, D.J., Hugelius, G., Koven, C.D., Kuhry, P., Lawrence, D.M. and Natali, S.M. (2015) 'Climate change and the permafrost carbon feedback', *Nature*, vol 520, no 7546, p.171–179.

Scott, A.C., W.G. Chaloner, C.M. Belcher, and C.I. Roos (2016) 'The interaction of fire and mankind', *Philosophical Transactions of the Royal Society of London B: Biological Sciences* vol 371, p. 1696.

Seddon, A.W., Macias-Fauria, M., Long, P.R., Benz, D. and Willis, K.J., 2016. Sensitivity of global terrestrial ecosystems to climate variability. *Nature*, 531(7593), pp.229.

Seidl, R., Schelhaas, M.J., Rammer, W., and Verkerk, P.J. (2014) 'Increasing forest disturbances in Europe and their impact on carbon storage', *Nature Climate Change*, 4, pp.806–810.

Seidl, R., Thom, D., Kautz, M., Martin-Benito, D., Peltoniemi, M., Vacchiano, G., Wild, J., Ascoli, D., Petr, M., Honkaniemi, J. and Lexer, M.J. (2017) 'Forest disturbances under climate change', *Nature Climate Change*, vol 7, no. 6, pp. 395–402.

Serreze, M.C., Walsh, J.E., Chapin, F.S., Osterkamp, T., Dyurgerov, M., Romanovsky, V., Oechel, W.C., Morison, J., Zhang, T. and Barry, R.G., 2000. Observational evidence of recent change in the northern high-latitude environment. *Climatic Change*, 46(1–2), pp.159–207.

Sherriff, R.L., Berg, E.E. and Miller, A.E. 2011. Effects of interannual and multidecadal climate variability on spruce beetle (*Dendroctonus rufipennis*) activity in south-central and southwest Alaska. *Ecology*, 92, pp.1459–70.

Shiyatov, S.G., Terent'ev, M.M. and Fomin, V.V. 2005. Spatial-temporal dynamics of forest tundra communities in the Polar Ural Mountains. *Russian Journal of Ecology* 36, 69–75.

Stephens, S.L., Agee, J.K., Fule, P.Z., North, M.P., Romme, W.H., Swetnam, T.W. and Turner, M.G., 2013. Managing forests and fire in changing climates. *Science*, 342(6154), pp.41–42.

Stevens-Rumann, C.S., Kemp, K.B., Higuera, P.E., Harvey, B.J., Rother, M.T., Donato, D.C., Morgan, P. and Veblen, T.T., 2018. Evidence for declining forest resilience to wildfires under climate change. *Ecology Letters*, 21(2), pp.243–252.

Stocker, T.F., Qin, D., Plattner, G.K., Tignor, M., Allen, S.K., Boschung, J., Nauels, A., Xia, Y., Bex, V., Midgley, P.M. *et al.* (2014) *Climate Change 2013: The Physical Science Basis.* Cambridge University Press, Cambridge.

Sturm, M., Racine, C. and Tape, K. (2001) 'Increasing shrub abundance in the Arctic', *Nature*, vol 411, no 6837, pp546–547.

Sturrock, R.N., Frankel, S.J., Brown, A.V., Hennon, P.E., Kliejunas, J.T., Lewis, K.J., Worrall, J.J. and Woods, A.J. (2011) 'Climate change and forest diseases', *Plant Pathology*, vol 60, no 1, pp133–149.

Styger, J., Marsden-Smedley, J. and Kirkpatrick, J. (2018) 'Changes in lightning fire incidence in the Tasmanian Wilderness World Heritage Area, 1980–2016', *Fire*, vol 1, no 3, art 38.

Swetnam, T.W., Farella, J., Roos, C.I., Liebmann, M.J., Falk, D.A. and Allen, C.D. (2016) 'Multiscale perspectives of fire, climate and humans in western North America and the Jemez Mountains, USA', *Philosophical Transactions of the Royal Society B: Biological Sciences*, vol 371, no 1696, art 20150168.

Syphard, A.D., Radeloff, V.C., Keeley, J.E., Hawbaker, T.J., Clayton, M.K., Stewart, S.I. and Hammer, R.B. (2007) 'Human influence on California fire regimes', *Ecological Applications*, vol 17, no 5, pp1388–1402.

Tanner, E.V.J., Rodriguez-Sanchez, F., Healey, J.R., Holdaway, R.J. and Bellingham, P.J. (2014) 'Long-term hurricane damage effects on tropical forest tree growth and mortality', *Ecology*, vol 95, no 10, pp2974–2983.

Tape, K.E.N., Sturm, M. and Racine, C. (2006) 'The evidence for shrub expansion in Northern Alaska and the Pan-Arctic', *Global Change Biology*, vol 12, no 4, pp686–702.

Temperli, C., Bugmann, H. and Elkin, C. (2013) 'Cross-scale interactions among bark beetles, climate change, and wind disturbances: A landscape modeling approach', *Ecological Monographs*, vol 83, no 3, pp383–402.

Terenzi, J., Jorgenson, M.T. and Ely, C.R. (2014) 'Storm-surge flooding on the Yukon-Kuskokwim Delta, Alaska', *Arctic*, vol 67, pp360–374.

Thackeray, S.J., Henrys, P.A., Hemming, D., Bell, J.R., Botham, M.S., Burthe, S., Helaouet, P., Johns, D.G., Jones, I.D., Leech, D.I. and Mackay, E.B. (2016) 'Phenological sensitivity to climate across taxa and trophic levels', *Nature*, vol 535, no 7611, p241–245.

Thom, D., Rammer, W. and Seidl, R. (2017) 'Disturbances catalyze the adaptation of forest ecosystems to changing climate conditions', *Global Change Biology*, vol 23, no 1, pp269–282.

Thomas, C.D., Cameron, A., Green, R.E., Bakkenes, M., Beaumont, L.J., Collingham, Y.C., Erasmus, B.F., De Siqueira, M.F., Grainger, A. and Hannah, L. (2004) 'Extinction risk from climate change', *Nature*, vol 427, pp145–148.

Thuiller, W., Albert, C., Araujo, M.B., Berry, P.M., Cabeza, M., Guisan, A., Hickler, T., Midgley, G.F., Paterson, J., Schurr, F.M. and Sykes, M.T. (2008) 'Predicting global change impacts on plant species' distributions: Future challenges', *Perspectives in Plant Ecology, Evolution and Systematics*, vol 9, no 3–4, pp137–152.

Trenberth, K.E. (2011) 'Changes in precipitation with climate change', *Climate Research*, vol 47, pp123–138.

Trumbore, S., Brando, P. and Hartmann, H. (2015) 'Forest health and global change', *Science*, vol 349, no 6250, pp814–818.

Turner, B.L. II., Clark, W.C., Kates, R.W., Richards, J.F., Mathews, J.T. and Meyer, W.B. (eds) (1991) *The Earth as Transformed by Human Action: Global and Regional Changes in the Biosphere over the Past 300 Years*. Cambridge University Press, Cambridge.

Turner, M.G. (2010) 'Disturbance and landscape dynamics in a changing world', *Ecology*, vol 91, no 10, pp2833–2849.

Vannière, B., Colombaroli, D., Chapron, E., Leroux, A., Tinner, W. and Magny, M. (2008) 'Climate versus human-driven fire regimes in Mediterranean landscapes: The Holocene record of Lago dell'Accesa (Tuscany, Italy)', *Quaternary Science Reviews*, vol 27, no 11–12, pp1181–1196.

Vermaire, J.C., Pisaric, M.F.J., Thienpont, J.R., Courtney Mustaphi, C.J., Kokelj, S.V. and Smol, J.P. (2013) 'Arctic climate warming and sea ice declines lead to increased storm surge activity', *Geophysical Research Letters*, vol 40, pp1386–1390.

Virtanen, R., Luoto, M., Rämä, T., Mikkola, K., Hjort, J., Grytnes, J.A. and Birks, H.J.B. (2010) 'Recent vegetation changes at the high-latitude tree line ecotone are controlled by geomorphological disturbance, productivity and diversity', *Global Ecology and Biogeography*, vol 19, no 6, pp810–821.

Vitousek, P.M., Mooney, H.A., Lubchenco, J. and Melillo, J.M. (1997) 'Human domination of Earth's ecosystems', *Science*, vol 277, no 5325, pp494–499.

Vörösmarty, C., Lettenmaier, D., Leveque, C., Meybeck, M., Pahl-Wostl, C., Alcamo, J., Cosgrove, W., Grassl, H., Hoff, H., Kabat, P. and Lansigan, F. (2004) 'Humans transforming the global water system', *Eos, Transactions American Geophysical Union*, vol 85, no 48, pp509–514.

Walther, G.-R., Post, E., Convey, P., Menzel, A., Parmesan, C., Beebee, T.J., Fromentin, J.-M., Hoegh-Guldberg, O. and Bairlein, F. (2002) 'Ecological responses to recent climate change', *Nature*, vol 416, pp389–395.

Wang, D., Heckathorn, S.A., Mainali, K. and Hamilton, E.W. (2008) 'Effects of N on plant response to heat-wave: A field study with prairie vegetation', *Journal of Integrative Plant Biology*, vol 50, no 11, pp1416–1425.

Wermelinger, B. (2004) 'Ecology and management of the spruce bark beetle *Ips typographus*: A review of recent research', *Forest Ecology and Management*, vol 202, pp67–82.

Westerling, A.L. and Swetnam, T.W. (2003) 'Interannual to decadal drought and wildfire in the western United States', *EOS, Transactions American Geophysical Union*, vol 84, no 49, pp545–555.

Westerling, A.L., Hidalgo, H.G., Cayan, D.R. and Swetnam, T.W. (2006) 'Warming and earlier spring increase western US forest wildfire activity', *Science*, vol 313, no 5789, pp940–943.

Williams, A.P., Allen, C.D., Millar, C.I., Swetnam, T.W., Michaelsen, J., Still, C.J. and Leavitt, S.W. (2010) 'Forest responses to increasing aridity and warmth in the southwestern United States', *Proceedings of the National Academy of Sciences*, vol 107, no 50, pp21289–21294.

Williams, A.P., Allen, C.D., Macalady, A.K., Griffin, D., Woodhouse, C.A., Meko, D.M., Swetnam, T.W., Rauscher, S.A., Seager, R., Grissino-Mayer, H.D. and Dean, J.S. (2013) 'Temperature as a potent driver of regional forest drought stress and tree mortality', *Nature Climate Change*, vol 3, no 3, pp292–297.

Wipf, S., Rixen, C. and Mulder, C.P.H. (2006) 'Advanced snowmelt causes a shift towards positive neighbor interactions in a subarctic tundra community', *Global Change Biology*, vol 12, no 8, pp1496–1506.

Wondzell, S.M. and King, J.G. (2003) 'Postfire erosional processes in the Pacific Northwest and Rocky Mountain regions', *Forest Ecology and Management*, vol 178, pp75–87.

Woodhouse, C.A., Meko, D.M., MacDonald, G.M., Stahle, D.W. and Cook, E.R. (2010) 'A 1,200-year perspective of 21st century drought in southwestern North America', *Proceedings of the National Academy of Sciences*, vol 107, no 50, pp21283–21288.

Xie, S.P., Deser, C., Vecchi, G.A., Collins, M., Delworth, T.L., Hall, A., Hawkins, E., Johnson, N.C., Cassou, C., Giannini, A. and Watanabe, M. (2015) 'Towards predictive understanding of regional climate change', *Nature Climate Change*, vol 5, no 10, pp921–930.

Yoshikawa, K., Bolton, W.R., Romanovsky, V.E., Fukuda, M. and Hinzman, L.D. (2002) 'Impacts of wildfire on the permafrost in the boreal forests of interior Alaska', *Journal of Geophysical Research: Atmospheres*, vol 107, no D1, art 8148.

Zardi, G.I., Nicastro, K.R., Serrão, E.A., Jacinto, R., Monteiro, C.A. and Pearson, G.A. (2015) 'Closer to the rear edge: Ecology and genetic diversity down the core-edge gradient of a marine macroalga', *Ecosphere*, vol 6, no 2, pp1–25.

Zhang, T., Barry, R.G., Knowles, K., Heginbottom, J.A. and Brown, J. (1999) 'Statistics and characteristics of permafrost and ground-ice distribution in the Northern Hemisphere', *Polar Geography*, vol 23, no 2, pp132–154.

Zhao, M. and Running, S.W. (2010) 'Drought-induced reduction in global terrestrial net primary production from 2000 through 2009', *Science*, vol 329, no 5994, pp940–943.

10

Change from within
Bottom-up disturbances of ecosystems

James M.R. Brock and Sarah V. Wyse

Introduction

The term *bottom-up disturbance* is used to describe disturbance events that originate within an ecosystem and the subsequent response or recovery processes. A bottom-up disturbance comprises a discrete event in time that disrupts an aspect of the ecosystem, be it an individual, a community, or a population, and changes the available resources, safe establishment sites, and nature of the physical environment (White and Pickett 1985; White and Jentsch 2001). A key aspect of bottom-up disturbance events is that they are generally relatively small in extent, such as the effects of the fall of a single tree, although they may also involve large stands of a tree species. A senescent tree dropping branches, or tree ferns and palms dropping senescent fronds, is evident at a small extent (area observed), and although a disturbance such as this diffuses at a landscape scale, it influences the regeneration of canopy species in a forest. The effects of a fire that removes the vegetation structure from 50% of a forested area will likely be evident at a range of extents, both temporally and spatially. An important element of a disturbance event is the response of the ecosystem to recovery, and the various likely pathways of different disturbance types are considered (Nimmo et al. 2015). In contrast to disturbance events themselves, which take discrete periods of time, the effects of these disturbance events can take much longer periods of time to resolve within the ecosystem.

Disturbance occurs along a continuum from those that originate from within a community (endogenous/bottom-up) to those that originate outside a community (exogenous/top-down). An example of a bottom-up disturbance is an insect outbreak, but of a species that is indigenous to the ecosystem rather than an invasive species. The literature on disturbance regimes has emphasized top-down disturbances (see Chapters 8 and 9): those that originate from outside an ecosystem, e.g. drought, flood, fire, hurricanes, or volcanism. Bottom-up disturbances interact with top-down disturbances and extrinsic pressures, and we will highlight elements of these important interactions. Overall, in this chapter, we will focus on those intrinsic events or processes that originate from within an ecosystem.

A key element of bottom-up disturbances is that they are largely biotic, i.e. relating to the organic components of an ecosystem rather than geological or meteorological processes. Bottom-up disturbances can include death of individuals or loss of populations, the abnormal increase in abundance of one component of a community, pathogens and disease, or even the actions of one species creating a micro-habitat. Biotic events such as these can be highly localized in spatial extent; however, scale is important when interpreting the effects of disturbance events such as these on an ecosystem. A canopy tree falling in the forest constitutes a large bottom-up disturbance in a terrestrial ecosystem. In forests where canopy trees can reach 30–40 meters in height, a mature and senescing tree could affect an area of up to 400 m² when it falls.

Disturbances such as landslides, avalanches, earthquakes, high winds, floods, etc. are all essentially extrinsic, from outside the ecosystem, and many of these are directly or indirectly triggered by climate or weather. These extrinsic disturbances are therefore top-down and likely to affect considerably larger areas than bottom-up processes. Furthermore, bottom-up disturbances typically result in secondary successions, i.e. not all elements of the local biota are destroyed or removed in the disturbance event, and the response process is not slowed by a lack of soil. The response processes, therefore, though dependent on residual propagule sources, have the potential to be quicker than many of those experienced in response to top-down disturbances; however, this is not always the case, e.g. the rapid re-forestation after the Mount Saint Helens eruption (Dale et al. 2005). For example, the return to a closed forest canopy from the death and collapse of a mature tree is likely to be quicker than the return to a closed forest after a major volcanic eruption, provided local seed sources remain available. Top-down events can trigger both secondary and primary succession processes depending on whether soil structures remain after the disturbance event. However, the distinction between top-down and bottom-up can also be subtle and scale-dependent, and top-down and bottom-up disturbances may interact, rendering a dichotomous differentiation between the two categories difficult in many instances.

When considering how a system responds to disturbance, it is important that succession, changes in ecosystem, or recovery of an ecosystem (i.e. the establishment of a functioning biotic community that may or may not have the same species composition as the pre-disturbance community) are thought of in terms of ecosystems being complex, dynamic systems. The basis of dynamic ecological systems is that heterogeneity, i.e. a mosaic of habitats in the landscape, is both the response to and an outcome of natural processes; the mix of abiotic and biotic conditions within an ecosystem that drives intrinsic variability are part of an ever-changing landscape. The processes described here differ from linear and predictable Clementsian succession, i.e. where disturbance is perceived as an unnatural process and all ecosystems will ultimately return to a prior, stable species assemblage (the 'climax') (Clements 1936). Regeneration of a forest, for example, describes the return of an area affected by a bottom-up disturbance event to an ecosystem with a comparable structure and function to that prior, if not in composition. Ecosystems have evolved in response to and alongside disturbance regimes, and their structure, composition, and function are a direct derivative of these regimes (Wyse et al. 2018; Chapter 8).

Top-down disturbance events, such as fire or hurricanes, are frequently considered, from a human perspective, to be catastrophic. Bottom-up disturbance events are rarely considered catastrophic; however, this is a perspective limited by human perception of spatial and temporal scales, as well as the scales usually addressed by ecological studies (Estes et al. 2018). The actions of individual species in an ecosystem can create an array of different micro-disturbances that fundamentally alter establishment opportunities for taxa, none of which are as obvious as a fire or hurricane. Take, for example, a beaver building a dam and flooding an area of forest. This event constitutes a suite of comparatively small (in relation to the forest as a whole), local (influencing the area immediately about the felled tree only) disturbances (i.e. felling of individual trees within a forested landscape) that not only have short recovery times but in the case of the flooding event, can fundamentally alter the local community and trajectory of the ecosystem (in this case, from forest to wetland; Turner et al. 1993).

Small-scale disturbances increase the heterogeneity of the landscape, creating micro-habitats and safe establishment sites for species. Increased variability of habitat in an ecosystem provides niche space for a larger range of species diversity. Turnover of species in a small area, a subset of an ecosystem, drives an increase in species richness, increasing the complexity and diversity of ecosystems over time. Multiple local bottom-up disturbance events across different structural layers create complex temporal and spatial mosaics in species and communities, successional trajectories, and the functional dynamics of ecosystems. This multi-facetted heterogeneity is also influenced by top-down shifts in climate alongside, for example, tectonics, and non-native plant and animal species.

Drivers of bottom-up disturbances

Bottom-up disturbances can take the form of press (slow changes in ecological factors, such as the construction of a beaver dam that floods a forest habitat), pulse (abrupt changes, e.g. a flood event), or ramp (continuously increasing pressure, e.g. a drought) disturbances (Lake 2000; Nimmo et al. 2015) (Figure 10.1). If we view a disturbance as a change in the availability of resources in an ecosystem, then we can identify a broad range of drivers of bottom-up disturbance (Jentsch and White 2019).

Forest gap-forming processes

In forest ecosystems, gap-forming events are among the most frequent bottom-up disturbance processes. These pulse disturbances are inherently intrinsic to the ecosystem when caused by factors such as senescence of aging individuals, and at their smallest extent, may be limited to the disturbance associated with the fall of a single individual or even a component of a single individual (e.g. a branch or frond). Where even-aged stands have formed following a previous disturbance, senescence of multiple individuals may occur during a relatively small timeframe (relative to the longevity of the dominant species), opening up larger gaps, while stand-scale disturbances may occur by wind-throw and snow-break events or small landslides. These disturbance events can facilitate the regeneration of canopy species and light-demanding sub-canopy species in forest ecosystems. For example, in New Zealand's Nothofagaceae-dominated forests, gap-forming events trigger the regeneration of more shade-tolerant canopy species such as silver beech (*Lophozonia menziesii*), while larger stand-level events initiate the release of seedling pools of light-demanding canopy species such as red beech (*Fuscospora fusca*) (Ogden 1988; Figure 10.2; Ogden et al. 1996).

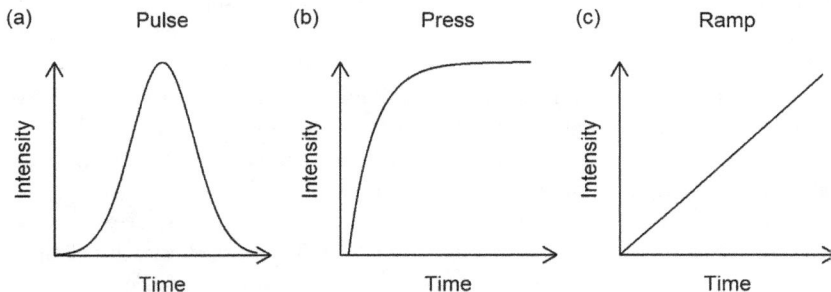

Figure 10.1   Schematic diagram depicting (a) pulse, (b) press, and (c) ramp disturbances with respect to the intensity of the effects on an ecosystem over time. Disturbance intensity can be defined as the strength of the effect (i.e. energy involved) per unit space and time.

Figure 10.2 Red beech, *Fuscospora fusca*, regenerating on a fallen silver beech, *Lophozonia menziesii*.

A mosaic of disturbance events across different spatial and temporal scales acts to maintain diversity in forest ecosystems, which is particularly evident when considering the second of the dominant New Zealand forest types: conifer-angiosperm forests (Ogden et al. 1996; McGlone et al. 2017). These forests can be regarded as 'two-component' systems, which respond to disturbances of different spatial extents and frequencies (temporal: period of return). The light-demanding and long-lived conifer components, dominated by species in the Podocarpaceae but also *Libocedrus* species (Cupressaceae) and the forest-dominant New Zealand kauri (*Agathis australis,* Araucariaceae), typically regenerate following large disturbances, which often include top-down events such as volcanism, earthquakes, or extreme weather events. Once the initial cohort reaches maturity, falling senescent individuals may create gaps of sufficient size for light to reach the forest floor and increase the growth and survival rates of conifer seedlings. Thus, as the initial cohort senesces, it is potentially replaced by a second and possibly subsequent cohorts. However, many gaps are captured by shade-tolerant broadleaved species, and thus, the successive conifer cohorts decrease in density (Figure 10.3). These more shade-tolerant broadleaved angiosperm species, which include members of the Lauraceae, Lamiaceae, and Meliaceae, typically regenerate in response to smaller, and more frequent, bottom-up disturbance events (Brock et al. 2018). Fine-scale pulse disturbance events are evident in the understory of forests, where taxa such as palms and tree ferns drop large, dead fronds (up to 25 × 3 m in species such as *Raphia regalis* in tropical moist forests of west Africa) that snap saplings and crush seedlings, creating gaps on the forest floor. Establishing seedlings in these micro-gaps (also caused by limbs from canopy trees and the senescence and fall of small trees) will not only experience reduced competition by the removal of established plants but may also benefit from increased light levels due to the reduction in biomass immediately above the site.

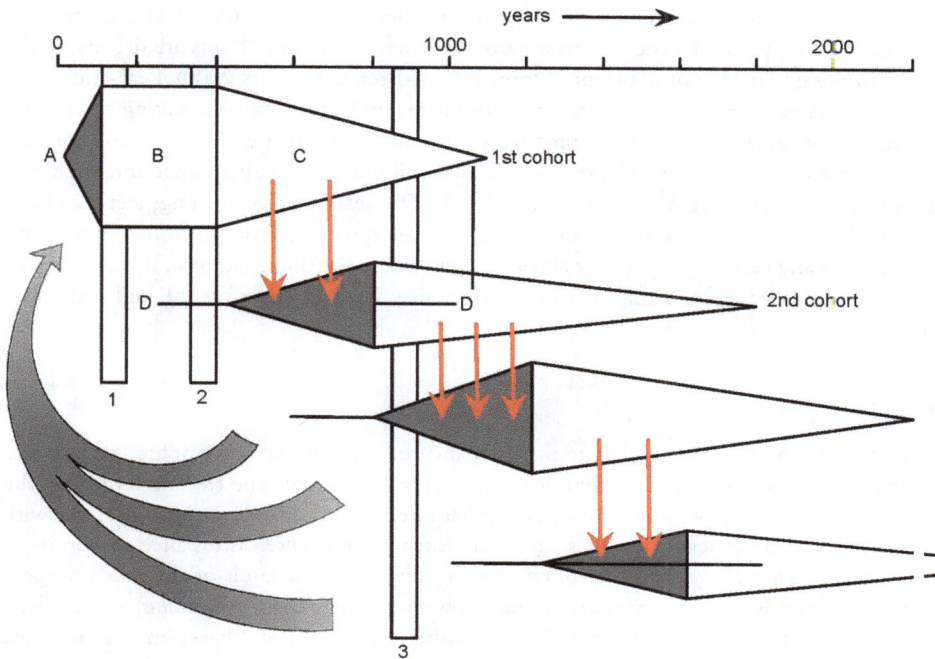

Figure 10.3 Species cohort regeneration in either top-down disturbance (grey arrows) or bottom-up disturbances (red arrows) such as canopy tree mortality. (Adapted from Ogden, J., *New Zealand Journal of Botany*, 23, 751, 1985.) The first cohort is divided into the recruitment phase during which seedling numbers and cohort biomass increases (A), thinning phase during which numbers of individuals decreases but biomass increases or remains constant (B), and a senescent phase during which biomass decreases (C). The second cohort (D) is only recruited when the first cohort begins to senesce. Columns 1, 2, and 3 are 50-year time spans during which observation of the population would reveal different population demographics.

Beyond senescence, biotic drivers of canopy tree death and disturbances to understory shrubs and small trees can include mega-fauna such as elephants (*Loxodonta* spp.). Elephants in the savannah systems of southern Africa can drive the mortality of trees through felling, debarking, and uprooting. The effects of elephant activity on tree mortality depend on the tree size classes affected, the density of elephant activity in time and space, and interactions with the weather (Chafota and Owen-Smith 2009). In Burkina Faso, elephants not only contribute to canopy tree mortality but also disperse arbuscular mycorrhizal fungi and are therefore likely to influence the soil microbiota and potentially have long-term effects on biogeochemical fluxes and the vegetation community (Paugy et al. 2004; Schmitz et al. 2018). In the northern Americas and parts of Europe, a common press disturbance in forests is caused by beaver activity (*Castor canadensis*) and other, human activities. The loss of disturbance events driven by species such as beaver can have significant long-term effects on the species present in ecosystems and the stability of the ecosystem itself (Brown et al. 2018). As with the tree ferns, smaller disturbances can occur from animal foraging that results in the loss of seedlings and occasionally, saplings. Examples of this dynamic can be seen with pigs (*Sus scrofa*) in forests, where they root for roots, shoots, and fungi, creating areas of disturbed soil clear of vegetation (Krull et al. 2013).

The extent to which the activities of a given species can be considered as a bottom-up disturbance is exemplified by the discussion around whether leaf-cutter ants are drivers of disturbance in the understory of neotropical forests (Farji-Brener and Illes 2000). Leaf-cutter ants collect and disperse leaf material and fruits of forest tree and shrub species, driving concentration gradients of organic material and propagules around their nest sites. A direct result of leaf-cutter ants moving into an area of forest is increased soil nutrition with a concomitant change in local shrub species (Farji-Brener and Silva 1995). The term 'ecological engineer' has been used to highlight the impacts of certain species with an apparent disproportionate impact on the function of any ecosystem (Jones et al. 1994). Identifying the threshold of significance when it seems that any species operating in an ecosystem may have an effect on the said ecosystem remains elusive.

Trophic cascades

Trophic cascades occur when a change in abundance of a species at one trophic level, be it a primary producer or apex predator, has flow-on effects up or down the trophic levels (i.e. the food-chain). An example of a disturbance event that directly impacts one trophic level but with flow-on effects to other levels can be seen in the impacts of a pulse disturbance to herbivory of grey willow (*Salix glauca*) and bog birch (*Betula glandulosa*) as a result of the predator–prey dynamics between lynx (*Lynx canadensis*) and snowshoe hares (*Lepus americanus*) in northern Canada. Snowshoe hares can reach population densities of up to 1,500 hares km^{-1} sometimes, providing abundant food supplies for lynx. In response to increased prey availability, the lynx population begins to swell, and a period of considerable predation pressure on snowshoe hares begins. As the hare populations decline, lynx health declines commensurately, and the lynx populations subsequently decline, reducing predation pressure on the snowshoe hares. During winter months, when vegetation is scarce or covered with snow, these hares predominantly graze on two species of shrub: grey willow and bog birch. If hare populations are high, there can be considerable impacts on the vegetation communities, with a selective effect impacting these shrub species. There are multiple effects of this over-grazing, as the reduction in food for hares causes starvation in the population and a reduced food source for other mammals. By reducing canopy cover, the localized loss of the shrub cover provides opportunities for grasses and mosses to establish, thereby driving successional processes. Other examples of a temporal release of predation driving herbivore disturbances in ecosystems can be seen where predation is reduced through disease in the predator population. Lions (*Panthera leo*) suffer a variety of diseases, such as canine distemper virus, across their distribution in Africa. When their populations decline or go locally extinct due to disease outbreaks, there is a reduction of predation pressure on herbivores such as zebra and buffalo. The response in the herbivore community to the loss of local predators is a significant increase in numbers that puts a strain on food resources.

Prey-switching can lead to different types of disturbance in ecosystems. Short-term or seasonal switching will drive a series of pulse disturbances (as with the lynxes and hares), whereas long-term change to predator–prey dynamics can drive press disturbances in ecosystems, such as with the loss of an apex predator through over-fishing or hunting. A long-term switching of prey species by a predator can be intrinsically driven, such as through disease in the prey population. Many examples of prey-switching are driven by local extinctions as a result of anthropogenic forces; therefore, localized intrinsic effects are the result of extrinsic processes (e.g. climate change). The main responses to prey-switching, as a bottom-up disturbance, would be as described previously, with increases in herbivory pressure driving gaps in vegetation structure, increase in fecal material driving changes in nutrient cycling, or vice versa, the loss of beaver (local extinction as a disturbance) driving a response that includes a collapse in freshwater components of an ecosystem with wetlands becoming forest.

Pest outbreaks

Mortality induced by outbreaks of plant pests and pathogens can be a significant cause of vegetation disturbance. These attacks are often species- or cohort-specific, and thus, entire stands are affected at once, opening up large areas of space. In the case of species-specific pathogens, a dominant canopy species may be removed from a forest ecosystem, often with far-reaching consequences for community change. An example of the loss of some species to pest or pathogen attack is eastern hemlock (*Tsuga canadensis*) in North American forests, where the hemlock woody adelgid (*Adelges tsugae*) has caused a shift in canopy composition that is altering understory composition, flora communities, and also hydrological regimes (Ellison et al. 2005). In some ecosystems, these outbreaks and the associated loss of canopy species are cyclical and associated with the age of the stands or exacerbated by weather conditions, while in others, the pests or pathogens may be introduced from other systems with devastating consequences. Cyclical outbreaks of bark beetles (Circulionidae: Scolytinae) in pine (*Pinus* spp.)-dominated forests in North America are associated with exogenous events, such as wind damage, that weaken host trees or threshold temperature conditions that influence over-wintering success in the beetle populations. There also appears to be an interaction between bark beetles and fire: over 10 years post outbreak, the increased abundance of standing dead trees drives increased fire severity (Harvey et al. 2014). Severe outbreaks (can) lead to widespread loss of host species and thus forest disturbance, but outbreaks cease as the forest thins. Conversely, pathogen attack can result in significant and ongoing disturbance to forest communities, with the potential to alter community composition indefinitely. In Australian jarrah forests, the dominant species, *Eucalyptus marginata*, can undergo synchronous collapse following waterlogging, which aids infection by the introduced *Phytophthora cinnamomi* (Burgess et al. 1999), while *Phytophthora ramorum* is responsible for causing sudden death of oak species in forest ecosystems in the western United States. American chestnut (*Castanea dentata*) was a dominant canopy species in North American forests, but an introduced chestnut blight saw their almost complete loss from the canopy by the mid-20th century, and this has not recovered (Ellison et al. 2005).

Fire

Fire is another potential bottom-up disturbance. Fire is a complex process, and the drivers of ignition are many; it is not necessarily the case that all wildfires are a result of extreme weather (i.e. drought), lightning, or human activity: sources of intrinsic ignition include spontaneous combustion and sparks from falling rocks, although these are rare. The impact of fire on an ecosystem will ultimately be in gap-formation (where gaps may span orders of magnitude in extent) but varies greatly with fuel load (an intrinsic property of the ecosystem), weather and climate, and the local fire history and other disturbance history. These factors influence the intensity (heat) and spread (extent of impact) of the fire, thus determining the effects of the fire on the ecosystem and communities therein (Keeley 2009). Fires rarely completely reset ecosystem structure and function, as some plants will re-sprout (such as tree ferns rapidly producing new fronds and *Eucalyptus* species sprouting from epicormic buds) and so persist through multiple fire events, while for many species, fire provides an opportunity to regenerate and capture space that was previously occupied by another species. In this manner, repeated fires over time and across different areas of forest, grassland, and wetland ecosystems contribute to ecosystem heterogeneity, driving secondary succession and increasing niche diversity. Ecosystems that are intrinsically linked with fire occur all over the world. Temporal predictability, in other words the variance around the mean return interval of fire, is important for the effects that fires have on ecosystems and the ecosystem's response. For example, in the fynbos heathland ecosystem in the Cape Province of South Africa, fires return after between 10 and 30 years, and plants' reproductive strategies may be cued to these intervals; likewise, in the Brazilian cerrado, fires are a regular event controlling biomass across vegetation communities. However, any changes to fire periodicity can drive homogenization effects in these communities (Enright et al. 2015). If there is a prolonged period without fire (up to 45 years) in the Kwongan Heathlands in Western Australia, there is a physiological response of structural and canopy senescence in the remaining plants (Bowman et al. 2016). However, in systems where fire has not been a component of the bottom-up disturbance regime, ecosystem resilience can be significantly affected by fire events. Where extrinsic pressures from climate change and other processes are driving novel fire events in naïve ecosystems, for example in New Zealand, where naturally occurring fires are a historical rarity, native species are losing out to competitive invaders like pine (*Pinus*) and wattle (*Acacia*) species (Perry et al. 2014). Additionally, the response in the ecosystem is not simply on the part of vegetation, as there can also be a clear change in trophic structures/food-webs in response to fire. For example, hares (*Lepus europaeus*) in Mediterranean regions are usually in greater abundance in fire scars that are several years old.

Anthropic disturbance

In many modern cultures, human activity has passed from intrinsic to extrinsic; humans are no longer an intrinsic functional component of an ecosystem, although through niche construction and modification of the landscape, we have driven the assembly of entirely novel ecosystems (such as indoor biomes). Many indigenous communities, where traditional cultures can still be expressed and practiced, function intrinsically within ecosystems, and historically at a global level, human activity was localized and interdependent of ecosystem resources. For example, in Europe, the traditional harvesting of timber products through coppicing opened up gaps in forest and woodland ecosystems that drove habitat heterogeneity and increased the available niches and subsequently, the diversity of flora and fauna in ecosystems. Aboriginal communities in Australia undertook controlled burning of the understory forests to manage resources, which maintained ecosystem structure in the landscape and, for example, likely contributed to the regeneration of the cypress pine (*Callitris intratropica*) (Trauernicht et al. 2016). All these behaviors enact disturbance events in an ecosystem that drive gap-forming events and heterogeneity in the landscape. In most traditional landscape practices, these disturbances would contribute to the maintenance of an increased range of niche space in ecosystems, driving positive effects on local biodiversity.

Discerning a threshold that defines when human activity should no longer be considered intrinsic to an ecosystem is complex. One such threshold that is used in the classification of plant introductions in the United Kingdom is the date of 1492; prior to this, human-introduced plant and animal species, such as walnut (*Juglans regia*) and rabbit (*Oryctolagus cuniculus*), are nominally native to the region (Preston et al. 2004). While this date is used not to specifically describe a change in human activity but to decide whether a tree species is native or alien, the implications are far-reaching. One of the reasons for this period being used as a threshold is that it was a period during which humans as a species were changing in their demography, industries, trade, and agricultural practices (trading and moving new species). As a result, there was a decreasing dependence on traditional practices and a move towards absolute (forest clearance for agriculture), impactful (driving extinctions), and extrinsic effects on ecosystems. Another aspect of changes in human behavior and social culture that drove extrinsic disturbances to ecosystems is migrations. Human migrations have occurred for millennia, and prior to any revolution in trade and industry, they drove extinctions and fundamental changes to ecosystems, for example the migration of people to the Pacific Islands (Kirch 2010). European colonization of countries and their indigenous cultures was also a key extrinsic disturbance, ecologically and culturally, and in colonized countries such as Australia, the arrival of colonists is used as a threshold for a shift from intrinsic to extrinsic influence of human activity (Kloot 1987). The change in socio-economic behavior coupled with, over longer time scales, the loss of traditional human activities and practices, as well as reduced ecological memory in the ecosystems (reduction of trait and propagule availability), has changed the bottom-up disturbance dynamic in many ecosystems (Johnstone et al. 2016). Changes such as a loss of coppicing and traditional burning practices have led to a loss of functional traits and diversity and an increased risk of large-scale landscape fire.

As with the loss of species from ecosystems, introduced vertebrate and invertebrate predators and herbivores also have substantial effects on ecosystem processes, including extinctions of plant and animal species and recruitment failures. However, as these species have been introduced by humans into these ecosystems, for example the introduction of deer (Cervinae and Capriolinae) and goats (*Capra hircus*) onto islands that lacked mammalian browsers prior to European colonization, we do not consider these as bottom-up sources of disturbance. Where these, or similar species, were introduced by Indigenous peoples with a similar deleterious outcome, it is not unreasonable to consider this a bottom-up dynamic; but such nuance suggests that a dichotomous definition (dividing top-down and bottom-up processes) is not always helpful. The effects of the activity and behavior of introduced animals can have significant effects on ecosystems when interacting with bottom-up disturbances. For example, the threat that feral cats (*Felis catus*; domestic cats breeding and living away from domestic associations with humans) present to global species and ecosystems can be seen in Australia, where feral cats will travel great distances to hunt in areas affected by natural wildfires, thus increasing predation in an area already affected by a pulse disturbance (McGregor et al. 2016). The extent and intensity of the fires occurring in the Catalina Mountains outside Tucson, Arizona at the time of writing (June 2020) are amplified by the presence of the invasive buffelgrass (*Cenchrus ciliaris*) – a species that was introduced to the United States to improve grazing for ranchers and has spread into indigenous ecosystems, altering their flammability. Therefore, invasive species can amplify the effects of disturbances on ecosystems that might drive affected ecosystems to tipping points: thresholds over which the composition, structure, and function of the ecosystem will change fundamentally (Gaertner et al. 2014).

Interactions among disturbance agents

Intrinsic and extrinsic ecosystem properties and disturbances can interact synergistically or in sequence to affect the frequency, intensity, and scale of disturbances, potentially resulting in profound effects on the ecosystem (Buma 2015). The composition of an ecosystem is an important intrinsic property that interacts with extrinsic processes such as climate and weather patterns to affect disturbance regimes, such as fire, hurricane, and ice-storm frequency and intensity. In many forest ecosystems, including conifer forests in North America or some *Eucalyptus* forests in Australia, the propensity of an ecosystem to burn increases with time following a fire, as there is a build-up of flammable biomass, especially dead fuels (Perry et al. 2014). Fire ignition is more likely to translate into a large disturbance event when forests are mature (i.e. there is abundant dead biomass). Species in these ecosystems have adapted to such regimes (not the disturbance per se), with many species having the ability to re-sprout epicormically or from lignotubers, or storing their seed in canopy seed banks in serotinous cones or capsules, which are released following the fire (Keeley et al. 2011). Conversely, in systems such as New Zealand, the propensity of a forest ecosystem to burn peaks in early stages of succession and then declines as the forest ages (Kitzberger et al. 2012; 2016). In these systems, the early successional species are the most flammable, while more mature forests contain fewer flammable species and have microclimates less conducive to fire spread. Thus, fire as a forest disturbance is more likely in early successional systems (Perry et al. 2014). In addition, positive feedback loops can occur that may see systems become 'stuck' in these early successional states. This may happen where fire frequency is increased by extrinsic factors that increase the likelihood of forest fire, such as human activities, dispersal failure, and climatic events (i.e. droughts), or introduced plant species that increase the flammability of early successional systems even further. In these New Zealand ecosystems, stalled successions may also occur as a result of introduced mammalian browsers that prevent the regeneration of palatable later-successional canopy species, again maintaining ecosystems in flammable, early successional communities (Richardson et al. 2014).

Top-down disturbances, such as fires and extreme weather events, also influence intrinsic sources of forest disturbances. A notable example is in North American conifer forests, where climate-forced drought events interact with bark beetle infestations to drive abnormally high levels of pine mortality. Increased canopy mortality has a cascade effect in forest ecosystems, where the loadings of deadwood can have significant effects on the intensity and extent of natural fires; increased intensity and increased extent means a significant decrease in the availability of conifer propagules, leading to an angiosperm-dominated ecosystem (Jewett et al. 2011). Changing climates and weather patterns can also influence forest disturbance through interactions with plant pathogens. Climate changes towards warmer and wetter conditions are associated with outbreaks of pine shoot and needle blights in Europe and North America, while flooding can increase both the spread of *Phytophthora* species and host susceptibility to the pathogens due to flooding stress in the host plants.

Response pathways

Ecosystem response is influenced by the extent, intensity, and return period (frequency) of the disturbance regime. After a disturbance, a system can:

1) Return to its previous state but at a different successional or temporal phase compared with adjacent unaffected areas; or
2) Shift across tipping points to an alternate stable state supporting both native species (e.g. a shift from forest to savannah ecosystems) and novel ecosystems (a type of alternate stable state) that support non-native species in a new assemblage (Figure 10.4) (Hobbs et al. 2012).

The emergence of novel ecosystems is a response to both bottom-up and top-down disturbances where exotic species are involved; i.e. the resulting species assemblage combines species that evolved in ecosystems in different biogeographic contexts and that come together in an introduced range alongside native species. In such cases, these novel assemblages represent the result of a perturbation. However, interactions with disturbance agents, such as fire, can reinforce the shift to this novel alternative stable state. The response element of the perturbation drives landscape patterns of heterogeneity.

In the following, we describe these two recovery pathways in more detail.

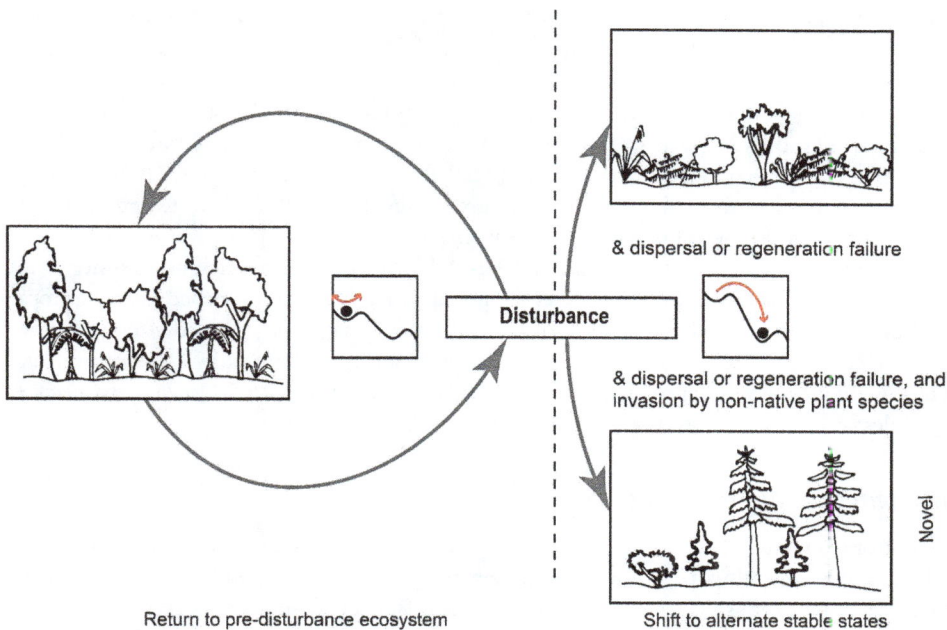

Figure 10.4 Post-disturbance response pathways highlighting the potential for novel communities to establish as an alternate stable state. The ball represents the ecosystem state and settles in stable states (the 'valleys'). 'Hilltops' are unstable states. A large perturbation is required to shift the ecosystem from one stable state to another (i.e. to get the ball over the hill to an adjacent valley).

Ecosystem returns to previous state

The most common response to a bottom–up disturbance is a secondary succession that returns the disturbed area to a structure, species composition, and function comparable to that present previously, and which is present in the landscape around the area affected. By this, we do not imply that there is a like-for-like replacement of individuals and species. Rather, species that were present prior to the disturbance would return to the disturbed area, and a similar structure would eventually re-establish (Bormann and Likens 1979). Over time and space in the landscape, bottom–up disturbance processes lead to differentially aged patches of the same ecosystem type: the patchwork or mosaic provides a range of nutrients and resources, heterogeneous niche space, that are available to flora and fauna present in the landscape. Each patch is, in essence, the same ecosystem type but at a different temporal stage of succession (the steady state shifting mosaic of Bormann and Likens 1979).

The extent of the individual patches across the landscape is determined by the magnitude of the disturbance. Annually, 100,000 to 200,000 km² of the Brazilian cerrado (wetland-savannah-forest) burns as part of a natural fire regime; however, individual fires range from 0.01 to 5 km² in extent, with several thousand fires occurring across the part of Brazil that supports cerrado (Miranda et al. 2009). At the other end of the spectrum, a falling canopy tree could impact up to 400 m² of forest, a falling tree fern frond a mere 12 m² (Delcourt et al. 1982; Brock et al. 2018). While falling fronds from understory tree ferns and palms cover an area orders of magnitude smaller than the extent of a cerrado fire, given the prominence of these guilds in forest understory's and that fronds drop annually (very different temporal scaling; Turner et al. 1993), the cumulative impact on the regeneration niche can be significant (Aguiar and Tabarelli 2010; Brock et al. 2018).

A key limiting factor to the ability of an ecosystem to return to a stable state is the availability and dispersal into the disturbed area of plant and fungal propagules. Dispersal can be influenced in multiple ways, either through reduced dispersal function (whether lack of available propagules to disperse into an area, or distances too great / dispersal vector absent for propagules to arrive) or through the loss of safe establishment sites when the propagules arrive in the system (propagules will arrive but not germinate/establish). If the extent, frequency, or intensity of the disturbance increases such that propagules cannot disperse into the disturbed area due to an interaction with extrinsic drivers, the disturbance response may derive an ecosystem dissimilar to that which was present before the event. Likewise, shifts in community composition may occur if the concentration of propagules on arrival at the disturbance site is low due to a lack of propagules being dispersed or high predation of propagules on dispersal.

Ecosystem develops to alternate condition

The bottom-up disturbances and recovery processes that we have described are generally local and create a patchwork or mosaic of different-aged vegetation stands with nutrients and resources heterogeneously distributed in the landscape. Bottom-up disturbances in combination with extrinsic forces, such as climate change and tectonic events, can lead to significant changes in ecosystem structure, composition, and function.

One possible outcome of a change in post-disturbance ecosystem is an 'alternate stable state' (Scheffer et al. 2001). For example, with a top-down (climactic) press disturbance, forest ecosystems can suffer from canopy die-off, and over time, a shift to a savannah ecosystem can occur (however, the presence of grasslands in the landscape should not be assumed to result from a collapsed closed forest). Trophic cascades after top-down disturbance can drive a series of collapses that will cause a shift to an alternate stable state (Estes et al. 2011). For example, in the 18th Century around islands in the Bering Sea, humans hunted both Steller's sea cow (*Hydrodamalis gigas*) (a 9-m, 9-tonne sirenian) and sea otter (*Enhydra lutris*) around the Commander Islands. The impacts to the sea otter, in particular, led to the release of predation pressure on sea urchins (Echinoidea); the increasing populations of sea urchins subsequently devastated the kelp (Laminariales) forests (pulse disturbance) around the islands, which were the main food resources for the sea cow (Estes et al. 2011; 2016). The lack of food, along with anthropic-driven mortality, drove the sea cow extinct, entirely and irreversibly altering the structure of the ecosystem. Coral reefs in the Pacific regularly experience top-down disturbance from cyclones; however, if this occurs alongside an outbreak of crown-of-thorn starfish (*Acanthaster planci*), significant areas of the reef can be destroyed. The areas of destroyed reef provide habitat for macro-algae, which provides a significantly increased food resource for herbivorous fishes: extrinsic and intrinsic forces combine to change a coral reef to an algal bed (Done 1992).

Novel ecosystems

If we consider the interaction of successional processes or alternate stable states in the context of climate change, anthropic effects (e.g. fire, nitrification), and in particular invasive species, the response pathway to a bottom-up disturbance might see the emergence of novel ecosystems. Where gap-forming processes previously saw the regeneration of the ecosystem and the maintenance of diversity through patch formation, the presence of new limiting factors driven by climate change, the presence of novel species in the ecosystem, and a shift in nutrient cycling due to extrinsic anthropogenic factors can result in entirely novel ecosystem structure, composition, and function. Novel ecosystems will arise, in particular, where key ecosystem processes are affected, such as pollination and dispersal, and where niches are made available to species that have evolved in a different biogeographic context through multiple processes at a local level, including decline in abundance, misalignment through disturbances, and extinction. Novel ecosystems are the product of iterations of extinction, invasion, and environmental change, a nexus of disturbance, both top-down and bottom-up. An example of a novel ecosystem can be seen in northern New Zealand in a landscape that has experienced anthropogenic fire (driving loss of nutrients and soils), agriculture (changing nutrient cycling), invasion (rodent seed predators), and extinction (loss of pollinators and dispersers of plant propagules). In these systems, this combination of pressures and altered disturbance regimes has led to the development of ecosystems dominated by plants from Australia, North and South America, and Europe that support a range of mammal and bird species from Europe and drive a regionally and historically uncommon disturbance regime (fire).

Disturbance in the face of global change

While threats to ecosystem resilience and function may be largely top-down, there are important questions around interactions between disturbance regimes and types in the future. Anthropogenic disturbances are driving changes to climate, species extinctions, and species invasion, and causing fragmentation of homogenizing ecosystems. The processes of extinction, invasion, shifts in biogeochemical cycles, and fragmentation will intrinsically alter bottom-up disturbances in ecosystems as the species composition, structure, and function of ecosystems change in response to these top-down pressures. Although homogenization is a product of both extinction and invasion, it is addressed here as a specific impacting factor on bottom-up events and processes.

Local species extinction (extinctions of species from an ecosystem) can fundamentally alter the disturbance regime. Imagine that the beaver is lost from a landscape: canals and pools will silt up, and no new flooding events will occur, which will change the hydrology of the ecosystem with the concomitant loss of wetland ecosystems. Furthermore, with no trees being felled to create dams and lodges, the canopy will mature and close, and the loss of local disturbances from felling will reduce the regeneration of light-demanding tree species. The loss of early successional terrestrial and freshwater habitat, created through disturbances caused by beaver activities, will reduce species richness in the ecosystem and in the longer term, change both the structure and the function of the ecosystem. In Yellowstone National Park, for example, over-grazing of riparian woody species by elk drove beaver from catchments; subsequent to the loss of beaver, and the collapse of dams, streams eroded gullies in the landscape, the water table dropped, and valleys that once supported pools and wet grasslands are now dry with channelized water courses.

An extreme example of a loss of bottom-up disturbance in ecosystems is the loss of megaherbivores such as mammoths and musk-ox from high-latitude ecosystems (Willerslev et al. 2014). The shift in nutrient cycling regimes due to the loss of disturbance associated with grazing animals is suggested to have caused a change in species composition in high-latitude grasslands, moving from grass and herbaceous species to a moss-dominated community (tundra). Changing from grass to moss meant a loss of root material in the topsoil, reducing the temperature of the soils and driving the development of permafrost. Whether or not this is the causal process for the development of permafrost during the late Pleistocene in these regions of the world is open for debate. In terms of the examples given in the forest gap-forming section, a further significant loss of disturbance can be seen in the southern Americas with the loss of tunnelling giant ground sloths. The excavations that these species undertook were vast and likely had a significant disturbance effect on the ecosystem.

One mechanism for managing the homogenization effects of anthropic activities on disturbances in ecosystems is to reintroduce, where possible, species lost from ecosystems. The purpose of undertaking reintroductions is to reinvigorate, or remind the landscape of, ecosystem functions that were lost when the species originally went locally extinct (Schweiger et al. 2019). The reintroduction of species of burrowing mammals into Australian landscapes increases disturbance rates in soils, which changes nutrient cycling, soil moisture, and seedling establishment. The reintroduction of rare horse breeds into grasslands in Sweden has increased not only plant species richness but also richness and pollination interactions from bees and butterflies. Returning to Yellowstone National Park, the success of introducing wolves as predatory controls on elk has in part driven localized regeneration of wet woodland alongside watercourses; however, without beaver returning and damming watercourses, many valleys in the park have not benefited from the return of the wolf. Restoration of one component of the disturbance regime of an ecosystem, particularly where changes to the historic ecosystem are significant, has not been powerful enough to rectify the loss of multiple important disturbance functions. Various re-wilding projects in the United Kingdom propose to reintroduce a number of species to regions of the country; however, the effects of the disturbance interactions caused by the activity of these species (lynx, wolf, beaver) are yet to be determined. The unknowable outcomes of reintroducing complex functions are in part derived from the absence of these species in the landscape for hundreds of years and the subsequent introduction of non-native species. A landscape-level experiment in Siberia that has been ongoing since 1996 (Zimov 2005) is attempting to answer the question as to whether the loss of disturbance from grazing mega-fauna drove the development of a species-poor tundra and permafrost by reintroducing as many large grazing animal species as possible, those species that remain extant.

Current rates of habitat loss and fragmentation are rescaling disturbance regimes; in other words, a recurrent fire of consistent extent in the same landscape will have an increased effect under increased fragmentation. In our discussion of the emergence of alternate states, we highlighted dispersal failure in recently disturbed areas as a key driver of regeneration failure. Fragmentation exacerbates this issue by amplifying the innate limitations of species dispersal – a reduction in propagule pressure through habitat loss (the loss of individuals of any species) reduces the ability of that species to disperse in the landscape, and where fragmentation is severe, propagules have a lower likelihood of reaching small and isolated fragments (Gowda et al. 2019). Bottom-up disturbance events, therefore, interact with extrinsic pressures to drive local extinctions and fundamental shifts in the composition, structure, and function of ecosystems.

A further example of an interaction is that between bottom-up disturbances and invasion of an ecosystem by non-native species. The Holocene (last c 10,000 years), and particularly the Anthropocene, has seen species moved from their native range to other regions by humans for cultural, agricultural, and ornamental purposes, as well as unintentional dispersal. Many of these species (~10%), although not all, will establish in their new landscape in the absence of predators/herbivores, adequate biotic competition, or in an available niche space and damage the ecological function and/or economic value of an ecosystem, causing extinctions of native flora and fauna along with fundamental disruption of ecosystems. Bottom-up disturbances will not introduce non-native species into areas per se, but where non-native species are present in the landscape, they may capture disturbed areas and displace those other species from dispersing into these same areas (Hobbs and Huenneke 1992). Invasion of non-native species, thus, continues a process of homogenization and species extinction. Species invasion has had a tragic impact on the biodiversity of islands (Tershy et al. 2015), where many thousands of indigenous species have been lost, for example the loss of endemic bird species on the Pacific island of Guam after the accidental introduction of the brown tree snake. The management of non-native species has come from two angles – both for preserving the intrinsic value of rare species and regional and culturally valued biodiversity, and to prevent damage to the local economy. Eradication of species such as these is challenging; however, eradication of rodents has been achieved in situations ranging from sub-Antarctic islands such as South Georgia (with significant benefits for the indigenous seabird population) to the extirpation of coypu (*Myocastor coypus*) in the United Kingdom.

Interactions between top-down and bottom-up disturbance can push ecosystems towards tipping points, thresholds whereby the availability of resources in ecosystems is so fundamentally changed that disturbance events will, rather than driving patch dynamics in a multi-aged community, drive a complete change in ecosystem type and function. Although tipping points have been observed and can be seen in some palaeoecological records (e.g. pollen from sediment cores), it is difficult to predict when such a threshold will be reached, and as such, how bottom-up disturbances will drive these events in the future.

References

Aguiar, A.V. and Tabarelli, M. (2010) 'Edge effects and seedling bank depletion: the role played by the early successional palm *Attalea oleifera* (Arecaceae) in the Atlantic Forest', *Biotropica*, vol 42 pp158–166.

Bormann, F.H. and Likens, G.E. (1979) 'Catastrophic disturbance and the steady state in northern hardwood forests: a new look at the role of disturbance in the development of forest ecosystems suggests important implications for land-use policies', *American Scientist*, vol 67, pp660–669.

Bowman, D.M.J.S., Perry, G.L.W., Higgins, S.I., Johnson, C.N., Fuhlendorf, S.D. and Murphy, B.P. (2016) 'Pyrodiversity is the coupling of biodiversity and fire regimes in food webs', *Philosophical Transactions of the Royal Society B: Biological Sciences*, vol 371, art 20150169.

Brock, J.M.R., Perry, G.L.W., Burkhardt, T. and Burns, B.R. (2018) 'Forest seedling community response to understorey filtering by tree ferns', *Journal of Vegetation Science*, vol 29, pp887–897.

Brown, A.G., Lespez, L., Sear, D.A., Macaire, J.-J., Houben, P., Klimek, K., Brazier, R.E., Van Oost, K. and Pears, B. (2018) 'Natural vs anthropogenic streams in Europe: history, ecology and implications for restoration, river-rewilding and riverine ecosystem services', *Earth-Science Reviews*, vol 180, pp185–205.

Buma, B. (2015) 'Disturbance interactions: characterization, prediction, and the potential for cascading effects', *Ecosphere*, vol 6, art 70.

Burgess, T., McComb, J.A., Colquhoun, I. and Hardy, G.E. St J. (1999) 'Increased susceptibility of *Eucalyptus marginata* to stem infection by *Phytophthora cinnamomi* resulting from root hypoxia', *Plant Pathology*, vol 48 pp797–806.

Chafota, J. and Owen-Smith, N. (2009) 'Episodic severe damage to canopy trees by elephants: interactions with fire, frost and rain', *Journal of Tropical Ecology*, vol 25, pp341–345.

Clements, F.E. (1936) 'Nature and structure of the climax', *Journal of Ecology*, vol 24, pp252–284.

Dale, V.H., Swanson, F.J. and Crisafulli, C.M. (eds) (2005) *Ecological Responses to the 1980 Eruption of Mount St. Helens*. Springer, New York.

Delcourt, H.R., Delcourt, P.A. and Webb, T. (1982) 'Dynamic plant ecology: the spectrum of vegetational change in space and time', *Quaternary Science Reviews*, vol 1, pp153–175.

Done, T.J. (1992) 'Phase shifts in coral reef communities and their ecological significance', *Hydrobiologia*, vol 247, pp121–132.

Ellison, A.M., Bank, M.S., Clinton, B.D., Colburn, E.A., Elliott, K., Ford, C.R., Foster, D.R., Kloeppel, B.D., Knoepp, J.D., Lovett, G.M., Mohan, J., Orwig, D.A., Rodenhouse, N.L., Sobczak, W.V., Stinson, K.A., Stone, J.K., Swan, C.M., Thompson, J., Von Holle, B. and Webster, J.R. (2005) 'Loss of foundation species: consequences for the structure and dynamics of forested ecosystems', *Frontiers in Ecology and the Environment*, vol 3, pp479–486.

Enright, N.J., Fontaine, J.B., Bowman, D.M., Bradstock, R.A. and Williams, R.J. (2015) 'Interval squeeze: altered fire regimes and demographic responses interact to threaten woody species persistence as climate changes', *Frontiers in Ecology and the Environment*, vol 13, pp265–272.

Estes, J.A., Terborgh, J., Brashares, J.S., Power, M.E., Berger, J., Bond, W.J., Carpenter, S.R., Essington, T.E., Holt, R.D., Jackson, J.B.C., Marquis, R.J., Oksanen, L., Oksanen, T., Paine, R.T., Pikitch, E.K., Ripple, W.J., Sandin, S.A., Scheffer, M., Schoener, T.W., Shurin, J.B., Sinclair, A.R.E., Soulé, M.E., Virtanen, R. and Wardle, D.A. (2011) 'Trophic downgrading of planet Earth', *Science*, vol 333, no 6040, pp301–306.

Estes, J.A., Burdin, A. and Doak, D.F. (2016) 'Sea otters, kelp forests, and the extinction of Steller's sea cow', *Proceedings of the National Academy of Sciences*, vol 113, pp880–885.

Estes, L., Elsen, P.R., Treuer, T., Ahmed, L., Caylor, K., Chang, J., Choi, J.J. and Ellis, E.C. (2018) 'The spatial and temporal domains of modern ecology', *Nature Ecology & Evolution*, vol 2, pp819–826.

Farji-Brener, A.G. and Illes, A.E. (2000) 'Do leaf-cutting ant nests make "bottom–up" gaps in neotropical rain forests?: a critical review of the evidence', *Ecology Letters*, vol 3, pp219–227.

Farji-Brener, A.G. and Silva, J.F. (1995) 'Leaf-cutting ant nests and soil fertility in a well-drained savanna in western Venezuela', *Biotropica*, vol 27, pp250–254.

Gaertner, M., Biggs, R., Te Beest, M., Hui, C., Molofsky, J. and Richardson, D.M. (2014) 'Invasive plants as drivers of regime shifts: identifying high-priority invaders that alter feedback relationships', *Diversity and Distributions*, vol 20, pp733–744.

Gowda, J.H., Tiribelli, F., Mermoz, M., Kitzberger, T. and Morales, J.M. (2019) 'Fragmentation modulates the response of dichotomous landscapes to fire and seed dispersal', *Ecological Modelling*, vol 392, pp22–30.

Harvey, B.J., Donato, D.C. and Turner, M.G. (2014) 'Recent mountain pine beetle outbreaks, wildfire severity, and postfire tree regeneration in the US Northern Rockies', *Proceedings of the National Academy of Sciences*, vol 111, art 15120.

Hobbs, R.J. and Huenneke, L.F. (1992) 'Disturbance, diversity, and invasion: implications for conservation', *Conservation Biology*, vol 6, pp324–337.

Hobbs, W.O., Hobbs, J.M.R., LaFrançois, T., Zimmer, K.D., Theissen, K.M., Edlund, M.B., Michelutti, N., Butler, M.G., Hanson, M.A. and Carlson, T.J. (2012) 'A 200-year perspective on alternative stable state theory and lake management from a biomanipulated shallow lake', *Ecological Applications*, vol 22, pp1483–1496.

Jentsch, A. and White, P. (2019) 'A theory of pulse dynamics and disturbance in ecology', *Ecology*, vol 100, art e02734.

Jewett, J.T., Lawrence, R.L., Marshall, L.A., Gessler, P.E., Powell, S.L. and Savage, S.L. (2011) 'Spatiotemporal relationships between climate and whitebark pine mortality in the greater Yellowstone ecosystem', *Forest Science*, vol 57, pp320–335.

Johnstone, J.F., Allen, C.D., Franklin, J.F., Frelich, L.E., Harvey, B.J., Higuera, P.E., Mack, M.C., Meentemeyer, R.K., Metz, M.R., Perry, G.L., Schoennagel, T. and Turner, M.G. (2016) 'Changing disturbance regimes, ecological memory, and forest resilience', *Frontiers in Ecology and the Environment*, vol 14, pp369–378.

Jones, C.G., Lawton, J.H. and Shachak, M. (1994) 'Organisms as ecosystem engineers', *Oikos* vol 69, pp373–386.

Keeley, J. (2009) 'Fire intensity, fire severity and burn severity: a brief review and suggested usage', *International Journal of Wildland Fire*, vol 18, pp116–126.

Keeley, J., Pausas, J.G., Rundel, P.W., Bond, W.J. and Bradstock, R.A. (2011) 'Fire as an evolutionary pressure shaping plant traits', *Trends in Plant Science*, vol 18, pp406–411.

Kirch, P.V. (2010) 'Peopling of the Pacific: a holistic anthropological perspective', *Annual Review of Anthropology*, vol 39, pp131–148.

Kitzberger, T., Aráoz, E., Gowda, J.H., Mermoz, M. and Morales, J.M. (2012) 'Decreases in fire spread probability with forest age promotes alternative community states, reduced resilience to climate variability and large fire regime shifts', *Ecosystems*, vol 15, pp97–112.

Kitzberger, T., Perry, G., Paritsis, J., Gowda, J., Tepley, A., Holz, A. and Veblen, T. (2016) 'Fire–vegetation feedbacks and alternative states: common mechanisms of temperate forest vulnerability to fire in southern South America and New Zealand', *New Zealand Journal of Botany*, vol 54, pp247–272.

Kloot, P.M. (1987) 'The naturalised flora of South Australia 1. The documentation of its development', *Journal of the Adelaide Botanic Garden*, vol 10, pp81–90.

Krull, C.R., Choquenot, D., Burns, B.R. and Stanley, M.C. (2013) 'Feral pigs in a temperate rainforest ecosystem: disturbance and ecological impacts', *Biological Invasions*, vol 15, pp2193–2204.

Lake, P.S. (2000) 'Disturbance, patchiness, and diversity in streams', *Journal of the North American Benthological Society*, vol 19, pp573–592.

McGlone, M.S., Richardson, S.J., Burge, O.R., Perry, G.L.W. and Wilmshurst, J.M. (2017) 'Palynology and the ecology of the New Zealand conifers', *Frontiers in Earth Science*, vol 5, art 94.

McGregor, H.W., Legge, S., Jones, M.E. and Johnson, C.N. (2016) 'Extraterritorial hunting expeditions to intense fire scars by feral cats', *Scientific Reports*, vol 6, art 22559.

Miranda, H.S., Sato, M.N., Neto, W.N. and Aires, F.S. (2009) 'Fires in the cerrado, the Brazilian savanna', in M.A. Cochrane (ed) *Tropical Fire Ecology: Climate Change, Land Use, and Ecosystem Dynamics*. Springer, Heidelberg.

Nimmo, D.G., Mac Nally, R., Cunningham, S.C., Haslem, A. and Bennett, A.F. (2015) 'Vive la résistance: reviving resistance for 21st century conservation', *Trends in Ecology & Evolution*, vol 30, pp516–523.

Ogden, J. (1985) 'An introduction to plant demography with special reference to New Zealand trees', *New Zealand Journal of Botany*, vol 23, pp751–772.

Ogden, J. (1988) 'Forest dynamics and stand-level dieback in New Zealand's *Nothofagus* forests', *GeoJournal*, vol 17, pp225–230.

Ogden, J., Stewart, G.H. and Allen, R.B. (1996) 'Ecology of New Zealand *Nothofagus* forests , in T.T. Veblen, R.S. Hill and J. Read (eds) *The Ecology and Biogeography of Nothofagus Forests*. Yale University Press, New Haven.

Paugy, M., Baillon, F., Chevalier, D. and Duponnois, R. (2004) 'Elephants as dispersal agents of mycorrhizal spores in Burkina Faso', *African Journal of Ecology*, vol 42, pp225–227.

Perry, G.L.W., Wilmshurst, J.M. and McGlone, M.S. (2014) 'Ecology and long-term history of fire in New Zealand', *New Zealand Journal of Ecology*, vol 38, pp157–176.

Preston, C.D., Pearman, D.A. and Hall, A.R. (2004) 'Archaeophytes in Britain', *Botanical Journal of the Linnean Society*, vol 145, pp257–294.

Richardson, S.J., Holdaway, R.J. and Carswell, F.E. (2014) 'Evidence for arrested successional processes after fire in the Waikare River catchment, Te Urewera', *New Zealand Journal of Ecology*, vol 38, pp221–229.

Scheffer, M., Carpenter, S., Foley, J.A., Folke, C. and Walker, B. (2001) 'Catastrophic shifts in ecosystems', *Nature*, vol 413, pp591–596.

Schmitz, O.J., Wilmers, C.C., Leroux, S.J., Doughty, C.E., Atwood, T.B., Galetti, M., Davies, A.B. and Goetz, S.J. (2018) 'Animals and the zoogeochemistry of the carbon cycle', *Science*, vol 362, no 6419, art eaar3213.

Schweiger, A.H., Boulangeat, I., Conradi, T., Davis, M. and Svenning, J.-C. (2019) 'The importance of ecological memory for trophic rewilding as an ecosystem restoration approach', *Biological Reviews*, vol 94, pp1–15.

Tershy, B.R., Shen, K.-W., Newton, K.M., Holmes, N.D. and Croll, D.A. (2015) 'The Importance of islands for the protection of biological and linguistic diversity', *BioScience*, vol 65, pp592–597.

Trauernicht, C., Murphy, B.P., Prior, L.D., Lawes, M.J. and Bowman, D.M.J.S. (2016) 'Human-imposed, fine-grained patch burning explains the population stability of a fire-sensitive conifer in a frequently burnt northern Australia savanna', *Ecosystems*, vol 19, pp896–909.

Turner, M.G., Romme, W.H., Gardner, R.H., O'Neill, R.V. and Kratz, T.K. (1993) 'A revised concept of landscape equilibrium: disturbance and stability on scaled landscapes', *Landscape Ecology*, vol 8, pp213–227.

White, P.S. and Jentsch, A. (2001) 'The search for generality in studies of disturbance and ecosystems dynamics', in K. Esser, U. Lüttge, J.W. Kadereit and W. Beyschlag (eds) *Progress in Botany*. Springer, Heidelberg.

White, P.S. and Pickett, S.T.A. (1985) 'Natural disturbance and patch dymanics: an introduction', in S.T.A. Pickett and P.S. White (eds) *The Ecology of Natural Disturbance and Patch Dynamics*. Academic Press, London.

Willerslev, E., Davison, J., Moora, M., Zobel, M., Coissac, E., Edwards, M.E., Lorenzen, E.D., Vestergård, M., Gussarova, G., Haile, J., Craine, J., Gielly, L., Boessenkool, S., Epp, L.S., Pearman, P.B., Cheddadi, R., Murray, D., Bråthen, K.A., Yoccoz, N., Binney, H., Cruaud, C., Wincker, P., Goslar, T., Alsos, I.G., Bellemain, E., Brysting, A.K., Elven, R., Sønstebø, J.H., Murton, J., Sher, A., Rasmussen M., Rønn, R., Mourier, T., Cooper, A., Austin, J., Möller, P., Froese, D., Zazula, G., Pompanon, F., Rioux, D., Niderkorn, V., Tikhonov, A., Savvinov, G., Roberts, R.G., MacPhee, R.D.E., Gilbert, M.T.P., Kjær, K.H., Orlando, L., Brochmann, C. and Taberlet, P. (2014) 'Fifty thousand years of Arctic vegetation and megafaunal diet', *Nature*, vol 506, pp47–51.

Wyse, S.V., Wilmshurst, J.M., Burns, B.R. and Perry, G.L.W. (2018) 'New Zealand forest dynamics: a review of past and present vegetation responses to disturbance, and development of conceptual forest models', *New Zealand Journal of Ecology*, vol 42, pp87–106.

Zimov, S.A. (2005) 'Pleistocene Park: return of the mammoth's ecosystem', *Science*, vol 308, no 5723, pp796–798.

Part III
Methods and tools for landscape ecology

11

Fieldwork in landscape ecology

Jesse E.D. Miller, Carly D. Ziter, and Michael J. Koontz

Introduction

Landscape ecology explores the biological and societal causes and consequences of landscape heterogeneity. Landscape ecologists often seek to synthesize patterns and processes across multiple spatial scales, and fieldwork is an indispensable and central technique for accomplishing this. Here, we define fieldwork as personal, in situ observations of biological and societal patterns. Fieldwork and data from other sources, such as remotely sensed landscape imagery, often play complementary roles in landscape ecology research. For example, fieldwork allows the relatively precise quantification of biological and social patterns and processes at typically fine spatial scales, while remotely sensed data facilitates often coarser quantifications of landscape variables across broad spatial extents. Fieldwork has remained a critical component of much of landscape ecology research for decades because it continues to provide unique information, despite rapid advances in technology for characterizing ecological patterns and processes.

Fieldwork plays several roles in landscape ecology research. Perhaps most prominently, fieldwork is used to characterize biological or social patterns and processes so that they can be related to landscape context. Fieldwork is also frequently used to field-calibrate remotely sensed data, such as landscape imagery and sensor data. While these two roles often involve different methods, they both typically attempt to capture data at a finer scale than remotely sensed imagery can provide and often make types of measurements that are outside the capabilities of remote sensing. In one example of relating field measurements of biological patterns to landscape variables, researchers explored the role of landscape context in plant community restoration project outcomes (Grman et al. 2013). A large body of field research has also explored the influence of landscape spatial configuration on plant community diversity and functional traits (e.g., Marini et al. 2012, Auffret et al. 2016, Miller et al. 2018). Other examples of influential field studies in landscape ecology include research showing that forest fragmentation can influence host–parasitoid relationships (Roland and Taylor 1997) and research estimating the effects of wildfire on nutrient cycling (Walker et al. 2018).

In experimental landscape ecology, fieldwork may also involve manipulating landscapes either in microcosms or at broad spatial scales. One large, landscape-scale manipulative experiment, the Corridor Project, has been used to show that habitat connectivity affects numerous taxonomic groups, such as plants (Damschen et al. 2006), butterflies (Haddad and Baum 1999), and arthropods (Orrock et al. 2011). Fieldwork is also sometimes used to quantify landscape patterns themselves, especially for fine-scale landscapes; researchers in Newfoundland have used landscape ecology methods to explore the dynamics of lichen patches on tree trunks as micro-landscapes (Wiersma et al. 2019). Contemporary landscape ecology fieldwork may also involve deploying drones (discussed further later) or other sensors (Chapter 13) in the field to create customized landscape imagery and other data.

The broad spatial extent at which many landscape ecology studies operate leads to conceptual and logistical challenges that differ from those faced in locally focused work. Because landscape studies often focus on large study regions, fieldwork can involve significant travel, and individual studies may span multiple land ownership boundaries and major geographical gradients. Establishing sufficient independent landscape replicates for inference while avoiding pseudoreplication is another frequent challenge in landscape ecology study design. The successful navigation of such challenges, however, can lead to broad and meaningful insights into natural and social patterns. Indeed, when well designed, field-based landscape ecology research can develop inference that combines the depth of field-based natural history expertise and the breath that contemporary technological approaches (e.g., remote sensing and other big data) can confer.

Fieldwork and the development of landscape ecology

Fieldwork has played a substantial role in the development of landscape ecology, and many of the early foundational papers of the nascent discipline half a century ago made extensive use of fieldwork. In one influential field study, Wright (1974) used pollen deposition and tree ring analyses to show that fire should be used as a management tool, a then-controversial perspective in many circles of forest ecologists that has since entered the mainstream. Wiens (1976) drew on hundreds of field-based studies in his foundational review of population responses to patchy environments, which highlights the importance of scale in the relationship between species and their environment. Bormann et al. (1968) conducted one of the first landscape-scale vegetation manipulation experiments, demonstrating that timber harvest can cause large-scale nitrogen loss from ecosystems.

A diversity of techniques and approaches are needed to address the mounting global challenges that ecologists face in the contemporary era of unprecedented global change. As technological advances yield novel tools for inference in ecology, such as greatly expanded remote sensing capabilities and advances in modeling and computational approaches, some ecologists have expressed concern that fieldwork and natural history may be falling by the wayside (Ríos-Saldaña et al. 2018), and empirical landscape ecology studies in the United Kingdom are believed to have declined recently (Young et al. 2019). One common concern is that ecologists who lack field-based experience and natural history skills may not be able to meaningfully interpret 'big data' sets that they were not personally involved in collecting. We agree that fieldwork and natural history continue to be a critical tool for the advancement of landscape ecology, but we also recognize that these approaches are often complemented by non-field-based approaches. When done well, fieldwork and non-field-based approaches can be mutually reinforcing.

Best practices in landscape ecology fieldwork

Independence is often described as the most important assumption of parametric statistics. As the first rule of geography establishes, things that are spatially closer together tend to be more similar, and this may be true of landscape variables of interest, such as environmental, biological, or socioeconomic variables. Samples that are independent are not correlated, in the sense that a given observation of a variable does not depend on values of other observations as a function of time or space. The term *pseudoreplication* refers to non-independent samples being treated as independent for purposes of analysis. Designing landscape ecology studies while avoiding pseudoreplication can be challenging given the broad scale at which landscape ecology studies are often conducted. Establishing independent landscape replicates (i.e., multiple study landscapes that do not overlap or that overlap only minimally) is important in many landscape ecology studies but may require large study regions with well-dispersed study sites.

For landscape ecology studies to be meaningful, they must span substantial variation in predictor variables of interest (Eigenbrod et al. 2011). This is especially important for detecting non-linear patterns, such as a saturating relationship between variables, where no relationship will be detected if only part of the range of the predictor variable is sampled. Landscape ecology studies must also carefully consider the grain and extent at which research is conducted, since ecological relationships may be scale-dependent (Chapter 3). In one example of scale-dependence, Fricker et al. (2019) found that topography became an increasingly important driver of tree height relative to climate at finer scales.

Thoughtful study design is an essential precursor to successful landscape field studies. Developing specific research questions and goals is an important first step in designing a field study, since it is difficult to choose an appropriate sampling strategy when the research goals are vague (Sutherland 2006). Once goals are defined, a specific sampling protocol can be established; sampling should occur at a scale that will capture heterogeneity in variables of interest. Study plots, sometimes with nested quadrats, are often used to measure community diversity or estimate species abundance. Linear transects or belt transects, along which data are collected periodically at points or nested quadrats, are another approach that may be especially useful in landscapes characterized by clinal or hierarchical environmental heterogeneity (Sutherland 2006). Another important consideration in study design is the tradeoff between sampling effort at each plot and the total number of plots in the study; in general, replicating at the highest level (e.g., collecting samples from more sites with less intensive effort per site) will produce the best results (Karban and Huntzinger 2006).

Choosing sampling locations within study regions is another important consideration. Randomly locating plots or transects is often considered an ideal approach, though there can also be advantages to using an evenly spaced sampling layout (e.g., grid designs; Elzinga et al. 1998, Sutherland 2006). Subjectively selecting 'representative' study locations or arbitrarily choosing sampling locations in the field may lead to bias in site selection and should be avoided (Sutherland 2006). Stratified random sampling may be a useful approach in heterogeneous landscapes that can be blocked into multiple discrete categories based on ecological differences or other factors such as site accessibility. Under a stratified random sampling scheme, a predetermined number of sampling locations are randomly chosen within two or more discrete regions of the study area; this can be useful when the abundance of habitats or organisms of interest varies substantially between regions or when different regions cannot be sampled with equal intensity.

Designing studies that accurately capture variables of interest while avoiding bias and meeting the requirements of independence may be very challenging at times, especially in systems where potential study sites are limited. However, advance planning can help ameliorate these challenges. Carefully examining maps or landscape imagery and scouting potential field sites can help researchers anticipate and control for unexpected complications such as confounding variables. Conducting small pilot studies may also help researchers identify potential problems such as correlated predictor variables before a great deal of time and resources have been invested in study sites that may lead to problematic data sets. Finally, researchers should also keep in mind that no field study is perfect, and the guidelines we mention here are not absolute. For example, small amounts of overlap (e.g., pseudoreplication) in study landscapes may pose minimal problems for inference (Eigenbrod et al. 2011).

The logistical challenges of sampling study sites across large study regions can be substantial. Significant amounts of travel time are often required for a single researcher or team to conduct such studies, which may reduce time for actual sampling. Establishing networks of researchers who follow similar protocols at widely dispersed study sites or regions can be one effective approach to this challenge. Some examples of such networks in North America include the Long Term Ecological Research Network, the Nutrient Network, the National Phenology Network, the National Ecological Observatory Network, and the Global Observation Research Initiative in Alpine Environments (GLORIA). Similarly, collaborations with community members and tools such as iNaturalist can be useful in landscape ecology (see further discussion of community-based science later).

Adapting fieldwork to different socioeconomic contexts

Landscape ecology fieldwork takes place in landscapes that can be geographically and socially diverse. Navigating this heterogeneity is generally easiest when researchers conduct careful advance planning but also maintain an adaptable and flexible attitude. Perhaps the only universal rule of fieldwork is that it rarely goes exactly as planned. As experienced fieldworkers, we have learned to accept and even sometimes enjoy some of the unexpected occurrences that seem to characterize fieldwork, which can range from a down tree blocking a road to encounters with community members. An important first step in embarking on any field project is planning in advance for anticipated logistical and safety concerns, which vary with the geographic context of studies, as we describe later.

As scientists and fieldworkers, we may be outsiders to some extent, working in areas where we do not live (even if we have come to know them well). As such, we are guests in landscapes where other people live and work, and sensitivity to the needs and concerns of local communities and other stakeholders is important. Led primarily by our colleagues in the social sciences and humanities (e.g., human geographers, political economists, and historians of science), there has been growing recognition of the need for anti-colonial, feminist, or 'decolonized' methods of conducting research that meaningfully engage with local communities and value traditional ecological knowledge (e.g., the Civic Laboratory for Environmental Action Research; https://civiclaboratory.nl/). Regardless of our specific study system or location, it is critical that ecologists examine the ways in which our research agendas and fieldwork practices may relate to or reinforce inequalities in the areas in which we work (Baker et al. 2019). In this spirit, we (the authors of this chapter) acknowledge that our perspective and fieldwork experience are biased to that of researchers from the Global North (North America) conducting fieldwork within this region. We also highlight that not all researchers experience the same risks in the field, with minoritized researchers more vulnerable to conflict and violence in the field as result of identity prejudice. We encourage all researchers (and particularly those is supervisory positions) to discuss and implement safe fieldwork strategies prior to any fieldwork (Demery and Pipkin 2020).

One way in which researchers have engaged with local communities in their study areas is through the co-production of research, where scientists actively partner with the people affected by the research to shape how projects are conceived, supported, conducted, and disseminated (Hickey et al. 2018). While this is a challenging and often time-consuming process when done in an intentional and meaningful way, it can lead to outcomes that are better aligned with the values and needs of society. In the United States, one productive boundary-spanning organization is the U.S. Forest Service Regional Ecology Program, which facilitates collaborations between academic researchers and government agency land managers to address critical management challenges (Safford et al. 2017). In Canada, collaborative research between academics and Indigenous communities has led to more effective conservation plans for both communities and scientists; for example, combined use of genetic analysis of caribou scat and place-based traditional knowledge has broadened understanding of caribou population dynamics in northern landscapes (Polfus et al. 2016).

Even when not engaging fully in a co-production model, fieldwork typically requires receiving permission from third parties to access study sites. Landscape ecology fieldwork in particular often involves establishing study sites across political and/or land ownership boundaries due to the need to capture broad extents and spatial heterogeneity. Practically, this often means that permits must be obtained from multiple land managers, such as government agencies, non-government organizations, or private landowners. In the United States, numerous government agencies control vast land holdings, and it is our experience that agencies vary substantially in their approaches towards permitting researchers. Some agencies have very straightforward application processes and issue permits quickly, while others require labyrinthine processes to receive a permit even for benign activities such as observational studies. In the latter cases, one thing we have learned is that identifying and contacting the person who ultimately processes research applications is often the surest way to accelerate the process of permit approval, especially when permit applications are submitted via an opaque online portal. One of us once had a permit application stagnate unapproved for over a year until we determined who to contact to ask for assistance, after which it was approved within days. In our experience, individual agency personnel are often friendly and interested in providing assistance with navigating what can be complex bureaucratic processes.

In contrast to working with agencies, gaining permission to access private land may require a less formal approach. We have found (in North America) that the easiest method is typically to contact the landowner or manager directly with a request. In the case of businesses or corporate landowners, researchers may be directed to a permit process. However, in the case of individual landowners (e.g., farmers or homeowners), it is often sufficient to establish permission directly either over the phone, by email, or in person. In this case, the most difficult part of the process is often establishing initial contact. It is important to leave adequate time to navigate this process, as it is not uncommon for studies to require permission from numerous individual landowners in heterogeneous landscapes; e.g., in agricultural, ex-urban, or urban areas. Once permission has been granted, maintaining communication with property owners for the duration of the research can be important for a successful field season and can leave the door open to future research collaborations (Hilty and Merenlender 2003, Dyson et al. 2019).

In some cases, researchers may also choose not to request permits for research even when they may technically be required. For example, one colleague of ours successfully completed a large observational study that took place mostly along roadsides where the agency that controlled rights-of-way was uninterested in facilitating research. However, penalties for conducting illegal research are extremely severe in some parts of the world, and we advise that researchers comply with local laws. When beginning any new project, it is important to be aware of the local risks and repercussions that your presence and fieldwork presents and particularly, how these may differ across different governance contexts (e.g., public vs. private land). Communicating with other researchers or practitioners familiar with the social norms and legal obligations of the landscape of interest can be an informative step in fieldwork planning.

Urban fieldwork

Urban areas are complex mosaics of land covers characterized by different histories, vegetation, management, and climate (Cadenasso et al. 2007). This high spatial heterogeneity and

223

frequent temporal change makes cities powerful systems for exploring landscape ecology questions. Given the differences in biodiversity, ecosystem structure, and function between urban ecosystems and their wildland counterparts, and the comparatively recent consideration of cities as ecosystems within mainstream ecology (Wu 2014), fieldwork still plays a fundamental role in understanding the ecological fabric of our cities. It is tempting to see urban fieldwork as an 'easy' option. Field sites are often close to home, and access to amenities throughout the field season (the lab, hardware stores, repair shops, etc.) can reduce some of the preparation stress compared with a wildland expedition. However, there are additional logistical challenges to contend with in an urban context that many classically trained ecologists have little experience of navigating (Dyson et al. 2019).

One of the most obvious differences in urban fieldwork is the extent to which human interactions pervade the work. The same heterogeneity that often makes urban areas attractive study regions also means that numerous different permissions may be required to access sites, for example. One author of this chapter recently conducted an urban study requiring permission from 70 different individuals (and field encounters with numerous additional individuals), which necessitated a substantial investment in relationship building and communication before, during, and after the field season (Ziter and Turner 2018). As in other fieldwork contexts, positionality (race, gender, sexual orientation, class) will influence the way a researcher is perceived by project partners and members of the local community. Speaking with someone familiar with the neighborhoods in which you hope to work, and forming relationships with community leaders, can be an important step in building trust within the community as well as offsetting safety concerns.

In addition to planning for extensive interactions with community members, research and sampling designs may also need to be adapted for working in a heavily human-dominated context. Modifying the timing of fieldwork activities to suit property owners or managers, adapting to frequent interruption, and reducing the impact of invasive sampling methods are all common experiences in urban areas (Dyson et al. 2019). Patterns of species occurrence and behavior may also differ in urban areas compared with nearby rural or natural habitat, such that natural history knowledge developed outside urban settings may be less reliable (Kowarik 2011, Johnson and Munshi-South 2017). As mentioned earlier, pilot studies can help to identify and address pitfalls in research design and make for an ultimately more successful field season.

Working landscapes and agroecosystems

Much of the earth's surface in temperate and tropical regions is used for agriculture and ranching, and a large body of landscape ecology research focuses on these ecosystems (Kremen and Merenlender 2018, Ellis 2019). Working in agroecosystems involves a unique set of considerations and challenges. There can be a tendency for people to distrust science and scientists in rural areas, and building trust can be key to working safely and efficiently. Co-production of research, as described earlier, can be useful to this end. Finding common ground with local people can also be useful. For example, in rural regions where local people may be skeptical of research on climate change, describing research in terms of tangible effects such as flooding may make the research more meaningful. We have also found that identifying as part of the agricultural college within our university – rather than the university as a whole – can help form a connection with community members. As previously mentioned, it is common for researchers' identity (e.g., race, gender, and sexual orientation) to influence how they are perceived in the field in rural areas as well. Keeping related safety concerns in mind is important, and researchers overseeing students or technicians should ensure that they are prepared and supported during fieldwork.

Rural residents, including land managers and farmers, are often some of the most knowledgeable people regarding the ecology of their lands. Engaging respectfully with local people can be a useful way for researchers to gain insight into their study systems and appreciate multiple ways of knowing. On the same note, we encourage researchers to be wary of the attitude that they are there to discover something 'new'. One component of engaging respectfully with local people includes planning research with awareness of land uses such as planting/harvest schedules, hunting, seasons, etc. It is important to recognize that your research will rarely take priority over people's livelihoods (nor should it). For researchers from urban areas in particular, developing awareness of local customs in rural places can be useful. One of us has worked in an isolated rural region where the pace of life is slow and local people commonly engage in conversation with anyone passing by (Miller et al. 2015); failing to recognize and participate in this friendly ritual could make it difficult to engage with and gain the trust of the local community.

Wildlands

Much landscape ecology research takes place in relatively wild landscapes outside urban and agricultural areas (e.g., Turner et al. 2010, Miller et al. 2015, Tingley et al. 2016). Working in wildlands typically involves fewer human encounters than work in human-inhabited areas, though those that do occur may be similar to those mentioned in the agroecosystems section. Maintaining inclusive and collegial work and living environments is another particularly important consideration for fieldwork in remote places; reports of harassment and assault in remote field environments are distressingly common (Clancy et al. 2014). Principal investigators and crew leaders should consider how they can develop policies and procedures to help create safe fieldwork environments for their teams.

Wilderness safety and being prepared for the isolation of remote places are important considerations for planning wildlands fieldwork. We have learned that it is good to be prepared for field trips to remote areas to last longer than planned. Extra food and water are important safety measures, since surprise changes of itinerary or weather are common during fieldwork, especially in remote places. While satellite navigation devices (e.g., geographic positioning systems) have made navigating in the back country easier, we always pack paper maps and compasses as well. Satellite communication devices can also add an extra layer of safety for areas where cell phone reception is unreliable. Accidents are most likely when researchers are tired or in a hurry, so we recommend working at a moderate pace and eating and sleeping well as accident prevention measures. Fieldwork will always carry some inherent risk, and conscientious management of that risk is the surest route to a successful field campaign.

Frontiers in landscape ecology fieldwork

Community-based science

Community-based science, also known as citizen science, is an approach to field data collection that is gaining traction among landscape ecologists. Community-based science generally involves many individuals or groups collecting data independently with varying degrees of central coordination. Such collaborative efforts to collect geographically disparate field data may succeed where they would prove impossible if undertaken by a single research group and can often be undertaken at minimal cost. Community-based science may take the form of professionally led efforts with participation from non-science professionals (e.g., bioblitzes, GLORIA), or decentralized, entirely non-professional efforts (e.g., eBird, iNaturalist). While they are manifold in form, the success of these community-based efforts for transforming landscape ecology research is fundamentally tied to a philosophy of open science: the idea that knowledge production should be reproducible, transparent, and accessible (Hampton et al. 2015, Bahlai et al. 2019). The benefits of crowd-sourced fieldwork are greatly reduced if collected data don't adhere to FAIR principles (data should be findable, accessible, interoperable, and reusable; Wilkinson et al. 2016). Data from community-based science may have more quality issues than data collected by professionals if data collectors are untrained. Nonetheless, the potential for the combination of community building and open science to revolutionize science as a whole is great, and landscape ecology stands to benefit in particular.

Drones

Advances in portability, accessibility, and capability of instrumentation have blurred the distinction between 'fieldwork' and 'remote sensing' approaches to landscape ecology. For instance, small unhumanned aerial systems (colloquially referred to as 'drones') can fit in a backpack, are inexpensive enough to be purchased on a small grant, are relatively easy to fly manually or on a pre-programmed flight path, and often come equipped with a capable camera. This makes them ideal tools for capturing fine-grain detail at relatively broad spatial extents. However, the range (i.e., flight distance) of these systems is limited by a need for a direct radio link to a ground-based controller, and thus, field visits are usually needed in order to use these tools. The dual identity of drone-based approaches (bridging fieldwork and remote sensing) adds some additional considerations for fieldwork, even if much of the data is collected from the air.

Drone-based approaches have an impact beyond the footprint of a strictly ground-based operation. This is a key strength of drones as a tool for ecology but also presents ethical challenges. For instance, Indigenous scientists and scientists engaged in knowledge co-production with Indigenous people may choose to use drone-based sampling as a means to preserve sovereignty and ensure data ownership (Martínez 2015, Haney 2016, Smith 2017, 'Decolonizing Digital: Empowering Indigeneity Through Data Sovereignty', 2019). On the other side of the coin, drone data (e.g., imagery) may extend beyond the area that is being intentionally surveyed, which may impose on the privacy of nearby people. Thus, special care should be employed when using drones to ensure that stakeholders have the agency to opt in or opt out of data collection depending on their needs. The extensive footprint of drones (e.g., the aircraft itself can be seen and heard from a distance) also requires consideration of wildlife beyond the area directly impacted by humans on the ground (Mulero-Pázmány et al. 2017). Based in part on these considerations, laws governing the use of drones abound in a global patchwork of regulatory frameworks (Stöcker et al. 2017). Because the rules regarding flights over wilderness, wildlife, and society are so variable in space and time, we suggest due diligence in learning the current relevant restrictions on the use of drones well in advance of fieldwork not only to ensure compliance with the law but also to avoid extralegal constraints. Beyond legal compliance, we also recommend exercising restraint and an abundance of caution when it comes to deploying these tools over areas that may cause harm or that may have their own influence on the very phenomenon being studied.

Other sensor technologies

Alongside developments in technology such as drones, other advanced sensors have also become more common in landscape ecology studies (Chapter 13). Further, the interconnectedness of sensors with each other and with the Internet (i.e., the 'Internet of Things') facilitates finely resolved, real-time ecological data collection at unprecedented scales (Guo et al. 2015, Bakker and Ritts 2018). One specific development in sensor-enabled fieldwork capitalizes on advances in sensor mobility. Recent studies have combined traditional fieldwork techniques such as environmental transects with mobile sensor technology (e.g., environmental sensors mounted on bicycles, cars, or boats) to collect fine-scale data on a range of environmental variables. Mobile sampling is a relatively affordable, efficient, and flexible way to capture fine-scale spatial data over large extents, and it is replicable over time. In urban landscapes, mobile sampling techniques have been used to investigate and map spatial patterns in air temperature (Ziter et al. 2019) and air pollution (Adams and Kanaroglou 2016). Mobile sampling has also been used in freshwater environments. For example, researchers in Wisconsin have developed 'FLAMe' (Fast Limnological Automated Measurements) technology to generate spatially explicit, real-time observations of surface water quality (Crawford et al. 2015, Loken et al. 2018). This approach has advanced 'landscape limnology' – the spatially explicit study of lakes, streams, and wetlands – a field now closely aligned with landscape ecology (Soranno et al. 2010).

Conclusion

Fieldwork has played a critical role in the development of landscape ecology, and it remains essential for addressing contemporary challenges such as understanding the landscape ecology of global change. Advances in technology have expanded the scope of fieldwork to include the deployment of drones and other sensors, and in recent years, researchers have expressed concerns that traditional fieldwork (e.g., organismal observation) may be declining. Continuing to train the next generation of researchers in field methods should be a priority for landscape ecologists. Indeed, there is great potential for combining fieldwork with modern sensor data and computational approaches to advance the field of landscape ecology.

References

Adams, M.D. and Kanaroglou, P.S. (2016) 'Mapping real-time air pollution health risk for environmental management: Combining mobile and stationary air pollution monitoring with neural network models', *Journal of Environmental Management*, vol 168, pp133–141.

Auffret, A.G., Aggemyr, E., Plue, J. and Cousins, S.A.O. (2016) 'Spatial scale and specialization affect how biogeography and functional traits predict long-term patterns of community turnover', *Functional Ecology*, vol 31, pp436–443.

Bahlai, C., Bartlett, L.J., Burgio, K.R., Fournier, A.M.V. and Keiser, C.N., Poisot, T. and Whitney, K.S. (2019) 'Open science isn't always open to all scientists', *American Scientist*, vol 107, pp78–82.

Baker, K., Eichhorn, M.P., Griffiths, M. (2019) 'Decolonizing field ecology', *Biotropica* vol 51, pp288– 292. https://doi.org/10.1111/btp.12663

Bakker, K. and Ritts, M. (2018) 'Smart Earth: A meta-review and implications for environmental governance', *Global Environmental Change*, vol 52, pp201–211.

Bormann, F.H., Likens, G.E., Fisher, G.W. and Pierce, R.S. (1968) 'Nutrient loss accelerated by clear-cutting of a forest ecosystem', *Science*, vol 159, no 3817, pp882–884.

Cadenasso, M.L., Pickett, S.T.A. and Schwarz, K. (2007) 'Spatial heterogeneity in urban ecosystems: Reconceptualizing land cover and a framework for classification', *Frontiers in Ecology and the Environment*, vol 5, no 2, pp80–88.

Clancy, K.B.H., Nelson, R.G., Rutherford, J.N. and Hinde, K. (2014) 'Survey of Academic Field Experiences (SAFE): Trainees report harassment and assault', *PLoS ONE*, vol 9, no 7, pp1–9.

Crawford, J.T., Loken, L.C., Casson, N.J., Smith, C., Stone, A.G. and Winslow, L.A. (2015) 'High-speed limnology: Using advanced sensors to investigate spatial variability in biogeochemistry and hydrology', *Environmental Science & Technology*, vol 49, no 1, pp442–450.

Damschen, E.I., Haddad, N.M., Orrock, J.L., Tewksbury, J.J. and Levey, D.J. (2006) 'Corridors increase plant species richness at large scales', *Science*, vol 313, no 5791, pp1284–1286.

Demery, AJ.C. and Pipkin, M.A. (2021) 'Safe fieldwork strategies for at-risk individuals, their supervisors and institutions', *Nat Ecol Evol*, vol 5, pp5–9. https://doi.org/10.1038/s41559-020-01328-5

Dyson, K., Ziter, C., Fuentes, T.L. and Patterson, M.S. (2019) 'Conducting urban ecology research on private property: Advice for new urban ecologists', *Journal of Urban Ecology*, vol 5, no1, pp1–10.

Eigenbrod, F., Hecnar, S.J. and Fahrig, L. (2011) 'Sub-optimal study design has major impacts on landscape-scale inference', *Biological Conservation*, vol 144, no 1, pp 298–305.

Ellis, E.C. (2019) 'Sharing the land between nature and people', *Science*, vol 364, no 6447, pp1226–1228.

Elzinga, C.L.; Salzer, D.W. and Willoughby, J. (1998) Measuring and monitoring plant populations. Technical Reference 1730–1. Denver, CO: U.S. Department of the Interior, Bureau of Land Management. 477 p.

Fricker, G.A., Synes, N.W., Serra-Diaz, J.M., North, M.P., Davis, F.W. and Franklin, J. (2019) 'More than climate? Predictors of tree canopy height vary with scale in complex terrain, Sierra Nevada, CA (USA)', *Forest Ecology and Management*, vol 434, pp142–153.

Grman, E., Bassett, T. and Brudvig, L.A. (2013) 'Confronting contingency in restoration : Management and site history determine outcomes of assembling prairies, but site characteristics and landscape context have little effect', *Journal of Applied Ecology*, vol 50, pp1234–1243.

Guo, S., Qiang, M., Luan, X., Xu, P., He, G., Yin, X., Xi, L., Jin, X, Shao, J., Chen, X., Fang, D. and Li, B. (2015) 'The application of the Internet of Things to animal ecology', *Integrative Zoology*, vol 10, no 6, pp572–578.

Haddad, N.M. and Baum, K.A. (1999) 'An experimental test of corridor effects on butterfly densities', *Ecological Applications*, vol 9, no 2, pp623–633.

Hampton, S.E., Anderson, S.S., Bagby, S.C., Gries, C., Han, X., Hart, E.M., Jones, M.B., Lenhardt, W.C., MacDonald, A., Michener, W.K., Mudge, J., Pourmokhtarian, A., Schildhauer, M.P., Woo, K.H. and Zimmerman, N. (2015) 'The Tao of open science for ecology', *Ecosphere*, vol 6, no 7, art 120.

Haney, W.M. (2016) 'Protecting tribal skies: Why Indian tribes possess the sovereign authority to regulate tribal airspace', *American Indian Law Review*, vol 40, no 1, pp1–40.

Hickey, G., Richards, T. and Sheehy, J. (2018) 'Co-production from proposal to paper', *Nature*, vol 562, pp29–31.

Hilty, J. and Merenlender, A.M. (2003) 'Conservation in practice studying biodiversity on private lands', *Conservation Biology*, vol 17, no 1, pp132–137.

Indigenous Innovation (2019) 'Decolonizing digital: Empowering indigeneity through data sovereignty', https://www.animikii.com/news/decolonizing-digital-empowering-indigeneity-through-data-sovereignty, accessed 6 Dec 2020.

Johnson, M.T.J. and Munshi-South, J. (2017) 'Evolution of life in urban environments', *Science*, vol 358, no 6363, art eaam8327.

Karban, Richard, Huntzinger, Mikaela, and Pearse, Ian S. *How to Do Ecology: A Concise Handbook* (2nd ed.). Princeton, NJ: Princeton University Press.

Kowarik, I. (2011) 'Novel urban ecosystems, biodiversity, and conservation', *Environmental Pollution*, vol 159, no 8, pp1974–1983.

Kremen, C. and Merenlender, A.M. (2018) 'Landscapes that work for biodiversity and people', *Science*, vol 362, no 6412, art eaau6020.

Loken, L.C., J Crawford, J.T., Dornblaser, M.M., Striegl, R.G., Houser, J.N., Turner, P.A. and Stanley, E.H. (2018) 'Limited nitrate retention capacity in the Upper Mississippi River', *Environmental Research Letters*, vol 13, no 7, art 74030.

Marini, L., Bruun, H.H., Heikkinen, R.K., Helm, A., Honnay, O., Krauss, J., Kühn, I., Lindborg, R., Pärtel, M. and Bommarco, R. (2012) 'Traits related to species persistence and dispersal explain changes in plant communities subjected to habitat loss', *Diversity and Distributions*, vol 18, pp898–908.

Martínez, C. (2015) *Tecno-Sovereignty: An Indigenous Theory and Praxis of Media Articulated Through Art, Technology, and Learning.* PhD Dissertation, Arizona State University, https://repository.asu.edu/items/29790, accessed 6 Dec 2020.

Miller, J.E.D., Damschen, E.I., Harrison, S.P. and Grace, J.B. (2015) 'Landscape spatial structure affects specialists but not generalists in naturally fragmented grasslands', *Ecology*, vol 96, pp3323–3331.

Miller, J.E.D., Ives, A.R., Harrison, S.P. and Damschen, E.I. (2018) 'Early- and late-flowering guilds respond differently to landscape spatial structure', *Journal of Ecology*, vol 106, pp1033–1045.

Mulero-Pázmány, M., Jenni-Eiermann, S., Strebel, N., Sattler, T., Negro, J.J. and Tablado, Z. (2017) 'Unmanned aircraft systems as a new source of disturbance for wildlife: A systematic review', *PLoS ONE*, vol 12, no 6, pp1–14.

Orrock, J.L., Curler, G.R., Danielson, B.J. and Coyle, D.R. (2011) 'Large-scale experimental landscapes reveal distinctive effects of patch shape and connectivity on arthropod communities', *Landscape Ecology*, vol 26, no 10, pp 1361–1372.

Polfus, J.L., Manseau, M., Simmons, D., Neyelle, M., Bayha, W., Andrew, F., Andrew, L., Klütsch, C.F.C, Rice, K. and Wilson, P. (2016) 'Łeghágots'enete (learning together): The importance of Indigenous perspectives in the identification of biological variation', *Ecology and Society*, vol 2, no 2, art 18.

Ríos-Saldaña, C.A., Delibes-Mateos, M. and Ferreira, C.C. (2018) 'Are fieldwork studies being relegated to second place in conservation science?', *Global Ecology and Conservation*, vol 14, art e00389.

Roland, J. and Taylor, P.D. (1997) 'Insect parasitoid species respond to forest structure at different spatial scales', *Nature*, vol 386, no 6626, pp710–713.

Safford, H.D., Sawyer, S.C., Kocher, S.D., Hiers, J.K. and Cross, M. (2017) 'Linking knowledge to action: The role of boundary spanners in translating ecology', *Frontiers in Ecology and the Environment*, vol 15, no 10, pp560–568.

Smith, K.N. (2017) 'Indigenous people are deploying drones to preserve land and traditions', *Discover*, Dec 11, 2017, https://www.discovermagazine.com/environment/indigenous-people-are-deploying-drones-to-preserve-land-and-traditions, accessed 6 Dec 2020.

Soranno, P.A., Cheruvelil, K.S., Webster, K.E., Bremigan, M.T., Wagner, T. and Stow, C.A. (2010) 'Using landscape limnology to classify freshwater ecosystems for multi-ecosystem management and conservation', *BioScience*, vol 60, no 6, pp440–454.

Stöcker, C., Bennett, R., Nex, F., Gerke, M., and Zevenbergen, J. (2017) 'Review of the current state of UAV regulations', *Remote Sensing*, vol 9, no 5, art 459.

Sutherland, W. (Ed.). (2006). *Ecological Census Techniques: A Handbook* (2nd ed.). Cambridge: Cambridge University Press. doi:10.1017/CBO9780511790508

Tingley, M.W., Ruiz-Gutiérrez, V., Wilkerson, R.L., Howell, C.A. and Siegel, R.B. (2016) 'Pyrodiversity promotes avian diversity over the decade following forest fire', *Proceedings of the Royal Society B: Biological Sciences*, vol 283, no 1840, art 20161703.

Turner, M.G., Romme, W.H., Gardner, R.H. and Hargrove, W.W. (2010) 'Effects of fire size and pattern on early succession in Yellowstone National Park', *Ecological Monographs*, vol 67, no 4, pp411–433.

Walker, X.J., Rogers, B.M., Baltzer, J.L., Cumming, S.G., Day, N.J., Goetz, S.J., Johnstone, J.F., Schuur, E.A.G., Turetsky, M.R. and Mack, M.C. (2018) 'Cross-scale controls on carbon emissions from boreal forest megafires', *Global Change Biology*, vol 24, no 9, pp4251–4265.

Wiens, J.A. (1976) 'Population responses to patchy environments', *Annual Review of Ecology and Systematics*, vol 7, no 1, pp81–120.

Wiersma, Y.F., Wigle, R.D. and McMullin, R.T. (2019) 'A proposed microcosm for landscape ecology: beyond the binary to the patch-mosaic model', *bioRxiv*, 542985, https://doi.org/10.1101/542985.

Wilkinson, M.D. *et al.* (2016) 'Comment: The FAIR guiding principles for scientific data management and stewardship', *Scientific Data*, vol 3, pp1–9.

Wright, H.E. (1974) 'Landscape development, forest fires, and wilderness management', *Science*, vol 186, no 4163, pp487–495.

Wu, J. (2014) 'Urban ecology and sustainability: The state-of-the-science and future directions', *Landscape and Urban Planning*, vol 125, pp209–221.

Young, C., Bellamy, C., Burton, V., Griffiths, G., Metzger, M.J., Neumann, J., Porter, J. and Millington, J.D.A. (2019) 'UK landscape ecology: Trends and perspectives from the first 25 years of ialeUK', *Landscape Ecology*, vol 35, pp11–22.

Ziter, C. and Turner, M.G. (2018) 'Current and historical land use influence soil-based ecosystem services in an urban landscape', *Ecological Applications*, vol 28, no 3, pp643–654.

Ziter, C.D., Pedersen, E.J., Kucharik, C.J. and Turner, M.G. (2019) 'Scale-dependent interactions between tree canopy cover and impervious surfaces reduce daytime urban heat during summer', *Proceedings of the National Academy of Sciences of the United States of America*, vol 116, no 15, pp7575–7580.

12

Remote sensing and landscape mapping

Nathalie Pettorelli, Jennifer E. Smith,
Mailys Lopes, and Henrike Schulte to Bühne

Introduction

As discussed in Chapter 1, landscape ecology is concerned with the study of landscapes, which can be defined in many different ways (Wu 2013). All these definitions, however, commonly refer to large spatial scales and extents, with landscapes generally perceived as entities of several square kilometers with structural elements of patch, mosaic, and corridor, reflecting a mix of ecosystems and habitats. As acknowledged in the seminal review by Turner (2005), the use of remotely sensed data is the norm in landscape studies, with the significant association between remote sensing and landscape ecology due in large part to the strong spatial focus within landscape ecology. The strength of the association between these fields was well captured by Blaschke (2003), who noted that 'aerial photography and its interpretation was the starting point for Carl Troll to coin the term landscape ecology'.

Remote sensing, like landscape ecology, is a constantly evolving discipline, which focuses on the development of a wide array of technologies, from sensors onboard planes and satellites to ground-based devices and unmanned aerial vehicles (UAVs). Developments in techniques and algorithms to process remotely sensed images are also relatively rapid and continuous. For example, new combinations of remotely sensed data with methodologies such as machine learning algorithms (e.g., support vector machines, random forests, artificial neural networks) and multisensor image fusion (which aims to leverage the complementarity of different types of sensors) are regularly tested for application, while investigations for developing new indices to monitor vegetation dynamics have been occurring for decades (Pettorelli et al. 2014b). Similarly, research questions underpinning the development of landscape ecology are evolving as ecological knowledge grows, itself potentially influencing the development of new methodologies to capture information about our planet. This means that opportunities and challenges for landscape ecology to benefit from monitoring options offered by remote sensing are constantly changing, with the synergetic relationship between these fields having an incredible potential to move both disciplines forward.

In this chapter, we will explore the nature and strength of the linkages between remote sensing and landscape ecology. Specifically, we will first offer a quick introduction to remote sensing and then detail the current and future opportunities these linkages may provide. We will conclude this chapter by articulating some general thoughts about the challenges interdisciplinarity poses and the need for better communication between both disciplines for landscape ecology to fully benefit from remote sensing approaches.

Remote sensing: what is it and how does it work?

Remote sensing refers to the acquisition of information about an object or phenomenon through a device that is not in physical contact with the object (Jones and Vaughan 2010). Ecologists routinely use a large variety of remote sensing approaches, from acoustic sensors to camera traps and airborne and satellite-borne sensors (Marvin et al. 2016). In this chapter, we will primarily focus on remote sensing techniques that acquire imagery of the Earth's surface via sensors that are deployed on drones, planes, or satellites (which we will collectively refer to as Earth observation). This choice is motivated by the fact that such technologies are currently routinely used in landscape ecology.

Understanding what sensors are and what they do

Remote sensors used in Earth observation provide information about the Earth's surface by measuring the electromagnetic energy it emits or reflects. Because the reflectance or emission spectrum varies depending on the type of ground cover – such as the type, structure, and vigor of vegetation, the distribution of water, or the presence of human-made, sealed surfaces such as roads and buildings – this data can be used to derive information about the spatial distribution of Earth surface properties (Jackson and Huete 1991, Figure 12.1). Earth observation sensors are either active or passive (Figure 12.2): active sensors have their own source of radiation and

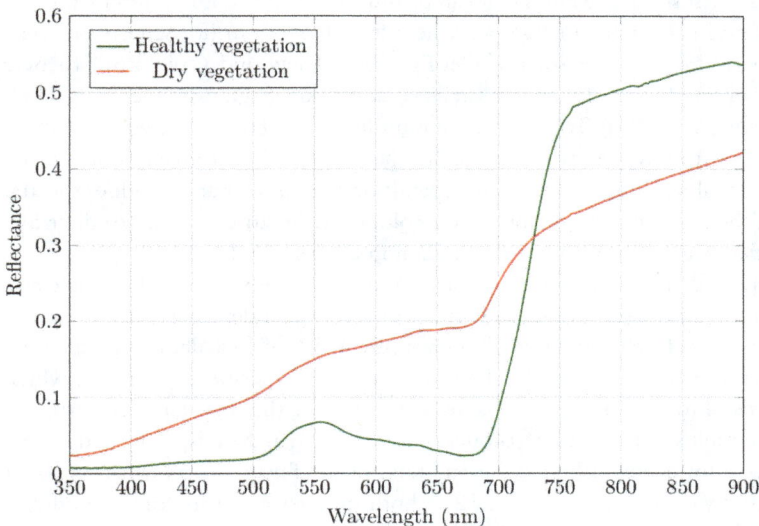

Figure 12.1　Examples of reflectance spectra of healthy vegetation (green) and of dry vegetation (red) acquired in a grassland field with a hyperspectral sensor. Remote sensors collect information about the reflectance over dozens or hundreds of sections, or bands, of the electromagnetic spectrum (multi- and hyperspectral sensors, respectively). Different surfaces can be distinguished if these bands cover those parts of the spectrum where their reflectance differs significantly.

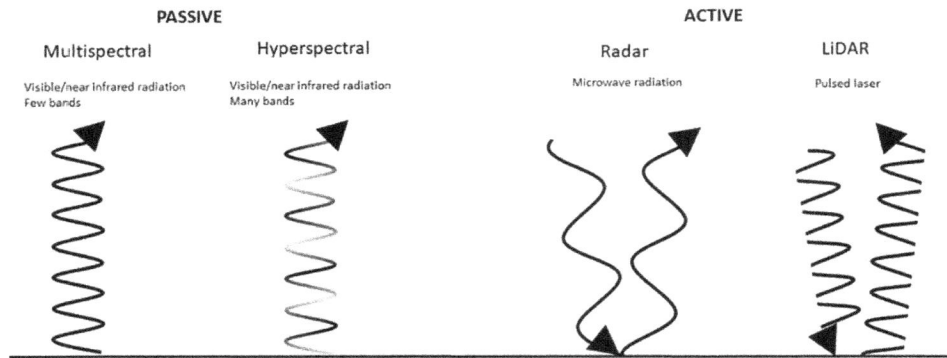

PASSIVE		ACTIVE	
Multispectral	Hyperspectral	Radar	LiDAR
Visible/near infrared radiation Few bands	Visible/near infrared radiation Many bands	Microwave radiation	Pulsed laser

Figure 12.2 Overview of some of the most commonly used sensors in Earth observation. Multispectral and hyperspectral sensors are both passive and acquire information about the reflectance in the visible and infrared spectrum. Hyperspectral sensors have a much higher spectral resolution, meaning that they can distinguish between many more wavelengths than multispectral sensors. Active sensors include radar sensors, which send out a microwave signal and measure the returning backscatter, and LiDAR sensors, which emit a pulsed laser signal.

measure the returning signal, whereas passive sensors measure the energy reflected or emitted by the Earth's surface that originates from external sources (such as the sun).

Sensors are characterized by their spatial, temporal, spectral, and radiometric resolutions; these characteristics in turn affect for which applications the sensors are appropriate. Spatial resolution limits the smallest surface features that can be distinguished and is commonly quantified by the size of a single pixel, which ranges anywhere from 0.4–0.5 m (for sensors on commercial, very high-resolution satellites such as GeoEye, WorldView, and QuickBird) to over tens of meters (e.g., Sentinel 1 and 2, Landsat satellites) and 1 km (e.g., Advanced Very High Resolution Radiometer [AVHRR]). Temporal resolution limits the fastest change that can be detected and depends on the frequency with which a platform revisits a given point on the Earth's surface. The spectral resolution of a sensor depends on the number and width of spectral bands that it distinguishes, whereas the radiometric resolution refers to the ability to discriminate different intensities of emitted or reflected radiation in a given band.

One important type of Earth observation sensor measures the reflectance in the visible and near infrared (VNIR) part of the electromagnetic spectrum. This includes multispectral sensors, which measure the VNIR reflectance in a handful (typically 4–30) of well-defined parts of the spectrum (called spectral bands or channels). This also includes hyperspectral sensors, which typically have hundreds of (often contiguous) bands and so resolve the reflectance spectrum in much more detail than multispectral sensors (Bioucas-Dias et al. 2013). Multispectral and hyperspectral sensors are good for distinguishing color and texture and for giving information about the photosynthetic activity of vegetation (Figure 12.1); both are passive sensors, for the most part only working in daylight (but see Elvidge et al. 2017).

Another important type of Earth observation sensor is sensitive to microwave radiation. Microwave radiation has a much longer wavelength than VNIR radiation and, as a result, interacts very differently with the Earth's surface. Microwave imagery can sometimes be hard to interpret intuitively: it provides information about the moisture content, surface roughness, volume, and spatial arrangement of objects rather than their color (Oh et al. 1992, Purkis and

Klemas 2011). There are both passive and active microwave sensors; the latter are called radar. Radar sensors typically emit a microwave signal in a single or very few bands, but they often have different polarizations. This means that the electric field lines of the electromagnetic waves they emit or receive are contained within a single plane – typically horizontal or vertical. Light Detection And Ranging (LiDAR) is another type of active sensor, emitting a laser beam and measuring the timing and intensity of the returning signal (Lim et al. 2003). LiDAR sensors are known to be the tool of choice for building up a high-resolution three-dimensional image of the environment (Côté et al. 2012, Melin et al. 2017).

Introducing common remote sensing platforms in landscape ecology

Earth observation sensors have been deployed on a wide range of platforms (Figure 12.3), which affects the spatial and temporal coverage as well as resolution of the data. In principle, any type of sensor can be deployed on any platform. In practice, LiDAR and hyperspectral sensors have so far been mounted predominantly on airborne systems, whereas multispectral and radar sensors are now commonly deployed on satellites (Toth and Józków 2016).

Sensors deployed on crewed airplanes are the oldest type of Earth observation. Airborne remote sensing can be very expensive, meaning that aerial observation campaigns are often carried out over a limited area and seldom repeated. However, planes have been an important platform for remote sensing data acquisition because they can provide imagery at high spatial resolution (Toth and Józków 2016). Data collected by sensors onboard these platforms thus allow mapping subtle changes in the Earth's surface, for instance vegetation growth and height, small-scale land or glacier deformation, and soil moisture (Reigber et al. 2013). These data also enable subtle differences in vegetation type to be detected, leading to different species (and even genotypes) being distinguished (Madritch et al. 2014).

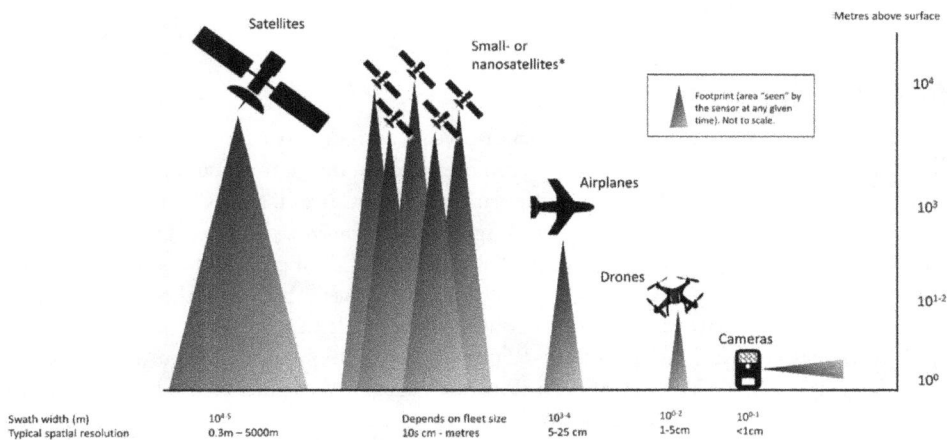

Figure 12.3 Overview of the common platforms used to deploy sensors for Earth observation. Platforms can observe the Earth's surface from a large range of heights. Typically, as this height decreases, the footprint of a sensor (i.e., its 'field of view') becomes smaller, so more images are required to cover the same area. See Toth and Józków (2016). *Nanosatellites or small satellites are an emerging technology for Earth Observations and not yet commonly used operationally.

Satellites are currently deployed by about 50 countries around the world for civilian use (Toth and Jóźków 2016). They often have a long life, running for several years or even decades. The Landsat program, for example, has provided multispectral imagery continuously since 1972 (Wulder et al. 2016). There are two different kinds of orbit that satellites can have: polar and geostationary. Polar satellites follow a route roughly from the south pole to the north pole and back again; geostationary satellites follow a path parallel to the equator at the same speed as the Earth's rotation, which means they appear to be fixed over a single point on the Earth's surface. Modern satellites typically acquire imagery over every point on the Earth's surface, with return times varying from once per day (e.g., MODIS Terra and Aqua) to more than 40 days (e.g., JERS-1 and ALOS PALSAR).

UAVs, or drones, are a recent addition to the range of platforms commonly used in remote sensing (Wich and Koh 2018). UAVs can enable access to data with higher spatial resolution than sensors onboard crewed airplanes because the distance between them and the ground is smaller. However, this means that their revisit time is typically lower than for satellites or even crewed planes (Anderson and Gaston 2013). UAVs are much cheaper to deploy than airplanes, but their range is often legally limited to the line of sight of the person controlling them, meaning that it can be time intensive to cover large areas using UAVs. Additionally, the rules regarding permission to fly drones for civilian use are only starting to emerge, with different countries having different regulatory frameworks (Stöcker et al. 2017).

Current opportunities

Remote sensing of the Earth's surface enables the spatial heterogeneity of the abiotic and biotic environment to be mapped at large spatial scales (up to global); it also allows the ecological processes that are caused by and that maintain these spatial patterns to be elucidated (Table 12.1). Remote sensing data are typically available for the recent past (a few years to a few decades), which allows retrospective analyses, tracking past as well as current landscape-level response to anthropogenic environmental change.

Vegetation and ecosystem mapping

An important use of remote sensing data is to map vegetation or ecosystem types at large spatial scales. This involves assigning each pixel in a remotely sensed image to a land cover type, either manually or (usually) using a classification algorithm. This automated approach relies on identified land cover types having different spectral signatures. In *supervised* land cover classifications, classification algorithms characterize the spectral signature of each pixel and compare it with a 'reference library' of the spectral signatures of land cover classes supplied by the user (Horning et al. 2016). These reference signatures are derived from pixels that are known to belong to a given land cover class, e.g., based on field data or from very high-resolution imagery (such as aerial photography). The classification algorithm then assigns each pixel to the land cover class whose reference spectrum is the most similar. Where the spatial heterogeneity of a landscape varies at spatial scales smaller than a single pixel, it is possible to estimate the relative coverage of several distinct land cover classes in each pixel (Meyer and Okin 2015). This involves characterizing the reflectance spectrum of each land cover type and then deriving their proportional coverage in each pixel via so-called spectral unmixing. *Unsupervised* classification approaches, by contrast, divide the pixels into groups with similar reflectance spectra without a priori knowledge of the classes – here, the user only specifies the number of land cover classes (Horning et al. 2016). In a second step, the groups of pixels are assigned to a thematic land cover class

Table 12.1 Non-exhaustive list of examples illustrating how remote sensing approaches can inform landscape ecology

Landscape feature	Sensors	Platform	References
Land cover types			
Tree cover	ETM+ (multispectral)	Landsat 7 satellite	Hansen et al. 2013
	AVIRIS (hyperspectral)	Aircraft	Huang et al. 2009
Water bodies	PALSAR (radar)	ALOS satellite	Chapman et al. 2015
Urban areas	ETM+ (multispectral	Landsat 7 satellite	Verpoorter et al. 2012
	OLS (night lights)	DMSP satellite	Zhou et al. 2014
Landscape structure			
Biomass	AVHRR (multispectral)	NOAA (satellite)	Hame et al. 1997
Vertical canopy structure	LiDAR	Aircraft	Skowronski et al. 2014
(e.g. LAI, height)	Hyperspectral	Handheld	Darvishzadeh et al. 2008
	LiDAR	Aircraft	Khosravipour et al. 2014
Landscape processes			
Primary productivity	AVHRR (multispectral)	NOAA (satellite)	Nayak et al. 2010
dynamics	Radarsat, JERS-1 (radar)	Satellite	Costa 2005
Herbivory	ETM+ (multispectral)	Landsat 7 satellite	Senf et al. 2015
	LiDAR	Aircraft	Asner et al. 2009
Disturbances			
Fire	MODIS (multispectral, thermal)	Terra/Aqua satellite	Hempson et al. 2018, Hantson et al. (2015)
Droughts	Radar	ENVISAT satellite	AghaKouchak et al. 2015
Floods	Radar	Sentinel-1 satellite	Twele et al. 2016
	Radar	ENVISAT satellite	Kuenzer et al. 2013
	Digital camera (RGB)	UAV	Perks et al. 2016
Anthropogenic impacts			
Roads	Multispectral	Aircraft	Brooks et al. 2017
Invasive species	Multispectral	QuickBird (satellite)	Christophe & Inglada 2007
	TM, ETM+ (multispectral)	Landsat 4-7 satellites	Gavier-Pizarro et al. 2012
	AVIRIS (hyperspectral), LiDAR	Aircraft	Asner et al. 2008
Eutrophication	Hyperion (hyperspectral)	Satellite	Kutser 2004
	ETM+ (multispectral)	Landsat 7 satellite	Huo et al. 2014
	Multispectral	UAV (drone)	Su & Chou 2015

ALOS: Advanced Land Observing Satellite. AVHRR: Advanced Very-High-Resolution Radiometer. AVIRIS: Airborne Visible/Infrared Imaging Spectrometer. DMSP: Defence Meteorological Satellite Program. ENVISAT: Environmental Satellite. ETM+: Enhanced Thematic Mapper Plus. MODIS: Moderate Resolution Imaging Spectroradiometer. NOAA: National Oceanic and Atmospheric Administration. JERS-1: Japan Earth Resources Satellite. LiDAR: Light Detection and Ranging. OLS: Operational Line-Scan System. PALSAR: Phased Array type L-band Synthetic Aperture Radar. TM: Thematic Mapper. UAV: unmanned aerial vehicle.

(either manually or by using a classification algorithm). Most classification algorithms used to produce categorical land cover maps are machine learning algorithms such as decision trees, random forests, support vector machines, and artificial neural networks. Which algorithm is most appropriate for a given application varies from case to case (Lawrence and Moran 2015). Once categorical land cover maps are derived, they can be validated with independent land cover information to quantify their accuracy (mainly the reliability and sensitivity; Strahler et al. 2006).

As the availability of satellite imagery and computing capacities have both increased, the volume of data used to carry out land cover classifications has also expanded. It is now common to use not only several bands from the same satellite acquisition but also several satellite acquisitions over the same location at different points in time. This can increase the ability of the classification algorithm to distinguish between land cover classes if they have characteristic and distinct temporal changes in reflectance signatures (Franklin et al. 2015) (Figure 12.4). For vegetation classes, this could be, e.g., differences in the time or magnitude of green-up after the winter or the dry season. Finally, combining imagery from different sensor types (e.g., optical and radar) can improve categorical land cover classification by taking advantage of the complementary information provided by different types of sensors (Schulte to Bühne and Pettorelli 2018, Figure 12.5).

Spatial structure

Categorical land cover maps have been used extensively to characterize the spatial structure of landscapes. The only global land cover products that are frequently being updated, to our knowledge, are the MODIS Land Cover products, which provide information about the distribution of 17 land cover classes each year at a spatial resolution of either 500 m or 1 km (Friedl et al. 2010). Regional and national-level products also exist. For instance, the European Union's CORINE land cover project has produced a European land cover map every 6 years since 2000 (Feranec et al. 2016), while the United Kingdom's land cover has been mapped several times since 1990 (Centre for Ecology & Hydrology [CEH] 2017).

In contrast to these examples, other land cover products focus on a single land cover type of interest. For instance, Hansen and colleagues (2013) used an automated classification approach on Landsat imagery to map changes in global tree cover extent between 2000 and 2012, thereby

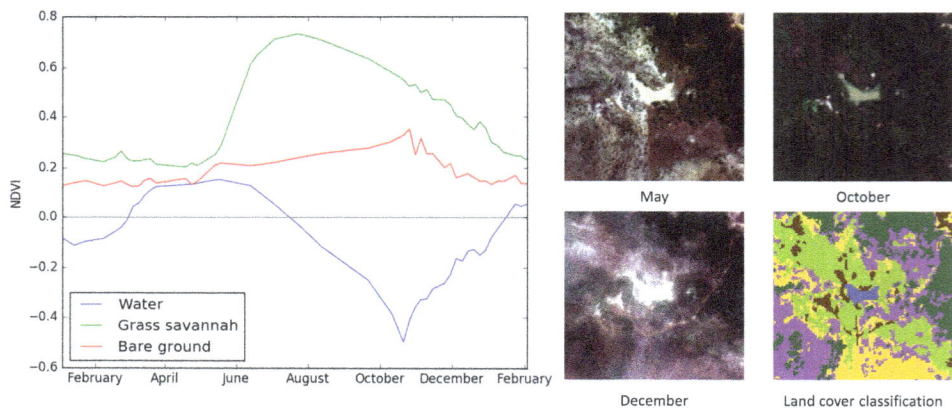

Figure 12.4 NDVI temporal signatures of three types of land cover (water, grass savannah, bare ground) in a savannah landscape measured with Sentinel-2 multispectral sensor. The temporal signature of bare ground remains stable over the year, whereas the temporal signature of grass savannah is marked by its seasonality. At the end of the dry season (April–May), pixels of water can have similar NDVI values to bare ground because of the drought. The images on the right are the Sentinel-2 color images acquired at different times of the year and the resulting classification map (blue: water, light green: grass savannah, brown: bare ground, yellow: shrub savannah, purple: tree savannah, dark green: woodland savannah). Satellite data courtesy of the European Space Agency.

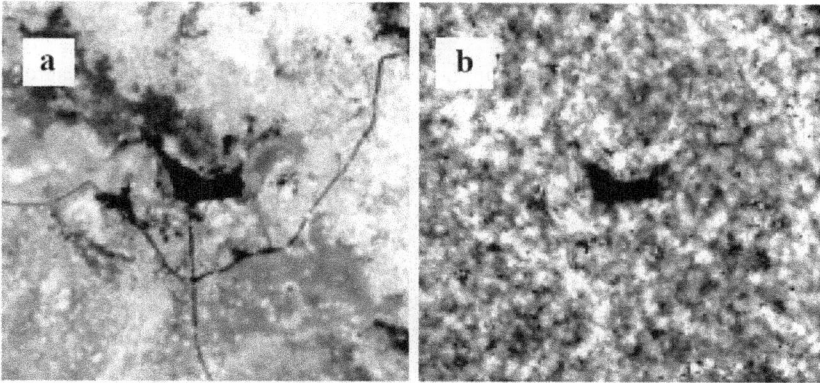

Figure 12.5 Greyscale images of the same landscape acquired with an optical sensor (a) and a radar sensor (b). Specifically, these images were acquired by the Sentinel-2 (NDVI) and Sentinel-1 (radar backscattering coefficient in Vertical-Horizontal polarization) satellites, respectively. Satellite data courtesy of the European Space Agency.

identifying hotspots of losses and gains. Pekel and colleagues (2016) used the same satellite suite to create a global map of surface water and its dynamics. Like the aforementioned multiclass land cover products, these maps are open access, meaning that users can directly download the finished land cover product. However, depending on the landscape and features of interest, a custom-made land cover map might be more appropriate, since the local accuracy of such maps can be low (Congalton et al. 2014). Examples for smaller-scale applications include mapping inundated vegetation across South America (Chapman et al. 2015), heterogenous vegetation cover in savannah landscapes (Marston et al. 2017), and human-dominated rural landscapes (Mack et al. 2017).

The resulting categorical land cover maps can then be used to assess the horizontal structure of a landscape, for instance by estimating the fragmentation and configuration of different land cover types (Kayiranga et al. 2016). There is a trade-off between the spatial resolution of the land cover map, its ability to capture small fragments and to correctly delineate irregular fragment boundaries, on the one hand, and ease of satellite data processing and data availability, on the other hand (Boyle et al. 2014). Satellite imagery with very high spatial resolution is often not freely available, and it takes much more processing power to generate land cover maps from it. However, small-scale variability in landscape horizontal structure can be ecologically important (Rösch et al. 2015, de Camargo Barbosa et al. 2017), so this trade-off needs to be resolved based on the context of the landscape in question.

Characterizing horizontal landscape structure can be a useful approach for quantifying structural landscape connectivity (sensu Fletcher et al. 2016). Structural landscape connectivity is typically derived from the spatial arrangement of patches of native vegetation classes, for instance by quantifying classic fragmentation statistics such as patch size distribution and geographical distances between patches (Pettorelli 2018), although more complicated indices of spatial arrangement exist (Saura et al. 2011, McGarigal 2015). Such information can be used to inform land use planning, including protected area gazetting (Magris et al. 2016, Pirnat and Hladnik 2016). Land cover maps can also be used to quantify the diversity of vegetation classes or habitats at different spatial scales, which has been shown to predict species-level biodiversity in some circumstances (Maskell et al. 2019).

Remote sensing can also help to derive continuous environmental information; this often involves relating reflectance values to ground measurements using a statistical model ('upscaling'). For instance, hyperspectral imagery has been combined with field measurements of chemical foliage composition to map plant functional trait proxies across large areas (Asner et al. 2015). Similarly, radar imagery has been used to scale up field observations of woody biomass and tree cover in a savannah landscape (Naidoo et al. 2015); optical data (specifically Landsat) have, moreover, been used to successfully upscale plot-scale measurements of primary productivity (Tebbs et al. 2017). Additionally, remotely sensed variables can be used to derive ecosystem characteristics using radiative transfer models, which predict how electromagnetic radiation interacts with the Earth's surface. This allows, for instance, derivation of biophysical parameters such as the leaf area index and the chlorophyll content from remote measurements of canopy spectra (Darvishzadeh et al. 2008).

There have also been attempts to use raw imagery values directly as proxies for landscape-level diversity. The spectral diversity hypothesis indeed posits that areas with more species will be more spectrally diverse than areas with a lower number of species, suggesting a way to estimate variation in alpha biodiversity across a landscape from space. However, this approach has had mixed success; although spectral diversity captures significant amounts of species richness or other diversity measures in some landscapes (e.g., Rocchini et al. 2004, Heumann et al. 2015), this relationship varies over time and in space and is often absent (Schmidtlein and Fassnacht 2017). As a result, it is difficult to derive meaningful insights into landscape spatial structure from raw spectral reflectance values alone.

Active remote sensing is a particularly valuable source of information when it comes to assessing the vertical structure of a landscape. For example, airborne LiDAR is a key tool for mapping canopy height (Khosravipour et al. 2014) and other canopy vertical structural properties such as leaf area index (Korhonen et al. 2011). There are now spaceborne radar sensors that allow forest biomass to be measured (Lu et al. 2016). Passive sensors (e.g., multispectral imagery) have also been used to describe vertical landscape structure. This is achieved either by relating field measurements of vertical structure (such as biomass) to remotely sensed vegetation or texture indices (Lu et al. 2016) or by taking stereo images of a study site and deriving a three-dimensional model of the landscape (Poli et al. 2015).

Landscape processes

Remote sensing makes it possible to monitor the processes that shape landscape structures at large spatial scales. Such processes include ecosystem functions and ecosystem processes, such as primary productivity and herbivory (Pettorelli et al. 2018), which are controlled by the biotic communities in a landscape. They also include natural disturbances, which can sometimes be driven by factors external to the landscape (such as climate variability).

Earth observation has been identified as a key methodology for the monitoring of a range of ecosystem functions (Pettorelli et al. 2018). Primary productivity is routinely measured remotely, for instance through vegetation indices such as the Normalized Difference Vegetation Index (NDVI; Pettorelli 2013). The NDVI is an established indicator of the amount of photosynthetically active vegetation (which is closely related to primary productivity) in a given area. Primary productivity can also be estimated from mapping biomass remotely (Zolkos et al. 2013), which involves linking remotely sensed spectral properties to field measurements of biomass or similar vegetation parameters. Secondary productivity is (as of now) difficult to estimate directly from Earth observation (Hollings et al. 2018), although there have been promising examples of directly detecting large mammals using spaceborne sensors with very high spatial resolution

(Yang et al. 2014); aerial surveys are also routinely used for wildlife counts (e.g., Thrash et al. 1995). The impact of herbivory at the landscape scale can be sensed remotely via its effects on foliage quantity and structure. Vegetation indices such as the NDVI can be used to quantify the effect of herbivory on vegetation if other confounding factors (e.g., rainfall, soil type, and vegetation type) can be controlled for (Rickbeil et al. 2015). Outbreaks of defoliator insects can sometimes lead to a dramatic loss of vegetation over large areas: these outbreaks can be detected using multispectral imagery (see Senf et al. 2015).

Fire is another important force driving the dynamics of many ecosystems that can be monitored remotely, with current approaches able to detect both active and recent fires. Active fires are characterized by high temperatures, which can be detected using infrared sensors, while recently burned areas typically show a rapid change in vegetation structure as well as charcoal and ash. Hempson and colleagues (2018) recently used satellite-derived information on the occurrence of both types of fires to characterize the diversity of fire regimes across Africa and then linked the observed patterns to variables such as rainfall and human land use. Similarly, Alencar and colleagues (2015) used a time series of remotely sensed fires in different forest types in the Amazon to suggest that fire regimes might be intensified by forest fragmentation.

Finally, extreme climatic conditions such as droughts and floods can significantly impact the structure and functioning of a landscape. Droughts can be detected by a range of remotely sensed indices, including precipitation indices, soil moisture indices, and evapotranspiration (mainly using spaceborne infrared, thermal, and microwave sensors; AghaKouchak et al. 2015). Floods can be detected using radar sensors, since very little of the original microwave signal is scattered back to the sensor by water, giving rise to characteristically 'dark' areas (Kuenzer et al. 2013). Floods have also successfully been detected with hyperspectral sensors (Ip et al. 2006) and multispectral sensors (Klemas 2014), both spaceborne and airborne.

Anthropogenic drivers of landscape change

Anthropogenic drivers comprise a wide range of stressors and disturbances, often disrupting landscape structure and processes in undesirable ways. Remote sensing allows the tracking of many of these drivers, sometimes in near-real time. The growth of urban areas (and subsequent loss or displacement of less intensive land uses) can, for example, be detected using spaceborne radar sensors (Ban et al. 2015). Light pollution (which is associated with urban areas) has been mapped using highly sensitive optical sensors that collect images at night (Zhou et al. 2014). Roads divide landscapes, alter connectivity, change vegetation communities, and are a hazard for wildlife. The development of road networks can be mapped remotely using Earth observations from multispectral, radar, or multiple data sources to extract linear features (Brooks et al. 2017). Increased road developments can sometimes indicate increased extraction of natural resources through, e.g., mining activities. In some circumstances, subsurface mining can be directly detected, for example where it is associated with changes in vegetation (e.g., Swenson et al. 2011). Alternatively, open-pit mines can be mapped using active sensors such as LiDAR, which can detect the depressions in the Earth's surface (Tong et al. 2015).

Water bodies can be significantly impacted by anthropogenic activity through eutrophication processes. Because this process alters the optical depth and color of the water, eutrophication can be detected using multispectral remote sensors. For instance, water quality indices such as Secchi disk depth or chlorophyll-α content can be derived from Earth observation sensors when ground measurements are available (Zhu et al. 2014). Since freshwater bodies often have a small size compared with the spatial resolution of many spaceborne sensors, such methods have predominantly been developed for coastal areas (Harvey et al. 2015, but see Matthews et al. 2010).

An indirect outcome of the growing levels of human activity is the increased introduction of invasive species. Invasive plants can be mapped remotely if their spectral signature, texture, or structure differs significantly from that of native vegetation (Bradley and Mustard 2006, He et al. 2011). Stand-forming invasive plants, in particular, have been successfully detected using multispectral sensors (Hoyos et al. 2010, Gavier-Pizarro et al. 2012). Though such detections are mostly limited to cases in which invasive plants reach the top of the canopy (and are thus visible to an airborne or spaceborne sensor), it is possible to track understory invasives if they significantly alter the overall spectral signature of the canopy, e.g., if they are green outside the growing season of native plants (Tuanmu et al. 2010).

Upcoming opportunities

Spaceborne missions

Hyperspectral missions: EnMAP (Environmental Monitoring and Analysis Program) and HyspIRI (Hyperspectral Infrared Imager) are two upcoming hyperspectral satellite missions aiming to open new monitoring opportunities for environmental parameters related to vegetation extent, vegetation health, carbon stock, and environmental disturbances. They are due to launch in 2021 and 2023, respectively. From a landscape ecology perspective, these missions will provide global, standardized, repeated information on changes in landscape structure, land use, and disturbance occurrence. The short revisit times for both satellites (5–19 days for HypsIRI and 4–27 days for EnMAP), in particular, will be important for monitoring climate change–driven changes in plant chemistry and phenology.

Spaceborne LiDAR missions: Notable spaceborne LiDAR missions, designed to focus on the Earth's vertical vegetation structure and sea ice thickness, include GEDI (Global Ecosystem Dynamics Investigation) and ICESat-2 (Ice, Cloud and land Elevation Satellite-2). By monitoring topography and vegetation structure with unprecedented precision (Melin et al. 2017), GEDI (launched in December 2018) will help assess how deforestation has contributed to increased atmospheric CO_2 or how much carbon forests will absorb in the future. ICESat-2 launched on 15 September 2018. It is an advanced version of the original ICESat, which was in action from 2003 to 2009. The upgrade to six lasers, compared with ICESat's single one, gives the ICESat-2 mission a much denser coverage of Earth's surface. As well as detecting changes in Earth's polar ice caps, ICESat-2 will survey the heights of forests, lakes, urban areas, and cloud cover. Thanks to this new dimension, specific amounts of vegetation that make up an area's biomass (e.g., trunks, leaves, branches) can be calculated, ultimately supporting the monitoring of wider landscape processes.

Radar missions: Biomass mission is a new mission carrying the first P-band (30 cm to 1 m wavelength) synthetic aperture radar (SAR) sensor that will provide measurements of global forest biomass currently not obtainable from techniques based on ground measurements (Quegan et al. 2019). The main objectives of this mission are to quantify and monitor the amount of carbon stored in the forests as well as forest disturbance and recovery. The launch is planned for 2021.

Nanosatellites: CubeSats and SmallSats (also known as nanosatellites) are small, relatively inexpensive satellites that are increasingly being discussed as options for targeted, short-term, very high–spatial resolution monitoring. These satellites are generally released in constellations (multiple orbitally synchronized satellites) working in parallel with each other to provide high spatial and temporal resolution imagery. Many would argue that we are currently going through a 'CubeSat revolution' with rapidly advancing technology and decreasing costs providing the potential for highly technical sensors and larger constellations (Marvin et al. 2016). Nanosatellites are thus expected to significantly advance opportunities to track changes in landscapes at very fine spatial and temporal resolution, potentially being a game changer for developing near-real time monitoring options.

Citizen science

The use of citizen science in Earth observation is seen by many as an important solution to the data validation bottlenecks that occur in remote sensing due to the enormous amounts of data being created at a rapid pace. Because many projects are volunteer sourced, the important but time-consuming ground-truthing data can indeed be provided for free. Increased numbers of Earth observation studies now contain citizen science; a non-exhaustive list of recent projects can be found in Fritz and colleagues (2017). Predominantly, citizens can assist with interpretation of images or collection of data in situ, normally by taking photographs. Examples of such projects include 'Season spotter', which required citizens to identify features in a selection of over 50,000 images taken from cameras across North America to speed up data collection on vegetation phenology (Kosmala et al. 2016), and 'Nature's notebook', which has non-specialists collect in situ data on vegetation phenology in North America, which are then used to validate satellite NDVI data (Wallace et al. 2016).

Citizen scientist data validation also proved valuable in the FotoQuest Austria campaign (which lasted from 2015 to 2018). Citizens take photos on a mobile app in given sample locations where officials normally collect ground data every 3 years for the Land Use/Cover Area frame Survey (LUCAS) survey (Laso Bayas et al. 2016). Periodically, a money-based reward system is used to encourage users. This increased the density of photos and reduced the time between repeat measurements, reducing the cost of data collection but increasing the value of the dataset. There are challenges with citizen science, such as the quality of the data from non-experts and how to engage and retain citizen participation in the longer term. Improvements are being made to increase integrated training and feedback during the data collection stage, and 'gamification' is being used to motivate users by making data collection more fun without the cost of providing rewards (Fritz et al. 2017).

Cameras and camera traps

Camera traps (which are remote sensors on stationary, often ground-level platforms) have predominantly been used to track elusive animal species. However, they are starting to be used for Earth observation as well. For instance, monitoring vegetation phenology through traffic webcams has proved an efficient method for data collection that fills the scale-gap between other approaches to measuring phenology (Morris et al. 2013). The webcams, alongside automated image interpretation, provide a low-cost, low-effort alternative to field-based data collection while maintaining a plot-size scale to complement large-scale satellite imagery. The high density and widespread network of these cameras across England allowed automated tracking of phenology on a landscape scale. The opportunity to expand this method to use other non-scientific camera networks (e.g., urban CCTV monitoring) could further utilize the huge amount of data already being collected on a landscape scale.

Near-surface remote sensing uses repeat photos from digital cameras based on the ground or mounted on towers. Seasonal changes in vegetation can be identified from the red, green, and blue channels extracted from digital photographs, and plant structure, phenology, and primary production can be recorded over time. For landscape ecology, useful networks of cameras can include camera traps used for wild animal tracking, which also collect information on the vegetation in the background of images, and those installed specifically to monitor changes in seasonal plant biomass (e.g., Inoue et al. 2015). These can provide valuable ground-truthing data to validate the data collected by satellite imagery.

Big Data analysis and data integration

Big Data refers to amounts of data that are so large that they overwhelm the 'usual' analysis tools and techniques (Hampton et al. 2013). In the context of Earth observation, we are seeing the beginning of 'Big Data ecology' as the amount of available remote sensing imagery, as well as its spatial and temporal resolution, continues to increase. This provides the opportunity to observe landscapes at high spatial and temporal detail without compromising spatial scale. An example of Big Data application for landscape ecology is the use of dense remote sensing time series to identify sudden shifts and more subtle but persistent trends in vegetation (Jamali et al. 2015) or the mapping of changes in tree cover over the global land surface (Hansen et al. 2013). Big Data approaches are also used to fuse different kinds of remote sensing imagery, e.g., multispectral and radar (Schulte to Bühne and Pettorelli 2018), to take advantage of the complementary information they provide. There are already tools and techniques available that can handle Big Data in Earth observation, including parallel processing and more efficient data storage (Ma et al. 2015). These approaches also require a lot of computational power, which has made the use of cloud computing in remote sensing more attractive. Examples for this include the open access Google Earth Engine platform and commercial options (such as Microsoft Azure or Amazon Web Services). Finally, given the significant increases in Earth observation and computing capacity, Big Data ecology could transform biodiversity monitoring by linking spatially sparse in situ data (e.g., on genetic or species biodiversity) to environmental variables derived from Earth observation (Bush et al. 2017).

Current challenges and ways forward

The potential for remote sensing to support landscape ecology has likely never been greater; however, several issues may limit this potential. First, remote sensing approaches can be expensive given the logistical requirements (i.e., qualified staff and training) for the acquisition, processing, and analysis of these sometimes large datasets. Second, integration of in situ data from local ecologists and expert knowledge from remote sensing analysts is still relatively limited, leading to remote sensing approaches frequently being underused and undervalued. The use of remote sensing data by landscape ecologists necessitates experts in both fields to work at the interface of their disciplines, but in many cases, that dialogue can be hard. All parties need to be able to understand the constraints shaping the datasets considered as well as the possible errors propagated in subsequent analyses. Yet, ecology trainees are rarely exposed to words such as 'multispectral', 'radiometer', or 'atmospheric corrections', while remote sensing experts often struggle when confronted with concepts such as 'ecological community', 'life history', or 'species evenness'. This lack of common reference frames can seriously hamper collaborative work.

The accessibility of data and algorithms might be a particularly important reason preventing the emergence of more collaborative projects between the remote sensing and ecology communities (Pettorelli et al. 2014a). Access to several satellite remote sensing products is, for example, still not free of charge (Boyle et al. 2014), while only a small fraction of ecological data ever collected is readily discoverable and accessible, much less usable. There are also several logistical and technical challenges associated with the storage, sharing, manipulation, and analysis of the datasets of common interest to both communities. Each community indeed stores its information in different databases and archives, and the data stored therein may exist in different formats, subjecting the integration of datasets across communities to significant processing and information technology hurdles. Combining remote sensing and ecological datasets for visual inspection or analysis can, moreover, require sophisticated algorithms and software, which might not be equally accessible to either community. Altogether, this can create a situation where partners do not have the same level of opportunities to contribute to the project, sometimes leading to partners' disengagement and the failure of the collaboration.

Given these considerations, training opportunities in remote sensing tailored for the landscape ecology community are of paramount importance to facilitate the emergence of a new generation of scientists able to carry out integrated, multi-disciplinary approaches. The number of universities providing tailored remote sensing courses or using 'service modules' to provide remote sensing skills to professionals is increasing, but recent analyses suggest that more needs to be done (Clark et al. 2017). Similarly, better and stronger communication channels need to be established between the remote sensing community and landscape ecologists to increase collaborative work and develop a coordinated, effective research agenda (Pettorelli et al. 2014a). Specific platforms facilitating exchanges of information and networking opportunities between these communities, such as the Group on Earth Observations Biodiversity Observation Network (GEO BON) or CRSNET (an online remote sensing conservation network), need to continue flourishing, while user-friendly, intuitive, and centralized data portals need to be developed to further enhance communication and exchange of experiences about remote sensing products and analyses.

Conclusions

Remote sensing is at the core of landscape ecology, having shaped its developments and research agenda from the start. As sensors are becoming more diversified and platforms more versatile, there is a real opportunity to broaden the scope of landscape ecology and increase the breadth of research questions that can be tackled with the support of this technology. Because landscape ecology is ultimately a discipline that sits at the interface between ecology and Earth observations, its progress will always be linked to the ability of its researchers to successfully integrate information and knowledge from both disciplines. Yet, this integration is becoming increasingly challenging due to the current pace of technological and analytical progress. As such, one could argue that we are moving from an area where progress was limited by technology to an area where progress may be limited by communication efficiency and collaborative skills. To fully capitalize on current and future opportunities, we believe that the new generation of landscape ecologists need to be equipped with tools and skills that allow them to identify feasible options and key partnerships for plugging the known knowledge gaps.

References

AghaKouchak, A., Farahmand, A., Melton, F.S., Teixeira, J., Anderson, M.C., Wardlow, B.D. and Hain, C.R. (2015) 'Remote sensing of drought: Progress, challenges and opportunities', *Reviews of Geophysics*, vol 53, pp452–480.

Alencar, A.A., Brando, P.M., Asner, G.P. and Putz, F.E. (2015) 'Landscape fragmentation, severe drought, and the new Amazon forest fire regime', *Ecological Applications*, vol 25, pp1493–1505.

Anderson, K. and Gaston, K.J. (2013) 'Lightweight unmanned aerial vehicles will revolutionize spatial ecology', *Frontiers in Ecology and the Environment*, vol 11, pp138–146.

Asner, G.P., Knapp, D.E., Kennedy-Bowdoin, T., Jones, M.O., Martin, R.E., Boardman, J. and Hughes, R.F. (2008) 'Invasive species detection in Hawaiian rainforests using airborne imaging spectroscopy and LiDAR', *Remote Sensing of Environment*, vol 112, pp1942–1955.

Asner, G.P., Levick, S.R., Kennedy-Bowdoin, T., Knapp, D.E., Emerson, R., Jacobson, J., Colgan, M.S. and Martin, R.E. (2009) 'Large-scale impacts of herbivores on the structural diversity of African savannas', *Proceedings of the National Academy of Sciences*, vol 106, pp4947–4952.

Asner, G.P., Martin, R.E., Anderson, C.B. and Knapp, D.E. (2015) 'Quantifying forest canopy traits: Imaging spectroscopy versus field survey', *Remote Sensing of Environment*, vol 158, pp15–27.

Ban, Y., Jacob, A. and Gamba, P. (2015) 'Spaceborne SAR data for global urban mapping at 30 m resolution using a robust urban extractor', *ISPRS Journal of Photogrammetry and Remote Sensing*, vol 103, pp28–37.

Bioucas-Dias, J.M., Plaza, A., Camps-Valls, G., Scheunders, P., Nasrabadi, N. and Chanussot, J. (2013) 'Hyperspectral remote sensing data analysis and future challenges', *IEEE Geoscience and Remote Sensing Magazine*, vol 1, pp6–36.

Blaschke T. (2003) 'Continuity, complexity and change: A hierarchical geoinformation-based approach to exploring patterns of change in a cultural landscape', in U. Mander and M. Antrop (eds) *Multifunctional Landscapes Vol. III: Continuity and Change*. WIT Press, Southampton.

Boyle, S.A., Kennedy, C.M., Torres, J., Colman, K., Perez-Estigarribia, P.E. and Noé, U. (2014) 'High-resolution satellite imagery is an important yet underutilized resource in conservation biology', *PLoS One*, vol 9, art e86908.

Bradley, B.A. and Mustard, J.F. (2006) 'Characterizing the landscape dynamics of an invasive plant and risk of invasion using remote sensing', *Ecological Applications*, vol 16, pp1132–1147

Brooks, C.N., Dean, D.B., Dobson, R.J., Roussi, C., Carter, J.F., VanderWoude, A.J., Colling, T. and Banach, D.M. (2017) 'Identification of unpaved roads in a regional road network using remote sensing', *Photogrammetric Engineering & Remote Sensing*, vol 83, pp377–383.

Bush, A., Sollmann, R., Wilting, A., Bohmann, K., Cole, B., Balzter, H., Martius, C., Zlinszky, A., Calvignac-Spencer, S., Cobbold, C.A. and Dawson, T.P. (2017) 'Connecting Earth observation to high-throughput biodiversity data', *Nature Ecology & Evolution*, vol 1, art 0176.

Centre for Ecology & Hydrology (CEH) (2017) 'Land cover map 2015: Dataset documentation', https://www.ceh.ac.uk/sites/default/files/LCM2015_Dataset_Documentation_22May2017.pdf, accessed 13 Sep 2019.

Chapman, B., McDonald, K., Shimada, M., Rosenqvist, A., Schroeder, R. and Hess, L. (2015) 'Mapping regional inundation with spaceborne L-Band SAR', *Remote Sensing*, vol 7, pp5440–5470.

Christophe, E. and Inglada, J. (2007) 'Robust road extraction for high resolution satellite images', IEEE International Conference on Image Processing, San Antonio, Texas, USA, vol 5, art V–437.

Clark, B.L., Bevanda, M., Aspillaga, E. and Jørgensen, N.H. (2017) 'Bridging disciplines with training in remote sensing for animal movement: An attendee perspective', *Remote Sensing in Ecology and Conservation*, vol 3, pp30–37.

Congalton, R., Gu, J., Yadav, K., Thenkabail, P. and Ozdogan, M. (2014) 'Global land cover mapping: A review and uncertainty analysis', *Remote Sensing*, vol 6, no 12, 12070–12093.

Costa, M. (2005) 'Estimate of net primary productivity of aquatic vegetation of the Amazon floodplain using Radarsat and JERS-1', *International Journal of Remote Sensing*, vol 26, pp4527–4536.

Côté, J.F., Fournier, R.A., Frazer, G.W. and Niemann, K.O. (2012) 'A fine-scale architectural model of trees to enhance LiDAR-derived measurements of forest canopy structure', *Agricultural and Forest Meteorology*, vol 166, pp72–85.

Darvishzadeh, R., Skidmore, A., Schlerf, M. and Atzberger, C. (2008) 'Inversion of a radiative transfer model for estimating vegetation LAI and chlorophyll in a heterogeneous grassland', *Remote Sensing of Environment*, vol 112, pp2592–2604.

de Camargo Barbosa, K.V., Knogge, C., Develey, P.F., Jenkins, C.N. and Uezu, A. (2017) 'Use of small Atlantic Forest fragments by birds in Southeast Brazil', *Perspectives in Ecology and Conservation*, vol 15, pp42–46.

Elvidge, C.D., Baugh, K., Zhizhin, M., Hsu, F.C. and Ghosh, T. (2017) 'VIIRS night-time lights', *International Journal of Remote Sensing*, vol 38, pp5860–5879.

Feranec, J., Soukup, T., Hazeu, G. and Jaffrain, G. (eds) (2016) *European Landscape Dynamics: CORINE Land Cover Data*. CRC Press, Boca Raton.

Fletcher, R.J., Burrell, N.S., Reichert, B.E., Vasudev, D. and Austin, J.D. (2016). 'Divergent perspectives on landscape connectivity reveal consistent effects from genes to communities', *Current Landscape Ecology Reports*, vol 1, pp67–79.

Franklin, S.E., Ahmed, O.S., Wulder, M.A., White, J.C., Hermosilla, T. and Coops, N.C. (2015) 'Large area mapping of annual land cover dynamics using multitemporal change detection and classification of Landsat time series data', *Canadian Journal of Remote Sensing*, vol 41, no 4, pp293–314.

Friedl, M.A., Sulla-Menashe, D., Tan, B., Schneider, A., Ramankutty, N., Sibley, A. and Huang, X. (2010) 'MODIS Collection 5 global land cover: Algorithm refinements and characterization of new datasets', *Remote sensing of Environment*, vol 114, pp168–182.

Fritz, S., Fonte, C.C. and See, L. (2017) 'The role of citizen science in Earth Observation', *Remote Sensing*, vol 9, art 357.

Gavier-Pizarro, G.I., Kuemmerle, T., Hoyos, L.E., Stewart, S.I., Huebner, C.D., Keuler, N.S. and Radeloff, V.C. (2012) 'Monitoring the invasion of an exotic tree (*Ligustrum lucidum*) from 1983 to 2006 with Landsat TM/ETM+ satellite data and Support Vector Machines in Córdoba, Argentina', *Remote Sensing of Environment*, vol 122, pp134–145.

Hame, T., Salli, A., Andersson, K. and Lohi, A. (1997) 'A new methodology for the estimation of biomass of conifer dominated boreal forest using NOAA AVHRR data', *International Journal of Remote Sensing*, vol 18, pp3211–3243.

Hampton, S.E., Strasser, C.A., Tewksbury, J.J., Gram, W.K., Budden, A.E., Batcheller, A.L., Duke, C.S. and Porter, J.H. (2013) 'Big data and the future of ecology', *Frontiers in Ecology and the Environment*, vol 11, pp156–162.

Hansen, M.C., Potapov, P.V., Moore, R., Hancher, M., Turubanova, S.A.A., Tyukavina, A., Thau, D., Stehman, S.V., Goetz, S.J., Loveland, T.R. and Kommareddy, A. (2013) 'High-resolution global maps of 21st-century forest cover change', *Science*, vol 342, pp850–853.

Hantson, S., Pueyo, S. and Chuvieco, E. (2015) 'Global fire size distribution is driven by human impact and climate', *Global Ecology and Biogeography*, vol 24, pp77–86.

Harvey, E.T., Kratzer, S. and Philipson, P. (2015) 'Satellite-based water quality monitoring for improved spatial and temporal retrieval of chlorophyll-a in coastal waters', *Remote Sensing of Environment*, vol 158, pp417–430.

He, K.S., Rocchini, D., Neteler, M., and Nagendra, H. (2011) 'Benefits of hyperspectral remote sensing for tracking plant invasions', *Diversity and Distributions*, vol 17, pp381–392.

Hempson, G.P., Parr, C.L., Archibald, S., Anderson, T.M., Mustaphi, C.J.C., Dobson, A.P., Donaldson, J.E., Morrison, T.A., Probert, J. and Beale, C.M. (2018) 'Continent-level drivers of African pyrodiversity', *Ecography*, vol 41, pp889–899.

Heumann, B.W., Hackett, R.A. and Monfils, A.K. (2015) 'Testing the spectral diversity hypothesis using spectroscopy data in a simulated wetland community', *Ecological Informatics*, vol 25, pp29–34.

Hollings, T., Burgman, M., van Andel, M., Gilbert, M., Robinson, T. and Robinson, A. (2018) 'How do you find the green sheep? A critical review of the use of remotely sensed imagery to detect and count animals', *Methods in Ecology and Evolution*, vol 9, pp881–892.

Horning, N., Leutner, B. and Wegmann, M. (2016) 'Land cover or image classification approaches', in M. Wegmann, B. Leutner, and S. Dech (eds) *Remote Sensing and GIS for Ecologists: Using Open Source Software*. Pelagic Publishing, Exeter.

Hoyos, L.E., Gavier-Pizarro, G.I., Kuemmerle, T., Bucher, E.H., Radeloff, V.C. and Tecco, P.A. (2010) 'Invasion of glossy privet (*Ligustrum lucidum*) and native forest loss in the Sierras Chicas of Córdoba, Argentina', *Biological Invasions*, vol 12, pp3261–3275.

Huang, C.Y., Asner, G.P., Martin, R.E., Barger, N.N. and Neff, J.C. (2009) 'Multiscale analysis of tree cover and aboveground carbon stocks in pinyon–juniper woodlands', *Ecological Applications*, vol 19, pp668–681.

Huo, A., Zhang, J., Qiao, C., Li, C., Xie, J., Wang, J., and Zhang, X. (2014) 'Multispectral remote sensing inversion for city landscape water eutrophication based on Genetic Algorithm-Support Vector Machine', *Water Quality Research Journal*, vol 49, pp285–293.

Inoue, T., Nagai, S., Kobayashi, H. and Koizumi, H. (2015) 'Utilization of ground-based digital photography for the evaluation of seasonal changes in the aboveground green biomass and foliage phenology in a grassland ecosystem', *Ecological Informatics*, vol 25, pp1–9.

Ip, F., Dohm, J.M., Baker, V.R., Doggett, T., Davies, A.G., Castano, R., Chien, S., Cichy, B., Greeley, R., Sherwood, R. and Tran, D. (2006) 'Flood detection and monitoring with the Autonomous Sciencecraft Experiment onboard EO-1', *Remote Sensing of Environment*, vol 101, pp463–481.

Jackson, R.D. and Huete, A.R. (1991) 'Interpreting vegetation indices', *Preventive Veterinary Medicine*, vol 11, pp185–200.

Jamali, S., Jönsson, P., Eklundh, L., Ardö, J. and Seaquist, J. (2015) 'Detecting changes in vegetation trends using time series segmentation', *Remote Sensing of Environment*, vol 156, pp182–195.

Jones, H.G. and Vaughan, R.A. (2010) *Remote Sensing of Vegetation: Principles, Techniques and Applications*. Oxford University Press, Oxford.

Kayiranga, A., Kurban, A., Ndayisaba, F., Nahayo, L., Karamage, F., Ablekim, A., Li, H. and Ilniyaz, O. (2016) 'Monitoring forest cover change and fragmentation using remote sensing and landscape metrics in Nyungwe-Kibira park', *Journal of Geoscience and Environment Protection*, vol 4, pp13–33.

Khosravipour, A., Skidmore, A.K., Isenburg, M., Wang, T. and Hussin, Y.A. (2014) 'Generating pit-free canopy height models from airborne lidar', *Photogrammetric Engineering & Remote Sensing*, vol 80, pp863–872.

Klemas, V. (2014) 'Remote sensing of floods and flood-prone areas: An overview', *Journal of Coastal Research*, vol 31, pp1005–1013.

Korhonen, L., Korpela, I., Heiskanen, J. and Maltamo, M. (2011) 'Airborne discrete-return LIDAR data in the estimation of vertical canopy cover, angular canopy closure and leaf area index', *Remote Sensing of Environment*, vol 115, pp1065–1080.

Kosmala, M., Crall, A., Cheng, R., Hufkens, K., Henderson, S. and Richardson, A.D. (2016) 'Season Spotter: Using citizen science to validate and scale plant phenology from near-surface remote sensing', *Remote Sensing*, vol 8, 726.

Kuenzer, C., Guo, H., Huth, J., Leinenkugel, P., Li, X. and Dech, S. (2013) 'Flood mapping and flood dynamics of the Mekong Delta: ENVISAT-ASAR-WSM based time series analyses', *Remote Sensing*, vol 5, pp687–715.

Kutser, T. (2004) 'Quantitative detection of chlorophyll in cyanobacterial blooms by satellite remote sensing', *Limnology and Oceanography*, vol 49, pp2179–2189.

Laso Bayas, J.C., See, L., Fritz, S., Sturn, T., Perger, C., Dürauer, M., Karner, M., Moorthy, I., Schepaschenko, D., Domian, D. and McCallum, I. (2016) 'Crowdsourcing in-situ data on land cover and land use using gamification and mobile technology', *Remote Sensing*, vol 8, art 905.

Lawrence, R.L., & Moran, C.J. (2015) 'The AmericaView classification methods accuracy comparison project: A rigorous approach for model selection', *Remote Sensing of Environment*, vol 170, pp115–120.

Lim, K., Treitz, P., Wulder, M., St-Onge, B. and Flood, M. (2003) 'LiDAR remote sensing of forest structure', *Progress in Physical Geography*, vol 27, pp88–106.

Lu, D., Chen, Q., Wang, G., Liu, L., Li, G. and Moran, E. (2016) 'A survey of remote sensing-based aboveground biomass estimation methods in forest ecosystems', *International Journal of Digital Earth*, vol 9, pp63–105.

Ma, Y., Wu, H., Wang, L., Huang, B., Ranjan, R., Zomaya, A. and Jie, W. (2015) 'Remote sensing big data computing: Challenges and opportunities', *Future Generation Computer Systems*, vol 51, pp47–60.

Mack, B., Leinenkugel, P., Kuenzer, C. and Dech, S. (2017) 'A semi-automated approach for the generation of a new land use and land cover product for Germany based on Landsat time-series and Lucas in-situ data', *Remote Sensing Letters*, vol 8, pp244–253.

Madritch, M.D., Kingdon, C.C., Singh, A., Mock, K.E., Lindroth, R.L. and Townsend, P.A. (2014) 'Imaging spectroscopy links aspen genotype with below-ground processes at landscape scales', *Philosophical Transactions of the Royal Society of London B: Biological Sciences*, vol 369, art 20130194.

Magris, R.A., Treml, E.A., Pressey, R.L. and Weeks, R. (2016) 'Integrating multiple species connectivity and habitat quality into conservation planning for coral reefs', *Ecography*, vol 39, pp649–664.

Marston, C.G., Aplin, P., Wilkinson, D.M., Field, R. and O'Regan, H.J. (2017) 'Scrubbing up: Multi-scale investigation of woody encroachment in a southern African savannah', *Remote Sensing*, vol 9, art 419.

Maskell, L.C., Botham, M., Henrys, P., Jarvis, S., Maxwell, D., Robinson, D.A., Rowland, C.S., Siriwardena, G., Smart, S., Skates, J. and Tebbs, E.J. (2019) 'Exploring relationships between land use intensity, habitat heterogeneity and biodiversity to identify and monitor areas of High Nature Value farming', *Biological Conservation*, vol 231, pp30–38.

Marvin, D.C., Koh, L.P., Lynam, A.J., Wich, S., Davies, A.B., Krishnamurthy, R., Stokes, E., Starkey, R. and Asner, G.P. (2016) 'Integrating technologies for scalable ecology and conservation', *Global Ecology and Conservation*, vol 7, pp262–275.

Matthews, M.W., Bernard, S. and Winter, K. (2010) 'Remote sensing of cyanobacteriadominant algal blooms and water quality parameters in Zeekoevlei, a small hypertrophic lake, using MERIS', *Remote Sensing of Environment*, vol 114, pp2070–2087.

McGarigal, K. (2015) FRAGSTATS help. University of Massachusetts, Amherst.

Melin, M., Shapiro, A. and Glover-Kapfer, P. (2017) *LiDAR for ecology and conservation*. WWF Conservation Technology Series (3). WWF-UK, Woking.

Meyer, T. and Okin, G.S. (2015) 'Evaluation of spectral unmixing techniques using MODIS in a structurally complex savanna environment for retrieval of green vegetation, nonphotosynthetic vegetation, and soil fractional cover', *Remote Sensing of Environment*, vol 161, pp122–130.

Morris, D.E., Boyd, D.S., Crowe, J.A., Johnson, C.S. and Smith, K.L. (2013) 'Exploring the potential for automatic extraction of vegetation phenological metrics from traffic webcams', *Remote Sensing*, vol 5, pp2200–2218.

Naidoo, L., Mathieu, R., Main, R., Kleynhans, W., Wessels, K., Asner, G. and Leblon, B. (2015) 'Savannah woody structure modelling and mapping using multi-frequency (X-, C-and L-band) Synthetic Aperture Radar data', *ISPRS Journal of Photogrammetry and Remote Sensing*, vol 105, pp234–250.

Nayak, R.K., Patel, N.R. and Dadhwal, V.K. (2010) 'Estimation and analysis of terrestrial net primary productivity over India by remote-sensing-driven terrestrial biosphere model', *Environmental Monitoring and Assessment*, vol 170, pp195–213.

Oh, Y., Sarabandi, K. and Ulaby, F.T. (1992) 'An empirical model and an inversion technique for radar scattering from bare soil surfaces', *IEEE transactions on Geoscience and Remote Sensing*, vol 30, pp370–381.

Pekel, J.F., Cottam, A., Gorelick, N. and Belward, A.S. (2016) 'High-resolution mapping of global surface water and its long-term changes', *Nature*, vol 540, pp418–422.

Perks, M.T., Russell, A.J. and Large, A.R. (2016) 'Advances in flash flood monitoring using unmanned aerial vehicles (UAVs)', *Hydrology and Earth System Sciences*, vol 20, pp4005–4015.

Pettorelli, N. (2013) *The Normalized Difference Vegetation Index*. Oxford University Press, Oxford.

Pettorelli, N. (2018) *Satellite Remote Sensing and the Management of Natural Resources*. Oxford University press, Oxford.

Pettorelli, N., Kamran S. and Turner (2014a) 'Satellite remote sensing, biodiversity research and conservation of the future', *Philosophical Transactions of the Royal Society B*, vol 369, art 20130190.

Pettorelli N., Laurance B., O'Brien T., Wegmann M., Harini N. and Turner W. (2014b) 'Satellite remote sensing for applied ecologists: Opportunities and challenges', *Journal of Applied Ecology*, vol 51, pp839–848.

Pettorelli, N., Schulte to Bühne, H., Tulloch, A., Dubois, G., Macinnis-Ng, C., Queirós, A.M., Keith, D.A., Wegmann, M., Schrodt, F., Stellmes, M., Sonnenschein, R., Geller, G.N., Roy, S., Somers, B., Murray, N., Bland, L., Geijzendorffer, I., Kerr, J.T., Broszeit, S., Leitão, P.J., Duncan, C., El Serafy, G., He, K.H., Blanchard, J.L., Lucas, R., Mairota, P., Webb, T.J. and Nicholson, E. (2018) 'Satellite remote sensing of ecosystem functions: Opportunities, challenges and way forward', *Remote Sensing in Ecology and Conservation*, vol 4, pp71–93.

Pirnat, J. and Hladnik, D. (2016) 'Connectivity as a tool in the prioritization and protection of sub-urban forest patches in landscape conservation planning', *Landscape and Urban Planning*, vol 153, pp129–139.

Poli, D., Remondino, F., Angiuli, E. and Agugiaro, G. (2015) 'Radiometric and geometric evaluation of GeoEye-1, worldview-2 and Pleiades-1A stereo images for 3D information extraction', ISPRS Journal of Photogrammetry and Remote Sensing, vol 100, pp35–47.

Purkis, S.J. and Klemas, V.V. (2011) *Remote Sensing and Global Environmental Change*. Wiley, Hoboken.

Quegan, S., Le Toan, T., Chave, J., Dall, J., Exbrayat, J.-F., Ho Tong Minh, D., Lomas, M., Mariotti D'Alessandro, M., Paillou, P., Papathanassiou, K., Rocca, F., Saatchi, S., Scipal, K., Shugart, H., Smallman, L., Soja, M., Tebaldini, S., Ulander, L., Villard, L. and Williams, M. (2019) 'The European Space Agency BIOMASS mission: Measuring forest above-ground biomass from space', *Remote Sensing of Environment*, vol 227, pp44–60.

Reigber, A., Scheiber, R., Jager, M., Prats-Iraola, P., Hajnsek, I., Jagdhuber, T., Papathanassiou, K.P., Nannini, M., Aguilera, E., Baumgartner, S. and Horn, R. (2013) 'Very-high-resolution airborne synthetic aperture radar imaging: Signal processing and applications', *Proceedings of the IEEE*, vol 101, pp759–783.

Rickbeil, G.J.M., Coops, N.C. and Adamczewski, J. (2015) 'The grazing impacts of four barren ground caribou herds (*Rangifer tarandus groenlandicus*) on their summer ranges: An application of archived remotely sensed vegetation productivity data', *Remote Sensing of Environment*, vol 164, pp314–323.

Rocchini, D., Chiarucci, A. and Loiselle, S.A. (2004) 'Testing the spectral variation hypothesis by using satellite multispectral images', *Acta Oecologica*, vol 26, pp117–120.

Rösch, V., Tscharntke, T., Scherber, C. and Batáry, P. (2015) 'Biodiversity conservation across taxa and landscapes requires many small as well as single large habitat fragments', *Oecologia*, vol 179, pp209–222.

Saura, S., Estreguil, C., Mouton, C. and Rodríguez-Freire, M. (2011) 'Network analysis to assess landscape connectivity trends: Application to European forests (1990–2000)', *Ecological Indicators*, vol 11, pp407–416.

Schmidtlein, S. and Fassnacht, F.E. (2017) 'The spectral variability hypothesis does not hold across landscapes', *Remote Sensing of Environment*, vol 192, pp114–125.

Schulte to Bühne, H. and Pettorelli, N. (2018) 'Better together: Integrating and fusing multispectral and radar satellite imagery to inform biodiversity monitoring, ecological research and conservation science', *Methods in Ecology and Evolution*, vol 9, pp849–865.

Senf, C., Pflugmacher, D., Wulder, M.A. and Hostert, P. (2015) 'Characterizing spectraltemporal patterns of defoliator and bark beetle disturbances using Landsat time series', *Remote Sensing of Environment*, vol 170, pp166–177.

Skowronski, N.S., Clark, K.L., Gallagher, M., Birdsey, R.A., and Hom, J.L. (2014) 'Airborne laser scanner-assisted estimation of aboveground biomass change in a temperate oak–pine forest', *Remote Sensing of Environment*, vol 151, pp166–174.

Stöcker, C., Bennett, R., Nex, F., Gerke, M. and Zevenbergen, J. (2017) 'Review of the current state of UAV regulations', *Remote Sensing*, vol 9, art 459.

Strahler, A.H., Boschetti, L., Foody, G.M., Friedl, M.A., Hansen, M.C., Herold, M., Mayaux, P., Morisette, J.T., Stehman, S.V. and Woodcock, C.E. (2006) 'Global land cover validation: Recommendations for evaluation and accuracy assessment of global land cover maps', GOFC-GOLD Report No. 25. European Communities, Luxembourg, https://gofcgold.org/sites/default/files/docs/ReportSeries/GOLD_25.pdf, accessed 11 Dec 2020.

Su, T.C., and Chou, H.T. (2015) 'Application of multispectral sensors carried on unmanned aerial vehicle (UAV) to trophic state mapping of small reservoirs: a case study of Tain-Pu reservoir in Kinmen, Taiwan', *Remote Sensing*, vol 7, no. 8, pp10078–10097.

Swenson, J.J., Carter, C.E., Domec, J.C. and Delgado, C.I. (2011) 'Gold mining in the peruvian amazon: Global prices, deforestation, and mercury imports', *PLoS One*, vol 6, art e18875.

Tebbs, E., Rowland, C., Smart, S., Maskell, L. and Norton, L. (2017) 'Regional-scale high spatial resolution mapping of aboveground net primary productivity (ANPP) from field survey and Landsat data: A case study for the country of Wales', *Remote Sensing*, vol 9, art 801.

Thrash, I., Theron, G.K. and Bothma, J.D.P. (1995) 'Dry season herbivore densities around drinking troughs in the Kruger National Park', *Journal of Arid Environments*, vol 29, pp213–219.

Tong, X., Liu, X., Chen, P., Liu, S., Luan, K., Li, L., Liu, S., Liu, X., Xie, H., Jin, Y. and Hong, Z. (2015) 'Integration of UAV-based photogrammetry and terrestrial laser scanning for the three-dimensional mapping and monitoring of open-pit mine areas', *Remote Sensing*, vol 7, pp6635–6662.

Toth, C. and Jóźków, G. (2016) 'Remote sensing platforms and sensors: A survey', *ISPRS Journal of Photogrammetry and Remote Sensing*, vol 115, pp22–36.

Tuanmu, M.N., Viña, A., Bearer, S., Xu, W., Ouyang, Z., Zhang, H. and Liu, J. (2010) 'Mapping understory vegetation using phenological characteristics derived from remotely sensed data', *Remote Sensing of Environment*, vol 114, pp1833–1844.

Turner M.G. (2005) 'Landscape ecology: What is the state of science?', Annual Review of Ecology, Evolution, and Systematics, vol 36, pp319–344.

Twele, A., Cao, W., Plank, S. and Martinis, S. (2016) 'Sentinel-1-based flood mapping: A fully automated processing chain', *International Journal of Remote Sensing*, vol 37, pp2990–3004.

Verpoorter, C., Kutser, T. and Tranvik, L. (2012) 'Automated mapping of water bodies using Landsat multispectral data', *Limnology and Oceanography: Methods*, vol 10, pp1037–1050.

Wallace, C.S., Walker, J.J., Skirvin, S.M., Patrick-Birdwell, C., Weltzin, J.F. and Raichle, H. (2016) 'Mapping presence and predicting phenological status of invasive buffelgrass in Southern Arizona using MODIS, climate and citizen science observation data', *Remote Sensing*, vol 8, art 524.

Wich S.A. and Koh L.P. (2018) *Conservation Drones - Mapping and Monitoring Biodiversity*. Oxford University Press, Oxford.

Wu, J. (2013) 'Key concepts and research topics in landscape ecology revisited: 30 years after the Allerton Park workshop', *Landscape Ecology*, vol 28, pp1–11.

Wulder, M.A., White, J.C., Loveland, T.R., Woodcock, C.E., Belward, A.S., Cohen, W.B., Fosnight, E.A., Shaw, J., Masek, J.G. and Roy, D.P. (2016) 'The global Landsat archive: Status, consolidation, and direction', *Remote Sensing of Environment*, vol 185, pp271–283.

Yang, Z., Wang, T., Skidmore, A.K., De Leeuw, J., Said, M.Y. and Freer, J. (2014) 'Spotting East African mammals in open savannah from space', *PLoS One*, vol 9, art e115989.

Zhou, Y., Smith, S.J., Elvidge, C.D., Zhao, K., Thomson, A. and Imhoff, M. (2014) 'A cluster-based method to map urban area from DMSP/OLS nightlights', *Remote Sensing of Environment*, vol 147, pp173–185.

Zhu, W., Yu, Q., Tian, Y.Q., Becker, B.L., Zheng, T. and Carrick, H.J. (2014) 'An assessment of remote sensing algorithms for colored dissolved organic matter in complex freshwater environments', *Remote Sensing of Environment*, vol 140, pp766–778.

Zolkos, S.G., Goetz, S.J. and Dubayah, R. (2013) 'A meta-analysis of terrestrial aboveground biomass estimation using lidar remote sensing', Remote Sensing of Environment, vol 128, pp289–298.

13

Sensors in the landscape

John H. Porter

Introduction

Sensors play an important role in ecology and science in general. They provide much of the time-series data needed to test theories and to propose new ones. Although 'hands on' measurements remain important, detailed environmental measurements, especially those taken frequently, are typically sensor-based (Yao et al. 2003, Szewczyk et al. 2004, Porter et al. 2005, Hart and Martinez 2006). Some environmental sensors are in situ, in actual contact with the entity being sensed, whereas others use remote sensing (non-contact) to infer the properties of an entity (Jensen 1996, Lillesand et al. 2015). Still others sample flowing streams of water or air at a point to adduce properties of a region surrounding the sensors (Xiao et al. 2010, Thompson et al. 2011).

In most areas of ecology, in situ sensors are most commonly used (Szewczyk et al. 2004, Porter et al. 2005, Porter et al. 2009). For example, many automated meteorological stations capture air temperature, humidity, and precipitation data at a frequency of once an hour or higher. Similarly, soil moisture, water levels, and even chemical properties (e.g., nitrate and phosphate concentrations) can also be monitored on a near-continuous basis. Sensor nodes comprised of a suite of different sensors can be linked to form sensor networks that measure comparable data across a range of locations. Some sensor networks include high-speed communications so that video and audio data can be captured in near real time.

In contrast, the use of sensors in landscape ecology is dominated by satellite and aerial remote sensing. Here, the sensor is not actually in contact with the environmental entities being sampled. Instead, the sensor detects radiation emitted or reflected from the landscape and infers properties of landscape elements based on the characteristics of that electromagnetic radiation (Lillesand et al. 2015). In situ sensor systems can be used to calibrate and refine remote sensing products (Bailey and Werdell 2006, Vogeler et al. 2016, Valente et al. 2019). However, even this can be challenging because of the differences in the pixel size of remote sensing platforms and the scale of in situ measurements, often requiring the use of models to reconcile the different scales (Turner et al. 2006). Given the wide-spatial extent of many remote sensing platforms, remote sensing is a good match for the landscape scale. Nonetheless, it is restricted to directly sensing what is available on the vegetation or land surface. There are still some potentially important environmental parameters that need to be measured with in situ sensors.

Here, I will focus on the in situ sensors rather than on remote sensing (see Chapter 12). As noted, they have not been widely used at the landscape scale. There are some good reasons for this, but there are also some new developments that will make these in-place sensor systems more feasible and valuable in the future.

A sensor network primer

The increasing use of in situ sensors in ecology has been driven by the development of new sensors based on microelectronics and has been coupled to advances in computing technology and communication technologies (Arzberger et al. 2005). There are three major elements in a sensor network (Szewczyk et al. 2004, Collins et al. 2006, Porter et al. 2012). The first is the sensor itself. At the most basic level, a sensor takes some aspect of the environment and reacts to it in a way that generates or modulates an electrical voltage or current. In many cases, the relationship is direct. For example, the electrical current flowing through a thermistor (a temperature-sensitive variable resistor) will vary as the air temperature changes. Similarly, in a tipping-bucket-style rain gauge, a funnel fills a chamber within the rain gauge. When the chamber is full, the weight of the water causes an arm to move, bringing a new chamber into play and incidentally, making or breaking an electrical connection, thus generating an electrical pulse, which when counted, will indicate how many times the bucket has moved. Some sensors actually generate electricity themselves. Wind monitors may incorporate propeller-driven generators whose output voltages are related to the wind speed, just as solar sensors using photovoltaic panels change their output based on light levels. The relationship between the sensor and the electrical signal produced may, or may not, be a linear one, but most sensor components are selected to ensure as linear as possible a relationship over the intended range of use.

In other cases, especially for chemical or photosynthesis sensors, the measurements may be indirect (Daly et al. 2004, Milani et al. 2015, Nightingale et al. 2015, Bagshaw et al. 2016, Rode et al. 2016). For example, in a Pulse-Amplitude-Modulation (PAM) fluorimetry system, specific wavelengths of light are used to induce chlorophyll to fluoresce. The brightness of that fluorescence can be measured using a light sensor, with the brightness being related to the amount and activity of chlorophyll (Schreiber 2004). For chemical measurements, colorimetric or spectroscopic measurements may be used to measure an array of biologically important molecules such as nitrate, phosphate, silicate, and ammonia (Daly et al. 2004). Similarly, some chemical and biochemical properties can be assessed using fiber-optic-based sensors (Wang and Wolfbeis 2015).

The next step in the process is to convert the electrical change into digital data through the use of an analog-to-digital converter. Historically, this conversion was most frequently done inside a data logger or other specialized hardware. However, the availability of low-cost microelectronics has led to an increasing integration of the analog-to-digital converter into the sensor itself. Thus, the integrated sensor generates a digital data stream rather than a raw current or voltage. Moreover, internally, the sensor may use a processor to address issues of linearity and scaling, so that the digital result is already corrected (Alberts et al. 2013). A variety of digital protocols (e.g., SDI-12, 1-Wire, I²C, RS-422, RS-232) have been developed to allow these digitally enabled sensors to communicate with computers. In some cases, the sensors may even include communications systems allowing them to transmit their readings in the absence of a physical connection.

A sensor node consists of one or more sensors connected to a central controller that provides additional processing, storage, and communications capabilities (Figure 13.1). The controller may take the form of a data logger (Connolly 2010), a micro-computer (e.g., Arduino, Raspberry Pi) (Ferdoush and Li 2014), or even a full-blown computer. The controller may log data internally, or it may use its communications capabilities to immediately transfer data to another system. In some cases, the sensor, controller, and storage systems may be bundled together into a single package (e.g., iButton, Hobo Pendant).

Power is a critical element for any sensor node. Indeed, it is often the availability of electrical power that dictates the capabilities of the sensor system (Anastasi et al. 2009,

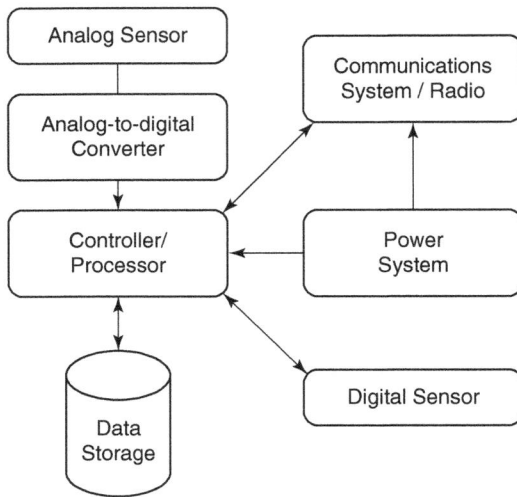

Figure 13.1 Components in a typical sensor node. Some nodes may exclude specific components (e.g., storage or radio) or have only analog or digital sensors.

Abdul-Salaam et al. 2016). For field researchers, with no ready access to commercial line power, it is typically necessary to depend on stored power (e.g., batteries) or solar, wind, water, or wave power. Typically the solar, wind, water, or wave power needs to be coupled to a battery system to provide power during the night or during calm winds or seas. Solar power has the advantage that it is available in most places (although forest and arctic researchers may disagree), has no moving parts, is extensible, and provides reliable power, so it tends to be the most frequently used in ecological applications. However, large solar arrays are large, conspicuous, and difficult to install, so in some cases, wind or water turbines may provide a better power source.

A communications system is the final component that needs to be incorporated into a sensor node. In many cases, especially for small single sensor nodes, communication of data may be via a wire, using a universal serial bus (USB) or serial connection to a personal computer or phone. For sealed, stand-alone nodes designed for rigorous conditions where providing an opening for a wire might compromise physical integrity, flashing lights or low-power radio signals (e.g., Bluetooth radio) may be used to convey data. However, if a sensor node is to be integrated into a wireless sensor network, more powerful (and power-consuming) methods need to be applied.

The advent of spread-spectrum radio technology has revolutionized our ability to transfer data wirelessly. Previous systems wherein a high-power radio signal was used to convey data using a single radio frequency were slow and extremely consumptive of power. In contrast, spread-spectrum radios use multiple radio frequencies at low power levels to transfer data at a high rate. Examples of systems using spread-spectrum technologies include WiFi networks, cellular phone data networks, and Bluetooth radios. Nonetheless, they do have some limitations. First, most of them operate at high frequencies (in the microwave range of the radio spectrum) that require line-of-sight communications; thus, spacing between nodes is limited by the curvature of the Earth or more likely, by local topography. Even worse, many of the available frequencies (e.g., 2.4 and 5.8 GHz) are easily blocked by vegetation. Additionally, most of the less expensive and 'commodity' radio systems (e.g., WiFi) have high duty cycles (i.e., they don't power down when they are idle), so they tend to have high power requirements relative to the other elements in a sensor node (albeit much lower than the old-style single frequency radios).

A large number of sensors may be required to adequately cover an area. Although it is possible to manage sensor nodes individually (e.g., manually going to each location and dumping the data from a logger), as the number of sensors increases, it is more efficient to integrate the sensor nodes into a sensor network, typically using wireless communications to knit the network together. One way of doing this is shown in Figure 13.2a. It depicts a 'star' topology wherein each sensor node communicates with a central element. It has the advantage of simplicity – each sensor node only needs to be configured to communicate with a single central node. Moreover, the central node can be programmed to successively query each of the sensor nodes, ensuring an orderly flow of data. However, it also has some practical disadvantages. The first is that it may not be possible to find a location for the central node that can be reached by radio from each of the desired sensor node locations. Some of the desired locations may be in places that are not in the line-of-sight of the central node. Additionally, there is a single point of failure. If each node is simply reporting data, rather than storing it, failure of the central node will lead to a total loss of all the data from the sensor network.

An alternative topology is a 'mesh' topology (Figure 13.2b). The central node is gone. Each sensor node communicates with one or more neighbors in the mesh. A mesh topology requires more sophisticated processing in each sensor node. Protocols need to control how data will flow

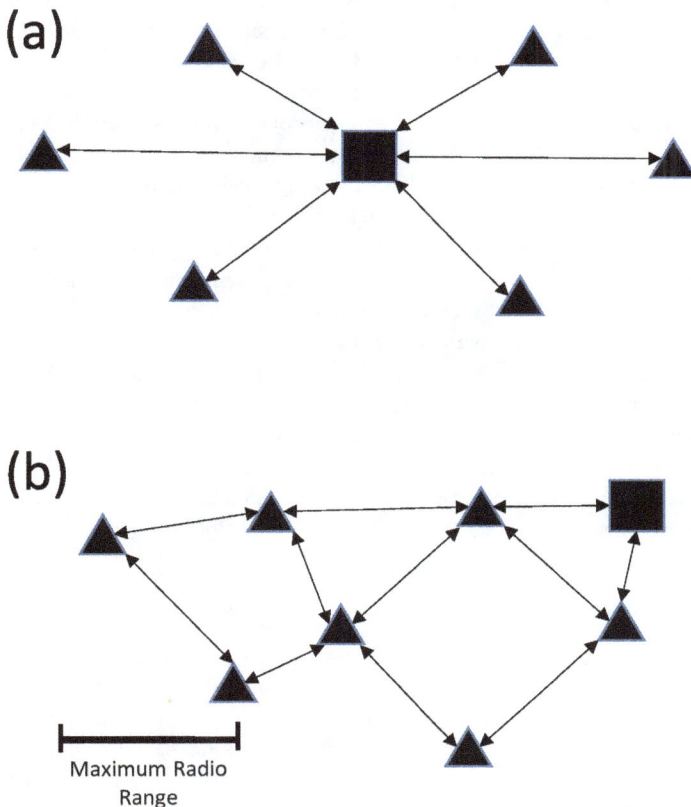

Figure 13.2   (a) Star and (b) mesh topologies for sensor network design. Triangles represent sensor nodes, and squares represent 'sinks'. Sinks are nodes where data are extracted from the network.

through the network, so each node needs to recognize which neighboring nodes it should send data to or receive data from. Increased power will be required for each node, because at times, a node will be transmitting data that has been forwarded by other nodes as well as its own data. To some degree, this increase will be mitigated by allowing reduced power output on the radio signals due to the decreased distance to be traversed by the signal relative to what is required in a star topology. It is this latter feature that makes a mesh attractive. For a sensor to be part of the network, it only needs to be able to talk to one or more neighboring sensor nodes. If it can communicate with more than one node, there will be redundant transmission paths, so that failure of a single sensor node in the mesh will not prevent transmission of data, since other paths are available.

When considering communications, there is an important relationship between rate of data transmission, transmission distance, and the power required. Frequently, the power used for communications is higher than the power used by either the sensors or the on-board processor (Anastasi et al. 2009). Large data volumes necessarily require more power. Similarly, if longer distances are to be traversed, if there are elements of the landscape that impede the signal, or if there is radio interference, then a more powerful radio signal may be required, or different radio frequencies need to be used.

Sensors networks are not immune to failures that can lead to data loss or corruption (Campbell et al. 2013). At the node level, in my experience, the most frequent cause of failure is loss of power, followed by issues with sensors, radios, and microprocessors. Power system problems can be caused by battery failures, issues with charging systems (e.g., damage to solar panels or solar controllers), blown fuses, corroded connectors, or all the above. These tend to be the most serious, along with microprocessor failures, because they compromise the entire data stream, not just data from a single sensor. Sensors and sensor nodes can also be damaged by intrusions by insects, water leakage into sensitive components, corrosion, corrosion-induced heating, arcing or power surges, simple component failures, and fouling of sensors. These problems are most likely when sensor nodes are deployed for extended periods, because waterproof seals take time to deteriorate, and exposed elements require time to corrode. Similarly, battery capacity typically declines over time, and microelectronics can be stressed by wide environmental changes (e.g., temperature, humidity). Additionally, the chance of a node failure increases with the number of nodes deployed. So, a large network is likely to have one or more nodes requiring repair at any specific time, and as the spatial extent of the network increases, repairs become more difficult to perform for logistical reasons.

Human actions can cause damage almost immediately. Sensors and sensor nodes can be stolen, vandalized, shot, and even, in one case, blown up by police, who thought it was a terrorist-planted device (Margaret O'Brien, personal communication). While inclusion of a node in a sensor network does not guarantee its survival or proper functioning, rapid access to the data can facilitate rapid detection of problems and their remediation (Porter et al. 2005).

Examples of sensor networks

Porter et al. (2005) identified five major reasons why ecologists use wireless sensor networks. The one most applicable to landscape ecology is the need to collect data over wide areas. Many landscape projects may also need to collect data at high frequencies. Rarer are experimental systems where real or near-real time data are needed, along with bidirectional data connections to allow robotic manipulations. The reason least likely for landscape studies is the need to be unobtrusive, which is most critical for organismal or behavioral studies.

Most frequently encountered are site-based sensor networks wherein sensor nodes are deployed over tens to hundreds of meters, but typically not more than a kilometer. For example, the 'Sensor Web' system deployed by Collins et al. (2006) consisted of 14 nodes covering a

300-meter transect with data being transferred node-to-node across the sensor network. These networks typically have high densities of nodes, often required by the limited transmission range of each node, but cover relatively small areas from the perspective of the landscape.

Even more common are networks consisting of only a limited number of nodes, which typically do not interact. For example, a survey of the Organization of Biological Field Stations found that of 105 sensor systems in use, only 46% included more than 3 sensor nodes, and only 9% had more than 25 nodes (Porter et al. 2009). Typically, these would be meteorological (~55%) or water sampling stations (~30%) that would be separated from one another by up to tens of kilometers. For example, the Lienhuachih Experimental Forest in Taiwan has three instrumented towers, each equipped with meteorological sensors, a video camera, and a microphone, that are separated by 1 to 2 km. Similarly, The Virginia Coast Reserve Long-Term Ecological Research Project operates three meteorological stations and three tide monitoring stations separated by approximately 20 km (Porter et al. 2009). However, these are only loosely networked with little or no inter-node communication.

There are some large, multi-watershed, forest sites that have been instrumented to a higher degree and that constitute landscape studies. For example the H.J. Andrews Experimental Forest developed a network of 14 monitoring stations across a variety of landscapes with inter-station distances varying from less than 100 m to 12 km (Henshaw et al. 2008). Pypker et al. (2007) used some of the stations, supplemented by an additional 30 temperature sensors along a 1-km transect, to discover diurnal 'rivers of air' transporting carbon through one of the watersheds. Daly et al. (2010) used 11 of the stations to elucidate patterns in air drainage and the resulting temperature changes in the same watersheds.

At the national to international scale, there are also networks of sites, such as the National Ecological Observatory Network (NEON), Ameriflux, and the Global Lake Ecological Observatory Network (GLEON) (Thompson et al. 2011, Kao et al. 2012, Weathers et al. 2013, Hanson et al. 2016), as well as government monitoring programs such as the National Oceanic and Atmospheric Administration (NOAA) Climate Reference Network (CRN) and the United States Geological Survey (USGS) stream flow gauge network (Hirsch and Costa 2004, Diamond et al. 2013). Here, individual stations are often widely spaced by tens to hundreds to thousands of kilometers. Flux towers within NEON and Ameriflux, although located at particular points, provide estimates of atmospheric fluxes in the vicinity of the stations, albeit with footprints much smaller than a few kilometers (Xiao et al. 2010). Similarly, stream gauges can integrate information on hydrology across their associated watersheds. However, none of the national and international networks provide high-density, distributed measurements at what typically would be considered the landscape scale.

Calibration and validation of remote sensing products has been accomplished using published data assembled from formal or informal networks of sensing stations (Bailey and Werdell 2006, Albergel et al. 2012, Vogeler et al. 2016, Valente et al. 2019). For example, Turner et al. (2006) assembled ground-based measurements of leaf area index and net primary production, along with measurements of gross primary production and meteorological data obtained from carbon flux towers, to compare with satellite-based estimates. They needed to scale the ground-based measurements both spatially and temporally to match the granularity of the satellite-derived products. To do this, they used the ground-based measurements to drive an ecological model (BIOME-BGC) to scale measurements from 25-m cells to 1-km cells and aggregated data temporally across 8 days to match the temporal resolution of the satellite-based products.

In addition to fixed or robotic sensors, sensors attached to mobile animals, typically as collars, have been widely used in landscape studies of populations and migration (White and Garrott 2012). Radio tracking goes back many decades (Cochran and Lord 1963), but sensing units that incorporate global positioning system capabilities, sometimes associated with satellite transmitters, provide more detail than was previously available (Pépin et al. 2004, Bridge et al. 2011).

There are some individual sensors that can collect data at the landscape scale. Ground-based LiDAR has the ability to map, in three dimensions, vegetation surfaces over wide areas (Hopkinson et al. 2004, Stovall et al. 2017, Stovall et al. 2018, Disney et al. 2019). This terrestrially based remote sensing technology is typically used to collect snapshots of forest structure rather than continuous data, in part because of the cost of the sensors, but mostly because of the high volumes of data produced for each survey and the concomitant processing time required. However, these are individual sensors (albeit sensors with a wide reach) rather than a sensor network.

Design of sensor networks

The design of sensor networks is driven by three major issues: the science requirements for the data a network will generate, the range of the radios used to communicate data, and the topology of the sensor nodes relative to data 'sinks' (a 'sink' is a node where data may exit the sensor network). The science requirements are dictated by traditional statistical spatial sampling issues (Fortin et al. 2006, Gelfand et al. 2010). Issues of sample size and statistical power can be similar to those for conventional, non-spatial statistics except for the addition of spatial sampling challenges. A minimum distance for the spacing between sensor nodes can be set by the degree and scale of spatial autocorrelation. Nodes that are located too closely together may be effectively measuring the same thing, leading to the lack of statistical independence, redundancy, and a concomitant loss of sampling efficiency (Fotheringham 2009). Nodes that are more widely separated will not have this issue but also may not be useful for helping to correct data based on data from surrounding nodes (Dereszynski and Dietterich 2011, Shahid et al. 2015).

The range of radios used for data communication may require that the maximum distance between sensor nodes is much shorter than would be dictated by science needs alone. This is especially true for landscape ecology where large areas are to be considered. For example, in Figure 13.3a, a mesh network is shown along with the area of high spatial autocorrelation (dotted lines) associated with each node. This is a near ideal situation, where the communication distance between nodes is comparable to the domain in which each node is statistically independent of its neighbors. More likely, given current technologies, is Figure 13.3b. Here, sensor nodes are more concentrated than would normally be scientifically required, with autocorrelated stations providing redundant data. However, moving the sensor nodes further apart would eliminate their ability to intercommunicate due to radio range constraints. Such overly dense systems do have some advantages from the perspective of error detection and correction because adjacent sensors should be reporting similar results (Dereszynski and Dietterich 2011, Shahid et al. 2015). However, communication range limitations can greatly increase the required density of nodes and hence, the total number of nodes required and thus, the cost of a sensor network. This issue is especially serious for two- and three-dimensional study areas because the number of nodes required to maintain a given spacing goes up as the square or cube of the dimensions of a study area.

For sensor networks using a mesh topology, the number of nodes that must be traversed to reach a sink can have important implications for the data rates and power requirements of the nodes (Oyman and Ersoy 2004). Each downstream node is responsible for passing on the data from the associated upstream nodes. Thus, a downstream node with ten nodes upstream from it will need to transmit ten times as much data in addition to its own locally generated data. Thus, it would also require ten times the power for data transmission and have a maximum

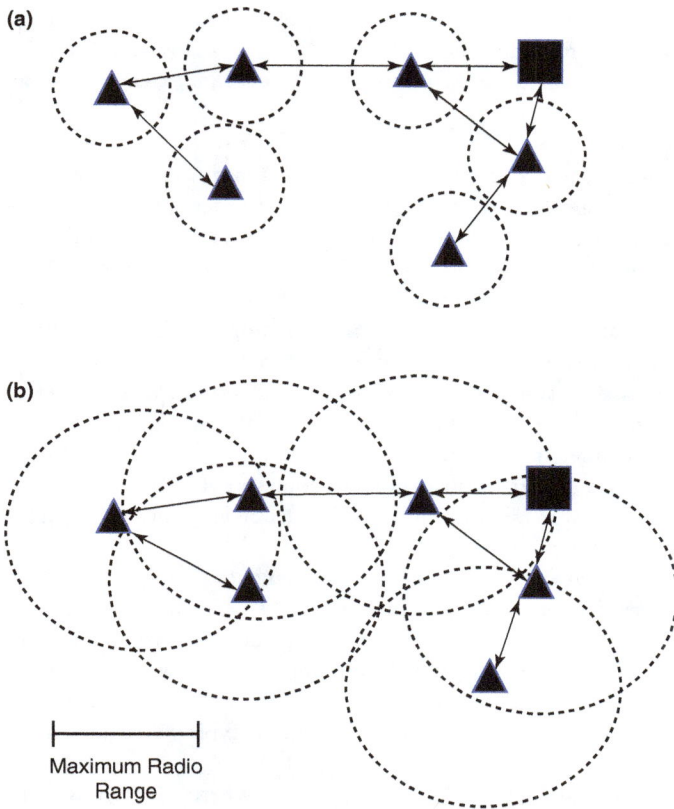

Figure 13.3 (a) A mesh sensor network where each sensor makes statistically independent measurements of the landscape. Here, the radio range is larger than the region of high spatial autocorrelation surrounding each node. (b) A mesh sensor network where the region of high spatial autocorrelation exceeds the range of the radios, leading to some level of redundancy in measurements.

data transmission speed for its own data reduced to roughly one-tenth of the normal link speed. Additionally, some network bandwidth is required for coordinating mesh nodes, further reducing the total throughput. It is therefore desirable to place some limits on the number of intercommunicating nodes and intervening links in a mesh network, perhaps by incorporating multiple sinks (Oyman and Ersoy 2004). Sensor network design and optimization remains an extremely active research area with nearly 29,000 articles and books since 2018 (Google Scholar search).

Challenges for sensor use at the landscape scale

As noted by Millington (Chapter 3), the landscape scale is not fixed. It varies with the size and characteristics of the organisms and processes of concern. Nonetheless, for large vertebrates (e.g., deer), the landscape scale is sufficiently extensive to make detailed monitoring using a network of static sensors nodes a challenge. For ground-based insects, the landscape scale is less extensive, but the physical size of sensor nodes may be an issue. For highly mobile organisms such as birds or flying insects, the appropriate scale for landscape studies may be particularly extensive. The mismatch between the landscape scale and most extant in situ sensor technologies remains an issue.

There are several reasons for this mismatch. First, there is the large spatial extent of most landscape studies for large organisms such as vertebrates or trees. This means that either a very large number of sensor nodes need to be deployed or individual sensor nodes need to be widely separated. This problem is especially acute because the number of sensors required to cover a rectangular area increases as the product of the dimensions. So, a study instrumenting a 10 × 10 km square requires 100 times as many sensor nodes as a study examining only a 1 × 1 km square while keeping the density of sensors constant. That many sensors would be expensive, both in terms of the initial cost and also in terms of the infrastructure required to manage the sensors and data flows while providing sufficient quality control and assurance (Porter et al. 2012). So, compromises need to be made between the desired density of sensor nodes and their cost. For this reason, studies with larger spatial extents, such as the H.J. Andrews Cyberforest (Henshaw et al. 2008), and national and international efforts (Thompson et al. 2011, Kao et al. 2012, Hanson et al. 2016) use a relatively modest number of nodes at very low density, thus limiting the spatial resolution of measurements.

The wide spacing of sensors in landscapes poses another problem: communications. Many of the communication modes used by commonly used sensor nodes such as Bluetooth, WiFi, and Zigbee (Ahamed 2009) either have limited ranges of 100 meters or less (a lot less for Bluetooth) or, in the case of WiFi, require line-of-sight communications, which are easily obstructed by vegetation or topography (Aust et al. 2015). Larger inter-sensor distances also pose challenges in the area of power. Longer-range communications typically require more power.

Commercial cellular networks can also be used, but only in areas where coverage is available (Wang and Fapojuwo 2017), and unfortunately, many ecological studies are conducted well away from urban areas where cellular data networks are most dense. In the case of NEON observatories, most observatories are connected to the Internet using a fiber-optic cable. Due to its expense and required level of environmental disruption, this is not an option that would work for a large number of sensor nodes.

Satellite communications have been used, but they typically are expensive and limited to relatively low bandwidth (slow data transfers). For example, the U.S. CRN (Diamond et al. 2013) uses the GOES weather satellite to collect hourly temperature data from its isolated stations. Similarly, robotic ocean gliders use commercial satellite links to communicate data and receive instructions (Rudnick 2016).

Thus, ecological sensor networks have been primarily deployed either over limited areas, with low-power sensor nodes with limited communication ranges (Delin 2002, Collins et al. 2006), or sparsely over large areas, with large sensor nodes with high power requirements (Hirsch and Costa 2004, Henshaw et al. 2008, Kao et al. 2012, Hanson et al. 2016). This leaves room for improvement in the development of landscape-scale sensor networks.

New opportunities

One solution to the scale issues for sensing of landscapes is the development of low-cost sensors that could be essentially 'scattered' across the landscape. There has been substantial discussion of extremely small, inexpensive sensor units referred to as 'Smart Dust' (Kahn et al. 1999), but there remain a variety of challenges in their development, including environmental, privacy, control, and cost concerns (Marr 2018). Additionally, the small sizes lead to limited communications distances, thus requiring a high density of sensor nodes.

One approach to overcoming communication gaps is to have roving nodes, referred to as 'mobile mules' (Shah et al. 2003, Tseng et al. 2013), that move through the sensor network collecting transmissions from sensors that might otherwise be isolated or require numerous hops (Abdul-Salaam et al. 2016). These mobile agents can include unmanned aerial vehicles (Jawhar et al. 2014). There has been a large amount of, largely theoretical, development of this approach with reference to discovery, data transfer, routing, and motion control (Di Francesco et al. 2011).

Mobility is also a potential answer for overcoming some of the scaling issues associated with landscape-scale studies. Mobile robotic sampling platforms are increasingly being used by researchers. For example, ocean gliders have been used for decades to measure ocean characteristics over wide areas (Rudnick 2016). These robotic platforms repeatedly submerge and surface by making small changes in volume that cause them to sink or rise. On-board processors implement course corrections based on input from global positioning systems while surfaced and send data, or accept new commands, via satellite phone (Sherman et al. 2001).

Implementing sensor mobility in a terrestrial context has been more difficult. However, the recent advent of low-cost unmanned aerial vehicles (i.e., 'drones') has made mobile sensing more practical (Anderson and Gaston 2013, Ivošević et al. 2015, Cruzan et al. 2016). Often, camera-equipped unmanned aerial vehicles (UAVs) play the role of low-cost remote sensing platforms (Woodget et al. 2017). However, they are also capable of transporting instruments for taking in situ measurements. For example, Palomaki et al. (2017) mounted a sonic anemometer on a UAV in order to assess winds at different altitudes. Potentially, UAVs could be used to automatically patrol study areas, carrying sensor payloads to specific spots in the landscape, landing to take measurements, and then flying off to the next location. However, there are legal restrictions regarding operation of UAVs without direct human oversight (Federal Aviation Administration 2016), along with a plethora of technical issues (e.g., automatic recharging of batteries, geolocation in forests where geographic positioning systems may be unreliable, unexpected weather, mechanical problems, and changes in the landscape).

The communications challenges associated with widely distributed sensors across a landscape are being, to some degree, addressed by the development of new communication systems associated with the 'Internet of Things' (IoT). Several of these systems increase the range over which sensor nodes can intercommunicate while maintaining relatively low power costs, albeit at low rates of data transmission relative to Bluetooth or WiFi (Sinha et al. 2017). Low-power, wide area (LPWA) devices have increased range relative to the Bluetooth and Zigbee systems previously available. They typically communicate at much lower speeds than Bluetooth or WiFi. For example, LoRa is a proprietary standard wherein individual nodes intercommunicate with one or more gateways in what is sometimes referred to as a 'star-of-stars' topology (similar to Figure 13.2a except that more than one gateway is likely to be in range). Unlike Zigbee, which has a nominal range of less than 100 meters, LoRa is designed to operate over many kilometers, although the actual range depends on where nodes are located and on environmental conditions, particularly temperature (Augustin et al. 2016, Kartakis et al. 2016). LPWA devices are also designed to be very efficient with respect to power use, with potentially years of operation off a single battery. However, power requirements are highly dependent on the amount of data being sent. A sensor node that only transmits data infrequently will have a much better power profile than a node that is constantly sending data.

Although in a whole different category with respect to power requirements, new high-density, low-altitude satellite constellations are being built with the goal of providing high-speed wireless Internet world-wide. For example, the proposed SpaceX Starlink system plans to have 12,000 satellites orbiting around 1,000 kilometers above the Earth by 2027 (Mosher 2019). Less ambitious is OneWeb, which plans to launch between 650 and 2,000 satellites and be operational by 2021 (Knapp 2019). Unlike geosynchronous satellites, which have significant latency as signals traverse the 35,000 kilometers between the satellites and the Earth, low-altitude satellites have low latency, enabling higher speeds. They also may require less radio power because smaller distances are being covered. However, because they are visible in the sky over a given spot less frequently, more satellites are required. There is substantial competition in the development of these satellite constellations, so it is likely that not all the proposed networks will come to fruition. But if even one is successful, it will help avoid the 'last mile' problem that makes connecting environmental sensors away from built environments so challenging.

Thus, although sensor networks have not been widely used in landscape ecological studies, improvements in sensor, processing, communication, and robotic technologies may make their use more attractive and valuable in the near future.

Summary

1. Sensors and sensor networks have proven to be valuable tools for most areas of ecology;
2. Apart from remote sensing, in situ sensor networks are uncommonly used in landscape ecology;
3. The large spatial extent required for most landscape studies of large organisms is a poor match for the existing communication technologies used in most ecological sensor networks;
4. New communications modes are being deployed that will match up better with the needs of landscape ecologists;
5. New mobile robotic sensors platforms may be useful for landscape studies.

References

Abdul-Salaam, G., Abdullah, A.H., Anisi, M.H., Gani, A. and Alelaiwi, A. (2016) 'A comparative analysis of energy conservation approaches in hybrid wireless sensor networks data collection protocols', *Telecommunication Systems*, vol 61, pp159–179.

Ahamed, S. (2009) 'The role of ZIGBEE technology in future data communication systems', *Journal of Theoretical & Applied Information Technology*, vol 5, pp129–135.

Albergel, C., De Rosnay, P., Gruhier, C., Muñoz-Sabater, J., Hasenauer, S., Isaksen, L., Kerr, Y. and Wagner, W. (2012) 'Evaluation of remotely sensed and modelled soil moisture products using global ground-based in situ observations', *Remote Sensing of Environment*, vol 118, pp215–226.

Alberts, M., Grinbergs, U., Kreismane, D., Kalejs, A., Dzerve, A., Jekabsons, V., Veselis, N., Zotovs, V., Brikmane, L. and Tikuma, B. (2013) 'New wireless sensor network technology for precision agriculture', in International Conference on Applied Information and Communication Technologies (AICT2013), 25–26 April, 2013, Jelgava, Latvia.

Anastasi, G., Conti, M., Di Francesco, M. and Passarella, A. (2009) 'Energy conservation in wireless sensor networks: A survey', *Ad Hoc Networks*, vol 7, no 3, pp537–568.

Anderson, K. and Gaston, K.J. (2013) 'Lightweight unmanned aerial vehicles will revolutionize spatial ecology', *Frontiers in Ecology and the Environment*, vol 11, pp138–146.

Arzberger, P., Bonner, J. Fries, D. and Sanderson, A. (2005) *Sensors for Environmental Observatories: Report of the NSF Sponsored Workshop December 2004.* World Technology Evaluation Center (WTEC), Inc., Baltimore, MD.

Augustin, A., Yi, J., Clausen, T. and Townsley, W. (2016) 'A study of LoRa: Long range & low power networks for the internet of things', *Sensors*, vol 16, pp1466.

Aust, S., Prasad, R.V. and Niemegeers, I.G. (2015) 'Outdoor long-range WLANs: A lesson for IEEE 802.11 ah', *IEEE Communications Surveys & Tutorials*, vol 17, pp1761–1775.

Bagshaw, E.A., Beaton, A., Wadham, J.L., Mowlem, M., Hawkings, J.R. and Tranter, M. (2016) 'Chemical sensors for in situ data collection in the cryosphere', *TrAC Trends in Analytical Chemistry*, vol 82, pp348–357.

Bailey, S.W. and Werdell, P.J. (2006) 'A multi-sensor approach for the on-orbit validation of ocean color satellite data products', *Remote Sensing of Environment*, vol 102, pp12–23.

Bridge, E.S., Thorup, K., Bowlin, M.S., Chilson, P.B., Diehl, R.H., Fléron, R.W., Hartl, P., Kays, R., Kelly, J.F. and Robinson, W.D. (2011) 'Technology on the move: Recent and forthcoming innovations for tracking migratory birds', *BioScience*, vol 61, pp689–698.

Campbell, J.L., Rustad, L.E., Porter, J.H., Taylor, J.R., Dereszynski, E.W., Shanley, J.B., Gries, C., Henshaw, D.L., Martin, M.E., Sheldon, W.M. and Boose, E.R. (2013) 'Quantity is nothing without quality: Automated QA/QC for streaming environmental sensor data', *BioScience*, vol 63, pp574–585.

Cochran, W.W. and Lord Jr., R.D. (1963) 'A radio-tracking system for wild animals', *Journal of Wildlife Management*, vol 27, no 1, pp9–24.

Collins, S.L., Bettencourt, L.M., Hagberg, A., Brown, R.F., Moore, D.I., Bonito, G., Delin, K.A., Jackson, S.P., Johnson, D.W. and Burleigh, S.C. (2006) 'New opportunities in ecological sensing using wireless sensor networks', *Frontiers in Ecology and the Environment*, vol 4, pp402–407.

Connolly, C. (2010) 'A review of data logging systems, software and applications', *Sensor Review*, vol 30, pp192–196.

Cruzan, M.B., Weinstein, B.G., Grasty, M.R., Kohrn, B.F., Hendrickson, E.C., Arredondo, T.M. and Thompson, P.G. (2016) 'Small unmanned aerial vehicles (micro-UAVs, drones) in plant ecology', *Applications in Plant Sciences*, vol 4, no 9, art 1600041.

Daly, C., Conklin, D.R. and Unsworth, M.H. (2010) 'Local atmospheric decoupling in complex topography alters climate change impacts', *International Journal of Climatology*, vol 30, pp1857–1864.

Daly, K.L., Byrne, R.H., Dickson, A.G., Gallager, S.M., Perry, M.J. and Tivey, M.K. (2004) 'Chemical and biological sensors for time-series research: Current status and new directions', *Marine Technology Society Journal*, vol 38, pp121–143.

Delin, K.A. (2002) 'The sensor web: A macro-instrument for coordinated sensing', *Sensors*, vol 2, pp270–285.

Dereszynski, E.W. and Dietterich, T.G. (2011) 'Spatiotemporal models for data-anomaly detection in dynamic environmental monitoring campaigns', *ACM Transactions on Sensor Networks*, vol 8, art 3.

Di Francesco, M., Das, S.K. and Anastasi, G. (2011) 'Data collection in wireless sensor networks with mobile elements: A survey', *ACM Transactions on Sensor Networks*, vol 8, art 7.

Diamond, H.J., Karl, T.R., Palecki, M.A., Baker, C.B., Bell, J.E., Leeper, R.D., Easterling, D.R., Lawrimore, J.H., Meyers, T.P. and Helfert, M.R. (2013) 'US climate reference network after one decade of operations: Status and assessment', *Bulletin of the American Meteorological Society*, vol 94, pp485–498.

Disney, M., Burt, A., Calders, K., Schaaf, C. and Stovall, A. (2019) 'Innovations in ground and airborne technologies as reference and for training and validation: Terrestrial laser scanning (TLS)', *Surveys in Geophysics*, vol 40, pp937–958.

Federal Aviation Administration (2016) 'Summary of small unmanned aircraft rule (part 107)', in *FAA News*. Federal Aviation Administration, Washington, DC.

Ferdoush, S. and Li, X. (2014) 'Wireless sensor network system design using Raspberry Pi and Arduino for environmental monitoring applications', *Procedia Computer Science*, vol 34, pp103–110.

Fortin, M.J., Dale, M.R. and Ver Hoef, J.M. (2006) 'Spatial analysis in ecology', in A.H. El-Shaarawi and W.W. Piegorsch (eds) *Encyclopedia of Environmetrics*. Wiley, Chichester.

Fotheringham, A.S. (2009) '"The problem of spatial autocorrelation" and local spatial statistics', *Geographical Analysis*, vol 41, pp398–403.

Gelfand, A.E., Diggle, P., Guttorp, P. and Fuentes M. (2010) *Handbook of Spatial Statistics*. CRC Press, Boca Raton.

Hanson, P.C., Weathers, K.C. and Kratz T.K. (2016) 'Networked lake science: How the Global Lake Ecological Observatory Network (GLEON) works to understand, predict, and communicate lake ecosystem response to global change', *Inland Waters*, vol 6, pp543–554.

Hart, J.K. and Martinez, K. (2006) 'Environmental sensor networks: A revolution in earth system science?', *Earth Science Reviews*, vol 78, pp177–191.

Henshaw, D.L., Bierlmaier, F., Bond, B.J. and O'Connell, K.B. (2008) 'Building a "cyber forest" in complex terrain at the andrews experimental forest', in C. Gries and M.B. Jones (eds) Environmental Information Management Conference 2008, Albuquerque, New Mexico.

Hirsch, R.M. and Costa, J.E. (2004) 'US stream flow measurement and data dissemination improve', *Eos, Transactions American Geophysical Union*, vol 85, pp197–203.

Hopkinson, C., Chasmer, L., Young-Pow, C. and Treitz, P. (2004) 'Assessing forest metrics with a ground-based scanning lidar', *Canadian Journal of Forest Research*, vol 34, pp573–583.

Ivošević, B., Han, Y.-G., Cho, Y. and Kwon, O. (2015) 'The use of conservation drones in ecology and wildlife research', *Ecology and Environment*, vol 38, pp113–188.

Jawhar, I., Mohamed, N., Al-Jaroodi, J. and Zhang, S. (2014) 'A framework for using unmanned aerial vehicles for data collection in linear wireless sensor networks', *Journal of Intelligent & Robotic Systems*, vol 74, pp437–453.

Jensen, J.R. (1996) *Introductory Digital Image Processing: A Remote Sensing Perspective*, 2nd edition. Prentice Hall, Upper Saddle River, NJ.

Kahn, J.M., Katz, R.H. and Pister, K.S. (1999) 'Next century challenges: Mobile networking for "smart dust"' in Proceedings of the 5th Annual ACM/IEEE International Conference on Mobile Computing and Networking. ACM, Seattle Washington USA August, 1999.

Kao, R.H., Gibson, C.M., Gallery, R.E., Meier, C.L., Barnett, D.T., Docherty, K.M., Blevins, K.K., Travers, P.D., Azuaje, E. and Springer, Y.P. (2012) 'NEON terrestrial field observations: Designing continental-scale, standardized sampling', *Ecosphere*, vol 3, pp1–17.

Kartakis, S., Choudhary, B.D., Gluhak, A.D., Lambrinos, L. and McCann, J.A. (2016) 'Demystifying low-power wide-area communications for city IoT applications', in Proceedings of the Tenth ACM International Workshop on Wireless Network Testbeds, Experimental Evaluation, and Characterization. ACM, New York City New York October, 2016.

Knapp, A. (2019) 'With a successful launch, OneWeb just joined SpaceX and others in the Satellite Internet Race', *Forbes*. 2019-02-27. https://www.forbes.com/sites/alexknapp/2019/02/27/with-a-successful-launch-oneweb-just-joined-spacex-and-others-in-the-internet-race, accessed 23 Dec 2020.

Lillesand, T., Kiefer, R.W. and Chipman, J. (2015) *Remote Sensing and Image Interpretation*. Wiley, Chichester.

Marr, B. (2018) 'Smart dust is coming. Are you ready? *Forbes*, www.forbes.com/sites/bernardmarr/2018/09/16/smart-dust-is-coming-are-you-ready/?sh=477f4ad95e41, accessed 23 Dec 2020.

Milani, A., Statham, P.J., Mowlem, M.C. and Connelly, D.P. (2015) 'Development and application of a microfluidic in-situ analyzer for dissolved Fe and Mn in natural waters', *Talanta*, vol 136, pp15–22.

Mosher, D. (2019) 'Elon Musk just revealed new details about Starlink, a plan to surround Earth with 12,000 high-speed internet satellites. Here's how it might work', *Business Insider*, https://www.businessinsider.com/spacex-starlink-satellite-internet-how-it-works-2019-5, accessed 23 Dec 2020.

Nightingale, A.M., Beaton, A.D. and Mowlem, M.C. (2015) 'Trends in microfluidic systems for in situ chemical analysis of natural waters', *Sensors and Actuators B: Chemical*, vol 221, pp1398–1405.

Oyman, E.I. and Ersoy, C. (2004) 'Multiple sink network design problem in large scale wireless sensor networks', in IEEE International Conference on Communications (IEEE Cat. No.04CH37577), Paris France, pp. 3663–3667 Vol.6, doi: 10.1109/ICC.2004.1313226.

Palomaki, R.T., Rose, N.T., van den Bossche, M., Sherman, T.J. and De Wekker, S.F. (2017) 'Wind estimation in the lower atmosphere using multirotor aircraft', *Journal of Atmospheric and Oceanic Technology*, vol 34, pp1183–1191.

Pépin, D., Adrados, C., Mann, C. and Janeau, G. (2004) 'Assessing real daily distance traveled by ungulates using differential GPS locations', *Journal of Mammalogy*, vol 85, pp774–780.

Porter, J.H., Arzberger, P., Braun, H.-W., Bryant, P., Gage, S., Hansen, T., Hanson, P., Lin, F.P., Lin, C.C., Kratz, T., Michener, W., Shapiro, S. and Williams, T. (2005) 'Wireless sensor networks for ecology', *BioScience*, vol 55, pp561–572.

Porter, J.H., Nagy, E., Kratz, T.K., Hanson, P., Collins, S.L. and Arzberger, P. (2009) 'New eyes on the world: Advanced sensors for ecology', *BioScience*, vol 59, pp385–397.

Porter, J.H., Hanson, P.C. and Lin, C.-C. (2012) 'Staying afloat in the sensor data deluge', *Trends in Ecology & Evolution*, vol 27, pp121–129.

Pypker, T., Unsworth, M.H., Lamb, B., Allwine, E., Edburg, S., Sulzman, E., Mix, A. and Bond, B. (2007) 'Cold air drainage in a forested valley: Investigating the feasibility of monitoring ecosystem metabolism', *Agricultural and Forest Meteorology*, vol 145, pp149–166.

Rode, M., Wade, A.J., Cohen, M.J., Hensley, R.T., Bowes, M.J., Kirchner, J.W., Arhonditsis, G.B., Jordan, P., Kronvang, B. and Halliday, S.J. (2016) 'Sensors in the stream: The high-frequency wave of the present', *Environmental Science and Technology*, vol 50, no 19, pp10297–10307.

Rudnick, D.L. (2016) 'Ocean research enabled by underwater gliders', *Annual Review of Marine Science*, vol 8, pp519–541.

Schreiber, U. (2004) 'Pulse-amplitude-modulation (PAM) fluorometry and saturation pulse method: An overview', in G.C. Papageorgiou and Govindjee (eds) *Advances in Photosynthesis and Respiration*. Springer, Dordrecht.

Shah, R.C., Roy, S., Jain, S. and Brunette, W. (2003) 'Data mules: Modeling and analysis of a three-tier architecture for sparse sensor networks', *Ad Hoc Networks*, vol 1, pp215–233.

Shahid, N., Naqvi, I.H. and Qaisar, S.B. (2015) 'Characteristics and classification of outlier detection techniques for wireless sensor networks in harsh environments: A survey', *Artificial Intelligence Review*, vol 43, pp193–228.

Sherman, J., Davis, R.E., Owens, W. and Valdes, J. (2001) 'The autonomous underwater glider "Spray"', *IEEE Journal of Oceanic Engineering*, vol 26, pp437–446.

Sinha, R.S., Wei, Y. and Hwang, S.-H. (2017) 'A survey on LPWA technology: LoRa and NB-IoT', *ICT Express*, vol 3, pp14–21.

Stovall, A.E., Vorster, A.G., Anderson, R.S., Evangelista, P.H. and Shugart, H.H. (2017) 'Non-destructive aboveground biomass estimation of coniferous trees using terrestrial LiDAR', *Remote Sensing of Environment*, vol 200, pp31–42.

Stovall, A.E., Anderson-Teixeira, K.J. and Shugart, H.H. (2018) 'Assessing terrestrial laser scanning for developing non-destructive biomass allometry', *Forest Ecology and Management*, vol 427, pp217–229.

Szewczyk, R., Osterweil, E., Polastre, J., Hamilton, M., Mainwaring, A. and Estrin, D. (2004) 'Habitat monitoring with sensor networks', *Communications of the ACM*, vol 47, pp34–40.

Thompson, S.E., Harman, C., Konings, A., Sivapalan, M., Neal, A. and Troch, P.A. (2011) 'Comparative hydrology across AmeriFlux sites: The variable roles of climate, vegetation, and groundwater', *Water Resources Research*, vol 47, art W00J07.

Tseng, Y.-C., Wu, F.-J. and Lai, W.-T. (2013) 'Opportunistic data collection for disconnected wireless sensor networks by mobile mules', *Ad Hoc Networks*, vol 11, pp1150–1164.

Turner, D.P., Ritts, W.D., Cohen, W.B., Gower, S.T., Running, S.W., Zhao, M., Costa, M.H., Kirschbaum, A.A., Ham, J.M., Saleska, S.R. and Ahl, D.E. (2006) 'Evaluation of MODIS NPP and GPP products across multiple biomes', *Remote Sensing of Environment*, vol 102, pp282–292.

Valente, A., Sathyendranath, S., Brotas, V., Groom, S., Grant, M. and Bracher, A. (2019) 'A compilation of global bio-optical in situ data for ocean-colour satellite applications: version two', *Earth System Science Data*, vol 11, pp1037–1068.

Vogeler, J.C., Yang, Z. and Cohen, W.B. (2016) 'Mapping post-fire habitat characteristics through the fusion of remote sensing tools', *Remote Sensing of Environment*, vol 173, pp294–303.

Wang, H. and Fapojuwo, A.O. (2017) 'A survey of enabling technologies of low power and long range machine-to-machine communications', *IEEE Communications Surveys & Tutorials*, vol 19, pp2621–2639.

Wang, X.-d. and Wolfbeis, O.S. (2015) 'Fiber-optic chemical sensors and biosensors (2013–2015)', *Analytical Chemistry*, vol 88, pp203–227.

Weathers, K.C., Hanson, P.C., Arzberger, P., Brentrup, J., Brookes, J., Carey, C.C., Gaiser, E., Gaiser, E., Hamilton, D.P. and Hong, G.S. (2013) 'The global lake ecological observatory network (GLEON): The evolution of grassroots network science', *Limnology and Oceanography Bulletin*, vol 22, pp71–73.

White, G.C. and Garrott, R.A. (2012) *Analysis of Wildlife Radio-tracking Data*. Academic Press, New York.

Woodget, A.S., Austrums, R., Maddock, I.P. and Habit, E. (2017) 'Drones and digital photogrammetry: From classifications to continuums for monitoring river habitat and hydromorphology', *WIREs Water*, vol 4, art e1222.

Xiao, J., Zhuang, Q., Law, B.E., Chen, J., Baldocchi, D.D., Cook, D.R., Oren, R., Richardson, A.D., Wharton, S. and Ma, S. (2010) 'A continuous measure of gross primary production for the conterminous United States derived from MODIS and AmeriFlux data', *Remote Sensing of Environment*, vol 114, pp576–591.

Yao, K., Estrin, D. and Hu, Y.H. (2003) 'Special issue on sensor networks', *Eurasip Journal on Applied Signal Processing*, vol 2003, pp319–320.

14

The role of paleoecology in understanding landscape-level ecosystem dynamics

George L.W. Perry, Richard E. Brazier, and Janet M. Wilmshurst

Introduction

Landscape ecology is the sub-discipline of ecology concerned with interactions between spatial patterns and ecological processes (Turner and Gardner 2015). During the early 1980s, as the North American flavor of landscape ecology was developed (Chapter 1), Risser et al. (1984) articulated four focal research questions for the field (reordered here from the source):

1. Which processes, past and present, generate and support landscape heterogeneity?
2. How does landscape pattern influence the spread of disturbance?
3. How does spatial pattern influence fluxes of energy, matter, organisms, information through the landscape?
4. What is the role of landscape ecology in 'natural resource management'?

As a conceptual model, the patch mosaic underpins these principles of landscape and spatial ecology (Wiens 1995; Wu and Loucks 1995). In the patch-mosaic view, landscape-level spatial heterogeneity emerges from the interplay between ecological processes, disturbances, and underlying abiotic patterns. The patch-mosaic model (Chapter 2) conceptualizes the landscape as repeatedly overwritten by disturbances, amplified and constrained by other sources of spatio-temporal heterogeneity; the landscape is a 'palimpsest'. As with conventional palimpsests, the information available on the surface (i.e., 'now') is a snapshot in time of a landscape's dynamics. Revealing the hidden components (i.e., the past) is the focus of paleoecology, and the benefits of closer integration between paleoecology and landscape ecology are the focus of this chapter. There has been a tendency among (some) neoecologists to see paleoecological data as merely a source of data for the parameterization and validation of ecological models (Anderson et al. 2006; Perry et al. 2016) or offering *ad hoc* explanations for contemporary patterns (McGlone 1996). However, paleoecological understanding and data offer much more for landscape ecology (and neoecology more widely). We start the chapter with a brief overview of the key questions and approaches that paleoecology is concerned with, before outlining the scaling challenges that confront the integration of paleoecological and neoecological approaches. We then review the potential contribution of paleoecological methods and understanding to the research foci articulated by Risser et al. (1984), considering their first and second foci concurrently. In each, we use case studies to demonstrate how paleoecology can contribute to our understanding of landscape ecological dynamics in ways that complement and extend those of neoecology.

What is 'paleoecology'?

Birks (2008, p. 2623) describes paleoecology as 'the ecology of the past. It is mainly concerned with reconstructing past biota, populations, communities, landscapes, environments, and ecosystems from available geological and biological (fossil) evidence'. The discipline spans time scales from the deep past (the domain of paleobiology; geological epochs) to the more recent. In this chapter, we focus on 'long-term ecology' (decades to millennia) with an emphasis on ecological reconstruction rather than the use of biological proxies to reconstruct facets of the environment such as climate.

Proxy information is used to reconstruct the ecosystems of the past, and these proxies may describe patterns and processes that span tens to millions of years (Willis and Birks 2006). Paleoecologists use a suite of biological, including plant and animal macrofossils (such as charcoal, leaf cuticles, seeds, leaves, and bones), microfossils (pollen, phytoliths, testate amoebae, charcoal, and diatoms), tree-rings, ancient DNA (aDNA), and geochemical and geophysical proxies (National Research Council et al. 2005). Different proxies complement each other, and their careful analysis and synthesis can reveal detailed pictures of long-term ecosystem change. Because they provide an indirect picture (e.g., pollen abundance is not linearly related to vegetation abundance), understanding how proxy records form (e.g., how far does pollen from different species travel? how rapidly do fossils decay? which species are preserved? how confidently can fossils be matched to specific species?) is crucial. Although such proxies can provide extraordinarily detailed pictures of the ecosystems of the past, there have been repeated calls for a more experimental, and less descriptive, paleoecology (Birks 1993). However, there are fundamental constraints in applying experimental approaches to the past; for example, how can we replicate and control history? Similar questions have plagued landscape-level ecology (how can we replicate and control large spatial extents? Hargrove and Pickering 1992). In short, there are limits to paleoecological data that arise from their fragmented nature in space and time and the uncertainties that accompany them. Thus, as a historical science, the emphasis of paleoecology is perhaps best focused on disentangling multiple competing hypotheses rather than strict falsification (see Cleland 2001).

A characteristic of paleoecological data is that they are usually based on fewer sites than a typical field-based neoecological study. For example, sediment records may be collected from a single site, a chronology developed from them (e.g., using radiocarbon dating), proxies described in detail, and inferences made about the processes that produced those patterns. By comparison, neoecological studies can usually draw on a richer source of contextual information to inform pre-investigation inference confidently. However, the data neoecologists collect and use do not usually span the temporal extents as those used by paleoecologists do (Willis and Birks 2006; Bennington et al. 2009). Long-term data sets (e.g., observational data sets of decades or longer) provide a link between paleoecology and neoecology (Kosnik and Kowalski 2016; Burge et al. 2020). Although there are many long-running ecological data sets, some extending up to 130 years, in a meta-analysis of the scale(s) at which ecological studies are conducted (nearly 350 studies from 2004 to 2014), Estes et al. (2018) observed that most ecological studies span less than 1 year. In short, and returning to the palimpsest analogy, paleoecologists reconstruct not the whole palimpsest but just parts of it, while neoecologists see the whole document but only the surface layer. Integrating long-term ecological understanding and data into landscape-scale (i.e., appropriate extent and granularity) ecological studies requires dialogue between these different types and scales of data, which in turn, requires addressing difficult scaling challenges.

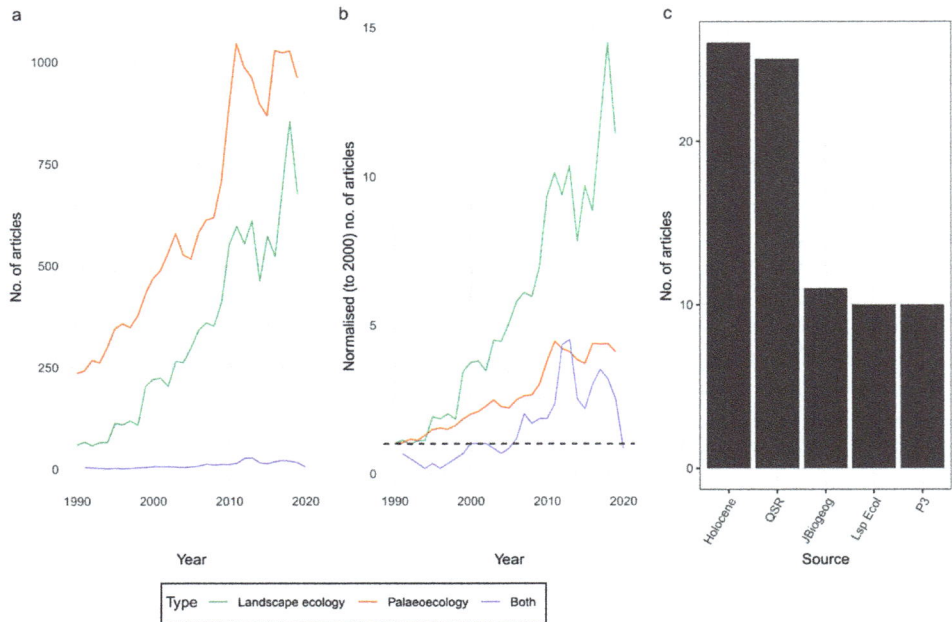

Figure 14.1 Bibliometric analysis of the interaction between paleoecological and landscape ecological research showing (a) publications vs. time, (b) publications vs. time normalized to 2000, and (b) journals with most publications (1980–2019). QSR = Quaternary Science Research, J Biogeog = Journal of Biogeography, Lsp Ecol = Landscape Ecology, P3 = Palaeogeography, Palaeoclimatology, Palaeoecology. Topic searches were conducted in Scopus for the title-abstract-keyword terms (i) 'landscape ecolog*', (ii) palaeoecolog* OR paleoecolog* and (iii) 'landscape ecolog*' AND (palaeoecolog* OR paleoecolog* OR 'long-term ecolog*' OR 'long term ecolog*'). Search was conducted on 25 August 2020, and the figure does not include 2020 citation information.

Links between landscape ecology and paleoecology

Because landscape dynamics involve processes at multiple scales in space and time, landscape ecology and paleoecology seem to be natural partners. However, as Turner (2005, p. 336) states, a 'compelling need for expanding the temporal horizon of landscape studies also exists. Paleoecological studies provide a critical context for understanding landscape dynamics, and historical dynamics shape current landscapes and may constrain future responses'. The call for closer integration between neo- and paleoecology is not new (West 1964; Huntley 1996; Froyd and Willis 2008), but despite the apparent potential, research in this area remains limited (Figure 14.1).

The challenges of scale and measurement

From fast to slow and big to small

Delcourt and Delcourt (1988, p. 29) outline the elements of an integrated paleolandscape ecology: the careful consideration of scale, the adoption of appropriate methods and data collection, and the use of paleoecological information to develop scenarios and test hypotheses.

Scale is central to Delcourt and Delcourt's (1988) integrated view. Ecosystem dynamics are scale-dependent; the patterns we observe are a function of the scale at which we view and measure the environment. Thus, selection of the scale at which to view a phenomenon of interest is critical to study design and inference (Wiens 1989; Levin 1992). A key challenge for integrating paleoecological and neoecological studies is bridging the disparate spatial and temporal scales they typically address. Conceptual tools such as hierarchy theory (see Chapter 3 for a detailed description) emphasize the cross-scale nature of ecological systems and so provide a way to think about integrating data from different scale (frequency) domains (King 1997; Wu 1999).

Spatializing paleoecological data and information

A second challenge for integrating paleoecological and neoecological data is translating the measures provided by paleoecological reconstructions to those used by neoecologists and those informative for ecosystem management (Jeffers et al. 2015; Grace et al. 2019). For example, the taxonomic resolution offered by palynological data may not match that used by conservation managers (i.e., species-level). Simulation and other forms of modelling offer a way to make spatial inferences about paleoecological records and to translate them to other measures. Model-based landscape reconstruction methods have been developed to infer land cover from palynological records. These methods use contemporary vegetation data and surface pollen samples alongside pollen production data to calibrate models of pollen deposition and movement (Figure 14.2). The landscape reconstruction algorithm (LRA) described by Sugita (2007a, b) is a widely adopted method. This approach incorporates regional (REVEALS) and local (LOVES) scale reconstructions to reconstruct land cover. Mariani et al. (2017) used the LRA to explore

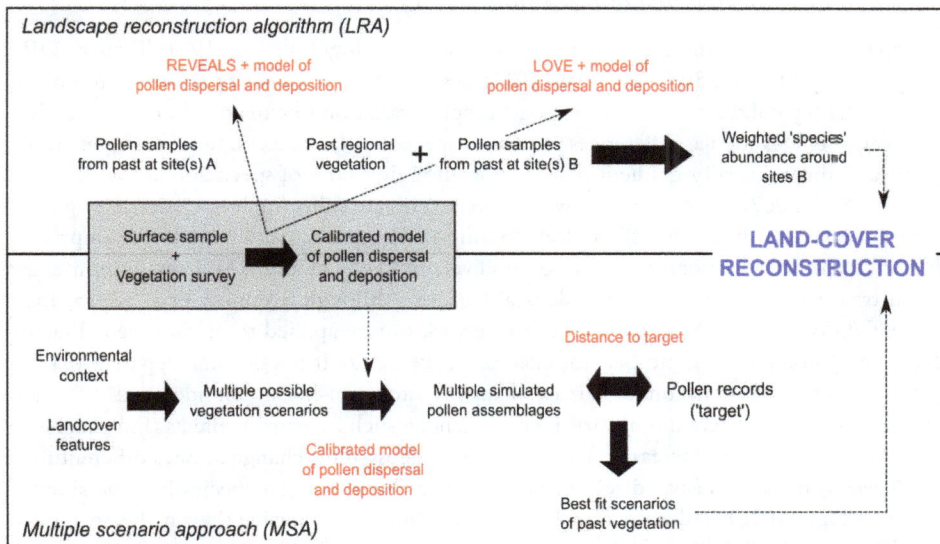

Figure 14.2 Schematic overview of the landscape reconstruction algorithm (LRA) and multiple scenario approach (MSA) methods for inferring land cover from pollen records. (Modified from Bunting et al. 2018; original figure © Bunting et al. 2018)

long-standing questions about the openness of the Tasmanian landscape, which are difficult to address using pollen records alone. A range of other approaches have been developed, such as the multiple scenario approach (MSA; Bunting and Middleton 2009), in which multiple land cover scenarios are simulated and assessed against pollen records, and the land cover scenarios within some distance of the pollen record are retained as plausible land covers. Bunting et al. (2018) used the MSA to test how Neolithization affected land cover at two sites in the United Kingdom (the Orkneys and the Somerset levels).

Another, but related, use of models is to make inferences about the mechanisms that generated paleoecological records (Perry et al. 2016). In this context, paleorecords are used in a pattern-oriented modelling framework (Grimm et al. 2005) to understand the conditions generating the patterns observed in the available data (similarly to the MSA). Thus, the modelling approaches developed by landscape ecologists become a powerful support for the hypothesis testing advocated by Delcourt and Delcourt (1988). As an example, Perry et al. (2012a, b) used machine-learning and spatial simulation to assess how small populations effected widespread deforestation following human settlement of New Zealand. Using spatial models informed by pollen and charcoal records, they showed how positive fire–vegetation feedbacks might have facilitated these changes; in the language of landscape ecology, this transformation emerged from reciprocal interactions between landscape pattern (vegetation structure and composition) and process (fire spread).

Long-term landscape dynamics and the role of disturbance

The first two of Risser et al.'s (1984) foci for landscape ecology were long-term landscape dynamics and the role of disturbance in generating landscape pattern; paleoecology can provide insights about both.

Long-term legacies: climate change and refugia

Landscapes and ecosystems are dynamic and ever-changing (Sprugel 1991; Turner 2010). Ricklefs et al. (1999) and Svenning et al. (2015) describe two biodiversity responses to climate change. First, they consider that long-term geographic variation in climatic conditions can lead to cumulative effects such as differences in rates of speciation (as an example, it has been argued that the latitudinal diversity gradient may arise from higher rates of speciation at low latitudes; Mittelbach et al. 2007). Second, relatively short-term climate shifts (such as Quaternary glacial-interglacial cycles or disturbances) can lead to long-lasting effects (for example, the imprint of glacial refugia on contemporary patterns of biodiversity; Gavin et al. 2014). These responses can result in transient and/or persistent biological legacies. Although Svenning et al. (2015) focus on the effects of climate change, the same framework can be applied to environmental factors such as the hydrological regime (see, for example, Lake 2000). It has become apparent that the reorganization of ecosystems under climate change is often non-linear (Blonder et al. 2017) and influenced by legacy effects and historical contingencies such as priority effects (Fukami 2015). As a result, there has been increasing recognition that ecosystems change at rates different from those expected from niche-based responses to climate alone and that species have persisted in areas (i.e., refugia) that, based on regional patterns in climate, are outside their niche. In short, it cannot be assumed that climate and ecosystems are at equilibrium (Blonder et al. 2017; Gaüzère et al. 2018).

Because the legacies of past climates are evident in contemporary landscapes, they are directly relevant to the questions landscape ecologists consider. Ecosystem response to climate change will be driven by biophysical processes at temporal extents from the millisecond (e.g., physiological) to the millennial and longer (e.g., speciation). However, the scales intersecting landscape ecology and paleoecology are likely to be between decades and millennia, including shifts in local population abundance and species ranges. Although studies of ecosystem reorganization have concentrated on range-shifts by individual taxa (and hence composition), the long-term legacies of past climates can also be seen in spatial patterns in genetic structures (Gavin et al. 2014) and trait expression (Battersby et al. 2017).

Climate refugia are an example of phenomena determining landscape heterogeneity across organizational levels. The term 'refugia' was coined by paleoecologists to describe climatically anomalous areas during past periods of change (e.g., thermal refugia during glacial periods) that provided safe havens for some taxa (Mosblech et al. 2011). These past refugia are visible in contemporary patterns of diversity (Stewart et al. 2010). Recognizing the role of refugia confirms the inherently spatial nature of the reorganization of ecosystems under climate change; for example, suitable climate space must be accessible via dispersal (McGuire et al. 2016; Littlefield et al. 2019). Although paleoecological data reveal the complex responses of ecosystems to past change (Williams et al. 2011; Svenning et al. 2015), they do not of themselves reveal the spatial processes responsible for these changes. Thus, understanding how ecosystems have responded / will respond to climate change requires integrating an understanding of paleoclimate with the focus on movement of energy and matter through heterogeneous systems emphasized by landscape ecology (Risser et al. 1984).

Disturbance dynamics and landscape stability

Landscape ecologists have developed a rich theoretical framework linking the disturbance regime to landscape structure (i.e., composition [what's there?] and configuration [where is it?]). As an early example, Jackson (1968) developed a conceptual model to explain the seemingly anomalous co-occurrence of forest and non-woody moorland in south-west Tasmania as a function of soil conditions (see also Bowman and Perry 2017). More generally, Turner et al. (1993) argued that if the extent (relative to total landscape area) and frequency (relative to recovery time) of disturbance events are known, landscape dynamics can be predicted (measured as variability in old vegetation in the landscape; Figure 14.4). Intuitively, where disturbance is infrequent and small, most of the landscape will be 'old' vegetation most of the time ('A' in Figure 14.3a); occasional large events result in stable systems but with high temporal variance ('E' in Figure 14.3a). However, where disturbances are relatively frequent and large, landscape dynamics may be difficult to predict with the potential for unstable and bifurcating systems ('F' in Figure 14.3a). The part of the frequency-extent parameter space where landscape dynamics are unpredictable and may bifurcate has attracted considerable recent attention through the lens of alternative stable states (Bowman et al. 2015; Johnstone et al. 2016). In systems where alternate stable states occur, contrasting states (e.g., forest vs. shrubland, turbid vs. clear lakes) arise due to changes in disturbance regime and once initiated, are maintained by feedback dynamics and difficult to reverse (Scheffer et al. 2001). As a recent example of the integration of paleoecological and long-term observational data to explore such dynamics, Burge et al. (2020) used pollen and charcoal from soil cores to evaluate whether taxonomic, functional, or baseline convergence had occurred since a large fire nearly 130 years ago.

These conceptual frameworks are challenging to evaluate because of the extended temporal extents they encompass (e.g., multiple generations of long-lived trees). Classic definitions of stability (e.g., Connell and Sousa 1983) emphasize the need to evaluate ecosystem change over multiple generations; for systems where long-lived organisms dominate, this temporal extent is

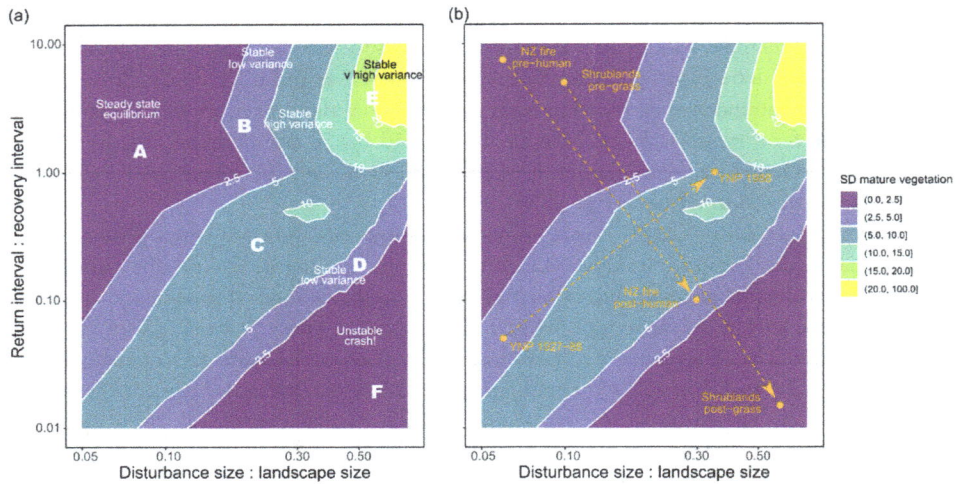

Figure 14.3 Conceptual model of landscape equilibrium developed by Turner et al. (1993) showing how landscape dynamics might be qualitatively predicted from the temporal and spatial scaling of recurrent disturbances. The filled contours in (a) show the SD of mature vegetation through tome with associated stability regimes. The orange circles in (b) show examples of three well-documented shifts in disturbance regime and associated change in landscape dynamics (arrows show direction of transition in time). YNP = Yellowstone National Park. (Figure reproduced using a reimplementation of the original model in NetLogo v. 6.1.1; Perry, G.L.W., http://dx.doi.org/10.17608/k6.auckland.11659230, 2020.)

impractical for direct observational (neoecological) approaches. As described earlier, during and after environmental change, ecological communities may remain disequilibrial for long periods of time and show complex lag dynamics (Blonder et al. 2017). In these settings, paleoecological data and understanding are invaluable.

Case study: The anthropic introduction of fire to Aotearoa-New Zealand's ecosystems

The archipelagos of remote Oceania were the last major landmasses settled by humans (Wilmshurst et al. 2011). Aotearoa-New Zealand (NZ) was settled in c. 1280 CE by Polynesian voyagers (Māori) with a relatively small population of humans (likely fewer than 500) as part of this diaspora. Paleoecological records, especially fossil pollen and charcoal, show that human settlement of NZ was associated with rapid loss of vegetation and increased fire activity (McGlone and Wilmshurst 1999). Two obvious questions about this transformation are: (i) what role did humans play in it? and (ii) if humans are responsible, *how* did they enact this change, given the small populations involved? Paleoecological approaches can help to resolve the first of these questions. Although many paleocharcoal records from across NZ show spikes in fire activity after human arrival, these are time transgressive, suggesting that regional climate change alone is unlikely to be responsible (it would be more synchronous). These records also suggest that fire activity was concentrated in a short window (c. 100–150 years) called the 'initial burning period' (McWethy et al. 2010). The patterns in the paleoecological record, however, do not reveal the behaviors responsible for the change, and it is here that the tools of landscape ecology become valuable. Perry et al. (2012b) used machine-learning approaches to identify the biophysical variables that best explained patterns of fire loss in NZ in the prehistoric (c. 1230–1840 CE) period. These empirical models identified climate predictors as by far the most important, despite the fact that the fire was human in origin, and confirm the seemingly paradoxical outcome that fire-driven forest loss was most prevalent in areas where human population densities were low.

The two commonest explanations for this fire-driven transformation have been the use of fire for hunting large animals such as moa and accidental ignitions. Given that humans are responsible for the change, the next question is the effort required to drive it; would humans have had to burn incessantly and ubiquitously? or would more targeted ignitions have had the same effect? The anthropological record suggests that indigenous peoples use fire to manage landscapes in a directed way, suggesting that a model where humans randomly 'flood' the landscape with ignitions does *not* explain the dynamics observed in the paleoecological record. Central to understanding how human fire could have transformed these landscapes so rapidly is that due to composition and microclimate differences, young vegetation is more flammable than older vegetation in these systems (Kitzberger et al. 2016). Perry *et al.* (2012a) used spatial simulation models grounded in landscape ecological theory to explore how this fire–vegetation relationship might have facilitated rapid forest loss. They demonstrated that positive feedbacks between fire and vegetation inherent in NZ's ecosystems mean that as long as ignitions tended to occur in younger (more flammable) vegetation, large-scale forest loss was possible with a low ignition frequency. Similar feedback-driven landscape transformations have been described in analogous ecosystems in South America (Kitzberger et al. 2016) and elsewhere in the Pacific (Perry and Enright 2002). Unravelling these sorts of changes requires a synthesis of the long-term information provided by paleoecological information and the tools and concepts developed by landscape ecologists.

The landscape ecology of fluxes of energy, matter, organisms, and information

Landscape ecology is not solely the study of spatial patterns in ecosystem composition. Indeed, the principles of the discipline as articulated by Risser et al. (1984) emphasize understanding flows of organisms, energy, and matter through spatially heterogeneous systems. These flows include the cycling of biogeochemicals and water, the movement of animals, and processes associated with animal movement (e.g., seed dispersal and pollination); these flows in turn, generate landscape heterogeneity. Changes in ecosystem composition, including vegetation shifts and faunal extinctions, have dramatically altered many of these processes (Dirzo et al. 2014). Restoring *function* to ecosystems and landscapes requires understanding how these functions have changed, which, in turn, requires adequate baselines; again, paleoecology is central to such reconstruction.

Reconstructing sediments and nutrient fluxes

Two signatures of human activity in ecosystems are altered sedimentation rates (Bennett and Buck 2016) and changes in biogeochemical fluxes, such as the overloading of sinks (e.g., cultural eutrophication; Carpenter et al. 1998). These shifts arise from processes such as deforestation, altered disturbance and hydrological regimes, and anthropogenic deposition or removal of materials (McLauchlan et al. 2014). Also central to the impacts of humans are the different structures of ecology, topography, and hydrology that impose profound functional changes on ecosystems. For examples of the interplay between structural and functional connectivity in controlling rates of soil erosion and nutrient loss from different landscapes, see Wainwright et al. (2011).

Radio-isotope dating underpins the development of chronologies for recent sediment records (e.g., 210Pb), and older sediments dating back 60,000 years (mostly using 14C). By dating materials at different depths in a sediment core, an 'age-depth' model can be developed, and sedimentation rates inferred (Blaauw and Heegaard 2012; Bennett and Buck 2016). Sedimentation rates vary through time and space, and the age–depth relationship allows estimation of age at a given depth and provides a chronology for paleoecological records developed from sediment cores. Temporal variations in sedimentation rate can arise from climate and other environmental change but are often also associated with human activity (Webb and Webb 1988). Thus, changes in sedimentation rate, alongside other paleoecological indicators, can be used to infer changes in landscape dynamics.

Shifts in sedimentation rates are often accompanied by changes in nutrient fluxes. McLauchlan et al. (2013) describe three ways to reconstruct nutrient fluxes using paleoecological information: (i) geochemical analyses (e.g., isotopic ratios in sediments), (ii) use of biological proxies, including environmental DNA, known to relate to nutrient conditions (e.g., diatoms), and (iii) quantitative reconstruction of nutrient budgets. Paleoecological approaches have been used to describe changes in lake conditions resulting from land-use and cover change in the surrounding catchment. For example, in Europe, assessment of reference conditions is required as part of the European Union's water framework directive (Bennion et al. 2011). In parallel, landscape ecologists have developed frameworks to understand the spatial organization of lakes in heterogeneous landscapes (Soranno 1999; Soranno et al. 2010). 'Landscape limnology' adopts the patch-mosaic framework (with the lakes as patches) and emphasizes the wider landscape context for understanding lake dynamics (Soranno et al. 2010). The concepts and tools of paleolimnology and landscape limnology offer an example of how landscape ecology and paleoecology could fruitfully be integrated. When combined with the long-term perspective provided by paleolimnology, such landscape ecological frameworks provide the opportunity to develop understanding of lake dynamics across broad spatio-temporal extents.

Reconstructing ecological interactions

Extinction involves the loss of ecological functions (Galetti et al. 2018) and ecosystem services (Jeffers et al. 2015). This loss of ecological function has important consequences for population and community dynamics and is relevant at the landscape level because processes such as seed dispersal and metaecosystem fluxes generate spatial heterogeneity (Gounand et al. 2018). At a fundamental level, paleoecological studies can identify and document unknown or unobserved interactions (for example, forgotten extinct seed predators; Carpenter et al. 2020). Wood et al. (2012) provide a serendipitous example of this knowledge generation in their observation of the pollen of the endangered endemic New Zealand root parasite *Dactylanthus taylorii* in coprolites of the endangered flightless kākāpō (*Strigops habroptilus*). Prior to Wood et al.'s observation, the short-tailed bat (*Mystacina tuberculata*) was the sole known pollinator of this plant species. Using such data, plant–animal interaction networks can be reconstructed, allowing the potential role of extinct taxa to be inferred (Pires et al. 2014; Wood et al. 2017) and the reconfiguration of interaction networks due to humans to be pieced together (Yeakel et al. 2014). While these studies reveal the interaction networks present in the ecosystems of the past, they depict them as static in time and space. Recently, Pires et al. (2018) used random walk models informed by allometric body-mass scaling relationships to reconstruct seed dispersal kernels of extinct South American megafauna. Their modelling extends the static reconstruction of interaction networks to further reveal the functional consequences of species loss. Various branches of neoecology, including movement ecology and landscape ecology, have paid considerable attention to interactions between animal movement, landscape structure, and the ecological services emerging from them. These offer the potential for an integrated 'paleo-movement ecology' drawing on advances in allometric scaling of animal movement (Hirt et al. 2018); such an integrated approach would provide a novel way to reconstruct changes in the ecosystem functions and services emerging from the movement of individual animals.

What is the role of landscape ecology in 'natural resource management'?

Risser et al.'s fourth focus concerns the role of landscape ecology in landscape and ecosystem management. However, as previous reviews (Opdam et al. 2002; Turner et al. 2002) have suggested, there remains a distance between landscape ecology and landscape/ecosystem management. Challenges of scale are central to the effective integration of a landscape perspective into ecosystem and natural resource management (Parrott and Meyer 2012). As Stevens et al. (2007) point out, landscape and ecosystem dynamics operate at scales potentially quite different from those at which management activities are targeted (e.g., centuries vs. years to decades; Christensen et al. 1996). They further argue that dialogue between disciplines in the natural and social sciences interested in landscape and ecosystem change is limited by the different scales various disciplines focus on. Here, we concentrate on the issues associated with 'ecological' versus 'management' scale domains, because it has been a recurrent theme in the literature (Turner 2005; Willis et al. 2005), and the potential role of paleoecology in resolving these challenges.

As the rate and extent of human change to the biosphere continue to increase, there has been growing interest in using paleoecological data to understand how ecosystems responded to environmental change in the past (Barnosky et al. 2017; Nogué et al. 2017). However, despite the potential for paleoecological data and understanding to anchor landscape management and restoration in an appropriately long-term perspective, concrete examples of its application are few (Davies et al. 2014; Barak et al. 2016). For example, Willis et al. (2005) note that of seven major biodiversity assessments published between 2000 and 2005, only one included records longer than 50 years (and it was 52 years long!). So, how can paleoecology contribute to landscape management and restoration? Jackson (2007) suggests three ways: (i) informing baselines, (ii) contextualizing ecosystem response to change, and (iii) identifying future conditions outside current frames of reference ('no-analogue conditions'). We consider each of these in turn.

Informing baselines

Ecosystems in the past have shown a range of responses to environmental change; this dynamism has led ecologists to question the concept of 'naturalness' and to ask how to develop baselines against which to contextualize ecosystem change (Sprugel 1991; Willis et al. 2010; Brown et al. 2018). This dynamism makes *static* historical targets inappropriate for restoration and conservation efforts (Jackson and Hobbs 2009). What is deemed 'natural' depends on how humans perceive change, which, in turn, is a function of the scale at which we perceive the environment (Sprugel 1991). For example, in northern NZ, forests on offshore islands dominated by pōhutakawa (*Meterosideros excelsa*; Myrtaceae) have been seen as the 'natural' pre-human condition; however, pollen and ancient DNA analyses of a soil core suggest that before human settlement, these forests were dominated by an assemblage of tree species with no current analogue (Wilmshurst et al. 2014). Likewise, paleoecological data can inform reference states for water quality in lake ecosystems (Bennion et al. 2011) and historical population conditions in marine systems (McClenachan et al. 2012). However, the contribution of paleoecology to informing baselines goes beyond descriptions of past ecosystem conditions. For example, questions of whether a species is 'native' or 'exotic' can be contested, and paleoecology can contribute to their resolution (van Leeuwen et al. 2008; Wilmshurst et al. 2015; and the following case study). Likewise, paleoecological studies can contribute to the development of baselines for disturbance regimes (Iglesias et al. 2015).

Case study: Beavers in Britain

Castor fiber, the Eurasian beaver, has not been present in Britain for at least 400 years, having been extirpated due to the value of its fur (most notably for hats), castoreum (believed to be widely used as a painkiller), and its meat (the beaver's tail and feet were defined by the church as a fish, enabling consumption on Lenten days) (Kitchener and Conroy 1997; Nolet and Rosell 1998; Coles 2006). Indeed, while the beaver may have been widespread across Eurasia, populations sank to near extinction on the continent due to hunting of this native species (Halley and Rosell 2002). In striving to meet the goals of the EU Habitats Directive (European Union 1992), which seeks to (Pillai and Heptinstall 2013) 'ensure the conservation of a wide range of rare, threatened or endemic animal and plant species', conservationists have argued for and indeed, delivered in most EU countries reintroduction of the Eurasian beaver. However, reintroduction of the beaver, a keystone species that delivers significant change to riparian landscapes as the very definition of an 'ecosystem engineer', has not been without controversy, and the role of paleoecological data describing its former extent and impact has been an important factor in its reintroduction (Kitchener and Conroy 1997).

The paleoecological and archaeological evidence base describing the extent and activity of beavers in Britain is best summarized in Coles (2006). Extensive fieldwork, often alongside archaeological digs, has been conducted in locations as diverse as the Somerset levels (a seasonally inundated wetland in southern England); Creswell Crags (a limestone valley in Derbyshire/Nottinghamshire); Starr Carr, in the Vale of Pickering, North Yorkshire; Middlestot's Bog, Berwickshire, south-east Scotland; and the Forest of Dean on the English/Welsh border. Archaeological/paleoenvironmental work across these sites reveals a plethora of evidence of beaver presence, including beaver-gnawed sticks, preserved in anaerobic soil conditions; beaver lodges, dams, and canals, preserved in sediment stratigraphies; and beaver remains, including teeth and skulls used as tools by humans across the Holocene. Alongside such direct evidence, Coles (2006) draws together a wealth of secondary evidence, including place-names (Beverley, Beaver dyke, Bewerly, Bardale Beck, Bar Brook – all in Yorkshire), which likely describe the presence of beavers in the paleolandscape for significant periods of time. Carvings are also presented, which still exist in churches such as St. Leonards, Ribbesford, Worcestershire; this example appears to show a beaver being hunted, possibly during Norman times. In addition, a 13th-century pew was carved in the shape of a beaver at Salisbury Cathedral from oak that was felled in the Royal Forest of Chippenham at the request of Henry III in 1236 CE. Translations of manuscripts also describe beavers in Britain, for example, Barber's (2006) translation of the *Bestiary*, which was written between 1220 and 1250 CE, based on previous texts spanning back to the 1st century CE (Coles 2006, p. 172). Finally, written records, primarily referring to beavers in terms of their value (probably for fur), date back to 940 CE, when the laws of Hywel Dda, the then king of Wales, stipulate that 'the King is to have the worth of Beavers, Martens and Ermines'. In total, Coles (2006) brings together more than 200 pieces of independent evidence describing the presence of beavers in Britain ranging from the Late Paleolithic to the 16th century.

Arguably, the collation of paleoecological and historical evidence describing the presence of beavers and the impacts that they delivered has enabled the contemporary reintroduction of this species as the first large mammal to be reintroduced to the British Isles. Without such an extraordinary body of evidence, the case for beaver reintroduction would be weaker, and whether the beaver is a 'native' species would remain contested, especially in light of the significant changes that have occurred to the ecology, hydrology, and biodiversity of modern Britain since the beaver was extirpated well before living memory. Ironically, it might be argued that many of the negative changes that we see in contemporary, intensively farmed, densely populated ecosystems (for example, reduced biodiversity due to loss of wetlands, lack of flow attenuation in over-deepened/straightened streams and rivers) are due to the trophic collapse that has occurred since the removal of beavers. Perhaps, following their widespread reintroduction, future paleoecological data sets will demonstrate the true value of the return of this native species.

Contextualizing environmental change

The ecological response to past climates provides a context for understanding how ecosystems may respond to future climate change. Particular attention has been given to understanding ecosystem dynamics under comparable ('analogue') climate conditions (Willis and MacDonald 2011) and the dynamics of regime shifts in the past (Williams et al. 2011). Attention has also focused on reconstructing ecosystems to inform conservation and restoration ('putting the dead to work', as Dietl and Flessa 2011 phrase it). These, and other associated applications of paleoecological data, align with the questions considered by landscape ecologists. As described earlier, numerous paleoecological studies have described species range-shifts, and associated community reorganization, in response to climate change. However, the extent of human modification of landscapes through habitat loss and fragmentation means that the ability of species to move their ranges cannot be evaluated by considering climate alone (Mayer et al. 2016). The approaches used by landscape ecologists have an important role to play in understanding the spatial dynamics of species movement by, for example, using frameworks such as metapopulation theory (Mosblech et al. 2011 and see Chapter 5) and supporting model-based exploration of species response to climate change under no-analogue conditions (Littlefield et al. 2019).

The challenges of 'no-analogue' conditions

As Hartley (1953) famously opened *The Go-Between*, 'The past is a foreign country; they do things differently there'. Nevertheless, the assumption underpinning the use of paleoecological information to forecast how ecosystems will respond to future change is that the past provides examples of conditions comparable to those expected in the future. However, rapid environmental change may create conditions without a parallel in the paleoecological record (Williams and Jackson 2007). In the past, such 'no-analogue conditions' have arisen from unique combinations of climate and environmental forcing. For example, using dung fungi, pollen, and charcoal, Gill et al. (2009) show that post-glacial climate change, megafaunal extinction, and altered fire regimes led to woodland communities with no known analogue in the paleoecological record. It seems likely that similar 'no-analogue' (or novel) conditions will arise under the stresses the environment is experiencing in the Anthropocene (Ordonez et al. 2016; Figure 14.4).

The potential for a no-analogue future led Williams and Jackson (2007, p. 475) to ask, 'how do you study an ecosystem that no ecologist has seen?' Paleoecologists have identified potential model systems to understand future change (e.g., changes during the last deglaciation may parallel some aspects of projected future climate; Willis and MacDonald 2011; Williams et al. 2013), so studying these transitions may be informative. Another solution that ecologists have used is simulation models such as those used by landscape ecologists to explore vegetation dynamics across large spatial extents. However, using simulation models to explore no-analogue conditions poses difficult challenges for model parametrization and evaluation. One solution to these problems is 'inverse modelling', which is a set of statistical approaches designed to infer parameter values for processes that are only partially observed (Hartig et al. 2012). For example in the case of a population model, aggregated information on population size might be available but estimates of vital rates (e.g., fecundity, growth, mortality) absent; in an inverse modelling approach, vital rates are estimated by comparing the population trajectories produced by *posterior* estimates against the observational data, with those performing adequately being retained (in the form of a *prior* distribution). Needham et al. (2016) used an inverse approach to estimate demographic rates for ash (*Fraxinus excelsior*) to evaluate the potential effects of ash dieback disease in the United Kingdom; in this case, the observational data available did not estimate fecundity, and the ecological implications of the disease were poorly understood. In a paleoclimatological context, Garreta et al. (2010) used a similar approach to estimate long-term (in this case, the entire Holocene) precipitation records from a dynamic vegetation model constrained by modern pollen surface samples and high-resolution sediment records. In developing an understanding of future no-analogue conditions, the role of paleoecological information is twofold: (i) to provide analogues from the past for future change and (ii) to contribute long-term information to support the inverse parameter estimation process (Williams et al. 2013).

Figure 14.4   Emerging 'no-analogue' plant communities on a recent fire site, Claris, Aotea-Great Barrier Island, northern New Zealand. Here, a range of native NZ species and invasives (e.g., *Pinus, Ulex, Hakea*) co-occur in a landscape shaped by a novel fire regime.

Towards a 'paleolandscape ecology'?

Despite repeated calls for closer integration between paleoecology and landscape ecology, there remains a distance between the two disciplines. As Birks (1993) points out, this separation is true of paleoecology and neoecology more generally. Nevertheless, paleoecology provides information relevant to many of the questions that interest landscape ecologists and has a value well beyond just climate reconstruction or model evaluation. We are optimistic that the environmental issues we face are so urgent that landscape ecology and paleoecology will by necessity become closer over the coming years, especially as interest in both disciplines moves from composition to encompass function. There are difficult unresolved challenges in bridging the scales that landscape ecology and paleoecology focus on, but their integration promises important insights with which to confront the challenges of projected and realized ecological change.

Acknowledgments

GP and JW acknowledge support from the New Zealand 'Biological Heritage' Science Challenge. The University of Exeter funded a visit by GP to RB at the Dept. of Geography, during which some of this chapter was conceptualized and drafted. Matt McGlone and James Millington provided useful comments on an earlier version of this chapter.

References

Anderson, N.J., Bugmann, H., Dearing, J.A. and Gaillard, M.-J. (2006) 'Linking palaeoenvironmental data and models to understand the past and to predict the future', *Trends in Ecology and Evolution*, vol 21, pp696–704.

Barak, R.S., Hipp, A.L., Cavender-Bares, J., Pearse, W.D., Hotchkiss, S.C., Lynch, E.A., et al. (2016). 'Taking the long view: integrating recorded, archeological, paleoecological, and evolutionary data into ecological restoration', *International Journal of Plant Sciences*, vol 177, pp90–102.

Barber, R.W. (ed) (2006) *Bestiary: Being an English Version of the Bodleian Library, Oxford, MS Bodley 764 with All the Original Miniatures Reproduced in Facsimile*. Reprinted in pbk. Boydell, Woodbridge.

Barnosky, A.D., Hadly, E.A., Gonzalez, P., Head, J., Polly, P.D., Lawing, A.M., et al. (2017) 'Merging paleobiology with conservation biology to guide the future of terrestrial ecosystems', *Science*, vol 355, art eaah4787.

Battersby, P.F., Wilmshurst, J.M., Curran, T.J., McGlone, M.S. and Perry, G.L.W. (2017) 'Exploring fire adaptation in a land with little fire: serotiny in *Leptospermum scoparium* (Myrtaceae)', *Journal of Biogeography*, vol 44, pp1306–1318.

Bennett, K. and Buck, C.E. (2016) 'Interpretation of lake sediment accumulation rates', *Holocene*, vol 26, pp1092–1102.

Bennington, J.B., Dimichele, W.A., Badgley, C., Bambach, R.K., Barrett, P.M., Behrensmeyer, A.K., et al. (2009) 'Critical issues of scale in paleoecology', *PALAIOS*, vol 24, pp1–4.

Bennion, H., Battarbee, R.W., Sayer, C.D., Simpson, G.L. and Davidson, T.A. (2011) 'Defining reference conditions and restoration targets for lake ecosystems using palaeolimnology: a synthesis', *Journal of Paleolimnology*, vol 45, pp533–544.

Birks, H.J.B. (1993) 'Quaternary palaeoecology and vegetation science: current contributions and possible future developments', *Review of Palaeobotany and Palynology*, vol 79, pp153–177.

Birks, H.J.B. (2008) 'Paleoecology', in S.E. Jørgensen and B.D. Fath (eds) *Encyclopedia of Ecology*. Elsevier, Amsterdam.

Blaauw, M. and Heegaard, E. (2012) 'Estimation of age-depth relationships', in H.J.B. Birks, A.F. Lotter, S. Juggins and J.P. Smol (eds) *Tracking Environmental Change Using Lake Sediments*. Springer, Dordrecht, Netherlands.

Blonder, B., Moulton, D.E., Blois, J., Enquist, B.J., Graae, B.J., Macias-Fauria, M., et al. (2017) 'Predictability in community dynamics', *Ecology Letters*, vol 20, pp293–306.

Bowman, D.M.J.S. and Perry, G.L.W. (2017) 'Soil or fire: what causes treeless sedgelands in Tasmanian wet forests?', *Plant and Soil*, vol 420, pp1–18.

Bowman, D.M.J.S., Perry, G.L.W. and Marston, J.B. (2015) 'Feedbacks and landscape-level vegetation dynamics', *Trends in Ecology and Evolution*, vol 30, pp255–260.

Brown, A.G., Lespez, L., Sear, D.A., Macaire, J.-J., Houben, P., Klimek, K., et al. (2018) 'Natural vs anthropogenic streams in Europe: history, ecology and implications for restoration, river-rewilding and riverine ecosystem services', *Earth-Science Reviews*, vol 180, pp185–205.

Bunting, M.J. and Middleton, R. (2009) 'Equifinality and uncertainty in the interpretation of pollen data: the multiple scenario approach to reconstruction of past vegetation mosaics', *Holocene*, vol 19, pp799–803.

Bunting, M.J., Farrell, M., Bayliss, A., Marshall, P. and Whittle, A. (2018) 'Maps from mud: using the multiple scenario approach to reconstruct land cover dynamics from pollen records: a case study of two neolithic landscapes', *Frontiers in Ecology and Evolution*, vol 6, art 36.

Burge, O.R., Bellingham, P.J., Arnst, E.A., Bonner, K.I., Burrows, L.E., Richardson, S.J., et al. (2020) 'Integrating permanent plot and palaeoecological data to determine subalpine post-fire succession, recovery and convergence over 128 years', *Journal of Vegetation Science*, vol 31, no 5, pp755–767.

Carpenter, J.K., Wilmshurst, J.M., McConkey, K.R., Hume, J.P., Wotton, D.M., Shiels, A.B., et al. (2020) 'The forgotten fauna: native vertebrate seed predators on islands', *Functional Ecology*, vol 34, no 9, pp1802–1813.

Carpenter, S.R., Caraco, N.F., Correll, D.L., Howarth, R.W., Sharpley, A.N. and Smith, V.H. (1998) 'Nonpoint pollution of surface waters with phosphorus and nitrogen', *Ecological Applications*, vol 8, pp559–568.

Christensen, N.L., Bartuska, A.M., Brown, J.H., Carpenter, S., D'Antonio, C., Francis, R., et al. (1996) 'The report of the ecological society of America committee on the scientific basis for ecosystem management', *Ecological Applications*, vol 6, pp665–691.

Cleland, C.E. (2001) 'Historical science, experimental science, and the scientific method', *Geology*, vol 29, pp987–990.

Coles, B. (2006) *Beavers in Britain's Past*. WARP occasional paper. Oxbow Books, WARP, Oxford.

Connell, J.H. and Sousa, W.P. (1983) 'On the evidence needed to judge ecological stability or persistence', *American Naturalist*, vol 121, pp789–824.

Davies, A.L., Colombo, S. and Hanley, N. (2014). 'Improving the application of long-term ecology in conservation and land management', *Journal of Applied Ecology*, vol 51, pp63–70.

Dietl, G.P. and Flessa, K.W. (2011) 'Conservation paleobiology: putting the dead to work', *Trends in Ecology and Evolution*, vol 26, pp30–37.

Dirzo, R., Young, H.S., Galetti, M., Ceballos, G., Isaac, N.J.B. and Collen, B. (2014) 'Defaunation in the Anthropocene', *Science*, vol 345, pp401–406.

Estes, L., Elsen, P.R., Treuer, T., Ahmed, L., Caylor, K., Chang, J., et al. (2018) 'The spatial and temporal domains of modern ecology', *Nature Ecology and Evolution*, vol 2, pp819–826.

European Union (1992) Council Directive 92/43/EEC of 21 May 1992 on the conservation of natural habitats and of wild fauna and flora [EUR-Lex - 31992L0043 - EN - EUR-Lex]. https://eur-lex.euro pa.eu/eli/dir/1992/43/oj.

Froyd, C.A. and Willis, K.J. (2008) 'Emerging issues in biodiversity & conservation management: the need for a palaeoecological perspective', *Quaternary Science Reviews*, vol 27, pp1723–1732.

Fukami, T. (2015) 'Historical contingency in community assembly: integrating niches, species pools, and priority effects', *Annual Review of Ecology, Evolution, and Systematics*, vol 46, pp1–23.

Galetti, M., Moleón, M., Jordano, P., Pires, M.M., Guimarães, P.R., Pape, T., et al. (2018) 'Ecological and evolutionary legacy of megafauna extinctions', *Biological Reviews*, vol 93, pp845–862.

Garreta, V., Miller, P.A., Guiot, J., Hély, C., Brewer, S., Sykes, M.T., et al. (2010) 'A method for climate and vegetation reconstruction through the inversion of a dynamic vegetation model', *Climate Dynamics*, vol 35, pp371–389.

Gaüzère, P., Iversen, L.L., Barnagaud, J.-Y., Svenning, J.-C. and Blonder, B. (2018) 'Empirical predictability of community responses to climate change', *Frontiers in Ecology and Evolution*, vol 6, art 186.

Gavin, D.G., Fitzpatrick, M.C., Gugger, P.F., Heath, K.D., Rodríguez-Sánchez, F., Dobrowski, S.Z., et al. (2014) 'Climate refugia: joint inference from fossil records, species distribution models and phylogeography', *New Phytologist*, vol 204, pp37–54.

Gill, J.L., Williams, J.W., Jackson, S.T., Lininger, K.B. and Robinson, G.S. (2009) 'Pleistocene megafaunal collapse, novel plant communities, and enhanced fire regimes in North America', *Science*, vol 326, pp1100–1103.

Gounand, I., Harvey, E., Little, C.J. and Altermatt, F. (2018) 'Meta-ecosystems 2.0: rooting the theory into the field', *Trends in Ecology and Evolution*, vol 33, pp36–46.

Grace, M., Akçakaya, H.R., Bennett, E., Hilton-Taylor, C., Long, B., Milner-Gulland, E.J., et al. (2019) 'Using historical and palaeoecological data to inform ambitious species recovery targets', *Philosophical Transaction of the Royal Society B*, vol 374, art 20190297.

Grimm, V., Revilla, E., Berger, U., Jeltsch, F., Mooij, W.M., Railsback, S.F., et al. (2005) 'Pattern-oriented modeling of agent-based complex systems: lessons from ecology', *Science*, vol 310, pp987–991.

Halley, D.J. and Rosell, F. (2002) 'The beaver's reconquest of Eurasia: status, population development and management of a conservation success', *Mammal Review*, vol 32, pp153–178.

Hargrove, W.W. and Pickering, J. (1992) 'Pseudoreplication: a *sine qua non* for regional ecology', *Landscape Ecology*, vol 6, pp251–258.

Hartig, F., Dyke, J., Hickler, T., Higgins, S.I., O'Hara, R.B., Scheiter, S., et al. (2012) 'Connecting dynamic vegetation models to data: an inverse perspective', *Journal of Biogeography*, vol 39, pp2240–2252.

Hartley, L.P. (1953) *The Go-Between*. New York Review Books Classics. Group West, New York, Berkeley.

Hirt, M.R., Grimm, V., Li, Y., Rall, B.C., Rosenbaum, B. and Brose, U. (2018) 'Bridging scales: allometric random walks link movement and biodiversity research', *Trends in Ecology and Evolution*, vol 33, pp701–712.

Huntley, B. (1996) 'Quaternary palaeoecology and ecology', *Quaternary Science Reviews*, vol 15, pp591–606.

Iglesias, V., Yospin, G.I. and Whitlock, C. (2015) 'Reconstruction of fire regimes through integrated paleoecological proxy data and ecological modeling', *Frontiers in Plant Science*, vol 5, art 785.

Jackson, S.T. (2007) 'Looking forward from the past: history, ecology, and conservation', *Frontiers in Ecology and the Environment*, vol 5, art 455.

Jackson, S.T. and Hobbs, R.J. (2009) 'Ecological restoration in the light of ecological history', *Science*, vol 325, pp567–569.

Jackson, W.D. (1968) 'Fire, air, water and earth – an elemental ecology of Tasmania', *Proceedings of the Australian Ecological Society*, vol 3, pp9–16.

Jeffers, E.S., Nogué, S. and Willis, K.J. (2015) 'The role of palaeoecological records in assessing ecosystem services', *Quaternary Science Reviews*, vol 112, pp17–32.

Johnstone, J.F., Allen, C.D., Franklin, J.F., Frelich, L.E., Harvey, B.J., Higuera, P.E., et al. (2016) 'Changing disturbance regimes, ecological memory, and forest resilience', *Frontiers in Ecology and the Environment*, vol 14, pp369–378.

King, A.W. (1997) 'Hierarchy theory: a guide to system structure for wildlife biologists', in J.A. Bissonette (ed) *Wildlife and Landscape Ecology*. Springer, New York.

Kitchener, A.C. and Conroy, J.W.H. (1997) 'The history of the Eurasian Beaver *Castor fiber* in Scotland', *Mammal Review*, vol 27, pp95–108.

Kitzberger, T., Perry, G.L.W., Paritsis, J., Gowda, J.H., Tepley, A.J., Holz, A., et al. (2016) 'Fire–vegetation feedbacks and alternative states: common mechanisms of temperate forest vulnerability to fire in southern South America and New Zealand', *New Zealand Journal of Botany*, vol 54, pp247–272.

Kosnik, M.A. and Kowalewski, M. (2016) 'Understanding modern extinctions in marine ecosystems: the role of palaeoecological data', *Biology Letters*, vol 12, art 20150951.

Lake, P.S. (2000). 'Disturbance, patchiness, and diversity in streams', *Journal of the North American Benthological Society*, vol 19, pp573–592.

van Leeuwen, J.F.N., Froyd, C.A., van der Knaap, W.O., Coffey, E.E., Tye, A. and Willis, K.J. (2008) 'Fossil pollen as a guide to conservation in the Galapagos', *Science*, vol 322, pp1206.

Levin, S.A. (1992) 'The problem of pattern and scale in ecology', *Ecology*, vol 73, pp1943–1967.

Littlefield, C.E., Krosby, M., Michalak, J.L. and Lawler, J.J. (2019) 'Connectivity for species on the move: supporting climate-driven range shifts', *Frontiers in Ecology and the Environment*, vol 17, no 5, pp270–278.

Mariani, M., Connor, S.E., Fletcher, M.-S., Theuerkauf, M., Kuneš, P., Jacobsen, G., et al. (2017). 'How old is the Tasmanian cultural landscape? A test of landscape openness using quantitative land cover reconstructions', *Journal of Biogeography*, vol 44, pp2410–2420.

Mayer, A.L., Buma, B., Davis, A., Gagné, S.A., Loudermilk, E.L., Scheller, R.M., et al. (2016) 'How landscape ecology informs global land-change science and policy', *BioScience*, vol 66, pp458–469.

McClenachan, L., Ferretti, F. and Baum, J.K. (2012) 'From archives to conservation: why historical data are needed to set baselines for marine animals and ecosystems: from archives to conservation', *Conservation Letters*, vol 5, pp349–359.

McGlone, M.S. (1996) 'When history matters: scale, time, climate and tree diversity', *Global Ecology and Biogeography Letters*, vol 5, pp309–314.

McGlone, M.S. and Wilmshurst, J.M. (1999) 'Dating initial Maori environmental impact in New Zealand', *Quaternary International*, vol 59, pp5–16.

McGuire, J.L., Lawler, J.J., McRae, B.H., Nuñez, T.A. and Theobald, D.M. (2016) 'Achieving climate connectivity in a fragmented landscape', *Proceedings of the National Academy of Sciences*, vol 113, pp7195–7200.

McLauchlan, K.K., Williams, J.J. and Engstrom, D.R. (2013) 'Nutrient cycling in the palaeorecord: fluxes from terrestrial to aquatic ecosystems', *Holocene*, vol 23, pp1635–1643.

McLauchlan, K.K., Higuera, P.E., Gavin, D.G., Perakis, S.S., Mack, M.C., Alexander, H., et al. (2014) 'Reconstructing disturbances and their biogeochemical consequences over multiple timescales', *BioScience*, vol 64, pp105–116.

McWethy, D.B., Whitlock, C., Wilmshurst, J.M., McGlone, M.S., Fromont, M., Li, X., et al. (2010). 'Rapid landscape transformation in South Island, New Zealand, following initial Polynesian settlement', *Proceedings of the National Academy of Sciences*, vol 107, pp21343–21348.

Mittelbach, G.G., Schemske, D.W., Cornell, H.V., Allen, A.P., Brown, J.M., Bush, M.B., et al. (2007) 'Evolution and the latitudinal diversity gradient: speciation, extinction and biogeography', *Ecology Letters*, vol 10, pp315–331.

Mosblech, N.A.S., Bush, M.B. and van Woesik, R. (2011) 'On metapopulations and microrefugia: palaeoecological insights', *Journal of Biogeography*, vol 38, pp419–429.

National Research Council, Division on Earth and Life Studies, Board on Life Sciences, Board on Earth Sciences and Resources, Committee on the Geologic Record of Biosphere Dynamics and National Academy of Sciences (2005) *The Geological Record of Ecological Dynamics: Understanding the Biotic Effects of Future Environmental Change*. National Academies Press, Washington.

Needham, J., Merow, C., Butt, N., Malhi, Y., Marthews, T.R., Morecroft, M., et al. (2016) 'Forest community response to invasive pathogens: the case of ash dieback in a British woodland', *Journal of Ecology*, vol 104, pp315–330.

Nogué, S., de Nascimento, L., Froyd, C.A., Wilmshurst, J.M., de Boer, E.J., Coffey, E.E.D., et al. (2017) 'Island biodiversity conservation needs palaeoecology', *Nature Ecology and Evolution*, vol 1, art 0181.

Nolet, B.A. and Rosell, F. (1998) 'Comeback of the beaver *Castor fiber*: an overview of old and new conservation problems', *Biological Conservation*, vol 83, pp165–173.

Opdam, P., Foppen, R. and Vos, C. (2002) 'Bridging the gap between ecology and spatial planning in landscape ecology', *Landscape Ecology*, vol 16, pp767–779.

Ordonez, A., Williams, J.W. and Svenning, J.-C. (2016) 'Mapping climatic mechanisms likely to favour the emergence of novel communities', *Nature Climate Change*, vol 6, pp1104–1109.

Parrott, L. and Meyer, W.S. (2012) 'Future landscapes: managing within complexity', *Frontiers in Ecology and the Environment*, vol 10, pp382–389.

Perry, G.L.W. (2020) 'NetLogo implementation of the Turner et al. 1993 landscape equilibrium model', http://dx.doi.org/10.17608/k6.auckland.11659230, accessed 7 Dec 2020.

Perry, G.L.W. and Enright, N.J. (2002) 'Humans, fire and landscape pattern: understanding a maquis-forest complex, Mont Do, New Caledonia using a spatial "state-and-transition" model', *Journal of Biogeography*, vol 29, pp1143–1159.

Perry, G.L.W., Wilmshurst, J.M., McGlone, M.S., McWethy, D.B. and Whitlock, C. (2012a) 'Explaining fire-driven landscape transformation during the initial burning period of New Zealand's prehistory', *Global Change Biology*, vol 18, pp1609–1621.

Perry, G.L.W., Wilmshurst, J.M., McGlone, M.S. and Napier, A. (2012b) 'Reconstructing spatial vulnerability to forest loss by fire in pre-historic New Zealand', *Global Ecology and Biogeography*, vol 21, pp1029–1041.

Perry, G.L.W., Wainwright, J., Etherington, T.R. and Wilmshurst, J.M. (2016) 'Experimental simulation: using generative modeling and palaeoecological data to understand human-environment interactions', *Frontiers in Ecology and Evolution*, vol 4, art 109.

Pillai, A. and Heptinstall, D. (2013) 'Twenty years of the habitats directive: a case study on species reintroduction, protection and management', *Environmental Law Review*, vol 15, pp27–46.

Pires, M.M., Galetti, M., Donatti, C.I., Pizo, M.A., Dirzo, R. and Guimarães, P.R., Jr. (2014) 'Reconstructing past ecological networks: the reconfiguration of seed-dispersal interactions after megafaunal extinction', *Oecologia*, vol 175, pp1247–1256.

Pires, M.M., Guimarães, P.R., Galetti, M. and Jordano, P. (2018) 'Pleistocene megafaunal extinctions and the functional loss of long-distance seed-dispersal services', *Ecography*, vol 41, pp153–163.

Ricklefs, R.E., Latham, R.E. and Qian, H. (1999) 'Global patterns of tree species richness in moist forests: distinguishing ecological influences and historical contingency', *Oikos*, vol 86, no 2, pp369–373.

Risser, P.G., Karr, J.R. and Forman, R.T.T. (1984) *Landscape Ecology: Directions and Approaches*. No. Special Publication Number 2. Illinois Natural History Survey, Champaign.

Scheffer, M., Carpenter, S., Foley, J.A. and Walker, B. (2001) 'Catastrophic shifts in ecosystems', *Nature*, vol 413, pp591–596.

Soranno, P.A. (1999) 'Spatial variation among lakes within landscapes: ecological organization along lake chains', *Ecosystems*, vol 2, pp395–410.

Soranno, P.A., Cheruvelil, K.S., Webster, K.E., Bremigan, M.T., Wagner, T. and Stow, C.A. (2010) 'Using landscape limnology to classify freshwater ecosystems for multi-ecosystem management and conservation', *BioScience*, vol 60, pp440–454.

Sprugel, D.G. (1991) 'Disturbance, equilibrium and environmental variability: what is "natural" vegetation in a changing environment', *Biological Conservation*, vol 58, pp1–18.

Stevens, C.J., Fraser, I., Mitchley, J. and Thomas, M.B. (2007) 'Making ecological science policy-relevant: issues of scale and disciplinary integration', *Landscape Ecology*, vol 22, pp799–809.

Stewart, J.R., Lister, A.M., Barnes, I. and Dalén, L. (2010) 'Refugia revisited: individualistic responses of species in space and time', *Proceedings of the Royal Society B: Biological Sciences*, vol 277, pp661–671.

Sugita, S. (2007a) 'Theory of quantitative reconstruction of vegetation I: pollen from large sites REVEALS regional vegetation composition', *Holocene*, vol 17, pp229–241.

Sugita, S. (2007b) 'Theory of quantitative reconstruction of vegetation II: all you need is LOVE', *Holocene*, vol 17, pp243–257.

Svenning, J.-C., Eiserhardt, W.L., Normand, S., Ordonez, A. and Sandel, B. (2015) 'The influence of paleoclimate on present-day patterns in biodiversity and ecosystems', *Annual Review of Ecology, Evolution, and Systematics*, vol 46, pp551–572.

Turner, M.G. (2005) 'Landscape ecology: what is the state of the science?', *Annual Review of Ecology and Systematics*, vol 36, pp319–344.

Turner, M.G. (2010) 'Disturbance and landscape dynamics in a changing world', *Ecology*, vol 91, pp2833–2849.

Turner, M.G. and Gardner, R.H. (2015) *Landscape Ecology in Theory and Practice: Pattern and Process*. Springer, New York.

Turner, M.G., Romme, W.H. and Gardner, R.H. (1993) 'A revised concept of landscape equilibrium: disturbance and stability on scaled landscapes', *Landscape Ecology*, vol 8, pp213–227.

Turner, M.G., Crow, T.R., Liu, J., Rabe, D., Rabeni, C.F., Soranno, P.A., et al. (2002) 'Bridging the gap between landscape ecology and natural resource management', in J. Liu and W.W. Taylor (eds) *Integrating Landscape Ecology and Natural Resource Management*. Cambridge University Press, Cambridge.

Wainwright, J., Turnbull, L., Ibrahim, T.G., Lexartza-Artza, I., Thornton, S.F. and Brazier, R.E. (2011). 'Linking environmental régimes, space and time: Interpretations of structural and functional connectivity', *Geomorphology*, vol 126, pp387–404.

Webb, R.S. and Webb, T. (1988) 'Rates of sediment accumulation in pollen cores from small lakes and mires of eastern North America', *Quaternary Research*, vol 30, pp284–297.

West, R.G. (1964) 'Inter-relations of ecology and quaternary palaeobotany', *Journal of Animal Ecology*, vol 33, pp47–57.

Wiens, J.A. (1989) 'Spatial scaling in ecology', *Functional Ecology*, vol 3, pp385–397.

Wiens, J.A. (1995) 'Landscape mosaics and ecological theory', in: L. Hansson, L. Fahrig and G. Merriam (eds) *Mosaic Landscapes and Ecological Processes*. Chapman and Hall, London.

Williams, J.W. and Jackson, S.T. (2007) 'Novel climates, no-analog communities, and ecological surprises', *Frontiers in Ecology and the Environment*, vol 5, pp475–482.

Williams, J.W., Blois, J.L. and Shuman, B.N. (2011) 'Extrinsic and intrinsic forcing of abrupt ecological change: case studies from the late Quaternary', *Journal of Ecology*, vol 99, pp664–677.

Williams, J.W., Blois, J.L., Gill, J.L., Gonzales, L.M., Grimm, E.C., Ordonez, A., et al. (2013) 'Model systems for a no-analog future: species associations and climates during the last deglaciation', *Annals of the New York Academy of Sciences*, vol 1297, pp29–43.

Willis, K.J. and Birks, H.J.B. (2006) 'What is natural? The need for a long-term perspective in biodiversity conservation', *Science*, vol 314, pp1261–1265.

Willis, K.J. and MacDonald, G.M. (2011) 'Long-term ecological records and their relevance to climate change predictions for a warmer world', *Annual Review of Ecology, Evolution, and Systematics*, vol 42, pp267–287.

Willis, K., Gillson, L., Brncic, T. and Figueroarangel, B. (2005) 'Providing baselines for biodiversity measurement', *Trends in Ecology and Evolution*, vol 20, pp107–108.

Willis, K.J., Bailey, R.M., Bhagwat, S.A. and Birks, H.J.B. (2010) 'Biodiversity baselines, thresholds and resilience: testing predictions and assumptions using palaeoecological data', *Trends in Ecology and Evolution*, vol 25, pp583–591.

Wilmshurst, J.M., Hunt, T.L., Lipo, C.P. and Anderson, A.J. (2011) 'High-precision radiocarbon dating shows recent and rapid initial human colonization of East Polynesia', *Proceedings of the National Academy of Sciences*, vol 108, pp1815–1820.

Wilmshurst, J.M., Moar, J.R., Wood, J.R., Bellingham, P.J., Findlater, A.M., Robinson, J.J., et al. (2014) 'Use of pollen and ancient DNA as conservation baselines for offshore islands in New Zealand', *Conservation Biology*, vol 28, pp202–212.

Wilmshurst, J.M., McGlone, M.S. and Turney, C.S.M. (2015) 'Long-term ecology resolves the timing, region of origin and process of establishment for a disputed alien tree', *AoB Plants*, 7, art plv104.

Wood, J.R., Wilmshurst, J.M., Worthy, T.H., Holzapfel, A.S. and Cooper, A. (2012) 'A lost link between a flightless parrot and a parasitic plant and the potential role of coprolites in conservation paleobiology', *Conservation Biology*, vol 26, pp1091–1099.

Wood, J.R., Perry, G.L.W. and Wilmshurst, J.M. (2017) 'Using palaeoecology to determine baseline ecological requirements and interaction networks for de-extinction candidate species', *Functional Ecology*, vol 31, pp1012–1020.

Wu, J. (1999) 'Hierarchy and scaling: extrapolating information along a scaling ladder', *Canadian Journal of Remote Sensing*, vol 25, pp367–380.

Wu, J. and Loucks, O.L. (1995) 'From balance of nature to hierarchical patch dynamics: a paradigm shift in ecology', *Quarterly Review of Biology*, vol 70, pp439–466.

Yeakel, J.D., Pires, M.M., Rudolf, L., Dominy, N.J., Koch, P.L., Guimarães, P.R., et al. (2014) 'Collapse of an ecological network in Ancient Egypt', *Proceedings of the National Academy of Sciences*, vol 111, pp14472–14477.

15

Landscape pattern analysis

Tarmo K. Remmel and Scott W. Mitchell

Introduction

Why landscape metrics?

Real landscapes are inherently complex and comprise biotic and abiotic elements of multiple sizes, shapes, complexities, and extents while having characteristics that interact and organize dynamically and hierarchically at, and across, multiple scales. This fundamental compositional and organizational complexity, ranging from atoms to organisms to structured landscape entities, produces systems of systems. The resulting landscape states can be considered instantaneously within a temporal continuum of change that results from interacting processes operating at varied scales of time and space, and intensities. Within this overwhelming realization of complexity, scientists continually seek order, relationships, correlations, and trends in order to better understand landscapes, environments, flows, and processes. For the purposes of this chapter, we concentrate on landscapes that have been represented as categorical maps, generally in raster format (although space is also dedicated to vector representations and continuous data).

The difficult question is how to make sense out of the inherent complexity of landscape patterns; what should be abstracted, retained, or ignored? While each study has unique drivers toward answering these queries, parallel tracks of inquiry seek to quantify those patterns for the sake of simplifying the detection, interpretation, and quantification of landscape changes or to link ecological processes with spatial patterns (Turner, 1990). The field of landscape ecology since the 1980s has been dominated by the patch mosaic model (PMM), a framework for describing and thinking about landscapes and their comprising elements, focusing on the structure, function, and change of landscapes (Forman and Godron, 1986). This logical ordering builds on the concept that landscapes are composed primarily of patches in a background matrix and that the patches may vary due to their origin and dynamics but can be described by their size, shape, edge characteristics, and spatial configuration (Forman and Godron, 1981).

Rapidly developing computational abilities in the 1980s meant that it was not long before geographic information systems (GIS) and other specialized software tools were developed to facilitate the quantification of landscape structure for both vector and raster datasets (Baker and Cai, 1992; McGarigal et al., 2012). The availability of tools that simplified complex spatial calculations led to a prolific introduction of landscape indices and metrics (O'Neill et al., 1988; Haines-Young and Chopping, 1996), and by about the year 2000, the term 'landscape metrics' (sometimes abbreviated as 'LM') vastly overtook the use of 'landscape indices' (Uuemaa et al., 2009) and has become the *de facto* standard in the literature. Within this context of a landscape mosaic of patches, assessments of landscape structure can happen at either (1) full landscape,

(2) thematic class, or (3) patch scales (McGarigal and Marks, 1995). Each scale of analysis yields differing views on the landscape state and produces different numbers of output values depending on the number of entities over which the metrics are computed. Thus, when we mention landscape metrics, we usually (but see the section on gradient surface metrics) mean specific quantifications of categorical map patterns that represent real landscapes.

With numerous metrics flooding the research arena, parsimonious metric selection became a concern; organizing metrics into groups through a multivariate factor analysis sought to identify redundancies among the multitude of related quantifications (Riitters et al., 1995). That work resulted in six broad dimensions for classifying metrics: (1) average patch compaction, (2) image texture, (3) average patch shape, (4) patch perimeter-area scaling, (5) the number of attribute classes, and (6) large patch density-area scaling. While this seminal work has been highly cited and used by many to set their bearings and methods within the complex domain of landscape pattern analysis, it was based on limited scope and the bending of assumptions for principal components analysis. Regardless, this and other similar approaches aim to reduce landscape complexity to simple numeric summaries that hopefully have intuitive or tangible interpretations if conducted at appropriate spatial scales and with valid assumptions.

In the broader inclusive domains of landscape ecology, ecology, conservation, land management, area protection, resource management, energy flows, hydrology, flooding, erosion, planning, and beyond, landscape metrics have been applied to quantify the generally unquantifiable (see the section on applications). There is a sense that making many measurements and having values that range from tangible to abstract is better than making decisions based on gut-feel and non-replicable decision-making tactics; thus, landscape metrics have become commonplace in studies in these cognate disciplinary areas. Challenges arise, however, when using traditional statistical tests for hypothesis testing or comparing groups that assume normal distributions, stationarity, and isotropy, all of which are often absent in spatial ecological data, thus raising questions about the validity of the tests.

In this chapter, we seek to set the development and use of landscape metrics into a broader context while highlighting areas of application, some important considerations for their use, and their known limitations, and summarizing some of the possible and viable solutions that have been proposed in the past decades to make the use and interpretation of these metrics as feasible as possible. Landscape metrics are incontrovertibly popular and unlikely to fade from popular use; the ability to represent the infinite complexity of a real landscape with a single number, or a suite of representative landscape metrics, is attractive from a simplicity perspective. This simplicity drives their prevalence in the research literature and applications, appealing to those seeking to cut through complexity in their analyses. The summaries can be easily tracked across geographic space or through time as indicators and measures upon which change direction and magnitude can be assessed more simply than if attempting to consider the full complexity of landscapes.

Metrics summary

Individual metrics and families of metrics

The peer-reviewed literature is loaded with papers that introduce, test, and implement landscape metrics (de Smith et al., 2018). Our informal count of metrics nears 300 across all possible vector, raster, and scope variations and derivatives (Baker and Cai, 1992; McGarigal and Marks, 1995; Haines-Young and Chopping, 1996; Uuemaa et al., 2009). This number bears little absolute meaning except to indicate that means for quantifying many aspects of landscape structure and pattern have been proposed, and there is a high likelihood of redundancy, or at least similarity, among subsets of published metrics. We alluded to attempts to organize metrics under broad umbrella terms in the section on Why landscape metrics? (based on McGarigal and Marks, 1995; Riitters et al., 1995) and elaborate on those here.

The quantification of landscape spatial pattern has traditionally been split to include non-spatial (composition) and spatial (configuration) properties (Gustafson, 1998); this important topic was recently revisited in Gustafson (2018) and Riitters (2018). While some efforts to separate composition and configuration effects exist (Machado et al., 2018; Bosco et al., 2019), such separation is even less frequent using a common measurement framework (Remmel and Csillag, 2006), and the two components have been shown to be inexorably linked (Remmel, 2009). The divisions can then be further subdivided, or informative crossovers among them may be determined. This intuitive division is also noted when humans assess and compare spatial patterns. People typically identify and more readily detect differences related to pattern composition (what thematic classes are present and the relative proportions of their cells) rather than configuration (how those cells are distributed in space) (Csillag and Boots, 2005). The widely popular FRAGSTATS software (McGarigal and Marks, 1995; McGarigal et al., 2012) organizes metrics into six families, and we preserve that classification in the following discussion; but keep in mind that the spatial scale of representation and the landscape extent will affect each landscape metric in individual ways (Saura and Martinez-Millan, 2001; Saura, 2002, 2004).

In an attempt to organize the realm of landscape metrics into a common framework for standardizing language, identifying options, and reminding users of key influencers, we produced Figure 15.1 to provide context for the remainder of this chapter and beyond. For any landscape metric, we use broad family groupings drawn from the literature to classify metrics

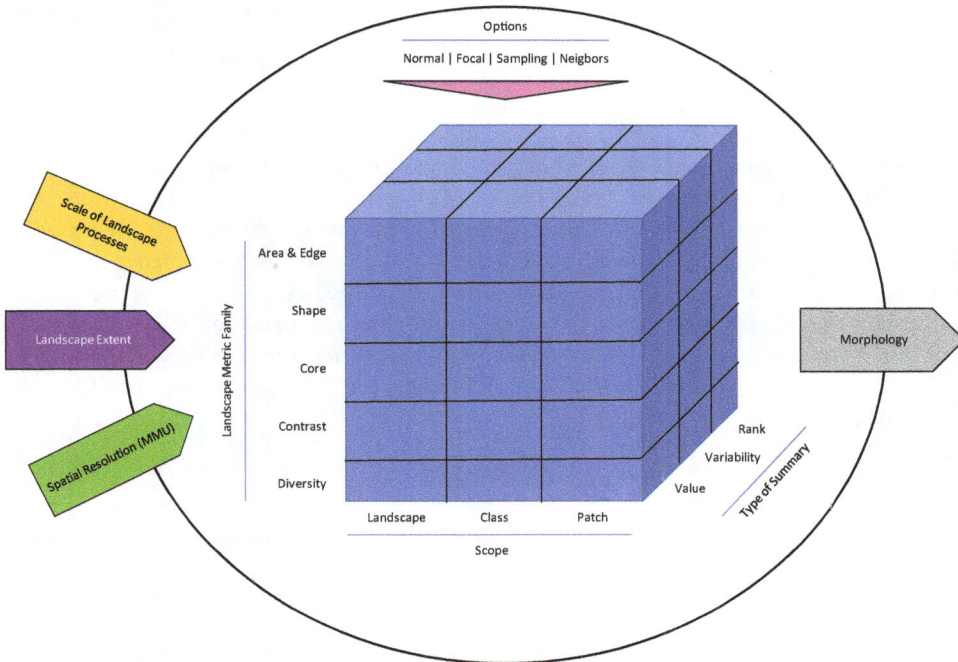

Figure 15.1 Conceptual mapping of landscape metric options and influencers.

that measure similar characteristics of landscape pattern. These include measures of area and edge, shape, core, contrast, or diversity, though other groupings have been presented (Giles and Trani, 1999). Each metric will exist in one or more scopes (landscape, class, or patch), identifying the specific subset of a landscape over which the metric will be computed. Some metrics by their definition cannot be computed in some scopes. Further, depending on specific metrics, it is possible to compute either the metric value or measures of that metric's variability across the landscape, or to rank the metric relative to others computed across the landscape. In terms of options, the landscape metrics can be computed in normal mode, i.e., as intended by the scope, but it is also possible to use focal (moving) windows (Hagen-Zanker, 2006) to compute localized values of these metrics or to rely on landscape sampling for computation (Hassett et al., 2012). The figure also identifies the spatial resolution, landscape extent, and the scale of the landscape processes measured as influencers on the computed metrics (see also Chapter 3). Finally, there is an evolving body measuring the mathematical morphology of landscape patterns that deviates from this core set of landscape tools.

Area and edge metrics

This comprehensive grouping deals with those metrics focused on various measures of patch size and the amount of edge present that demarcate the patches comprising the landscape of interest (e.g., Table 15.1). Measures of area invoke an understanding of absolute and relative composition, while measures of edge begin to indicate the degree of parceling or even fragmentation (Fahrig, 2003; see also Chapter 6) within the landscape. These metrics can be presented as first-order actual measures of area or length in map units or as rankings of specific landscape entities within the context of the larger landscape. Similarly, a suite of second-order metrics that capture the variability (or alternately, a ranking) of those measures is also available.

Shape metrics

Measures of shape rely on the assessment of patch boundary geometry and relate to the overall complexity of a landscape's structure (e.g., Table 15.2). For landscape processes that rely on edges or edge-effects, a quantification of edge complexity can provide insight into ecotone processes such as light penetration, seed dispersal, wildlife space utilization, foraging, and predation success. Shape is also often associated with the identification of anthropogenic activity (straighter and simpler boundaries) relative to natural processes (irregular, less geometric boundaries) (Lourenço et al., 2019). Attempts to quantify the abruptness of boundaries have also been presented but have been shown to be most successful if landscapes are already more patchy than not (Bowersox and Brown, 2001).

Table 15.1 Examples of area and edge metrics

Scope	Examples of metrics
Patch	Patch area, patch perimeter, radius of gyration
Class	Total class area, percentage of landscape, largest patch index, total edge, edge density
Landscape	Total area, largest patch index, total edge, edge density

Table 15.2 Examples of shape metrics

Scope	Examples of metrics
Patch	Perimeter-area ratio, shape index, fractal dimension index, ShrinkShape, porosity
Class	Perimeter-area fractal dimension, perimeter-area ratio distribution, shape index distribution, fractal index distribution, linearity index distribution
Landscape	Perimeter-area fractal dimension, perimeter-area ratio distribution, shape index distribution, fractal index distribution, related circumscribing square distribution, contiguity index distribution

Table 15.3 Examples of core area metrics

Scope	Examples of metrics
Patch	Core area, number of core areas, core area index
Class	Total core area, core area percentage of landscape, number of disjunct core areas, disjunct core area density, core area distribution, disjunct core area distribution, core area index distribution
Landscape	Total core area, number of disjunct core areas, disjunct core area density, core area distribution, disjunct core area distribution, core area index distribution

Core area metrics

Core area is a theoretical construct that is defined by the extent of a patch that is at least a specified distance from any point on the edge of that patch (e.g., Table 15.3). The delineation of core areas is perhaps best imagined as an internal setback from the patch perimeter. Where patches are narrow, core area may not exist, unlike in areas where a patch is wide. Depending on the shape of a patch, pinch-points may create the formation of multiple disconnected core areas for any given patch. Since the determination of core areas is predicated on the definition of an edge width (or distance from the patch perimeter), the resulting core area(s) and associated metrics will vary based on this distance.

Contrast metrics

Contrast metrics seek to quantify the magnitude of difference between adjacent patch types at a specific scale (Kotliar and Wiens, 1990). This requires a weighting of contrasts based on user input, since not all contrasts can be considered equal with respect to species distributions or ecological processes present on the landscape. Contrast metrics describe the abruptness, severity, or substantive changes across landscapes. Interpretation of results can relate to assessing barriers to movement or the ease with which resources flow across a landscape (e.g., Table 15.4).

Aggregation metrics

Aggregation measures the likelihood of certain patch types to be spatially clustered or to form contiguous groupings (e.g., Table 15.5). This umbrella group includes measures of dispersion, interspersion, subdivision, and isolation, each of which broadly characterizes the texture of a

Tarmo K. Remmel and Scott W. Mitchell

Table 15.4 Examples of contrast metrics

Scope	Examples of metrics
Patch	Edge contrast index
Class	Contrast-weighted contrast index, edge contrast index distribution
Landscape	Contrast-weighted edge density, total edge contrast index, edge contrast index distribution

Table 15.5 Examples of aggregation metrics

Scope	Examples of metrics
Patch	Euclidean nearest neighbor distance, proximity index, similarity index
Class	Interspersion and juxtaposition index, percentage of like adjacencies, aggregation index, clumpiness index, landscape shape index, normalized landscape shape index, patch cohesion index, landscape division index, splitting index, effective mesh size, Euclidean nearest neighbor distance distribution, connectance
Landscape	Contagion, interspersion and juxtaposition index, percentage of like adjacencies, aggregation index, landscape shape index, patch cohesion index, landscape division index, splitting index, effective mesh size, Euclidean nearest neighbor distance distribution, proximity index distribution, similarity index distribution, connectance

Table 15.6 Examples of diversity metrics

Scope	Examples of metrics
Landscape	Patch richness, patch richness density, relative patch richness, Shannon's diversity index, Simpson's diversity index, modified Simpson's diversity index, Shannon's evenness Index, Simpson's evenness index, modified Simpson's evenness index

landscape. Here, dispersion refers to the distribution of patches of a specific class without explicit regard to any other patch types present on the landscape. This concept measures how dispersed the class type is within a landscape. A related measure, interspersion, measures the spatial inter-mixing of specified patch types to characterize how certain landscape classes mix on landscapes. Subdivision is similar to the measure of dispersion but explicitly includes the degree to which patch types are fragmented. Isolation adds a measure of distance to other patches as an extension of subdivision. It is often difficult to distinguish between these subtle differences on real landscapes, since highly subdivided patches tend also to be more isolated. However, as indicators, high interspersion leads to the juxtaposition of more land cover types, and this could be related to increased landscape diversity.

Diversity metrics

These metrics attempt to describe the richness and evenness of landscapes, which again separate the compositional and structural components of diversity, respectively (e.g., Table 15.6). Richness is a measure that captures the number and variety of patch types within a landscape, while even-ness characterizes their spatial distribution. Two common indices that measure diversity are the

Shannon's diversity index (Shannon, 1948) and Simpson's diversity index (Simpson, 1949). The former is more sensitive to the presence of rare patch types, indicating that the appropriate diversity index needs to be selected depending on the purpose for computing the landscape metrics (Nagendra, 2002). The Shannon index is based on information theory and quantifies the amount of information, in a mathematical sense, attributed to individual patches; the index is most useful when used in a relative sense to compare the same landscape through time. The Simpson index quantifies the probability that two randomly selected cells would be from different patch types. Note that while landscapes may be identified as being diverse or rich, this is based purely on the land cover and does not necessarily translate to, for example, species diversity in a community. Attempts to quantify species richness in heterogeneous environments have led to measures of specificity (Wagner and Edwards, 2001), which estimate the proportion of observed species occurrences within a given spatial unit and can be scaled to larger units in ways that simple measures of richness cannot.

Gradient surface metrics

While the PMM and its categorical landscape representations are conceptually simple, there have been pushes to consider landscapes as more complex gradient surface models (GSM) to allow for spatially varying and transitional states (Frazier and Kedron, 2017; Kedron et al., 2018). Connectivity and graph theory provide an intermediate step, relying heavily on weighted connections among patches across edges (Baranyi et al., 2011), but the extension to fully specified surface metrics aims to consider landscapes as continuously varying and hence, may better connect observed patterns with landscape processes. Surface metrics, which originate from the assessment of industrial surfaces to evaluate wear and friction of bearings, while interesting and innovative, still require clear ecological reasoning and substantive software development to facilitate implementation in landscape ecology (McGarigal et al., 2009). It is not inherently obvious how to interpret such metrics, which is partially due to the large amounts of continuous data that they provide, as opposed to compact summaries, but also due to lack of standardization (Lausch et al., 2015). Initial progress has been made with specific studies (e.g., Abdel Moniem and Holland, 2013; Gadelmawla et al., 2002; Wu et al., 2017), but the substantially longer history of using category-based metrics means that surface metrics require a suitable duration of evolution to reach a common level of utilization and understanding, which will be partially driven by practitioner uptake once these metrics are better understood and tested.

Tools

When the calculation of landscape metrics entered the realm of personal computing, tools were packaged into software packages and toolsets for GIS systems (e.g., GRASS GIS). Most computing environments were based on Unix or MS-DOS, requiring command line calls for argumentation. However, with time, tools migrated into graphical- and menu-based systems, greatly facilitating access to non-specialist users and proliferating the computation and reliance on these quantifications of landscape pattern (Vogt, 2019). The rising popularity of free and open source (FOSS) software is further making access to, tailoring of, and inclusion into larger scripts and programs easier across different platforms. Table 15.7 provides a listing of some major software tools that provide means for computing landscape metrics and for the morphological segmentation of landscapes, with additional options and context provided by Kupfer (2012). A more enhanced overview of tools can be found in Steiniger and Hay (2009).

Table 15.7 Common tools for computing landscape metrics and morphological segmentation

Software	Reference and access
FRAGSTATS	www.umass.edu/landeco/research/fragstats/fragstats.html (McGarigal et al., 2012)
r.le	http://wgbis.ces.iisc.ernet.in/grass/gdp/landscape/r_le_manual5.pdf
MSPA (Morphological Spatial Pattern Analysis) + GIS plugins for ArcGIS, QGIS, and R	https://forest.jrc.ec.europa.eu/en/activities/lpa/mspa/
ArcView Patch Analyst	www.cnfer.on.ca/SEP/patchanalyst/Patch5_2_Install.htm
ArcGIS Toolbox – Landscape Analysis Tools	www.biology.ualberta.ca/facilities/GIS/uploads/instructions/AVVectorMetrics.pdf, www.umesc.usgs.gov/dss.html, www.arcgis.com/home/item.html?id=36f9728a895e4f5386bdec68be6d08ac
SAGA GIS	www.saga-gis.org/en/index.html
SAM	www.ecoevol.ufg.br/sam/
PASSaGE	www.passagesoftware.net/
Idrisi	https://clarklabs.org/
GeoDa	https://geodacenter.github.io/
LISA	www.spatialanalysisonline.com/HTML/index.html?local_indicators_of_spatial_as.htm, https://sgsup.asu.edu/geodacenter-redirect (Anselin, 1995)
Join-counts	(Upton and Fingleton, 1985; Kabos and Csillag, 2002)
spdep in R	https://cran.r-project.org/web/packages/spdep/index.html
PatternClass in R	https://cran.r-project.org/web/packages/PatternClass/PatternClass.pdf) (Remmel and Fortin, 2013)
hdeco in R	https://cran.r-project.org/src/contrib/Archive/hdeco/ (Remmel and Csillag, 2006)
ShapePattern in R	https://cran.r-project.org/web/packages/ShapePattern/index.html (Remmel, 2015)
SDMTools in R	https://cran.r-project.org/web/packages/SDMTools/index.html (VanDerWal et al., 2014)
landscapemetrics in R	https://cran.r-project.org/web/packages/landscapemetrics/index.html
Motif in R	https://doi.org/10.1007/s10980-020-01135-0

While all landscape metrics can be presented in tabular format, those that are computed at class or patch scope can also be well presented in spatial (map) output format. The spatial output provides additional information about the distribution of metric values in geographic space and may form input to further geoprocessing or modelling. Moving window analyses are inherently spatial and intended to characterize the varying nature of metrics through space but can be troubled by elements of multiple counting due to the overlapping nature of moving windows. Since the questions we ask of landscapes can vary infinitely, the limitations of considering a patch to be contiguous may be limiting, particularly if assessing fragmentation processes where multiple parts of a former shape may need to be assessed together or if multiple disconnected parts form a higher level of landscape organization. In this context, Remmel (2018) provides the ability to specifically modify 'patch' membership rules prior to computing landscape metrics.

With the ability to easily compute a large variety of landscape metrics came the desire to compare landscapes based on these values. These comparisons took the form of statistical comparisons using *t*-tests and analysis of variance (ANOVA) and also spatial map comparisons

(Hagen, 2003). While troubles with landscape metrics have been identified (Li and Wu, 2004), methods to appropriately compare metrics remain complex and challenging to implement (Fortin et al., 2003; Remmel and Csillag, 2003), as they require the simulation of landscapes that can serve as neutral models, but these are further limited to being binary, stationary and isotropic landscapes. This is complicated by ecologists' desire to use hexagonal grids rather than rectangular ones to make connectivity and representation less ambiguous (Birch et al., 2007; White and Kiester, 2008; Pond, 2016), but software tools and their utilization for hexagonal grids are relatively rare compared with those using regular square grids (Adamczyk and Tiede, 2017).

While numerous spatial point and area patterns with kernel methods, spatial autocorrelation, join-count statistics, and geographically weighted and spatial regression, among other geostatistical tools, they do not fall directly within the scope of this chapter; we refer the reader to some of the many excellent papers and books written on these topics for further reference (Cressie, 1993; Bailey and Gatrell, 1995; Kabos and Csillag, 2002; Getis, 2007; Paez et al., 2002a,b; Dale and Fortin, 2014).

Raster versus vector metrics

The GIS world is broadly carved into broad families according to decisions surrounding data representation. While several families exist (and variations on them), raster and vector representations account for the bulk of data formats and thus also primary or most popular options for landscape representation. Arguably, even with vector data being primarily tied to the creation of topographic maps and the representation of discrete entities, it is raster data that tend to form the basis of most landscape analytical work. The reasons for this can be debated, yet the simplicity of the underlying spatial support and regularity of raster grids provides several motivations. First, the raster structure ensures that there are no gaps or holes; in other words, all ground areas will have representation and pixel-level detail to characterize spatially varying conditions. Second, the geometry is considerably simpler, and that makes describing perimeters, areas, neighbor-structures, and all associated calculations much less intensive than dealing with vector coordinates and irregular polygons (although error can be introduced by rasterization processes, particularly if the spatial resolution is too coarse relative to the scale of the objects being represented). Third, many relevant data are primarily produced in grid format (e.g., classified satellite images, spatial models), and maintaining work in a consistent data representational framework makes intuitive sense and ensures efficiency. Fourth, the minimum mapping unit (MMU) remains consistent across a dataset, and thus, there is no need to control for variable scale or shifting precision or to deal with vertex spacing issues that introduce variable complexity within a single dataset.

Corresponding landscape metric calculations have been compared between raster and vector data formats (Wade et al., 2003; Ramezani and Holm, 2011), but some metrics are better suited to one data format or the other, while this suitability remains influenced by sampling, scale, and MMU. When complex shapes are represented in vector format, specific measures of shape (Remmel, 2015) that cross spatial scales can be useful for quantifying and comparing shape complexity and identifying pinch-points (where shape thickness reduces to points of constriction). Not all landscape metrics, however, are easily computed across spatial scales or lend themselves to either data representational format.

Scales: landscape, class, patch

Landscape metrics are generally computed at one of three set levels (often referred to as scales): landscape, class, or patch (see also Chapter 3); each provides a differing denominator (spatial extent or partitioning) over which a landscape's pattern is assessed (Figure 15.2). Not all metrics

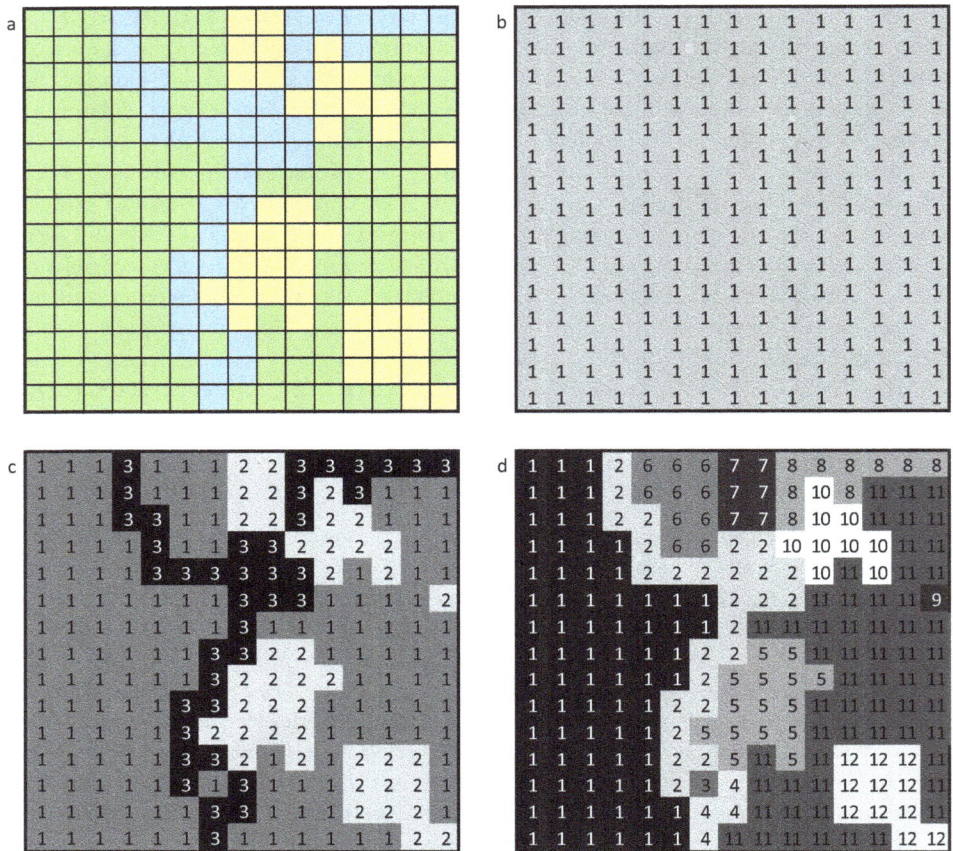

Figure 15.2 Depicting the scale of analysis for landscape metrics: (a) categorical three-class landscape, (b) landscape metrics produce a single value for the entire landscape, (c) class metrics produce one value for each aggregate class on the landscape, and (d) patch metrics produce a value for each contiguous patch, however defined, for the landscape. In this case, there are three classes in (c), and thus, there would be a metric value computed for each of the dark, medium, and light grey areas in aggregate. Twelve metric values would be produced at the patch level (d) assuming a 4-neighbor connectivity rule. See also discussion of moving windows and potential problems with them in section on tools.

can be computed at all levels, but many can be hierarchically nested or extended across these three levels (see Figure 15.1). The scale at which computations are focused should be a function of the research question asked or the reason for computing the selected metric(s). Landscape-level metrics assess the decided metric's value or state across an entire landscape extent (as defined by the extent of the dataset, excluding cells with no data). The result is a single value that summarizes the entirety of the provided landscape. While dramatically summarizing the complexity of landscapes, these broadest-level metrics can also mask trends, spatial variability, anisotropy, and non-stationarity, elements that may be crucial to understanding landscape processes.

Class-level metrics compute a metric for each represented class on the landscape. Thus, if the landscape comprises five land cover types (classes), a metric is computed for each of those five class types by aggregating all patches of a common class prior to computation. This method has the benefit of partitioning pattern quantifiers by class and thereby controlling for possible inequality among classes. For example, a landscape with many rivers may see the water class as relatively linear and sinuous compared with the other land cover classes.

Patch-level metrics compute an individual metric for each patch within a selected landscape. Obviously, as one moves from landscape- to class- and patch-level metrics, the number of results will generally increase (unless the landscape only comprises one large patch), and thus, the interpretation of results will become more arduous. These levels are hierarchically nested, and as one moves toward the coarser levels of assessment, values will represent greater averaging of values than at the individual patch level. Further, cell contiguity (i.e., neighborhood structures) can be defined by the 4- or 8-neighbor rules (the Rook and Queen case, respectively) and will adjust the rules used to determine whether two cells are to be considered spatial neighbors or not (Sawada, 1999).

Morphology

To move away from traditional and often abstract landscape metrics summaries of landscape spatial pattern, efforts arise to segment raster cells defining landscapes into mutually exclusive spatial feature classes. This concept of mathematical landscape morphology can be applied to binary landscape representations (Riitters et al., 2007; Soille and Vogt, 2009), where a class of interest is referred to as belonging to the foreground relative to background cells that are not the focus of the current assessment and can be regarded as the complementary area. A sequence of mathematical morphological operators processes the binary map to further characterize the foreground class into a fully spatial and intuitive result, where each cell representing the foreground is assigned to a single morphological class defined by its geometry and connectivity (Vogt et al., 2007; Ostapowicz et al., 2008; Soille and Vogt, 2009). A cell may be labelled as core, edge, perforation, branch, loop, bridge, or islet; all remaining cells are labelled as background. When the seven morphological classes are merged, they will exactly correspond to the foreground class that was processed (Figure 15.3).

The morphological elements represent intuitive and mappable spatial components of a landscape. They relate to a single land cover class (e.g., presence of some feature, class, or condition) and describe its component parts. The core defines all areas within a set edge distance from the nearest background pixel. Edges are the cells forming an outside boundary intermediate to the

Figure 15.3 Sample morphological segmentation of a binary landscape. The left panel is a binary input image with black representing the background and white the foreground. The right panel displays the MSPA output using a 1-pixel edge width and an 8-neighbor connectivity and transition turned on.

background and core cells; the widths of edges are specified a priori. Perforations are similar to edges but form the inside boundary to background enclosed by core pixels. Islets represent isolated patches that are not sufficiently wide enough to contain cores and hence, have no specific edge designations either. Branches, loops, and bridges are related in that they represent linear elements. Branches extend outward from cores or other linear features but do not connect anywhere. Bridges connect two or more core areas, while loops are connecting pathways emanating from a single core back to itself.

Fine-tuning the identification of the various morphological elements is possible by adjusting the four Morphological Spatial Pattern Analysis (MSPA) parameters: a) edge width (to control the width of patch boundaries); b) the connectivity rule (Sawada, 1999) to determine whether 4- or 8-neighbor connectivity is implemented; c) the transition rule (to control whether connecting features traverse edge or perforation classes to connect to cores or not), and finally d) intext, to add an additional layer of MSPA classes for features inside the holes existing within core areas.

Transition controls a purely visual state of the two connecting feature classes, bridge and loop, determining their pathways to core areas. Connector class pixels must traverse boundary classes to reach core areas, and hence, transition pixels are those boundary pixels that are also connecting pathway pixels. 'Transition ON' means showing those pixels as connecting pixels, while 'Transition OFF' will show them as respective boundary pixels. Figure 15.3 shows transition pixels loops and bridges. Switching to Transition OFF would provide the same numerical result but with visually pleasing closed edges at the expense of hiding some parts of the connecting pathways.

Fine-tuning the MSPA parameters may result in up to 25 individual classes (see the MSPA Guide[1]), providing the flexibility to mimic specific landscape processes as required. Influencing all segmentations is the original spatial resolution of the data, as this will affect the level of detail with which the foreground landscape class is mapped, the precision with which edge width can be defined, and the relative likelihood of elements existing (e.g., there is a higher likelihood of perforations as spatial resolution becomes finer; Ostapowicz et al., 2008).

From an ecological perspective, much attention has been given to ecotones (edges) and fragmentation (patchiness) and having means to better isolate and measure these landscape elements may improve assessments of ecotone abruptness (Bowersox and Brown, 2001; Arnot et al., 2004) since each cell contributing to being an edge is identified and mapped whereas landscape metrics summarize more generally over patches, or abstractions and statistical descriptions of what an edge is. Having morphological element layers that classify each cell into mutually exclusive element classes permits subsequent zonal analyses (Shan et al., 2003) to summarize related attribute values within specific morphological class areas. Such segmentations would facilitate studies that compare conditions constrained to certain morphological classes (Harper et al., 2015), allowing automated analyses to be scripted or to standardize the definition and detection of edges.

Using metrics to compare landscapes

Statistics versus indicators/diagnostics

One of the challenges with landscape metrics is the absoluteness with which they are often interpreted or used for assessing landscape changes. We believe that landscape metrics would be more effectively considered as diagnostic indicators rather than absolute quantifications of landscape state. The subtle difference here is critically important; at the heart, it is important to realize that multiple landscape patterns can yield identical values for one or more landscape metrics (Remmel and Csillag, 2003). Given this reality, failing to detect a change in a landscape metric value does not necessarily indicate that a change has not occurred, or vice versa, a landscape metric value difference may not equate to a substantive landscape alteration. The key point here is that while landscape metrics *may* indicate that something has happened, this is not always the case. Similarly, most landscape metrics do not scale linearly (Wu et al., 2002; Wu, 2004), and the evaluation of statistical significance is fraught with complexity, complicating interpretation. Furthermore, it is known that landscape metrics are sensitive to extent, spatial autocorrelation, and grain size; thus, if any of these elements differs between compared landscapes, then it becomes difficult to tease out whether these elements are responsible for the observed landscape metric changes or whether the pattern actually differs due to true changes on the ground.

Estimation of composition (e.g., relative class proportions) and configuration (e.g., spatial autocorrelation) from landscape patterns permits replicate landscapes to be simulated that have consistent constraints on the varied patterns that they portray. This approach allows the variability of landscape metrics to be assessed relative to fixed parameters. The challenges here are related to requiring binary, stationary, and isotropic landscapes (Fortin et al., 2003; Remmel and Fortin, 2013), but decisions regarding landscape modifications to control certain landscape metric value ranges become possible. The constraints and complexity of even three land cover classes make the number of compositional and configurational parameters unwieldy. However, composition and configuration are not purely independent (Remmel, 2009), and estimating these parameters is not necessarily a perfect science (e.g., as a landscape becomes increasingly dominated by one class, the number of possible configurations becomes increasingly constrained).

An often-overlooked aspect of using landscape metrics for quantifying landscape pattern or for assessing landscape changes is deciding what will form the focus of the analysis. Typically, it is land cover change that is measured or quantified in terms of both the form and the magnitude of changes. However, landscape persistence (Pontius et al., 2004) is the condition of land cover not changing, and this too may yield important insights into landscape processes, often representing the dominant state of an examined landscape (e.g., how much area remained unfragmented?). When thinking about landscape metrics of all sorts, it is critical to consider all aspects of how the pattern may be best quantified or assessed for changes. This is particularly important in attempts to link observed patterns with their underlying spatial processes (Fortin et al., 2003; Csillag and Boots, 2005; Turner, 2005).

Limitations and external influences on landscape metrics

The literature has identified several influences on the outcome of landscape metrics that have little to do with the actual pattern on the landscape (see Figure 15.1). The Modifiable Areal Unit Problem (MAUP) is known to influence most summaries of quantitative values as the partitioning of geographic space changes, and landscape metrics are not exempt from being sensitive to changes in areal partitioning (Jelinski and Wu, 1996). While this underlying condition impacts other influencers that follow, attempts have been made to examine the sensitivity of landscape metrics to these environmental changes in general (Baldwin et al., 2004) and also more specifically to landscape factors (Peng et al., 2010; Wei et al., 2017). The extent of landscapes is a key consideration (Saura and Martinez-Millan, 2001), since even an extra row or column of data can lead to varied landscape metric outcomes. As an extension of the extent, the spatial resolution (Saura, 2004), or more broadly the MMU or what is sometimes referred to as the grain size (Saura, 2002), will affect landscape metric outcomes (Wickham and Riitters, 2019). This stresses the importance of comparing landscapes with identical extent and spatial resolution to control for these basic conditions.

Given the numerous combinations of these influencers and the underlying complexity of processes leading to landscapes, the need to consider isotropy and stationarity in determining how landscape patterns will be simulated to produce replicates for assessing metric variabilities is critically important. The interplay between composition and configuration is complex (Remmel, 2009) and makes it necessary to develop statistically valid and stable null distributions against which landscape metrics can be compared (Remmel and Csillag, 2003). Such rigor allows questions to be asked that grapple with assessing whether a given landscape pattern could have arisen due to chance alone or whether two landscape patterns differ beyond a significant statistical margin. Neither question is trivial to answer without first building null models, since normal distributions and the use of t-tests and ANOVA are simply not sufficient in these cases.

The map comparison problem

One focus of landscape metric analysis is in map comparison, specifically, the assessment of spatial similarities and differences between or among categorical landscape patterns (Trani and Giles, 1999; Loran et al., 2018). In search of linkages among landscape metrics and spatial landscape processes, the role of stochastic simulation of landscape patterns has created opportunities for comparing landscape metrics against null models and treating observed (or simulated) landscapes as realizations of processes (Fortin et al., 2003). With enough understanding of landscape metric outcomes for given landscape structures (given composition and configuration), it becomes possible to create inferences about spatial processes leading to observed patterns. This is further understood to mean that observed changes in pattern could result from a changed spatial process, and thus, landscape patterns, processes, and map comparison are all interrelated and can be approached from pattern-based or process-based comparisons (Csillag and Boots, 2005; Boots and Csillag, 2006).

Primarily in remote sensing, but also in ecological studies, the reliance on Kappa statistics dominated for some time when assessing accuracy, or alternatively using pixel-to-pixel cross tabulations to compare maps, but limitations of these metrics have been identified by many, and the articulation of their definitive demise is documented in a study that identifies serious limitations of five different Kappa summary metrics (Pontius and Millones, 2011). Reliance on measures such as symbolic entropy (Ruiz et al., 2012) or feature-based comparisons and shape (Dungan, 2006) is an important alternative, as is the need to assess the statistical significance of map differences or changes through time (Boots and Csillag, 2006). This conceptual need spurred workshops and a hive of activity nearly a decade ago that culminated in a special issue of the *Journal of Geographical Systems* (e.g., Hagen-Zanker, 2006; Hargrove et al., 2006; Stehman, 2006; White, 2006). Related methods incorporate fuzziness (Hagen, 2003) and conditional entropy to measure aspects of composition and configuration in common units (Remmel and Csillag, 2003), but the perfect separation of composition and configuration is not simple or even fully possible. Still, it is possible to assess the variability of possible patterns given a coincidence matrix that enumerates all the joint pixel classes between two maps (Remmel, 2009), providing some insight into the link between composition and configuration. Overlap statistics have been explored for map comparison, too, and represent means for assessing spatial coincidence that can be compared against null models to permit statistical testing (Jacquez, 1995; Remmel and Perera, 2002).

Applications

Despite the challenges outlined in the previous section, there is a wide range of valuable applications using landscape metrics to compare landscapes or track changes in specific landscapes through time. Applications share an attempt to link observed spatial patterns to underlying processes, normally for some combination of monitoring change through time, linking pattern to desired states, or estimating something about the processes contributing to the observed patterns. For reasons described earlier, a general relationship between pattern and process is still elusive, but improved understanding of the behavior and limitations of landscape metrics has fueled their expanded use and improved their usefulness.

Perhaps not surprising given their roots in landscape ecology, many examples of the application of landscape metrics lie in the realm of ecological questions (e.g., how habitat availability or biodiversity is impacted by anthropogenic alterations to land cover patterns), but other areas of study have also emerged. Uuemaa et al. (2009) found 478 unique articles in the Web of Science from 1994 to October 2008, using either 'landscape metrics' or 'landscape indices' as a search term, and classified the applications into the following categories: biodiversity and habitat analysis; estimating water quality; evaluation of landscape pattern and change (in response to disturbance); urban landscape patterns and road networks; landscape aesthetics; and management and planning. Uuemaa et al. (2013) provide a more hierarchical classification of landscape analysis based on categories from De Groot (2006) to review trends in the use of metrics.

As a simple extension in time to Uuemaa et al. (2009)'s analysis, we conducted a similar search from 2009 to 2018 in the Web of Science Core Collection.[2] This yielded 1,480 articles, demonstrating a continued growth of interest in research activity using these tools. Several application areas in this updated review did not neatly fall into any category in the previous survey, including papers exploring relationships between landscape metrics and: climate/microclimate (Faye et al., 2016; Kong et al., 2014), air and noise pollution (Lee et al., 2014; Lin et al., 2018), ecosystem services (studied broadly, not focusing on a specific ecosystem service, e.g., Syrbe and Walz, 2012), and geomorphology and soil erosion (Kot, 2018; Ren et al., 2013; Szabo et al., 2017), and even application of landscape metrics in art restoration (Henriques and Goncalves, 2010). Based on this simple update, the following sub-sections discuss applications of landscape metrics in slightly broader and often overlapping classes:

- Human-dominated landscape processes and impacts;
- Biodiversity and habitat analysis;
- Ecosystem processes and services;
- Spatial aesthetics; and
- Management and planning.

Human-dominated landscape processes and impacts

Much of our interest in landscape pattern is due to the impact humans are having on our environment, part of which involves changes in landscape patterns and the interplay between landscape pattern and process. Our modifications to land cover and land use include those caused by urbanization, agriculture, forestry, mining, and energy production. Changing landscape patterns caused by these transitions are a dominant subject of research using landscape metrics. For example, the indicator Landscape Mosaic (Riitters et al., 2009) classifies a given location according to the relative proportions of the three land cover types, Agriculture, Natural, and Developed, in a neighborhood surrounding that location. By accounting for the interplay of different land cover classes, this indicator provides a measure of the human impact on the landscape.

Disturbance regimes have been recognized for at least two decades as part of natural ecological processes and contributing to the patch mosaic model of landscape heterogeneity (Turner and Gardner, 2015); however, disturbance regimes are shifted by human activities, extending the use of landscape metrics analysis from natural ecosystem processes to landscape changes caused by human activities. Given that disturbance occurs along a spectrum of causation from natural to human processes, this discussion could overlap with the section on biodiversity and habitat analysis, so here, we highlight examples that are most strongly associated with human activities.

Impacts of urbanization dominate this application category in the landscape metrics literature; for example, Uuemaa et al. (2013)'s review identified evaluation of land use/cover (LULC) patterns or changes in urban/urbanizing areas as the most common category (57/116) of non-generic LULC studies, and in our Web of Science search from 2009–2018, 24 out of the top 50 most highly cited papers had either urban, urbanizing, or urbanization in their titles (both highly conservative indicators of the degree to which urbanization is implicated in the research). Urban sprawl can remove and fragment natural habitat, which is often characterized using landscape metrics, and studying processes of urbanization through time has been made particularly accessible by the availability of open imagery archives (e.g., Taubenböck et al., 2009; Araya and Cabral, 2010; Ramachandra et al., 2012; Yue et al., 2013). Jaeger et al. (2010) note that studies of urban sprawl can be difficult to compare, due to variable definitions of sprawl, and they define 13 criteria to assess the consistency and reliability of particular indicators.

Pattern and texture within urban areas and their changes as we redevelop cities have also been studied with landscape metrics (Herold et al., 2003; Wu et al., 2011; Reis et al., 2016), as well as the implications of these patterns for non-human species. Many studies in the past decade have concentrated on how urban heat islands and/or areas of mediating green cover are structured in space and time within cities (e.g., Li et al., 2011; Zhou et al., 2011; Connors et al., 2013). Others examine how landscape pattern in urban areas impacts species of concern – i.e., the focus of the following subsection in this chapter, but within urban environments. Muderere et al. (2018) performed a systematic review to characterize all the urban ecology studies they could find in nine landscape ecology journals. Interestingly, they found that very few studies employed remote sensing data or techniques, whereas many of the studies of urban form or processes themselves, identified earlier, do employ remote sensing; the disconnect may invite further investigation or collaboration between the research communities. There have also been calls for an increasing emphasis on multi-taxa and cross-region studies (Muderere et al., 2018).

Forested regions generally involve larger areas and extensive distributions, often in remote areas; thus, research in this area benefits from the trend towards increasing availability of temporal archives of imagery at no cost to users (Wulder et al., 2012). Landscape pattern analysis in forested areas can examine general changes to the structure of forest patches (often with respect to forest habitat, further discussed in section titled Biodiversity and habitat analysis) or landscape changes caused by specific processes such as fire (Remmel, 2018), pest infestation (Lustig et al., 2017), reforestation (Sitzia et al., 2010) or deforestation (Ignacio Gasparri and Ricardo Grau, 2009), or infrastructure associated with logging and other human activity, such as roads (Kittle et al., 2015).

Uuemaa et al. (2013) noted a shortage of studies applying landscape pattern analysis in agricultural areas and called for increasing attention to the impacts of policies on changes in landscape heterogeneity. The review we conducted for this section found that 21 of the top 100 most-cited papers since 2013 assessed landscape pattern with at least a partial focus on agricultural areas, so perhaps that shortage has been addressed. This expanded interest seems to be partially due to concern about the expansion of agriculture in areas like the Amazon basin (e.g., Silva et al., 2018) and a recognition (Marull et al., 2018) that species conservation and ecosystem service preservation are unlikely to be achieved by protected areas alone. More specific areas that have attracted increased interest in recent years include examining relationships between landscape spatial patterns and biodiversity or ecosystem services (see sections on biodiversity and habitat analysis and spatial aesthetics).

Biodiversity and habitat analysis

There is a long history of studying species richness or habitat preferences with respect to landscape composition and/or configuration as measured with landscape metrics. These often aim to identify aspects of landscape heterogeneity which have been introduced by human changes to the landscape, as described in the previous section, but also how such findings, coupled with studies in areas with relatively little human impact, could be used in a management context to promote biodiversity or habitat conservation for species of concern (see also Chapter 19). Generally, for a given habitat, larger and more heterogeneous patches support more species and often greater abundances than smaller, less heterogeneous patches (Turner and Gardner, 2015). Beyond this relatively simple and intuitive observation, however, a wide variety of biodiversity responses have been observed with respect to edge versus interior habitat and the landscape context within which patches are situated. This involves debates on fundamentals of landscape ecology that are beyond the scope of this chapter, but some highlights are noted here with links to other sections of this volume.

Landscape heterogeneity can be measured structurally, without any reference to the needs or preferences of any specific species or group, or functionally, where land cover assessment includes identification of differing species or group dependencies on resources in the landscape (Fahrig et al., 2011). In some ways, a functional approach examining a single species' habitat preferences with respect to landscape pattern is the simplest approach, and there are many such studies. However, contrasting life histories and resource requirements mean that an ideal landscape for one species may not benefit overall biodiversity, and there are likewise very many structural studies testing for impacts of landscape patch characteristics and/or context on different measures of biodiversity. In either approach, finding generalities becomes the challenge.

Seemingly contradictory findings from individual studies may be explained by a combination of variable levels of species response to patch versus surrounding landscape characteristics (Thornton et al., 2011) and inherited assumptions from the adoption of island biogeography principles to conservation biology, including the adoption of a binary classification of land cover into habitat and uninhabitable areas (Prugh et al., 2008; Franklin and Lindenmayer, 2009). Prugh et al. (2008) looked at presence/absence data of 1,015 different populations across multiple taxonomic groups and found that patch area and isolation were generally poor predictors of species occupancy in a patch. They called for greater emphasis on examining the surrounding landscape matrix, which contributed to Thornton et al. (2011)'s study to seek generalities about the role of scales of heterogeneity on species distribution and abundance. They reviewed 122 focal patch studies incorporating data on 954 species, looking at heterogeneity measures within patches, ranges of patch sizes, and compositions of the surrounding landscape. Across diverse taxa, many species responded to measures at all three scales, and more than half of the studied species responded to at least one landscape context measure.

Analysis of negative impacts of human alteration of ecosystems on biodiversity often cites fragmentation of habitat as a primary concern, but this has elaborated into questions of the relative importance of habitat amount or configuration. Fahrig (2003) reviewed 100 studies and argued that many of the differing responses in that work could be caused by failing to separate the two components of heterogeneity, therefore overestimating the effect of habitat configuration with respect to amount of habitat. She showed that a large proportion of studies identifying fragmentation effects as influencing biodiversity failed to control for the amount of habitat; thereby, they extrapolated from patch size or isolation studies into conclusions about landscape-level habitat fragmentation. A later review (Fahrig, 2017) filtered out studies that did separate out habitat amount and found that most (76%) of the significant relationships found a (usually weak) positive impact of habitat fragmentation, and of the 24% that showed negative effects, no conditions (e.g., biomes/size of home range/degree of competition) could be found across which this remained consistent. These findings have led to a considerable amount of debate (Hadley and Betts, 2016; Fletcher et al., 2018; Fahrig, 2019), which for the purposes of this chapter emphasizes the importance of carefully and explicitly outlining what aspects of landscape heterogeneity are being measured, including the choice of individual metrics and the sampling scheme in which they are applied. Fragmentation literature also suffers from biases towards research in North America and Europe, a concentration on forest environments, and a taxonomic focus on birds, mammals, and vascular plants (Fardila et al., 2017).

There are many calls for caution in using landscape metrics within such biodiversity and habitat analysis, corresponding with general concerns summarized in previous sections of this chapter, and presentations of potential solutions. For example, scale-dependence of the relationships has been well established, leading to multi-scale approaches (e.g., Schindler et al., 2013). Eigenbrod et al. (2011) identified 'statistical pitfalls' able to reverse the direction of inferred relationships between species response and landscape structure. Pasher et al. (2013) outlined potential approaches to select landscapes for studying species responses to reduce the likelihood of such pitfalls. Those methods have been employed in several studies of relationships between farmland heterogeneity and biodiversity, which, among other things, have demonstrated a negative relationship between mean field size and multitrophic biodiversity, and a positive impact of the number of crop types on biodiversity, across 435 landscapes in 8 contrasting regions in Europe and Canada (Sirami et al., 2019).

Highly fragmented landscapes are often associated with invasive species, but the generality of this relationship is challenged by individual species showing contrary ecological relationships (Rodewald and Arcese, 2016). This variability can sometimes be exploited by identifying specific traits that impact how organisms react to landscape pattern – for example, a range of successful forest pest models include pest-specific parameters of climatic sensitivity and dispersal rates (Ferrenberg, 2016). Alternatively, to get away from the problems with species-specific responses, some studies elect to concentrate on functional traits (e.g., Barbaro and Halder, 2009; Kissick et al., 2018; van der Walt et al., 2015). The biodiversity literature contains many examples of research related to patch isolation and fragmentation (e.g., Tischendorf and Fahrig, 2000; Bender et al., 2003; Fahrig, 2003; UNCBD Biodiversity Source Book[3]).

The degree to which human activities cause differences in relative levels of heterogeneity within land cover categories may be an important factor in the success of attempts to link spatial pattern to processes. Landscapes with low human influences tend to have greater heterogeneity in time and space than those that have been heavily modified, and this has led to suggestions that metrics that retain this heterogeneity (e.g., gradient metrics – see the section on gradient surface metrics) may be more appropriate (McGarigal et al., 2009; Lausch et al., 2015). Biodiversity monitoring employing the Spectral Variation Hypothesis (Palmer et al., 2002; Rocchini et al., 2010) seems to be an independent but conceptually similar approach linking pattern and process, and it is interesting to note that Duro et al. (2014) reanalyzed the data from Fahrig et al. (2015), and found that there were stronger empirical relationships between the biodiversity measures and derivatives of the continuous spectral data than with the latter study's categorical land cover data.

Ecosystem processes and services

While ecosystem processes are inherent in most other application areas discussed previously and later in the chapter, there are also examples where landscape influences are explicitly studied with respect to specific processes of interest. The recognition of landscape influences in geomorphology and hydrology likely accounts for developments linking landscape metrics to research in these fields.

The combination of landscape ecology with microclimatology and climate change science is an emerging theme; the proliferation of urban ecology studies that have included spatial consideration of urban heat island effects and the mediating impacts of green space has already been mentioned (see the section on human-dominated landscape processes and impacts), but the broader field includes consideration of concepts such as thermal niches, refugia, thermal heterogeneity or continuity, climate change, and species range shifts. Challenges and possible future directions in this work include the mismatch between typical climate change datasets and the ecological processes of concern, including the fact that assumptions about the thermal habitat needs of species based on past distributions typically provide no information about the heterogeneity of thermal conditions within the ranges, whereas that heterogeneity is likely to be very important to how species can adapt to changing climatic conditions (Nowakowski et al., 2018).

More broadly, ecosystem services, the benefits that humans receive from functioning ecosystems, have become a well-established concept in environmental management, policy, and research, largely thanks to the influence (Daily et al., 2009; Turner and Gardner, 2015) of the Millenium Ecosystem Assessment (2005). The complexities of feasibly and defensibly understanding ecosystem dynamics in large complex systems to aid decision-making in a timely fashion are recognized, but so are the needs to consider such complexity to evaluate how human decisions change ecosystem services (Villa et al., 2014). There is a solid spatial framework to assessing ecosystem services, and landscape metrics have been advocated as relevant tools for characterizing landscape heterogeneity to develop landscape service indices and help delineate relevant place-based landscape units for services that have strong spatial structure (Syrbe and Walz, 2012). A recent meta-analysis on studies of effects of landscape metric analysis on several landscape services reiterated the need to consider the relevance of landscape structure (both composition and configuration) in management and decision-making, and also showed that the importance of the landscape effect varies widely by type of ecosystem service (Duarte et al., 2018).

Spatial aesthetics

The visual aspect of landscapes has been studied in the context of design and for the provision of visual ecosystem services, a specific sub-category of the work summarized earlier. The importance of landscape aesthetics as a cultural ecosystem service has been identified (Millenium Ecosystem Assessment, 2005). Common aspects to the visual and ecological characteristics of landscapes have been identified; conceptual visual landscape descriptors from Tveit et al. (2006) have been mapped to corresponding concepts in landscape ecology, with relevant landscape metrics identified (Fry et al., 2009). For example, the concept of landscape complexity in the visual realm includes the dimensions of land cover diversity, which is also an important part of habitat heterogeneity in ecology measured by landscape metrics such as the patch heterogeneity index. Uuemaa et al. (2013) identified a shortage of research objectively testing the potential of landscape metrics as visual indicators in different contexts and where landscape metrics are tested for specific aspects of visual character instead of overall attractiveness. At least one highly cited study has used a combination of objective evaluation employing landscape metrics alongside a validation test using a subjective study of individual users' perceptions of landscape photos, showing that there was good agreement between the subjective ratings of beauty and the landscape metrics, although this varied with the degree of professional accreditation and the spatial origin of the research subjects (Frank et al., 2013). Potential challenges to generalizable findings include the degree of human influence on a landscape and the degree to which humans living in that area have a subjective preference for that kind of landscape. As another example, studies of scenic beauty in the European Alps, where historically, human influences on landscape pattern have been strong, and abandonment of farmlands is seen as a negative influence (Schirpke et al., 2013), are in a very different context than more remote and/or pristine mountain landscapes elsewhere on the planet.

Management and planning

While many of the applications described can have planning or management in mind, some research explicitly develops tools or frameworks to support or plan for the management of human–environment systems. Frank et al. (2012) combined a cellular automaton-based landscape change model, a multi-criteria evaluation, ecosystem service estimates, and landscape metrics to build a planning tool to evaluate afforestation scenarios. They concluded that without the inclusion of landscape metrics (e.g., effective mesh size and hemeroby index) to assess spatial context, their system overestimated the delivery of ecosystem services. They also noted the sensitivity of the system to assumptions made about specific land cover – ecosystem service – landscape metric associations, and spatial and thematic resolution dependencies of the landscape evaluation, challenging the model's transferability to other regions. Other uses of landscape metrics within planning tools include evaluating fragmentation of agricultural lands as an indicator of fragmentation or differentiation of agricultural practices and cover through time (Giardino et al., 2010) and the simulation of forest fire spread in a model designed to test forest management plans (Gonzalez-Olabarria and Pukkala, 2011). Planning and management of protected areas is another area where landscape metrics are commonly employed to assess protected habitat quantity and quality. Limitations of landscape metric analyses, as discussed earlier, are frequently acknowledged in these studies, but this is balanced against the advantages of standardized analyses with available data and the absence of better alternatives (Townsend et al., 2009; Banks-Leite et al., 2011). Such work can be extended in space and time to assist with planning entire protected networks incorporating climate and land use change scenarios (Costanza and Terando, 2019). Ecosystem restoration may also benefit from landscape metric analysis; a review of empirical restoration studies between 1997 and 2011 found that 84% of the studies concluded that landscape context was an important influence on restoration processes (Leite et al., 2013).

Where are things headed?

As with most academic domains, it makes sense to occasionally step back and examine the larger field of developments, to take stock, and to assess where things are headed. The realm of landscape metrics, map comparison, and ecological assessments based on pattern quantification has been progressing with varied emphasis and vigor for decades, leading to a large stock of tools and disconnected techniques that still rely heavily on user context, experience, and best practices. We have yet to agree on consistent workflows and clear guidance surrounding the scale, extent, or appropriateness of specific minimum mappable units. Assumptions are generally ad hoc and site- or study-specific rather than being clearly defined as, for example, in approaches to non-spatial hypothesis testing. There is a deep and growing need, and hence an opportunity, to comb through the numerous metrics, null models, and landscape simulation approaches to seek consistency and robustness in how landscape patterns are quantified, compared, and more generally described.

Complicating matters is the ongoing advancement of technology that continually provides higher-resolution imagery (whether increased spatial, temporal, spectral, or radiometric resolution), which translates into larger datasets with greater precision and detail. Each of these aspects stands to complicate summaries and comparisons and to potentially extend too far in capturing local variation that is based less on land cover than on illumination and shadow effects or the natural variability of a land cover class. Such increases in detail emphasize a need to be even more diligent in defining what is being measured or compared. Parallel to this rapid increase in data resolution is the multiplicative reality of newer data types that are gaining popularity and ubiquity in acquisition (e.g., LiDAR point clouds, video, three-dimensional [3-D] surfaces).

Quantifying and comparing landscape patterns is likely to expand to include landscape structure that incorporates the third dimension and time as the fourth dimension, and with a greater uptake of metrics for gradient surfaces (Frazier and Kedron, 2017). Within this context, 3-D building structures and patterns (Liu et al., 2017) and the development of morphological definitions that extend beyond the planar dimension will become increasingly necessary. Similarly, the partitioning of 3-D space into variably sized voxels and then using that new spatial structure as the base on which new metrics will be developed seems like an intuitive future extension for landscape metrics. Some of these ideas dovetail with computer science developments, computer vision, object recognition, and the tracking of change through varied intervals of time. Recent developments by Nowosad and Stepinski (2021) introduces the integrated co-occurrence matrix (INCOMA) as a signature for representing multi-thematic categorical patterns which can be used to compare and search for patterns within datasets. This relates to the work of Remmel (2020) that characterizes and compares patterns based on empirical distributions of all first-order orthogonal neighbour configurations, permitting simplified pattern comparisons and boundary detections.

We foresee a great opportunity for artificial intelligence and extensions of models such as cellular automata (Li et al., 2013) to play a role in learning and classifying complex patterns across multiple metrics, scales, and data products.

While the preceding remarks point to technical improvements in how we measure things, there also exists a great body of work that needs to advance to connect patterns with processes; a formidable task that is still largely untapped (Turner, 2005). The ability to infer processes from observed and quantified patterns remains the 'golden goose' in landscape ecology, greatly adding value to image interpretation for ecological analyses. The current limitations are not likely to cease being constraints, but deeper understandings of metrics, environments, and the spatial patterns that operate within them will allow compartmentalized assessments, which if coupled with artificial intelligence and simulation, stand to provide targeted benefits.

Following the conceptual developments described by Boots (2003) and extending research described in the section on the map comparison problem, there is benefit from extending local statistics to categorical spatial data. The idea that configuration assessment can be conditioned on composition is presented for binary maps, but extension to multi-class maps of large size remains mathematically challenging as the combinations become numerous. The field needs methods for partitioning landscapes into binary representation and then robustly estimating composition and configuration parameters that can be used to simulate landscape replicates while autocorrelation, randomness, and structures can be compared with statistical expectations. Work on this aspect of pattern analysis seems to have stalled; a reboot would be welcome.

Recently, there has been a growing grass-roots call for the statistical community to do away with p-values and the cleanliness of binary hypothesis testing, as is common for classical t-tests and ANOVA, among others. The argument stems from the fact that uncertainty and variability are more interesting than an exact test outcome, particularly when the distributions of measured values are not normally distributed and may bias outcomes (as may large numbers of observations that are prevalent in spatial data). In this spirit, spatial data are well suited to focusing on variability and uncertainty rather than simply differences in means, as the uncertainty provides a broader context to both the range of variability and the expectation for pattern alternatives given a set of constraints (Remmel, 2009). We also see the continued development of gradient surface metrics as beneficial, but as alluded to earlier, these new metrics would benefit from the more than three decades of research and development that traditional landscape metrics have received (Frazier and Kedron, 2017).

Overall, landscape metrics provide a myriad of tools for quantifying aspects of shape, pattern, and related derivatives. For the most part, they aim to quantify that which is not easily quantified by a single value or a small suite of indicator values, as these mask the true complexity of the real world. This itself makes the reliance and assessment of those values difficult (like assigning the human genome a single representative number); a quantification that may be repeatable in one direction becomes meaningless in the opposing direction. That is, from a summary metric, it is virtually impossible to reconstitute the original pattern from which it was computed. Even knowing very detailed and precise constraints, the metric values summarize pattern but do not fully encode them; these are not reversible computations. We argue that in this scope, landscape metrics serve better as indicators than as absolute target values. Thus, exact landscape metric values should never form targets for remediation or landscape manipulation, nor should landscape changes be decided by recorded changes in landscape metric values alone. Landscape metrics may be usable to detect changes but not easily explain them, as the variability or uncertainty surrounding measured metrics can easily mask the actual signal.

Notes

1 http://ies-ows.jrc.ec.europa.eu/gtb/GTB/MSPA_Guide.pdf
2 We searched TOPIC: ('landscape metrics') OR TOPIC: ('landscape indices') and limited the results to 2009–2018 publication years in the Web of Science Core Collection. Abstracts were individually reviewed to evaluate fit into the application categories defined in Uuemaa et al. (2009).
3 UNCBD: A Sourcebook of Methods and Procedures for Monitoring Essential Biodiversity Variables in Tropical Forests with Remote Sensing. /www.gofcgold.wur.nl/sites/gofcgold-geobon_biodiversitys ourcebook.php

References

Abdel Moniem, H.E.M. and Holland, J.D. (2013) 'Habitat connectivity for pollinator beetles using surface metrics', *Landscape Ecology*, vol 28, no 7, pp1251–1267.
Adamczyk, J. and Tiede, D. (2017) 'ZonalMetrics: A Python toolbox for zonal landscape structure analysis', *Computers & Geosciences*, vol 99, pp91–99.
Anselin, L. (1995) 'Local indicators of spatial association: LISA', *Geographical Analysis*, vol 27, no 2, pp93–115.
Araya, Y.H. and Cabral, P. (2010) 'Analysis and modeling of urban land cover change in Setubal and Sesimbra, Portugal', *Remote Sensing*, vol 2, no 6, pp1549–1563.
Arnot, C., Fisher, P.F., Wadsworth, R. and Wellens, J. (2004) 'Landscape metrics with ecotones: Pattern under uncertainty', *Landscape Ecology* vol 19, no 2, pp181–195.
Bailey, T.C. and Gatrell, A.C. (1995) *Interactive Spatial Data Analysis*. Longman Group Ltd., London.
Baker, W.L. and Cai, Y.M. (1992) 'The r.le-programs for multiscale analysis of landscape structure using the GRASS geographical information-system', *Landscape Ecology*, vol 7, no 4, pp291–302.
Baldwin, D.J.B., Weaver, K., Schnekenburger, F. and Perera, A.H. (2004) 'Sensitivity of landscape pattern indices to input data characteristics on real landscapes: Implications for their use in natural disturbance emulation', *Landscape Ecology*, vol 19, no 3, pp255–271.

Banks-Leite, C., Ewers, R.M., Kapos, V., Martensen, A.C. and Metzger, J.P. (2011) 'Comparing species and measures of landscape structure as indicators of conservation importance: Species vs. landscape-based indicators', *Journal of Applied Ecology*, vol 48, no 3, pp706–714.

Baranyi, G., Saura, S., Podani, J. and Jordán, F. (2011) 'Contribution of habitat patches to network connectivity: Redundancy and uniqueness of topological indices', *Ecological Indicators*, vol 11, no 5, pp1301–1310.

Barbaro, L. and Halder, I.V. (2009) 'Linking bird, carabid beetle and butterfly life-history traits to habitat fragmentation in mosaic landscapes', *Ecography*, vol 32, no 2, pp321–333.

Bender, D.J., Tischendorf, L. and Fahrig, L. (2003) 'Using patch isolation metrics to predict animal movement in binary landscapes', *Landscape Ecology*, vol 18, no 1, pp17–39.

Birch, C.P.D., Oom, S.P. and Beecham, J.A. (2007) 'Rectangular and hexagonal grids used for observation, experiment and simulation in ecology', *Ecological Modelling*, vol, 206, no 3–4, pp347–359.

Boots, B. (2003) 'Developing local measures of spatial association for categorical data', *Journal of Geographical Systems*, vol 5, no 2, pp139–160.

Boots, B. and Csillag, F. (2006) 'Categorical maps, comparisons, and confidence', *Journal of Geographical Systems*, vol 8, no 2, pp109–118.

Bosco, L., Wan, H.Y., Cushman, S.A., Arlettaz, R. and Jacot, A. (2019) 'Separating the effects of habitat amount and fragmentation on invertebrate abundance using a multi-scale framework', *Landscape Ecology*, vol 34, no 1, pp105–117.

Bowersox, M.A. and Brown, D.G. (2001) 'Measuring the abruptness of patchy ecotones – a simulation-based comparison of landscape pattern statistics', *Plant Ecology*, vol 156, no 1, pp89–103.

Connors, J.P., Galletti, C.S. and Chow, W.T.L. (2013) 'Landscape configuration and urban heat island effects: Assessing the relationship between landscape characteristics and land surface temperature in Phoenix, Arizona', *Landscape Ecology*, vol 28, no 2, pp271–283.

Costanza, J.K. and Terando, A.J. (2019) 'Landscape connectivity planning for adaptation to future climate and land-use change', *Current Landscape Ecology Reports*, vol 4, no 1, pp1–13.

Cressie, N.A.C. (1993) *Statistics for Spatial Data*. Wiley, New York.

Csillag, F. and Boots, B. (2005) 'Toward comparing maps as spatial processes', in Fisher, P.F. (ed) *Developments in Spatial Data Handling*. Springer, Heidelberg.

Daily, G.C., Polasky, S., Goldstein, J., Kareiva, P.M., Mooney, H.A., Pejchar, L., Ricketts, T.H., Salzman, J. and Shallenberger, R. (2009) 'Ecosystem services in decision making: Time to deliver', *Frontiers in Ecology and the Environment*, vol 7, no 1, pp21–28.

Dale, M.R.T. and Fortin, M-J. (2014) *Spatial Analysis: A Guide for Ecologists*. 2nd ed. Cambridge University Press, Cambridge.

de Groot, R. (2006) 'Function-analysis and valuation as a tool to assess land use conflicts in planning for sustainable, multi-functional landscapes', *Landscape and Urban Planning*, vol 75, no 3–4, pp175–186.

de Smith, M.J., Goodchild, M.F. and Longley, P.A. (2018) *Geospatial Analysis: A Comprehensive Guide to Principles, Techniques, and Software Tools*. 6th ed. The Winchelsea Press, London.

de Souza Leite, M., Tambosi, L.R., Romitelli, I. and Metzger, J.P. (2013) 'Landscape ecology perspective in restoration projects for biodiversity conservation: A review', *Natureza & Conservacco*, vol 11, no 2, pp108–118.

Duarte, G.T., Santos, P.M., Cornelissen, T.G., Ribeiro, M.C. and Paglia, A.P. (2018) 'The effects of landscape patterns on ecosystem services: Meta-analyses of landscape services', *Landscape Ecology*, vol 33, no 8, pp1247–1257.

Dungan, J.L. (2006) 'Focusing on feature-based differences in map comparison', *Journal of Geographical Systems*, vol 8, no 2, pp131–143.

Duro, D.C., Girard, J., King, D.J., Fahrig, L., Mitchell, S., Lindsay, K. and Tischendorf, L. (2014) 'Predicting species diversity in agricultural environments using Landsat TM imagery', *Remote Sensing of Environment*, vol 144, pp214–225.

Eigenbrod, F., Hecnar, S.J. and Fahrig, L. (2011) 'Sub-optimal study design has major impacts on landscape-scale inference', *Biological Conservation* vol 144, pp298–305.

Fahrig, L. (2003) 'Effects of habitat fragmentation on biodiversity', *Annual Review of Ecology Evolution and Systematics*, vol 34, pp487–515.

Fahrig, L. (2017) 'Ecological responses to habitat fragmentation per se', *Annual Review of Ecology, Evolution, and Systematics*, vol 48, no 1, pp1–23.

Fahrig, L. (2019) 'Habitat fragmentation: A long and tangled tale', *Global Ecology and Biogeography*, vol 28, no 1, pp33–41.

Fahrig, L., Baudry, J., Brotons, L., Burel, F.G., Crist, T.O., Fuller, R.J., Sirami, C., Siriwardena, G.M. and Martin, J-L. (2011) 'Functional landscape heterogeneity and animal biodiversity in agricultural landscapes: Heterogeneity and biodiversity', *Ecology Letters*, vol 14, no 2, pp101–112.

Fahrig, L., Girard, J., Duro, D., Pasher, J., Smith, A., Javorek, S., King, D., Lindsay, K.F., Mitchell, S. and Tischendorf, L. (2015) 'Farmlands with smaller crop fields have higher within-field biodiversity', *Agriculture, Ecosystems & Environment*, vol 200, pp219–234.

Fardila, D., Kelly, L.T., Moore, J.L. and McCarthy, M.A. (2017) 'A systematic review reveals changes in where and how we have studied habitat loss and fragmentation over 20 years', *Biological Conservation*, vol 212, pp130–138.

Faye, E., Rebaudo, F., Yanez-Cajo, D., Cauvy-Fraunié, S. and Dangles, O. (2016) 'A toolbox for studying thermal heterogeneity across spatial scales: From unmanned aerial vehicle imagery to landscape metrics', *Methods in Ecology and Evolution*, vol 7, no 4, pp437–446.

Ferrenberg, S. (2016) 'Landscape features and processes influencing forest pest dynamics', *Current Landscape Ecology Reports*, vol 1, no 1, pp19–29.

Fletcher, R.J., Didham, R.K., Banks-Leite, C., *et al.* (2018) 'Is habitat fragmentation good for biodiversity?', *Biological Conservation*, vol 226, pp9–15.

Forman, R.T.T. and Godron, M. (1981) 'Patches and structural components for a landscape ecology', *Bioscience*, vol 31, no 10, pp733–740.

Forman, R.T.T. and Godron, M. (1986) *Landscape Ecology*. Wiley, New York.

Fortin, M-J., Boots, B., Csillag, F. and Remmel, T.K. (2003) 'On the role of spatial stochastic models in understanding landscape indices in ecology', *Oikos*, vol 102, no 1, pp203–212.

Frank, S., Fürst, C., Koschke, L. and Makeschin, F. (2012) 'A contribution towards a transfer of the ecosystem service concept to landscape planning using landscape metrics', *Ecological Indicators*, vol 21, pp30–38.

Frank, S., Fürst, C., Koschke, L., Witt, A. and Makeschin, F. (2013) 'Assessment of landscape aesthetics: Validation of a landscape metrics-based assessment by visual estimation of the scenic beauty', *Ecological Indicators*, vol 32, pp222–231.

Franklin, J.F. and Lindenmayer, D.B. (2009) 'Importance of matrix habitats in maintaining biological diversity', *Proceedings of the National Academy of Sciences*, vol 106, no 2, pp349–350.

Frazier, A.E. and Kedron, P. (2017) 'Landscape metrics: Past progress and future directions', *Current Landscape Ecology Reports*, vol 2, no 3, pp63–72.

Fry, G., Tveit, M.S., Ode, Å. and Velarde, M.D. (2009) 'The ecology of visual landscapes: Exploring the conceptual common ground of visual and ecological landscape indicators', *Ecological Indicators*, vol 9, no 5, 933–947.

Gadelmawla, E.S., Koura, M.M., Maksoud, T.M.A., Elewa, I.M. and Soliman, H.H. (2002) 'Roughness parameters', *Journal of Materials Processing Technology*, vol 123, no 1, pp133–145.

Getis, A. (2007) 'Reflections on spatial autocorrelation', *Regional Science and Urban Economics*, vol 37, no 4, pp491–496.

Giardino, C., Bresciani, M., Villa, P. and Martinelli, A. (2010) 'Application of remote sensing in water resource management: The case study of Lake Trasimeno, Italy', *Water Resources Management*, vol 24, no 14, pp3885–3899.

Giles, R.H. and Trani, M.K. (1999) 'Key elements of landscape pattern measures', *Environmental Management*, vol 23, no 4, pp477–481.

Gonzalez-Olabarria, J-R. and Pukkala, T. (2011) 'Integrating fire risk considerations in landscape-level forest planning', *Forest Ecology and Management*, vol 261, no 2, pp278–287.

Gustafson, E.J. (1998) 'Quantifying landscape spatial pattern: What is the state of the art?', *Ecosystems*, vol 1, no 2, pp143–156.

Gustafson, E.J. (2018) 'How has the state-of-the-art for quantification of landscape pattern advanced in the twenty-first century?', *Landscape Ecology*, vol 34, pp2065–2072.

Hadley, A.S. and Betts, M.G. (2016) 'Refocusing habitat fragmentation research using lessons from the last decade', *Current Landscape Ecology Reports*, vol 1, no 2, pp55–66.

Hagen, A. (2003) 'Fuzzy set approach to assessing similarity of categorical maps', *International Journal of Geographical Information Science*, vol 17, no 3, pp235–249.

Hagen-Zanker, A. (2006) 'Map comparison methods that simultaneously address overlap and structure', *Journal of Geographical Systems*, vol 8, no 2, pp165–185.

Haines-Young, R. and Chopping, M. (1996) 'Quantifying landscape structure: A review of landscape indices and their application to forested landscapes', *Progress in Physical Geography*, vol 20, no 4, pp418–445.

Hargrove, W.W., Hoffman, F.M. and Hessburg, P.F. (2006) 'Mapcurves: A quantitative method for comparing categorical maps', *Journal of Geographical Systems*, vol 8, no 2, pp187–208.

Harper, K.A., Macdonald, S.E., Mayerhofer, M.S., *et al.* (2015) 'Edge influence on vegetation at natural and anthropogenic edges of boreal forests in Canada and Fennoscandia', in P. Bellingham (ed) *Journal of Ecology*, vol 103, no 3, pp550–562.

Hassett, E.M., Stehman, S.V. and Wickham, J.D. (2012) 'Estimating landscape pattern metrics from a sample of land cover', *Landscape Ecology*, vol 27, no 1, pp133–149.

Henriques, F. and Goncalves, A. (2010) 'Analysis of lacunae and retouching areas in panel paintings using landscape metrics', in M. Ioannides, D. Fellner, A. Georgopoulos and D.G. Hadjimitsis (eds) *Digital Heritage*. Springer, London.

Herold, M., Liu, X.H. and Clarke, K.C. (2003) 'Spatial metrics and image texture for mapping urban land use', *Photogrammetric Engineering and Remote Sensing*, vol 69, no 9, pp991–1001.

Ignacio Gasparri, N. and Ricardo Grau, H. (2009) 'Deforestation and fragmentation of Chaco dry forest in NW Argentina (1972–2007)', *Forest Ecology and Management*, vol 258, no 6, pp913–921.

Jacquez, G.M. (1995) 'The map comparison problem: Tests for the overlap of geographic boundaries', *Statistics in Medicine*, vol 14, no 21–22, pp2343–2361.

Jaeger, J.A.G., Bertiller, R., Schwick, C. and Kienast, F. (2010) 'Suitability criteria for measures of urban sprawl', *Ecological Indicators*, vol 10, no 2, pp397–406.

Jelinski, D.E. and Wu, J.G. (1996) 'The modifiable areal unit problem and implications for landscape ecology', *Landscape Ecology*, vol 11, no 3, pp129–140.

Kabos, S. and Csillag, F. (2002) 'The analysis of spatial association on a regular lattice by join-count statistics without the assumption of first-order homogeneity', *Computers & Geosciences*, vol 2, no 8, pp901–910.

Kedron, P.J., Frazier, A.E., Ovando-Montejo, G.A. and Wang, J. (2018) 'Surface metrics for landscape ecology: A comparison of landscape models across ecoregions and scales', *Landscape Ecology*, vol 33, no 9, pp1489–1504.

Kissick, A.L., Dunning, Jr. J.B, Fernandez-Juricic, E. and Holland, J.D. (2018) 'Different responses of predator and prey functional diversity to fragmentation', *Ecological Applications*, vol 28, no 7, pp1853–1866.

Kittle, A.M., Anderson, M., Avgar, T., *et al.* (2015) 'Wolves adapt territory size, not pack size to local habitat quality', *Journal of Animal Ecology*, vol 84, no 5, pp1177–1186.

Kong, F., Yin, H., Wang, C., Cavan, G. and James, P. (2014) 'A satellite image-based analysis of factors contributing to the green-space cool island intensity on a city scale', *Urban Forestry & Urban Greening*, vol 13, no 4, pp846–853.

Kot, R. (2018) 'A comparison of results from geomorphological diversity evaluation methods in the Polish Lowland (Toru Basin and Chemno Lakeland)', *Geografisk Tidsskrift-Danish Journal of Geography*, vol 118, no 1, pp17–35.

Kotliar, N.B. and Wiens, J.A. (1990) 'Multiple scales of patchiness and patch structure: A hierarchical framework for the study of heterogeneity', *Oikos*, vol 59, no 2, pp253–260.

Kupfer, J.A. (2012) 'Landscape ecology and biogeography: Rethinking landscape metrics in a post-FRAGSTATS landscape', *Progress in Physical Geography: Earth and Environment*, vol 36, no3, pp400–420.

Lausch, A., Blaschke, T., Haase, D., Herzog, F., Syrbe, R-U., Tischendorf, L. and Walz, U. (2015) 'Understanding and quantifying landscape structure: A review on relevant process characteristics, data models and landscape metrics', *Ecological Modelling*, vol 295, pp31–41.

Lee, P.J., Hong, J.Y. and Jeon, J.Y. (2014) 'Assessment of rural soundscapes with high-speed train noise', *Science of the Total Environment*, vol 482, pp432–439.

Leite, M. de S., Tambosi, L.R., Romitelli, I, and Metzger, J.P. (2013) 'Landscape ecology perspective in restoration projects for biodiversity conservation: a review', *Natureza & Conservacao* vol 11, 108+.

Li, H. and Wu, J. (2004) 'Use and misuse of landscape indices', *Landscape Ecology*, vol 19, no 4, pp389–399.

Li, J., Song, C., Cao, L., Zhu, F., Meng, X. and Wu, J. (2011) 'Impacts of landscape structure on surface urban heat islands: A case study of Shanghai, China', *Remote Sensing of Environment*, vol 115, no 12, pp3249–3263.

Li, X., Lin, J., Chen, Y., Liu, X. and Ai, B. (2013) 'Calibrating cellular automata based on landscape metrics by using genetic algorithms', *International Journal of Geographical Information Science*, vol 27, no 3, pp594–613.

Lin, L., Yan, J., Chen, G., *et al.* (2018) 'Does magnification of SEM image influence quantification of particulate matters deposited on vegetation foliage', *Micron*, vol 115, pp7–16.

Liu, M., Hu, Y-M. and Li, C-L. (2017) 'Landscape metrics for three-dimensional urban building pattern recognition', *Applied Geography*, vol 87, pp66–72.

Loran, C., Haegi, S. and Ginzler, C. (2018) 'Comparing historical and contemporary maps – a methodological framework for a cartographic map comparison applied to Swiss maps', *International Journal of Geographical Information Science*, vol 32, no 11, pp2123–2139.

Lourenço, G.M., Soares, G.R., Santos, T.P., Dáttilo, W., Freitas, A.V.L. and Ribeiro, S.P. (2019) 'Equal but different: Natural ecotones are dissimilar to anthropic edges', *PLOS ONE*, vol 14, no 3, art e0213008.

Lustig, A., Stouffer, D.B., Doscher, C. and Worner, S.P. (2017) 'Landscape metrics as a framework to measure the effect of landscape structure on the spread of invasive insect species', *Landscape Ecology*, vol 32, no 12, pp2311–2325.

Machado, R., Godinho, S., Pirnat, J., Neves, N. and Santos, P. (2018) 'Assessment of landscape composition and configuration via spatial metrics combination: Conceptual framework proposal and method improvement', *Landscape Research*, vol 43, no 5, pp652–664.

Marull, J., Cunfer, G., Sylvester, K. and Tello, E. (2018) 'A landscape ecology assessment of land-use change on the Great Plains-Denver (CO, USA) metropolitan edge', *Regional Environmental Change*, vol 18, no 6, pp1765–1782.

McGarigal, K. and Marks, B.J. (1995) *FRAGSTATS: Spatial Pattern Analysis Program for Quantifying Landscape Structure*. U.S. Department of Agriculture, Forest Service, Pacific Northwest Research Station, Portland, OR, Gen. Tech. Rep. PNW-GTR-351, 122 p.

McGarigal, K., Tagil, S. and Cushman, S.A. (2009) 'Surface metrics: An alternative to patch metrics for the quantification of landscape structure', *Landscape Ecology*, vol 24, no 3, pp433–450.

McGarigal, K., Cushman, S.A. and Ene, E. (2012) *FRAGSTATS. Spatial Pattern Analysis Program for Categorical and Continuous Maps*. University of Massachusetts, Amherst, MA, www.umass.edu/landeco/research/fragstats/fragstats.html, accessed 7 Dec 2020.

Millenium Ecosystem Assessment (2005) *Ecosystems and Human Well-being: Synthesis*. Island Press, Washington, DC.

Muderere, T., Murwira, A. and Tagwireyi, P. (2018) 'An analysis of trends in urban landscape ecology research in spatial ecological literature between 1986 and 2016', *Current Landscape Ecology Reports*, vol 3, no 3, pp43–56.

Nagendra, H. (2002) 'Opposite trends in response for the Shannon and Simpson indices of landscape diversity', *Applied Geography*, vol 22, no 2, pp175–186.

Nowakowski, A.J., Frishkoff, L.O., Agha, M., Todd, B.D. and Scheffers, B.R. (2018) 'Changing thermal landscapes: Merging climate science and landscape ecology through thermal biology', *Current Landscape Ecology Reports*, vol 3, no 4, pp57–72.

Nowosad, J. and Stepinski, T.F. (2021) 'Pattern-based identification and mapping of landscape types using multi-thematic data', *International Journal of Geographical Information Science* 1–16 doi:10.1080/13658816.2021.1893324.

O'Neill, R.V., Krummel, J.R., Gardner, R.H., Sugihara, G., Jackson, B., DeAngelis, D.L., Milne, B.T., Turner, M.G., Zygmunt, B., Christensen, S.W., Dale, V.H. and Graham, R.L. (1988) 'Indices of landscape pattern', *Landscape Ecology*, vol 1, no 3, pp153–162.

Ostapowicz, K., Vogt, P., Riitters, K.H., Kozak, J. and Estreguil, C. (2008) 'Impact of scale on morphological spatial pattern of forest', *Landscape Ecology*, vol 23, no 9, pp1107–1117.

Paez, A., Uchida, T. and Miyamoto, K. (2002a) 'A general framework for estimation and inference of geographically weighted regression models: 1. Location-specific kernel bandwidths and a test for locational heterogeneity', *Environment and Planning A*, vol 34, no 4, pp733–754.

Paez, A., Uchida, T. and Miyamoto, K. (2002b) 'A general framework for estimation and inference of geographically weighted regression models: 2. Spatial association and model specification tests', *Environment and Planning A*, vol 34, no 5, pp883–904.

Palmer, M.W., Earls, P.G., Hoagland, B.W., White, P.S. and Wohlgemuth, T. (2002) 'Quantitative tools for perfecting species lists', *Environmetrics*, vol 13, no 2, pp121–137.

Pasher, J. et al. (2013) 'Optimizing landscape selection for estimating relative effects of landscape variables on ecological responses', *Landscape Ecology* vol 28, pp371–383.

Peng, J., Wang, Y.L., Zhang, Y., Wu, J., Li, W. and Li, Y. (2010) 'Evaluating the effectiveness of landscape metrics in quantifying spatial patterns', *Ecological Indicators*, vol 10, no 2, pp217–223.

Pond, B.A. (2016) 'Across the grain: Multi-scale map comparison and land change assessment', *Ecological Indicators*, vol 71, pp660–668.

Pontius, R.G. Jr. and Millones, M. (2011) 'Death to Kappa: Birth of quantity disagreement and allocation disagreement for accuracy assessment', *International Journal of Remote Sensing*, vol 32, no 15, pp4407–4429.

Pontius, R.G., Shusas, E. and Mceachern, M. (2004) 'Detecting important categorical land changes while accounting for persistence', *Agriculture Ecosystems & Environment*, vol 101, no 2–3, pp251–268.

Prugh, L.R., Hodges, K.E., Sinclair, A.R.E. and Brashares, J.S. (2008) 'Effect of habitat area and isolation on fragmented animal populations', *Proceedings of the National Academy of Sciences of the United States of America*, vol 105, no 52, pp20770–20775.

Ramachandra, T.V., Aithal, B.H. and Sanna, D.D. (2012) 'Insights to urban dynamics through landscape spatial pattern analysis', *International Journal of Applied Earth Observation and Geoinformation*, vol 18, pp329–343.

Ramezani, H. and Holm, S. (2011) 'A distance dependent contagion function for vector-based data', *Environmental and Ecological Statistics*, vol 19, pp 161–181.

Reis, J.P., Silva, E.A. and Pinho, P. (2016) 'Spatial metrics to study urban patterns in growing and shrinking cities', *Urban Geography*, vol 37, no 2, pp246–271.

Remmel, T.K. (2009) 'Investigating global and local categorical map configuration comparisons based on coincidence matrices', *Geographical Analysis*, vol 41, no 1, pp113–126.

Remmel, T.K. (2015) 'ShrinkShape2: A FOSS toolbox for computing rotation-invariant shape spectra for characterizing and comparing polygons', *Canadian Geographer*, vol 59, no 4, pp532–547.

Remmel, T.K. (2018) 'An incremental and philosophically different approach to measuring raster patch porosity', *Sustainability*, vol 10, no 10, art 3413.

Remmel, T.K. (2020) 'Distributions of hyper-local configuration elements to characterize, compare, and assess landscape-level spatial patterns', *Entropy* vol 22, p420.

Remmel, T.K. and Csillag, F. (2003) 'When are two landscape pattern indices significantly different?', *Journal of Geographical Systems*, vol 5, no 4, pp331–351.

Remmel, T.K. and Csillag, F. (2006) 'Mutual information spectra for comparing categorical maps', *International Journal of Remote Sensing*, vol 27, no 7, pp1425–1452.

Remmel, T.K. and Fortin, M-J. (2013) 'Categorical, class-focused map patterns: Characterization and comparison', *Landscape Ecology*, vol 28, no 8, pp1587–1599.

Remmel, T.K. and Perera, A.H. (2002) 'Accuracy of discontinuous binary surfaces: A case study using boreal forest fires', *International Journal of Geographical Information Science*, vol 16, no 3, pp287–298.

Ren, L., Huang, J., Huang, Q., Lei, G., Cui, W., Yuan, Y. and Liang, Y. (2018) 'A fractal and entropy-based model for selecting the optimum spatial scale of soil erosion', *Arabian Journal of Geosciences*, vol 11, no 8, art 161.

Riitters, K. (2018) 'Pattern metrics for a transdisciplinary landscape ecology', *Landscape Ecology*, vol 34, pp2057–2063.

Riitters, K.H., O'Neill, R.V., Hunsaker, C.T., Wickham, J.D., Yankee, D.H., Timmins, S.P., Jones, K.B. and Jackson, B.L. (1995) 'A factor analysis of landscape pattern and structure metrics', *Landscape Ecology*, vol 10, no 1, pp23–39.

Riitters, K.H., Vogt, P., Soille, P., Kozak, J. and Estreguil, C. (2007) 'Neutral model analysis of landscape patterns from mathematical morphology', *Landscape Ecology*, vol 22, no 7, pp1033–1043.

Riitters, K.H., Wickham, J.D. and Wade, T.G. (2009) 'An indicator of forest dynamics using a shifting landscape mosaic', *Ecological Indicators*, vol 9, no 1, pp107–117.

Rocchini, D., Balkenhol, N., Carter, G.A., *et al.* (2010) 'Remotely sensed spectral heterogeneity as a proxy of species diversity: Recent advances and open challenges', *Ecological Informatics*, vol 5, no 5, pp318–329.

Rodewald, A.D. and Arcese, P. (2016) 'Direct and indirect interactions between landscape structure and invasive or overabundant species', *Current Landscape Ecology Reports*, vol 1, no 1, pp30–39.

Ruiz, M., López, F. and Páez, A. (2012) 'Comparison of thematic maps using symbolic entropy', *International Journal of Geographical Information Science*, vol 26, no 3, pp413–439.

Saura, S. (2002) 'Effects of minimum mapping unit on land cover data spatial configuration and composition', *International Journal of Remote Sensing*, vol 23, no 22, pp4853–4880.

Saura, S. (2004) 'Effects of remote sensor spatial resolution and data aggregation on selected fragmentation indices', *Landscape Ecology*, vol 19, no 2, pp197–209.

Saura, S. and Martinez-Millan, J. (2001) 'Sensitivity of landscape pattern metrics to map spatial extent', *Photogrammetric Engineering and Remote Sensing*, vol 67, no 9, 1027–1036.

Sawada, M. (1999) 'ROOKCASE: An Excel 97/2000 Visual Basic (VB) add-in for exploring global and local spatial autocorrelation', *Bulletin of the Ecological Society of America*, vol 80, no 4, pp231–234.

Schindler, S., von Wehrden, H., Poirazidis, K., Wrbka, T. and Kati, V. (2013) 'Multiscale performance of landscape metrics as indicators of species richness of plants, insects and vertebrates', *Ecological Indicators*, vol 31, pp41–48.

Schirpke, U., Tasser, E. and Tappeiner, U. (2013) 'Predicting scenic beauty of mountain regions', *Landscape and Urban Planning*, vol 111, pp1–12.

Shan, J., Zaheer, M. and Hussain, E. (2003) 'Study on Accuracy of 1-Degree DEM Versus Topographic Complexity Using GIS Zonal Analysis', *Journal of Surveying Engineering*, vol 129, no 2, pp85–89.

Shannon, C.E. (1948) 'A mathematical theory of communication', *BELL System Technical Journal*, vol 27, no 4, pp623–656.

Silva, A.L., Alves, D.S. and Ferreira, M.P. (2018) 'Landsat-based land use change assessment in the Brazilian Atlantic Forest: Forest transition and sugarcane expansion', *Remote Sensing*, vol 10, no 7, art 996.

Simpson, E.H. (1949) 'Measurement of diversity', *Nature*, vol 163, no 4148, pp688–688.

Sirami, C., Gross, N., Baillod, A.B., et al. (2019) 'Increasing crop heterogeneity enhances multitrophic diversity across agricultural regions', *Proceedings of the National Academy of Sciences*, vol 116, no 33, pp 16442–16447.

Sitzia, T., Semenzato, P. and Trentanovi, G. (2010) 'Natural reforestation is changing spatial patterns of rural mountain and hill landscapes: A global overview', *Forest Ecology and Management*, vol 259, no 8, pp1354–1362.

Soille, P. and Vogt, P. (2009) 'Morphological segmentation of binary patterns', *Pattern Recognition Letters*, vol 30, no 4, pp456–459.

Stehman, S.V. (2006) 'Design, analysis, and inference for studies comparing thematic accuracy of classified remotely sensed data: A special case of map comparison', *Journal of Geographical Systems*, vol 8, no 2, pp209–226.

Steiniger, S. and Hay, G.J. (2009) 'Free and open source geographic information tools for landscape ecology', *Ecological Informatics*, vol 4, no 4, pp183–195.

Syrbe, R–U. and Walz, U. (2012) 'Spatial indicators for the assessment of ecosystem services: Providing, benefiting and connecting areas and landscape metrics', *Ecological Indicators*, vol 21, pp80–88.

Szabo, Z., Toth, C.A., Tomor, T. and Szabó, S. (2017) 'Airborne LiDAR point cloud in mapping of fluvial forms: A case study of a Hungarian floodplain', *Giscience & Remote Sensing*, vol 54, no 6, pp862–880.

Taubenböck, H., Wegmann, M., Roth, A., Mehl, H. and Dech, S. (2009) 'Urbanization in India – Spatiotemporal analysis using remote sensing data', *Computers, Environment and Urban Systems*, vol 33, no 3, pp179–188.

Thornton, D.H., Branch, L.C. and Sunquist, M.E. (2011) 'The influence of landscape, patch, and within-patch factors on species presence and abundance: A review of focal patch studies', *Landscape Ecology*, vol 26, no 1, pp7–18.

Tischendorf, L. and Fahrig, L. (2000) 'On the usage and measurement of landscape connectivity', *Oikos*, vol 90, no 1, pp7–19.

Townsend, P.A., Lookingbill, T.R., Kingdon, C.C. and Gardner, R.H. (2009) 'Spatial pattern analysis for monitoring protected areas', *Remote Sensing of Environment*, vol 113, no 7, pp1410–1420.

Trani, M.K. and Giles, R.H. (1999) 'An analysis of deforestation: Metrics used to describe pattern change', *Forest Ecology and Management*, vol 114, no 2–3, pp459–470.

Turner, M.G. (1990) 'Spatial and temporal analysis of landscape patterns', *Landscape Ecology*, vol 4, no 1, pp21–30.

Turner, M.G. (2005) 'Landscape ecology: What is the state of the science?', *Annual Review of Ecology, Evolution, and Systematics*, vol 36, pp319–344.

Turner, M.G. and Gardner, R.H. (2015) *Landscape Ecology in Theory and Practice: Pattern and Process*. 2nd ed. Springer, New York.

Tveit, M., Ode, Å. and Fry, G. (2006) 'Key concepts in a framework for analysing visual landscape character', *Landscape Research*, vol 31, no 3, pp229–255.

Upton, G.J.G. and Fingleton, B. (1985) *Spatial Data Analysis by Example*. Wiley, Toronto.

Uuemaa, E., Antrop, M., Roosaare, J., Marja, R. and Mander, U. (2009) 'Landscape metrics and indices: An overview of their use in landscape research', *Living Reviews in Landscape Research*, vol 3, no 1, pp1–28.

Uuemaa, E., Mander, Ü. and Marja, R. (2013) 'Trends in the use of landscape spatial metrics as landscape indicators: A review', *Ecological Indicators*, vol 28, pp100–106.

van der Walt, L., Cilliers, S.S., Du Toit, M.J. and Kellner, K. (2015) 'Conservation of fragmented grasslands as part of the urban green infrastructure: How important are species diversity, functional diversity and landscape functionality?', *Urban Ecosystems*, vol 18, no 1, pp87–113.

VanDerWal, J., Falconi, L., Januchowski, S., *et al.* (2014) *SDMTools*, https://cran.r-project.org/src/contrib/Archive/SDMTools/, accessed 7 Dec 2020.

Villa, F., Bagstad, K.J., Voigt, B., Johnson, G.W., Portela, R., Honzák, M. and Batker, D. (2014) 'A methodology for adaptable and robust ecosystem services assessment', *PLoS ONE*, vol 9, no 3, art e91001.

Vogt, P. (2019) 'Patterns in software design', *Landscape Ecology*, vol 34, pp 2083–2089.

Vogt, P., Riitters, K.H., Estreguil, C., Kozak, J., Wade, T.G. and Wickham, J.D. (2007) 'Mapping spatial patterns with morphological image processing', *Landscape Ecology*, vol 22, no 2, pp171–177.

Wade, T.G., Wickham, J.D., Nash, M.S., Neale, A.C., Riitters, K.H. and Jones, K.B. (2003) 'A comparison of vector and raster GIS methods for calculating landscape metrics used in environmental assessments', *Photogrammetric Engineering and Remote Sensing*, vol 69, no 12, pp1399–1405.

Wagner, H.H. and Edwards, P.J. (2001) 'Quantifying habitat specificity to assess the contribution of a patch to species richness at a landscape scale', *Landscape Ecology*, vol 16, pp121–131.

Wei, X., Xiao, Z., Li, Q., Li, P. and Xiang, C. (2017) 'Evaluating the effectiveness of landscape configuration metrics from landscape composition metrics', *Landscape and Ecological Engineering*, vol 13, no 1, pp169–181.

White, D. and Kiester, A.R. (2008) 'Topology matters: Network topology affects outcomes from community ecology neutral models', *Computers, Environment and Urban Systems*, vol 32, no 2, pp165–171.

White, R. (2006) 'Pattern based map comparisons', *Journal of Geographical Systems*, vol 8, no 2, pp145–164.

Wickham, J. and Riitters, K.H. (2019) 'Influence of high-resolution data on the assessment of forest fragmentation', *Landscape Ecology*, vol 34, pp2169–2182.

Wu, J. (2004) 'Effects of changing scale on landscape pattern analysis: Scaling relations', *Landscape Ecology*, vol 19, no 2, pp125–138.

Wu, J., Jenerette, G.D., Buyantuyev, A. and Redman, C.L. (2011) 'Quantifying spatiotemporal patterns of urbanization: The case of the two fastest growing metropolitan regions in the United States', *Ecological Complexity*, vol 8, no 1, pp1–8.

Wu, J.G., Shen, W.J., Sun, W.Z. and Tueller, P.T. (2002) 'Empirical patterns of the effects of changing scale on landscape metrics', *Landscape Ecology*, vol 17, no 8, pp761–782.

Wu, Q., Guo, F., Li, H. and Kang, J. (2017) 'Measuring landscape pattern in three dimensional space', *Landscape and Urban Planning*, vol 167, pp49–59.

Wulder, M.A., Masek, J.G., Cohen, W.B., Loveland, T.R. and Woodcock, C.E. (2012) 'Opening the archive: How free data has enabled the science and monitoring promise of Landsat', *Remote Sensing of Environment*, vol 122, pp2–10.

Yue, W., Liu, Y. and Fan, P. (2013) 'Measuring urban sprawl and its drivers in large Chinese cities: The case of Hangzhou', *Land Use Policy*, vol 31, pp358–370.

Zhou, W., Huang, G. and Cadenasso, M.L. (2011) 'Does spatial configuration matter? Understanding the effects of land cover pattern on land surface temperature in urban landscapes', *Landscape and Urban Planning*, vol 102, no 1, pp54–63.

16

Quantitative modelling and computer simulation

Calum Brown

Introduction

Landscape ecology is a particularly dynamic field that has often generated rapid methodological development. This innovative tendency has recently been strengthened by the need to account for landscape processes in the context of the whole Earth system and so to address pressing societal concerns about climatic, environmental, and social change. Quantitative modelling and computer simulation play key roles in this development because they allow complex system dynamics to be analyzed, explored, and projected with unparalleled detail. These methods comprise a wide range of techniques for simplifying and representing real-world systems through key relationships and processes and have gained power as computational resources improve. They have been used to establish a positive feedback in which novel techniques and research questions inform one another.

Nevertheless, modelling and simulation have their own challenges and complexities. Among the former, the need to achieve fully coherent representations of socio-ecological systems is particularly notable, not least because it requires integration of traditionally fragmented areas of research. While landscapes are widely conceptualized as key interfaces between social and ecological processes, they are still often represented in simulation models as discrete social or (usually) ecological systems. The design of models to bridge this gap can increase the ever-present risk of misinterpretation of results by adding complexity and reducing the ease with which models can be understood. Increasing scope also forces model users to consider the roles of factors that are hard to quantify or combine and to assess associated uncertainties. The contributions of models to landscape ecology depend upon repeated and careful engagement with these issues.

In this chapter, I review the development of modelling in landscape ecology to identify some of the most promising – and perilous – avenues for future research. I focus in particular on recent convergence between social and ecological modelling, which promises significant advances but must negotiate persistent disciplinary barriers. This brief survey of past and potential future research priorities suggests that landscape ecology is currently at a decisive threshold, beyond which substantial new findings are likely to be made through computational research.

Background

Landscape ecology is fundamentally interdisciplinary in its prioritization of the site of the interface between various processes, human and natural. As such, it has always represented a challenge to established approaches from any one scientific field. This basic mismatch between embedded concepts and available tools has effectively acted as a difference engine, driving the development of new perspectives and methods. Initially, development focused on the recognition and description of the role of the spatial environment in shaping ecological outcomes; a process that helped to create modern ecology (von Humboldt and Bonpland 1805; Nicolson 1996).

Perhaps less usefully, preoccupation with configuration also produced a dizzying abundance of metrics of landscape structure, all capturing slightly different (at best) aspects of studied systems (Cushman et al. 2008). Eventually, the sheer number of metrics began to obscure the insights they were intended to provide, replacing understanding with measurement. While certainly recognized as problematic in its own right (Lausch and Herzog 2002; Cushman et al. 2008), this excess of quantification was partly tamed by the recognition that ecological communities were temporally as well as spatially structured. The early concept of three-dimensional environmental niches (Grinnell 1917) evolved to one of (partially unmeasurable) multi-dimensional interactive or functional niches (Elton 1927; Hutchinson 1958), establishing a creative tension between dynamic conceptualizations and static descriptions of landscape ecology (Pimm 1984; Wu and Loucks 1995).

The resultant focus on process as well as pattern became – and remains – a defining characteristic of landscape ecology (Turner 1989). It also required methods that revealed not only observable structure but unobservable relationships, providing not only description but representation and if possible, explanation as well (Li and Wu 2004). Statistical (and later, computational) modelling was recognized as an ideal tool for this purpose (Turner 1990; Pickett and Cadenasso 1995). While incorporating a range of specific approaches, such modelling is united by its objectives of representing, studying, and often, extrapolating some aspects of complex system dynamics using mathematical or computational techniques. Not only were such models quickly able to utilize existing data, but rapid mathematical and computational advances allowed them to keep pace with novel research questions as well. Furthermore, in providing stochastic methods of generating patterns, they allowed assessment of uncertainties and therefore, far more robust inference about the roles of non-deterministic ecological processes (Fortin et al. 2003).

An important example is provided by the interactions of landscape ecology with the field of spatial statistics. Originally a source of metrics describing landscape structure, spatial statistics later developed to include a broad range of statistical models with considerable power for revealing the processes structuring ecological communities (overviews are provided by, e.g., Wiegand et al. 2003; Illian et al. 2008). Partly motivated by landscape ecology questions, the development of these models still represents a rich seam of research, allowing inference about underlying processes through detailed descriptions of spatial (and spatio-temporal) patterns. Spatial statistical methods have been used in forest ecology, for example, to investigate the associations between different species (Ogata and Tanemura 1985; Wiegand et al. 2007a), clustering in mortality (Sterner et al. 1986; Queenborough et al. 2007), and patterns of colonization (Salonen et al. 1992) and to try and separate the effects of local dispersal, species interactions, and environmental niche differentiation (Tuomisto et al. 2003; Hardy and Sonké 2004; Wiegand et al. 2007b).

Nevertheless, there remains a significant unresolved gap between pattern-based and process-based research in landscape ecology; a gap that modelling often straddles uneasily. Fundamentally, the lack of formal relationships between process and pattern makes inference of one from the other, whether via statistical models, computational models, or some other approach, risky at best and inappropriate at worst (Baddeley and Silverman 1984; Stoyan and Penttinen 2000; Fortin et al. 2003; Cushman and Landguth 2010; Brown et al. 2016). Lepš (1990, p. 9), for example, argued that 'mechanisms can be suggested on the basis of observed patterns, but they cannot be tested'.

The difficulty of linking pattern and process has perhaps been most apparent in the controversial use of models of simple (often first-order or non-spatial) patterns for inference about putative fundamental ecological laws (e.g. Andrewartha 1958; Hajnal 1958; Prigogine 1980; Cramer 1993). Macroecological patterns are often especially amenable to consistent statistical descriptions that resemble general laws, but considerable disagreement exists over whether these patterns contain any real ecological information (Lawton 1999; Turchin 2001; Colyvan and Ginzburg 2003). Whether they do or not, a wide variety of different models can (and do) replicate those patterns, sometimes using detailed ecological explanations and sometimes using no ecological explanations at all; a 'spectacular warning of the dangers of inferring process from pattern' (Chave and Leigh 2002, p. 164).

Conversely, over-reliance on metrics continues to hamper some areas of landscape ecology. In particular, poorly chosen or formulated metrics often fail to map onto the needs of policy and practice. This can be because they prioritize different types of information (as in a focus on landscape configuration rather than its impacts on policy-relevant outcomes) or because they prioritize entirely different sectors (as in a focus on environmental variables rather than policy-relevant social variables; Winkler et al. 2018; Scown et al. 2019). Indeed, it is widely recognized that landscape ecology retains a strong bias towards ecology even as landscapes become increasingly dominated by socially mediated (i.e. anthropogenic) processes (Nassauer 1995; Wu 2010). This bias is partially historical in nature but is undoubtedly exacerbated by the relative difficulty of 'measuring' and modelling the characteristics of social systems. To the extent that methods for dealing with the social aspects of landscape ecology have been developed, they remain largely separated by their distinct histories, data requirements, terminologies, and often, purposes (Figure 16.1 provides an outline of the development of social and ecological modelling in landscape ecology). Models also suffer from biases in the types of landscapes they represent, being primarily focused on agricultural landscapes in either highly developed or highly marginal contexts, with little coverage of traditional livelihoods or value systems (Malek et al. 2019).

Despite these difficulties, much of the momentum and motivation in present-day landscape ecology derives from deeply socio-ecological questions related to human impacts on natural systems and how these can be controlled to balance food production, biodiversity preservation,

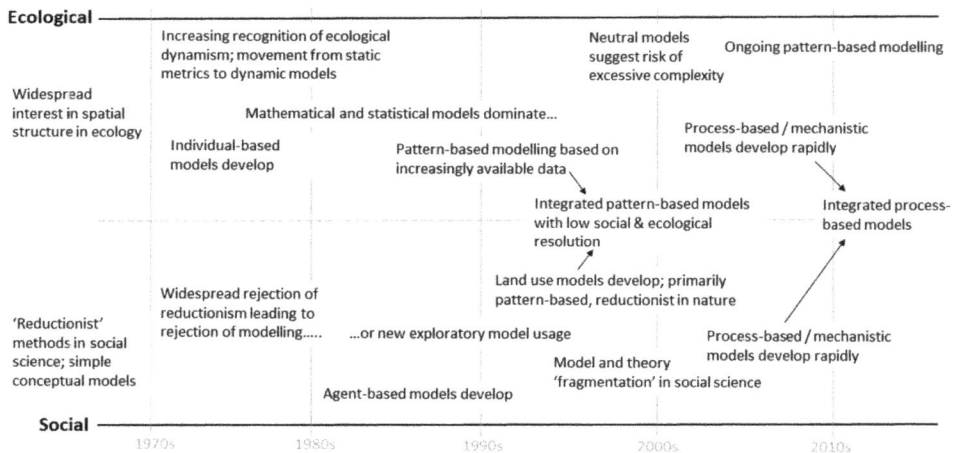

Figure 16.1 Schematic timeline of major developments in ecological, social, and socio-ecological modelling.

and climate change mitigation (Naveh 2007; Mayer et al. 2016). The scale and urgency of these questions demand novel, integrated methods, of which modelling is an indispensable component. In developing these methods, landscape ecologists must grapple with a number of issues that have so far remained unresolved. Particularly important are methods of bridging the long-term divide between ecological and social modelling, to represent and explore the deep landscape-scale links between the two areas, without generating excessive levels of uncertainty or data requirements.

Modelling ecological processes

Landscape ecology is often practiced and viewed simply as a branch of ecology, in which the landscape is an incidental setting rather than a dynamic entity (Bastian 2002; Wu 2006). This is especially apparent in modelling applications, which if not purely ecological in nature, are generally more advanced in ecological than in social terms. On the one hand, this clear focus has resulted in an invaluable array of advanced models that tell us a great deal about many ecological processes (Synes et al. 2016). On the other hand, it has allowed blind spots to persist not only where social processes are concerned but also where those processes make under-recognized ecological processes particularly important (Crane 2010; Schlüter et al. 2012).

Nevertheless, the landscape perspective has ensured that among the most modelled processes in landscape ecology are those involved in disturbance dynamics. At an early stage, this was prompted by arguments that ecological communities did not exist in or move towards equilibrium but rather, that frequently observed fluctuations in species abundances and community composition were fundamental aspects of those communities (Ayala 1972; Grubb 1977; Wiens 1977; Connell 1978; Levins 1979; Pickett 1980; Chesson and Warner 1981). This (still debated) perspective led to the development of diverse models of particular ecological processes that graduated from simple mathematical functions to complex simulation models as computational resources developed. In recent years, such models have been used to explore various aspects of processes, including reproduction, dispersal, predation, and mortality, in ecological communities (e.g. Temple and Cary 1988; Donovan and Thompson 2001; Nathan and Casagrandi 2004; Brown et al. 2011; Bocedi et al. 2014).

Widespread process modelling of this kind has been complemented and occasionally challenged by neutral models that replicate observed patterns without recourse to ecological mechanisms (Gardner et al. 1987; Hubbell 2001; O'Dwyer and Green 2010; Synes et al. 2016). These models have provided a valuable basis for comparison with other models, militating against the inclusion of unnecessary detail and complexity. Validation of neutral and non-neutral models using observational data has also proved useful in revealing the actions of particular ecological processes, as these can be individually introduced and tested from a neutral baseline (e.g. Brown et al. 2013; Püttker et al. 2015; Prevedello et al. 2016).

Recently, the cautionary brake applied by neutral models has been weakened by a new momentum towards an exploration of anthropogenic impacts on landscapes and ecology. The often novel and urgent nature of this exploration has driven the development of new forms and applications of models, many of which are still being evaluated. At a general level, and in the absence of precise knowledge of future conditions, this has resulted in a renewed emphasis on disturbances in ecological communities and the processes that determine their responses (Temple and Cary 1988; Pickett and Cadenasso 1995; Baker 1995; Bengtsson et al. 2000; Arroyo-Rodríguez et al. 2017). Models of forest dynamics and management have been particularly successful in including various forms of environmental change (e.g. Scheller et al. 2007; Seidl et al. 2012).

The evolution of issues that landscape ecology addresses has prompted the development of many different types of ecological model (reviewed at different stages by, e.g., Baker 1989; Perry and Enright 2006; Brown et al. 2006; Synes et al. 2016). Most of these model types remain in use, from mathematical or statistical models (e.g. O'Dwyer and Green 2010; Urban et al. 2019), to neutral models (e.g. Etherington et al. 2015; Sciaini et al. 2018), to detailed mechanistic and cross-sectoral integrated models (e.g. Voinov et al. 1999; Aspinall and Pearson 2000; Rudner et al. 2007; Vanwalleghem et al. 2013) (Table 16.1). Perceptible within this diversification has been a movement towards models that encode basic processes from which patterns emerge (Schröder and Seppelt 2006), even while new approaches such as Machine Learning reinforce mathematical or pattern-based relationships. Individual-based models (IBMs) that focus on the behavior and interactions of entities within ecological communities have been particularly popular, allowing virtual experiments to be conducted at community scale on the basis of well-understood individual characteristics (Grimm and Railsback 2013). Embedded within the exploration of different model types has been a movement towards transparent, open-source model design that allows better interpretation, evaluation, and re-use of models (Steiniger and Hay 2009; Grimm et al. 2010).

Models dealing with species, community, or ecosystem dynamics across or in the context of large spatial extents remain, for the most part, correlational, statistical, or pattern based, often in the guise of correlational Species Distribution Models (Urban 2015, 2019). These models can be characterized by their focus on reproducing observations without necessarily explaining how they arise, providing relatively simple, transferrable representations that do not necessarily hold in novel contexts. For instance, in tropical forests – some of the most diverse ecosystems in the world – models of ecological responses to climate change omit almost all the variation between tree species (van der Sande et al. 2017; Fisher et al. 2018). Furthermore, because the main focus of these models is on climate–ecosystem interactions, they have emphasized characteristics related to carbon, water, and nitrogen fluxes through plants and soils and neglected functions relating to the interactions that affect the composition and resilience of plant communities (Verheijen et al. 2015; van der Sande et al. 2017). A substantial disconnect therefore persists between the many detailed, mechanistic models of individual communities or processes and the relatively few generalized models of large-scale systems (Grimm and Railsback 2013; DeAngelis and Grimm 2014).

Not only does the focus on pattern-matching under projected climate change risk relying on correlations that may not hold under novel future environmental conditions; it also overshadows efforts to capture other impacts on ecological communities (Thuiller et al. 2019). For instance, studies that seek to identify impacts of human land use on ecological communities (including around half of meta-analyses) do so using simple metrics, such as total species richness and abundance, that carry little information about community composition, function, or stability (Blüthgen et al. 2016; Hekkala and Roberge 2018). This omission poses a significant challenge for efforts to anticipate future change, producing abstract, unrelated snapshots of landscape characteristics rather than dynamic, evolving explanations.

Nevertheless, recent years have seen the development of new types of large-scale mechanistic models that upscale knowledge of basic ecological processes. These model types range from representations of particular species across scales (as in the agent-based model of bumblebee colony development of Becher et al. 2018) to representations of ecosystem or habitat types (as with flexible, scalable forest ecology models such as *LANDIS-II*, developed by Scheller et al. 2007). At a more general level, Bocedi et al. (2014) developed the *RangeShifter* model of species eco-evolutionary dynamics, which has since been applied in a range of ecological settings related to connectivity and population dynamics (e.g. Heikkinen et al. 2015; Henry et al. 2017).

Table 16.1 Major model types with key references general reviews of model types are provided by Turner et al. (2001), Perry and Enright (2006), Houet et al. (2010), and Synes et al. (2016).

Model type	Description	Selected references
Statistical	Models intended to reveal causal relationships responsible for observations, as explanations in their own right or as a basis for further modelling	(Tischendorf 2001; Serneels and Lambin 2001; Veldkamp and Lambin 2001; Gutzwiller and Riffell 2007)
Spatial statistical	A special class of statistical models widely used in plant ecology, which allow stochastic simulation of spatial patterns as well as statistical inference about causes utilizing spatial information	(Wiegand and Moloney 2004; Páez and Scott 2005; Illian et al. 2008)
Neutral	Parsimonious or null models that attempt to replicate observations with a minimum (or absence) of mechanistic explanation	(Hubbell 2001; Gardner and Urban 2007; Wang and Malanson 2008; Hagen-Zanker and Lajoie 2008)
Pattern-based	Models that replicate some key features of observations ('patterns', though not necessarily spatial in nature), usually on the basis of statistical relationships with measurable variables, and often for the purpose of extrapolation into unobserved contexts	(Verburg et al. 2002; Hernandez et al. 2006; Phillips and Dudík 2008; Mas et al. 2014)
Process-based/mechanistic	Models that represent key processes or mechanisms, often at the level of individual entities (individual- and agent-based models), for the purposes of explanation or projection into unobserved contexts	(Matthews et al. 2007; Seidl et al. 2012; DeAngelis and Grimm 2014; Urban 2019)
Hybrid	Models that adopt different approaches for different parts of the study system depending on levels of knowledge available or detail required	(Lørup et al. 1998; Castella et al. 2007; Vincenot et al. 2011; Alexander et al. 2018)
Integrated	Combinations of models of different sub-systems, adapted to work together to represent real-world interactions between those sub-systems	(Harrison et al. 2012; Polhill et al. 2013; Guillem et al. 2015; Verkerk et al. 2018; Robinson et al. 2018)

The *Madingley* model is intended to represent entire ecosystems up to global extents (Purves et al. 2013) and has been used to examine the different effects of habitat loss and fragmentation (Bartlett et al. 2016) and the ecosystem impacts of bushmeat harvesting (Barychka et al. 2019). In adopting flexible and scalable approaches to modelling key ecological processes, these models are particularly well suited to integration with models of other components of land systems, allowing landscape ecology to be further embedded in a cross-disciplinary context.

Modelling social processes

The dominant cross-disciplinary feature of landscape ecology is, arguably, that it encompasses social processes. Indeed, landscapes often act as the main direct, tangible interfaces in socio-ecological systems and are the most appropriate settings in which to address a wide range of socio-cultural as well as ecological issues (Nassauer 1995; Matthews and Selman 2006; Houet et al. 2010). In landscape ecology, many of the most important social processes therefore relate to human land use. Often, however, social processes are included only in theory or as external drivers, with models generally focusing exclusively on ecological processes, even when they operate in response to socially mediated changes (Perry and Enright 2006; Elsawah et al. 2020). As a result, landscape-relevant social models have developed along largely separate, albeit parallel, lines to ecological models (Figure 16.1).

The divergence of ecological- and social-process modelling has been reinforced by the mismatch between the qualitative complexity of much of social science and the quantitative simplifications of modelling. In fact, much of the early development of social modelling was associated with an explicit rejection of this complexity and a search for general patterns and laws underlying social dynamics (Batty and Torrens 2005; Brown et al. 2016). These simple mathematical models, in avoiding much of the detail of the social systems, were also of limited relevance to ecology, focusing instead on large-scale social outcomes to which landscapes and ecosystems were incidental at best. However, just as recognition of the importance of dynamism in ecology prompted the development of new forms of process-based model, the obvious divergence of social systems from their idealized model counterparts prompted a reconsideration of the role of modelling in social science too (Bell 1964; Batty 1980; Martin et al. 2016). In some cases, this reconsideration involved a rejection of quantitative modelling as an inherently inappropriate technique (e.g. Harvey 1972), but in others, it led to new, exploratory models of the emergence of social outcomes from theorized or observed behaviors and ultimately, the development of modern computational social science (Lazer et al. 2009; Giles 2012; Conte et al. 2012).

The refocusing of social models on particular theories, cases, or processes has limited the scope for synthesis and generalization of the kind achieved in ecology. This may be attributable to the lack of 'neutral models' in social science and also to the fragmentation of social science between approaches and schools (Watts 2017). This fragmentation is apparent in the development of agent-based models that, like ecological IBMs, focus on how system characteristics develop from the behavior and interactions of individual entities. These models are frequently applied to land use change but have been criticized for being excessively diverse and context-specific without producing sufficient cumulative insight (O'Sullivan et al. 2016). At the same time, they share with other land system models a bias towards economic factors and neglect many important behavioral and environmental processes (Brown et al. 2017; Holman et al. 2018). Computational social science itself tends to simulate emergence and evolution in social systems rather than extrapolate observations to identify generalized system dynamics (Abbott 1997; Cederman 2005).

This points to a fundamental problem for models of landscape ecology: the fact that they must formalize inherently contingent relationships (in the sense that relationships are context-dependent and difficult, if not impossible, to establish as universal; e.g. Sayer 1982). Similarly to Lawton's (1999, p. 181) argument that '[ecological] rules are contingent in so many ways … as to make the search for patterns unworkable', Abbott (1997, p. 1152) warned that 'every social fact is situated, surrounded by other contextual facts, and brought into being by a process relating it to past contexts'. This holds true even of something as central as 'the environment'; a subjective and culturally mediated concept that is nonetheless an essential component of land system decision-making (Atran et al. 2005). This lack of basic 'ground truths' poses serious challenges for models that seek even to delineate, let alone simulate, social processes in landscape ecology.

These basic difficulties provide some explanation, if not justification, for the widespread use of assumptions and exogenous scenarios to represent human activity in landscape ecology models. Nevertheless, recent years have seen substantial progress in new techniques to make modelling more robust to such conceptual challenges. Especially prominent have been moves towards transparency, the use of clear conceptual models, and the participation of 'stakeholders' to explore different perspectives and social processes (e.g. Stephenson 2008; Holtz et al. 2015; Parrott 2017; Burton et al. 2018a; Keuschnigg et al. 2018; Pedde et al. 2019). Explicitly process-based approaches such as agent-based modelling have proved particularly amenable to uses of this kind, allowing behavioral information to be directly incorporated in some form (e.g. Castella et al. 2005; Prell et al. 2007; Millington et al. 2011; Robinson et al. 2012; Van Berkel and Verburg 2012). Agent-based modelling is also used to represent the individual-level human decision-making that mediates social and environmental interactions, including, recently, at large scales (e.g. Groeneveld et al. 2017; Schulze et al. 2017; Lippe et al. 2019; Brown et al. 2019).

A new, exploratory role is not limited to one form of modelling. Indeed, perhaps the most significant contribution of the development of agent-based land use models has been the stimulation of approaches that can be adopted more widely. These approaches include the identification and representation of high-level or cross-context processes (Parker et al. 2008) and their relative impacts on socio-ecological change (Brown et al. 2018). This often relies on representations of the functional roles of actors within land systems, which incorporate socially (and often, ecologically) meaningful information about system dynamics while also providing generic, flexible modelling frameworks (Arneth et al. 2014; Blüthgen et al. 2016; Hekkala and Roberge 2018; Grêt-Regamey et al. 2019). Informative simplifications of this kind can also be developed and used in a range of other modelling approaches, from statistical to process-based (e.g. Millington et al. 2007; Schößer et al. 2010; Rounsevell et al. 2014), and are highly compatible with parallel representations of ecological communities (Arneth et al. 2014). As such, attempts to reduce complexity while increasing accuracy in social models have converged with similar attempts in ecological models, allowing – in principle at least – the development of fully integrated landscape models.

Modelling integrated landscape processes

As a result of the separate and often internally fragmented nature of ecological and social models of landscape processes, by far the most common form of integrated landscape ecology models has involved acknowledgement of socio-ecological links, factors, or processes without their explicit representation. In this sense, neutral and other pattern-based models that make no explicit assumptions about processes represent integrated systems even if they can reveal little about their dynamics. Far more common, however, are specialized models that focus on one field but include some exogenous factors to account for the other.

The use of ecological models under social driving conditions, and vice versa, is a well-established and powerful technique for exploring some socio-ecological interactions. Particularly common are models that use pre-defined scenarios of land use change, human disturbance, or management regimes as inputs to ecological models to project changes in landscape systems (e.g. Schröder et al. 2008; Touza et al. 2013; Püttker et al. 2015; Prevedello et al. 2016). In these cases, ecological models vary from neutral to strongly process-based, although they rarely account for the spatial interactions or horizontal transfers of energy, water, nutrients, or other foundations of natural processes (Robinson et al. 2009). The social scenarios they utilize, meanwhile, tend to be relatively simplistic, often comprising only assumed landscape configurations. In some cases, however, far more detailed scenarios are used as a basis for deriving input parameters, as with the use of the Shared Socio-Economic Pathways (SSPs) developed by the Intergovernmental Panel on Climate Change (van Vuuren and Carter 2014; O'Neill et al. 2017). These provide a unified basis for socio-economic and land use projections even where specific model input requirements vary substantially (Saito et al. 2019; Nordström et al. 2019). Sometimes, land use scenarios have been far more context-specific, developed and assessed using participatory processes and local expertise (e.g. Thorn et al. 2017; Koo et al. 2018).

Ecological scenarios for use in land use and other social system models are generally less coherent or advanced. While land use models are an important part of environmental and climatic research from local to global scales, their predominantly economic focus means that ecological inputs are limited even where appropriate scenarios are available. For example, despite the global importance of tropical forests and high-profile international efforts to preserve them, few if any models capture the effects of global pressures on tropical landscape processes. Such interactions have only recently been explicitly treated in any form (Alexander et al. 2018), and then as broad processes with limited spatial resolution. Furthermore, forest dynamics in such models tend to emerge largely as a residual once area needs for other broad land use types (such as cropland, pasture, and urban land) have been fulfilled (Rosa et al. 2014; Lawrence et al. 2016; Brown et al. 2017). Decisions about these land uses in tropical and other areas are simulated not as the result of social, cultural, and behavioral factors but as outcomes of pre-defined transition matrices most sensitive to population sizes, profit levels, agricultural efficiency, and proximity to roads (Rosa et al. 2014; Stehfest et al. 2019). Even in agent-based models that prioritize individual decision-making, economic factors dominate (Groeneveld et al. 2017; Huber et al. 2018). Where ecological scenarios have been developed, they tend to be climate-focused and less detailed with respect to land use change (Titeux et al. 2016), although they have occasionally been used in more complete contexts that consider a range of ecological and social issues (e.g. Berman et al. 2004).

The most advanced method for modelling integrated landscape processes, then, is the combination of existing social and ecological models or to a lesser extent, the development of new socio-ecological models. Both of these are challenging tasks that cross disciplinary boundaries and require technical as well as field-specific expertise (Robinson et al. 2018). For these reasons, most integrated models have limited representation of one sub-system and detailed representation of another, depending on the field in which they originated. For example, an increasing number of ecologically focused models have incorporated economic components to allow simulation of land use changes (e.g. Voinov et al. 1999; Rudner et al. 2007). Meanwhile, land use models have in many cases been linked with species distribution or other correlational models of ecological change, usually to allow one-way effects of land use on ecological systems to be explored (e.g. Linderman et al. 2005; Rounsevell et al. 2006; Evans and Kelley 2008; Gibon et al. 2010; Matsumura et al. 2019). In some cases, process-based ecological modelling has been used in this way, for example, to explore the succession of natural vegetation on abandoned farmland (Verburg and Overmars 2009).

In a relatively small number of cases, process-based models of social and ecological subsystems have been combined to allow realistic, emergent dynamics to be simulated and to feed back to one another. This has been done, for example, using agent- and individual-based models at small scales to examine agricultural interactions with populations of birds and pollinators (Guillem et al. 2015; Synes et al. 2018) or using hybrid approaches at larger scales to examine land use–ecosystem interactions and policy impacts (Polhill et al. 2013; Alexander et al. 2018; Harrison et al. 2019). Spatially explicit process-based combinations of this kind are particularly useful as they can allow accurate treatment of spatial patterns, flows, and processes that underpin many of the dynamics of socio-ecological systems (Robinson et al. 2009). These studies also suggest that the dynamics produced by integrated models diverge from those produced by discrete models, particularly when feedbacks are strong or systems encounter sudden external changes (Filatova et al. 2016; Robinson et al. 2018; Synes et al. 2018). This, in turn, suggests a key role for integrated landscape modelling as processes of global change generate novel socio-ecological conditions.

Changing requirements and implications

The evolution of landscape ecology models from discrete ecological or social pattern-based models to integrated process-based models has generated very different data requirements as well as implications for model usage and interpretation. Broadly speaking, both data requirements and model complexity have increased, forcing a number of difficult choices on model developers and users. Key amongst these are the extent and nature of data required for acceptable calibration and validation of models, the point at which increasing complexity becomes counter-productive in terms of model usage, and the handling of model uncertainty. Without careful consideration of these issues, inappropriate usage can invalidate any benefits of advanced modelling approaches.

From an initially luxurious situation in which relatively simple models could utilize a small number of key data, landscape ecological models have developed to extend far beyond the boundaries of available data. Process-based models especially require detailed spatio-temporal data to build an accurate representation of processes across relevant contexts; data that are often unavailable for ecological or social systems (Boero and Squazzoni 2005; Urban 2019). When relevant data are available, they can impose particular perspectives or architectures on models, for instance in emphasizing economic over social conditions or in allowing only approximate calibration and therefore, exploration rather than prediction (Boero and Squazzoni 2005; Synes et al. 2016). Furthermore, data often only cover small spatial extents that are not necessarily representative of the processes or problems with which models are concerned, limiting the generality of model outputs (Matthews and Selman 2006; Cumming et al. 2006).

Nevertheless, major new data sources do now exist, and models are beginning to take advantage of these. Links are being made, for example, between large-scale Earth Observation data and models of ecosystem service provision (Hill et al. 2002; Ramirez-Reyes et al. 2019), while collections of field data provide foundations for more extensive modelling (e.g. Ellis 2012; Goetz et al. 2015; Smithsonian Institute 2018; Griffiths et al. 2019). Widely used databases include the global land use case-study database *GLOBE* (Ellis 2012; GLOBE 2016), open access data on socio-economic conditions and changes (e.g. World Bank Open Data and NASA's Socioeconomic Data and Applications Center; NASA 2018b; World Bank 2018), ecological census repositories such as that of the ForestGEO network (Anderson-Teixeira et al. 2015; Smithsonian Institute 2018), and high-resolution satellite observations of land use and ecosystem change from European Space Agency and NASA data repositories (e.g. SENTINEL II and MODIS; ESA 2018; NASA 2018a). Meanwhile, the development of long-term 'natural experiments' takes advantage of historical changes to understand socio-ecological dynamics (e.g. Watts et al. 2016; Damschen et al. 2019).

While these new data resources allow models to expand their purview greatly, they cannot necessarily support the representation of human decision-making in a landscape context; a process that is fundamentally psychological, social, and cultural in nature (Boero and Squazzoni 2005; Brown et al. 2016). However, techniques have been developed to partially address these difficulties. Social media data have been used to understand the links between landscape use and ecosystem services (e.g. García-Palomares et al. 2015; Yoshimura and Hiura 2017; Lee et al. 2019), and sufficient studies exist to allow tentative general conclusions to be drawn through reviews or meta-analyses about some of the socio-ecological processes responsible for those links (e.g. van Vliet et al. 2015; Díaz et al. 2018; Burton et al. 2018b; Feurer et al. 2019; Malek et al. 2019). A substantial body of research now exists on using social 'big data' to answer new questions and develop new models (e.g. Lazer et al. 2009; Giles 2012; Conte et al. 2012), and more active crowd-sourcing provides an alternative means of gathering and analyzing such data (e.g. Lesiv et al. 2019).

Despite these advances, the inherent unpredictability of human decisions, and their inconsistency with quantitative measures of landscape characteristics, require different approaches. As Martin et al. (2016) and Hofman et al. (2017) show, predictive accuracy remains elusive for landscape models, as indeed would be expected in principle (Brown et al. 2016). The strong cultural element of a landscape can only be properly elucidated by interaction with members of that culture, and models are not normally suited to use in such interactions (Matthews and Selman 2006). Traditionally, this requirement has forced process-based land use models to operate over small spatial extents in which stakeholder behavior can be explored in participatory settings (Janssen and Ostrom 2006; Robinson et al. 2007; Smajgl et al. 2011). Upscaling may be facilitated by techniques such as fuzzy cognitive mapping, which allows perceptions to be expressed in ways amenable to quantitative modelling (Gray et al. 2015; Pedde et al. 2019) – if not without some inevitable biases and simplifications (Gray et al. 2014). As well as making models hard to develop and calibrate, the qualitative, subjective nature of human processes makes true model validation (in the sense of identifying one 'correct' representation of a system) impossible, evaluation very complex, and calibration imprecise (Boero and Squazzoni 2005; Windrum et al. 2007; Synes et al. 2016; Brown et al. 2016).

As a result, good modelling practice includes a number of complementary approaches to ensure appropriate usage. Perhaps the most fundamental of these is transparency in model descriptions, giving full, accessible details of model purpose, design, and data usage. In order to facilitate such transparency, a number of schemes have been proposed to guide the content and even the format of model-based publications, ensuring minimum standards and comparability –– to the extent that these schemes are adopted (Schmolke et al. 2010; Grimm et al. 2010; Robinson et al. 2018). Another important component of good practice is making models (and underlying data) open-source, allowing them to be checked, used, and developed by others. Again, a number of schemes have been established to encourage open-source modelling and to provide suitable repositories, and a range of landscape ecological models – particularly agent-based models – are now available in this form (Steiniger and Hay 2009; Thiele and Grimm 2010; CoMSES Net 2019).

Nevertheless, openness and transparency are not in themselves guarantees of appropriate model usage. Indeed, many models are so complex that fully understanding their design and behavior is practically impossible for anyone not involved in model development. For this reason, the burden of ensuring model interpretability must fall more heavily on model developers than it does on model users and requires analysis as well as description. The most necessary type of analysis in this regard is uncertainty analysis; any one of a number of approaches that explore how a model responds to uncertainty in input data, parameterization, or design (Schlüter et al. 2012; Robinson et al. 2018; Elsawah et al. 2020).

While methods for uncertainty analysis are now well-developed (if less well-used, as suggested by Brown et al. 2017; Robinson et al. 2018; Elsawah et al. 2020), they inevitably leave some aspects of model selection to be decided in other ways. For example, simple models usually have fewer and smaller quantifiable uncertainties, requiring only limited setup to produce apparently definite results. But these models typically neglect important aspects of landscape ecological dynamics containing hidden uncertainties that can affect outputs in unrecognized ways (Levis 2010). Similarly, observed outcomes often fall within the uncertainty ranges of many different models, making uncertainty analysis of limited help in model selection even when used to generate likelihoods of agreement with observations (Fortin et al. 2003). Ranges of uncertainty are even greater when diverse future conditions are considered – even to the extent that uncertainty analysis must become an imaginative, qualitative exercise that uses models as exploratory tools rather than discriminating between them on the basis of certainty (Yusoff and Gabrys 2011).

Fundamentally, a range of theoretical and computational approaches are equally valid in landscape ecology, being both philosophically and practically impossible to choose between on the basis of accuracy. As a result, an overarching requirement is that models, while transparent, open-source, and with well-understood uncertainties, should also adopt a diversity of designs, parameterizations, and applications. In this respect, the development of modelling in landscape ecology has not been a process of gradual, directional improvement but one of the cumulative contributions to a multi-perspective, multi-disciplinary understanding of complex systems. This process can only benefit from the increasing scope, urgency, and methodological diversity of landscape ecology as it seeks to engage fully with anthropogenic global change.

Future prospects

From diverse and rapidly evolving current practices, quantitative modelling and computer simulation in landscape ecology are likely to advance substantially in the coming years. Nevertheless, hard-won progress towards greater model coverage, detail, and coupling must be balanced by improved model accessibility, evaluation, and generality, and all these technical development goals must be congruous with improved use of qualitative information and participatory, transdisciplinary approaches. Perhaps most challenging of all, landscape ecology is increasingly required to answer big, pressing questions about socio-ecological dynamics and reconcile contradictory societal objectives (Freeman et al. 2015).

One of the most important responses to these new challenges may be a renewed focus on basic research methods. The novel pressures on socio-ecological systems mean that new data and tools must be built up to complement established approaches; a process that requires time and rigor. Data biases and gaps must be corrected so that a wider range of social and ecological systems can be represented (Bird et al. 2014; GLOBE 2016; Smithsonian Institute 2018; Urban 2019; Malek et al. 2019). Models must also be freed of their dependence on historical patterns so that they can build on these data to explore the emergence of new systems under future conditions (Urban et al. 2016; Brown et al. 2019). New, untested, and poorly understood models have great potential to mislead in these conditions if not developed and applied with care.

That said, modelling does and must continue to push the boundaries of landscape ecology towards greater process accuracy, socio-ecological integration, and exploration of novel dynamics. The parallel development of process-based models in ecology and social science is of great value in this respect. Often, these models focus on particular geographical areas or types of process, but they have also been used to generalize context-dependent knowledge to simulate landscape processes at far larger scales (e.g. Purves et al. 2013; Bocedi et al. 2014;

Grêt-Regamey et al. 2019; Brown et al. 2019). Furthermore, because the processes those models represent are some of the key interfaces between human and natural systems, they provide a firm basis on which to build coupled models that more accurately reflect system dynamics (as, for example, in the dispersal of pollinating insects between changing habitat patches in Synes et al. 2018). Coupled, process-based models, then, allow investigation of new impacts, pressures, and interventions, including those related to thresholds, tipping points, or feedback loops (Liu et al. 2007; Lafuite and Loreau 2017; Cumming and Peterson 2017; Robinson et al. 2018).

Notwithstanding the utility of such models, it will remain necessary to identify new areas of knowledge to which models can contribute. While ecological models have often been used to test and develop theory (e.g. Lawton 1995; Van Nes and Scheffer 2005), social models (outside computational social science) have so far done less, generally adopting a data-driven and theory-neutral perspective (but see Magliocca et al. 2013, Brown et al. 2016, and Meyfroidt et al. 2018). Theoretically aligned or targeted models can do much to illuminate areas of landscape ecology where data are scarce, be those areas geographical, temporal, or thematic in nature. Indeed, modelling inevitably involves the adoption of a particular theoretical perspective, even if this is done implicitly, and a diversity of modelling approaches remains necessary simply in order to explore an appropriate range of such perspectives. Similar cumulative gains can be made by the application of models with a range of data requirements, levels of detail, and breadths of scope; models that when interpreted correctly, reveal different aspects of the systems under study.

This expansion of modelling clearly risks engendering model fatigue (O'Sullivan et al. 2016) but is a fundamentally informative process, even ignoring the practical (and political) impossibility of narrowing consideration down to particular data, theories, applications, or models. As such, modelling remains a necessary part of landscape ecology for the same reasons that it originally developed: because the complexity of real-world landscapes cannot be reduced to metrics, patterns, or laws. Instead, it demands active exploration from a variety of imaginative and technical outposts that lie beyond the reach of other empirical methods. In this respect at least, the age of landscape exploration is far from over.

Summary

1. Quantitative modelling and computer simulation have been highly productive methods in landscape ecology, generating many specific approaches and findings. They have been less successful, however, in terms of synthesizing knowledge and fostering interdisciplinarity;

2. The historical divide between social and ecological modelling remains particularly problematic, with methods in each field at different stages of development and with different foci;

3. Recently, convergence between different strands of landscape ecological modelling has shown substantial promise, allowing links between social and ecological sub-systems to be represented by similar mechanistic or process-based models that utilize large new data resources;

4. These integrated approaches are particularly suitable for addressing pressing questions about landscape ecology in a period of rapid global change that is strengthening the ties between human and natural systems;

5. To capitalize on emerging opportunities and rise to the challenge of urgent new research questions, modelling practice must improve to maximize interpretability, evaluation, and exploration of key issues and exploit new computational and data resources to better represent landscape processes.

References

Abbott, A. (1997) 'Of time and space: the contemporary relevance of the Chicago School', *Social Forces*, vol 75, no 4, pp 1149–1182.

Alexander, P., Rabin, S., Anthoni, P., Henry, R., Pugh, T.A.M., Rounsevell, M.D.A. and Arneth, A. (2018) 'Adaptation of global land use and management intensity to changes in climate and atmospheric carbon dioxide', *Global Change Biology*, vol 24, no 7, pp2791–2809.

Anderson-Teixeira, K.J., Davies, S.J., Bennett, A.C., *et al.* (2015) 'CTFS-ForestGEO: a worldwide network monitoring forests in an era of global change', *Global Change Biology*, vol 21, pp528–549.

Andrewartha, H.G. (1958) 'The use of conceptual models in population ecology', in Cold Spring Harbor Symposia on Quantitative Biology, Volume XXII. Long Island Biological Association, Cold Spring Harbor, Long Island, New York.

Arneth, A., Brown, C. and Rounsevell, M.D.A. (2014) 'Global models of human decision-making for land-based mitigation and adaptation assessment', *Nature Climate Change*, vol 4, pp550–557.

Arroyo-Rodríguez, V., Melo, F.P.L., Martínez-Ramos, M., Bongers, F., Chazdon, R.L., Meave, J.A., Norden, N., Santos, B.A., Leal, I.R. and Tabarelli, M. (2017) 'Multiple successional pathways in human-modified tropical landscapes: new insights from forest succession, forest fragmentation and landscape ecology research', *Biological Reviews*, vol 92, pp326–340.

Aspinall, R. and Pearson, D. (2000) 'Integrated geographical assessment of environmental condition in water catchments: linking landscape ecology, environmental modelling and GIS', *Journal of Environmental Management*, vol 59, pp299–319.

Atran, S., Medin, D.L. and Ross, N.O. (2005) 'The cultural mind: environmental decision making and cultural modeling within and across populations', *Psychological Review*, vol 112, pp744–776.

Ayala, F.J. (1972) 'Competition between species: the diversity of environments in which most organisms live permits the coexistence of many species, even when they compete for the same resources', *American Scientist*, vol 60, pp348–357.

Baddeley, A.J. and Silverman, B.W. (1984) 'A cautionary example on the use of second-order methods for analyzing point patterns', *Biometrics*, vol 40, pp1089–1093.

Baker, W.L. (1989) 'A review of models of landscape change', *Landscape Ecology*, vol 2, pp111–133.

Baker, W.L. (1995) 'Longterm response of disturbance landscapes to human intervention and global change', *Landscape Ecology*, vol 10, pp143–159.

Bartlett, L.J., Newbold, T., Purves, D.W., Tittensor, D.P. and Harfoot, M.B.J. (2016) 'Synergistic impacts of habitat loss and fragmentation on model ecosystems', *Proceedings of the Royal Society B: Biological Sciences*, vol 283, art 20161027.

Barychka, T., Mace, G.M. and Purves, D.W. (2019) 'Modelling variation in bushmeat harvesting among seven African ecosystems using the Madingley Model: yield, survival and ecosystem impacts', *bioRxiv*, art 695924. doi: 10.1101/695924.

Bastian, O. (2002) 'Landscape Ecology: towards a unified discipline?', *Landscape Ecology*, vol 16, pp757–766.

Batty, M. (1980) 'Limits to prediction in science and design science', *Design Studies*, vol 1, pp153–159.

Batty, M. and Torrens, P.M. (2005) 'Modelling and prediction in a complex world', *Futures*, vol 37, pp745–766.

Becher, M.A., Twiston-Davies, G., Penny, T.D., Goulson, D., Rotheray, E.L. and Osborne, J.L. (2018) 'Bumble-BEEHAVE: a systems model for exploring multifactorial causes of bumblebee decline at individual, colony, population and community level', *Journal of Applied Ecology*, vol 55, pp2790–2801.

Bell, D. (1964) 'Twelve modes of prediction: a preliminary sorting of approaches in the social sciences', *Daedalus*, vol 93, pp845–880.

Bengtsson, J., Nilsson, S.G., Franc, A. and Menozzi, P. (2000) 'Biodiversity, disturbances, ecosystem function and management of European forests', *Forest Ecology and Management*, vol 132, pp39–50.

Berman, M., Nicolson, C., Kofinas, G., Tetlichi, J. and Martin, S. (2004) 'Adaptation and sustainability in a small Arctic community: results of an agent-based simulation model', *Arctic*, vol 57, pp401–414.

Bird, T.J., Bates, A.E., Lefcheck, J.S., *et al.* (2014) 'Statistical solutions for error and bias in global citizen science datasets', *Biological Conservation*, vol 173, pp144–154.

Blüthgen, N., Simons, N.K., Jung, K., Prati, D., Renner, S.C., Boch, S., Fischer, M., Hölzel, N., Klaus, V.H., Kleinebecker, T., Tschapka, M., Weisser, W.W. and Gossner, M.M. (2016) 'Land use imperils plant and animal community stability through changes in asynchrony rather than diversity', *Nature Communications*, vol 7, art 10697.

Bocedi, G., Palmer, S.C.F., Pe'er, G., Heikkinen, R.K., Matsinos, Y.G., Watts, K. and Travis, J.M.J. (2014) 'RangeShifter: a platform for modelling spatial eco-evolutionary dynamics and species' responses to environmental changes', *Methods in Ecology and Evolution*, vol 5, pp388–396.

Boero, R. and Squazzoni, F. (2005) 'Does empirical embeddedness matter? Methodological issues on agent-based models for analytical social science', *Journal of Artificial Societies and Social Simulation*, vol 8, no 4, art 6.

Brown, C., Law, R., Illian, J.B. and Burslem, D.F.R.P. (2011) 'Linking ecological processes with spatial and non-spatial patterns in plant communities', *Journal of Ecology*, vol 99, no 6, pp1402–1414.

Brown, C., Burslem, D.F.R.P., Illian, J.B., et al. (2013) 'Multispecies coexistence of trees in tropical forests: spatial signals of topographic niche differentiation increase with environmental heterogeneity', *Proceedings of the Royal Society B: Biological Sciences*, vol 280, art 20130502.

Brown, C., Brown, K. and Rounsevell, M. (2016) 'A philosophical case for process-based modelling of land use change', *Modelling Earth Systems and Environment*, vol 2, art 50.

Brown, C., Alexander, P., Holzhauer, S. and Rounsevell, M.D.A. (2017) 'Behavioral models of climate change adaptation and mitigation in land-based sectors', *WIREs Climate Change*, vol 8, no 2, art e448.

Brown, C., Holzhauer, S., Metzger, M.J., Paterson, J.S. and Rounsevell, M. (2018) 'Land managers' behaviours modulate pathways to visions of future land systems', *Regional Environmental Change*, vol 18, pp831–845.

Brown, C., Seo, B. and Rounsevell, M. (2019) 'Societal breakdown as an emergent property of large-scale behavioural models of land use change', *Earth System Dynamics*, vol 10, pp809–845.

Brown, D.G., Aspinall, R. and Bennett, D.A. (2006) 'Landscape models and explanation in landscape ecology: a space for generative landscape science?', *Professional Geographer*, vol 58, no 4, 369–382.

Burton, V., Metzger, M.J., Brown, C. and Moseley, D. (2018a) 'Green Gold to Wild Woodlands; understanding stakeholder visions for woodland expansion in Scotland', *Landscape Ecology*, vol 100, no 1, pp104–115.

Burton, V., Moseley, D., Brown, C., Metzger, M.J. and Bellamy, P. (2018b) 'Reviewing the evidence base for the effects of woodland expansion on biodiversity and ecosystem services in the United Kingdom', *Forest Ecology and Management*, vol 430, pp366–379.

Castella, J.-C., Trung, T. and Boissau, S. (2005) 'Participatory simulation of land-use changes in the northern mountains of Vietnam: the combined use of an agent-based model, a role-playing game, and a Geographic Information System', *Ecology and Society*, vol 10, no 1, art 27.

Castella, J.-C., Pheng Kam, S., Dinh Quang, D., Verburg, P.H. and Thai Hoanh, C. (2007) 'Combining top-down and bottom-up modelling approaches of land use/cover change to support public policies: application to sustainable management of natural resources in northern Vietnam', *Land Use Policy*, vol 24, pp531–545.

Cederman, L. (2005) 'Computational models of social forms: advancing generative process theory', *American Journal of Sociology*, vol 110, pp864–893.

Chave, J. and Leigh, E.G. (2002) 'A spatially explicit neutral model of β-diversity in tropical forests', *Theoretical Population Biology*, vol 62, pp153–168.

Chesson, P.L. and Warner, R.R. (1981) 'Environmental variability promotes coexistence in lottery competitive systems', *American Naturalist*, vol 117, pp923–943.

Colyvan, M. and Ginzburg, L.R. (2003) 'Laws of nature and laws of ecology', *Oikos*, vol 101, pp649–653.

CoMSES Net (2019) 'CoMSES Net: OpenABM', https://www.comses.net/, accessed 16 May 2019.

Connell, J.H. (1978) 'Diversity in tropical rain forests and coral reefs', Science, vol 199, no 4335, pp1302–1310.

Conte, R., Gilbert, N., Bonelli, G., Cioffi-Revilla, C., Deffuant, G., Kertesz, J., Loreto, V., Moat, S., Nadal, J.-P., Sanchez, A., Nowak, A., Flache, A., San Miguel, M. and Helbing, D. (2012) 'Manifesto of computational social science', *European Physical Journal Special Topics*, vol 214, pp325–346.

Cramer, F. (1993) *Chaos and Order*. VCH Verlagsgesellschaft, Weinheim.

Crane, T. (2010) 'Of models and meanings: cultural resilience in social-ecological systems', *Ecology and Society*, vol 15, no 4, art 19.

Cumming, G.S. and Peterson, G.D. (2017) 'Unifying research on social–ecological resilience and collapse', *Trends in Ecology and Evolution*, vol 32, pp695–713.

Cumming, G.S., Cumming, D.H.M. and Redman, C.L. (2006) 'Scale mismatches in social-ecological systems: causes, consequences, and solutions', *Ecology and Society*, vol 11, no 1, art 14.

Cushman, S.A. and Landguth, E.L. (2010) 'Spurious correlations and inference in landscape genetics', *Molecular Ecology*, vol 19, pp3592–3602.

Cushman, S.A., McGarigal, K. and Neel, M.C. (2008) 'Parsimony in landscape metrics: strength, universality, and consistency', *Ecological Indicators*, vol 8, pp691–703.

Damschen, E.I., Brudvig, L.A., Burt, M.A., Fletcher Jr., R.J., Haddad, N.M., Levey, D.J., Orrock, J.L., Resasco, J. and Tewksbury, J.J. (2019) 'Ongoing accumulation of plant diversity through habitat connectivity in an 18-year experiment', *Science*, vol 365, no 6460, pp1478–1480.

DeAngelis, D.L. and Grimm, V. (2014) 'Individual-based models in ecology after four decades', *F1000Prime Rep*, vol 6, art 39.

Díaz, S., Pascual, U., Stenseke, M., *et al.* (2018) 'Assessing nature's contributions to people', *Science*, vol 359, no 6373, pp270–272.

Donovan, T.M. and Thompson, F.R. (2001) 'Modeling the ecological trap hypothesis: a habitat and demographic analysis for migrant songbirds', *Ecological Applications*, vol 11, pp871–882.

Ellis, E.C. (2012) 'The GLOBE project: accelerating global synthesis of local studies in land change science', *Newsletter of the Global Land Project*, vol 8, pp5–6.

Elsawah, S., Filatova, T., Jakeman, A.J., Kettne, A.J., Zellner, M.L., Athanasiadis, I.N., Hamilton, S.H., Axtell, R.L., Brown, D.G., Gilligan, J.M., Janssen, M.A., Robinson, D.T., Rozenberg, J., Ullah, I.I.T. and Lade, S.J. (2020) 'Eight grand challenges in socio-environmental systems modeling', *Socio-Environmental Systems Modelling*, vol 2, art 16226.

Elton, C.S. (1927) *Animal Ecology*. Sidgwick & Jackson Ltd, London.

European Space Agency (ESA) (2018) 'SENTINEL II', https://www.esa.int/Our_Activities/Observing_the_Earth/Copernicus/Sentinel-2, accessed 26 Apr 2018.

Etherington, T.R., Holland, E.P. and O'Sullivan, D. (2015) 'NLMpy: a python software package for the creation of neutral landscape models within a general numerical framework', *Methods in Ecology and Evolution*, vol 6, pp164–168.

Evans, T.P. and Kelley, H. (2008) 'Assessing the transition from deforestation to forest regrowth with an agent-based model of land cover change for south-central Indiana (USA)', *Geoforum*, vol 39 pp819–832.

Feurer, M., Heinimann, A., Schneider, F., Jurt, C., Myint, W. and Zaehringer, J.G. (2019) 'Local perspectives on ecosystem service trade-offs in a forest frontier landscape in Myanmar', *Land*, vol 8, art 45.

Filatova, T., Polhill, J.G. and van Ewijk, S. (2016) 'Regime shifts in coupled socio-environmental systems: review of modelling challenges and approaches', *Environmental Modelling and Software*, vol 75, pp333–347.

Fisher, R.A., Koven, C.D., Anderegg, W.R.L., *et al.* (2018) 'Vegetation demographics in Earth System Models: a review of progress and priorities', *Global Change Biology*, vol 24, pp35–54.

Fortin, M-J., Boots, B., Csillag, F. and Remmel, T.K. (2003) 'On the role of spatial stochastic models in understanding landscape indices in ecology', *Oikos*, vol 102, pp203–212.

Freeman, O.E., Duguma, L.A. and Minang, P.A. (2015) 'Operationalizing the integrated landscape approach in practice', *Ecology and Society*, vol 20, art 24.

García-Palomares, J.C., Gutiérrez, J. and Mínguez, C. (2015) 'Identification of tourist hot spots based on social networks: a comparative analysis of European metropolises using photo-sharing services and GIS', *Applied Geography*, vol 63, pp408–417.

Gardner, R.H. and Urban, D.L. (2007) 'Neutral models for testing landscape hypotheses', *Landscape Ecology*, vol 22, pp15–29.

Gardner, R.H., Milne, B.T., Turner, M.G. and O'Neill, R.V. (1987) 'Neutral models for the analysis of broad-scale landscape pattern', *Landscape Ecology*, vol 1, pp19–28.

Gibon, A., Sheeren, D., Monteil, C., Ladet, S. and Balent, G. (2010) 'Modelling and simulating change in reforesting mountain landscapes using a social-ecological framework', *Landscape Ecology*, vol 25, pp267–285.

Giles, J. (2012) 'Computational social science: making the links', *Nature*, vol 488, pp448–450.

GLOBE (2016) 'GLOBE project', http://ecotope.org/projects/globe/, accessed 26 Apr 2018.

Goetz, S.J., Hansen, M., Houghton, R.A., Walker, W., Laporte, N. and Busch, J. (2015) 'Measurement and monitoring needs, capabilities and potential for addressing reduced emissions from deforestation and forest degradation under REDD+', *Environmental Research Letters*, vol 10, art 123001.

Gray, S.A., Zanre, E. and Gray, S.R.J. (2014) 'Fuzzy cognitive maps as representations of mental models and group beliefs', in E.I. Papageorgiou (ed) *Fuzzy Cognitive Maps for Applied Sciences and Engineering: From Fundamentals to Extensions and Learning Algorithms*. Springer, Heidelberg.

Gray, S.A., Gray, S., de Kok, J.L., Helfgott, A.E.R., O'Dwyer, B., Jordan, R. and Nyaki, A. (2015) 'Using fuzzy cognitive mapping as a participatory approach to analyze change, preferred states, and perceived resilience of social-ecological systems', *Ecology and Society*, vol 20, no 2, art 11.

Grêt-Regamey, A., Huber, S.H. and Huber, R. (2019) 'Actors' diversity and the resilience of social-ecological systems to global change', *Nature Sustainability*, vol 2, pp 290–297.

Griffiths, P., Nendel, C., Pickert, J. and Hostert, P. (2019) 'Towards national-scale characterization of grassland use intensity from integrated Sentinel-2 and Landsat time series', *Remote Sensing of Environment*, vol 238, art 111124.

Grimm, V. and Railsback, S.F. (2013) *Individual-based Modeling and Ecology*. Princeton University Press, Princeton.

Grimm, V., Berger, U., DeAngelis, D.L., Polhill, J.G., Giske, J. and Railsback, S.F. (2010) 'The ODD protocol: a review and first update', *Ecological Modelling*, vol 221, pp2760–2768.

Grinnell, J. (1917) 'Field tests of theories concerning distributional control', *American Naturalist*, vol 51, pp115–128.

Groeneveld, J., Müller, B., Buchmann, C.M., *et al.* (2017) 'Theoretical foundations of human decision-making in agent-based land use models: a review', *Environmental Modelling and Software*, vol 87, pp39–48.

Grubb, P.J. (1977) 'The maintenance of species-richness in plant communities: the importance of the regeneration niche', *Biological Reviews*, vol 52, pp107–145.

Guillem, E.E., Murray-Rust, D., Robinson, D.T., Barnes, A. and Rounsevell, M.D.A. (2015) 'Modelling farmer decision-making to anticipate tradeoffs between provisioning ecosystem services and biodiversity', *Agricultural Systems*, vol 137, pp12–23.

Gutzwiller, K.J. and Riffell, S.K. (2007) 'Using statistical models to study temporal dynamics of animal-landscape relations', in I. Storch and J.A. Bissonette (eds) *Temporal Dimensions of Landscape Ecology: Wildlife Responses to Variable Resources*. Springer, Boston.

Hagen-Zanker, A. and Lajoie, G. (2008) 'Neutral models of landscape change as benchmarks in the assessment of model performance', *Landscape and Urban Planning*, vol 86, pp284–296.

Hajnal, J. (1958) 'Mathematical models in demography', in Cold Spring Harbor Symposia on Quantitative Biology, Volume 22 – Population Studies: Animal Ecology and Demography. The Biological Laboratory, Long Island Biological Association, Inc., Cold Spring Harbor, Long Island, New York.

Hardy, O.J. and Sonké, B. (2004) 'Spatial pattern analysis of tree species distribution in a tropical rain forest of Cameroon: assessing the role of limited dispersal and niche differentiation', *Forest Ecology and Management*, vol 197, pp191–202.

Harrison, P.A., Holman, I.P., Cojocaru, G., Kok, K., Kontogianni, A., Metzger, M.J. and Gramberger, M. (2012) 'Combining qualitative and quantitative understanding for exploring cross-sectoral climate change impacts, adaptation and vulnerability in Europe', *Regional Environmental Change*, vol 13, pp761–780.

Harrison, P.A., Dunford, R.W., Holman, I.P., Cojocaru, G., Madsen, M.S., Chen, P-Y., Pedde, S. and Sandars, D. (2019) 'Differences between low-end and high-end climate change impacts in Europe across multiple sectors', *Regional Environmental Change*, vol 19, pp695–709.

Harvey, D. (1972) 'Revolutionary and counter revolutionary theory in geography and the problem of ghetto formation', *Antipode*, vol 4, pp1–13.

Heikkinen, R.K., Pöyry, J., Virkkala, R., Bocedi, G., Kuussaari, M., Schweiger, O., Settele, J. and Travis, J.M.J. (2015) 'Modelling potential success of conservation translocations of a specialist grassland butterfly', *Biological Conservation*, vol 192, pp200–206.

Hekkala, A-M. and Roberge, J-M. (2018) 'The use of response measures in meta-analyses of land-use impacts on ecological communities: a review and the way forward', *Biodiversity and Conservation*, vol 27, pp2989–3005.

Henry, R.C., Palmer, S.C.F., Watts, K., Mitchell, R.J., Atkinson, N. and Travis, J.M.J. (2017) 'Tree loss impacts on ecological connectivity: developing models for assessment', *Ecological Informatics*, vol 42, pp90–99.

Hernandez, P.A., Graham, C.H., Master, L.L. and Albert, D.L. (2006) 'The effect of sample size and species characteristics on performance of different species distribution modeling methods', *Ecography*, vol 29, pp773–785.

Hill, R.A., Smith, G.M., Fuller, R.M. and Veitch, N. (2002) 'Landscape modelling using integrated airborne multi-spectral and laser scanning data', *International Journal of Remote Sensing*, vol 23, pp2327–2334.

Hofman, J.M., Sharma, A. and Watts, D.J. (2017) 'Prediction and explanation in social systems', *Science*, vol 355, pp486–488.

Holman, I.P., Brown, C., Carter, T.R., Harrison, P.A. and Rounsevell, M. (2018) 'Improving the representation of adaptation in climate change impact models', *Regional Environmental Change*, vol 19, pp711–721.

Holtz G, Alkemade F, de Haan F, *et al.* (2015) Prospects of modelling societal transitions: position paper of an emerging community. *Environmental Innovation and Societal Transitions* vol 17, pp41–58. doi: 10.1016/J. EIST.2015.05.006.

Houet T, Verburg PH, Loveland TR (2010) Monitoring and modelling landscape dynamics. Landsc Ecol 25:163–167. doi: 10.1007/s10980-009-9417-x.

Hubbell, S.P. (2001) *The Unified Neutral Theory of Biodiversity and Biogeography*. Princeton University Press, Princeton.

Huber, R., Bakker, M., Balmann, A., *et al.* (2018) 'Representation of decision-making in European agricultural agent-based models', *Agricultural Systems*, vol 167, pp143–160.

Hutchinson, G.E. (1958) 'Concluding remarks', *Cold Spring Harbor Symposia on Quantitative Biology*, vol 22, pp415–427.

Illian, J.B., Penttinen, A., Stoyan, H. and Stoyan, D. (2008) *Statistical Analysis and Modelling of Spatial Point Patterns*. Wiley, Chichester.

Janssen, M. and Ostrom, E. (2006) 'Empirically based, agent-based models', *Ecology and Society*, vol 11, no 2, art 37.

Keuschnigg, M., Lovsjö, N. and Hedström, P. (2018) 'Analytical sociology and computational social science', *Journal of Computational Social Science*, vol 1, pp3–14.

Koo, H., Kleemann, J. and Fürst, C. (2018) 'Land use scenario modeling based on local knowledge for the provision of ecosystem services in Northern Ghana', *Land*, vol 7, art 59.

Lafuite, A.S. and Loreau, M. (2017) 'Time-delayed biodiversity feedbacks and the sustainability of social-ecological systems', *Ecological Modelling*, vol 351, pp96–108.

Lausch, A. and Herzog, F. (2002) 'Applicability of landscape metrics for the monitoring of landscape change: issues of scale, resolution and interpretability', *Ecological Indicators*, vol 2, pp3–15.

Lawrence, D.M., Hurtt, G.C., Arneth, A., *et al.* (2016) 'The land use model intercomparison project (LUMIP) contribution to CMIP6: rationale and experimental design', *Geoscientific Model Development*, vol 9, pp2973–2998.

Lawton, J.H. (1995) 'Ecological experiments with model systems', *Science*, vol 269, no 5222, pp328–331.

Lawton, J.H. (1999) 'Are there general laws in ecology?', *Oikos*, vol 84, pp177–192.

Lazer, D., Pentland, A., Adamic, L., Aral, S., Barabasi, A.L., Brewer, D., Christakis, N., Contractor, N., Fowler, J., Gutmann, M., Jebara, T., King, G., Macy, M., Roy, D. and Van Alstyne, M. (2009) 'Computational social science', *Science*, vol 323, no 5915, pp721–723.

Lee, H., Seo, B., Koellner, T. and Lautenbach, S. (2019) 'Mapping cultural ecosystem services 2.0: potential and shortcomings from unlabeled crowd sourced images', *Ecological Indicators*, vol 96, pp505–515.

Lepš, J. (1990) 'Can underlying mechanisms be deduced from observed patterns?', in F. Krahulec, A.D.Q. Agnew and J.H. Willems (eds) *Spatial Processes in Plant Communities*. Academia, Prague.

Lesiv, M., Laso Bayas, J.C., See, L., *et al.* (2019) 'Estimating the global distribution of field size using crowdsourcing', *Global Change Biology*, vol 25, pp174–186.

Levins, R. (1979) 'Coexistence in a variable environment', *American Naturalist*, vol 114, pp765–783.

Levis, S. (2010) 'Modeling vegetation and land use in models of the Earth System', *WIREs Climate Change*, vol 1, pp840–856.

Li, H. and Wu, J. (2004) 'Use and misuse of landscape indices', *Landscape Ecology*, vol 19, pp389–399.

Linderman, M.A., An, L., Bearer, S., Guangming, H., Ouyang, Z. and Liu, J. (2005) 'Modeling the spatio-temporal dynamics and interactions of households, landscapes, and giant panda habitat', *Ecological Modelling*, vol 183, pp47–65.

Lippe, M., Bithell, M., Gotts, N., *et al.* (2019) 'Using agent-based modelling to simulate social-ecological systems across scales', *Geoinformatica*, vol 23, pp269–298.

Liu, J., Dietz, T., Carpenter, S.R., *et al.* (2007) 'Complexity of coupled human and natural systems', *Science*, vol 317, no 5844, pp1513–1516.

Lørup, J.K., Refsgaard, J.C. and Mazvimavi, D. (1998) 'Assessing the effect of land use change on catchment runoff by combined use of statistical tests and hydrological modelling: case studies from Zimbabwe', *Journal of Hydrology*, vol 205, pp147–163.

Magliocca, N.R., Brown, D.G. and Ellis, E.C. (2013) 'Exploring agricultural livelihood transitions with an agent-based virtual laboratory: global forces to local decision-making', *PLoS One*, vol 8, art e73241.

Malek, Ž., Douw, B., Van Vliet, J., Van Der Zanden, E.H. and Verburg, P.H. (2019) 'Local land-use decision-making in a global context', *Environmental Research Letters*, vol 14, art 083006.

Martin, T., Hofman, J.M., Sharma, A., *et al.* (2016) 'Exploring limits to prediction in complex social systems', in *Proceedings of the 25th International Conference on World Wide Web – WWW '16*. ACM Press, New York.

Mas, J-F., Kolb, M., Paegelow, M., Olmedo, M.T.C. and Houet, T. (2014) 'Inductive pattern-based land use/cover change models: a comparison of four software packages', *Environmental Modelling and Software*, vol 51, pp94–111.

Matsumura, S., Beardmore, B., Haider, W., Dieckmann, U. and Arlinghaus, R. (2019) 'Ecological, angler, and spatial heterogeneity drive social and ecological outcomes in an integrated landscape model of freshwater recreational fisheries', *Reviews in Fisheries Science and Aquaculture*, vol 27, pp170–197.

Matthews, R. and Selman, P. (2006) 'Landscape as a focus for integrating human and environmental processes', *Journal of Agricultural Economics*, vol 57, pp199–212.

Matthews, R.B., Gilbert, N.G., Roach, A., Polhill, J.G. and Gotts, N.M. (2007) 'Agent-based land-use models: a review of applications', *Landscape Ecology*, vol 22, pp1447–1459.

Mayer, A.L., Buma, B., Davis, A., Gagné, S.A., Loudermilk, E.L., Scheller, R.M., Schmiegelow, F.K.A., Wiersma, Y.F. and Franklin, J. (2016) 'How landscape ecology informs global land-change science and policy', Bioscience, vol 66, pp458–469.

Meyfroidt, P., Roy Chowdhury, R., de Bremond, A., *et al.* (2018) 'Middle-range theories of land system change', *Global Environmental Change*, vol 53, pp52–67.

Millington, J.D.A., Perry, G.L.W. and Romero-Calcerrada, R. (2007) 'Regression techniques for examining land use/cover change: a case study of a Mediterranean landscape', *Ecosystems*, vol 10, pp562–578.

Millington, J.D.A., Demeritt, D. and Romero-Calcerrada, R. (2011) 'Participatory evaluation of agent-based land-use models', *Journal of Land Use Science*, vol 6, pp195–210.

NASA (2018a) 'NASA: MODIS', https://modis.gsfc.nasa.gov/data/, accessed 26 Apr 2018.

NASA (2018b) 'Socioeconomic Data and Applications Center (SEDAC)', http://sedac.ciesin.columbia.edu/, accessed 26 Apr 2018.

Nassauer, J.I. (1995) 'Culture and changing landscape structure', *Landscape Ecology*, vol 10, pp229–237.

Nathan, R. and Casagrandi, R. (2004) 'A simple mechanistic model of seed dispersal, predation and plant establishment: Janzen-Connell and beyond', *Journal of Ecology*, vol 92, pp733–746.

Naveh, Z. (2007) 'Landscape ecology and sustainability', *Landscape Ecology*, vol 22, pp1437–1440.

Nicolson, M. (1996) 'Humboldtian plant geography after Humboldt: the link to ecology', *British Journal for the History of Science*, vol 29, pp289–310.

Nordström, E-M., Nieuwenhuis, M., Başkent, E.Z., *et al.* (2019) 'Forest decision support systems for the analysis of ecosystem services provisioning at the landscape scale under global climate and market change scenarios', *European Journal of Forest Research*, vol 138, pp561–581.

O'Dwyer, J.P. and Green, J.L. (2010) 'Field theory for biogeography: a spatially explicit model for predicting patterns of biodiversity', *Ecology Letters*, vol 13, pp87–95.

Ogata, Y. and Tanemura, M. (1985) 'Estimation of interaction potentials of marked spatial point patterns through the maximum likelihood method', *Biometrics*, vol 41, pp421–433.

O'Neill, B.C., Kriegler, E., Ebi, K.L., *et al.* (2017) 'The roads ahead: narratives for shared socioeconomic pathways describing world futures in the 21st century', *Global Environmental Change*, vol 42, pp169–180.

O'Sullivan, D., Evans, T., Manson, S., Metcalf, S., Ligmann-Zielinska, A. and Bone, C. (2016) 'Strategic directions for agent-based modeling: avoiding the YAAWN syndrome', *Journal of Land Use Science*, vol 11, pp177–187.

Páez, A. and Scott, D.M. (2005) 'Spatial statistics for urban analysis: a review of techniques with examples', *GeoJournal*, vol 61, pp53–67.

Parker, D.C., Entwisle, B., Rindfuss, R.R., *et al.* (2008) 'Case studies, cross-site comparisons, and the challenge of generalization: comparing agent-based models of land-use change in frontier regions', *Journal of Land Use Science*, vol 3, pp41–72.

Parrott, L. (2017) 'The modelling spiral for solving 'wicked' environmental problems: guidance for stakeholder involvement and collaborative model development', *Methods in Ecology and Evolution*, vol 8, pp1005–1011.

Pedde, S., Kok, K., Onigkeit, J., Brown, C., Holman, I. and Harrison, P.A. (2019) 'Bridging uncertainty concepts across narratives and simulations in environmental scenarios', *Regional Environmental Change*, vol 19, pp655–666.

Perry, G.L.W. and Enright, N.J. (2006) 'Spatial modelling of vegetation change in dynamic landscapes: a review of methods and applications', *Progress in Physical Geography, Earth and Environment*, vol 30, pp47–72.

Phillips, S.J. and Dudík, M. (2008) 'Modeling of species distributions with Maxent: new extensions and a comprehensive evaluation', *Ecography*, vol 31, pp161–175.

Pickett, S.T.A. (1980) 'Non-equilibrium coexistence of plants', *Bulletin of the Torrey Botanical Club*, vol 107, pp238–248.

Pickett, S.T.A. and Cadenasso, M.L. (1995) 'Landscape ecology: spatial heterogeneity in ecological systems', *Science*, vol 269, no 5222, pp331–334.

Pimm, S.L. (1984) 'The complexity and stability of ecosystems', *Nature*, vol 307, pp321–326.

Polhill, J.G., Gimona, A. and Gotts, N.M. (2013) 'Nonlinearities in biodiversity incentive schemes: a study using an integrated agent-based and metacommunity model', Environmental Modelling and Software, vol 45, pp74–91.

Prell, C., Hubacek, K., Reed, M., Quinn, C., Jin, N., Holden, J., Burt, T., Kirby, M. and Sendzimir, J. (2007) 'If you have a hammer everything looks like a nail: traditional versus participatory model building', *Interdiscipinary Science Reviews*, vol 32, pp263–282.

Prevedello, J.A., Gotelli, N.J. and Metzger, J.P. (2016) 'A stochastic model for landscape patterns of biodiversity', *Ecological Monographs*, vol 86, pp462–479.

Prigogine, I. (1980) *From Being to Becoming: Time and Complexity in the Physical Sciences.* WH Freeman and Company, San Francisco.

Purves, D., Scharlemann, J.P.W., Harfoot, M., Newbold, T., Tittensor, D.P., Hutton, J. and Emmott, S. (2013) 'Time to model all life on Earth', *Nature*, vol 493, pp295–297.

Püttker, T., de Arruda Bueno, A., Prado, P.I. and Pardini, R. (2015) 'Ecological filtering or random extinction? Beta-diversity patterns and the importance of niche-based and neutral processes following habitat loss', *Oikos*, vol 124, pp206–215.

Queenborough, S.A., Burslem, D.F.R.P., Garwood, N.C. and Valencia, R. (2007) 'Neighborhood and community interactions deteremine the spatial pattern of tropical tree seedling survival', *Ecology*, vol 88, pp2248–2258.

Ramirez-Reyes, C., Brauman, K.A., Chaplin-Kramer, R., *et al.* (2019) 'Reimagining the potential of Earth observations for ecosystem service assessments', *Science of the Total Environment*, vol 665, pp1053–1063.

Robinson, D.T., Brown, D.G., Parker, D.C., Schreinemachers, P., Janssen, M.A., Huigen, M.., Wittmer, H., Gotts, N., Promburom, P., Irwin, E., Berger, T., Gatzweiler, F. and Barnaud, C. (2007) 'Comparison of empirical methods for building agent-based models in land use science', *Journal of Land Use Science*, vol 2, pp31–55.

Robinson, D.T., Brown, D.G. and Currie, W.S. (2009) 'Modelling carbon storage in highly fragmented and human-dominated landscapes: linking land-cover patterns and ecosystem models', *Ecological Modelling*, vol 220, no 9–10, pp1325–1338.

Robinson, D.T., Murray-Rust, D., Rieser, V., Milicic, V. and Rounsevell, M. (2012) 'Modelling the impacts of land system dynamics on human well-being: using an agent-based approach to cope with data limitations in Koper, Slovenia', *Computers, Environment and Urban Systems*, vol 36, pp164–176.

Robinson, D.T., Di Vittorio, A., Alexander, P., Arneth, A., Michael Barton, C. Brown, D.G., Kettner, A., Lemmen, C., O'Neill, B.C., Janssen, M., Pugh, T.A.M., Rabin, S.S., Rounsevell, M., Syvitski, J.P., Ullah, I. and Verburg, P.H. (2018) 'Modelling feedbacks between human and natural processes in the land system', *Earth System Dynamics*, vol 9, pp895–914.

Rosa, I.M.D., Ahmed, S.E. and Ewers, R.M. (2014) 'The transparency, reliability and utility of tropical rainforest land-use and land-cover change models', *Global Change Biology*, vol 20, pp1707–1722.

Rounsevell, M.D.A., Berry, P.M. and Harrison, P.A. (2006) 'Future environmental change impacts on rural land use and biodiversity: a synthesis of the ACCELERATES project', *Environmental Science and Policy*, vol 9, pp93–100.

Rounsevell, M.D.A., Arneth, A., Alexander, P., Brown, D.G., de Noblet-Ducoudré, N., Ellis, E., Finnigan, J., Galvin, K., Grigg, N., Harman, I., Lennox, J., Magliocca, N., Parker, D., O'Neill, B.C., Verburg, P.H. and Young, O. (2014) 'Towards decision-based global land use models for improved understanding of the Earth system', *Earth System Dynamics*, vol 5, pp117–137.

Rudner, M., Biedermann, R., Schröder, B. and Kleyer, M. (2007) 'Integrated Grid Based Ecological and Economic (INGRID) landscape model – A tool to support landscape management decisions', *Environmental Modelling and Software*, vol 22, pp177–187.

Saito, O., Hashimoto, S., Managi, S., Aiba, M., Yamakita, T., DasGupta, R. and Takeuchi, K. (2019) 'Future scenarios for socio-ecological production landscape and seascape', *Sustainability Science*, vol 14, pp1–4.

Salonen, V., Penttinen, A. and Sarkka, A. (1992) 'Plant colonization of a bare peat surface: population changes and spatial patterns', *Journal of Vegetation Science*, vol 3, pp113–118.

Sayer, A. (1982) 'Explanation in economic geography: abstraction versus generalization', *Progress in Human Geography*, vol 6, pp68–88.

Scheller, R.M., Domingo, J.B., Sturtevant, B.R., Williams, J.S., Rudy, A., Gustafson, E.J. and Mladenoff, D.J. (2007) 'Design, development, and application of LANDIS-II, a spatial landscape simulation model with flexible temporal and spatial resolution', *Ecological Modelling*, vol 201, pp409–419.

Schlüter, M., McAllister, R.R.J., Arlinghaus, R., Bunnefeld, N., Eisenack, K., Hölker, F., Milner-Gulland, E.J., Müller, B., Nicholson, E., Quaas, M. and Stöven, M. (2012) 'New horizons for managing the environment: a review of coupled social-ecological systems modeling', *Natural Resource Modeling*, vol 25, pp219–272.

Schmolke, A., Thorbek, P., DeAngelis, D.L. and Grimm, V. (2010) 'Ecological models supporting environmental decision making: a strategy for the future', *Trends in Ecology and Evolution*, vol 25, pp479–486.

Schößer, B., Helming, K. and Wiggering, H. (2010) 'Assessing land use change impacts: a comparison of the SENSOR land use function approach with other frameworks', *Journal of Land Use Science*, vol 5, pp159–178.

Schröder, B. and Seppelt, R. (2006) 'Analysis of pattern–process interactions based on landscape models: overview, general concepts, and methodological issues', *Ecological Modelling*, vol 199, pp505–516.

Schröder, B., Rudner, M., Biedermann, R., Kögl, H. and Kleyer, M. (2008) 'A landscape model for quantifying the trade-off between conservation needs and economic constraints in the management of a semi-natural grassland community', *Biological Conservation*, vol 141, pp719–732.

Schulze, J., Müller, B., Groeneveld, J. and Grimm, V. (2017) 'Agent-Based Modelling of social-ecological systems: achievements, challenges, and a way forward', *Journal of Artificial Societies and Social Simulation*, vol 20, art 8.

Sciaini, M., Fritsch, M., Scherer, C. and Simpkins, C.E. (2018) 'NLMR and landscapetools: an integrated environment for simulating and modifying neutral landscape models in R', *Methods in Ecology and Evolution*, vol 9, pp2240–2248.

Scown, M.W., Winkler, K.J. and Nicholas, K.A. (2019) 'Aligning research with policy and practice for sustainable agricultural land systems in Europe', *Proceedings of the National Academy of Sciences of the USA*, vol 116, pp4911–4916.

Seidl, R., Rammer, W., Scheller, R.M. and Spies, T.A. (2012) 'An individual-based process model to simulate landscape-scale forest ecosystem dynamics', *Ecological Modelling*, vol 231, pp87–100.

Serneels, S. and Lambin, E.F. (2001) 'Proximate causes of land-use change in Narok District, Kenya: a spatial statistical model', *Agriculture, Ecosystems and Environment*, vol 85, pp65–81.

Smajgl, A., Brown, D.G., Valbuena, D. and Huigen, M.G.A. (2011) 'Empirical characterisation of agent behaviours in socio-ecological systems', *Environmental Modelling and Software*, vol 26, pp837–844.

Smithsonian Institute (2018) 'ForestGEO: global earth observatory network', https://forestgeo.si.edu/, accessed 25 Apr 2018.

Stehfest, E., van Zeist, W.-J., Valin, H., Havlik, P., Popp, A., Kyle, P., Tabeau, A., Mason-D'Croz, D., Hasegawa, T., Bodirsky, B.L., Calvin, K., Doelman, J.C., Fujimori, S., Humpenöder, F., Lotze-Campen, H., van Meijl, H. and Wiebe, K. (2019) 'Key determinants of global land-use projections', *Nature Communications*, vol 10, art 2166.

Steiniger, S. and Hay, G.J. (2009) 'Free and open source geographic information tools for landscape ecology', *Ecological Informatics*, vol 4, pp183–195.

Stephenson, J. (2008) 'The cultural values model: an integrated approach to values in landscapes', *Landscape and Urban Planning*, vol 84, pp127–139.

Sterner, R.W., Ribic, C.A. and Schatz, G.E. (1986) 'Testing for life historical changes in spatial patterns of four tropical tree species', *Journal of Ecology*, vol 74, pp621–633.

Stoyan, D. and Penttinen, A. (2000) 'Recent applications of point process methods in forestry statistics', *Statistical Science*, vol 15, pp61–78.

Synes, N.W., Brown, C., Watts, K., White, S.M., Gilbert, M.A. and Travis, J.M.J. (2016) 'Emerging opportunities for landscape ecological modelling', *Current Landscape Ecology Reports*, vol 1, pp146–167.

Synes, N.W., Brown, C., Palmer, S.C.F., Bocedi, G., Osborne, P.E., Watts, K., Franklin, J. and Travis, J.M.J. (2018) 'Coupled land use and ecological models reveal emergence and feedbacks in socio-ecological systems', *Ecography*, vol 42, no 4, pp814–825.

Temple, S.A. and Cary, J.R. (1988) 'Modeling dynamics of habitat-interior bird populations in fragmented landscapes', *Conservation Biology*, vol 2, pp340–347.

Thiele, J.C. and Grimm, V. (2010) 'NetLogo meets R: linking agent-based models with a toolbox for their analysis', *Environmental Modelling and Software*, vol 25, pp972–974.

Thorn, A.M., Wake, C.P., Grimm, C.D., Mitchell, C.R., Mineau, M.M. and Ollinger, S.V. (2017) 'Development of scenarios for land cover, population density, impervious cover, and conservation in New Hampshire, 2010–2100', *Ecology and Society*, vol 22, no 4, art 19.

Thuiller, W., Guéguen, M., Renaud, J., Karger, D.N. and Zimmermann, N.E. (2019) 'Uncertainty in ensembles of global biodiversity scenarios', *Nature Communications*, vol 10, art 1446.

Tischendorf, L. (2001) 'Can landscape indices predict ecological processes consistently?', *Landscape Ecology*, vol 16, pp235–254.

Titeux, N., Henle, K., Mihoub, J-B., Regos, A., Geijzendorffer, I.R., Cramer, W., Verburg, P.H. and Brotons, L. (2016) 'Biodiversity scenarios neglect future land-use changes', *Global Change Biology*, vol 22, pp2505–2515.

Touza, J., Drechsler, M., Smart, J.C.R. and Termansen, M. (2013) 'Emergence of cooperative behaviours in the management of mobile ecological resources', *Environmental Modelling and Software*, vol 45, pp52–63.

Tuomisto, H., Ruokolainen, K. and Yli-Halla, M. (2003) 'Dispersal, environment, and floristic variation of Western Amazonian forests', *Science*, vol 299, no 5604, pp241–244.

Turchin, P. (2001) 'Does population ecology have general laws?', *Oikos*, vol 94, pp17–26.

Turner, M.G. (1989) 'Landscape ecology: the effect of pattern on process', *Annual Review of Ecology and Systematics*, vol 20, pp171–197.

Turner, M.G. (1990) 'Spatial and temporal analysis of landscape patterns', *Landscape Ecology*, vol 4, pp21–30.

Turner, M.G., Gardner, R.H. and O'Neill, R.V. (2001) *Landscape Ecology in Theory and Practice: Pattern and Process*. Springer, Heidelberg.

Urban, M.C. (2015) 'Accelerating extinction risk from climate change', *Science*, vol 348, no 6234, pp571–573.

Urban, M.C. (2019) 'Projecting biological impacts from climate change like a climate scientist', *WIREs Climate Change*, vol 10, no 4, art e585.

Urban, M.C., Bocedi, G., Hendry, A.P., *et al.* (2016) 'Improving the forecast for biodiversity under climate change', *Science*, vol 353, no 6304, art aad8466.

Urban, M.C., Scarpa, A., Travis, J.M.J. and Bocedi, G. (2019) 'Maladapted prey subsidize predators and facilitate range expansion', *American Naturalist*, vol 194, pp590–612.

Van Berkel, D.B. and Verburg, P.H. (2012) 'Combining exploratory scenarios and participatory backcasting: using an agent-based model in participatory policy design for a multi-functional landscape', *Landscape Ecology*, vol 27, pp641–658.

van der Sande, M.T., Poorter, L., Balvanera, P., *et al.* (2017) 'The integration of empirical, remote sensing and modelling approaches enhances insight in the role of biodiversity in climate change mitigation by tropical forests', *Current Opinion in Environmental Sustainability*, 26–27, 69–76.

Van Nes, E.H. and Scheffer, M. (2005) 'A strategy to improve the contribution of complex simulation models to ecological theory', *Ecological Modelling*, vol 185, pp153–164.

van Vliet, J., de Groot, H.L.F., Rietveld, P. and Verburg, P.H. (2015) 'Manifestations and underlying drivers of agricultural land use change in Europe', *Landscape and Urban Planning*, vol 133, pp24–36.

van Vuuren, D.P. and Carter, T.R. (2014) 'Climate and socio-economic scenarios for climate change research and assessment: reconciling the new with the old', *Climatic Change*, vol 122, pp415–429.

Vanwalleghem, T., Stockmann, U., Minasny, B. and McBratney, A.B. (2013) 'A quantitative model for integrating landscape evolution and soil formation', *Journal of Geophysical Research: Earth Surface*, vol 118, pp331–347.

Veldkamp, A. and Lambin, E. (2001) 'Predicting land-use change', *Agriculture, Ecosystems and Environment*, vol 85, pp1–6.

Verburg, P.H. and Overmars, K.P. (2009) 'Combining top-down and bottom-up dynamics in land use modeling: exploring the future of abandoned farmlands in Europe with the Dyna-CLUE model', *Landscape Ecology*, vol 24, pp1167–1181.

Verburg, P.H., Soepboer, W., Veldkamp, A., Limpiada, R., Espaldon, V. and Mastura, S.S.A. (2002) 'Modeling the spatial dynamics of regional land use: the CLUE-S model', *Environmental Management*, vol 30, pp391–405.

Verheijen, L.M., Aerts, R., Brovkin, V., Cavender-Bares, J., Cornelissen, J.H.C., Kattge, J. and van Bodegom, P.M. (2015) 'Inclusion of ecologically based trait variation in plant functional types reduces the projected land carbon sink in an earth system model', *Global Change Biology*, vol 21, pp3074–3086.

Verkerk, P.J., Lindner, M., Pérez-Soba, M., *et al.* (2018) 'Identifying pathways to visions of future land use in Europe', *Regional Environmental Change*, vol 18, pp817–830.

Vincenot, C.E., Giannino, F., Rietkerk, M., Moriya, K. and Mazzoleni, S. (2011) 'Theoretical considerations on the combined use of system dynamics and individual-based modeling in ecology', *Ecological Modelling*, vol 222 pp210–218.

Voinov, A., Costanza, R., Wainger, L., Boumans, R., Villa, F., Maxwell, T. and Voinov, H. (1999) 'Patuxent landscape model: integrated ecological economic modeling of a watershed', *Environmental Modelling and Software*, vol 14, pp473–491.

von Humboldt, A. and Bonpland, A. (1805) *Essai sur la géographie des plantes*. Levrault, Schoell & Company, Paris.

Wang, Q. and Malanson, G.P. (2008) 'Neutral landscapes: bases for exploration in landscape ecology', *Geography Compass*, vol 2, pp319–339.

Watts, D.J. (2017) 'Should social science be more solution-oriented?', *Nature Human Behaviour*, vol 1, art 0015.

Watts, K., Fuentes-Montemayor, E., Macgregor, N.A., Peredo-Alvarez, V., Ferryman, M., Bellamy, C., Brown, N. and Park, K.J. (2016) 'Using historical woodland creation to construct a long-term, large-scale natural experiment: the WrEN project', *Ecology and Evolution*, vol 6, pp3012–3025.

Wiegand, T. and Moloney, K.A. (2004) 'Rings, circles, and null-models for point pattern analysis in ecology', *Oikos*, vol 104, pp209–229.

Wiegand, T., Jeltsch, F., Hanski, I. and Grimm, R. (2003) 'Using pattern-oriented modelling for revealing hidden information: a key for reconciling ecological theory and application', *Oikos*, vol 100, pp209–222.

Wiegand, T., Gunatilleke, C.V.S., Gunatilleke, I.A.U.N. and Huth, A. (2007a) 'How individual species structure diversity in tropical forests', *Proceedings of the National Academy of Sciences*, vol 104, pp19029–19033.

Wiegand, T., Gunatilleke, S. and Gunatilleke, N. (2007b) 'Species associations in a hetergeneous Sri Lankan dipterocarp forest', *American Naturalist*, vol 170, ppE77–E95.

Wiens, J.A. (1977) 'On competition and variable environments: populations may experience "ecological crunches" in variable climates, nullifying the assumptions of competition theory and limiting the usefulness of short-term studies of population patterns', *American Scientist*, vol 65, pp590–597.

Windrum, P., Fagiolo, G. and Moneta, A. (2007) 'Empirical validation of agent-based models: alternatives and prospects, *Journal of Artificial Societies and Social Simulation*, vol 10, no 2, art 8.

Winkler, K.J., Scown, M.W. and Nicholas, K.A. (2018) 'A classification to align social-ecological land systems research with policy in Europe', *Land Use Policy*, vol 79, pp137–145.

World Bank (2018) 'World bank open data|data', https://data.worldbank.org/, accessed 26 Apr 2018.

Wu, J. (2006) 'Landscape ecology, cross-disciplinarity, and sustainability science', *Landscape Ecology*, vol 21, pp1–4.

Wu, J. (2010) 'Landscape of culture and culture of landscape: does landscape ecology need culture?', *Landscape Ecology*, vol 25, pp1147–1150.

Wu, J. and Loucks, O.L. (1995) 'From balance of nature to hierarchical patch dynamics- a paradigm shift in ecology', *Quarterly Review of Biology*, vol 70, pp441–466.

Yoshimura, N., and Hiura, T. (2017) 'Demand and supply of cultural ecosystem services: use of geotagged photos to map the aesthetic value of landscapes in Hokkaido', *Ecosystem Services*, vol 24, pp68–78.

Yusoff, K. and Gabrys, J. (2011) 'Climate change and the imagination', *WIREs Climate Change*, vol 2, pp516–534.

17

Landscape character assessment and participatory approaches

Andrew Butler and Ingrid Sarlöv Herlin

Introduction

Landscape characterization (or landscape character assessment [LCA]) is a process of identifying what makes a landscape unique. It is considered a judgment-free process: not assigning value but rather, developing an understanding through describing the elements that make up the character of the landscape. Characterization approaches attempt to capture the diverse complexity of a landscape through a process of determining values attributed to a landscape, where each element plays its part in defining character and creating the landscape as a whole. Landscape character assessment provides an image of the state of the landscape, or a baseline for managing change, ensuring that transformations are planned with consideration of their wider surroundings and ideally, contribute positively to the landscape (Butler and Berglund, 2014). Consequently, the knowledge attained through a landscape assessment informs future planning and identification of landscape resources.

Numerous LCA approaches have emerged, based on national and disciplinary traditions and policies, yet it is the LCA approach developed in England and Scotland towards the end of the 20th century that has gained prominence (Fairclough et al., 2018b, Sarlöv Herlin, 2016). The United Kingdom (UK) approach has since been widely utilized in Europe and beyond (Käyhkö et al., 2018, Loupa-Ramos and Pinto-Correia, 2018, Caspersen, 2009) and now represents the dominant approach to addressing landscapes in planning issues (Selman, 2010, Fairclough et al., 2018b). The LCA approach is the focus of this chapter.

The LCA approach promotes a specific view of the landscape as 'the relationship between people and place. It provides the setting for our day-to-day lives' and emphasizes that '[p]eople's perceptions turn land into the concept of landscape' (Swanwick and Land Use Consultants, 2002, Tudor, 2014). This view of the landscape moves away from the visual and physical aspects, which have dominated planning, and places those who directly experience the landscape as central to its formation.

Through this chapter, we explore the relevance of the European Landscape Convention (ELC) as a support and guide for public participation in LCA. We follow this by drawing out factors that influence meaningful participation in the assessment of a landscape, using vignettes to highlight specific aspects of participation. Finally, we reflect on the shortfalls in participation in landscape assessment and also how landscape assessments can be beneficial to democratic systems.

European Landscape Convention

The academic discussion and practical endeavors that spawned the LCA approach also informed the production of the ELC, which has become an influential tool for forwarding landscapes in planning issues. The ELC is the first international treaty to directly address landscapes, providing a geographical dimension to the Council of Europe (CoE) aims of promoting democracy and protecting human rights by realizing and legitimizing the surroundings of citizens' everyday life. Conventions of the CoE do not constitute European law, representing the moral voice of Europe (Olwig, 2007), yet the ELC requires signatories to recognize landscapes in national laws of the signatory states and implement the convention in line with existing legal systems and policy frameworks.

Accordingly, the ELC does not dictate how landscapes should be handled. Yet, it defines how landscapes should be recognized and hence, what is brought into law. A landscape becomes defined and legitimized as

> an area, as perceived by people, whose character is the result of the action and interaction of natural and/or human factors' (Chapter 1, Art 1).

This makes a landscape reliant on the perceptions of the people who experience it, moving it away from being solely a professional arena for engagement. Landscapes become seen as being informed and created by those who inhabit them; they are where people engage with everyday life and are central for defining individual and group identities. Landscapes thus become the all-encompassing surroundings to life, understood as holistic entities. Recognition that landscapes are of benefit to all and of value for individuals and society is central to the ELC, as is the realization that an understanding of landscapes is reliant on the individuals and groups who experience it.

European Landscape Convention (ELC) and participation

The ELC can be seen as a bastion for public participation in landscape issues, providing moral authority and a basis for member states to develop inclusive strategies. Participation in landscape issues opens up a space for participants to express differences and provides the opportunity to question dominant discourses and values. This reflects a growing recognition that the whole population, and not just experts, should be involved in planning processes (Stephenson, 2010, Selman, 2012). Participation within the ELC also has the possibility to develop democracy and legitimacy and provide social justice (Jones, 2007, Calderon and Butler, 2019). As such, participation should be seen as more than just handing over knowledge and hoping that it will be utilized effectively. Yet, even if participation is encouraged, numerous factors effect individuals' willingness to participate in landscape issues, including attachment to the local area, confidence in tangible outcomes, and an interest in the development of their local landscape. Subsequently, participatory processes should focus on these elements to provide motivation for engagement in landscape issues (Höppner et al., 2008).

The democratic values of the CoE are taken up in the general measures of the convention as the need to establish procedures for the participation of the general public, local and regional authorities, and other parties with an interest in the definition and implementation of the landscape policies (Chapter 2, Art 5c).

The legal framework of the ELC also builds on the United Nations Economic Commission for Europe's 'Aarhus Convention on Public Participation in Decision-Making and Access to Justice on Environmental Matters'. The Aarhus convention recognizes that 'public participation in decision-making enhance[s] the quality and the implementation of decisions, contribute[s] to public awareness of environmental issues, give[s] the public the opportunity to express its concerns and enable[s] public authorities to take due account of such concern' (UNECE, 1998: 2).

The views of the CoE are further bolstered through the Nafpion declaration of 2015 that public participation 'is an exercise in democracy and consolidates the legitimacy of a shared decision-making process and the sustainability of its outcomes' and 'enables communities to develop and pursue a shared vision for their territory and enhances their sense of belonging' (Council of Europe, 2014 2.A).

A general approach to landscape assessment

A landscape assessment presents a description of the landscape for others to argue for its values and to provide insight and understanding of place (Stahlschmidt et al., 2017). In essence, a landscape assessment can be seen as providing a solid knowledge base prior to planning interventions in the landscape. Landscape assessments are central for implementation of the ELC; they are seen as a means for a ratifying nation to the convention to 'identify its own landscapes', 'analyse their characteristics and the forces and pressures transforming them', and 'take note of changes' (CoE, 2000a, Ch II, Art. 6). Such action must be based on detailed knowledge of the characteristics of each landscape, the evolutionary processes affecting it, and values assigned to the landscape by the population concerned (Jones, 2007). The methods used to represent the landscape are instrumental in defining how the landscape is understood (Geelmuyden, 2016, Brunetta and Voghera, 2008). Yet, traditional approaches to landscape assessment tend to build on natural science ideals of objectivity, approaches that struggle to deal with the subjective and ephemeral nature that is an essential aspect of a landscape (Geelmuyden, 2016, Dakin, 2003, Brunetta and Voghera, 2008). As society's and individuals' perceptions are fundamental in the formation of the concept of landscape, their multiple and varied values and knowledge must be considered (Jørgensen et al., 2016). The recognition that people's perceptions are central to understanding landscapes places participation as central to attaining true insight into what a landscape means and how a landscape is identified (Council of Europe, 2000).

Landscape character assessment approach

Since the turn of the century, the LCA approach from the UK has gained recognition across Europe due in part to the freely accessible and comprehensive online landscape character assessment guidelines (Swanwick and Land Use Consultants, 2002). LCA has since been recognized as a tool for contributing to the implementation of the ELC (Dower, 2007). A partial update of the 2002 guidelines was produced in 2014 (Tudor, 2014) after the post-financial crisis public sector cuts and was considered by many consultants as a step backwards.

The LCA approach is broadly divided into two stages: characterization and judgment-making. Characterization is seen as a relatively value-free stage, defining and describing the landscape character. Although it is the judgment stage that often demands public engagement,

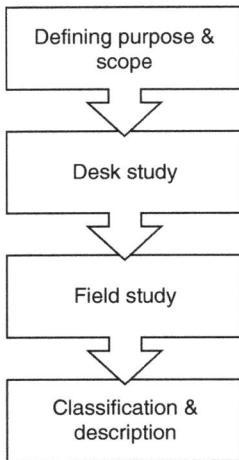

Figure 17.1 Simplified overview of the LCA characterization process.

it is the characterization stage, when the landscape is framed and defined, that is the focus of this chapter (Figure 17.1). Landscape characterization involves the purposeful identification and assessment of the social, cultural, economic, and environmental interactions and relationships that constitute the landscape. Characterization has a central role in informing, framing, and directing landscape decision-making (Butler, 2016, Dalglish and Leslie, 2016). The process identifies certain characteristics as relevant and legitimate while marginalizing others. The resulting division into distinct character areas is not political or administrative; the areas represent living spaces, meeting places, and vessels of collective memories (Nogué and Sala 2018). However, while the LCA approach raises the importance of the public for understanding a landscape, the landscape characterizations themselves are predominantly prepared by and used by professionals (Conrad et al., 2011b, Butler and Berglund, 2014).

The initial stage of the characterization process is to define the purpose and scope of the assessment, the resources that are available, the method and scale of the assessment, and consequently, how landscape is recognized (Figure 17.1). The next stage is the desk study, where an initial understanding of the landscape is attained primarily through geographic information system (GIS) analysis, providing the context for draft landscape character areas. This is followed by a field survey, which involves identifying sensory aspects in the landscape, refining boundaries, assessing the condition of the landscape, and corroborating desk study information. Finally, the data are drawn together to provide a communicable document of the landscape's character. The characterization phase in turn informs the subsequent stage of making judgments and decisions made relating to the landscape. How these judgments are determined should be transparent and traceable from the results of the characterization process.

Landscape character assessment and participation

The LCA guidelines espouse the significance of individuals with attachment to a place for understanding a landscape, making public perceptions central to any genuine landscape assessment. The LCA provides the opportunity to bring public involvement into the planning process at a very early stage, allowing the public to be included in defining and identifying landscapes as outlined in the ELC (Butler and Berglund, 2014).

Engaging the public in LCAs can provide numerous benefits: (1) help individuals to understand and be aware of the landscape, (2) provide valuable information, (3) result in engaged individuals who are more likely to be committed to the outcome, (4) help build consensus, and (5) help mobilize a broader spectrum of individuals and organizations for implementing subsequent goals (Swanwick et al., 2002, Caspersen, 2009). The LCA guidelines provide minimal information on how to engage and incorporate the public in the landscape assessment. However, the benefit of some stakeholder involvement, especially at local scale and in the later judgment stages, is recommended (Tudor, 2014).

A supplementary topic paper to the 2002 LCA guidelines (Topic paper 3: Landscape Character Assessment – How Stakeholders Can Help) elaborated further: 'Stakeholders play a vital role in Landscape Character Assessment. Their involvement can produce a more informed assessment, greater ownership of applications, and establish valuable partnerships for future work' (Swanwick et al., 2002: 1). Although supplementary, this topic paper is seen as an important document that continues to be used for practical suggestions for involving the public.

While calls for participation have increased, evidence from the literature reveals a gap between participation rhetoric in policy and participation as practice (Scott, 2011, Conrad et al., 2011b, Butler and Berglund, 2014). Whereas academia has engaged critically with participation in landscape issues, practice has driven numerous ways of operationalizing participation. Endeavors from practitioners have been framed in academic discussion as a means to enhance human rights and democratic processes (Fairclough et al., 2018a, Conrad et al., 2011b, Butler and Berglund, 2014). Consequently, scholarly attempts have been made to define aspects of 'good practice' to ensure that the involvement of the public produces real and tangible benefits. In this chapter, we utilize these aspects of participation as an analytical means for addressing participation. We broadly divide these aspects into *practicalities* of the process and the *product* of the process. Practicalities relate to what (scope), when (timing), who (representativeness), and how (convenience) of participation; product refers to influence and knowledge transfer in the participatory process (Butler and Berglund, 2014, Conrad et al., 2011b, Butler, 2016). It is these operationalized norms that the rest of the chapter builds on. The vignettes presented through the following sections highlight examples where participation has been included in the approach to LCA and illustrate aspects of 'good practice'.

Scope of participation

The scope of participation considers the degree to which stakeholders or the public are included (see Table 17.1). Since Arnstein's 'ladder of participation' (Arnstein, 1969), various 'levels' of participation have been recognized as relevant for different processes. The International Association of Public Participation (IAP2) devised a framework that includes five levels of participation. The lowest level of meaningful participation is 'Inform', where the public is helped to understand; the next level, 'Consult', aims at gaining feedback from the public; 'Involve' requires the consideration of public values and aspirations; 'Collaborate' necessitates partnerships with the public to identify solutions; and finally, 'Empower' places final decision-making in the hands of the public. The latter three categories can be seen as more 'active participation' (IAP2, 2007). We utilize the IAP2 categories to represent levels of participation in LCA (Table 17.1)

In practice, participation in most landscape assessments is limited to informing the public or verifying professional studies, or (if ambitious) consultation on draft LCA documents prepared by experts and/or the provision of public input into results (Butler and Berglund, 2014). Few processes involve the public as active participants in the process of compiling data for the LCA

Table 17.1 Levels of participation in a landscape character assessment, based on International Association of Public Participation (IAP2). The large circle represents the practitioners who compile the characterization, the small circles represent the individuals or groups whose knowledge is incorporated in the LCA, and the arrows represent the flow of information between the different actors during the characterization process

Inform	Providing the public information with information about their landscape, the issues affecting their landscape, or the assessment being used to understand the landscape.	
Verify	The public provides verification of the professional-based understanding of the landscape or professionals' interpretation of the public's view.	
Consult	The public feeds information into the assessment through consultation.	
Involve	The public are engaged in discussing their own landscape values and the issues affecting them in the landscape. Approaches seek two-way exchange and empowerment of participants.	
Collaborate	The public co-create the assessment and the meaning of the landscape in a multidirectional flow of knowledge with practitioners.	
Empower	The public own the process and the final product. They define their landscape on their terms (see Vignette 17.1).	

(Conrad et al., 2011b). In professional-led processes, it is unrealistic to think that empowerment of the public can be attained. However, community-driven initiatives can come close to achieving such lofty aims (James and Gittins, 2007) (Vignette 17.1).

Vignette 17.1: Community landscape character assessments (UK)

At a local scale, there is an increasing number of community landscape character assessments (CLCAs) being undertaken in the UK as support for statutory Neighbourhood Plans. It is easier to draw the public into landscape issues at a local scale. CLCAs are basically the same as professionally-driven LCAs but undertaken by the community. They allow the community to highlight what they value. Such approaches allow the local community to better understand their landscape and at the same time, express what makes the landscape special for them. There are two distinct alternatives with CLCAs: one is that the community defines the character areas and undertakes the entire analysis; the other is that the community further develops professionally defined character areas with identified boundaries.

While this work can be driven entirely by a local community, there is often a need for help with technical solutions, e.g. accessing maps and data collection. This work can only really be successfully undertaken where there is an active community, yet representation issues can arise if the process is driven by a single group. Several toolkits or guides exist for this type of work: Unlocking the Landscape (England), Tirlun Lleol (Wales), Talking About Our Place (Scotland), and Landscape Identification (ECOVAST, European initiative).[1]

Timing of participation

There is a widespread recognition that participants should be included as early as possible in the process, when there is the potential to influence scope and methods (Rowe and Frewer, 2000, Conrad et al., 2011b, Cox, 2006). Landscape assessment signifies the early stage of the landscape planning process, the point where the landscape is framed and diverse landscape values can be recognized. Thus, LCA represents a significant point at which involvement of the general public can be instigated.

Timing is inextricably linked to scope of participation. Numerous stages of the characterization process are considered potential for participation: during desk study, while undertaking draft characterization, in the classification and description phase, or later through judgment-making (Figure 17.2) (Swanwick and Land Use Consultants, 2002). Due to practicalities relating to finances and resources, there is no definitive consensus on when is most feasible.

Although the importance of early inclusion of the public is widely recognized and is linked to the scope of participation, public engagement is characteristically restricted to the final stages of the assessment process. At this point, the public either validate documents or provide limited input into the characterization of the landscape. However, there are examples where the public have been involved at an earlier stage, including establishing the nature of the LCA process, framing the landscape characterization process, and gathering relevant data for setting the agenda of the assessment (Vignette 17.2).

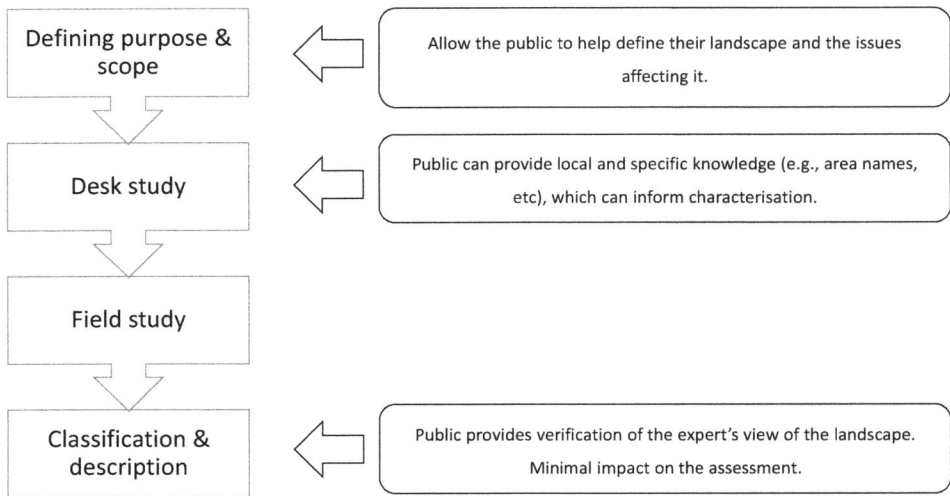

Figure 17.2 When and how public participation can be incorporated in the LCA process.

Vignette 17.2: East Lindsey District LCA (UK) (A consultant-driven assessment where participation was used to strengthen the evidence base for the district's local development framework)

Lying on the east coast of England, East Lindsey covers an area of 1,765 km² with a population of approximately 139,700. Participation in the East Lindsay landscape assessment was undertaken prior to the field study, thus providing an opportunity to influence the actual assessment process, rather than simplifying affirming the view of the consultants. Participants completed questionnaires about valued or disliked features in the landscape and their concerns about changes affecting the landscape character. At workshops, the purpose and nature of LCA and how the participants could contribute and help was outlined. Participants placed Post-It notes on maps with their comments and spoke to landscape architects and planning officers present at the workshops. The information provided on field survey sheets and maps used during the participation process helped to frame how experts engaged with the landscape.

Representation of participation

It is accepted that those involved in a participatory process should comprise a broadly representative sample of the affected population (Carr and Halvorsen, 2001, Rowe and Frewer, 2000, Butler and Berglund, 2014). Results from participatory processes are generated by amalgamations of arguments (Sager, 2005). It is this plurality that provides the diversity that should be central to participatory planning and provides the catalyst for generating creativity (Sager, 2005, Healey, 1996). Consequently, a balance needs to be struck between an in-depth approach with few participants and a shallow understanding from many. A small number of LCAs have made efforts to ensure representativeness through, for example, citizen panels (Vignette 17.3). However, practical constraints often make this unviable.

Vignette 17.3: Hertfordshire County Council LCA (UK) (undertaken by the County Council as evidence base for informing local plans)

The county of Hertfordshire lies north of London and covers an area of 1,643 km² with a population of 1,107,600. The LCA approach aimed to get a representative overview of the perception of the landscape from those who live in the county. A questionnaire went out to 2,500–3,000 individuals from a citizens' panel representing the social composition of the county. The questionnaire was circulated prior to draft landscape character areas being identified.

Both qualitative and quantitative responses were requested from the respondents The quantitative responses were expressed by a text- and map-based questionnaire. Respondents marked on maps and described areas of the landscape that they considered to be 'distinctive'. The responses provided information about perceived area boundaries, preferences, key features, and local expert advice, which helped inform the LCA. Respondents also enhanced the area descriptions; this was very subjective, as the individual descriptions became an amalgamation of several voices. Multiple-choice questions helped the consultants attain a deeper understanding of what is valued by the public in the landscape.

Although the study looked at the county as a whole, individual responses related to a local scale, based on maps of their local area at 1:50,000. This allowed respondents to comment on the landscape that they understood: their everyday landscape. A consistent approach across the county that is 'representative, yields results, and can be used at different scales' was seen as the main benefit of undertaking a questionnaire approach in this study.

The supplementary LCA Topic Paper 3 (Landscape Character Assessment – How Stakeholders Can Help) refers to 'stakeholders', who are seen as individuals and groups with an interest in the landscape. For LCA, the range of stakeholders is wide and often divided into two broad categories: 'communities of place' and 'communities of interest' (Swanwick et al., 2002:1). Communities of interest are identified as governmental bodies and non-governmental organizations, etc., who can provide important baseline information. 'Communities of place' are seen as those who live, visit, work, or have businesses in the area. Local communities are considered to have the greatest stake, but they also have the greatest diversity of issues and perspectives; communities cannot be seen as homogeneous, static, and harmonious units with entirely shared visions and interests that need to be listened to and understood (Cooke and Kothari, 2001). This dichotomy between knowledge of place and interest has been drawn into question. Raymond et al. (2010, 2014) summarize that local knowledge also constitutes local experts and therefore, represents different points on a continuum.

A second aspect of representation is geographical, questioning how inclusive participation is for the entirety of the area. LCA operates at various scales nesting within each other and informing each other, from national or even European down to regional and local. The scale influences what detail comes out and affects the efficiency and ability of public involvement. The importance and relevance of a participatory process increase when LCA is adapted for use at the local scale due to the citizens' local knowledge of the landscape in which they find themselves and their knowledge of local narratives (Caspersen, 2009). It is at this scale that people can relate to the landscape and understand how the landscape relates to their lives; this is the scale at which they 'dwell' (Ingold, 2000). Yet, municipal- or even county-level landscape assessments can gain representative coverage through utilizing the local by aggregating the fine grain of multiple everyday landscapes (Vignettes 17.3 and 17.4).

Vignette 17.4: Havant Borough LCA workshops (UK) (Participation was undertaken by consultants to ensure that views of communities of both place and interest were considered in the landscape planning)

Sitting on the south coast of England, Havant Borough covers an area of 55.3 km² with a population of 116,800. The Havant community consultation workshop engaged the public once landscape character areas had been defined and described. The exercise consisted of short workshops using an adaption of the DesignWays approach, specifically designed for the LCA by Countryscape.[2] Participants were divided into groups, representing different locations in the municipality, to look at their local landscape. A structured mindmap was provided to all groups to develop 'brainstorming'. The mindmap was structured with the main topic in the center (Havant LCA) with six 'branches' radiating from it; the branches represented ecology and nature, history and stories, activities, local economy, experience of the landscape, and landscape patterns and features. On these branches, the participants placed 'leaves' corresponding to the mindmap topic and relating to their local landscape. Each group then decided on the three most important values and visions and translated them into geographically located actions, communicated on a large-format map marked with the Landscape Character Area boundaries.

Convenience of participation

This criterion reflects the potential logistics of participation for the public (Halvorsen, 2001). Relevant considerations include notice (e.g., does the onus for finding out about participation opportunities rest predominantly with the public?), timing and location of participation events, and methods used for engaging the public. These factors influence the ease with which the public can participate and influence the public's motivation to engage in the process (Conrad et al., 2011b, Butler et al., 2018).

Communication needs to make potential participants aware of their skills and capabilities, providing them with the sense that their knowledge is of value (Höppner et al., 2008). Planners should make specific efforts to inform the public of participation possibilities (Vignette 17.5) beyond general advertising in the media. However, the onus of participating tends to lie with the public. In some cases, the process of participation is facilitated through mailing out documents and questionnaires or offering multiple possibilities for public involvement (see Vignette 17.3) (Conrad et al., 2011b).

Vignette 17.5: Participatory landscape assessment Lindö, Sweden (Participation was part of an academic project, addressing how participation can, with minimal effort, provide meaningful input into a landscape assessment)

Lindö is a residential area on the eastern outskirts of Norrköping in the east of Sweden, covering 321 hectares, with a population of 4,900. The aim of participation in the assessment was

Figure 17.3   The pin mapping process in Lindö. Photo by the author.

to attain, by as simple means as possible, mapped data on values that the local community attached to their landscape. Individuals placed pins on a map of Lindö and its surrounding landscape relating to different landscape values. Two categories of valued places were defined: individual and collective. Factors for individuals valuing a place consisted of physical activities, foraging and consumption, aesthetic view points, and experiencing the environment; collective reasons for valuing places were organized activities, special events, and significant landmarks (Figure 17.3).

The mapped data were used to inform the creation of landscape character areas. The data collection was undertaken outside the local supermarket, a site selected to capture a broad spectrum of the population in a neutral place on their own terms. Posters were put up around the area 2 weeks prior to the data collection to raise awareness of the event.

Influence of participation

Influence is a measure of the outcome, reflecting whether the output of the participatory process influences policy (Rowe and Frewer, 2000, Conrad et al., 2011b). The degree of influence the public will have on policy is an aspect that needs to be considered from the inception of the landscape assessment. One of the central aims of public participation should be a better-informed landscape assessment with greater ownership of the results, providing the opportunity to develop partnerships, develop commitment to outcomes, build consensus, increase local awareness, and validate character areas (Butler and Berglund, 2014). Influence comes about in many ways, yet important considerations are (1) the degree to which the public participates in the process (relating to the scope), (2) how transparently public views are expressed in the final output, and (3) the effectiveness of the process in terms of outcomes, performance, and sustainability (IAP2, 2007).

Influence is often limited to fine-tuning of results or incorporation of public views within the experts' characterization. Yet, even seemingly minor input, such as naming of areas by locals, can have a significant impact by making descriptions more familiar and relevant (Nogué and Sala 2018), while names and descriptions inflicted on a community can alienate them from the process and their landscape (Lee, 2018).

Knowledge transfer of participation

This criterion pertains to both the type of knowledge sought and the way it has been transferred. Various types of knowledge can be attained through the landscape assessment: scientific knowledge to supplement professional understanding, local knowledge of and perspective on physicality of a landscape, perception as a sensed phenomenon, or landscape perception as a lived experience. The knowledge sought raises the fundamental question of what constitutes the landscape. Different platforms for discussion are needed to make participants with different requirements aware of their importance to the process and to attain a diversity of knowledge (Höppner et al., 2008) (Vignette 17.6).

Vignette 17.6: Cornwall and Isles of Scilly landscape character study (UK) (Participation was undertaken as a pilot study, examining different approaches for 'deep' participation)

The county of Cornwall lies in the southwest of England, covering 3,563 km² with a population of 537,400. However, participation was undertaken at local parish level. Workshops attended by parish-level representatives and planners were undertaken after character areas had been defined. The process incorporated several tools with the aim of gaining a deep understanding and drawing out diverse knowledge of what the landscape means to those who dwell in the area. One tool was a *creative writing workshop*, in which participants could describe a sense of place, their perceptions of landscape character, and their personal experience. The writing workshop generated good-quality information in a short time. A second tool was a *rich picture exercise*, in which mind mapping was used to combine images and text. Participants worked in small groups to draw out shared values and understandings of the landscape. A third tool was a *photographic exercise*, in which attendees provided photos to examine what is significant and special about the landscape. The workshop was rounded off with a *plenary discussion*, providing a forum for articulating issues not taken up in the other sections of the workshop.

The responses and output were framed within expert-defined areas and influenced by a professional description of the area. It was seen as easier for these approaches to be utilized, and gave 'greater value', when work covered a well-defined spatial area. Although overly ambitious for a county-wide study, this approach provides a deep understanding of the local landscape.

How knowledge is transferred relates to the flow of information within the characterization process and to raising awareness with the public. Awareness-raising can take many forms: informing the public (e.g., through public presentation or printed information), gaining knowledge from the public (e.g., through public meetings or simple workshops), weak two-way communication (e.g., using public meetings to provide feedback on the assessment document), strong two-way communication (e.g., using public meetings to discuss the landscape), and multidirectional knowledge transfer (e.g., through focus group discussions or workshops).

With an entity such as a landscape, where all have both a stake and relevant knowledge, awareness-raising needs to be based on multidirectional knowledge transfer. The importance of awareness-raising is stated and clarified in the 2008 ELC guidelines, recognized as being 'a knowledge spreading process operating in all directions' (Council of Europe, 2008). Transfer of knowledge moves from a traditional, top-down process towards awareness-raising as a process for handling landscapes as dynamic entities constructed by those who encounter them (Butler and Åkerskog, 2014). Knowledge transfer has the potential to develop co-construction of the meaning of the landscape as well as help to develop social networks that can support landscape values, both of which are crucial for appropriation of the landscape (Le Du-Blayo, 2018).

Awareness-raising should be a way of clarifying the link between people, the activities they pursue during their daily lives, and the characteristics of their surroundings (Council of Europe, 2000). If awareness-raising is viewed in this light, then it can be seen as developing the understanding that all have a stake in the landscape (Sevenant and Antrop, 2010), and awareness-raising becomes a significant part of the assessment process (Butler and Åkerskog 2014). As well as being central to the participation process, awareness is also central to the product, the assessment document. The relevance of this lies in which values are conveyed and how they are communicated, as this will define future discourses on the landscape (Le Dû-Blayo, 2018) (Vignette 17.7).

Vignette 17.7: Landskabsatlas Langeland Kommune (Denmark) (Landscape atlas)

The landscape atlas for the island of Langeland conveys the landscape character of the place. While the process of this landscape assessment did not focus on participation, the characterization document acts as a tool with many functions: to increase interest in and understanding of the landscape for individuals and groups who experience the landscape on a daily basis, to serve as a guide for visitors to the island, and to provide a means for devising policies and political decisions relating to the landscape.

Although the landscape atlas presents an expert view of the landscape, the document also acts as a basis for further discussion about the landscape. Consequently, through raising awareness of a landscape and lifting specific values, the assessment document develops a continuous discussion on what the landscape means or could mean for people. The Ministry of the Environment and Food, together with Langeland Municipality,[3] has prepared Landlaskaatlas for Langeland Municipality (Figure 17.4).

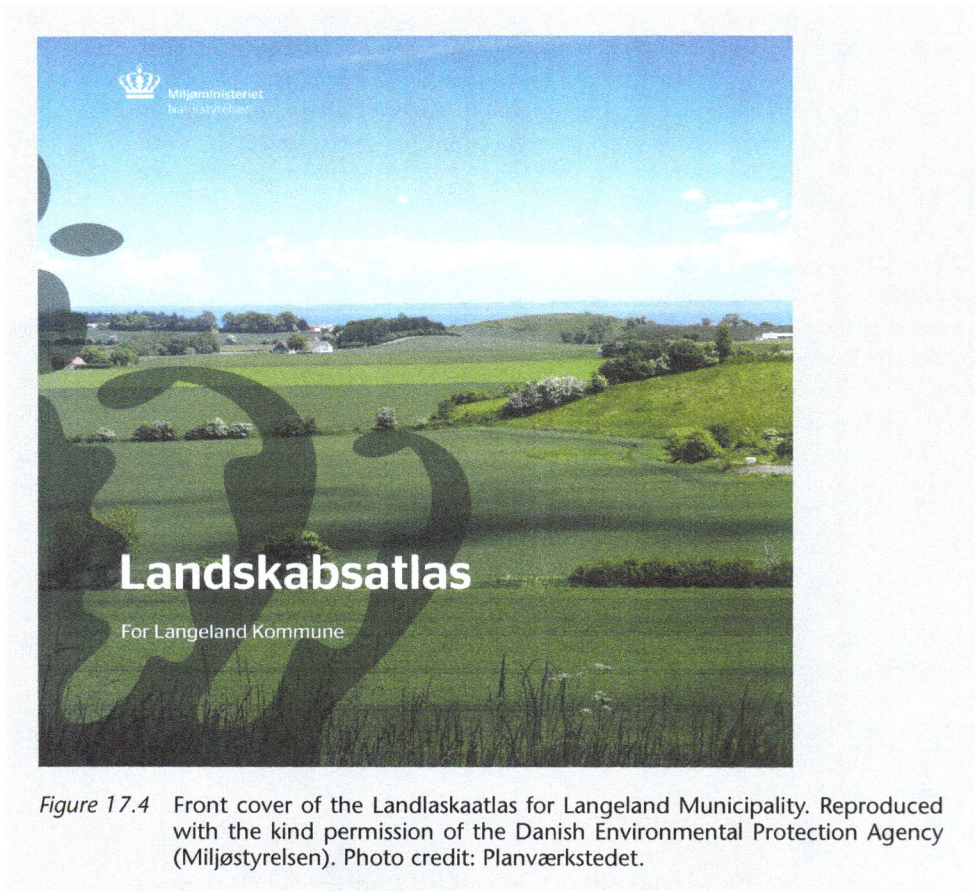

Figure 17.4   Front cover of the Landlaskaatlas for Langeland Municipality. Reproduced with the kind permission of the Danish Environmental Protection Agency (Miljøstyrelsen). Photo credit: Planværkstedet.

Discussion

LCA is a process whereby individuals can embrace or rationalize complexity, raising the question of whether we characterize to understand or to direct future change of the landscape. Is the focus to define what the landscape is or to question what the landscape could be (Dobson, 2018)? If all are indeed considered stakeholders in the landscape, then all have the right to have their views and values on the landscape heard (Jones, 2007). All interventions or even discourses on the landscape will inevitably lead to inclusion and exclusion of certain values, experiences, or interests, and there will always be winners and losers in the landscape. Consequently, conflicts are inevitable in all landscapes and at all scales. Expressing views and values early on creates space for conflicting issues to be recognized and legitimized early in the planning process. When the alternative is to reduce differences to counterproductive NIMBY (Not in My Backyard) issues once decisions have been made, allowing voices to be aired through the landscape assessment seems a more productive way forward. The landscape assessment becomes an arena for discussing the landscape.

Accordingly, participatory processes should continue to aim to be as genuine as possible and make explicit who was excluded from discussions and who won or lost with decisions. The process should make visible the conflicts, power mechanisms, contestation of unjust or unsustainable hegemonies, and the possible institutionalization of new counter-hegemonic projects that can alter them (Dobson, 2018). The process should demonstrate how social capacity can be developed to enable communities to gather around (Conrad et al., 2011a).

LCAs have the opportunity to legitimize local values, bringing them into an official discourse (Butler and Åkerskog, 2014). To close the gap between the rhetoric of participation and how it is practiced in landscape planning (Conrad et al., 2011b, Butler and Berglund, 2014), tools that provide the opportunity to identify the diversity of landscape values need to be advanced. These tools can include both simple-to-deploy approaches, such as placing pins on maps, which provide a shallow yet broad overview of landscape values, and more in-depth pseudo-anthropological, literature-based, or sense-of-place approaches and strategies (Edwards, 2018). The six aspects described in this chapter (scope, timing, representation, convenience, influence, and knowledge transfer), while in no way definitive, can help inform future developments in LCA by allowing practitioners and other actors to be reflective and question their actions.

Notes

1 Unlocking the landscape: www.cpre.org.uk/resources/countryside/landscapes/item/1929-
Tirlun Lleol: www.monmouthshiregreenweb.co.uk/locallandscapes/index.html

Ecovast: www.ecovast.org/papers/good_guid_e.pdf

Talking about our place: https://www.nature.scot/sites/default/files/2018-11/Talking%20About%20
Our%20Place%20Toolkit.pdf
2 www.havant.gov.uk/sites/default/files/documents/0605%20Report.pdf
3 Landskabsatlas Langeland: www2.nst.dk/download/planlaegning/landskabsatlas.pdf

References

Arnstein, S.R. (1969) 'A ladder of citizen participation', *Journal of the American Institute of Planners*, vol 34, pp216–225.

Brunetta, G. and Voghera, A. (2008) 'Evaluating landscape for shared values: tools, principles, and methods', *Landscape Research*, vol 33, pp71–87.

Butler, A. (2016) 'Dynamics of integrating landscape values in landscape character assessment: the hidden dominance of the objective outsider', *Landscape Research*, vol 41, pp239–252.

Butler, A. and Åkerskog, A. (2014) 'Awareness-raising of landscape in practice: an analysis of landscape character assessments in England', *Land Use Policy*, vol 36, pp441–449.

Butler, A. and Berglund, U. (2014) 'Landscape character assessment as an approach to understanding public interests within the European landscape convention', *Landscape Research*, vol 39, pp219–236.

Butler, A., Eriksson, M. and Berglund, U. (2018) 'Can simple tools for mapping landscape convey insider perspectives?', *Nordic Journal of Architectural Research*, vol 3, pp57–80.

Calderon, C. and Butler, A. (2019) 'Politicising the landscape: a theoretical contribution towards the development of participation in landscape planning', *Landscape Research*, vol 45, no 2, pp152–163.

Carr, D. and Halvorsen, K. (2001) 'An evaluation of three democratic, community-based approaches to citizen participation: surveys, conversations with community groups, and community dinners', *Society & Natural Resources*, vol 14, pp107–126.

Caspersen, O.H. (2009) 'Public participation in strengthening cultural heritage: the role of landscape character assessment in Denmark', *Geografisk Tidsskrift-Danish Journal of Geography*, vol 109, pp33–45.

Conrad, E., Cassar, L.F, Christie, M. and Fazey, I. (2011a) 'Hearing but not listening? A participatory assessment of public participation in planning', *Environment and Planning C: Government and Policy*, vol 29, pp761–782.

Conrad, E., Cassar, L. F., Jones, M., Eiter, S., Izaovičová, Z., Barankova, Z., Christie, M. and Fazey, I. (2011b) 'Rhetoric and reporting of public participation in landscape policy', *Journal of Environmental Policy & Planning*, vol 13, pp23–47.

Cooke, B. and Kothari, U. (2001) 'The case for participation as tyranny', in B. Cooke and U. Kothari (eds) *Participation: The New Tyranny?* Zed Books Ltd., London.

Council of Europe (2000) *European Landscape Convention.* Council of Europe, Florence and Strasbourg.

Council of Europe (2008) *Recommendation of the Committee of Ministers to Member States on the Guidelines for the Implementation of the European Landscape Convention.* Council of Europe, Florence and Strasbourg.

Council of Europe (2014) *Napflion Declaration: Promoting Territorial Democracy in Spatial Planning.* Council of Europe Conference of Ministers responsible for Spatial/Regional Planning, Napflion.

Cox, R. (2006) *Environmental Communication and the Public Sphere.* Sage, Thousand Oaks, CA.

Dakin, S. (2003) 'There's more to landscape than meets the eye: towards inclusive landscape assessment in resource and environmental management', *Canadian Geographer/Le Géographe canadien*, vol 47, pp185–200.

Dalglish, C. and Leslie, A. (2016) 'A question of what matters: landscape characterisation as a process of situated, problem-orientated public discourse', *Landscape Research*, vol 41, pp212–226.

Dobson, S. (2018) 'The embodied city and metropolitan landscape', in G. Fairclough, I Sarlöv Herlin and C. Swanwick (eds) *Routledge Handbook of Landscape Character Assessment.* Routledge, Abingdon.

Dower, M. (2007) 'The European landscape convention: its origins, focus and relevance at European level to land use and landscape planning', *Landscape Character Network News*, vol 27, pp10–15.

Edwards, J. (2018) 'Literature and sense of place in UK landscape strategy', *Landscape Research*, vol 44, no 6, pp659–670.

Fairclough, G., Sarlöv Herlin, I. and Swanwick, C. (2018a) 'Landscape character approaches in global, disciplinary and policy context: an introduction', in G. Fairclough, I Sarlöv Herlin and C. Swanwick (eds) *Routledge Handbook of Landscape Character Assessment.* Routledge, Abingdon.

Fairclough, G., Sarlöv Herlin, I. and Swanwick, C. (eds) (2018b) *Routledge Handbook of Landscape Character Assessment.* Routledge, Abingdon.

Geelmuyden, A.K. (2016) 'Landscape assessments as imaginative (poetic) landscape narratives', in K. Jørgensen, M. Clemetsen, K. Halvorsen Thorén and T. Richardson (eds) *Mainstreaming Landscape through the European Landscape Convention.* Routledge, London and New York.

Healey, P. (1996) 'The communicative turn in planning theory and its implications for spatial strategy formation', in S.F.S. Campbell (ed) *Readings in Planning Theory*, 2nd ed. Blackwell, Oxford.

Höppner, C., Frick, J. and Buchecker, M. (2008) 'What drives people's willingness to discuss local landscape development?', *Landscape Research*, vol 33, pp605–622.

IAP2 (2007) *IAP2 Spectrum of Public Participation*, www.iap2.org/associations/4748/files/IAP2%20Spectr um_vertical.pdf [Online], accessed 15 June 2009.

Ingold. T. (2000) *The Perception of the Environment: Essays on Livelihood, Dwelling and Skill.* Routledge, London.

James, P. and Gittins, J. (2007) 'Local landscape character assessment: an evaluation of community-led schemes in cheshire', *Landscape Research* vol 32, no 4, pp423–442.

Jones, M. (2007) 'The European landscape convention and the question of public participation', *Landscape Research*, vol 32, pp613–633.

Jørgensen, K., Clemetsen, M., Halvorsen Thorén, K. and Richardson, T. (2016) *Mainstreaming Landscape through the European Landscape Convention.* Routledge, London and New York.

Käyhkö, N., Fagerholm, N., Khamis, M., Hamdan, S. and Juma, M. (2018) 'The collaborative participatory process of landscape character mapping for land and forest planning in Zanzibar, Tanzania', in G. Fairclough, I Sarlöv Herlin and C. Swanwick (eds) *Routledge Handbook of Landscape Character Assessment.* Routledge, Abingdon.

Le Du-Blayo, L. (2018) 'Atlas du paysage: landscape atlases in France and Wallonia', in G. Fairclough, I Sarlöv Herlin and C. Swanwick (eds) *Routledge Handbook of Landscape Character Assessment.* Routledge, Abingdon.

Lee, G. (2018) 'On calling place: language, naming and the understanding of landscape character attributes of cultural places in the Asia_pasific region', in G. Fairclough, I Sarlöv Herlin and C. Swanwick (eds) *Routledge Handbook of Landscape Character Assessment.* Routledge, Abingdon.

Loupa-Ramos, I. and Pinto-Correia, T. (2018) 'Landscape character assesment across scales: insight from portugese experience of policy and planning', in G. Fairclough, I. Sarlöv Herlin and C. Swanwick (eds) *Routledge Handbook of Landscape Character Assessment.* Routledge, Abingdon.

Nogué, J. and Sala, P. (2018) 'Landscape, local knowledge and democracy: the work of the Landscape Observatory of Catalonia', in G. Fairclough, I. Sarlöv Herlin and C. Swanwick (eds) *Routledge Handbook of Landscape Character Assessment.* Routledge, Abingdon, 265–278.

Olwig, K. (2007) 'Multiple interfaces of the European landscape convention: the European landscape convention as "interface"', *Norsk Geografisk Tidsskrift*, vol 61, no 4, pp214–215.

Raymond, C., Fazey, I., Reed, M., Stringer, L., Robinson, G. and Evely, A. (2010) 'Integrating local and scientific knowledge for environmental management', *Journal of Environmental Management*, vol 91, pp1766–1777.

Raymond, C., Kenter, J., Plieninger, T., Turner, N. and Alexander, K. (2014) 'Comparing instrumental and deliberative paradigms underpinning the assessment of social values for cultural ecosystem services', *Ecological Economics*, 107, pp145–156.

Rowe, G. and Frewer, L. (2000) 'Public participation methods: a framework for evaluation', *Science, Technology, & Human Values*, vol 25, pp3–29.

Sager, T. (2005) 'Planning through inclusive dialogue: no escape from social choice dilemmas', *Economic Affairs*, vol 25, pp32–35.

Sarlöv Herlin, I. (2016) 'Exploring the national contexts and cultural ideas that preceded the landscape character assessment method in England', *Landscape Research*, vol 41, pp175–185.

Scott, A. (2011) 'Beyond the conventional: meeting the challenges of landscape governance within the European landscape convention?', *Journal of Environmental Management*, vol 92, pp2754–2762.

Selman, P. (2010) 'Landscape planning: preservation, conservation and sustainable development', *Town Planning Review*, vol 81, pp382–406.

Selman, P. (2012) *Sustainable Landscape Planning: the Reconnection Agenda*. Routledge, London.

Sevenant, M. and Antrop, M. (2010) 'Transdisciplinary landscape planning: does the public have aspirations? Experiences from a case study in Ghent (Flanders, Belgium)', *Land Use Policy*, vol 27, pp373–386.

Stahlschmidt, P., Swaffield, S., Primdahl, J. and Nellemann, V. (2017) *Landscape Analysis: Investigating the Potentials of Space and Place*. Routledge, London.

Stephenson, J. (2010) 'People and place', *Planning Theory & Practice*, vol 11, pp9–21.

Swanwick, C. and Land Use Consultants (2002) *Landscape Character Assessment: Guidance for England and Scotland*. Cheltenham Countryside Agency and Scottish Natural Heritage, Edinburgh.

Swanwick, C., Bingham, L. and Parfitt, A. (2002) 'Landscape character assessment guidance topic paper 3: how stakeholders can help', *Landscape Character Assessment Guidance*. Countryside Agency, Cheltenham, and Scottish Natural Heritage, Edinburgh.

Tudor, C. (2014) *An Approach to Landscape Character Assessment*. Natural England, York.

United Nations Economic Commision for Europe (UNECE) (1998) 'Aarhus convention', *Convention on Access to Information, Public Participation in Decision-making and Access to Justice on Environmental Matters*, http://www.unece.org/env/pp/ UNECE, United Nations Economic Commision for Europe, Aarhus.

18

Experimentation in landscape ecology

G. Darrel Jenerette

Overview of experimentation in landscape ecology: definitions, challenges, opportunities

Across scientific disciplines, experimentation has diverse meanings. Within ecological sciences, experimentation is most commonly defined as 'a deliberate manipulation conducted such that other variables are held constant or rendered unimportant through suitable design' (Tilman 1989). Essential to the definition, experiments exclude studies that are observational or model-based. Although observational studies can frequently be designed to take advantage of natural events, as in 'natural experiments', or extensively evaluate simulated dynamics in 'computational experiments', both are recognized as more limited in inferential power than experimentation with controls, replication, and deliberate manipulation. Ecological experiments include activities that take place using model systems in the laboratory, constructed outdoor mesocosms, and full field manipulations. Such experimental work has had large impacts in physiological, population, community, and ecosystem ecological research, yet a pervading misconception has been that experimental work has limited possibilities and relevance to landscape ecology (Hargrove and Pickering 1992; Nabe-Nielsen et al. 2010). Nevertheless, landscape ecology has a rich history of experiments, and many opportunities for improving landscape theory can be achieved through experimentation (Cunningham and Lindenmayer 2017; Ims 2005; Jenerette and Shen 2012).

While less frequently used than correlative or modeling approaches, landscape ecology theory and predictive models have benefited from direct experimentation. Landscape experiments have been directed to a broad range of questions, from the effects of landscape structure on ecological processes to thevaluation of the processes governing landscape formation. Similarly, landscape ecologists have conducted experiments at a broad range of scales – from less than 1 m² model landscapes for some studies of microbial metapopulation dynamics (Kurkjian 2019) to more than 7000 ha for studies of landscape fragmentation (Ewers et al. 2011). Compared with other ecological experiments, landscape experiments are inherently complicated by the necessity to include spatial heterogeneity and frequently incorporate multiple scales in their design (McGarigal and Cushman 2002). As in other fields of ecology, experimentation can provide key information for testing hypotheses and advancing theory. Landscape experiments are directed to testing hypotheses of landscape structure, functioning, or dynamics of both biotic and abiotic components. Landscape structure includes descriptions of composition, the presence of differing components, and configuration, the arrangement of components. Experiments can be used to directly assess the structure as well as to identify processes that have led to the formation of structure (Ewers et al. 2011; Pangle et al. 2015). Alternatively, experiments can assess how landscape structure, both composition and configuration, in turn, affects landscape functioning (Sponseller and Fisher 2008; Silva et al. 2012). Functioning can include ecological processes occurring within a landscape as well as aggregate whole-landscape processes.

The design and creation of landscape experiments are frequently considered more challenging than in other subdisciplines of ecology (Cunningham and Lindenmayer 2017; Ims 2005; Jenerette and Shen 2012). Extending ecological experiments to landscapes can present major logistical impediments. Landscapes are multiple-scale distributions of ecological heterogeneity with variable and dynamic connectivity among locations that interact with ecological processes (Forman and Godron 1981; Wu and Loucks 1995). Compared with human experience, landscapes can be relatively small, for example in the context of a microbiome, or large and extending to continents, for example in the context of animal species range distributions. Manipulations at large scales often require extensive coordination among many participants and landowners. Nevertheless, experimentation has led to many important discoveries and remains a strong component of landscape ecological research. The remaining sections of this chapter summarize and evaluate the different types of experiments used in landscape ecology and look toward a future of expanded opportunities for landscape experiments.

Three major types of landscape experiments

Landscape ecology has used a variety of experimental approaches that address diverse scientific challenges. Experiments range in size from benchtop to thousands of hectares and include a wide range of specific approaches tailored to individual research questions. Frequently, experiments are integrated with other approaches as part of a larger science program. Here, I follow and update the classification described in Jenerette and Shen (2012), which originally identified 15 types of experiments organized into 4 distinct groups. Two of the previous groups were combined based on research goals, which allowed a more general categorization of experiments into three main categories – identify landscape structure, evaluate effects of landscape structure on ecological processes, and evaluate processes causing landscape change. In some cases, experiments look to achieve multiple goals. In evaluating experiments, one key criterion is identifying what is being manipulated, with an important distinction being whether the landscape itself is being manipulated or whether manipulations are occurring in different landscape components. This issue can be critical for the kinds of inferences that are possible from different landscape experiments.

One large class of experiments is directed to evaluate landscape structure. While much landscape structure is directly observable from the land surface, many important landscape components cannot be directly evaluated or are dependent on organismal characteristics. Experimental evaluations of how the landscape may be perceived have been conducted for both humans (Larsen and Harlan 2006) and other species (Knowlton and Graham 2010), where preferences are assessed through exposure to manipulated differences in an aspect of landscape pattern. For example, Knowlton and Graham (2010) summarized an extensive number of behavioral landscape ecological experiments and grouped these into six categories: translocation, playback, landscape alteration, manipulation of food resources, perceived predation risk, and reproductive success. Across all these behavioral experiments, the goal was to identify organism preferences for different landscape components and configurations.

For belowground landscapes, experimental approaches can be uniquely valuable for uncovering landscape structure. Many key landscape components are distributed belowground, and identifying their distributions and connectivity is a major challenge. However, even in aboveground conditions, the movement pathways and connectivity of a landscape may still be challenging to quantify. Tracer experiments, which track the movement of an inert material tracer, can provide a functional assessment of connectivity and flow paths (Neuman 2005). Tracer experiments have been particularly valuable for identifying hydrological flow paths belowground (Drummond et al. 2015) and also for pathways connected by wind (Allen et al. 1989) and animal (Hershey et al. 1993) movement vectors. These pathways may be associated with ecosystem fluxes of materials or energy as well as population and community fluxes of propagules and food resources. The use of experimentation to evaluate the structure of the landscape has been an important approach for identifying key components of landscape structure and connectivity among patches.

Another large class of experiments looks to evaluate how landscape functioning responds to changes in landscape structure. Key examples of these experiments are distributed experiments, which replicate protocols throughout a landscape to evaluate how experimental results change depending on location (Jenerette and Chatterjee 2012). Distributed experiments replicate many individual experiments whose placement differs in landscape composition, configuration, and connectivity. The effects of landscape pattern can be assessed by either field experiments throughout a landscape (Jenerette and Chatterjee 2012; Sponseller and Fisher 2008) or laboratory experiments conducted on samples collected from different locations in a landscape (Frank and Groffman 1998; Richardson et al. 2012). While individual nodes of a distributed experiment need not necessarily evaluate landscape phenomena themselves, the network of experimental nodes can address landscape-level questions. In an alternate experimental approach, translocation experiments can be used to evaluate the effect of local and regional landscape pattern on processes (Betts et al. 2015; Hart 2006; Schooley and Wiens 2004). As influencing connectivity is a key mechanism through which landscape structure influences functioning, some landscape experiments have been directed to influence connectivity directly by either removing or enhancing exchanges between landscape components (Kawaguchi et al. 2003; Sayer 2006). For example, Kawaguchi et al. (2003) enclosed a stream in netting to evaluate how the movement of aquatic insects within streams to adjacent riparian lands provides a resource subsidy between these two patches. Using a variety of different approaches, landscape experiments can provide valuable information on how components of landscape structure influence many important ecological processes.

Box 18.1 Deep Canyon Transect Pulse Experiments

The Deep Canyon Transect is a 2200-m elevation transect within only ~20 km horizontal distance in southern California (Zabriskie 1979). The transect varies from desert scrub at the bottom of the canyon to Pinyon-Juniper woodland, chaparral, and sub-alpine coniferous forest at the upper elevations. This transect has been extensively used to evaluate drivers of landscape variation in ecosystem functioning, with early work conducted by Robert Whittaker (Hanawalt and Whittaker 1976). More recently, the Deep Canyon Transect Pulse Experiments were conducted to evaluate how soil responses to wetting varied throughout the Deep Canyon landscape. Here, a series of connected distributed experiments were conducted to evaluate how ecosystem metabolic rates and their sensitivity to short-term water additions varied across this landscape. These experiments were all directed to evaluating how soils varied spatially within a site as well as across the entire transect. Several experiments were conducted by collecting soils throughout the transect and performing laboratory incubations of soil mesocosms under varying experimental treatments (Chatterjee and Jenerette 2011a, b; Richardson et al. 2012). In one of these experiments, soils were collected from local transects and were then experimentally wetted in laboratory conditions. The results from laboratory experiments were then used to develop semi-variograms of wetting responses for sites across the transect (Chatterjee and Jenerette 2011b). These experiments showed an increasing strength but shorter distance of autocorrelation at the lower elevations compared with higher elevations – that is, with increasing elevation, patches became less distinct but larger. These initial laboratory experiments served as a basis for a subsequent series of distributed field experiments evaluating the spatial and temporal responses of ecosystems to wetting events (Jenerette and Chatterjee 2012).

The field experiment identified an important scale discontinuity in drivers of ecosystem responses to wetting. Within sites, increasing soil organic matter was correlated with increased metabolic activity following wetting, while across sites, organic matter was negatively correlated. That is, the lowest elevation sites had the lowest amount of organic matter but responded the most to wetting, and within these sites, more organic matter was associated with greater responses. In contrast, the highest elevation site, which had the highest amounts of organic matter, showed the lowest responses to wetting – although here as well, increased organic matter locally was associated with greater responses to wetting. Through applications of distributed experiments, both in the laboratory and in the field, the project was able to show how landscape position influences responses to wetting at both the site and the transect scale. These experiments further pointed to new questions about cross-scale relationships in sensitivity to wetting and the roles of local climate, organic matter, and microbial communities.

A large sub-class of experiments evaluating the effects of landscape structure on functioning has been directed to evaluating the effects of landscape configuration on many ecological processes. Experiments of this type have included manipulations of individual patch characteristics, corridors between patches, and the construction of entire landscapes. These experiments have been conducted in field (Laurance et al. 2011), mesocosm (Silva et al. 2012), and laboratory settings that use existing landscapes or construct entirely new landscapes (Huxman et al. 2009). Applications of these experiments have been directed broadly to population (Schmitz 2005), community (Collins et al. 2009), and ecosystem dynamics (Maestre and Reynolds 2007). These experiments have examined the effects of properties of individual patches on connectivity and fragmentation across a landscape (Haddad and Baum 1999) and have shown that the effects of configuration can continue to accumulate through 18 years of monitoring (Damschen et al. 2019). Some of the largest landscape experiments fall in this category and include several fragmentation experiments covering thousands of hectares (Ewers et al. 2011; Haddad et al. 2015).

Box 18.2 Biological Dynamics of Forest Fragments Project

The Biological Dynamics of Forest Fragments Project (BDFFP; Figure 18.1; Laurance et al. 2018; Laurance et al. 2002), located in the central Amazon basin, has 'had the single greatest impact on the general understanding of the ecological impact of forest fragmentation and is routinely cited in the conservation literature around the world' (Ewers et al. 2011). The experiment is ~1000 km²

Figure 18.1 Biological Dynamics of Forest Fragments Project. (Credit Robert Johnson.)

in area, spanning three large cattle ranches (~5000 ha each), and includes 11 forest fragments (5 of 1 ha, 4 of 10 ha, and 2 of 100 ha). An additional large area of nearby continuous forest serves as a control for the experiment. Prior to initiating the experiment, extensive surveys for animals and plants were conducted at the site. As originally conceived in 1979, it is the longest-running fragmentation experiment. Now in its 40th year, the BDFFP has experienced extensive changes in the matrix of the fragmented plots, climate, and extreme events (Laurance et al. 2018). In the late 1980s, ranching was largely abandoned at the site, and regrowth occurred around the original fragments. Isolation was maintained by establishing 100 m-wide strips around each of the original fragments. The landscape changes pose challenges and opportunities to evaluate long-term dynamics associated with patches and their larger landscape context. Coincidentally with local land cover changes and global climate changes, the climate of the site has also been altered. Repeated disturbances have affected several of the fragments.

The research conducted as part of the project has made extensive scientific contributions – as of October 2010, BDFFP research was featured in 562 manuscripts and 143 graduate theses (Laurance et al. 2011). The experiment has shown extensive physical effects of fragmentation, notably the increased rate of drying in fragmented patches as convection from forests to cleared land is increased by warming of the adjacent cleared lands. These whole-patch changes are juxtaposed on a large suite of edge effects that differ in their persistence from only a few meters, for effects including species invasions, vapor pressure, and soil moisture, to other effects that persist for 100 m or more, including canopy height, invertebrates in leaf litter, and elevated tree mortality (Laurance et al. 2018). Importantly, the edge effects were found to differ through time – newly established fragments are much more physically open than later stages when lianas and lateral stems 'seal' the edge. As an overarching finding, the experimental results suggest that fragmentation introduces 'hyperdynamism' in population, community, and ecosystem characteristics, meaning that the fragmented plots exhibited much greater variability in forest dynamics than control plots. The research enabled by the BDFFP experiment is a testament to the power of long-term landscape experiments to test landscape hypotheses and provide a foundation for extensive research in directions outside the original plans of the study.

A third set of experiments are directed to evaluating dynamics of landscape structure and the formation of pattern. Several approaches have been considered that influence alternate components of landscape formation. One prominent approach has been to alter disturbance regimes and evaluate the consequences for landscape dynamics (Andersen et al. 1998; Renofalt and Nilsson 2008). Alternatively, experiments have altered rates of key vectors including wind (Ravi et al. 2009) and water (Konrad et al. 2011). Finally, processes connected to self-organization of landscape dynamics have been experimentally evaluated with regard to both top-down (Weerman et al. 2011) and bottom-up (Van De Koppel et al. 2001) constraints. These experiments are less frequent than other landscape experiments and can present extreme logistical challenges.

Box 18.3 Landscape Evolution Observatory

The Biosphere 2 facility, located in Oracle, AZ (United States), has a dynamic history, from its development as a test-case for long-term space-colonization technology in 1986 to its current incarnation as a home for the Landscape Evolution Observatory (LEO; Figure 18.2). LEO is an experiment designed to test controls on geomorphic and biologic dynamics associated with the development of landscape structure (Pangle et al. 2015). LEO consists of three artificial landscapes (330 m² surface area; 330 m³ volume; 10° average slope) that imitate a zero-order basin. The three landscapes are designed as replicates with a consistent structure and parent material. Among landscapes, meteorological conditions can be experimentally manipulated. In the future, vegetation will be introduced to the landscapes, and this introduces another experimental opportunity. The design of the experiment was informed by extensive preliminary modeling activities (Huxman et al. 2009). As part of installation, the landscapes were extensively instrumented

Figure 18.2 Landscape Evolution Observatory. (Photo published in Pangle et al. 2015.)

with 496 individual sampling locations and more than 1800 individual sensors. The monitoring capabilities of the experiment provide opportunities to evaluate atmosphere, soil, and surface dynamics at high temporal and spatial resolutions. The ability of models to use these data and the potential losses as instruments age were also explicitly assessed (Pasetto et al. 2015). LEO has a unique role in the context of previously developed landscape evolution experiments. Unlike open field experiments that are dependent on ambient meteorological conditions, in LEO, the temperature, humidity, and precipitation are all controlled. Similarly, while other field experiments exist in the context of geologic legacies already in place, with LEO, a consistent, homogeneous substrate is used, which minimizes the effects of historic contingencies. As a constructed landscape, LEO had an opportunity to install monitoring instrumentation as part of the construction, which is not possible in field settings.

While it is a relatively new experiment, the benefits of LEO are still accruing. Already, research from the experiments has enabled improved models of rainfall-runoff dynamics, CO_2 production, and weathering rates (Pasetto et al. 2015). Microbial sampling within each landscape has shown separation of bacteria within the surface profile with surprisingly larger numbers of autotrophic cyanobacteria at deeper soil layers (Volkman et al. 2018). With future interest in community use of the experimental facility, there is an expectation that increasing numbers of science applications will be identified. In particular, future inclusion of vegetation in the experiment will offer many new opportunities for examining the dynamics of the experimental landscapes. Nonetheless, already, LEO serves as a unique example of a constructed landscape experiment that is being used to answer questions about the formation of landscape structure.

Regardless of the type of experiment, their applications all emphasize hypothesis testing through the application of replication, control, and planned manipulation. Among experimental approaches, a key distinction is whether experiments are directed to manipulating the landscape itself or manipulating conditions within and throughout a landscape. In some cases, multiple experiments have been combined to overcome the limitations of any single approach. Examples include combinations of laboratory and field tracer experiments (Drummond et al. 2015), landscape structure and perception experiments (Nachappa et al. 2011), and distributed and transplant experiments (Warren 2010). Designing and evaluating experimental approaches remains a key challenge to the deployment of effective experiments.

In maximizing the benefits of landscape experimental approaches, trade-offs become apparent in the usefulness of different experiments for different goals. Costs of an experiment versus scope are a prominent trade-off, but as many experiments have shown, costs are not always limiting in designs. In designing a landscape experiment, another important trade-off reflects how experiments are used for predicting dynamics in specific landscapes versus generalization across landscapes (Beier and Noss 1998; Haddad et al. 2000). At larger scales of experiments, the local ecological context becomes a key component of ecological processes, and the goals of experimentation tend to reflect the importance of local context. In contrast, smaller experiments, especially those using constructed landscapes, allow designs that can be more separated from the environmental context. Such trade-offs highlight differences in design and the capabilities for achieving experimental goals.

Combining experiments with other approaches

While many landscape experiments are designed as self-contained research projects, landscape experiments are frequently conducted in the context of multiple approaches that may include extensive observational and/or modeling studies. A combination of approaches provides a valuable opportunity for a 'triangulation' (Plowright et al. 2008) of research that takes advantage of contrasting trade-offs between individual methods. Within landscape ecology, experiments are only one approach that will always present major trade-offs in what can be learned. Integrating experimentation more broadly within landscape ecology will help researchers understand the underlying mechanisms responsible for landscape dynamics and improve the capability of models to forecast future dynamics. To achieve this integration, approaches must facilitate the evaluation of landscapes across multiple scales.

From regional to patch scales, landscape experiments can be integrated with observations that complement the scale of experimentation. Sensor networks and imagery platforms can provide finer-resolution and broader-extent data for evaluating responses to experimental manipulations. Increasingly, unmanned aerial systems (UAS) can generate imagery data at centimeter or lower resolution over hectares of extent for ecological research (Anderson and Gaston 2013). Sensor networks are increasingly distributed, with greater reliance on wireless sensors that communicate across sensor networks to central data repositories. From regional to continental or even global scale, landscape experiments can themselves be combined into networks of experiments and associated networks of sensor networks. Networks of networks are a compelling feature of the National Ecological Observatory Network (NEON) as well as the Long Term Ecological Research (LTER) programs within the United States, with similar programs in other parts of the world. At the same time, satellite observations provide a large suite of alternative imaging systems that can measure land, aquatic, and atmospheric conditions at global scales with varying spatial and temporal resolution.

Observational approaches can be useful in contextualizing and extending the results of experiments (Greenwood and McIntosh 2008). Observational data within an experiment extend from direct field observations (Ewers et al. 2011) to embedded sensor networks (Pasetto et al. 2015) and applications of remotely sensed data resources (Laurance et al. 2018). Data from larger contexts and embedded within experiments can facilitate extrapolations as well as improve the description of regional conditions within which an experiment is embedded. Sensor data from networks distributed within an experiment can provide high-resolution description of experimental conditions. Parallel observations to an experiment, such as with a natural experiment or an environmental gradient, can allow robust tests of experimental results and improve the understanding of processes.

Similarly to observations, landscape experiments have opportunities for integration with models. Models can be deployed across all ecological scales, and increasingly, connecting models across scales has become an appreciated research need (Allen 2001; Knowlton and Graham 2010; Wiens 2001). As one notable example, biogeochemical models are increasingly directed to extending microbial-scale dynamics to whole Earth system models (Wieder et al. 2013). At finer scales than an experiment, models can represent the functioning of individual organismal dynamics, and in the case of vegetation, even specific organs, including leaves, stems, and roots. Whole-landscape models are increasingly developed from finer-scale models, including individual organisms that are aggregated into patches (Moorcroft et al. 2001). Movement and connectivity across landscapes are modeled through both active mobility in animals as well as exchanges driven by hydrologic or atmospheric vectors. At coarse scales covering the entire planet, Earth system models are increasingly simulating dynamics of biogeochemical cycling and energy balances as well as population and community dynamics. In the context of landscape experiments, models are used to help design experimental protocols and provide quantitative predictions that can be tested by the experiments. Following an experiment, modeling the results can be used to better constrain uncertainty in model parameters and equations as well as extrapolating results from experiments.

Combinations of multiple approaches, including experimentation, can be an especially valuable approach for ecological research. The LEO experiment (Huxman et al. 2009) and the SAFE fragmentation experiment (Ewers et al. 2011) provide examples of experiments conducted in the context of extensive monitoring and modeling. A multiple-pronged approach that includes mechanistic detail achievable through experimentation along with broader context achievable through observations provides a strong foundation for ecological research with multiple scales of causality.

Synthesis and conclusions

In looking toward integrative and cross-scale landscape ecology science, experiments play a key role. While experiments can provide key information for advancing theory, they are also certainly not the only approach for achieving strong inferences about how the world works. An over reliance on the requirement for experimentation can even impede the advancement of science. For example, in cases where experimentation would be logistically and ethically impossible, for example in the context of air pollution impacts on human health, robust analyses of observational data can lead to improved hypotheses (Goldman and Dominici 2019). Landscape ecology in particular needs a balanced and nuanced view in which experimentation provides one of many approaches for learning. Individual experiments may be useful in evaluating targeted processes within a landscape, and they may also be placed in the context of observational analyses.

In conclusion, experiments have historically played, and will continue to play an important role in landscape science. While landscape experiments cannot provide the only approach for addressing all landscape questions, landscape experiments provide many benefits and have been widely used. Experiments provide a powerful approach for testing hypotheses of pattern–process relationships. They also can take advantage of expanded collaborations and cyberinfrastructure. At the same time, because of the large investments required for landscape experiments, they often provide extensive opportunities to leverage the experimental infrastructure for additional research. While experiments are not always the answer, they can and should be part of the answers to uncertainties in landscape ecology.

Acknowledgments

I thank Emily Minor for the invitation to contribute and help with reviewing the manuscript. This work was supported by the National Science Foundation (DEB −1656062; CBE − 1444758).

References

Allen, M.F., Hipps, L.E. and Wooldridge, G.L. (1989) 'Wind dispersal and subsequent establishment of VA mycorrhizal fungi across a successional arid landscape', *Landscape Ecology*, vol 2, pp165–171.

Allen, T.F.H. (2001) 'The nature of the scale issue in experimentation', in Gardner, R.H., Kemp, W.M., Kennedy, V.S. and Petersen, J.E. (eds) *Scaling Relations in Experimental Ecology*. Columbia University Press, New York.

Andersen, A.N., Braithwaite, R.W., Cook, G.D., Corbett, L.K., Williams, R.J., Douglas, M.M., Gill, A.M., Setterfield, S.A. and Muller, W.J. (1998) 'Fire research for conservation management in tropical savannas: Introducing the Kapalga fire experiment', *Australian Journal of Ecology*, vol 23, pp95–110.

Anderson, K. and Gaston, K.J. (2013) 'Lightweight unmanned aerial vehicles will revolutionize spatial ecology', *Frontiers in Ecology and the Environment*, vol 11, pp138–146.

Beier, P. and Noss, R.F. (1998) 'Do habitat corridors provide connectivity?', *Conservation Biology*, vol 12, pp1241–1252.

Betts, M.G., Gutzwiller, K.J., Smith, M.J., Robinson, W.D. and Hadley, A.S. (2015) 'Improving inferences about functional connectivity from animal translocation experiments', *Landscape Ecology*, vol 30, pp585–593.

Chatterjee, A. and Jenerette, G. (2011a) 'Changes in sol respiration temperature sensitivity during drying-rewetting along a semi-arid elevation gradient', *Geoderma*, vol 163, pp171–177.

Chatterjee, A. and Jenerette, G.D. (2011b) 'Spatial variability of soil metabolic rate along a dryland elevation gradient', *Landscape Ecology*, vol 26, pp1111–1123.

Collins, C.D., Holt, R.D. and Foster, B.L. (2009) 'Patch size effects on plant species decline in an experimentally fragmented landscape', *Ecology*, vol 90, pp2577–2588.

Cunningham, R.B. and Lindenmayer, D.B. (2017) 'Approaches to landscape scale inference and study design', *Current Landscape Ecology Reports*, vol 2, pp42–50.

Damschen, E.I., Brudvig, L.A., Burt, M.A., Fletcher, R.J., Haddad, N.M., Levey, D.J., Orrock, J.L., Resasco, J. and Tewksbury, J.J. (2019) 'Ongoing accumulation of plant diversity through habitat connectivity in an 18-year experiment', *Science*, vol 365, pp1478–1480.

Drummond, J.D., Davies-Colley, R.J., Stott, R., Sukias, J.P., Nagels, J.W., Sharp, A. and Packman, A.I. (2015) 'Microbial transport, retention, and inactivation in streams: A combined experimental and stochastic modeling approach', *Environmental Science & Technology*, vol 49, pp7825–7833.

Ewers, R.M., Didham, R.K., Fahrig, L., Ferraz, G., Hector, A., Holt, R.D., Kapos, V., Reynolds, G., Sinun, W., Snaddon, J.L. and Turner, E.C. (2011) 'A large-scale forest fragmentation experiment: The Stability of Altered Forest Ecosystems Project', *Philosophical Transactions of the Royal Society B-Biological Sciences*, vol 366, pp3292–3302.

Forman, R.T.T. and Godron, M. (1981) 'Patches and structural components for a landscape ecology', *Bioscience*, vol 31, pp733–740.

Frank, D.A. and Groffman, P.M. (1998) 'Denitrification in a semi-arid grazing ecosystem', *Oecologia*, vol 117, pp564–569.

Goldman, G.T. and Dominici, F. (2019) 'Don't abandon evidence and process on air pollution policy', *Science*, vol 363, no 6434, pp1398–1400.

Greenwood, M.J. and McIntosh, A.R. (2008) 'Flooding impacts on responses of a riparian consumer to cross-ecosystem subsidies', *Ecology*, vol 89, pp1489–1496.

Haddad, N.M. and Baum, K.A. (1999) 'An experimental test of corridor effects on butterfly densities', *Ecological Applications*, vol 9, pp623–633.

Haddad, N.M., Rosenberg, D.K. and Noon, B.R. (2000) 'On experimentation and the study of corridors: Response to Beier and Noss', *Conservation Biology*, vol 14, pp1543–1545.

Haddad, N.M., Brudvig, L.A., Clobert, J., Davies, K.F., Gonzalez, A., Holt, R.D., Lovejoy, T.E., Sexton, J.O., Austin, M.P., Collins, C.D., Cook, W.M., Damschen, E.I., Ewers, R.M., Foster, B.L., Jenkins, C.N., King, A.J., Laurance, W.F., Levey, D.J., Margules, C.R., Melbourne, B.A., Nicholls, A.O., Orrock, J.L., Song, D.X. and Townshend, J.R. (2015) 'Habitat fragmentation and its lasting impact on Earth's ecosystems', *Science Advances*, vol 1, art e1500052.

Hanawalt, R.B. and Whittaker, R.H. (1976) 'Altitudnally coordinated patterns of soils and vegetation in San Jacinto Mountains, California', *Soil Science*, vol 121, pp114–124.

Hargrove, W.W. and Pickering, J. (1992) 'Pseudoreplication: A *sine-qua-non* for regional ecology', *Landscape Ecology*, vol 6, pp251–258.

Hart, S.C. (2006) 'Potential impacts of climate change on nitrogen transformations and greenhouse gas fluxes in forests: A soil transfer study', *Global Change Biology*, vol 12, pp1032–1046.

Hershey, A.E., Pastor, J., Peterson, B.J., Kling, G.W. (1993) 'Stable isotopes resolve the drift paradox for Baetis mayflies in an arctic river', *Ecology*, vol 74, pp2315–2325.

Huxman, T., Troch, P., Chorover, J., Breshears, D.D., Saleska, S., Pelletier, J. and Zeng, X. (2009) 'The hills are alive: Earth science in a controlled environment', *Eos, Transactions, American Geophysical Union*, vol 90, no 14, pp120–120.

Ims, R.A. (2005) 'The role of experiments in landscape ecology', in J.A. Wiens and M.R. Moss (eds) *Issues and Perspectives in Landscape Ecology*. Cambridge University Press, Cambridge.

Jenerette, G.D. and Chatterjee, A. (2012) 'Soil metabolic pulses: Water, substrate, and biological regulation', *Ecology*, vol 93, pp959–966.

Jenerette, G.D. and Shen, W.J. (2012) 'Experimental landscape ecology', *Landscape Ecology*, vol 27, pp1237–1248.

Kawaguchi, Y., Taniguchi, Y. and Nakano, S. (2003) 'Terrestrial invertebrate inputs determine the local abundance of stream fishes in a forested stream', *Ecology*, vol 84, pp701–708.

Knowlton, J.L. and Graham, C.H. (2010) 'Using behavioral landscape ecology to predict species' responses to land-use and climate change', *Biological Conservation*, vol 143, pp1342–1354.

Konrad, C.P., Olden, J.D., Lytle, D.A., Melis, T.S., Schmidt, J.C., Bray, E.N., Freeman, M.C., Gido, K.B., Hemphill, N.P., Kennard, M.J., McMullen, L.E., Mims, M.C., Pyron, M., Robinson, C.T. and Williams, J.G. (2011) 'Large-scale flow experiments for managing river systems', *Bioscience*, vol 61, pp948–959.

Kurkjian, H.M. (2019) 'The metapopulation microcosm plate: A modified 96-well plate for use in microbial metapopulation experiments', *Methods in Ecology and Evolution*, vol 10, pp162–168.

Larsen, L. and Harlan, S.L. (2006) 'Desert dreamscapes: Residential landscape preference and behavior', *Landscape and Urban Planning*, vol 78, pp85–100.

Laurance, W.F., Lovejoy, T.E., Vasconcelos, H.L., Bruna, E.M., Didham, R.K., Stouffer, P.C., Gascon, C., Bierregaard, R.O., Laurance, S.G. and Sampaio, E. (2002) 'Ecosystem decay of Amazonian forest fragments: A 22-year investigation', *Conservation Biology*, vol 16, pp605–618.

Laurance, W.F., Camargo, J.L.C., Luizao, R.C.C., Laurance, S.G., Pimm, S.L., Bruna, E.M., Stouffer, P.C., Williamson, G.B., Benitez-Malvido, J., Vasconcelos, H.L., Van Houtan, K.S., Zartman, C.E., Boyle, S.A., Didham, R.K., Andrade, A. and Lovejoy, T.E. (2011) 'The fate of Amazonian forest fragments: A 32-year investigation', *Biological Conservation*, vol 144, pp56–67.

Laurance, W.F., Camargo, J.L.C., Fearnside, P.M., Lovejoy, T.E., Williamson, G.B., Mesquita, R.C.G., Meyer, C.F.J., Bobrowiec, P.E.D. and Laurance, S.G.W. (2018) 'An Amazonian rainforest and its fragments as a laboratory of global change', *Biological Reviews*, vol 93, pp223–247.

Maestre, F.T. and Reynolds, J.F. (2007) 'Amount or pattern? Grassland responses to the heterogeneity and availability of two key resources', *Ecology*, vol 88, pp501–511.

McGarigal, K. and Cushman, S.A. (2002) 'Comparative evaluation of experimental approaches to the study of habitat fragmentation effects', *Ecological Applications*, vol 12, pp335–345.

Moorcroft, P.R., Hurtt, G.C. and Pacala, S.W. (2001) 'A method for scaling vegetation dynamics: The ecosystem demography model (ED)', *Ecological Monographs*, vol 71, pp557–585.

Nabe-Nielsen, J., Sibly, R.M., Forchhammer, M.C., Forbes, V.E. and Topping, C.J. (2010) 'The effects of landscape modifications on the long-term persistence of animal populations', *PLoS One*, vol 5, no 1, art e8932.

Nachappa, P., Margolies, D.C., Nechols, J.R. and Campbell, J.F. (2011) 'Variation in predator foraging behaviour changes predator-prey spatio-temporal dynamics', *Functional Ecology*, vol 25, pp1309–1317.

Neuman, S.P. (2005) 'Trends, prospects and challenges in quantifying flow and transport through fractured rocks', *Hydrogeology Journal*, vol 13, pp124–147.

Pangle, L.A., DeLong, S.B., Abramson, N., Adams, J., Barron-Gafford, G.A., Breshears, D.D., Brooks, P.D., Chorover, J., Dietrich, W.E., Dontsova, K., Durcik, M., Espeleta, J., Ferre, T.P.A., Ferriere, R., Henderson, W., Hunt, E.A., Huxman, T.E., Millar, D., Murphy, B., Niu, G.Y., Pavao-Zuckerman, M., Pelletier, J.D., Rasmussen, C., Ruiz, J., Saleska, S., Schaap, M., Sibayan, M., Troch, P.A., Tuller, M., van Haren, J. and Zeng, X.B. (2015) 'The landscape evolution observatory: A large-scale controllable infrastructure to study coupled Earth-surface processes', *Geomorphology*, vol 244, pp190–203.

Pasetto, D., Niu, G.Y., Pangle, L., Paniconi, C., Putti, M. and Troch, P.A. (2015) 'Impact of sensor failure on the observability of flow dynamics at the Biosphere 2 LEO hillslopes', *Advances in Water Resources*, vol 86, pp327–339.

Plowright, R.K., Sokolow, S.H., Gorman, M.E., Daszak, P. and Foley, J.E. (2008) 'Causal inference in disease ecology: Investigating ecological drivers of disease emergence', *Frontiers in Ecology and the Environment*, vol 6, pp420–429.

Ravi, S., D'Odorico, P., Zobeck, T.M. and Over, T.M. (2009) 'The effect of fire-induced soil hydrophobicity on wind erosion in a semiarid grassland: Experimental observations and theoretical framework', *Geomorphology*, vol 105, pp80–86.

Renofalt, B.M. and Nilsson, C. (2008) 'Landscape scale effects of disturbance on riparian vegetation', *Freshwater Biology*, vol 53, pp2244–2255.

Richardson, J., Chatterjee, A. and Jenerette, G.D. (2012) 'Optimum temperatures for soil respiration along a semi-arid elevation gradient in southern California', *Soil Biology & Biochemistry*, vol 46, pp89–95.

Sayer, E.J. (2006) 'Using experimental manipulation to assess the roles of leaf litter in the functioning of forest ecosystems', *Biological Reviews*, vol 81, pp1–31.

Schmitz, O.J. (2005) 'Scaling from plot experiments to landscapes: Studying grasshoppers to inform forest ecosystem management', *Oecologia*, vol 145, pp225–234.

Schooley, R.L. and Wiens, J.A. (2004) 'Movements of cactus bugs: Patch transfers, matrix resistance, and edge permeability', *Landscape Ecology*, vol 19, pp801–810.

Silva, F.R., Oliveira, T.A.L., Gibbs, J.P. and Rossa-Feres, D.C. (2012) 'An experimental assessment of landscape configuration effects on frog and toad abundance and diversity in tropical agro-savannah landscapes of southeastern Brazil', *Landscape Ecology*, vol 27, pp87–96.

Sponseller, R.A. and Fisher, S.G. (2008) 'The influence of drainage networks on patterns of soil respiration in a desert catchment', *Ecology*, vol 89, pp1089–1100.

Tilman, D. (1989) 'Ecological experimentation: Strengths and conceptual problems', in G. Likens (ed) *Long-Term Studies in Ecology: Approaches and Alternatives*. Springer, New York.

Van De Koppel, J., Herman, P.M.J., Thoolen, P. and Heip, C.H.R. (2001) 'Do alternate stable states occur in natural ecosystems? Evidence from a tidal flat', *Ecology*, vol 82, pp3449–3461.

Volkmann, T.H.M., Sengupta, A., Pangle, L.A., Dontsova, K., Barron-Gafford, G.A., Harman, C.J., Niu, G., Meredith, L.K., Abramson, N., Meira Neto, A.A., Wang, Y., Adams, J.R., Breshears, D.D., Bugaj, A., Chorover, J., Cueva, A., DeLong, S.B., Durcik, M., Ferre, T.P.A., Hunt, E.A., Huxman, T.E., Kim, M., Maier, R.M., Monson, R.K., Pelletier, J.D., Pohlmann, M., Rasmussen, C., Ruiz, J., Saleska, S.R., Schaap, M.G., Sibayan, M., Tuller, M., van Haren, J.L.M., Zeng X. and Troch, P.A. (2018) 'Landscape evolution observatory: Toward prediction of coupled hydrological, biogeochemical, and ecological change', in J. Liu and W. Gu (eds) *Hydrology of Artificial and Controlled Experiments*. IntechOpen, London. https://www.intechopen.com/books/hydrology-of-artificial-and-controlled-experiments/controlled-experiments-of-hillslope-coevolution-at-the-biosphere-2-landscape-evolution-observatory-t

Warren, R.J. (2010) 'An experimental test of well-described vegetation patterns across slope aspects using woodland herb transplants and manipulated abiotic drivers', *New Phytologist*, vol 185, pp1038–1049.

Weerman, E.J., Herman, P.M.J. and Van de Koppel, J. (2011) 'Top-down control inhibits spatial self-organization of a patterned landscape', *Ecology*, vol 92, pp487–495.

Wieder, W.R., Bonan, G.B. and Allison, S.D. (2013) 'Global soil carbon projections are improved by modelling microbial processes', *Nature Climate Change*, vol 3, pp909–912.

Wiens, J.A. (2001) 'Understanding the problem of scale in experimental ecology', in R.H. Gardner, W.M. Kemp, V.S. Kennedy and J.E. Petersen (eds) *Scaling Relations in Experimental Ecology*. Columbia University Press, New York.

Wu, J.G. and Loucks, O.L. (1995) 'From balance of nature to hierarchical patch dynamics: A paradigm shift in ecology', *Quarterly Review of Biology*, vol 70, pp439–466.

Zabriskie, J.G. (1979) *Plants of Deep Canyon and the Central Coachella Valley, California Plants of Deep Canyon and the Central Coachella Valley, California*. University of California, Riverside.

Part IV
Landscape ecology frontiers

19

Landscape ecology contributions to biodiversity conservation

Robert F. Baldwin, R. Daniel Hanks, and Jeremy S. Dertien

Introduction

Among the environmental issues about which the public appears to be the least concerned is the decline of biodiversity. Rather, the trend appears to be towards prioritizing topics perceived as critical to our own well-being (Novacek 2008, Kusmanoff et al. 2017, DellaSala 2020). Despite the perception that biodiversity loss is not going to compete with climate change, disease, or pollution for our attention, it is clear that it poses a threat, as biodiversity contributes to psychological health, regulates infectious diseases, and is the source of many medicines and foods (Fuller et al. 2007, Chivian and Bernstein 2008). Past conservation efforts have been substantial in the form of globally extensive protected areas, law and policy, university research and teaching programs, and actions by government agencies and non-governmental organizations, but these efforts have been inadequate to meet the extent of biodiversity declines (Dirzo et al. 2014, Baldwin and Beazley 2019) and the reduction in biodiversity may threaten society's ability to keep within planetary boundaries (Steffen et al. 2015). Landscape ecology is an integrative, spatial discipline and offers a means of understanding links between land use and land cover change and biodiversity, forming a basis for planning for a sustainable future (Opdam et al. 2002, Forman 1995).

Landscape-level patterns and processes are directly related to the diversity of species, genes, and ecosystems (Clark 1991, Forman 1995, Hanski 1998). Humans, by influencing habitat patterns in time and space, influence spatial distributions of populations. For example, by creating small patches of agricultural land and their edges in an otherwise forested region, humans introduce heterogeneity that may actually increase diversity for a period of time in the early stages of conversion (Hunter and Gibbs 2007, Hepinstall et al. 2008). As the degree of human influence increases and more habitat is converted, the heterogeneity will decrease, resulting in a more homogeneous, primarily anthropogenic landscape and declines in the diversity of native species (Laurance et al. 2002, McKinney 2002, Hendrickx et al. 2007). By predicting how landscape conditions at varying spatial scales (resolution and extent) influence populations, landscape ecologists provide fundamental understanding for conservation models, such as the location of new protected areas and corridors, and/or management regimes.

The field of systematic conservation planning is an applied spatial discipline with the specific goal of mapping landscapes for biodiversity conservation (Margules and Pressey 2000,

Groves et al. 2002). Advances in recent decades have been related to improvements in theory, data repositories, and analytical software, some of which will be discussed later (Moilanen et al. 2009, Dickson et al. 2019). Data and software improvements for landscape analysis include remote sensing products measuring land use and habitat distribution, species location data, protected areas coverage, and spatial software to map species distributions, reserve site selection, and habitat corridors (Ball et al. 2009, Elith et al. 2011, Dickson et al. 2019). These advances are driven by scientists' desire to make robust predictions about spatial aspects of biodiversity conservation given an uncertain future (Hunter et al. 1988, Pressey et al. 2007, Newbold et al. 2016).

The purpose of this chapter is to provide an overview of a nexus between landscape ecology and biodiversity conservation. We have chosen to focus on systematic conservation planning as the nexus. Our goals are to touch upon general topics and then give two special cases: aquatic-terrestrial integration and species distribution modeling. We conclude with some recommendations.

Scale

Landscape ecology explicitly considers 'scale', which incorporates both grain size (resolution) and spatial extent (area, distance, length of time) and both temporal and spatial scale (Forman 1995, Turner et al. 2001). To understand composition, structure, and function at a particular site or community, it is necessary to incorporate events and conditions occurring at different distances and time periods (Foster 2000, Foster and Aber 2004, Semlitsch 2008). Thus, many studies have incorporated historical effects of land use (land use legacies) on biodiversity, and effects of human activities at different distances from ecological responses that are being measured (Foster et al. 2003, Homan et al. 2004, Scott 2005, Hepinstall et al. 2008). Conservation efforts must consider both spatial and temporal scale.

Dynamic landscapes

Human activities dominate ecosystem form and function, changing landscape pattern and process (Theobald 2004, Haddad et al. 2015, Venter et al. 2016). The theory of island biogeography formed a basis for understanding how biogeographic patterns on oceanic islands of varying areas and distances to source populations relate to fragments of natural ecosystems such as remnant forest patches on land (MacArthur and Wilson 1967). Since then, there have been many advances in how to map, measure, and categorize habitat fragments and test hypotheses relative to their effects on biodiversity (McGarigal and Cushman 2002, Fahrig 2003).

The *Biological Dynamics of Forest Fragments Project* spanned 22 years and found that size of fragment, edge effects, and matrix quality influence diversity in the Amazon Basin (Laurance et al. 2002). Recent analyses suggest an increase in habitat fragmentation across the planet with a 9% increase in overall human footprint; in the past 16 years, fragmentation has reduced biodiversity 13–75%, with more than 50% of the habitat of area-sensitive species in fragmented environments (Crooks et al. 2011, Haddad et al. 2015, Venter et al. 2016).

A number of approaches have been developed to detect and monitor changes in natural landscape condition at the global level. They measure fragmentation, degradation, or conversion of natural land cover using remotely sensed data products such as land use and land cover data (Sanderson et al. 2002a, Theobald 2010). Examples include the Human Footprint, which has been used to assess global changes in human domination over time (Venter et al. 2016); the Human Modification Index, which has been used in habitat connectivity analyses as a permeability or resistance layer, to assess the condition of a protected area, and help select new conservation areas (Theobald 2013); and the Index of Ecological Integrity (McGarigal et al. 2018). Developing repeatable, transparent metrics that can be deployed to track changes over time as new remote sensing data are released, and at multiple scales, is a major contribution of landscape ecology to biodiversity conservation.

Protected area coverage and targets

Protected areas are variable in how well they represent and protect biodiversity (Woodroffe and Ginsberg 1998, Gaston et al. 2006, DeFries et al. 2010). There are many types of protected areas – a relatively small portion is managed specifically for biodiversity, while greater areas are open to multiple uses. The governmental systems for categorizing them according to the degree of protection vary considerably and are based on imperfect information, meaning that there are many areas mapped as 'protected' whose status is lower or uncertain (Boitani et al. 2008, Dudley et al. 2010).

Global areal coverage of protected areas is far below the levels predicted to be needed to capture biodiversity; 25–75% of the extent of each biome is expected to be needed, and yet only 15% of terrestrial habitats, and 8% of marine habitats, currently are within protected areas of any status (Brooks et al. 2004, Baldwin and Beazley 2019). A troubling problem is that some protected areas have been found to be disjunct from and not inclusive of the ecological hotspots they are expected to defend (Margules and Pressey 2000, Jenkins et al. 2015). Such 'mismatches' provide a reason for ongoing and rigorous landscape-scale conservation planning projects that seek to strategically place new areas in the network that would more effectively meet conservation needs.

Management practices and land use policies in the landscape matrix in which protected areas are embedded illustrate the power of landscape ecology as a spatial discipline that integrates human and natural pattern and process. The landscape matrix surrounding protected areas influences conditions within the protected area. People often live and work outside the boundaries of protected areas and in many parts of the world, inside them as well. Increased human activity in the adjacent areas may ecologically isolate protected areas, and without connectivity with other core areas – which can essentially increase the effective size of all reserves – the individual, isolated reserves can lose species (Newmark 1995, Parks and Harcourt 2002). Local human populations may need to extract resources from the protected areas to survive; thus, understanding their needs as well as natural ecosystems may both improve human life and also make biodiversity conservation efforts more effective (deFries et al. 2007). The integrative quality of landscape ecology is that human and natural features are considered interacting parts of the same system. Emerging analyses focus on the dynamic between humans living near, and often inside, protected areas and biodiversity conservation.

Systematic conservation planning

History

Computing in general, and mapping software in particular, has become an essential tool used by ecologists and landscape ecology and has made large contributions to the spatial aspects of biodiversity conservation (McHarg 1969, Opdam et al. 2002). Concurrently, mapped conservation-relevant data is available at scales ranging from highly local to global. For example, the United States Geological Survey (USGS) Land Use/Land Cover program maintains a large spatial dataset used in many biodiversity conservation applications in the United

States (Theobald et al. 2005, Theobald 2010). Other examples are NatureServe and Global Biodiversity Information Facility, which serve species distribution data (occurrences, models) that are used in conservation planning. Data on protected areas are compiled and served globally by the International Union for Conservation of Nature (IUCN)-Protected Planet (World Database on Protected Areas) and for the United States, by the USGS (Protected Areas Database of the United States). Technological advances have facilitated the increasing ability to predict which areas are most critical for maintaining biogeographic pattern and process, beginning with simple co-occurrence and layering and moving to complex spatial computing (Scott et al. 1993).

The science of predicting where and when to add new protected areas to systematically represent the biota took shape in the 1980s and 1990s (Kirkpatrick 1983). Projects such as the USGS Gap Analysis project used early geographic information system (GIS) software and biogeographic data to map 'gaps' in protection for species richness and rare species occurrence (Scott et al. 1993). As computers became more capable of handling spatial data, and as more spatial data became digitized, the questions that were asked became more complex. Optimization-based algorithms were designed into software such as Marxan and Zonation, whose goal it was to develop spatially explicit designs capturing the most biodiversity targets while minimizing cost (Moilanen et al. 2009).

Advances in habitat connectivity modeling enabled mapping of corridors and linkages, identifying 'pinch points', where gene flow may be constricted, and flows across very large areas, leading to a new way of viewing landscapes as interconnected networks (Crooks and Sanjayan 2006, Dickson et al. 2019). The practical result of these advances has been a plethora of conservation plans made by organizations attempting to capture and plan for conservation targets and goals at multiple spatial and temporal scales (Sinclair et al. 2018). Major conservation organizations such as The Nature Conservancy, government agencies, and citizen groups have used landscape ecological tools for biodiversity conservation applications, as can be seen on websites such as Conservation Gateway (www.conservationgateway.org) and DataBasin (https://databasin.org/).

Systematic approach

The biased distribution and incomplete coverage of protected areas drove the development of systematic conservation planning (Margules and Pressey 2000). Systematic conservation planning attempts to make placement of new reserves more effective by solving the problem of representation of biogeographic patterns in reserve systems and creating a system of reserves that functions as a network and is resilient to change. Rather than following the historical pattern of establishing reserves based on opportunity, the effort is to make each desired ecological outcome an explicit target or goal, weighted, and included in an algorithm that tests various combinations (Watts et al. 2009).

Systematic conservation planning approaches have been thoroughly reviewed elsewhere (Moilanen et al. 2009). The uptake of these methodologies by conservation groups provides evidence that the science–implementation gap has been bridged to a great degree, and there are numerous conservation planning case studies available (Trombulak and Baldwin 2010, Hilty et al. 2012)

The ideal representative reserve network includes core areas selected because they meet multiple and specific goals, linkages among those core areas, and a relatively permeable landscape matrix (Noss 1983, Margules and Pressey 2000, Beier et al. 2011). Well-designed reserve networks are resilient to change: to the degree possible, future conditions (such as those due to climate or land use change) have been considered (Hilty et al. 2012). Put simply, if landscapes are relatively permeable, then populations of plants and animals facing inhospitable, new climate regimes will be more likely to move to places that have their preferred climate regimes (Beier and Brost 2010, Nuñez et al. 2013, McGuire et al. 2016).

The limitations of systematic conservation planning are many. These include (1) using a static plan to protect a dynamic world, (2) subjectivity involved in setting conservation targets and goals based on the values of the participants, (3) scale dependency in data availability, especially the widespread use of relatively coarse-scale, broad-extent data for more local applications, (4) the relative lack of consideration of human needs and the relative paucity of socioeconomic and especially cultural data, and (5) a disconnect between the scientific engine of landscape ecology and actual decision making to protect biodiversity (Theobald et al. 2000, Pressey et al. 2007, Knight et al. 2008, Woolmer et al. 2008, Imperial et al. 2016).

Habitat connectivity

An aspect of conservation planning that has recently received significant attention, due partly to conceptual and technological advances, is habitat connectivity. Linking protected areas into a network increases the functional size of the protected areas and may facilitate metapopulation processes (Carroll et al. 2003, Wikramanayake et al. 2004). Migration, such as seasonal movements among resource patches, is also facilitated. Yet, predicting movement is complex (Zeller et al. 2012).

A creative period in landscape ecology and conservation biology related to habitat connectivity occurred over the past decade. Perhaps the most innovative thinking came from Brad McCrae, who adapted electrical circuit theory to model spatial patterns of gene flow (McRae et al. 2008). He and his colleagues built software to help conservation organizations predict where on the landscape genes flow easily (high permeability) and where the flow becomes most restricted. Circuitscape, the software McRae developed, and related programs have been used thousands of times (Dickson et al. 2019). Circuit theory applications are intuitively easy, have been applied by conservation groups in numerous ways around the world, and are undergoing continual improvements despite the untimely passing of Dr McRae, a landscape ecology and conservation innovator (Lawler et al. 2019).

One trend has been a re-imagining of the ideal conservation landscape as being less binary and focused on discrete habitat polygons, and more a managed gradient (Cushman et al. 2010, Theobald et al. 2012, Zeller et al. 2012). With human domination, managed activities in the matrix between core habitats – such as sustainable agriculture – can have defining effects on the overall levels of biodiversity at landscape scales (Vandermeer and Perfecto 2007). Gradient analyses show a continuous surface over which organisms are more (or less) likely to travel given the conditions, whereas corridors are generally depicted as landscape features with 'hard boundaries'. While the corridor approach is concrete and fits more easily into the 'protected area' paradigm, and may be more easily managed because its area is clearly defined, the gradient approach may more closely reflect biological reality. Even so, managing a gradient is more difficult because it is more diffuse.

Planning for climate change

To accommodate range shifts and new migration pathways, a number of spatial approaches have been explored. These models include predicting pathways to new environments such that the pathways themselves are permeable. Examples include (1) models that do not expect shifts of mountaintop species to occur through intervening much lower areas or inhospitable slopes and aspects (Beier and Brost 2010, Nuñez et al. 2013), (2) models that consider the velocity of changes (Burrows et al. 2014), and (3) models that anticipate salt water intrusion as a form of fragmentation (Leonard et al. 2017b).

The driving factor in successful movement is current and future fragmentation caused by land use and climate change (McGuire et al. 2016). Given uncertainties, conservation biologists have advocated for various coarse-filter responses that take into account paleoecological trajectories in order to conserve adequate space and diversity of microenvironments. Such plans attempt to maintain ecological pattern and process even if the individual taxa making up present communities change (Hunter et al. 1988). If enough area with enough heterogeneity and connectivity is set aside, ecological communities will assemble and reassemble but potentially remain diverse (Anderson and Ferree 2010).

Socioeconomic and cultural dimensions

Conservation is a broadly integrative problem, but socioeconomic and cultural aspects of conservation have only recently been treated systematically as a research topic (Mascia et al. 2003, Powell et al. 2009, Armsworth et al. 2011). Mapping cultural resources for conservation plans (Ogletree et al. 2019), including stakeholder objectives in marine conservation plans (Gurney et al. 2015), and assessing how natural areas contribute to human well-being (Larson et al. 2016) and how the management of protected areas may influence poverty (Ferraro and Hanauer 2014) are a few examples from a rapidly growing field. It is possible that the future of biodiversity conservation lies in new set-aside areas in which natural pattern and process prevail, and in understanding and managing human activities outside protected areas with an eye to improving both the human and the natural condition across a gradient of land uses from heavily industrial to wild (Theobald 2004, Baldwin and Beazley 2019).

Technological fundamentalism

Just before the information superhighway exploded, philosopher David Orr warned conservation biologists to be skeptical of computers (Orr 1994). Experience with landscape conservation planning projects tells us it is best not to fall into the trap of 'newer technology is better' but rather, to understand exactly what each data product and processing step does and compare it with the projects' assumptions and conservation goals. In some situations, earlier approaches such as co-occurrence 'layering' using readily available data (Scott et al. 1993) sufficiently accomplish goals, and complex modeling and time-intensive development of new input data may be detrimental. Anecdotal evidence from many conservation planning exercises leads us to conjecture that overly complex modeling processes and expensive data generation may lead to implementation gaps. We have heard important stakeholders complain about 'black box' modeling and expensive and protracted data generation steps. Stakeholders who feel alienated from the science may not be compelled to follow its recommendations. Conversely, investing the effort into making more complex algorithms and acquiring higher-quality data can result in meaningful improvements in conservation modeling (Leonard et al. 2016). Even basic components of a study such as extent of the analysis (e.g., county versus region), and grain size-resolution of the dataset can have meaningful impacts on results (Woolmer et al. 2008, Huber et al. 2010, Leonard et al. 2016)

Planning–implementation gaps

Despite efforts to conduct studies that have meaning and relevance to real-world situations, implementation gaps – gaps between doing applied science and applying the science – are a persistent problem (Theobald et al. 2000, Knight et al. 2008). Spatial planning processes are integrative and iterative, engaging multiple stakeholders, each with differing perspectives and goals. The object is to include both end users (such as conservation practitioners) and scientists (including landscape ecologists, taxonomic specialists, etc.) in the process of making the plan, with feedback loops between engagement and product and between implementation and modeling (Knight et al. 2009). This is often framed in an adaptive management context (Sayer et al. 2013). Implementation gaps between spatial conservation plans and how they are used in policy, land use, and management context may be narrowing; a recent survey found that 74% of landscape-level prioritizations intended for implementation had translated into on-the-ground action (Sinclair et al. 2018).

The special challenge of conservation planning in the freshwater realm

Systematic conservation planning has traditionally been evaluated and performed within the context of terrestrial and marine ecosystems (Vance-Borland et al. 2008, Linke et al. 2011), with little attention directed towards freshwater ecosystems and particularly freshwater lotic systems (Ban et al. 2013). This is in part due to the challenges associated with these linear and networked systems (Hermoso et al. 2015). However, due to the poor conservation status of freshwater systems worldwide (Vörösmarty et al. 2010, Collen et al. 2014), the call for effective management of freshwater systems (Hermoso et al. 2016), the lack of inclusion in systematic conservation planning (Nel et al. 2009), particularly the inclusion of freshwater protected areas (Hermoso et al. 2016), and recent novel approaches, systematic conservation planning in freshwater aquatic ecosystems has begun to grow (Hermoso et al. 2015) and is quickly becoming a fertile discipline with potential yet to be fully understood and implemented.

Freshwater taxa are the earth's most threatened group of organisms (IUCN 2019). Of those taxa assessed, 31% of amphibians and 13% of fish are listed as endangered, vulnerable, or near threatened on the IUCN red list (IUCN 2019). The increased risk to freshwater biodiversity may in part be due to the fact that unlike terrestrial systems, riverine systems are interconnected by flowing waters, and therefore, conservation efforts must consider upstream and downstream threats as well as those adjacent to a given site (Linke et al. 2011). The need for conservation scientists to develop systematic conservation plans that provide the essential core habitat and connectivity for aquatic species to continue to persist under a changing and uncertain future (e.g., due to land conversion and climate change) is paramount.

Challenges for freshwater biodiversity

Data availability and the uncertainty associated with modeled species distributions (see following section) for freshwater systems have been somewhat limiting for systematic conservation plans. Fortunately, GIS data have become increasingly available, as have contemporary modeling techniques that recognize the unique topological aspects of hydrologic networks (Peterson et al. 2013, Hanks et al. 2020), e.g., boosted regression trees (Elith and Leathwick 2015, Hanks et al. 2020), or valid spatial models (Isaak et al. 2014). Therefore, data limitations have become less of an issue (McManamay and Derolph 2019), even in data-poor regions (Linke et al. 2011).

At times when the limitations in data availability cannot be overcome (Thieme et al. 2007), surrogates that represent ecosystem pattern and/or process may be employed. Surrogates can include biological surrogates, combined biological and abiotic surrogates, or abiotic surrogates alone. Care must be taken, however, particularly because of the limitation on conservation resources, to ensure that surrogates provide for the protection of the highest-valued priority sites. Community-level surrogates can represent various diversity measures (Ferrier 2002) and account for unknown or under-surveyed species (Ferrier and Guisan 2006) and are therefore preferred (Linke et al. 2011).

Difficulties in freshwater conservation planning arise largely due to the linear structure of freshwater ecosystems and due to the importance of river connectivity to the maintenance of key ecological processes such as migration of aquatic fauna and fluxes of energy and nutrients (Fausch et al. 2002, Hermoso et al. 2015). The implementation of conservation recommendations is difficult because of the extensive geographies needed to meet the conservation objectives (i.e., targets and goals) within entire catchments (Abell et al. 2007, Hermoso et al. 2015). Additionally, freshwater ecosystems are particularly vulnerable to exogenous threats such as chemical transport via drainage and irrigation (Roche et al. 2002), altered hydrology (Barendregt et al. 1995), flow modification structures (Schindler 2000, Grill et al. 2019), and lack of, or partial, inclusion in current protected areas (Driver et al. 2005). While good land management provides ecological uplift, particularly within protected areas (Driver et al. 2005), land cover changes and other landscape activities outside protected area boundaries affect freshwater systems within protected areas due to their permeability to biophysical processes (Peres and Terborgh 1995, Pringle 2001).

Integration of freshwater conservation plans with those of the terrestrial and marine realms has become increasingly recognized as the best way forward for systematic conservation planning (Kukkala and Moilanen 2013). Doing so allows conservation plans to minimize cross-realm threats (Nel et al. 2009) while maintaining cross-realm interconnected ecological processes (Naiman et al. 2002), thereby solidifying their interdependence (Fairbanks et al. 2001, Fox and Beckley 2005, Strange et al. 2007). However, distinct differences exist between freshwater and terrestrial systems, which therefore require, to some extent, separate analyses that should be followed by vertical integration for systematic conservation planning (Amis et al. 2009).

Overcoming complexities for freshwater biodiversity

Abell et al. (2007) proposed a hierarchical approach to freshwater conservation planning due to the implication that 'protected areas' indicate sweeping and restrictive regulations on the landscape, which are often seen as intractable with current land use practices and the status of coupled human–natural systems. They proposed a three-level freshwater conservation system, where each subsequent zone becomes more extensive in area but less restrictive in management and land use practices: freshwater focal zones, critical management zones, and catchment management zones. Freshwater focal zones are managed to provide localized protection for key areas of biological diversity (similarly to protected areas in marine and terrestrial plans) and have the most restrictive regulations and management, which typically aim to prevent disturbance to the identified conservation target. Critical management zones are managed to provide for the maintenance of ecological functionality to and between focal areas (e.g., connectivity, protection of spawning and nursery areas, etc.). Land use practices that do not interfere with the focal area's conservation targets are allowed. Catchment management zones are managed to link the entire upstream catchment with the critical management zone. In these zones, land use activities are typically not constrained, but best practices are encouraged (Abell et al. 2008, Hermoso et al. 2015). Hermoso et al. (2015) tested this framework using Marxan with Zones in the Daly River catchment in northern Australia and found that it had the potential to make conservation recommendations for freshwater ecosystems more informative and explicit regarding how different management regimes could be allocated to meet conservation goals.

An integrated freshwater and terrestrial ecosystem approach was proposed by Amis et al. (2009), in which freshwater systems are assessed first, followed by a terrestrial system assessment that is driven by freshwater priority areas. The authors successfully used Marxan to identify a near-optimal set of planning units that achieved both aquatic and terrestrial goals. However, the authors were unable to overcome some of the major complications of systematic conservation planning in the freshwater realm, namely connectivity along riverine networks. In a systematic conservation design for the southern and central Appalachian Mountain region, used the framework developed by Amis et al. (2009) and further extended the model by including a conservation design phase where design elements for the systematic conservation plan were created (Figure 19.1) (Amis et al. 2009, Leonard et al. 2017a). Additionally, they modeled aquatic connectivity using density of dams and road crossings at the catchment and watershed scales, which was used to inform boundary relations so that highly connected catchments were selected during Marxan optimization (Leonard et al. 2017a). Neither Amis et al. (2009) nor Leonard et al. (2017) performed their aquatic conservation plan in the mode of Abell et al. (2007), nor

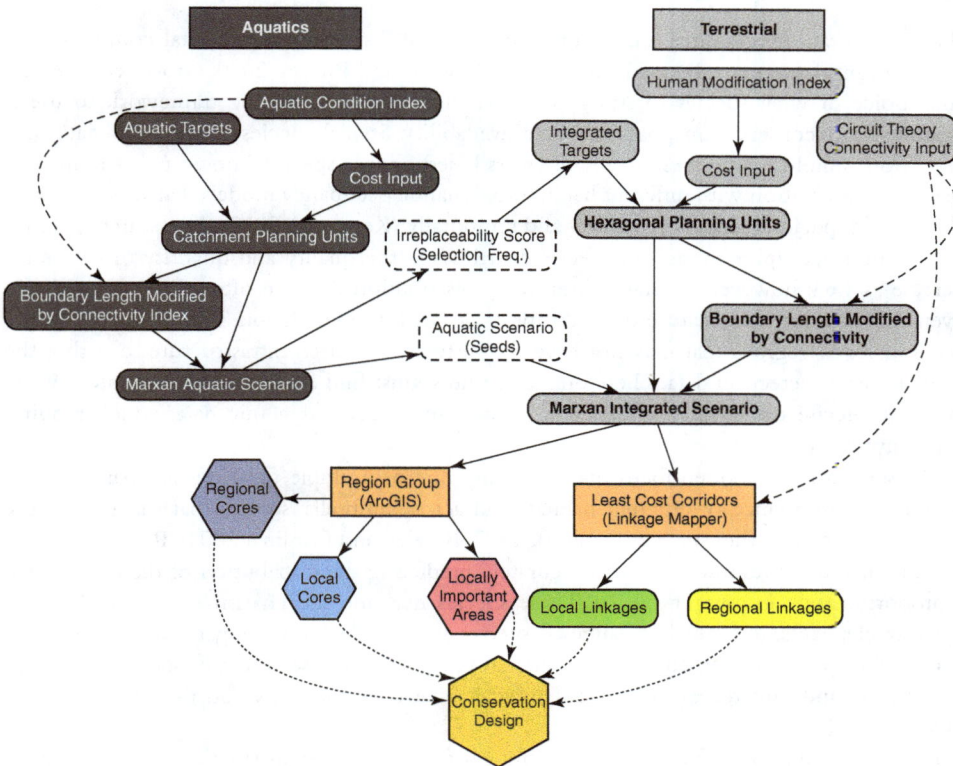

Figure 19.1 Schematic that shows a framework proposed by Leonard et al. (2017a) for the integration of cross-realm planning, where freshwater planning is represented by black boxes, and terrestrial planning is represented by grey boxes. Colored boxes show the integrated scenario of the author's landscape conservation design across the central and southern Appalachian Mountains.

did they use the method proposed by Hermoso et al. (2015) to overcome issues associated with freshwater ecosystem connectivity. The inclusion of these methods should continue to improve systematic conservation plans in the future and further develop the science of systematic conservation planning by integration of freshwater and terrestrial realms.

For systematic conservation planning to realize its full potential, much work is needed to integrate conservation science and planning with additional disciplines such as economics, sociology, landscape planning, and others (Abell et al. 2008, Álvarez-Romero et al. 2015). Using a fully integrated and holistic model will provide insight for decision makers that will enable reduced cross-realm threats and provide for the maintenance of healthy links between socioecological systems (Álvarez-Romero et al. 2015). The technical challenges of cross-realm integration have been a hindrance to systematic conservation planning, and while progress has been made towards overcoming those barriers, potentially the greater hurdle and threat to conservation is convincing conservation scientists, practitioners, and policy makers that siloed work (e.g., conservation [separate freshwater, marine, and terrestrial plans], economic, sociology, etc.) towards conservation goals is inefficient and unrealistic (Linke et al. 2011).

Fundamental importance of species distributions

The incorporation of animal and plant species' spatial distributions is a vital component of conservation planning at the landscape scale (Margules and Pressey 2000, Groves et al. 2002). The choice of what species to study and what distribution modeling framework to use is dependent on conservation goals and data availability. Spatial species distribution data may range from simple binary present/absent layers based upon a species' known (or at minimum presumed) association with different habitats to dynamic occupancy models that vary the probabilities of a species occupancy spatially and temporally (Kery et al. 2013). As researchers move towards more complex models of species distribution, the quality and quantity of data must increase as well; however, accurate biodiversity distribution data are often difficult to obtain given the time-intensive nature of landscape-scale field data collection. This is especially true in countries or regions that may not have the historical research infrastructure to gather the needed species detection data. Therefore, researchers must find a parsimonious point between coarse, imprecise data, which is easier to obtain, and precise, dynamic data, which requires heavy investment.

Conservation plans often focus on identifying one or multiple focal species that represent the relative importance of different habitats and act as 'umbrella' species that can lead to the protection of many other species (Lambeck 1997, Beazley and Cardinal 2004). Researchers can direct their limited resources towards accurately predicting the distribution of these presumed disproportionately important species. These species may include charismatic species such as tigers or elephants, imperiled/endangered species such as the snow leopard or red-cockaded woodpecker, presumed keystone species such as prairie dog and sea otter, or species known to specialize in and thus represent a well-defined habitat type, such as wood frogs in ephemeral ponds.

Binary habitat association maps are the simplest form of species distribution data and require, at a minimum, expert opinion as a source of the likely distribution of a select species. The choice of experts can be rather broad (e.g., professional biologists, professors, volunteer naturalists) if they have personal experience with the landscape and the habitat use of the species (Burgman et al. 2011, Krueger et al. 2012). Expert opinion, if collected in a systematic fashion (Burgman et al. 2001, Burgman et al. 2011, Martin et al. 2012), can provide important species distribution information; however, if the data are not collected in a systematic way, they can be biased and qualitative and lack estimates of sampling error (Johnson and Gillingham 2004, Krueger et al. 2012). In addition, expert opinion is based on the past association of a species with the biotic and abiotic structure of a specific location. While it can be presumed that a species will continue to persist in that habitat, expert opinion does not provide predictive spatial data related to environmental variables that can be applied to predict future species distributions. Predictive spatial data are especially important for conservation planning during the global ecological change of the Anthropocene, since conservation planning should be focused on present and potential future conditions.

The software *Maxent* (short for maximum entropy) is an increasingly popular species distribution modeling program used globally by researchers to produce predictive spatial data of species distribution using presence-only wildlife detections (Phillips et al. 2006, Radosavljevic and Anderson 2014). While there is a growing body of species distribution modeling techniques available, such as generalized linear models and boosted regression trees, Maxent provides a helpful user interface that can use multiple layers of environmental variables to model probabilities of wildlife habitat use. Presence-only data can include systematically collected records from camera traps and point counts or anecdotal data from roadkill, museum specimens, or observations by citizen scientists. Maxent results produce a probabilistic raster layer of species habitat associations and plots of the estimated relationships between environmental variables and species habitat association. While this computer program is popular and relatively easy to use, results should be viewed cautiously, as with many modeling methods, data assumptions are often violated without acknowledgment (Yackulic et al. 2013).

There are several biases associated with anecdotally collected presence-only data, including that detections may be disproportionally higher from places where humans travel frequently, and data may be lacking on places where a species was surveyed for but not detected. Some of this bias can be accounted for by estimating the level of effort used in collecting the biological data (Zuckerberg et al. 2016). The online repository eBird attempts to account for this effort by asking citizen scientists to include estimates of the time spent birding, distance traveled, and type of birding (Sullivan et al. 2009). Furthermore, systematic surveying programs, such as the U.S. Geological Survey's North American Amphibian Monitoring Program and Breeding Bird Survey, control for survey effort by establishing rigorous survey protocols.

Beyond estimates of survey effort, issues with variability in detecting different species can be directly accounted for through the use of occupancy models that directly estimate a species' detection probability (MacKenzie et al. 2002, 2003). Whereas other modeling methods can use presence-only data, occupancy models require that researchers establish multiple sampling sites and then take multiple temporally independent observations for the species of interest at each site during one survey period (i.e., season). Each of these temporally independent observations (i.e., capture occasions) show when that species was either detected (1) or not detected (0) during a survey period. With this detection–non-detection data, the occupancy model can estimate the probability of detecting a species given its presence, which is an important consideration, as most species are not detected every time they are present. If this procedure is repeated for multiple survey periods, one can model the dynamic probabilities of colonization and extinction of a species from a site (Dertien et al. 2017). Process and environmental covariates can be included to model all four of these probabilities, which allows spatially and temporally varying raster layers of species occupancy across a landscape. With a representative sample of sites, the results of these models should be closer to the true distribution of the species of interest and therefore, a better landscape-scale representation to input into a systematic conservation planning program (MacKenzie et al. 2002, 2003).

Of the discussed options, occupancy models likely compute the least biased estimate of how a species is distributed across a landscape. The degree to which occupancy models are an improvement over binary habitat association models and models derived from presence-only data is debatable and is dependent on a suite of factors. Regarding conservation planning, there is likely a parsimonious space between simple, coarse data and detailed, fine spatial grain biodiversity data that most efficiently uses research resources while providing accurate estimates of species distributions. As we have explored, simpler and coarser data may introduce biased information to a conservation plan, while the collection of detailed dynamic data may be investing unnecessary resources for the ultimate needs of the conservation plan. When possible, research groups should use the modeling option that achieves project goals and that seeks a balance between precise, accurate model results and finite project resources.

Conclusions

Landscape ecology has been a central field addressing one of the main causes of the biodiversity crisis – human land cover transformation. It explicitly considers spatial and temporal scale, which are both necessary for understanding pattern and process, and integrates human geography with biogeography in a planning context, which is necessary for implementation.

Systematic conservation planning is a map-based discipline addressing biodiversity conservation. The field is evolving rapidly with new software and data products emerging on an annual basis. For those entering the field for the first time and hoping to make a difference by mapping priority areas for biodiversity conservation, it is important to choose the appropriate technology for the conservation situation. Choosing the right tool for the job may speed implementation and lead to more effective conservation.

Following this overview of landscape ecology and biodiversity conservation, there remain significant questions. Those interested in a deeper understanding should consult the reviews we cited and foundational papers. We tried to make the case that systematic conservation planning is a direct application of landscape ecology principles to biodiversity conservation. We explored two areas of active research: conservation design and planning for aquatic landscapes, and modeling of species distributions. There is much we did not cover, and there are many unanswered questions in the field. Much work is needed in understanding and planning for synergies among climate, land use, and biodiversity changes; integration of biological, cultural, and socioeconomic theory and data; and barriers to translating conservation models into implemented plans.

Lliterature cited

Abell, R., David Allan, J. and Lehner, B. (2007) 'Unlocking the potential of protected areas for freshwaters', *Biological Conservation*, vol 134, pp48–63. https://doi.org/10.1016/j.biocon.2006.08.017.

Abell, R., Thieme, M.L., Revenga, C., Bryer, M., Kottelat, M., Bogutskaya, N., Coad, B., Mandrak, N., Balderas, S.C., Bussing, W., Stiassny, M.L.J., Skelton, P., Allen, G.R., Unmack, P., Naseka, A., Ng, R., Sindorf, N., Robertson, J. Armijo, E., Higgins, J.V., Heibel, T.J., Wikramanayake, E., Olson, D., López, H.L., Reis, R.E., Lundberg, J.G., Sabaj Pérez, M.H. and Petry, P. (2008) 'Freshwater ecoregions of the world: a new map of biogeographic units for freshwater biodiversity conservation', *Bioscience*, vol 58, pp403–414.

Álvarez-Romero, J.G., Adams, V.M., Pressey, R.L., Douglas, M., Dale, A.P., Augé, A.A., Ball, D., Childs, J., Digby, M., Dobbs, R., Gobius, N., Hinchley, D., Lancaster, I., Maughan M. and Perdrisat, I. (2015) 'Integrated cross-realm planning: a decision-makers' perspective', *Biological Conservation*, vol 191, pp799–808.

Amis, M.A., Rouget, M., Lotter, M. and Day, J. (2009) 'Integrating freshwater and terrestrial priorities in conservation planning', *Biological Conservation*, vol 142, pp2217–2226.

Anderson, M.G. and Ferree, C.E. (2010) 'Conserving the stage: climate change and the geophysical underpinnings of species diversity', *PLoS One*, vol 5, no 7, p e11554. https://doi.org/10.1371/journal.pone.0011554.

Armsworth, P.R., Cantú-Salazar, L., Parnell, M., Davies, Z.G. and Stoneman, R. (2011) 'Management costs for small protected areas and economies of scale in habitat conservation', *Biological Conservation*, vol 144, pp423–429.

Baldwin, R.F. and Beazley, K.F. (2019) 'Emerging paradigms for biodiversity and protected areas', *Land* vol 8, no 3, art 43.

Ball, I.J., Possingham, H.P. and Watts, M.E.J. (2009) 'Marxon and relatives: software for conservation prioritization', in A. Moilanen and K.A. Wilson (eds). *Spatial Conservation Prioritization: Quantitative Methods and Computational Tools*. Oxford University Press, Oxford.

Ban, N.C., Januchowski-Hartley, S., Alvarez-Romero, J., Mills, M., Pressey, R.L., Linke, S. and de Freitas, D. (2013) 'Marine and freshwater conservation planning: from representation to persistence', in L.F. Craighead and C.L.J. Convis (eds). *Conservation Planning: Shaping the Future*. Esri Press, Redlands, CA.

Barendregt, A., Wassen, M.J. and Schot, P.P. (1995) 'Hydrological systems beyond a nature reserve, the major problem in wetland conservation of Naardermeer (The Netherlands)', *Biological Conservation*, vol 72, pp393–405.

Beazley, K. and Cardinal, N. (2004) 'A systematic approach for selecting focal species for conservation in the forests of Nova Scotia and Maine', *Environmental Conservation*, vol 31, pp91–101.

Beier, P. and Brost, B. (2010) 'Use of land facets to plan for climate change: conserving the arenas, not the actors', *Conservation Biology*, vol 24, pp701–710.

Beier, P., Spencer, W.D., Baldwin, R. F and McRae, B.H. (2011) 'Toward best practices for developing regional connectivity maps', *Conservation Biology*, vol 25, pp879–892.

Boitani, L., Cowling, R.M., Dublin, H.T., Mace, G.M., Parrish, J., Possingham, H.P., et al. (2008) 'Change the IUCN protected area categories to reflect biodiversity outcomes', *PLoS Biology*, vol 6, no 3, p e66. https://doi.org/10.1371/journal.pbio.0060066.

Brooks, T.M., Bakarr, M.I., Boucher, T.M. da Fonseca, G.A.B., Hilton-Taylor, C., Hoekstra, J.M., Moritz, T., Olivieri, S., Parrish, J., Pressey, R.L., Rodrigues, A.S.L., Sechrest, W., Stattersfield, A., Strahm, W. and Stuart, S.N. (2004) 'Coverage provided by the global protected-area system: is it enough?', *Bioscience*, vol 54, pp1083–1091.

Burgman, M., Carr, A., Godden, L., Gregory, R., McBride, M., Flander, L. and Maguire, L. (2011) 'Redefining expertise and improving ecological judgment', *Conservation Letters*, vol 4, pp81–87.

Burgman, M.A., Possingham, H.P., Lynch, A.J.J., Keith, D.A., McCarthy, M.A., Hopper, S.D., Drury, W.L., Passioura, J.A. and Devries, R.J. (2001) 'A method for setting the size of plant conservation target areas', *Conservation Biology*, vol 15, pp603–616.

Burrows, M.T., Schoeman, D.S., Richardson, A.J., Molinos, J.G., Hoffmann, A., Buckley, L.B., Moore, P.J., Brown, C.J., Bruno, J.F., Duarte, C.M., Halpern, B.S., Hoegh-Guldberg, O., Kappel, C.V., Kiessling, W., O'Connor, M.I., Pandolfi, J.M., Parmesan, C., Sydeman, W.J., Ferrier, S., Williams, K.J. and Poloczanska, E.S. (2014) 'Geographical limits to species-range shifts are suggested by climate velocity', *Nature*, vol 507, pp492–495.

Carroll, C., Noss, R.F., Paquet, P.C. and Schumaker, N.H. (2003) 'Use of population viability analysis and reserve selection algorithms in regional conservation plans', *Ecological Applications*, vol 13 pp1771–1789.

Chivian, E. and Bernstein, A. (eds) (2008) *Sustaining Life: How Human Health Depends on Biodiversity*. Oxford University Press, New York.

Clark, J.S. (1991) 'Disturbance and population structure on the shifting mosaic landscape', *Ecology*, vol 72, pp1119–1137.

Collen, B., Whitton, F., Dyer, E.E., Baillie, J.E.M., Cumberlidge, N., Darwall, W.R.T., Pollock, C., Richman, N.I., Soulsby, A.M. and Böhm, M. (2014) 'Global patterns of freshwater species diversity, threat and endemism', *Global Ecology and Biogeography*, vol 23, pp40–51.

Crooks, K.R. and Sanjayan, M. (eds) (2006) *Connectivity Conservation*. Cambridge University Press, Cambridge.

Crooks, K.R., Burdett, C.L., Theobald, D.M., Rondinini, C. and Boitani, L. (2011) 'Global patterns of fragmentation and connectivity of mammalian carnivore habitat', *Philosophical Transactions of the Royal Society B: Biological Sciences*, vol 36, pp2642–2651.

Cushman, S.A., Gutzweiler, K., Evans, J.S. and McGarigal, K. (2010) 'The gradient paradigm: a conceptual and analytical framework for landscape ecology', in Cushman, S.A., Huettmann, F. (eds.), *Spatial Complexity, Informatics, and Wildlife Conservation*. Springer, Tokyo. https://doi.org/10.1007/978-4-431-87771-4_5.

deFries, R., Hanson, A., Turner, B.L., Reid, R. and Liu, J. (2007) 'Land use change around protected areas: management to balance human needs and ecological function', *Ecological Applications*, vol 17, pp1031–1038.

DeFries, R., Karanth, K.K. and Pareeth, S. (2010) 'Interactions between protected areas and their surroundings in human-dominated tropical landscapes', *Biological Conservation*, vol 143, pp2870–2880.

DellaSala, D.A. (2020) 'Has Anthropocentrism replaced Ecocentrism in conservation?', in H. Kopnina and H. Washington (eds) *Conservation: Integrating Social and Ecological Justice*. Springer International Publishing, Cham.

Dertien, J.S., Doherty, P.F., Bagley, C.F., Haddix, J.A., Brinkman, A.R. and Neipert, E.S. (2017) 'Evaluating dall's sheep habitat use via camera traps', *Journal of Wildlife Management*, vol 81, pp1457–1467.

Dickson, B.G., Albano, C.M., Anantharaman, R., Beier, P., Fargione, J., Graves, T.A., Gray, M.E., Hall, K.R., Lawler, J.J., Leonard, P.B., Littlefield, C.E., McClure, M.L., Novembre, J., Schloss, C.A., Schumaker, N.H., Shah, V.B. and Theobald, D.M. (2019) 'Circuit-theory applications to connectivity science and conservation', *Conservation Biology*, vol 33, pp239–249.

Dirzo, R., Young, H.S., Galetti, M., Ceballos, G., Isaac, N.J.B. and Collen, B. (2014) 'Defaunation in the Anthropocene', *Science*, vol 345 pp401–406.

Driver, A., Maze, K., Rouget, M., Lombard, A.T., Nel, J., Turpie, J.K., Cowling, R.M., Desmet, P., Goodman, P., Harris, J., Jfinas, Z., Reyers, B., Sink, K. and Strauss, T. (2005) 'National spatial biodiversity assessment priorities for biodiversity conservation in South Africa', *Strelitzia*, vol 17, South African National Biodiversity Institute, Pretoria.

Dudley, N., Parrish, J., Redford, K. and Stolton, S. (2010) 'The revised IUCN protected area management categories: the debate and ways forward', *Oryx*, vol 44, no 4, pp485–490. https://doi.org/10.1017/S 0030605310000566.

Elith, J. and Leathwick, J. (2016) 'Boosted regression trees for ecological modeling', https://rspatial.org/rast er/sdm/9_sdm_brt.html, accessed 12 Dec 2020.

Elith, J., Phillips, S.J., Hastie, T., Dudik, M., Chee, Y.E. and Yates, C.J. (2011) 'A statistical explanation of MaxEnt for ecologists', *Diversity and Distributions*, vol 17, no 1, pp43–57.

Fahrig, L. (2003) 'Effects of habitat fragmentation on biodiversity', *Annual Review of Ecology, Evolution, and Systematics*, vol 34, pp487–515.

Fairbanks, D.H.K., Reyers, B. and Van Jaarsveld, A.S. (2001) 'Species and environment representation: selecting reserves for the retention of avian diversity in KwaZulu-Natal, South Africa', *Biological Conservation*, vol 98, no 3, pp365–379.

Fausch, K.D., Torgersen, C.E., Baxter, C.V. and Li, H.W. (2002) 'Landscapes to riverscapes: the advent of a landscape ecology for stream fishes', *Bioscience*, vol 52, pp483–438.

Ferraro, P.J. and Hanauer, M.M. (2014) 'Quantifying causal mechanisms to determine how protected areas affect poverty through changes in ecosystem services and infrastructure', *Proceedings of the National Academy of Sciences*, vol 111, pp4332–4337.

Ferrier, S. (2002) 'Mapping spatial pattern in biodiversity for regional conservation planning: where to from here?', *Systematic Biology*, vol 51, no 2, pp331–363.

Ferrier, S. and Guisan, A. (2006) 'Spatial modelling of biodiversity at the community level', *Journal of Applied Ecology*, vol 43, pp393–404.

Forman, R.T.T. (1995) *Land Mosaics: the Ecology of Landscapes and Regions*. Cambridge Univesity Press, Cambridge.

Foster, D.R. (2000) 'From bobolinks to bears: interjecting geographical history into ecological studies, environmental interpretation, and conservation planning', *Journal of Biogeography*, vol 27, pp27–30.

Foster, D.R. and Aber, J.D. (2004) *Forests in Time: 1,000 Years of Change in New England*. Yale University Press, New Haven.

Foster, D.R., Swanson, F.J., Aber, J.D., Burke, I., Brokaw, N., Tilman, D. and Knapp, A. (2003) 'The importance of land-use legacies to ecology and conservation', *Bioscience*, vol 53, pp77–88.

Fox, N.J. and Beckley, L.E. (2005) 'Priority areas for conservation of Western Australian coastal fishes: a comparison of hotspot, biogeographical and complementarity approaches', *Biological Conservation*, vol 125, no 4, pp399–410.

Fuller, R.A., Irvine, K.N., Devine-Wright, P., Warren, P.H. and Gaston, K.J. (2007) 'Psychological benefits of greenspace increase with biodiversity', *Biology Letters*, vol 3, pp390–394.

Gaston, K.J., Charman, K., Jackson, S.F., Armsworth, P.R., Bonn, A., Briers, R.A., Callaghan, C.S.Q., Catchpole, R., Hopkins, J., Kunin, W.E., Latham, J., Opdam, P., Stoneman, R., Stroud, D.A. and Tratt, R. (2006) 'The ecological effectiveness of protected areas: the United Kingdom', *Biological Conservation*, vol 132, pp76–87.

Grill, G., Lehner, B., Thieme, M., Geenen, B., Tickner, D., Antonelli, F., Babu, S., Borrelli, P., Cheng, L., Crochetiere, H., Ehalt Macedo, H., Filgueiras, R., Goichot, M., Higgins, J., Hogan, Z., Lip, B., McClain, M.E., Meng, J., Mulligan, M., Nilsson, C., Olden, J.D., Opperman, J.J., Petry, P., Reidy Liermann, C., Sáenz, L., Salinas-Rodríguez, S., Schelle, P., Schmitt, R.J.P., Snider, J., Tan, F., Tockner, K., Valdujo, P.H., van Soesbergen, A. and Zarfl, C. (2019) 'Mapping the world's free-flowing rivers', *Nature*, vol 569, pp215–221.

Groves, C., Jensen, D., Valutis, L.L., Redford, K.H., Shaffer, M., Scot, J.M., Baumgartner, J.V., Higgins, J.V., Beck, M.W. and Anderson, M.G. (2002) 'Planning for biodiversity conservation: putting conservation science into practice', *Bioscience*, vol 52, pp499–512.

Gurney, G.G., Pressey, R.L., Ban, N.C., Álvarez-Romero, J.G., Jupiter, S. and Adams, V.M. (2015) 'Efficient and equitable design of marine protected areas in Fiji through inclusion of stakeholder-specific objectives in conservation planning', *Conservation Biology*, vol 29, pp1378–1389.

Haddad, N.M., Brudvig, L.A., Clobert, J., Davies, K.F., Gonzalez, A., Holt, R.D., Lovejoy, T.E., Sexton, J.O., Austin, M.P., Collins, C.D., Cook, W.M., Damschen, E.I., Ewers, R.M., Foster, B.L., Jenkins, C.N., King, A.J., Laurance, W.F., Levey, D.J., Margules, C.R., Melbourne, B.A., Nicholls, A.O., Orrock, J.L., Song, D.-X. and Townshend, J.R. (2015) 'Habitat fragmentation and its lasting impact on Earth's ecosystems', *Science Advances*, vol 1, no 2, art e1500052.

Hanks, R.D., Leonard, P.B. and Baldwin, R.F. (2020) 'Understanding landscape influences on aquatic fauna across the Central and Southern Appalachians', *Land*, vol 9, no 1, art 16.

Hanski, I. (1998) 'Metapopulation dynamics', *Nature*, vol 396, pp41–49.

Hendrickx, F., Maelfait, J.-P., Van Wingerden, W., Schweiger, O., Speelmans, M., Aviron, S., Augenstein, I., Billeter, R., Bailey, D., Bukacek, R., Burel, F., Diekotter, T., Dirksen, J., Herzog, F., Liira, J., Roubalova, M., Vandomme, V. and Bugter, R. (2007) 'How landscape structure, land-use intensity and habitat diversity affect components of total arthropod diversity in agricultural landscapes', *Journal of Applied Ecology*, vol 44, pp340–351.

Hepinstall, J.A., Alberti, M. and Marzluff, J.M. (2008) 'Predicting land cover change and avian community responses in rapidly urbanizing environments', *Landscape Ecology*, vol 23, pp1257–1276.

Hermoso, V., Cattarino, L., Kennard, M.J., Watts, M. and Linke, S. (2015) 'Catchment zoning for freshwater conservation: refining plans to enhance action on the ground', *Journal of Applied Ecology*, vol 52, pp940–949.

Hermoso, V., Filipe, A.F., Segurado, P. and Beja, P. (2016) 'Catchment zoning to unlock freshwater conservation opportunities in the Iberian Peninsula', *Diversity and Distributions*, vol 22, pp960–969.

Hilty, J.A., Chester, C.C. and Cross, M.S. (eds) (2012) *Climate and Conservation: Landscape and Seascape Science, Planning, and Action.* Island Press, Washington, DC.

Homan, R.N., Windmiller, B.S. and Reed, J.M. (2004) 'Critical thresholds associated with habitat loss for two vernal pool-breeding amphibians', *Ecological Applications*, vol 14, pp1547–1553.

Huber, P.R., Greco, S.E. and Thorne, J.H. (2010) 'Spatial scale effects on conservation network design: trade-offs and omissions in regional versus local scale planning', *Landscape Ecology*, vol 25, pp683–695.

Hunter, M.L. and Gibbs, J.P. (2007) *Fundamentals of Conservation Biology*, 3rd edition. Blackwell, Oxford.

Hunter Jr., M.L., Jacobson, G.L. and Webb, T.I. (1988) 'Paleoecology and the coarse-filter approach to maintaining biological diversity', *Conservation Biology*, vol 2, pp375–385.

Imperial, M.T., Ospina, S., Johnston, E., O'Leary, R., Thomsen, J., Williams, P. and Johnson, S. (2016) 'Understanding leadership in a world of shared problems: advancing network governance in large landscape conservation', *Frontiers in Ecology and the Environment*, vol 14, pp126–134.

Isaak, D.J., Peterson, E.E., Ver Hoef, J.M., Wenger, S.J., Falke, J.A., Torgersen, C.E., Sowder, C., Steel, E.A., Fortin, M.-J., Jordan, C.E., Ruesch, A.S., Som, N. and Monestiez, P. (2014) 'Applications of spatial statistical network models to stream data', *WIREs Water*, vol 1, pp277–294.

IUCN (International Union for Conservation of Nature) (2019) 'IUCN red list of endangered species', https://www.iucnredlist.org/, accessed 12 Dec 2020.

Jenkins, C.N., Van Houtan, K.S., Pimm, S.L. and Sexton, J.O. (2015) 'US protected lands mismatch biodiversity priorities', *Proceedings of the National Academy of Sciences*, vol 112, pp5081–5086.

Johnson, C.J. and Gillingham, M.P. (2004) 'Mapping uncertainty: sensitivity of wildlife habitat ratings to expert opinion', *Journal of Applied Ecology*, vol 41, pp1032–1041.

Kery, M., Guillera-Arroita, G. and Lahoz-Monfort, J.J. (2013) 'Analysing and mapping species range dynamics using occupancy models', *Journal of Biogeography*, vol 40, pp1463–1474.

Kirkpatrick, J.B. (1983) 'An iterative method for establishing priorities for the selection of nature reserves: an example from Tasmania', *Biological Conservation*, vol 25, pp127–134.

Knight, A.T., Cowling, R.M., Rouget, M., Balmford, A., Lombard, A.T. and Cambell, B.M. (2008) 'Knowing but not doing: selecting priority conservation areas and the research–implementation gap', *Conservation Biology*, vol 22, pp610–617.

Knight, A.T., Cowling, R.M., Possingham, H.P. and Wilson, K.A. (2009) 'From theory to practice: designing and situating spatial prioritization approaches to better implement conservation action', in A. Moilanen, K.A. Wilson and H.P. Possingham (eds) *Spatial Conservation Prioritization: Quantitative Methods and Computational Tools*. Oxford University Press, Oxford.

Krueger, T., Page, T., Hubacek, K., Smith, L. and Hiscock, K. (2012) 'The role of expert opinion in environmental modelling', *Environmental Modelling & Software*, vol 36, pp4–18.

Kukkala, A.S. and Moilanen, A. (2013) 'Core concepts of spatial prioritisation in systematic conservation planning', *Biological Reviews*, vol 88, pp443–464.

Kusmanoff, A.M., Fidler, F., Gordon, A. and Bekessy, S.A. (2017) 'Decline of 'biodiversity' in conservation policy discourse in Australia', *Environmental Science & Policy*, vol 77, pp160–165.

Lambeck, R.J. (1997) 'Focal species: a multi-species umbrella for nature conservation', *Conservation Biology*, vol 11, pp849–856.

Larson, L.R., Jennings, V. and Cloutier, S.A. (2016) 'Public parks and wellbeing in urban areas of the United States', *PloS ONE*, vol 11, art e0153211.

Laurance, W.F., Lovejoy, T.E., Vasconcelos, H.L., Bruna, E.M., Didham, R.K., Stouffer, P.C., Gascon, C., Bierregaard, R.O., Laurance, S.G. and Sampaio, E. (2002) 'Ecosystem decay of Amazonian forest fragments: a 22-year investigation', *Conservation Biology*, vol 16, pp605–618.

Lawler, J., Beier, P., Dickson, B., Fargione, J., Novembre, J. and Theobald, D. (2019) 'A tribute to a true conservation innovator, Brad McRae, 1966–2017', *Conservation Biology*, vol 33, 480–482.

Leonard, P., Baldwin, R.F. and Hanks, D. (2017a) 'Landscape-scale conservation design across biotic realms: sequential integration of aquatic and terrestrial landscapes', *Scientific Reports*, vol 7, art 14556.

Leonard, P.B., Duffy, E.B., Baldwin, R.F., McRae, B.H., Shah, V.B. and Mohapatra, T.K. (2016) 'gflow: software for modelling circuit theory-based connectivity at any scale', *Methods in Ecology and Evolution*, vol 8, no 4, pp519–526.

Leonard, P.B., Sutherland, R.W., Baldwin, R.F., Fedak, D.A., Carnes, R.G. and Montgomery, A.P. (2017b) 'Landscape connectivity losses due to sea level rise and land use change', *Animal Conservation*, vol 20, pp80–90.

Linke, S., Turak, E. and Nel, J. (2011) 'Freshwater conservation planning: the case for systematic approaches', *Freshwater Biology*, vol 56, pp6–20.

MacArthur, R.H. and Wilson, E.O. (1967) *The Theory of Island Biogeography*. Princeton University Press, Princeton.

MacKenzie, D.I., Nichols, J.D., Lachman, G.B., Droege, S., Andrew, J. and Langtimm, C.A. (2002) 'Estimating site occupancy rates when detection probabilities are less than one', *Ecology*, vol 83, pp2248–2255.

MacKenzie, D.I., Nichols, J.D., Hines, J.E., Knutson, M.G. and Franklin, A.B. (2003) 'Estimating site occupancy, colonization, and local extinction when a species is detected imperfectly', *Ecology*, vol 84, pp2200–2207.

Margules, C.R. and Pressey, R.L. (2000) 'Systematic conservation planning', *Nature*, vol 405, pp243–253.

Martin, G.I., Kirkman, L.K. and Hepinstall-Cymerman, J. (2012) 'Mapping geographically isolated wetlands in the Dougherty Plain, Georgia, USA', *Wetlands*, vol 32, pp149–160.

Mascia, M.B., Brosius, J.P., Dobson, T.A., Forbes, B.C., Horowitz, L., McKean, M.A. and Turner, N.J. (2003) 'Conservation and the social sciences', *Conservation Biology*, vol 17, pp649–650.

McGarigal, K. and Cushman, S.A. (2002) 'Comparative evaluation of experimental approaches to the study of habitat fragmentation effects', *Ecological Applications*, vol 12, pp335–345.

McGarigal, K., Compton, B.W., Plunkett, E.B., DeLuca, W.V., Grand, J., Ene, E. and Jackson, S.D. (2018) 'A landscape index of ecological integrity to inform landscape conservation', *Landscape Ecology*, vol 33, pp1029–1048.

McGuire, J.L., Lawler, J.J., McRae, B.H., Nuñez, T.A. and Theobald, D.M. (2016) 'Achieving climate connectivity in a fragmented landscape', *Proceedings of the National Academy of Sciences*, vol 113, pp7195–7200.

McHarg, I.L. (1969) *Design with Nature*. Natural History Press, Garden City, NY.

McKinney, M.L. (2002) 'Urbanization, biodiversity, and conservation', *Bioscience*, vol 52, pp883–890.

McManamay, R.A., and Derolph, C.R. (2019) 'Data descriptor: a stream classification system for the conterminous United States', *Scientific Data*, vol 6, pp1–18.

McRae, B.H., Dickson, B.G., Keitt, T.H. and Shah, V.B. (2008) 'Using circuit theory to model connectivity in ecology, evolution, and conservation', *Ecology*, vol 89, pp2712–2724.

Moilanen, A., Wilson, K.A. and Possingham, H.P. (eds) (2009) *Spatial Conservation Prioritization: Quantitative Methods and Computational Tools*. Oxford University Press, Oxford.

Naiman, R.J., Bilby, R.E., Schindler, D.E. and Helfield, J.M. (2002) 'Pacific salmon, nutrients, and the dynamics of freshwater and riparian ecosystems', *Ecosystems*, vol 5, pp399–417.

Nel, J.L., Roux, D.J., Abell, R., Ashton, P.J., Cowling, R.M., Higgins, J.V., Thieme, M. and Viers, J.H. (2009) 'Progress and challenges in freshwater conservation planning', *Aquatic Conservation: Marine and Freshwater Ecosystems*, vol 19, pp474–485.

Newbold, T., Hudson, L.N., Arnell, A.P., Contu, S., De Palma, A., Ferrier, S., Hill, S.L.L., Hoskins, A.J., Lysenko, I., Phillips, H.R.P., Burton, V.J., Chng, C.W.T., Emerson, S., Gao, D., Pask-Hale, G., Hutton, J., Jung, M., Sanchez-Ortiz, K., Simmons, B.I., Whitmee, S., Zhang, H., Scharlemann, J.P.W. and Purvis, A. (2016). 'Has land use pushed terrestrial biodiversity beyond the planetary boundary? A global assessment', *Science*, vol 353, pp288–291.

Newmark, W.D. (1995) 'Extinction of mammal populations in Western North American National Parks', *Conservation Biology*, vol 9, pp512–526.

Noss, R.F. (1983) 'A regional landscape approach to maintain diversity', *Bioscience*, vol 33, pp700–706.

Novacek, M.J. (2008). 'Engaging the public in biodiversity issues', *Proceedings of the National Academy of Sciences*, vol 105, no. Supplement 1, pp11571–11578.

Nuñez, T., Lawler, J.J., McRae, B.H., Pierce, D.J., Krosby, M.R., Kavanagh, D.M., Singleton, P.H. and Tewksbury, J.J. (2013) 'Connectivity planning to facilitate species movements in response to climate change', *Conservation Biology*, vol 27, pp407–416.

Ogletree, S.S., Powell, R.B., Baldwin, R.F. and Leonard, P.B. (2019) 'A framework for mapping cultural resources in landscape conservation planning', *Conservation Science and Practice*, vol 1, art e41.

Opdam, P., Foppen, R. and Vos, C. (2002) 'Bridging the gap between ecology and spatial planning in landscape ecology', *Landscape Ecology*, vol 16, pp767–779.

Orr, D.W. (1994) 'Technological fundamentalism', *Conservation Biology*, vol 8, pp335–337.

Parks, S.A. and Harcourt, A.H. (2002) 'Reserve size, local human density, and mammalian extinctions in U.S. protected areas', *Conservation Biology*, vol 16, pp800–808.

Peres, C.A. and Terborgh, J.W. (1995) 'Amazonian nature reserves: an analysis of the defensibility status of existing conservation units and design criteria for the future', *Conservation Biology*, vol 9, no 1, pp34–46.

Peterson, E.E., Ver Hoef, J.M., Isaak, D.J., Falke, J.A., Fortin, M.J., Jordan, C.E., McNyset, K., Monestiez, P., Ruesch, A.S., Sengupta, A., Som, N., Steel, E.A., Theobald, D.M., Torgersen, C.E. and Wenger, S.J. (2013) 'Modelling dendritic ecological networks in space: an integrated network perspective', *Ecology Letters*, vol 16, pp707–719.

Phillips, S.J., Anderson, R.P. and Schapire, R.E. (2006) 'Maximum entropy modeling of species geographic distributions', *Ecological Modelling*, vol 190, pp231–259.

Powell, R.B., Cuschnir, A. and Peiris, P. (2009) 'Overcoming governance and institutional barriers to integrated coastal zone, marine protected area, and tourism management in Sri Lanka', *Coastal Management*, vol 37, pp633–655.

Pressey, R.L., Cabeza, M., Watts, M.E.J., Cowling, R.M. and Wilson, K.A. (2007) 'Conservation planning in a changing world', *Trends in Ecology and Evolution*, vol 22, pp583–592.

Pringle, C.M. (2001) 'Hydrologic connectivity and the management of biological reserves: a global perspective', *Ecological Applications*, vol 11, pp981–998.

Radosavljevic, A. and Anderson, R.P. (2014) 'Making better Maxent models of species distributions: complexity, overfitting and evaluation', *Journal of Biogeography*, vol 41, pp629–643.

Roche, H., Buet, A. and Ramadé, F. (2002) 'Relationships between persistent organic chemicals residues and biochemical constituents in fish from a protected area: the French National Nature Reserve of Camargue', *Comparative Biochemistry and Physiology*, vol 133, pp393–410.

Sayer, J., Sunderland, T., Ghazoul, J., Pfund, J.-L., Sheil, D., Meijaard, E., Venter, M., Boedhihartono, A.K., Day, M., Garcia, C., van Oosten, C. and Buck, L.E. (2013) 'Ten principles for a landscape approach to reconciling agriculture, conservation, and other competing land uses', *Proceedings of the National Academy of Sciences*, vol 110, pp8349–8356.

Schindler, D.W. (2000) 'Aquatic problems caused by human activities in Banff National Park, Alberta, Canada', *Ambio*, vol 29, pp401–407.

Scott, J.M., Davis, F., Csuti, F., Noss, R., Butterfield, B., Groves, C., Anderson, H., Caicco, S., D'Erchia, F., Edwards, T.C.J., Ulliman, J. and Wright, R.G. (1993). 'Gap analysis: a geographic approach to protection of biological diversity', *Wildlife Monographs*, vol 57, pp5–41.

Scott, M.C. (2005) 'Winners and losers among stream fishes in relation to land use legacies and urban development', *Biological Conservation*, vol 127, pp301–309.

Semlitsch, R.D. (2008) 'Differentiating migration and dispersal processes for pond-breeding amphibians', *Journal of Wildlife Management*, vol 72, pp260–267.

Sinclair, S.P., Milner-Gulland, E.J., Smith, R.J., McIntosh, E.J., Possingham, H.P., Vercammen, A. and Knight, A.T. (2018) 'The use, and usefulness, of spatial conservation prioritizations', *Conservation Letters*, vol 11, no 6, art e12459.

Steffen, W., Richardson, K., Rockström, J., Cornell, S.E., Fetzer, I., Bennett, E.M., Biggs, R., Carpenter, S.R., de Vries, W., de Wit, C.A., Folke, C., Gerten, D., Heinke, J., Mace, G.M., Persson, L.M., Ramanathan, V., Reyers, B. and Sörlin, S. (2015) 'Planetary boundaries: guiding human development on a changing planet', *Science*, vol 347, no 6223, art 1259855.

Strange, N., Theilade, I., Thea, S., Sloth, A. and Helles, F. (2007) 'Integration of species persistence, costs and conflicts: an evaluation of tree conservation strategies in Cambodia', *Biological Conservation*, vol 137, no 2, pp223–236.

Sullivan, B.L., Wood, C.L., Iliff, M.J., Bonney, R.E., Fink, D. and Kelling, S. (2009) 'eBird: a citizen-based bird observation network in the biological sciences', *Biological Conservation*, vol 142, pp2282–2292. https ://doi.org/10.1016/j.biocon.2009.05.006.

Theobald, D.M. (2004) 'Placing exurban land-use change in a human modification framework', *Frontiers in Ecology and the Environment*, vol 2, pp139–144.

Theobald, D.M. (2010) 'Estimating natural landscape changes from 1992 to 2030 in the conterminous US', *Landscape Ecology*, vol 25, pp999–1011.

Theobald, D.M. (2013) 'A general model to quantify ecological integrity for landscape assessments and US application', *Landscape Ecology*, vol 28, pp1859–1874. https://doi.org/10.1007/s10980-013-9941-6.

Theobald, D.M., Hobbs, R.J., Bearly, T., Zack, J.A., Shenk, T. and Riebsame, W.E. (2000) 'Incorporating biological information in local land-use decision-making: designing a system for conservation planning', *Landscape Ecology*, vol 15, pp35–45.

Theobald, D.M., Reed, S.E., Fields, K. and Soulé, M. (2012) 'Connecting natural landscapes using a landscape permeability model to prioritize conservation activities in the United States', *Conservation Letters*, vol 5, pp123–133. https://doi.org/10.1111/j.1755-263X.2011.00218.x.

Theobald, D.M., Spies, T., Kline, J.D., Maxwell, B., Hobbs, N.T. and Dale, V.H. (2005) 'Ecological support for rural land-use planning', *Ecological Applications*, vol 15, pp1906–1914.

Thieme, M., Lehner, B., Abell, R., Hamilton, S.K., Kellndorfer, J., Powell, G. and Riveros, J.C. (2007) 'Freshwater conservation planning in data-poor areas: an example from a remote Amazonian basin (Madre de Dios River, Peru and Bolivia)', *Biological Conservation*, vol 135, no 4, pp484–501.

Trombulak, S.C. and Baldwin, R.F. (eds) (2010) *Landscape-scale Conservation Planning*. Springer, New York.

Turner, M.G., Gardner, R.H. and O'Neill, R.V. (2001) *Landscape Ecology in Theory and Practice*. Springer, New York.

Vance-Borland, K., Roux, D., Nel, J. and Pressey, B. (2008) 'From the mountains to the sea: where Is freshwater conservation in the SCB Agenda?', *Conservation Biology*, vol 22, pp505–507.

Vandermeer, J. and Perfecto, I. (2007) 'The agricultural matrix and a future paradigm for conservation', *Conservation Biology*, vol 21, pp274–277. https://doi.org/10.1111/j.1523-1739.2006.00582.x.

Venter, O., Sanderson, E.W., Magrach, A., Allan, J.R., Beher, J., Jones, K.R., Possingham, H.P., Laurance, W.F., Wood, P., Fekete, B.M., Levy, M.A. and Watson, J.E.M. (2016) 'Sixteen years of change in the global terrestrial human footprint and implications for biodiversity conservation', *Nature Communications*, vol 7, art 12558.

Vörösmarty, C.J., Mcintyre, P.B., Gessner, M.O., Dudgeon, D., Prusevich, A., Green, P., Glidden, S., Bunn, S.E., Sullivan, C.A., Liermann, C.R. and Davies, P.M. (2010) 'Global threats to human water security and river biodiversity', *Nature*, vol 467, pp555–561.

Watts, M.E.J., Ball, I.R., Stewart, R.S., Klein, C.J., Wilson, K.A., Steinback, C., Lourival, R., Kircher, L. and Possingham, H.P. (2009) 'Marxan with Zones: software for optimal conservation based land and sea-use zoning', *Environmental Modeling and Software*, vol 24, pp1513–1521.

Wikramanayake, E., McKnight, M., Dinerstein, E., Joshi, A., Gurung, B. and Smith, D. (2004) 'Designing a conservation landscape for tigers in human-dominated environments', *Conservation Biology*, vol 18, no 3, pp839–844.

Woodroffe, R. and Ginsberg, J.R. (1998) 'Edge effects and the extinction of populations inside protected areas', *Science*, vol 280, no 5371, pp2126–2128.

Woolmer, G., Trombulak, S.C., Ray, J.C., Doran, P.J., Anderson, M.G., Baldwin, R.F., Morgan, A. and Sanderson, E.W. (2008) 'Rescaling the human footprint: a tool for conservation planning at an ecoregional scale', *Landscape and Urban Planning*, vol 87, pp42–53.

Yackulic, C.B., Chandler, R., Zipkin, E.F., Royle, J.A., Nichols, J.D., Grant, E.H.C. and Veran, S. (2013) 'Presence-only modelling using MAXENT: when can we trust the inferences?', *Methods in Ecology and Evolution*, vol 4, pp236–243.

Zeller, K.A., McGarigal, K. and Whiteley, A.R. (2012) 'Estimating landscape resistance to movement: a review', *Landscape Ecology*, vol 27, pp777–797.

Zuckerberg, B., Fink, D., La Sorte, F.A., Hochachka, W.M. and Kelling, S. (2016) 'Novel seasonal land cover associations for eastern North American forest birds identified through dynamic species distribution modelling', *Diversity and Distributions*, vol 22, pp717–730.

20

Ecosystem services in the landscape

Matthew Mitchell

Ecosystem services at landscape scales

How people affect, manage, and structure landscapes has important effects on ecosystem services – the benefits that people derive from natural and semi-natural ecosystems (Millennium Ecosystem Assessment 2005). As human impacts on ecosystems continue to increase and expand to broader spatial scales, it is critical that we improve our ability to manage landscapes simultaneously for different ecosystem services. Without knowledge to understand how landscapes supply different services, our capacity to create sustainable and resilient landscapes will be diminished, with real consequences for both ecosystems and human wellbeing (Wu 2013).

This chapter describes how ecosystem services are produced at landscape scales and the key ways in which landscape structure affects how and where ecosystem services are provided across landscapes. Rather than describing the key landscape processes for each possible service (up to 100 services according to some current classifications), this chapter takes a more theoretical approach. First, it presents and develops a conceptual framework centered on the co-production of ecosystem services by natural and human systems that integrates ideas of ecosystem service supply, demand, and flow. Next, it identifies key ways in which landscape structure – the composition, configuration, and connectivity of landscapes – affects these different components of ecosystem service provision, with a focus on ecosystem service supply. This section also highlights specific services and case study examples that demonstrate the usefulness of landscape-scale approaches for managing ecosystem services. The next section focuses on landscape multifunctionality – the ability of landscapes to provide multiple ecosystem service benefits – and ideas around ecosystem service relationships and bundles at landscape scales. Finally, the chapter closes with a forward-looking discussion of the challenges around landscape planning for ecosystem services, with a focus on key knowledge gaps.

What are ecosystem services, and why are they useful?

Ecosystem services are the benefits that people derive from natural and semi-natural ecosystems that maintain or improve their wellbeing. This concept, described first by Daily et al. (1997), was comprehensively assessed at a global scale by the Millennium Assessment (MA) in 2005, an international endeavor of hundreds of scientists to describe and evaluate the status of ecosystem services around the world (Millennium Ecosystem Assessment 2005). The MA reports, along with the dual realization that most ecosystem services were being degraded by human actions and that there was a lack of scientific knowledge to understand and manage most services, led to an explosion of ecosystem services research over the past 15 years (Mulder et al. 2015). The MA was followed by a number of similar international initiatives, including The Economics of Ecosystem Services and Biodiversity (TEEB), which described the linkages between our economic systems and ecosystem services and biodiversity, and more recently, the Intergovernmental Science-Policy Panel on Biodiversity and Biodiversity and Ecosystem Services (IPBES). The goal of IPBES is to provide independent scientific information 'to strengthen the science-policy interface for biodiversity and ecosystem services for the conservation and sustainable use of biodiversity, long-term human well-being and sustainable development' (Díaz et al. 2015, pg. 3). Ecosystem services are now central to a wide variety of sustainability and conservation initiatives, including the United Nations' Sustainable Development Goals (Wood et al. 2018) and the Aichi Biodiversity Targets (CBD 2010), and have been incorporated into the work of global conservation organizations such as The Nature Conservancy and the World Wildlife Fund (Tallis et al. 2009).

A wide variety of classification systems have been developed to help organize and understand the variety of benefits that ecosystems provide to people. The best known is that of the MA, which includes four broad categories:

- *Provisioning*: the material products from ecosystems; includes food, water, fiber, and fuel;
- *Regulating*: ecological processes that contribute to favorable conditions for human life, such as climate, air quality, or water quality regulation;
- *Cultural*: non-material benefits from ecosystems that include recreational, aesthetic, and spiritual benefits;
- *Supporting*: The essential ecosystem processes that underlie all other ecosystem services, including photosynthesis, soil formation, and nutrient cycling.

The MA classification has been further developed and improved, specifically, to disentangle intermediate from final services (e.g., pollination from crop provisioning); distinguish between the benefits of nature (ecosystem services) and outcomes of ecological processes that negatively affect people (disservices) (Saunders and Luck 2016); establish standardized methods for defining, quantifying, and accounting for ecosystem services; and explicitly recognize the diversity of cultural views around humans and nature. In general, these new classifications also drop the supporting services category, viewing these more as ecosystem functions that underlie ecosystem service provision than as independent services themselves. Current and widely used classification schemes include those of TEEB (UNEP 2010), the Common International Classification of Ecosystem Services (CICES) (Haines-Young and Potschin-Young 2018), and the IPBES concept of 'nature's contribution to people' (Diaz et al. 2018) and its associated categorization system. Classification systems specific to service provision at landscape scales that explicitly include spatial patterns are also present (e.g., Vallés-Planells et al. 2014) but currently are not widely used. A related term, 'landscape services', has been developed to emphasize these social and ecological connections and the impact that spatial patterns across landscapes have on ecosystem services (Termorshuizen and Opdam 2009) and is often used in the field of landscape ecology.

The usefulness of the ecosystem services concept lies in its ability to separate and differentiate the different ways that ecosystems benefit people and its power to communicate these benefits to different stakeholders. Viewing landscapes through an ecosystem services lens is particularly valuable as it facilitates understanding of the ecological, social, and socio-ecological processes that structure landscapes. Developing the ecosystem services concept has also led to the ability to map ecosystem services and their economic values across broad scales (e.g., Costanza et al. 1997; Naidoo et al. 2008) and identify key areas where services are supplied or benefit people (e.g., Chan et al. 2006; Burkhard et al. 2012; Ala-Hulkko et al. 2019). These capacities are of particular use at landscape scales, where management decisions are likely to affect multiple ecosystem services and stakeholder groups at the same time. In particular, the ecosystem services framework has allowed a more quantitative understanding of landscape multifunctionality – the ability of landscapes to provide multiple ecosystem service benefits at the same time. Landscape multifunctionality is widely seen as a key property of many human-dominated landscapes and one that if achieved, should lead to more sustainable and resilient landscapes. This concept is discussed in more detail later.

A socio-ecological and spatially explicit understanding of ecosystem service provision

Ecosystem services are provided when people and ecosystems interact. Early ecosystem service frameworks focused on a cascade model, which emphasized a somewhat passive flow of ecosystem services from biophysical structures to ecological processes to services, which then resulted in benefits to people and influenced human values (Potschin and Haines-Young 2011). However, more recent approaches explicitly consider the spatial dynamics of the locations where ecosystem services are supplied, versus where they are used or provide benefits to people, and how these types of areas are connected (e.g., Syrbe and Walz 2012; Bagstad et al. 2013; Villamagna et al. 2013; Burkhard et al. 2014; Serna-Chavez et al. 2014). Here, I adopt an explicitly socio-ecological framework in which ecosystem service provision arises actively when ecosystems with the *capacity* to provide a service interact with human systems that have a *demand* for that service (Figure 20.1). When capacity and demand connect, then an ecosystem service *flow* results that leads to the *provision* of ecosystem services and benefits to people. Framing ecosystems in this way makes it easier to understand how capacity, demand, and flow interact at landscape scales to provide ecosystem services and how changes to landscape structure can influence these processes, sometimes in contrasting and diverging ways. I use this framework in the following to explore how ecosystem services are provided across different types of landscapes.

Ecosystem service capacity: The biophysical environment of a landscape determines its ability to provide ecosystem services. Capacity refers to the ability of natural or semi-natural ecosystems or natural capital (the stock of ecosystems, including their biological and physical features, that have the capacity to provide ecosystem services to people) to generate a potential benefit for people irrespective of it being realized or used (Tallis et al. 2012). For example, a natural landscape of forests, lakes, and rivers has the capacity to provide freshwater and recreational opportunities to people regardless of there being communities downstream to consume the water or the ability of people to access the area for hiking and canoeing. In other words, while capacity is necessary for ecosystem service provision, it is not sufficient. At a landscape scale, capacity largely depends on the types of ecosystems present and the ecosystem functions that they produce. For example, a landscape consisting of agricultural fields and small forest patches has the capacity to provide a much different suite of ecosystem services (food, pollination, pest control, aesthetic value) than a coastal wetland (carbon storage, storm protection, water quality regulation). In addition to the types of ecosystems present, the level of species diversity also matters for service capacity.

There is widespread evidence that ecosystem service capacity is strongly influenced by the level of diversity present in an area (Cardinale et al. 2012) and that biodiversity has a positive influence on most services (Balvanera et al. 2006; Isbell et al. 2011; Gamfeldt et al. 2013), especially with respect to important service-providing species (Ricketts et al. 2016). This is especially

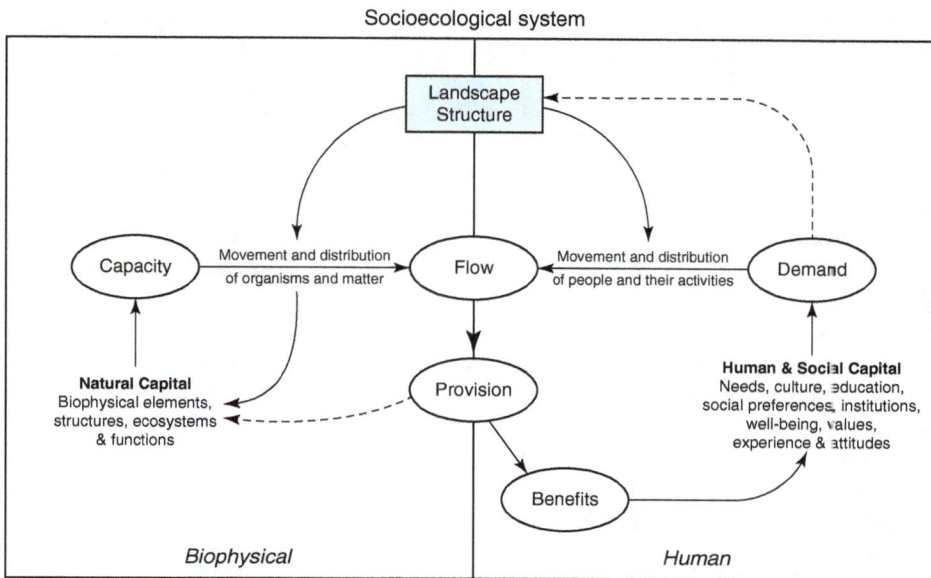

Socioecological system

Figure 20.1 Conceptual diagram of the effects of landscape structure on the provision of ecosystem services. Landscape structure affects ecosystem service capacity by affecting natural capital. Landscape structure also affects patterns of human distribution, activities, and movement across the landscape and ecosystem service demand. Thus, ecosystem service flows and ultimately, service provision depend on how landscape structure affects the movement and distribution of both organisms and people. In turn, the benefit derived from an ecosystem service affects service demand by altering human well-being and needs, which in turn drives human activities that alter landscape structure (broken arrow). Ecosystem service provision can also directly affect natural capital (broken arrow) through overexploitation. (Adapted from Tallis et al. 2012 and Mitchell et al. 2015.)

true for services that depend on specific species groups, such as pollination, pest regulation, and genetic resources (Tscharntke et al. 2005). Similarly, as more ecosystem services are considered, the number of species required to provide these services also increases. For example, an increasing number of ecosystem functions or services requires an increasing number of plant species (Lefcheck et al. 2015; Hertzog et al. 2019) especially as more locations or longer time periods are considered (Reich et al. 2012). In other words, the capacity of a landscape to provide multiple services through time is significantly influenced by the species diversity it contains and the ecosystem functions that this diversity contributes to and provides. However, whether that capacity actually provides a benefit to people depends on the presence of service demand and flows to connect capacity to demand.

Ecosystem service demand: The level of ecosystem service provision required or desired by people in a given landscape is ecosystem service *demand* (Villamagna et al. 2013; Mitchell et al. 2015b). While the vast majority of early ecosystem service studies focused on service capacity, there has been an increasing realization that service demand is also a critical component. Ecosystem service demand, in part, depends on the number of people or beneficiaries present in a landscape (Watson et al. 2019), as each individual has a basic need for food, water, and shelter. Ecosystem service demand is usually greatest in urban areas, where population densities are high, and decreases along urban–rural gradients (while service capacity often shows the opposite response) (Kroll et al. 2012), although there is considerable variation in this pattern among cities (Larondelle and Haase 2013).

Service demand is also affected by a wide suite of social and economic variables, although these have not been widely studied (Castro et al. 2014). These include economic status (Wolff et al. 2017), culture (Wilkerson et al. 2018), education, human values (Watson et al. 2019), institutions (Wolff et al. 2017), technological alternatives to ecosystem services (Luck et al. 2012), and the level of development of the built environment (Watson et al. 2019). For example, an increase in socioeconomic status may increase demand for recreational and aesthetic services (Wilkerson et al. 2018). Similarly, technology can influence demand by substituting for the services provided by ecosystems, although not always in cost-effective ways. For instance, water purification technology can replace natural water purification services (Brauman et al. 2007), and air conditioners can lessen the need for cooling by trees in urban areas (Cavan et al. 2014). While demand is clearly important for ecosystem service provision, its relationship with different socioeconomic drivers and the development of methods to quantify and map ecosystem service demand across and between landscapes are only now beginning to be explored.

Ecosystem service flow: The capacity to provide an ecosystem service and the presence of demand for that service in the same landscape do not guarantee that an ecosystem service will be provided or that it will be supplied at its maximum level. The actual delivery of an ecosystem service to people to be used or enjoyed results when capacity connects to demand and an ecosystem service flow results (Mitchell et al. 2015b). For most services, these flows occur through the movement of organisms, matter, people, or information (Bagstad et al. 2013). For example, pollination occurs when pollinators move between non-crop habitat and fields with pollinator-dependent crops (Kremen et al. 2007). Similarly, water provision results when above- and below-ground water flows to areas where people collect or consume water (Brauman et al. 2007), fisheries and recreation depend on the movement of people to lakes, rivers, marine environments, or protected areas (Bagstad et al. 2014), and aesthetic experiences depend on the flow of visual information from natural areas to places where people visit for recreation and enjoy landscapes (Egarter Vigl et al. 2017). Even food provision depends on flows, although in this case, it can be the movement of food products via transportation networks that is important (Ala-Hulkko et al. 2019). Contrastingly, some services depend on the restriction of movement across landscapes. Examples include flood regulation, storm protection, erosion control, and disease regulation (Bagstad et al. 2013). Finally, ecosystem service flows also depend on human-derived capital; human, social, cultural, and financial capital can be necessary for a flow to occur (Jones et al. 2016). In particular, ecosystem service flows between landscapes, such as those that occur through trade, can help a landscape with limited capacity to provide a service meet local demand and are strongly dependent on human capital (Palacios-Agundez et al. 2015).

Critically, ecosystem service flows are spatially explicit in that they result when movement from areas of ecosystem service capacity to areas of ecosystem service demand occurs. Thus, the location and magnitude of ecosystem service flows across landscapes should be strongly influenced by landscape structure. This is especially true for landscape fragmentation – the breaking apart of land cover into smaller and more isolated patches independently of change in the area of that land cover (Fahrig 2003). Fragmentation can increase the interspersion of areas of ecosystem service supply through a landscape, potentially bringing areas of supply and demand closer together and increasing service flow. Conversely, fragmentation also affects the spatial arrangement, size, and isolation of patches, possibly reducing ecosystem service flows. Mitchell et al. (2015b) identified four distinct positive and negative ways that landscape fragmentation can affect ecosystem service flows (Figure 20.2). Importantly, the varied ways that

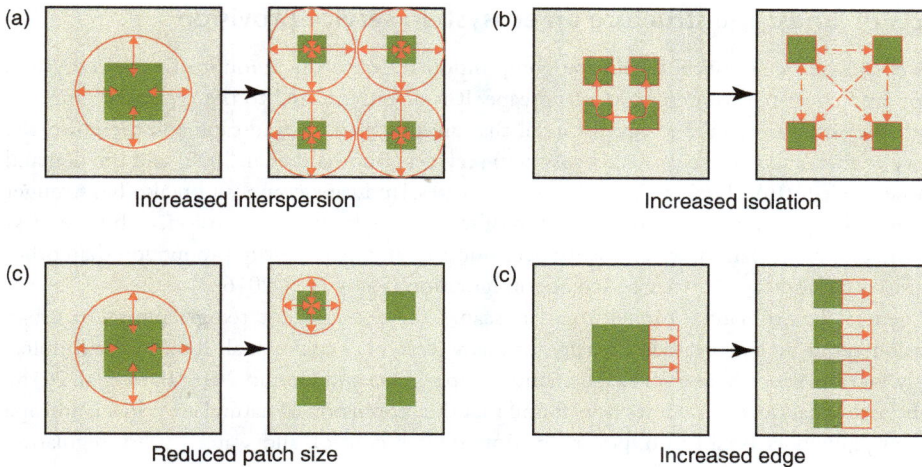

(a) Increased interspersion

(b) Increased isolation

(c) Reduced patch size

(c) Increased edge

Figure 20.2 The mechanisms by which landscape fragmentation, independently of a change in the area of natural land cover, can affect ecosystem service flow. Locations of natural land cover and ecosystem service capacity (green areas) provide ecosystem service flows (red arrows) and benefits (red areas) to the human-dominated matrix (light brown areas) that is affected by landscape fragmentation. Ecosystem service flows of organisms and people (arrows) can depend on proximity to natural areas (a) and will therefore be influenced by the interspersion of natural and anthropogenic land cover across the landscape (e.g., recreation, pollination, waste treatment, pest regulation). Simultaneously, increased isolation of patches and reduced connectivity (b), as well as decreased patch size (c), can decrease service flow in fragmented landscapes (e.g., pollination, seed dispersal, cultural services, watercourse recreation, water provision and regulation). Finally, for services that depend on restricting movement across landscapes, increased edge amounts with fragmentation (d) can help restrict the movement of water, waves, air or water pollutants, and soil, resulting in positive (e.g., storm protection, air quality regulation) or negative (e.g., water quality or soil erosion regulation) effects on ecosystem service flow. In each panel, the area of natural land cover and ecosystem service supply is unchanged between intact and fragmented landscapes. (Adapted from Fisher et al. 2009 and Mitchell et al. 2015.)

landscape fragmentation can affect service supply and flow means that the resulting patterns of service provision can be varied and could possibly change direction as landscape fragmentation proceeds. Unfortunately, the presence of these patterns in real landscapes has yet to be widely investigated.

Ecosystem service provision: Ecosystem services are provided when they produce a benefit to people that contributes to or improves their wellbeing (Villamagna et al. 2013). In the framework presented here, this occurs when ecosystem service flows connect areas of supply to areas of demand. However, how common measures or indicators of ecosystem service provision (e.g., freshwater consumed, crop production, recreation levels) actually relate to the physical or mental wellbeing of individuals (Bratman et al. 2019), or what the interrelationships are between ecosystem services, values, and wellbeing, is not well understood. In practice, a wide and often inconsistent suite of indicators or proxies for actual ecosystem service benefits are measured (Liss et al. 2013), and it is assumed that these are positively related to human wellbeing.

Effects of landscape structure on ecosystem service provision

Effects of landscape composition: Landscape composition refers to the amount of each ecosystem type or land cover that is present in a landscape. It is directly related to the types and amounts of both the natural capital and social capital that are present in a landscape and therefore, the capacity of that landscape to provide ecosystem services (Burkhard et al. 2009) and the demand for those services (Burkhard et al. 2012). In many cases, landscape composition also has stronger and more widespread effects on the provision of services such as food provision, carbon sequestration, nature appreciation, water quality, deer hunting, and maple syrup production than other dimensions of landscape structure such as configuration (Lamy et al. 2016).

In general, loss of natural capital from landscapes, for example, due to agricultural or urban expansion, results in a decrease in capacity for many services (Tratalos et al. 2007) while simultaneously increasing ecosystem service demand (Felipe-Lucia and Comín 2015; Baró et al. 2015). Recent meta-analyses and reviews have found that the proportion of natural areas in a landscape positively influences pollination, pest regulation, disease control, and water quality regulation (Duarte et al. 2018) and that there are widespread positive effects of semi-natural habitats on pest control and pollination (Holland et al. 2017). Similarly, a Europe-wide analysis of arable and agricultural landscapes identified strong positive relationships between semi-natural vegetation and regulating services (García-Feced et al. 2014). Changes in land cover from forests to pasture in Brazil and Colombia (Grimaldi et al. 2014) and from forests to plantations in China (Liu et al. 2017) have also shown strong negative effects on above-ground carbon stocks, water infiltration, and water provision, while crop production is often and unsurprisingly strongly correlated with the amount of converted land (Laterra et al. 2012).

The impacts of landscape structure, and in particular landscape composition, on ecosystem services provided by mobile organisms, including pollination, pest regulation, and seed dispersal, have been the most comprehensively studied (e.g., Kremen et al. 2007; Shackelford et al. 2013; Rusch et al. 2016; Landis 2017). In particular, the importance of natural and semi-natural habitats for these services in agricultural and urban landscapes is well known (Holzschuh et al. 2010; Kennedy et al. 2013; García-Feced et al. 2014). For example, the presence of at least 40% forest cover in coffee plantation landscapes results in reduced coffee berry borer presence, infestation, and damage due to the presence of ants (Aristizábal and Metzger 2019). Similarly, for pollination, an increase in perennial grasslands around cropland sites often results in increased pollinator abundance, diversity, and pollination services (Bennett and Isaacs 2014), although in some cases, crops can still be pollen-limited despite high natural forest cover (Samnegård et al. 2016). A recent meta-analysis found similar positive effects of native habitat cover on pest regulation by birds (Boesing et al. 2017), and comparable effects for seed-dispersing birds are also common (Moran and Catterall 2014) but not universal (Schäckermann et al. 2015). Positive effects of the amount of surrounding natural habitat (or conversely, negative effects of agricultural fields) have also been found for a variety of pollinator–crop systems (Kremen et al. 2004; Martins et al. 2015; Chatterjee et al. 2020) and control of a range of insect pests (Woltz et al. 2012; Rusch et al. 2013; Martin et al. 2013; Jonsson et al. 2014) by arthropod predators (Schmidt et al. 2008; Chaplin-Kramer et al. 2013; Power et al. 2016), vertebrates (Rodríguez-San Pedro et al. 2019), and parasitoids (Rusch et al. 2011; Pak et al. 2015; Tamburini et al. 2016). However, in some cases, cultivated lands in the surrounding landscape, especially if they represent important foraging resources for pollinators, can improve or support pollination services (Brittain et al. 2010; Zou et al. 2017), as can urban areas if they include private gardens with high floral resources (Kratschmer et al. 2018; Liu et al. 2018). There is also some evidence that increasing semi-natural habitats can lead to higher crop yields (Galpern et al. 2020), although the mechanisms behind this have yet to be identified. It is also important to note that the majority of these studies quantify service provision over single or short time scales. Service stability is a measure that is rarely quantified, although there is some evidence from ecological models that landscape composition can also have key positive effects on the stability of services such as pollination (Montoya et al. 2019).

The broad trends presented here can, however, mask complexities important for understanding how to manage landscape composition for different ecosystem services. First, the effects of changes in landscape composition can shift across spatial scales. For example, in Wisconsin potato fields, pest control of potato beetle increased primarily with local-scale field margin area, while aphid control increased with the amount of non-crop habitat at the 1.5-km scale (Werling and Gratton 2010), and in Brazilian coffee plantations, increased forest and coffee cover at a 300-m scale increased the presence of coffee berry borers despite the opposite trend at 2 km (Aristizábal and Metzger 2019). Second, changes in landscape composition can affect pests and predators in corresponding or contrasting ways. A study of cabbage pests found that the presence of meadows in the landscape decreased caterpillar numbers via increased parasitism but increased the abundance of other key pests (Perez-Alvarez et al. 2018). Similarly, in Israeli almond orchards, increasing proportions of semi-natural habitat increased wasp predation of almonds but did not affect parasitism of pests or seed predation by birds (Schäckermann et al. 2015). Conversely, increased non-crop habitat in Serbian agricultural landscapes increased both aphid densities as well as rates of parasitism (Plećaš et al. 2014), and the predation of both detrimental organisms (insect pests and weed seeds) and beneficial species (predators and crop seeds) in cereal fields can increase as the proportion of surrounding semi-natural grasslands increases (Tschumi et al. 2018). These last two results call into question whether there are overall benefits from increasing grasslands in all types of agricultural landscapes. Antagonistic interactions between biocontrol species, for example between insects and birds, can also increase with an increased proportion of natural habitat in the surrounding landscape, constraining pest regulation (Martin et al. 2013). Finally, loss of species diversity and abundance with changes in landscape composition does not always result in loss of ecosystem service provision. In central German agricultural landscapes, seed dispersal by birds and pollination by bees did not vary across a land use gradient despite significant decreases in bird abundance and diversity (Breitbach et al. 2010) and increases in pollinator species richness and visitation (Breitbach et al. 2012) along the gradient. In this case, changes in bird and bee movement and behavior across the gradient were able to maintain seed dispersal and pollination services.

There is also evidence that the effects of landscape composition (and configuration; see later) can interact with management practices. For example, pesticide use in agricultural landscapes can limit or counteract the positive effects of landscape structure on pest regulation (Ricci et al. 2019). Likewise, a study in Italian cereal fields (Tamburini et al. 2016) found that tillage interacted with landscape composition to affect water quality regulation and weed control. For weeds, a greater proportion of arable land increased the number of weeds (cover and species richness) when conventional tillage was present but decreased the number of weeds with conservation tillage; while for water quality regulation, when conventional tillage was present, an increase in arable land proportion led to a decrease in P saturation and predicted improvements in water quality.

Beyond landscape composition and changes in the proportion of natural or semi-natural habitats, the diversity of habitats present across landscapes can also be important for ecosystem services. For instance, compared with landscapes dominated by corn and soybean crops, an increased diversity of crop and non-crop habitats can increase pest regulation of aphids (Lansing et al. 2009; Liere et al. 2015) and activity levels of insectivorous bats (Rodríguez-San

Pedro et al. 2019), although these landscape diversity effects can vary across predator species (Liu et al. 2018) or be negative for parasitoids (Pak et al. 2015). Landscape diversity can also increase seed predation in agricultural landscapes, although pesticide use can help limit this disservice (Trichard et al. 2013). Landscape diversity effects are also not only limited to ecological effects for regulating services; cultural services can also be influenced. For example, an analysis using photos from the social media sites Flikr and Panoramio found that while certain cultural services depended on specific land cover types (e.g., terrestrial recreation was related to mountains; cultural heritage to landscapes with wood pastures and grasslands), landscape heterogeneity was positively related to the diversity of cultural ecosystem services provided by a landscape (Oteros-Rozas et al. 2018).

Landscape composition effects are also not limited to services provided by mobile organisms but can affect water regulation services as well. Qiu and Turner (2015), working in the Yahara watershed of Wisconsin, found that landscape composition consistently influenced freshwater supply and quality along with groundwater quality. Specifically, an increased proportion of cropland negatively influenced surface and groundwater quality, while water supply was negatively related to wetland and urban cover. At a continental scale across Europe, landscape composition is also predicted to influence flood regulation, with river catchments dominated by natural vegetation or agriculture providing the most flood regulation due to increased evapotranspiration from these vegetation types (Stürck et al. 2014). Finally, the proportions of both agriculture and roads in the landscapes of southern Quebec affected water quality regulation and water purification, although in different ways; agriculture primarily influenced suspended solids and phosphorus, while roads were associated with elevated nitrogen (Terrado et al. 2015).

Effects of landscape configuration: Changes in the spatial distribution of habitat and land cover patches across a landscape impact landscape configuration. Often, changes in landscape configuration are reflected in changes to the mean size and shape of patches in a landscape independently of any changes to landscape composition. These changes can have strong positive effects on a diversity of services. For instance, increased landscape complexity and patch aggregation have been found to have widespread positive effects on water quality regulation, pest regulation, pollination, and aesthetic value (Duarte et al. 2018). Changes to configuration can also have effects over and above changes to landscape composition for services where spatial structure is important, such as water quality regulation, recreation, and pollination (Lautenbach et al. 2011). Composition and configuration can also interact, influencing the heterogeneity of landscapes and affecting landscape-scale patterns of ecosystem functions and associated ecosystem services (Turner et al. 2013). In some cases, changes in configuration can have stronger effects on certain services (e.g., regulating services [Laterra et al. 2012] or summer home values [Lamy et al. 2016]) than changes in landscape composition.

One of the most widespread effects of landscape configuration on ecosystem services is a change in distance between key areas of ecosystem service capacity and demand. Key ecosystem service–providing organisms, whose movement results in a key function, often 'spill over' between habitats or land cover types (Rand et al. 2006; Blitzer et al. 2012). Thus, proximity to areas of ecosystem service capacity and the density of edges in a landscape can be key factors that influence the ability of mobile organisms to provide services (Martin et al. 2019). For example, pollinators typically forage from a central nesting location in natural or semi-natural habitat. Therefore, the distance to these types of habitats can influence pollinator abundance and diversity and pollination services across a landscape (Ricketts et al. 2008). These effects have been observed for pollinators in diverse crops (Blanche et al. 2006; Farwig et al. 2009; Carvalheiro et al. 2010) and spider and bird species that provide important pest regulation services (Boesing et al. 2017; Gallé et al. 2019). However, in some cases, these spillover effects may not be strong (Ferrante et al. 2017) or may not change dramatically with the distance from habitat if the agricultural matrix contains high-quality pollinator habitat (Jauker et al. 2009), pollinators do not transport high-quality pollen (Chacoff et al. 2008), or increased spillover near natural habitats is balanced by increased predation pressure (Herrmann et al. 2010).

These distance effects are not only possible for services provided by pollinators and insect predators but may also be present for other services. In the agricultural landscapes of southern Quebec, Mitchell et al. (2014b) found significant effects of distance-from-forest for crop production, pest regulation, and decomposition. Importantly, the patterns for each service were unique, such that no single landscape configuration could maximize all ecosystem services.

While these spillover and distance effects for pollinators appear to be especially common, those for arthropod predators are much more varied (Jauker et al. 2009; Karp et al 2018) or can even be nonexistent (Plećaš et al. 2014). In part, this can be due to similar effects of landscape configuration on pests and predators. For example, in Quebec soybean fields, increasing field widths and distance for hedgerows decreased the abundance and diversity of arthropod predators but also the abundance of insect herbivores, therefore resulting in an overall decrease in crop damage (Mitchell et al. 2014a). The scale of these effects can also vary across services and disservices. In South African macadamia fields, crop raiding by monkeys (a disservice) and biocontrol by birds and bats (a service) both decrease with distance from natural vegetation (Linden et al. 2019). However, crop raiding only occurred at small distances, while biocontrol occurred up to 500 m from vegetation, resulting in disproportionate economic gains from the presence of patches of natural vegetation in the landscapes.

Given these landscape effects of proximity to key ecosystem service capacity patches, loss of important patches, even those that are small, can result in a disproportionate loss of ecosystem services. In Madagascar, Bodin et al. (2006) found that the loss of even small patches of forest disproportionately affected the provision of pollination and seed dispersal across the landscape. Thus, the location of these small patches was more important to ecosystem service provision than their size. Empirical work in southern Quebec also observed that even small forest fragments (less than 5 ha in size) affected ecosystem services in the surrounding agricultural matrix in similar ways to much larger (greater than 100 ha) forest patches (Mitchell et al. 2014b). Similarly, modelling of agricultural landscapes to determine their optimal design for pollination has found that the presence of small pollination reservoirs scattered across the landscape at intermediate levels of fragmentation produces the best and most even pollination services (Brosi et al. 2008; Cong et al. 2016; Gunton et al. 2017). This can be especially true when mobile ecosystem service providers or ecosystem flows are limited spatially and do not diffuse widely across a landscape from areas of capacity. In this case, small but scattered areas of service capacity may lead to higher and more evenly distributed service provision (Mitchell et al. 2015a).

Edges not only affect the movement of key ecosystem service providers but can also result in changes to environmental conditions that result in edge effects: changes to forest growth and structure, reductions in tree biomass, and resulting shifts in climate-regulating services (e.g., carbon storage) near forest edges. For example, edge effects in tropical forests worldwide are estimated to have increased carbon emissions from forest loss by 10.3 Gt (Brinck et al. 2017). Similarly, a modelling study of agricultural expansion in Brazil found that cropland growth that maximized forest fragmentation, due to increased edge effects, resulted in the greatest decrease in carbon storage compared with scenarios that targeted core or edge forest areas for expansion (Chaplin-Kramer et al. 2015). These effects may also be prevalent in urban areas, as clumped forest fragments across the landscape can have significantly higher above-ground carbon densities per unit area than those that are more isolated by urban development (Mitchell et al. 2018).

For other services, landscape configuration can be a critical factor for determining the ecosystem service flows that control service provision. Using the widely used InVEST (Integrated Valuation of Ecosystem Services and Tradeoffs) model, Chaplin-Kramer et al. (2016) found, across the same amount of land conversion to agriculture, that the proximity of land use change to streams was the key driver of sediment export. Importantly, landscape configuration determined soil export and water purification services to a much greater degree than variables such as soil type, slope, and climate in this study. Similarly, including landscape configuration effects in ecosystem service models in Scotland substantially affected pollination, landscape aesthetics, nutrient retention and most intensely, flood control and sediment retention (Verhagen et al. 2016). A similar study of virtual landscapes and the ability of fragmented woodland patches to mitigate runoff, sediment loss, and diffuse pollution in surface water found that increased fragmentation, and in particular edge density, was the best predictor of the size of the mitigated area in modelled landscapes (Thomas et al. 2020). Wetland patch density and urban edge density have also been seen to positively affect surface water quality and freshwater supply, respectively (Qiu and Turner 2015), and landscape configuration can be a key driver of water quality regulation and water purification (Terrado et al. 2015). However, for some services, changes in landscape configuration may have little or no effect. In southern Sweden, changes in landscape heterogeneity (the amount and distribution of permanent pasture and field borders) had no effect on soil organic carbon, total nitrogen, water holding capacity, or plant-available phosphorus (Williams and Hedlund 2013)

Changes in landscape configuration can also reflect changes in patch size and shifts in the succession of plant communities that are important for ecosystem service provision via changes in dispersal, colonization, establishment, and extirpation. In English heathlands, increased patch size and changes in plant succession have been seen to influence a variety of ecosystem services, including decreasing carbon storage and aesthetic value but increasing recreational value (Cordingley et al. 2015). Despite this, there is increasing evidence that even small patches of habitat, and especially forest patches, still deliver critical ecosystem services to people, including timber, wild game, food, disturbance regulation, pest control, pollination, disease regulation, water regulation, and soil erosion control (Decocq et al. 2016). In some cases, these small forest patches can even provide higher levels of multiple services per unit than larger patches (Valdés et al. 2020) and often do so in locations where more people can benefit.

Effects of landscape connectivity and fragmentation: Landscape fragmentation and a corresponding loss of landscape connectivity can have both direct and indirect effects on ecosystem service provision. Direct effects stem from the impacts of fragmentation on ecosystem service flows – fragmentation can prevent the important flows that are crucial for service provision (Mitchell et al. 2015b). Indirect effects result from the impacts of fragmentation on biodiversity and the size and persistence of ecosystem service provider populations. Fragmentation can have strong negative effects on both species diversity and population abundances (Mitchell et al. 2013) as well as ecosystem functions (Haddad et al. 2015), which can, in turn, influence ecosystem service provision. Including these types of fragmentation effects in the modelling of ecosystem service values across landscapes can substantially change our assessment of the level of ecosystem service provision and the key locations for services (Ng et al. 2013). However, studies that focus directly on the effects of landscape connectivity, beyond simply changes in landscape configuration, are not common.

In agricultural landscapes, even small landscape elements, such as hedgerows and field margins, can provide important corridors and increase landscape connectivity for organisms important for pest regulation (Haenke et al. 2014; Puech et al. 2015) and pollination (Garratt et al. 2017), especially in landscapes with intermediate levels of natural habitat cover (Grab et al. 2018).

For example, Holzschuh et al. (2009) found that grass strips connected to forest increased the abundance of caterpillar-hunting wasps by over six times compared with isolated grass strips. Similarly, hedgerows can counteract cherry tree isolation from forested habitats, increasing predator abundance and decreasing aphid numbers (Stutz and Entling 2011). Landscape connectivity across vegetation types (e.g., crop, pasture, native vegetation) can also affect pest dynamics in agricultural landscapes (Macfadyen et al. 2015). It is important to keep in mind that it is difficult to separate connectivity/fragmentation effects from those of composition in real landscapes. In most cases, fragmentation takes place through the loss of natural habitats. Therefore, some of these effects of fragmentation on service provision might be due to loss of habitat area rather than fragmentation effects per se (Fahrig 2003). For example, a meta-analysis of fragmentation effects on herbivory found nonsignificant effects of fragmentation per se on herbivory but strong effects of fragment isolation on the abundance and richness of herbivores (Rossetti et al. 2017).

Connectivity between forest patches can also affect how they influence ecosystem services in the agricultural matrix around them. In Quebec agricultural landscapes, forest connectivity affected pest regulation and decomposition in soybean fields, with the highest levels of herbivore control near isolated forest patches and the highest levels of decomposition near forest patches with intermediate connectivity (Mitchell et al. 2014b). Schüepp et al. (2011) also found that woody habitat isolation decreased the abundance of natural enemies, resulting in substantially lower parasitism rates of insect pests (Schüepp et al. 2011). These connectivity effects can also be important within fragmented forest patches, as edge effects on sugar maple trees have been observed to vary according to the broader-scale connectivity of the forest patches (Maguire et al. 2016). While much theory suggests that connectivity is critical to maintaining many services, for pest regulation, increased connectivity may facilitate the spread of herbivorous insects, especially under outbreak conditions (Maguire et al. 2015), which could then affect services such as timber production, carbon sequestration, and soil or water quality. There is also evidence that connectivity influences carbon storage in fragmented forest patches; Ziter et al. (2013) found stronger positive relationships between tree functional diversity and carbon in unmanaged forest fragments in the agricultural landscapes of southern Quebec, suggesting that increased connectivity could lead to increased biodiversity and carbon storage in this landscape.

Landscape connectivity can also be important for services that involve the movement of water, people, and information. In Melbourne, Australia, fragmentation of urban forests is a better predictor of service provision than overall vegetation cover and resulted in an increased potential for recreation but decreased flooding regulation (Dobbs et al. 2014). Fragmentation can also have an important influence on water quality regulation. Modelled nitrogen loads in Florida were explained equally well by models of landscape fragmentation (number of patches) as by models of landscape composition (proportion of landscape for each land use class) (Yang 2012). Connectivity is also important for how people view and experience landscapes and in turn, the cultural ecosystem services they receive. For example, visitors to the Dutch Winterswijk region prefer landscapes with connected forest patches and hedgerows and mosaic landscapes, all of which increase the delivery of recreation, aesthetic beauty, cultural heritage, inspiration, and spirituality (Van Berkel and Verburg 2014).

Ecosystem service multifunctionality, relationships, and bundles

Ecosystem service multifunctionality: Landscape multifunctionality with respect to ecosystem services is the capacity of landscapes to simultaneously provide multiple benefits to society (Mastrangelo et al. 2014; Hölting et al. 2019). In the face of increasing human demand for most ecosystem services, a key aspect of ecosystem service research, especially at landscape scales, is understanding how landscapes provide multiple services and how to manage them to increase their multifunctionality. Three different but related research questions are prevalent in the ecosystem services literature:

- What are the relationships between ecosystem services at landscape scales?
- Are there groups or 'bundles' of ecosystem services that repeatedly occur or are provided across different landscapes?
- How can we best measure landscape multifunctionality and what drives it? (Manning et al. 2018)

Answering these questions should lead to better management decisions at a variety of spatial scales, including landscape scales, for multiple ecosystem services at the same time.

Ecosystem service relationships: Understanding how ecosystem services relate to and affect each other across landscapes is key to developing more multifunctional landscapes. A great deal of research has focused on identifying synergies (positive relationships) or trade-offs (negative relationships) between ecosystem services (e.g., Qiu and Turner 2013) or between ecosystem services and biodiversity (e.g., Nelson et al. 2009), and priority locations that provide multiple ecosystem services (e.g., Chan et al. 2006; Egoh et al. 2011). In general, this involves measuring ecosystem service supply across a set of landscapes and calculating the pairwise correlations between different services (Dade et al. 2018). Analyses of relationships between services that incorporate service supply, demand, and flow are rare, although they are becoming more common. For example, Castro et al. (2014) used surveys of local stakeholders' socio-cultural and economic values around ecosystem services along with biophysical evaluation of service supply to evaluate ecosystem service trade-offs in the landscapes of eastern Andalusia, Spain. They found mismatches between the biophysical supply and socioeconomic value of ecosystem services, such that in some cases, strong trade-offs from the biophysical analysis did not correspond to similar socio-cultural or economic trade-offs. Similar mismatches have also been found in central Belgian agricultural landscapes (De Vreese et al. 2016). However, this type of correlation approach largely ignores the actual processes or mechanisms that underlie these relationships. Pairs of ecosystem services can respond in similar ways due to direct interactions between the services or because they share a common driver (Bennett et al. 2009). For example, carbon sequestration and crop production can be negatively correlated when forest restoration that increases carbon storage leads to a loss of cropland. Conversely, they can be positively correlated if the forest restoration also leads to improved soil quality and pollinator abundances that help increase crop production. In particular, the question of whether landscape structure is important for determining relationships between services has not been widely investigated.

Ecosystem service bundles: Instead of looking at pairs of ecosystem services, other studies have set out to identify sets of ecosystem services that occur together repeatedly across space or that respond together in similar ways to socio-ecological drivers (e.g., Raudsepp-Hearne et al. 2010; Crouzat et al. 2015; Renard et al. 2015; Spake et al. 2017). Identifying ecosystem service bundles across landscapes is seen as a way of understanding how multiple ecosystem services interact with each other and identifying syndromes of service provision that may reflect common landscape structures and land uses. In general, these studies have shown trade-offs between provisioning and regulating services, but they have yet to be widely used to identify the key drivers of this bundling (Spake et al. 2017) or understand how multiple services and bundles change through time (but see Renard et al. 2015; Sutherland et al. 2016)

Effects of landscape structure on multiple ecosystem services and multifunctionality: While a number of studies of landscape multifunctionality with respect to ecosystem services have been published, analyses of these patterns with landscape structure are limited. One example is from the French Alps, where high ecosystem service richness did not always correspond with high landscape heterogeneity (Crouzat et al. 2015). While heterogeneous areas with artificial surfaces generally provided fewer services, and those with grasslands and pastures provided increased service diversity, homogeneous forest landscapes also had the capacity to provide multiple services. A similar Europe-wide study found a similar lack of association between land use diversity and ecosystem service multifunctionality (Stürck and Verburg 2017). Contrastingly, a modelling approach predicted that increased compositional diversity of agricultural landscapes (e.g., a greater number of land cover types) will lead to increased provision of multiple services as well as decreased uncertainty in reaching ecosystem service provision targets (Knoke et al. 2016). Similarly, modelled and empirical landscapes from the German Biodiversity Exploratories project predicted and showed that heterogeneity in land use intensity consistently led to higher landscape multifunctionality (van der Plas et al. 2019).

There is also some evidence that bundles of ecosystem services may be influenced or associated with specific landscape pattern types: an analysis of multiple ecosystem services in southern Quebec identified three distinct service bundle types that were associated with landscape composition, specifically, changes in forest composition and type of agriculture (Lamy et al. 2016). A similar study of Andean mountain regions indicated that bundles of land use change (i.e., clusters of similar changes in landscape composition) are associated with corresponding bundles of ecosystem service change (Madrigal-Martínez and Miralles i García 2019). Specifically, agricultural de-intensification led to increases in water, soil, and climate regulation, while agricultural expansion led to decreases in the same bundle of services. Finally, landscape composition may also help drive multifunctionality and hotspots of ecosystem service provision. Qiu et al. (2013) found that most service hotspots in the Yahara watershed, Wisconsin, were associated with areas of natural land cover such as parks and riparian areas, while coldspots were often cropland or urban areas.

Ecosystem service planning at landscape scales

Planning for ecosystem services, and especially multiple ecosystem services at landscape scales, is challenging. Even for single services, areas of supply and demand may be spatially separate, demanding cooperation between landowners, actors, and decision-makers at broad spatial scales to ensure effective management of service provision. When multiple ecosystem services are considered, the situation becomes even more complex; not only do trade-offs and synergies between services need to be accounted for, but also differing ecosystem service values and goals for their management across different groups that are present in the landscape. In addition, while there can be substantial gains for ecosystem services from managing landscape-level configuration for services like pollination, individuals can also gain by cheating and not engaging with landscape-level management actions when the majority of landowners do participate (Cong et al. 2016). A final layer of complexity involves ecosystem disservices – ecosystem processes that negatively impact human wellbeing (Lyytimäki et al. 2008), where management actions and changes at landscape scales can increase both services and disservices simultaneously (Dobbs et al. 2014; Tschumi et al. 2018), and how to balance these benefits and costs is unclear.

While the ecosystem services concept has been relatively quickly integrated into landscape planning approaches in a number of countries (Mascarenhas et al. 2014), how comprehensive this integration is, or how well initiatives such as the MA, TEEB, and IPBES are contributing to planning at the landscape level, is more uncertain. In other words, while planners are able to define ecosystem services and can use the ecosystem services concept to advance planning initiatives and lend support to the protection of natural capital, how well more complex understanding of ecosystem services (such as multifunctionality) is informing landscape-level planning is more uncertain. For example, how management decisions might influence bundles of ecosystem services and trade-offs between services (de Groot et al. 2010; Raudsepp-Hearne et al. 2010) and how to integrate these considerations into planning (Goldstein et al. 2012), how services and the relationships between them change through time (Holland et al. 2011; Renard et al. 2015), what the effects of historical legacies might be on current decisions and future conditions (Bürgi et al. 2015), how the spatial scale of ecosystem service data affects relationships between services (Anderson et al. 2009), and how to integrate disservices that might result from management decisions and increases in natural capital (Babí Almenar et al. 2018) are all questions that have currently been little studied. There is also interest in developing common landscape metrics as indicators of ecosystem service provision (Feld et al. 2009), using common approaches to quantifying and mapping ecosystem services (Englund et al. 2017), and standardizing measures of multifunctionality (Manning et al. 2018). However, landscape metric selection and how different indicators reflect service provision, especially given landscape-specific contexts, is not well known (Babí Almenar et al. 2018). How to best integrate measures of capacity, demand, flow, and human values into measures of multifunctionality is also not well known. While the development of remote-sensing approaches for ecosystem services is a promising way forward, how exactly to use these approaches to quantify supply, demand, and flow is an open question (Cord et al. 2017).

How to integrate different knowledge and values into landscape planning for services is also a critical component that needs to be further developed. How services are quantified and mapped, and whether this process can itself help integrate ecosystem service considerations into planning through processes such as public participatory geographic information systems (PPGIS), is not currently well understood. For example, combining traditional mapping of ecosystem service benefits via land cover and 'expert-based' assessment of service values could help integrate different knowledge systems around services and improve landscape-scale planning (Koschke et al. 2012). Integrating cultural services into landscape planning also has the potential to possibly facilitate the inclusion of different values and improve landscape multifunctionality (Plieninger et al. 2015) by making explicit different values and the utilitarian and non-utilitarian uses that landscapes provide. However, our ability to measure and understand where and why cultural ecosystem services are provided at landscape scales is currently not well developed. Finally, there is a widespread lack of assessments that investigate how successful specific policies, plans, or designs across landscapes have been for maintaining or improving ecosystem service provision (Jones et al. 2013) and linking this to changes in land use or landscape structure.

A potentially fruitful way forward for planning and managing ecosystem services at landscape scales is to use network approaches (QUINTESSENCE Consortium 2016). Ecosystem services result from the interaction of networks of organisms, matter, and people across landscapes. In addition, social and economic networks influence how people interact with ecosystems and value different ecosystem services. Quantifying these different networks and understanding how they interact across landscapes at different spatial and temporal scales offers a promising way forward for linking landscape structure to ecosystem services provision. They may also be well suited to dealing with power dynamics between stakeholders and understanding how these dynamics influence access to ecosystem services and associated service flows (Felipe-Lucia et al. 2015). Of course, the challenge is that few examples of successful approaches to building these sorts of multi-level networks exist, as they require long-term ecological, social, and economic data and likely need transdisciplinary approaches that involve scientists, decision-makers, and other stakeholders on the landscape.

Conclusions

Landscapes are a key spatial scale across which to understand and manage ecosystem services. Landscapes incorporate many of the biophysical, social, and economic processes that determine the provision of ecosystem services and represent a scale upon which human actions and decisions are managed for the creation of multifunctional landscapes. At this scale, how we structure and manage landscapes has important impacts on ecosystem service capacity, demand, and flow as well as ecosystem disservices that negatively impact humans. While our understanding of the effects of landscape composition, configuration, and connectivity on services is growing rapidly, it still remains focused on the capacity of landscapes to provide regulating services, especially those provided by mobile organisms. Expanding knowledge to explicitly encompass ecosystem service demand and flows; the full diversity of ecosystem services and disservices that landscapes provide, especially cultural services; and understanding how to manage landscapes for multiple services has the potential to transform our ability to create multifunctional and sustainable landscapes.

References

Ala-Hulkko, T., Kotavaara, O., Alahuhta, J., and Hjort, J. (2019) 'Mapping supply and demand of a provisioning ecosystem service across Europe', *Ecological Indicators*, vol 103, pp520–9.

Anderson, B.J., Armsworth, P.R., Eigenbrod F, Thomas, C.D., Gillings, S., Heinemeyer, A., Roy, D.B. and Gaston, K.J. (2009) 'Spatial covariance between biodiversity and other ecosystem service priorities', *Journal of Appled Ecology*, vol 46, pp888–96.

Aristizábal, N. and Metzger, J.P. (2019) 'Landscape structure regulates pest control provided by ants in sun coffee farms', *Journal of Applied Ecology*, vol 56, pp21–30.

Babí Almenar, J., Rugani, B., Geneletti, D. and Brewer, T. (2018) 'Integration of ecosystem services into a conceptual spatial planning framework based on a landscape ecology perspective', *Landscape Ecology*, vol 33, pp2047–59.

Bagstad, K.J., Johnson, G.W., Voigt, B. and Villa, F. (2013) 'Spatial dynamics of ecosystem service flows: A comprehensive approach to quantifying actual services', *Ecosystem Services*, vol 4, pp117–25.

Bagstad, K.J., Villa, F., Batker, D., Harrison-Cox, J., Voigt, B. and Johnson, G.W. (2014) 'From theoretical to actual ecosystem services: Mapping beneficiaries and spatial flows in ecosystem service assessments', *Ecology and Society*, vol 19, art 64.

Balvanera P, Pfisterer AB, Buchmann N, He, J-S., Nakashizuka, T., Raffaelli, D. and Schmid, B. (2006) 'Quantifying the evidence for biodiversity effects on ecosystem functioning and services', *Ecology Letters*, vol 9, pp1146–1156.

Baró, F., Haase, D., Gómez-Baggethun, E. and Frantzeskaki, N. (2015) 'Mismatches between ecosystem services supply and demand in urban areas: A quantitative assessment in five European cities', *Ecological Indicators*, vol 55, pp146–58.

Bennett, A.B. and Isaacs, R. (2014) 'Landscape composition influences pollinators and pollination services in perennial biofuel plantings', *Agriculture, Ecosystems and Environment*, vol 193, pp1–8.

Bennett, E.M., Peterson, G.D. and Gordon, L.J. (2009) 'Understanding relationships among multiple ecosystem services', *Ecology Letters*, vol 12, pp1394–404.

Blanche, K.R., Ludwig, J.A. and Cunningham, S.A. (2006) 'Proximity to rainforest enhances pollination and fruit set in orchards', *Journal of Applied Ecology*, vol 43, pp1182–1187.

Blitzer, E.J., Dormann, C.F., Holzschuh, A., Klein, A-M., Rand, T.A. and Tscharntke, T. (2012) 'Spillover of functionally important organisms between managed and natural habitats', *Agriculture, Ecosystems and Environment*, vol 146, pp34–43.

Bodin, O., Tengö, M., Norman, A., Lundberg, J. and Elmqvist, T. (2006) 'The value of small size: Loss of forest patches and ecological thresholds in southern Madagascar', *Ecological Applications*, vol 16, pp440–51.

Boesing, A.L., Nichols, E. and Metzger, J.P. (2017) 'Effects of landscape structure on avian-mediated insect pest control services: A review', *Landscape Ecology*, vol 32, pp931–44.

Bratman, G.N., Anderson, C.B., Berman, M.G., Cochran, B., de Vries, S., Flanders, J., Folke, C., Frumkin, H., Gross, J.J., Hartig, T., Kahn, P.H. Jr., Kuo, M., Lawler, J.J., Levin, P.S., Lindahl, T., Meyer-Lindenberg,

A., Mitchell, R., Ouyang, Z., Roe, J., Scarlett, L., Smith, J.R., van den Bosch, M., Wheeler, B.W., White, M.P., Zheng, H. and Daily, G.C. (2019) 'Nature and mental health: An ecosystem service perspective', *Science Advances*, vol 5, art eaax0903.

Brauman, K.A., Daily, G.C., Duarte, T.K. and Mooney, H.A. (2007) 'The nature and value of ecosystem services: An overview highlighting hydrologic services', *Annual Review of Environment and Resources*, vol 32, pp67–98.

Breitbach, N., Laube, I., Steffan-Dewenter, I. and Boehning-Gaese, K. (2010) 'Bird diversity and seed dispersal along a human land-use gradient: High seed removal in structurally simple farmland', *Oecologia*, vol 162, pp965–76.

Breitbach, N., Tillmann, S., Schleuning, M., Grünewald, C., Laube, I., Steffan-Dewenter, I. and Böhning-Gaese, K. (2012) 'Influence of habitat complexity and landscape configuration on pollination and seed-dispersal interactions of wild cherry trees', *Oecologia* vol 168, pp425–37.

Brinck K, Fischer R, Groeneveld J, Lehmann, S., De Paula, M.D., Pütz, S., Sexton, J.O., Song, D. and Huth, A. (2017) 'High resolution analysis of tropical forest fragmentation and its impact on the global carbon cycle', *Nature Communications*, vol 8, art 14855.

Brittain, C., Bommarco, R., Vighi, M., Settele, J. and Pottsa, S.G. (2010) 'Organic farming in isolated landscapes does not benefit flower-visiting insects and pollination', *Biological Conservation*, vol 143, pp1860–1867.

Brosi, B.J., Armsworth, P.R. and Daily, G.C. (2008) 'Optimal design of agricultural landscapes for pollination services', *Conservation Letters*, vol 1, pp27–36.

Bürgi, M., Silbernagel, J., Wu, J. and Kienast, F. (2015) 'Linking ecosystem services with landscape history', *Landscape Ecology*, vol 30, pp11–20.

Burkhard, B., Kroll, F., Müller, F. and Windhorst, W. (2009) 'Landscapes' capacities to provide ecosystem services – A concept for land-cover based assessments', *Landscape Online*, vol 15, pp1–22.

Burkhard, B., Kroll, F., Nedkov, S. and Müller, F. (2012) 'Mapping ecosystem service supply, demand and budgets', *Ecologcal Indicators*, vol 21, pp17–29.

Burkhard, B., Kandziora, M., Hou, Y. and Müller, F. (2014) 'Ecosystem service potentials, flows and demands – concepts for spatial localisation, indication and quantification', *Landscape Online*, vol 34, pp1–32.

Cardinale, B.J., Duffy, J.E., Gonzalez, A., Hooper, D.U., Perrings, C., Venail, P., Narwani, A., Mace, G.M., Tilman, D., Wardle, D.A., Kinzig, A.P., Daily, G.C., Loreau, M., Grace, J.B., Larigauderie, A., Srivastava, D.S. and Naeem, S. (2012) 'Biodiversity loss and its impact on humanity', *Nature*, vol 486, pp59–67.

Carvalheiro, L.G., Seymour, C.L., Veldtman, R. and Nicolson, S.W. (2010) 'Pollination services decline with distance from natural habitat even in biodiversity-rich areas', *Journal of Applied Ecology*, vol 47, pp810–820.

Castro, A.J., Verburg, P.H., Martín-López, B., Garcia-Llorente, M., Cabello, J., Vaughn, C.C. and López, E. (2014) 'Ecosystem service trade-offs from supply to social demand: A landscape-scale spatial analysis', *Landscape and Urban Planning*, vol 132, pp102–110.

Cavan, G., Lindley, S., Jalayer, F., Yeshitela, K., Pauleit, S., Renner, F., Gill, S., Capuano, P., Nebebe, A., Woldegerima, T., Kibassa, D. and Shemdoe, R. (2014) 'Urban morphological determinants of temperature regulating ecosystem services in two African cities', *Ecological Indicators*, vol 42, pp43–57.

Convention on Biological Diversity (2010) 'Strategic plan for biodiversity 2011–2020, including Aichi Biodiversity Targets', www.cbd.int/sp/.

Chacoff, N.P., Aizen, M.A. and Aschero, V. (2008) 'Proximity to forest edge does not affect crop production despite pollen limitation', *Proceedings of the Royal Society B Biological Sciences*, vol 275, pp907–913.

Chan KMA, Shaw MR, Cameron DR, Underwood, E.C. and Daily, G.C. (2006) 'Conservation planning for ecosystem services', *PLoS Biology*, vol 4, pp2138–2152.

Chaplin-Kramer, R., Valpine P. de, Mills N.J. and Kremen C. (2013) 'Detecting pest control services across spatial and temporal scales', *Agriculture, Ecosystems and the Environment*, vol 181, pp206–212.

Chaplin-Kramer, R., Sharp, R.P., Mandle, L., Sim, S., Johnson, J., Butnar, I., Milà i Canals, L., Eichelberger, B.A., Ramler, I., Mueller, C., McLachlan, N., Yousefi, A., King, H. and Kareiva, P.M. (2015) 'Spatial patterns of agricultural expansion determine impacts on biodiversity and carbon storage', *Proceedings of the National Academy of Sciences,* vol 112, pp7402–7407.

Chaplin-Kramer, R., Hamel, P., Sharp, R., Kowal, V., Wolny, S., Sim, S. and Mueller, C. (2016) 'Landscape configuration is the primary driver of impacts on water quality associated with agricultural expansion', *Environmental Research Letters*, vol 11, pp1–11.

Chatterjee, A., Chatterjee, S., Smith, B., Cresswell, J.E. and Basua, P. (2020) 'Predicted thresholds for natural vegetation cover to safeguard pollinator services in agricultural landscapes', *Agriculture Ecosystems and the Environment*, vol 290, art 106785.

Cong, R.G., Ekroos, J., Smith, H.G. and Brady, M.V. (2016) 'Optimizing intermediate ecosystem services in agriculture using rules based on landscape composition and configuration indices', *Ecological Economics*, vol 128, pp214–223.

Cord, A.F., Brauman, K.A., Chaplin-Kramer, R., Huth, A., Ziv, G. and Seppelt, R. (2017) 'Priorities to advance monitoring of ecosystem services using earth observation', *Trends in Ecology and Evolution*, vol 32, pp416–428.

Cordingley, J.E., Newton, A.C., Rose, R.J., Clarke, R.T. and Bullock, J.M. (2015) 'Habitat fragmentation intensifies trade-offs between biodiversity and ecosystem services in a heathland ecosystem in southern England', *PLoS One*, vol 10, art e0130004.

Costanza, R., d'Arge, R., deGroot, R., Farber, S., Grasso, M., Hannon, B., Limburg, K., Naeem, S., O'Neill, R.V., Paruelo, J., Raskin, R.G., Sutton, P. van den Belt, M. (1997) 'The value of the world's ecosystem services and natural capital', *Nature*, vol 387, pp253–260.

Crouzat, E., Mouchet, M., Turkelboom, F., Byczek, C., Meersmans, J., Berger, F., Verkerk, P.J. and Lavorel, S. (2015) 'Assessing bundles of ecosystem services from regional to landscape scale: Insights from the French Alps', *Journal of Applied Ecology*, vol 52, pp1145–1155.

Dade, M.C., Mitchell, M.G.E., McAlpine, C.A. and Rhodes, J.R. (2018) 'Assessing ecosystem service trade-offs and synergies: The need for a more mechanistic approach', *Ambio*, vol 9, pp1–13.

Daily GC, Alexander S, Ehrlich PR, Goulder, Larry, Lubchenco, J. Matson, P.A., Mooney, H.A., Postel, S., Schneider, S.H., Tilman, David and Woodwell, G.M (1997) 'Ecosystem services: Benefits supplied to human societies by natural ecosystems', *Issues in Ecology*, vol 2, pp1–16.

Decocq, G., Andrieu, E., Brunet J, Chabrerie, O., De Frenne, P., De Smedt, P., Deconchat, M., Diekmann, M., Ehrmann, S., Giffard, B., Gorriz Mifsud, E., Hansen, K., Hermy, M., Kolb, A., Lenoir, J., Liira, J., Moldan, F., Prokofieva, I., Rosenqvist, L., Varela, E., Valdés, A., Verheyen, K. and Wulf, M. (2016) 'Ecosystem services from small forest patches in agricultural landscapes', *Current Forestry Reports*, vol 2, pp30–44.

de Groot, R.S., Alkemade, R., Braat, L., Hein, L. and Willemen, L. (2010) 'Challenges in integrating the concept of ecosystem services and values in landscape planning, management and decision making', *Ecological Complexity*, vol 7, pp260–272.

DeVreese, R., Leys, M., Fontaine, C.M. and Dendoncker, N. (2016) 'Social mapping of perceived ecosystem services supply-The role of social landscape metrics and social hotspots for integrated ecosystem services assessment, landscape planning and management', *Ecological Indicators*, vol 66, pp517–533.

Díaz S, Demissew S, Carabias J, Joly, C., Lonsdale, M., Ash, N., Larigauderie, A., Adhikari, J.R., Arico, S., Báldi, A., Bartuska, A., Baste, I.A., Bilgin, A., Brondizio, E., Chan, K.M.A., Figueroa, V.E., Duraiappah, A., Fischer, M. and Zlatanova, D. (2015) 'The IPBES Conceptual Framework – connecting nature and people', *Current Opinion in Environmental Sustainability*, vol 14, pp1–16.

Diaz, S., Pascual, U., Stenseke, M., Martín-López, B., Watson, R.T., Molnár, Z., Hill, R., Chan, K.M.A., Baste, I.A., Brauman, K.A., Polasky, S., Church, A., Lonsdale, M., Larigauderie, A., Leadley, P.W., van Oudenhoven, A.P.E., van der Plaat, F., Schröter, M., Lavorel, S., Aumeeruddy-Thomas, Y., Bukvareva, E., Davies, K., Demissew, S., Erpul, G., Failler, P., Guerra, C.A., Hewitt, C.L., Keune, H. Lindley, S. and Shirayama, Y. (2018) 'Assessing nature's contributions to people', *Science*, vol 359, pp270–272.

Dobbs, C., Kendal, D. and Nitschke, C.R. (2014) 'Multiple ecosystem services and disservices of the urban forest establishing their connections with landscape structure and sociodemographics', *Ecological Indicators*, vol 43, pp44–55.

Duarte, G.T., Santos, P.M., Cornelissen, T.G., Ribeiro, M.C. and Paglia, A.P. (2018) 'The effects of landscape patterns on ecosystem services: Meta-analyses of landscape services', *Landscape Ecology*, vol 33, pp1247–1257.

Egarter Vigl, L., Depellegrin, D., Pereira, P., Groot, R. and Tappeiner, U. (2017) 'Mapping the ecosystem service delivery chain: Capacity, flow, and demand pertaining to aesthetic experiences in mountain landscapes', *Science of the Total Environment*, vol 574, pp422–36.

Egoh, B.N., Reyers, B., Rouget, M. and Richardson, D.M. (2011) 'Identifying priority areas for ecosystem service management in South African grasslands', *Journal of Environmental Management*, vol 92, pp1642–50.

Englund, O., Berndes, G. and Cederberg, C. (2017) 'How to analyse ecosystem services in landscapes – A systematic review', *Ecological Indicators*, vol 73, pp492–504.

Fahrig L. (2003) 'Effects of habitat fragmentation on biodiversity', *Annual Review of Ecology, Evolution and Systematics*, vol 34, pp487–515.

Farwig, N., Bailey, D., Bochud, E., Herrmann, J.D., Kindler, E., Reusser, N., Schüepp, C. and Schmidt-Entling, M.H. (2009) 'Isolation from forest reduces pollination, seed predation and insect scavenging in Swiss farmland', *Landscape Ecology*, vol 24, pp919–27.

Feld, C.K., da Silva, P.M., Sousa J.P., De Bello, F., Bugter, R., Grandin, U., Hering, D., Lavorel, S., Mountford, O., Pardo, I., Pärtel, M., Römbke, J., Sandin, L., Jones, K.B. and Harrison, P. (2009) 'Indicators of biodiversity and ecosystem services: A synthesis across ecosystems and spatial scales', *Oikos*, vol 118, pp1862–1871.

Felipe-Lucia, M.R. and Comín, F.A. (2015) 'Ecosystem services-biodiversity relationships depend on land use type in floodplain agroecosystems', *Land Use Policy*, vol 46, pp201–210.

Felipe-Lucia, M.R., Martín-López, B., Lavorel, S., Berraquero-Díaz, L., Escalera-Reyes, J. and Comín, F.A. (2015) 'Ecosystem services flows: Why stakeholders' power relationships matter', *PLoS One*, vol 10, art e0132232.

Ferrante, M., González, E. and Lövei, G.L. (2017) 'Predators do not spill over from forest fragments to maize fields in a landscape mosaic in central Argentina', *Ecology and Evolution*, vol 7, pp7699–7707.

Fisher, B., Turner, R.K. and Morling, P. (2009) 'Defining and classifying ecosystem services for decision making', *Ecological Economics*, vol 68, pp643–53.

Gallé R., Happe, A.K., Baillod, A.B., Tscharntke, T. and Batáry, P. (2019) 'Landscape configuration, organic management, and within-field position drive functional diversity of spiders and carabids', *Journal of Applied Ecology*, vol 56, pp63–72.

Galpern, P., Vickruck, J., Devries, J.H. and Gavin, M.P. (2020) 'Landscape complexity is associated with crop yields across a large temperate grassland region', *Agricicultre, Ecosystems and Environment*, vol 290, art 106724.

Gamfeldt L, Snäll T, Bagchi R, Jonsson, M., Gustafsson, L., Kjellander, P., Ruiz-Jaen, M.C., Fröberg, M., Stendahl, J., Philipson, C.D., Mikusiński, G., Andersson, E., Westerlund, B., Andrén, H., Moberg, F., Moen, J. and Bengtsson, J. (2013) 'Higher levels of multiple ecosystem services are found in forests with more tree species', *Nature Communications*, vol 4, art 1340.

García-Feced, C., Weissteiner, C.J., Baraldi, A., Paracchini, M.L., Maes, J., Zulian, G., Kempen, M., Elbersen, B. and Pérez-Soba, M. (2014) 'Semi-natural vegetation in agricultural land: European map and links to ecosystem service supply', *Agronomy for Sustainable Development*, vol 35, pp273–283.

Garratt, M.P.D., Senapathi, D., Coston, D.J., Mortimer, S.R. and Potts, S.G. (2017) 'The benefits of hedgerows for pollinators and natural enemies depends on hedge quality and landscape context', *Agriculture, Ecosystems and Environment*, vol 247, pp363–70.

Goldstein, J.H., Caldarone, G., Duarte, T.K., Ennaanay, D., Hannahs, N., Mendoza, G., Polasky, S., Wolny, S. and Daily, G.C. (2012) 'Integrating ecosystem-service tradeoffs into land-use decisions', *Proceedings of the National Academy of Sciences*, vol 109, pp7565–7570.

Grab, H., Poveda, K., Danforth, B. and Loeb, G. (2018) 'Landscape context shifts the balance of costs and benefits from wildflower borders on multiple ecosystem services', *Proceedings of the Royal Society B*, vol 285, art 20181102.

Grimaldi, M., Oszwald, J., Dolédec, S., del Pilar Hurtado, M., de Souza Miranda, I., de Sartre, X.A., de Assis, W.S., Castañeda, E., Desjardins, T., Dubs, F., Guevara, E., Gond, V., Lima, T.T.S., Marichal, R., Michelotti, F., Mitja, D., Noronha, N.C., Oliveira, M.N.D., Ramirez, B., Rodriguez, G., Sarrazin, M., da Silva, M.L. Jr., Costa, L.G.S., de Souza, S.L., Veiga, I., Velasquez, E. and Lavelle, P. (2014) 'Ecosystem services of regulation and support in Amazonian pioneer fronts: Searching for landscape drivers', *Landscape Ecology*, vol 29, pp311–28.

Gunton, R.M., Marsh, C.J., Moulherat, S., Malchow, A-K., Bocedi, G., Klenke, R.A. and Kunin, W.E. (2017) 'Multicriterion trade-offs and synergies for spatial conservation planning', *Journal of Applied Ecology*, vol 54, pp903–13.

Haddad, N.M., Brudvig, L.A., Clobert, J., Davies, K.F., Gonzalez, A., Holt, R.D, Lovejoy, T.E., Sexton, J.O., Austin, M.P., Collins, C.D., Cook, W.M., Damschen, E.I., Ewers, R.M., Foster, B.L., Jenkins, C.N., King, A.J., Laurance, W.F., Levey, D.G., Margules, C.R., Melbourne, B.A., Nicholls, A.O., Orrock, J.L., Song, D-X. and Townshend, J.R. (2015) 'Habitat fragmentation and its lasting impact on Earth's ecosystems', *Science Advances*, vol 1, art e1500052.

Haenke, S., Kovács-Hostyánszki, A., Fründ, J., Batáry, P., Jauker, B., Tscharntke, T. and Holzschuh, A. (2014) 'Landscape configuration of crops and hedgerows drives local syrphid fly abundance', *Journal of Applied Ecology*, vol 51, pp505–13.

Haines-Young, R. and Potschin-Young, M.B. (2018) 'Revision of the common international classification for ecosystem services (CICES V5.1): A policy brief', *One Ecosystem*, vol 3, pp1–6.

Herrmann, J.D., Bailey, D., Hofer, G., Herzog, F. and Schmidt-Entling, M.H. (2010) 'Spiders associated with the meadow and tree canopies of orchards respond differently to habitat fragmentation', *Landscape Ecology*, vol 25, pp1375–1384.

Hertzog, L.R., Boonyarittichaikij, R., Dekeukeleire, D., de Groote, S.R.E., van Schrojenstein Lantman, I.M., Sercu, B.K., Smith, H.K., de la Peña, E., Vandegehuchte, M.L., Bonte, D., Martel, A., Verheyen, K., Lens, L and Baeten, L. (2019) 'Forest fragmentation modulates effects of tree species richness and composition on ecosystem multifunctionality', *Ecology*, vol 100, pp1–9.

Holland, J.M., Douma, J.C., Crowley, L., James, L., Kor, L., Stevenson, D.R.W and Smith, B.M. (2017) 'Semi-natural habitats support biological control, pollination and soil conservation in Europe. A review', *Agronomy for Sustainable Development*, vol 37, art 31.

Holland, R.A., Eigenbrod, F., Armsworth, P.R., Anderson, B.J., Thomas, C.D. and Gaston, K.J. (2011) 'The influence of temporal variation on relationships between ecosystem services', *Biodiversity Conservation*, vol 20, pp3285–3294.

Hölting, L., Beckmann, M., Volk, M. and Cord, A.F. (2019) 'Multifunctionality assessments – more than assessing multiple ecosystem functions and services? A quantitative literature review', *Ecological Indicators*, vol 103, pp226–235.

Holzschuh, A., Steffan-Dewenter, I. and Tscharntke, T.T. (2009) 'Grass strip corridors in agricultural landscapes enhance nest-site colonization by solitary wasps', *Ecological Applications*, vol 19, pp123–32.

Holzschuh, A., Steffan-Dewenter, I. and Tscharntke, T. (2010) 'How do landscape composition and configuration, organic farming and fallow strips affect the diversity of bees, wasps and their parasitoids?', *Journal of Animal Ecology*, vol 79, pp491–500.

Isbell, F., Calcagno, V., Hector A, Connolly, J., Harpole, W.S., Reich, P.B., Scherer-Lorenzen, M., Schmid, B., Tilman, D., van Ruijven, J., Weigelt, A., Wilsey, B.J., Zavaleta, E.S. and Loreau, M. (2011) 'High plant diversity is needed to maintain ecosystem services', *Nature*, vol 477, pp199–202.

Jauker, F., Diekoetter, T., Schwarzbach, F. and Wolters, V. (2009) 'Pollinator dispersal in an agricultural matrix: Opposing responses of wild bees and hoverflies to landscape structure and distance from main habitat', *Landscape Ecology*, vol 24, pp547–55.

Jones, K.B., Zurlini, G., Kienast, F., Petrosillo, I., Edwards, T., Wade, T.G., Li, B. and Zaccarelli, N. (2013) 'Informing landscape planning and design for sustaining ecosystem services from existing spatial patterns and knowledge', *Landscape Ecology*, vol 28, pp1175–1192.

Jones, L., Norton, L., Austin, Z., Browne, A.L., Donovan, D., Emmett, B.A., Grabowski, Z.J., Howard, D.C., Jones, J.P.G., Kenter, J.O., Manley, W., Morris, C., Robinson, D.A., Short, C., Siriwardena, G.M., Stevens, C.J., Storkey, J., Waters, R.D. and Willis, G.F. (2016) 'Stocks and flows of natural and human-derived capital in ecosystem services', *Land Use Policy*, vol 52, pp151–162.

Jonsson, M., Bommarco, R., Ekbom, B., Smith, H.G., Bengtsson, J., Caballero-Lopez, B., Winqvist, C. and Olsson, O. (2014) 'Ecological production functions for biological control services in agricultural landscapes', *Methods in Ecology and Evolution*, vol 5, pp243–252.

Karp, D.S., Chaplin-Kramer, R., Meehan, T.D., et al. (2018) 'Crop pests and predators exhibit inconsistent responses to surrounding landscape composition', *Proceedings of the National Academy of Sciences*, vol 115, E7863–E7870.

Kennedy, C.M., Lonsdorf, E., Neel, M.C., et al. (2013) 'A global quantitative synthesis of local and landscape effects on wild bee pollinators in agroecosystems', *Ecology Letters*, vol 16, pp584–99.

Knoke, T., Paul, C., Hildebrandt, P., Calvas, B., Castro, L.M, Härtl, F., Döllerer, M., Hamer, U., Windhorst, D., Wiersma, Y.F., Curatola Fernández, G.F., Obermeier, W.A., Adams, J., Breuer, L., Mosandl, R., Beck, E., Weber, M., Stimm, B., Haber, W., Fürst, C. and Bendix, J. (2016) 'Compositional diversity of rehabilitated tropical lands supports multiple ecosystem services and buffers uncertainties', *Nature Communications*, vol 7, art 11877.

Koschke, L., Fürst, C., Frank, S. and Makeschin, F. (2012) 'A multi-criteria approach for an integrated land-cover-based assessment of ecosystem services provision to support landscape planning', *Ecological Indicators*, vol 21, pp54–66.

Kratschmer, S., Pachinger, B., Schwantzer, M., Paredes, D., Guernion, M., Burel, F., Nicolai, A., Strauss, P., Bauer, T., Kriechbaum, M., Zaller, Y.G and Winter, S. (2018) 'Tillage intensity or landscape features: What matters most for wild bee diversity in vineyards?', *Agriculture, Ecosystems and Environment*, vol 266, pp142–152.

Kremen, C., Williams, N.M., Bugg, R.L., Fay, J.P. and Thorp, R.W. (2004) 'The area requirements of an ecosystem service: Crop pollination by native bee communities in California', *Ecology Letters*, vol 7, pp1109–1119.

Kremen, C., Williams, N.M., Aizen, M.A.A., Gemmill-Herren, B., LeBuhn, G., Minckley, R., Packer, L., Potts, S.G., Roulston, T., Steffan-Dewenter, I., Vázquez, D.P., Winfree, R., Adams, L., Crone, E.E., Greenleaf, S.S., Keitt, T.H., Klein, A-M., Regetz, J. and Ricketts, T.H. (2007) 'Pollination and other ecosystem services produced by mobile organisms: A conceptual framework for the effects of land-use change', *Ecology Letters*, vol 10, pp299–314.

Kroll, F., Müller, F., Haase, D. and Fohrer, N. (2012) 'Rural-urban gradient analysis of ecosystem services supply and demand dynamics', *Land Use Policy*, vol 29, pp521–35.

Lamy, T., Liss, K.N., Gonzalez, A. and Bennett, E.M. (2016) 'Landscape structure affects the provision of multiple ecosystem services', *Environmetal Ressearch Letters*, vol 11, pp1–9.

Landis, D.A. (2017) 'Designing agricultural landscapes for biodiversity-based ecosystem services', *Basic and Applied Ecology*, vol 18, pp1–12.

Lansing, E., Gardiner, M.M., Landis, D.A., DiFonzo, C.D., O'Neal, M., Chacon, J.M., Wayo, M.T., Schmidt, N.P., Mueller, E.E. and Heimpel, G.E. (2009) 'Landscape diversity enhances biological control of an introduced crop pest in the north-central USA', *Ecolgcal Applications*, vol 19, pp143–54.

Larondelle, N. and Haase, D. (2013) 'Urban ecosystem services assessment along a rural-urban gradient: A cross-analysis of European cities', *Ecological Indicators*, vol 29, pp179–90.

Laterra, P., Orúe, M.E. and Booman, G.C. (2012) 'Spatial complexity and ecosystem services in rural landscapes', *Agriculture, Ecosystems and Environment*, vol 154, pp56–67.

Lautenbach, S., Kugel, C., Lausch, A. and Seppelt, R. (2011) 'Analysis of historic changes in regional ecosystem service provisioning using land use data', *Ecological Indicators*, vol 11, pp676–687.

Lefcheck JS, Byrnes JEK, Isbell F, Gamfeldt, L., Griffin, J.N., Eisenhauer, N., Hensel, M.J.S., Hector, A., Cardinale, B.J. and Duffy, J.E. (2015) 'Biodiversity enhances ecosystem multifunctionality across trophic levels and habitats', *Nature Communications*, vol 6, art 6936.

Liere, H., Kim, T.N., Werling, B.P., Meehan, T.D., Landis, D.A. and Gratton, C. (2015) 'Trophic cascades in agricultural landscapes: Indirect effects of landscape composition on crop yield', *Ecological Applications*, vol 25, pp652–661.

Linden, V.M.G., Grass, I., Joubert, E., Tscharntke, T., Weier, S.M., Taylor, P.J. (2019) 'Ecosystem services and disservices by birds, bats and monkeys change with macadamia landscape heterogeneity', *Journal of Applied Ecology*, vol 51, pp510–527.

Liss, K.N., Mitchell, M.G.E., MacDonald, G.K., Mahajan, S.L., Méthot, J., Jacob, A.L., Maguire, D.Y., Metson, G.S., Ziter, C., Dancose, K., Martins, K., Terrado, M. and Bennett, E.M. (2013) 'Variability in ecosystem service measurement: A pollination service case study', *Frontiers in Ecology and the Environment*, vol 11, pp414–422.

Liu B, Yang L, Zeng Y, Yang, F., Yang, Y. and Lu, Y. (2018) 'Secondary crops and non-crop habitats within landscapes enhance the abundance and diversity of generalist predators', *Agriculture, Ecosystems and Environment*, vol 258, pp30–39.

Liu S, Yin Y, Liu X, Cheng, F., Yang, J., Li, J., Dong, S. and Zhu, A. (2017) 'Ecosystem Services and landscape change associated with plantation expansion in a tropical rainforest region of Southwest China', *Ecological Modelling*, vol 353, pp129–138.

Luck, G.W., Chan, K.M.A.A.M., Klein, C. and Klien, C.J. (2012) 'Identifying spatial priorities for protecting ecosystem services', *F1000Research* vol 1, pp1–17.

Lyytimäki, J., Petersen, L., Normander, B. and Bezák, P. (2008) 'Nature as a nuisance? Ecosystem services and disservices to urban lifestyle', *Environmental Sciences* vol 5, pp161–172.

Macfadyen, S., Kramer, E.A., Parry, H.R. and Schellhorn, N.A. (2015) 'Temporal change in vegetation productivity in grain production landscapes: Linking landscape complexity with pest and natural enemy communities', *Ecological Entomology*, vol 40, pp56–69.

Madrigal-Martínez, S. and García, J.L.M. (2019) 'Land-change dynamics and ecosystem service trends across the central high-Andean Puna', *Scientific Reports*, vol 9, art 9688.

Maguire, D.Y., James, P.M.A., Buddle, C.M. and Bennett, E.M. (2015) 'Landscape connectivity and insect herbivory: A framework for understanding tradeoffs among ecosystem services', *Global Ecology and Conservation*, vol 4, pp73–84.

Maguire, D.Y., Buddle, C.M. and Bennett, E.M. (2016). 'Within and among patch variability in patterns of insect herbivory across a fragmented forest landscape', *PLoS One*, vol 11, art e0150843.

Manning, P., Plas, F., Soliveres, S., Allan, E., Maestre, F.T., Mace, G., Whittingham, M.J. and Fischer, M. (2018) 'Redefining ecosystem multifunctionality', *Nature Ecology and Evolution*, vol 2, pp427–436.

Martin, E.A, Reineking, B., Seo, B. and Steffan-Dewenter, I. (2013) 'Natural enemy interactions constrain pest control in complex agricultural landscapes', *Proceedings of the National Academy of Sciences*, vol 110, pp5534–5539.

Martin, E.A., Dainese, M., Clough, Y., *et al.* (2019) 'The interplay of landscape composition and configuration: New pathways to manage functional biodiversity and agroecosystem services across Europe', *Ecology Letters*, vol 22, pp1083–1094.

Martins, K.T., Gonzalez, A. and Lechowicz, M.J. (2015) 'Pollination services are mediated by bee functional diversity and landscape context', *Agriculture, Ecosystems and Environment*, vol 200, pp12–20.

Mascarenhas, A., Ramos, T.B., Haase, D. and Santos, R. (2014) 'Integration of ecosystem services in spatial planning: A survey on regional planners' views', *Landscape Ecology*, vol 29, pp1287–1300.

Mastrangelo, M.E., Weyland, F., Villarino, S.H., Barral, M.P., Nahuelhual, L. and Laterra, P. (2014) 'Concepts and methods for landscape multifunctionality and a unifying framework based on ecosystem services', *Landscape Ecology*, vol 29, pp345–358.

Millennium Ecosystem Assessment (2005) 'Ecosystems and human well-being', http://www.bioquest.org/wp-content/blogs.dir/files/2009/06/ecosystems-and-health.pdf.

Mitchell, M.G.E., Bennett, E.M. and Gonzalez, A. (2013) 'Linking landscape connectivity and ecosystem service provision: Current knowledge and research gaps', *Ecosystems*, vol 16, pp894–908.

Mitchell, M.G.E., Bennett, E.M. and Gonzalez, A. (2014a) 'Agricultural landscape structure affects arthropod diversity and arthropod-derived ecosystem services', *Agriculture, Ecosystems and Environment*, vol 192, pp144–151.

Mitchell, M.G.E., Bennett, E.M. and Gonzalez, A. (2014b) 'Forest fragments modulate the provision of multiple ecosystem services', *Journal of Applied Ecology*, vol 51, pp909–918.

Mitchell, M.G.E., Bennett, E.M. and Gonzalez, A. (2015a) 'Strong and nonlinear effects of fragmentation on ecosystem service provision at multiple scales', *Environmental Research Letters*, vol 10, art 94014.

Mitchell, M.G.E., Suarez-Castro, A.F., Martinez-Harms, M., Maron, M., McAlpine, C., Gaston, K.J., Johansen, K. and Rhodes, J.R. (2015b) 'Reframing landscape fragmentation's effects on ecosystem services', *Trends in Ecology and Evolution*, vol 30, pp190–198.

Mitchell, M.G.E., Johansen, K., Maron, M., McAlpine, C.A., Wu, D. and Rhodes, J.R. (2018) 'Identification of fine scale and landscape scale drivers of urban aboveground carbon stocks using high-resolution modeling and mapping', *Science of the Total Environment*, 622–623, pp57–70.

Montoya, D., Haegeman, B., Gaba, S., de Mazancourt, C., Bretagnolle, V. and Loreau, M. (2019) 'Trade-offs in the provisioning and stability of ecosystem services in agroecosystems', *Ecological Applications*, vol 29, art e01853.

Moran, C. and Catterall, C.P. (2014) 'Responses of seed-dispersing birds to amount of rainforest in the landscape around fragments', *Conservation Biology*, vol 28, pp551–560.

Mulder, C., Bennett, E.M., Bohan, D.A., Bonkowski, M., Carpenter, S.R., Chalmers, R., Cramer, W., Durance, I., Eisenhauer, N., Fontaine, C., Haughton, A.J., Hettelingh, J-P., Hines, J., Ibanez, S., Jeppesen, E., Krumins, J.A., Ma, a., Mancinelli, G. and Woodward, G. (2015) '10 years later: Revisiting priorities for science and society a decade after the Millennium Ecosystem Assessment', in Woodward, G. and Bohan, D.A. (eds) *Advances in Ecological Research*. Oxford Academic Press, Oxford.

Naidoo, R., Balmford, A., Costanza, R., Fisher, B., Green, R.E., Lehner, B., Malcolm, T.R. and Ricketts, T.H. (2008) 'Global mapping of ecosystem services and conservation priorities', *Proceedings of the National Academy of Sciences*, vol 105, pp9495–9500.

Nelson, E., Mendoza, G., Regetz, J., Polasky, S., Tallis, H., Cameron, D.R., Chan, K.M.A., Daily, G.C., Goldstein, J., Kareiva, P.M., Lonsdorf, E., Naidoo, R., Ricketts, T.H. and Shaw, M.R. (2009) 'Modeling multiple ecosystem services, biodiversity conservation, commodity production, and tradeoffs at landscape scales', *Frontiers in Ecology and the Environment*, vol 7, pp4–11.

Ng, C.N., Xie, Y.J. and Yu, X.J. (2013) 'Integrating landscape connectivity into the evaluation of ecosystem services for biodiversity conservation and its implications for landscape planning', *Applied Geography*, vol 42, pp1–12.

Oteros-Rozas, E., Martín-López, B., Fagerholm, N., Bieling, C. and Plieninger, T. (2018) 'Using social media photos to explore the relation between cultural ecosystem services and landscape features across five European sites', *Ecological Indicators*, vol 94, pp74–86.

Pak, D., Iverson, A.L., Ennis, K.K., Gonthier, D.J. and Vandermeer, J.H. (2015) 'Parasitoid wasps benefit from shade tree size and landscape complexity in Mexican coffee agroecosystems', *Agriculture Ecosystems and Environment*, vol 206, pp21–32.

Palacios-Agundez, I., Onaindia, M., Barraqueta, P. and Madariaga, I. (2015) 'Provisioning ecosystem services supply and demand: The role of landscape management to reinforce supply and promote synergies with other ecosystem services', *Land Use Policy*, vol 47, pp145–155.

Perez-Alvarez, R., Nault, B.A. and Poveda, K. (2018) 'Contrasting effects of landscape composition on crop yield mediated by specialist herbivores', *Ecological Applications*, vol 28, pp842–853.

Plas F van der, Allan E, Fischer M, *et al.* (2019) 'Towards the development of general rules describing landscape heterogeneity–multifunctionality relationships', *Journal of Appled Ecology*, vol 56, pp168–179.

Plećaš, M., Gagić, V., Janković, M., Petrović-Obradović, O., Kavallieratos, N.G., Tomanović, Z., Thiese, C., Tscharntke, T. and Ćetković, A. (2014) 'Landscape composition and configuration influence cereal aphid-parasitoid-hyperparasitoid interactions and biological control differentially across years', *Agriculture, Ecosystems and Environment*, vol 183, pp1–10.

Plieninger, T., Bieling, C., Fagerholm, N., Byg, A., Hartel, T., Hurley, P., López-Santiago, C.A., Nagabhatla, N., Oteros-Rozas, E., Raymond, C.M., van der Horst, D. and Huntsinger, L. (2015) 'The role of cultural ecosystem services in landscape management and planning', *Current Opinion in Environmental Sustainability*, vol 14, pp28–33.

Potschin, M.B. and Haines-Young, R.H. (2011) 'Ecosystem services: Exploring a geographical perspective', *Progress in Physical Geography*, vol 35, pp575–94.

Power, E.F., Jackson, Z. and Stout, J.C. (2016) 'Organic farming and landscape factors affect abundance and richness of hoverflies (Diptera, Syrphidae) in grasslands', *Insect Conservation and Diversity*, vol 9, pp244–253.

Programme UNE (2010) *The Economics of Ecosystems and Biodiversity: Ecological and Economic Foundations.* Routledge, New York.

Puech, C., Poggi, S., Baudry, J. and Aviron, S. (2015) 'Do farming practices affect natural enemies at the landscape scale?', *Landscape Ecology*, vol 30, pp125–40.

Qiu, J. and Turner, M.G. (2013) 'Spatial interactions among ecosystem services in an urbanizing agricultural watershed', *Proceedings of the National Academy of Sciences*, vol 110, pp12149–12154.

Qiu, J. and Turner, M.G. (2015) 'Importance of landscape heterogeneity in sustaining hydrologic ecosystem services in an agricultural watershed', *Ecosphere*, vol 6, pp1–19.

QUINTESSENCE Consortium (2016) 'Networking our way to better ecosystem service provision', *Trends in Ecology and Evolution*, vol 31, pp105–115.

Rand, T.A., Tylianakis, J.M. and Tscharntke, T. (2006) 'Spillover edge effects: The dispersal of agriculturally subsidized insect natural enemies into adjacent natural habitats', *Ecology Letters*, vol 9, pp603–614.

Raudsepp-Hearne, C., Peterson, G.D. and Bennett, E.M. (2010) 'Ecosystem service bundles for analyzing tradeoffs in diverse landscapes', *Proceedings of the National Academy of Sciences*, vol 107, pp5242–5247.

Reich, P.B., Tilman, D., Isbell, F., Mueller, K., Hobbie, S.E., Flynn, D.F.B. and Eisenhauer, N. (2012) 'Impacts of biodiversity loss escalate through time as redundancy fades', *Science*, vol 336, pp589–592.

Renard, D., Rhemtulla, J.M. and Bennett, E.M. (2015) 'Historical dynamics in ecosystem service bundles', *Proceedings of the National Academy of Sciences*, vol 112, pp13411–13416.

Ricci, B., Lavigne, C., Alignier, A., Aviron, S., Biju-Duval, L., Bouvier, J.C., Choisis, J-P., Franck, P., Joannon, A., Ladet, S., Mezerette, F., Plantegenest, M., Savary, G., Thomas, C., Vialatte, A. and Petit, S. (2019) 'Local pesticide use intensity conditions landscape effects on biological pest control', *Proceedings of the Royal Society B*, vol 286, art 20182898.

Ricketts, T.H., Regetz, J., Steffan-Dewenter, I., Cunningham, S.A., Kremen, C., Bogdanski, A., Gemmill-Herren, B., Greenleaf, S.S., Klein, A.M., Mayfield, M.M., Morandin, L.A., Ochieng', A. and Viana, B.F. (2008) 'Landscape effects on crop pollination services: Are there general patterns?', *Ecology Letters*, vol 11, pp499–515.

Ricketts, T.H., Watson, K.B., Koh, I., Ellis, A.M., Nicholson, C.C., Posner, S., Richardson, L.L. and Sonter, L.J. (2016) 'Disaggregating the evidence linking biodiversity and ecosystem services', *Nature Communications*, vol 7, art 13106.

Rodríguez-San Pedro, A., Rodríguez-Herbach, C., Allendes, J.L., Chaperon, P.N., Beltrán, C.A. and Grez, A.A. (2019) 'Responses of aerial insectivorous bats to landscape composition and heterogeneity in organic vineyards', *Agriculture, Ecosystems and Environment*, vol 277, pp74–82.

Rossetti, M.R., Tscharntke, T., Aguilar, R. and Batáry, P. (2017) 'Responses of insect herbivores and herbivory to habitat fragmentation: A hierarchical meta-analysis', *Ecology Letters*, vol 20, pp264–272.

Rusch, A., Valantin-Morison, M., Sarthou, J-P. and Roger-Estrade, J. (2011) 'Multi-scale effects of landscape complexity and crop management on pollen beetle parasitism rate', *Landscape Ecology*, vol 26, pp473–486.

Rusch, A., Bommarco, R., Jonsson, M., Smith, H.G. and Ekbom, B. (2013) 'Flow and stability of natural pest control services depend on complexity and crop rotation at the landscape scale', *Journal of Applied Ecology*, vol 50, pp345–354.

Rusch, A., Chaplin-Kramer, R., Gardiner, M.M., Hawro, V., Holland, J., Landis, D., Thies, C, Tscharntke, T., Weisser, W.W., Winqvist, C., Woltz, M. and Bommarco, R. (2016) 'Agricultural landscape simplification reduces natural pest control: A quantitative synthesis', *Agriculture, Ecosystems and Environment*, vol 221, pp198–204.

Samnegård, U., Hambäck, P.A., Lemessa, D., Nemomissa, S. and Hylander, K. (2016) 'A heterogeneous landscape does not guarantee high crop pollination', *Proceedings of the Royal Society B*, vol 283, art 20161472–9.

Saunders, M.E. and Luck, G.W. (2016) 'Limitations of the ecosystem services versus disservices dichotomy', *Conservation Biology*, vol 30, pp1363–1365.

Schäckermann, J., Pufal, G., Mandelik, Y. and Klein, A.M. (2015) 'Agro-ecosystem services and dis-services in almond orchards are differentially influenced by the surrounding landscape', *Ecological Entomology*, vol 40, pp12–21.

Schmidt, M.H., Thies, C., Nentwig, W. and Tscharntke, T. (2008) 'Contrasting responses of arable spiders to the landscape matrix at different spatial scales', *Journal of Biogeography*, vol 35, pp157–166.

Schüepp, C., Herrmann, J.D., Herzog, F. and Schmidt-Entling, M.H. (2011) 'Differential effects of habitat isolation and landscape composition on wasps, bees, and their enemies', *Oecologia*, vol 155, pp713–721.

Serna-Chavez, H.M., Schulp, C.J.E., van Bodegom, P.M., Bouten, W., Verburg, P.H. and Davidson, M.D. (2014) 'A quantitative framework for assessing spatial flows of ecosystem services', *Ecological Indicators*, vol 39, pp24–33.

Shackelford, G., Steward, P.R., Benton, T.G., Kunin, W.E., Potts, S.G., Biesmeijer, J.C. and Sait, S.M. (2013) 'Comparison of pollinators and natural enemies: A meta-analysis of landscape and local effects on abundance and richness in crops', *Biological Reviews*, vol 88, pp1002–1021.

Spake, R., Lasseur, R., Crouzat, E., Bullock, J.M., Lavorel, S., Parks, K.E., Schaafsma, M., Bennett, E.M., Maes, J., Mulligan, M., Mouchet, M., Peterson, G.D., Schulp, C.J.E., Thuiller, W., Turner, M.G., Verburg, P.H. and Eigenbrod, F. (2017) 'Unpacking ecosystem service bundles: Towards predictive mapping of synergies and trade-offs between ecosystem services', *Global Environmental Change*, vol 47, pp37–50.

Stürck, J., Poortinga, A. and Verburg, P.H. (2014) 'Mapping ecosystem services: The supply and demand of flood regulation services in Europe', *Ecological Indicators*, vol 38, pp198–211.

Stürck, J. and Verburg, P.H. (2017) 'Multifunctionality at what scale? A landscape multifunctionality assessment for the European Union under conditions of land use change', *Landscape Ecology*, vol 32, pp481–500.

Stutz, S. and Entling, M.H. (2011) 'Effects of the landscape context on aphid-ant-predator interactions on cherry trees', *Biological Control*, vol 57, pp37–43.

Sutherland, I.J., Bennett, E.M. and Gergel, S.E. (2016) 'Recovery trends for multiple ecosystem services reveal non-linear responses and long-term tradeoffs from temperate forest harvesting', *Forest Ecology and Management*, vol 374, pp61–70.

Syrbe, R-U. and Walz, U. (2012) 'Spatial indicators for the assessment of ecosystem services: Providing, benefiting and connecting areas and landscape metrics', *Ecological Indicators*, vol 21, pp80–88.

Tallis, H., Goldman, R.L., Uhl, M. and Brosi, B. (2009) 'Integrating conservation and development in the field: Implementing ecosystem service projects', *Frontiers in Ecology and the Environment*, vol 7, pp12–20.

Tallis, H., Mooney, H., Andelman, S., Balvanera, P., Cramer, W., Karp, D., Polasky, S., Reyers, B., Ricketts, T., Running, S., Thonicke, K., Tietjen, B. and Walz, A. (2012). 'A global system for monitoring ecosystem service change', *Bioscience*, vol 62, pp977–986.

Tamburini, G., de Simone, S., Sigura, M., Boscutti, F. and Marini, L. (2016) 'Soil management shapes ecosystem service provision and trade-offs in agricultural landscapes', *Proceedings of the Royal Society B*, vol 283, art 20161369.

Termorshuizen, J.W. and Opdam, P. (2009) 'Landscape services as a bridge between landscape ecology and sustainable development', *Landscape Ecology*, vol 24, pp1037–1052.

Terrado, M., Tauler, R. and Bennett, E.M. (2015) 'Landscape and local factors influence water purification in the Monteregian agroecosystem in Québec, Canada', *Regional Environmental Change*, vol 15, pp1743–1755.

Thomas, A., Masante, D., Jackson, B., Cosby, B., Emmett, B. and Jones, L. (2020) 'Fragmentation and thresholds in hydrological flow-based ecosystem services', *Ecological Applications*, vol 30, art e02046.

Tratalos, J., Fuller, R.A., Warren, P.H., Davies, R.G. and Gaston, K.J. (2007) 'Urban form, biodiversity potential and ecosystem services', *Landscape and Urban Planning*, vol 83, pp308–317.

Trichard, A., Alignier, A., Biju-Duval, L. and Petit, S. (2013) 'The relative effects of local management and landscape context on weed seed predation and carabid functional groups', *Basic and Applied Ecology*, vol 14, pp235–245.

Tscharntke, T., Klein, A.M, Kruess A, Steffan-Dewenter, I. and Thies, C. (2005) 'Landscape perspectives on agricultural intensification and biodiversity: Ecosystem service management', *Ecology Letters*, vol 8, pp857–874.

Tschumi, M., Ekroos, J., Hjort, C., Smith, H.G. and Birkhofer, K. (2018) 'Predation-mediated ecosystem services and disservices in agricultural landscapes', *Ecological Applications*, vol 28, pp2109–2118.

Turner, M.G., Donato, D.C. and Romme, W.H. (2013) 'Consequences of spatial heterogeneity for ecosystem services in changing forest landscapes: Priorities for future research', *Landscape Ecology*, vol 28, 1081–1097.

Valdés A, Lenoir J, Frenne P De, *et al.* (2020) 'High ecosystem service delivery potential of small woodlands in agricultural landscapes', *Journal of Applied Ecology*, vol 57, pp4–16.

Vallés-Planells, M., Galiana, F. and van Eetvelde, V. (2014) 'A classification of landscape services to support local landscape planning', *Ecology and Society*, vol 19, art 44.

Van Berkel, D.B. and Verburg, P.H. (2014) 'Spatial quantification and valuation of cultural ecosystem services in an agricultural landscape', *Ecological Indicators*, vol 37, pp163–74.

Verhagen, W., van Teeffelen, A.J.A.A., Compagnucci, A.B, Poggio, L., Gimona, A. and Verburg, P.H. (2016) 'Effects of landscape configuration on mapping ecosystem service capacity: A review of evidence and a case study in Scotland', *Landscape Ecology*, vol 31, pp1457–1479.

Villamagna, A.M., Angermeier, P.L. and Bennett, E.M. (2013) 'Capacity, pressure, demand, and flow: A conceptual framework for analyzing ecosystem service provision and delivery', *Ecological Complexity*, vol 15, pp114–121.

Watson, K.B., Galford, G.L., Sonter, L.J., Koh, I. and Ricketts, T.H. (2019) 'Effects of human demand on conservation planning for biodiversity and ecosystem services', *Conservation Biology*, vol 33, pp942–952.

Werling, B.P. and Gratton, C. (2010) 'Local and broadscale landscape structure differentially impact predation of two potato pests', *Ecological Applications*, vol 20, pp1114–1125.

Wilkerson, M.L., Mitchell, M.G.E., Shanahan, D., Wilson, K.A., Ives, C.D., Lovelock, C.E. and Rhodes, J.R. (2018) 'The role of socio-economic factors in planning and managing urban ecosystem services', *Ecosystem Services*, vol 31, pp102–110.

Williams, A. and Hedlund, K. (2013) 'Indicators of soil ecosystem services in conventional and organic arable fields along a gradient of landscape heterogeneity in southern Sweden', *Applied Soil Ecology*, vol 65, pp1–7.

Wolff, S., Schulp, C.J.E., Kastner, T. and Verburg, P.H. (2017) 'Quantifying Spatial Variation in Ecosystem Services Demand: A Global Mapping Approach', *Ecological Economics*, vol 136, pp14–29.

Woltz, J.M., Isaacs, R. and Landis, D.A. (2012) 'Landscape structure and habitat management differentially influence insect natural enemies in an agricultural landscape', *Agriculture, Ecosystems and Environment*, vol 152, 40–49.

Wood, S.L.R., Jones, S.K., Johnson, J.A., *et al.* (2018) 'Distilling the role of ecosystem services in the Sustainable Development Goals', *Ecosystem Services*, vol 29, pp70–82.

Wu, J. (2013) 'Landscape sustainability science: Ecosystem services and human well-being in changing landscapes', *Landscape Ecology*, vol 2, pp999–1023.

Yang, X. (2012) 'An assessment of landscape characteristics affecting estuarine nitrogen loading in an urban watershed', *Journal of Environmental Management*, vol 94, pp50–60.

Ziter, C., Bennett, E.M. and Gonzalez, A. (2013) 'Functional diversity and management mediate aboveground carbon stocks in small forest fragments', *Ecosphere*, vol 4, pp1–21.

Zou, Y., Bianchi, F.J.J.A., Jauker, F., Xiao, H., Chen, J., Cresswell, J., Luo, S., Huang, J., Deng, X., Hou, L. and van der Werf, W. (2017) 'Landscape effects on pollinator communities and pollination services in small-holder agroecosystems', *Agricultre, Ecosystems and Environment*, vol 246, pp109–116.

21

Riverscapes

Todd Lookingbill, Kimberly Meitzen, and Jason P. Julian

Background

From its beginnings, landscape ecology has emphasized the classification of terrestrial landscapes into distinct units (Christian 1958, Moss 1975, Naveh and Lieberman 1994). The recognition and characterization of the heterogeneity of these landscapes has been a cornerstone of the science. In contrast, rivers have historically been viewed within landscape ecology as more homogeneous systems than their terrestrial counterparts. Their spatial complexity and temporal variability have often been grossly oversimplified. Only within the last couple of decades have landscape ecologists begun applying a more spatially explicit lens to lotic systems (Poole 2002, Fausch et al. 2002, Wiens 2002), though the foundations for such a framework emerged much earlier within the arts and through scientific applications in stream ecology and fluvial geomorphology.

Riverscapes were imagined more than 500 years ago in both Western and Eastern civilizations in the visual arts; paintings captured their functional and aesthetic qualities. The term *riverscape* was appropriated from art to science by Ward (1998) to describe riverine landscapes and the structural and functional interactions between landscape pattern, stream ecology, and scale. Earlier studies by Frissell et al. (1986), Ward (1989), and Schlosser (1991) laid the groundwork for this perspective by considering the effect of both small-scale factors (channel cross-section to reach-scale dynamics) and large-scale factors (network-scale, longitudinal dynamics) on the spatial heterogeneity and hierarchy of functional stream habitats. These studies also described important links between stream ecology and fluvial geomorphology with a focus on connectivity and the transfer/movement of energy, matter, and biota within watersheds. Schlosser (1991) made key observations of the importance of human–environment factors, and Fausch et al. (2002) helped tie these ideas with Ward (1989) into a framework for using hierarchy theory to manage and conserve stream ecosystems. These original ideas continue to motivate modern riverscape-focused research (Wellemeyer et al. 2019, Perkin et al. 2019a).

While many early riverscape research applications focused on stream ecology, riverscapes are also a focus in geomorphology (Belletti et al. 2014), remote sensing (Carbonneau et al. 2012), and social-ecological management of river systems (Dunham et al. 2018). Reviewing traditional riverscape theory and incorporating modern advances in riverscape applications provides an opportunity to combine methods and technologies across river-related disciplines and serves as a catalyst for collaborative efforts aimed at riverine management and conservation (Erős and Lowe 2019).

In this chapter, we begin with a brief review of recent publication trends within riverscapes and provide a summary of the term's use following Ward (1998). We next consider two different organizational frameworks to describe riverscape heterogeneity (Figure 21.1). We first consider riverscapes from a hierarchical perspective, increasing in scale from patches to watersheds. Within this framework, we focus on cross-scale interactions (CSIs) as an area of research in which landscape ecology continues to make significant contributions. We then provide a longitudinal overview of riverscapes from headwater regions to large, coastal deposition zones. The importance of (dis)continuums and disruptions is highlighted here.

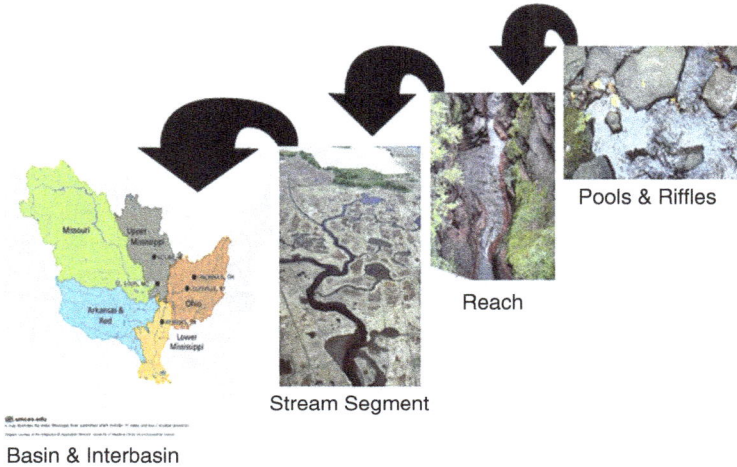

Pools & Riffles

Reach

Stream Segment

Basin & Interbasin

Hierarchical Model of Riverscapes

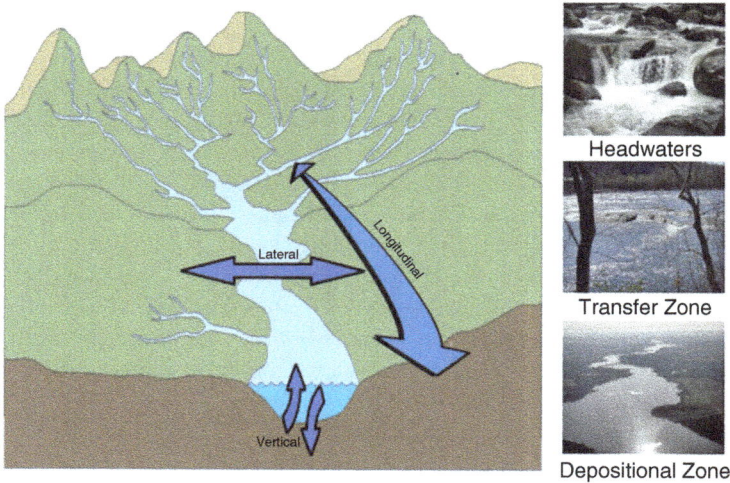

Headwaters

Longitudinal

Lateral

Transfer Zone

Vertical

Depositional Zone

Longitudinal Model of Riverscapes

Figure 21.1 Common organization frameworks for studying riverscapes include hierarchical (e.g., Frissell et al. 1986) and longitudinal (e.g., Schumm 1977) models. This review emphasizes the cross-scale interactions and cumulative downstream impacts that can be extrapolated from these models. (Symbols and images courtesy of the Integration and Application Network, University of Maryland Center for Environmental Science; ian.umces.edu/symbols/)

We end with a discussion of the future of riverscape ecology that ties theory to practice by providing examples of riverscape management and conservation that cross scales and transcend traditional boundaries. In this conclusion, we emphasize new directions, including the emerging field of using riverscapes in a social-ecological context for managing river resources.

Recent trends in riverscape research

Following Ward's (1998) use of the term *riverscape*, it has slowly become a buzzword, appearing in a wide variety of river-related research and management applications. A June 2020 EBSCO search for the use of 'riverscape' within a paper's title yielded 116 peer-reviewed articles. We grouped these papers relative to nine different themes and plotted their frequency over time (Figure 21.2). Although this analysis only captures the term's use within a title, it provides a good representation of the diversity of applications for the riverscape perspective.

Fausch et al. (2002) conceivably initiated the widespread use of the term within contemporary stream ecology literature following their foundational paper on how key concepts from landscape ecology could be applied to river systems and specifically, the conservation of aquatic biota and their habitats. This paper elucidated the importance of considering a hierarchy of scales from micro-habitat to drainage basin when examining intermediate-scale dynamics – which is recognized as a critical scale for many biotic interactions. These cross-scale concepts have since become a basis for examining population, community, and habitat dynamics (Curtis et al. 2018, Perkin et al. 2019a), organism movement (Dyer and Brewer 2020), genetics (Lapointe, 2011), nutrient cycling (Becker et al. 2017), and pollution distributions (Merriam et al. 2015). This perspective also motivated the development of stream classification systems (McManamay et al. 2018) and computational models that can be applied at multiple scales to quantify spatially explicit stream ecology functions involving connectivity, heterogeneity, and habitat matrix and patch dynamics (Erős and Grant 2015, Erős and Lowe 2019).

A common thread within geomorphology is using remote sensing to map riverscapes at a variety of spatial scales. Carbonneau et al. (2012) applied the concepts from Fausch et al. (2002) into a remote-sensing derived 'Fluvial Information System' to link high-resolution (sub-meter spatial scale) lidar and imagery-derived hydraulic variables to reach and segment-scale habitat metrics including connectivity and patchiness. Deitrich (2016) used structure-from-motion

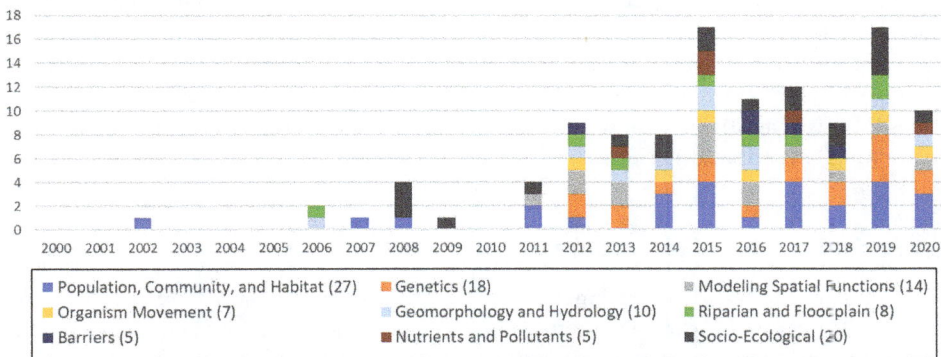

Figure 21.2 Frequency of the use of riverscapes in peer-reviewed journal titles categorized by research focus since Ward (1998). The number in parenthesis indicates the number for each color-coded category.

photogrammetry to map river morphology at the spatial scale of centimeters, while Demarchi et al. (2020) used ~2.5-m spatial scale imagery to map geomorphic-based riverscapes for multiple rivers at regional scales across Europe. Riparian and floodplain studies of riverscapes have largely focused on flood dynamics, resource exchanges, restoration, and management applications (Wohl 2014).

Common to most of the papers identified in our query is a persistent link between human impacts and altered watershed, stream, and ecological conditions (Allan 2004, Perkin et al. 2019b). This connection was especially apparent in the barrier-themed papers, which included studies on multi-stakeholder decision-making processes in dam removal (Grabowski et al. 2017), barrier removal prioritization planning models (Erős et al. 2017), and dam-altered habitats (Brenkman et al. 2012, Aguiar et al. 2016). About a decade after riverscapes became a mainstay within stream ecology, the term emerged in the socio-ecological literature and has become a conceptual basis for including dimensions of human behavior and perception into river management (Dunham et al. 2018, Hand et al. 2018) and for urban river park planning (Levin-Keitel 2014, Milinkovic et al. 2019). Urban planners, however, might suggest that this application was initiated first by Pomeroy et al. (1983) with regard to their cultural perceptions of riverscapes and aesthetics.

While riverscape applications and perspectives expanded amongst a variety of river-related interests, the term is minimally encountered within the landscape ecology–specific literature. A similar search for 'riverscape' in the title or keywords of the journal *Landscape Ecology* yielded only 21 results over the period of publication (July 1987 – June 2020). A recent topical collection under the section 'Landscape Ecology of Aquatic Systems' within the journal *Current Landscape Ecology Reports* included only one paper focused on rivers, while the others addressed marine or lentic systems. The one paper dedicated to rivers (Erős and Lowe 2019) summarized the history of riverscape applications in stream ecology and proposed re-evaluating the complex process-pattern hierarchy and scale-dependent relationships using newer, more sophisticated geostatistical modeling techniques, including spatially explicit graph theory and statistical stream network models. Erős and Lowe (2019) contend that these methods can elucidate new relationships in spatial and temporal heterogeneity and habitat connectivity and better inform management decisions relevant at both patch scales and entire basin scales.

In addition to this summary of peer-reviewed papers that contain the term 'riverscape' within their titles, there are numerous other studies that include these same concepts as core to the paper's content. Most notably, these include Wiens (2002) and Poole (2002), who – in combination with Fausch et al. (2002) – set the stage for the abundance of contemporary references over the past two decades. Others include the River Styles Framework (Brierly and Fryirs, 2005), Riverine Ecosystem Synthesis (Thorp et al. 2006), and reviews of riverscape-based environmental flow applications (Meitzen et al. 2013) and other ecology-influenced stream classification schemes (Melles et al. 2012). Additional studies are referenced through this chapter to provide a more thorough overview of the larger body of literature on riverscapes.

Hierarchical model of riverscapes

Rivers are generally scaled to the amount of water and sediment they carry (Leopold and Maddock 1953); however, channel vegetation (riparian and in-stream) and anthropogenic disturbances (watershed and in-stream) can influence river form and function (Julian et al. 2016, Manning et al. 2020). Accordingly, rivers can be considered the funnels and retainers of a landscape's history, reflecting all that has happened within the watershed over both space and time. Because river systems are shaped by interactions among geomorphic processes, biological communities, biogeochemical processes, weather patterns, and anthropogenic land uses over multiple spatiotemporal scales (Brierley et al. 2010), they are an excellent case study to investigate CSIs with climate and land use changes. CSIs occur when processes at one spatial or temporal scale interact with processes at finer or broader scales (Soranno et al. 2014). In systems where CSIs are connected, a change in an environmental driver (e.g., climate or land use) can result in positive feedbacks and cascading events that lead to dramatic and widespread changes in system dynamics (Carpenter and Turner 2000, Peters et al. 2007).

Through a combination of studies, Julian and colleagues (2008, 2011) showed how land use changes at the watershed scale (conversion from forest to agriculture) and the reach scale (removal of riparian trees) interacted to change two fluvial processes (aquatic light transmission and sediment runoff) that also interact at patch and reach scales. Given that rivers form highly connected networks across the landscape, these positive feedbacks ultimately led to changes in water quality and primary productivity that propagated all the way up to the watershed scale.

These hierarchical interactions and cascades are too often ignored in policy decision-making. For example, water policy targeted at localized point and non-point source reductions and reach-scale restoration have had some success in reducing local contaminant loading; however, these successes generally have not translated into improved conditions for large rivers and estuarine receiving bodies (Bernhardt et al. 2005, Lookingbill et al. 2009). Instead, whole-watershed approaches that consider CSIs are needed to achieve meaningful and lasting restoration of riverine resources (Palmer 2009, McCluney et al. 2014). These approaches include stormwater management in upper parts of the watershed, especially within urban core areas, where sources of nitrogen loads include lawns and impervious surfaces (Paul and Meyer 2001). The complexity of this problem also requires an interdisciplinary team of researchers to span the myriad social and physical connections that dictate land use practices, nutrient fluxes, and policy decisions within urban watersheds.

We illustrate a hierarchical approach to river management with two recent, community-based initiatives undertaken at drastically different scales and show how river restoration efforts can be strategically nested. The first is the Envision the James (ETJ) initiative, which is an interdisciplinary effort, undertaken at the large river basin level, to study, conserve, and restore one of America's most iconic watersheds (Figure 21.3). The initiative is guided by the community's multiple objectives to improve water quality, enhance wildlife habitat, protect important landscapes and viewsheds, improve heritage-based tourism, and increase public access and linkage to recreational resources (Chesapeake Conservancy 2013). To accomplish these diverse goals, the initiative mobilized dozens of academic, government, private business, and non-profit organizations with unique knowledge, expertise, and interests in the watershed. The ETJ infrastructure provides the network for these groups to interact to tackle projects at multiple spatial scales.

Nested within large initiatives like the ETJ are many local projects undertaken at the reach scale, such as the Gambles Mill Eco-Corridor (GMEC) restoration in Richmond, Virginia. While the long-term benefits of stream restoration projects are still uncertain given the many individual and combined stressors confronting urban streams (Palmer et al. 2010), the likelihood of project success is increased when the goals are broadened beyond a single chemical, physical, or biological objective. At the local level, the GMEC project is aligned with the City of Richmond's Clean Water Plan, a bold initiative, begun in 2014, to become one of the first cities in the nation to create a comprehensive watershed management strategy that integrates what are typically separate processes for managing wastewater, stormwater, and combined sewer overflows (RVAH2O 2017). The plan addresses water quality and related quality of life issues associated with recreational opportunities, economic prosperity, and public health. Stream and riparian area restoration are two of the primary strategies recommended for accomplishing plan

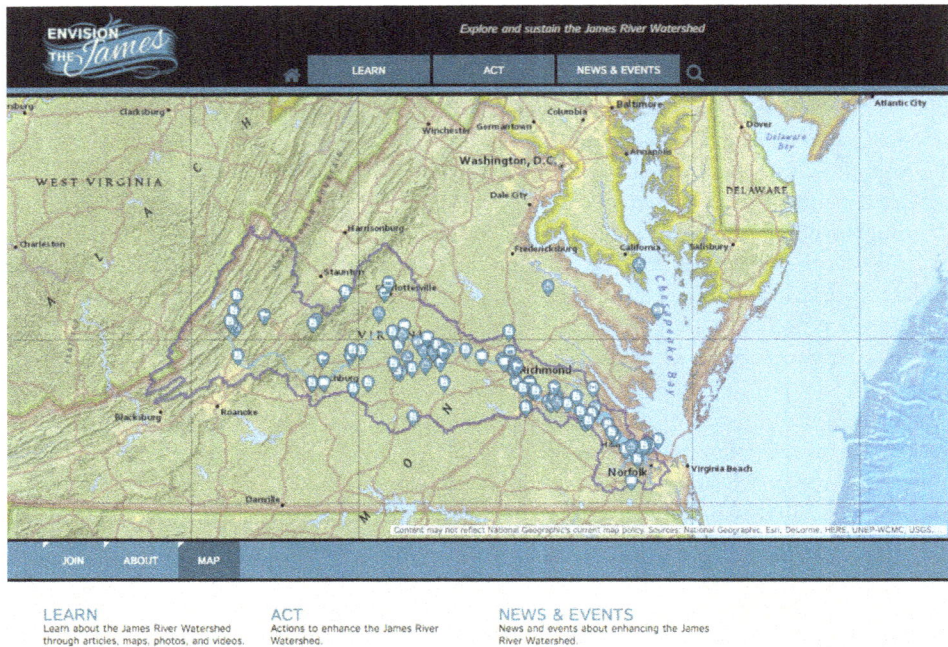

Figure 21.3 Envision the James provides an example of a collaborative initiative created to develop a regionally and locally endorsed community action agenda for the James River watershed. The many state, local, and federal government, non-profit, academic, and business organizations engaged in the initiative are represented on the map (www.envisionthejames.org).

goals, but projects like the GMEC also include social goals such as the co-creation of a multi-use recreational trail to improve public access to and awareness of the creek and ultimately, the James River and its management concerns (Figure 21.4).

Scaling back up to the whole basin, Courtenay and Lookingbill (2014) have identified similar opportunities for co-locating trails and riparian conservation and restoration efforts within the densely settled James River watershed using principles of circuit theory, a spatial analysis tool that builds upon least cost path modeling to identify potential ecological and recreational flowpaths across the landscape. A CSI approach to these types of reach-scale management efforts would consider the many interactions and constraints among projects at multiple levels (Rios and Bailey 2006, Burke et al. 2009).

Longitudinal model of riverscapes

In switching perspectives to a longitudinal view of riverscapes, Hawkes (1975) characterized rivers into discrete and predictable zones from headwater to coastal regions. These zones are arrayed along continuous morphological and ecological gradients according to the River Continuum Concept (RCC) (Vannote et al. 1980). The longitudinal impacts of disruptions to these gradients associated with dams has been broadly characterized, for example, by the Serial Discontinuity Concept (Ward and Stanford, 1983). More recent riverine classification systems invoke a range of hydrological, geomorphic, and ecological factors that draw heavily

Figure 21.4 The Gambles Mill Eco-Corridor project creates a multi-functional riverscape as part of the restoration of Little Westham Creek. The project includes a restored stream, a pollinator meadow, an enhanced community garden, outdoor classrooms, gathering spaces, and reflection areas. The project is designed to complement and reinforce clean water and associated goals at both local and watershed scales as articulated by the City of Richmond's Clean Water Plan and the Envision the James Watershed Initiative.

from principles of landscape ecology and recognize the importance of landscape connectivity and network position (Elosegi et al. 2010, Melles et al. 2012, Rinaldi et al. 2016, Julian et al. 2016, McManamay et al. 2018).

Headwaters

One of the earliest, but still enduring, riverine classification systems categorizes streams based on network position (Strahler 1957). Landscape ecology has played an important role in identifying the spatial heterogeneity and conservation value of different components of the river network. For example, first- and second-order (headwater) streams contain high levels of habitat heterogeneity (Lowe and Likens 2005) and are therefore targets in many protected areas strategies (Saunders et al. 2002). As described by the RCC, the functions provided by mountain headwater streams are notably valuable. These include the regulation of sediment export, the retention of nutrients, the processing of terrestrial organic matter, and the establishment of the chemical signature for water quality in the landscape (Lowe and Likens 2005). Intact headwater streams are important to maintaining downstream energy fluxes and water quality, as first-order streams contribute between 43% and 87% of the loadings to second-order streams and up to 40% of the loadings to sixth-order streams (Alexander et al. 2007).

In source zones, mountain headwater streams are vulnerable to forest clearing, mining, and other anthropogenic disturbances. Linear anthropogenic disturbances like pipelines can be particularly impactful to riverscapes in these regions. As oil and gas pipelines become increasingly common, the impacts of these features on riverscape attributes such as connectivity need to be quickly assessed and if necessary, mitigated (Richardson et al. 2017). As an example, the number of first- and second-order headwater stream crossings is one simple measure of the potential impact of a natural gas pipeline proposed for construction across a mixture of private and public lands in the eastern United States (Figure 21.5). The analysis indicates the large number of stream crossings that would be required within the dendritic stream networks of the Monongahela National Forest. Concerns about the dredge-and-fill activities at these crossings ultimately contributed to the cancellation of the $8-billion project after 6 years of planning, permitting, and court battles. Additional hydrologic concerns were tied to the prevalence of karst geomorphology and the intermixing of surface and groundwaters in the soluble limestone of the Valley and Ridge Physiographic Province within Highland and Augusta Counties. Other disturbances associated with mining and agricultural activities in the basin were another important consideration, as the cumulative effects of multiple disturbances can have profound effects on downstream hydrology when aggregated to the watershed scale (Ferrari et al. 2009).

Of course, not all headwater streams are located within mountain contexts, and headwater streams embedded within urban and agricultural settings are prone to intense disturbance (Roy et al. 2009, Julian et al. 2015). Some landscapes have lost virtually all their streams to land use change, creating 'stream deserts' (Napieralski and Carvalhaes 2016). New methods in landscape ecology that combine remote sensing and hydrologic modeling have resulted in vastly improved understanding of these changes, including the ubiquity of stream burial (Stammler et al. 2013, Weitzell et al. 2016). In one study of Baltimore City, Elmore and Kaushal (2008) estimated that 66% of all streams within the urbanized watershed were buried beneath impervious surfaces, with smaller headwater streams more likely to be buried than larger streams. The chemical, physical, and biological changes created by burying headwater streams are significant enough to warrant serious questions about whether buried streams should still be described as streams at all (Doyle and Bernhardt 2011). Impacts of burial include reduced nitrogen uptake by streams, altered flooding regime, lower gross primary production, and changes in composition of dissolved organic matter (Hope et al. 2014, Aguilar et al. 2019, Roebuck et al. 2020). Yet, the recent 2020 Navigable Waters Protection Rule that defined Waters of the United States (WOTUS) excluded all ephemeral streams (and some intermittent streams and floodplain wetlands) from environmental regulation and compliance. This new rule, if not changed, will have an unprecedented effect on arid and semi-arid regions where the majority of all streams are ephemeral (Manning et al. 2020).

Transfer zone

Moving downstream through the longitudinal profile, the major tributaries provide important intermediate conduits for the transport of materials and energy, along with longitudinal movements of organisms. This zone is often characterized by reaches alternating between confined bedrock and highly mobile, poorly sorted sediments, which create a variety of geomorphic forms and heterogeneous habitats that can be extremely sensitive to changes in sediment supply (Church 2002). This transition zone creates an important geomorphic threshold between headwater channels more coupled with their hillslopes and lowland mainstem channels more influenced by floodplain storage and base-level controls (Church 2002). Depending on regional physiography, this zone will often contain a range of stream gradients and bedforms controlled by geology, geomorphology, and localized sediment and flow inputs from tributaries.

Stream ecologists recognize the heterogeneity of the intermediate transport zone as important for supporting a high diversity of habitats and biota as predicted by the RCC (Vannote et al. 1980). Curtis et al. (2018) tested the RCC hypothesis on the Roaring River in Tennessee and found that intermediate-sized third- to fifth-order streams contained the greatest diversity of

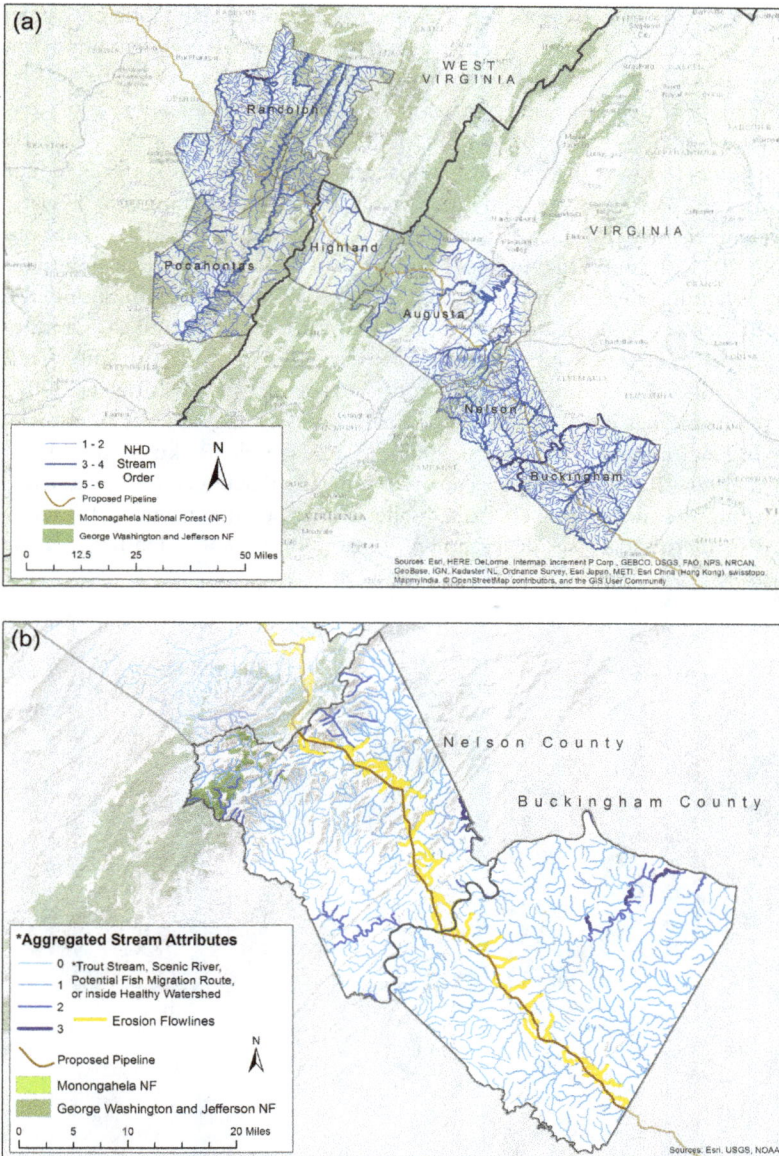

Figure 21.5 Drainage network patterns by county in western Virginia and West Virginia. The distinct differences between the dendritic drainage seen for Nelson and Buckingham Counties in western Virginia and the trellis drainage of Augusta and Highland counties are due to differences in the geomorphology between the Piedmont and the Valley and Ridge Physiographic Provinces. Once atop the Appalachian Plateau in Randolph County, the dendritic pattern of streams is present once again. The map illustrates the greater number of headwater streams that would be crossed by a proposed regional pipeline project on public lands with more dendritic networks. A close-up of Nelson and Buckingham Counties illustrates the potential erosion flowpaths created by the pipeline construction and their impact on streams with desired attributes. (Sources: NHDPLUS, NRCS SSURGO, VA DGIF, VA DCR, VCU INSTAR.)

prey assemblages for a generalist insectivorous fish. Transport zones also enable biotic exchanges of drifters and organism movement between tributaries and mainstems, making them important for community-level ecosystem functions (Bentley et al. 2015, Dyer and Brewer 2020).

Human development tends to increase in intensity within the transfer zone, and satellite data can be used to quantify these influences in urban riverscapes (Figure 21.6). By preventing water from seeping into the ground, compacted and impervious surfaces reduce the capacity of landscapes to slow, store, and filter water, thereby increasing storm water runoff into stream channels, which leads to increased peak flow rates (Arnold and Gibbons 1996, Boggs and Sun 2011). Impervious surfaces also facilitate the conveyance of pollutants into waterways by preventing water from percolating into the soil, where it would undergo natural processing (Brabec et al. 2002). By reducing groundwater recharge and thus, reducing base flow conditions in streams, impervious surface cover can have significant effects on macroinvertebrates and other sensitive aquatic and riparian species (Walsh et al. 2007). The urban stream syndrome has been a useful construct for describing this degradation of river systems in urban environments (Walsh et al. 2005). As an example, Perkin et al. (2019b) reported a long-term (1976–2016) loss in fish diversity within the Blackburn Fork watershed in Tennessee using a multi-scaled approach from its headwaters to mainstem and directly related diversity loss to increased urban development and the diagnostics identified by the urban stream syndrome.

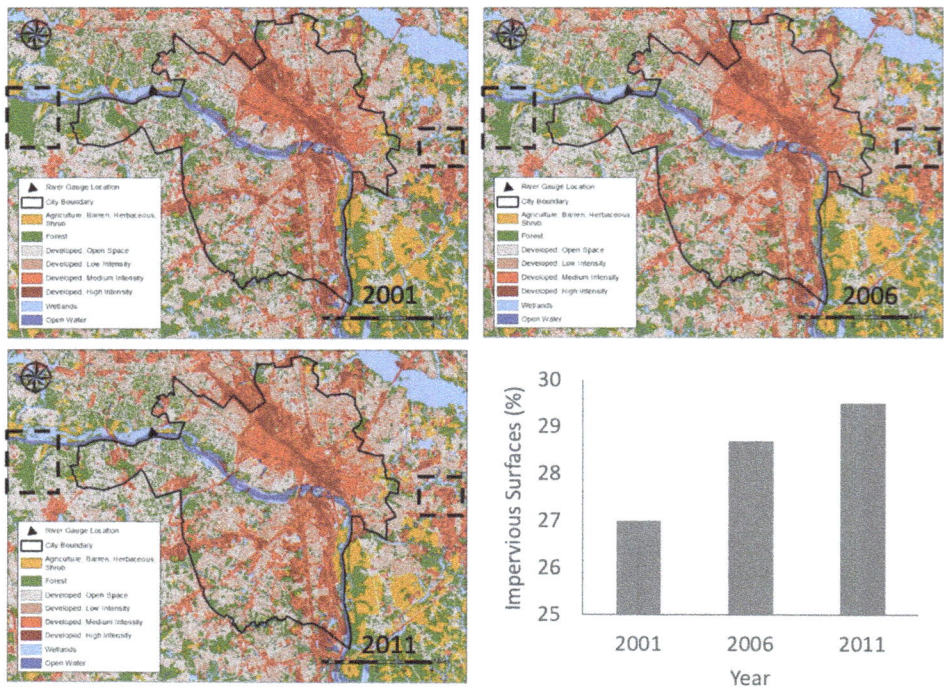

Figure 21.6 Land use changes and associated changes in percent impervious surface (bar graph) for the City of Richmond taken at three time points: 2001, 2006, and 2011. Green represents forest; yellow represents agriculture/barren/shrub/grass; and the four shades of red represent urban development of increasing density. The two dashed boxes highlight areas of change along rivers. (Source: Data from National Land Cover Database.)

Percent impervious surface has proved to be a useful, if imperfect, indicator of riverscape condition that incorporates the important concept of lateral fluxes into mainstem channels (Gergel et al. 2002). Ten percent impervious cover has become a common threshold of concern for watersheds, dating back to early studies by Schueler (1994). However, other studies have found impacts at impervious surface covers ranging from 3% to 15%, including increases in nutrients (Schoonover and Lockaby 2006), changes in streamflow (Yang et al. 2010), and declines in species diversity, including losses of sensitive aquatic invertebrates (Utz et al. 2009), fish (Uphoff et al. 2011), and riparian vegetation (Guida-Johnson et al. 2017). King and colleagues (2011) found that impervious surface cover as low as 0.5–2% resulted in the decline of the majority (80%) of stream taxa, while 2–25% cover showed a decline in 100% of taxa.

Depositional zone

Continuing further downstream, the transition from the intermediate transport zone into floodplain and depositional-dominated segments can be abrupt in valleys controlled by subsequent geologic constraints. In other cases, the transition occurs over a gradual decline in stream gradient via consequent semi-uniform erosion. The Fall Line separating the Piedmont (or foothills, or continental shield) from the Coastal Plain – examples include the eastern United States, Balcones Escarpment in Texas, Italy, western Africa, Brazil, and India – creates fairly abrupt mainstem transitions between the transport-dominated intermediate zones and depositional-dominated floodplain zones. In consequent controlled transitions, such as those in the Great Plains and much of the greater Mississippi River Valley, the transition is much less abrupt. The channel gradient gradually decreases, the valley widens, and the mainstem river decouples from the hillslopes and becomes dominated by alluvial floodplain processes. The different geomorphic zones along the longitudinal dimension of rivers provide different biological functions (Allan and Castillo 2007). The riverscape perspective offers a framework for describing, studying, and managing these transitions.

Regardless of the type of transition, a key ecosystem function taking place within these lower segments of the river network involves flood-pulses and the bi-directional lateral exchanges of water, sediments, materials, and organisms between the river and the floodplain (Junk et al. 1989, Pettit et al. 2017). The bulk of riverscape research in these environments focuses on measuring and modeling lateral connectivity and quantifying abiotic (e.g., sediment, nutrients, carbon) and biotic (e.g., living and dead organisms, wood) exchanges dependent on these lateral processes (Kupfer et al. 2015, Chanut et al. 2018). This research encompasses a variety of environments, including tropical (Jardine et al. 2015), temperate (Meitzen et al. 2018b), semi-arid (Rajesh et al. 2016), montane (Wohl 2013), and permafrost regions (Lininger and Wohl 2019). Quantifying river–floodplain dynamics is increasingly important for understanding how watershed-scale variables such as the density and volume of upstream dam storage, climate change (increased extremes in drought and rainfall intensity), disturbance, and land cover and use changes affect the discharge and sediment regimes that create the physical template and drive these lateral exchanges.

The heterogeneity of floodplain features (patch and matrix mosaic) and flow paths (connectivity and source conduits) provides numerous riverscape functions and supports a diversity of spawning and nursery habitats. Floodplain-dependent spawning fish require access to floodplain lakes, flooded riparian margins, or inundated wetland flats to meet their reproductive life history needs. Reduced flows (from dams and withdraws), floodplain land cover conversions (to agriculture, timber, urban development), and physical barriers (livestock fences and weir gates), among other factors, prevent access and disrupt reproductive success. As an example, these anthropogenic factors, along with over-harvesting, are of concern to the conservation of the Alligator gar (*Atroctosteus spatula*), native to Gulf Coast rivers from Mexico to Florida. Spatially explicit, landscape ecology–based habitat suitability models provide valuable tools for linking flood-pulse variables (e.g., depth, duration, temperature) to suitable habitat patches and quantifying connectivity pathways between the river and the floodplain spawning sites. Conservation management for Alligator gar benefits from these types of models with applications on the Trinity River, Texas (Robertson et al. 2018), the Guadalupe River, Texas (Meitzen et al. 2018a), and the greater Mississippi Alluvial Valley (Allen et al. 2020). A combination of public and private floodplain land protection, improving longitudinal connectivity and watershed conditions at the basin scale, and learning more about specific flow–ecology relationships of floodplain-dependent species will be necessary for sustaining the long-term integrity of these lowland ecosystems.

Reimagining riverscapes

Approaching riverscape management and conservation using a combination of the hierarchical structure and dynamic longitudinal process perspectives provides a useful framework for identifying socio-ecological problems and proposing sustainable solutions. At the heart of landscape ecology is the reciprocal relationship between pattern (structure) and process (function). The physical structure of rivers has long been used for understanding and comparing different riverscapes, and more recently, structural attributes have been used for assessing vulnerabilities to anthropogenic disturbances or selecting different management strategies (McManamay et al. 2018). We have described how different geomorphic zones along the longitudinal dimension of rivers provide different biological functions. These zones are linked along the river continuum. In addition, the lateral, and three-dimensional hyporheic (e.g., Dugdale et al. 2015), transport of matter and energy has been increasingly recognized within riverscapes. Small streams are well connected to their local riparian communities, which can give rise to diverse and specialized biotic assemblages (Meyer et al. 2007). Network analysis using graph theory has provided a valuable tool in recent years for analyzing both the longitudinal (upstream–downstream) and lateral (channel–riparian) fluxes of biological functions for heterogeneous riverscapes (Erős et al. 2012, Erős and Lowe 2019).

These interactions between riverscape structure and function occur across multiple spatial scales. One of the more promising future directions of riverscape research involves an expansion in perspective from the watershed scale to the global scale, which has been made possible with advances in remote sensing. Using Landsat-derived data products, Allen and Pavelsky (2018) showed that we have been underestimating the global surface area of streams by about a third. Satellite data analyses have also revealed global changes in river water quality (Gardner et al. 2021). With these advanced tools and panoramas of riverscapes, we will gain novel insight into global biogeochemistry, particularly as it relates to land use change and climate change. Regionally, riverscape research has been largely biased to temperate climate zones in relatively wealthy countries (Gao et al. 2013, Ramírez et al. 2014, Capps et al. 2016). New studies are needed in other parts of the world.

Atop the spatial complexities generated by the interactions of river structure and functions, riverscapes are dynamic systems. Many of these changes are part of natural short- and long-term cycles. However, humans have induced novel changes in most river ecosystems during the Anthropocene (Wohl 2019). These anthropogenic impacts include reductions in channel complexity that can have profound impacts on hydrology and the ability of rivers to support a diverse biota (Tracy-Smith et al. 2012). The diversion of headwater streams to meet human demands through pipes, impoundment, and burial, for example, may scale up to the regional level to eliminate specific habitats and alter the flow of organisms, energy, and matter within the river network (Freeman et al. 2007). The regulation of rivers through dams alters flow and sediment transport, reducing the heterogeneity of structural geomorphic habitat elements (Graf 2006) and the complexity of riparian vegetation (Aguiar et al. 2016). Whether the habitat alterations are from water diversions or dams, sensitive, specialist species of birds, fish, and invertebrates often cannot survive such changes, while tolerant, generalist species can become increasingly abundant (Agostinho et al. 2012). Future management and scientific understanding of riverscapes would benefit from a deeper understanding of the historical and ongoing influences of human actions on these systems.

Allan (2004) emphasized the primacy of relationships between land use and river condition in his early article on riverscapes and landscapes. He further advocated for the continued development of landscape metrics of greater diagnostic value than aggregated measures like total impervious surface area. This research has continued to evolve. Jantz et al. (2010) and others have demonstrated how the spatial patterning of impervious surface, not just the total amount, can be incorporated into hydrologic models. In other modeling work, Shields and Tague (2015) demonstrated how the amount of impervious surfaces with direct hydrologic connections to the stream network can be predictive of performance measures like ecosystem productivity. The research supporting the use of effective/connected impervious area to assess urban stream condition is compelling (Walsh and Kunapo 2009, Vietz et al. 2014, Ebrahimian et al. 2016). Even these more spatial measures do not account for the large amount of water in cities that gets transported directly to streams in concrete channels or pipes (Roy and Shuster 2009) or that gets mitigated by storm control measures (Bell et al. 2019). Nevertheless, the data are clear. Not just the total amount of impervious surface but the spatial configuration of those surfaces is a critical driver of their impact on river systems, and more research on these spatial relationships is needed.

After more than 500 years of imagining and studying riverscapes, we have learned that rivers are not merely features of the landscape, and humans are not just disturbances to the riverscape. In reality, humans and riverscapes are integrated into a social-ecological system where (1) they regularly interact across multiple spatial, temporal, and organizational scales; (2) their function and health are dependent on an assemblage of critical resources – natural, socioeconomic, and cultural; and (3) their relationship is perpetually dynamic with continuous adaptation (Burch and DeLuca 1984, Machlis et al. 1997, Redman et al. 2004). The river ahead must go beyond hierarchical and longitudinal natural process–based studies and now assess people's perceptions, values, and life experiences in riverscapes (Julian et al. 2018). Adding this social demand of blue spaces will reveal their true value, ensuring the protection of riverscapes for future generations of flora, fauna, and people.

Acknowledgments

Cassandra Troy, Taylor Holden, and Kimberley Browne assisted with the creation of figures for the manuscript.

References

Agostinho, A.A., Agostinho, C.S., Pelicice, F.M and Marques, E.E. (2012) 'Fish ladders: safe fish passage or hotspot for predation?', *Neotropical Ichthyology*, vol 10, no 4, pp687–696.

Aguiar, F.C., Martins, M.J., Silva, P.C. and Fernandes, M.R. (2016). 'Riverscapes downstream of hydropower dams: effects of altered flows and historical land-use change', *Landscape and Urban Planning*, vol 153, pp83–98.

Aguilar, M.F., Dymond, R.L. and Cooper, D.R. (2019) 'History, mapping, and hydraulic monitoring of a buried stream under a central business district', *Journal of Water Resources Planning and Management*, vol 145, no 12, p10.

Alexander, R.B., Boyer, E.W., Smith, R.A., Schwarz, G.E. and Moore, R.B. (2007) 'The role of headwater streams in downstream water quality', *Journal of the American Water Resources Association*, vol 43, no 1, pp41–59.

Allan, J.D. (2004). 'Landscapes and riverscapes: the influence of land use on stream ecosystems', *Annual Review of Ecology, Evolution, and Systematics*, vol 35, no 1, pp257–284.

Allan, J.D. and Castillo, M.M. (2007) *Stream Ecology: Structure and Function of Running Waters*. Springer, Dordrecht.

Allen, G.H. and Pavelsky, T.M. (2018) 'Global extent of rivers and streams', *Science*, vol 361, pp585.

Allen, Y., Kimmel, K. and Constant, G. (2020) 'Using remote sensing to assess Alligator gar spawning habitat suitability in the Lower Mississippi River', *North American Journal of Fisheries Management*, vol 40, pp580–594.

Arnold, C.L. and Gibbons, C.J. (1996) 'Impervious surface coverage: the emergence of a key environmental indicator', *Journal of the American Planning Association*, vol 62, no 2, pp243–258.

Becker, J.C. Rodibaugh, K.J., Hahn, D. and Nowlin, W. (2017) 'Bacterial community composition and carbon metabolism in a subtropical riverscape', *Hydrobiologia*, vol 792, pp2019–2226.

Bell, C.D., Tague, C.L. and McMillan, S.K. (2019) 'Modeling runoff and nitrogen loads from a watershed at different levels of impervious surface coverage and connectivity to storm water control measures', *Water Resources Research*, vol 55, no 4, pp2690–2707.

Belletti, B., Dufour, S., Piegay, H. (2014) 'Regional assessment of the multi-decadal changes in braided riverscapes following large floods (examples from 12 reaches in south east France)', *Advances in Geosciences*, vol 37, pp57–71.

Bentley, K.T., Schindler, D.E., Armstrong, J.B., Cline, T.J. and Brooks, G.T. (2015) 'Inter-tributary movements by resident salmonids across a boreal riverscape', *PLoS ONE* vol 10, no 9, art e0136985.

Bernhardt, E.S., M.A. Palmer, J.D. Allan, G. Alexander, K. Barnas, S. Brooks, J. Carr, S. Clayton, C. Dahm, J. Follstad-Shah, D. Galat, S. Gloss, P. Goodwin, D. Hart, B. Hassett, R. Jenkinson, S. Katz, G.M. Kondolf, P.S. Lake, R. Lave, J.L. Meyer, T.K. O'Donnell, L. Pagano, B. Powell, and E. Sudduth. (2005) 'Synthesizing U.S. river restoration efforts', *Science*, vol 308, pp636–637.

Boggs J.L. and Sun, G. (2011) 'Urbanization alters watershed hydrology in the piedmont of North Carolina', *Ecohydrology*, vol 4, pp256–264.

Brabec, E., Schulte, S. and Richards, P.L. (2002) 'Impervious surfaces and water quality: a review of current literature and its implications for watershed planning', *Journal of Planning Literature*, vol 16, no 4, pp499–514.

Brenkman, S.J., Duda, J.J., Weltey, E., Torgersen, C.E., Pess, G.R., Peters, R. and McHenry, M.L. (2012) 'A riverscape perspective of Pacific salmonids and aquatic habitats prior to large-scale dam removal in the Elwha River, Washington, USA', *Fisheries Management and Ecology*, vol 19, pp36–53.

Brierly, G.J. and Fryirs, K.A. (2005) *Geomorphology and River Management: Applications of the River Styles Framework*. Blackwell Publications, Oxford.

Brierley, G., Reid, H., Fryirs, K. and Trahan, N. (2010) 'What are we monitoring and why? Using geomorphic principles to frame eco-hydrological assessments of river condition', *Science of the Total Environment*, vol 408, pp2025–2033.

Burch, W.R. Jr. and DeLuca, D.R. (1984) *Measuring the Social Impact of Natural Resource Policies*. University of New Mexico Press, Albuquerque.

Burke, M., Jorde, K. and Buffington, J.M. (2009) 'Application of a hierarchical framework for assessing environmental impacts of dam operation: changes in streamflow, bed mobility and recruitment of riparian trees in a western North American river', *Journal of Environmental Management*, vol 90 Supplement 3, ppS224–S236.

Capps, K.A., Bentsen, C.N. and Ramírez, A. (2016) 'Poverty, urbanization, and environmental degradation: urban streams in the developing world', *Freshwater Science*, vol 35, no 1, pp429–435.

Carbonneau, P., Fonstad, M.A., Marcus, W.A. and Dugdale S.J. (2012) 'Making riverscapes real', *Geomorphology*, vol 137, no 1, pp74–86.

Carpenter, S.R. and Turner, M.G. (2000) 'Hares and tortoises: interactions of fast and slow variables in ecosystems', *Ecosystems*, vol 3, pp495–497.

Chanut, P.C.M., Datry, T., Gabbud, C. and Robinson, C. (2019) 'Direct and indirect effects of flood regime on macroinvertebrate assemblages in a floodplain riverscape', *Ecohydrology*, vol 12, no 5, art e2095.

Chesapeake Conservancy (2013) 'Envision the James: a vision for the James River Watershed', www .envisionthejames.org/, accessed 22 November 2020.

Christian, C.S. (1958) 'The concept of land units and land systems', in Proceedings of Ninth Pacific Science Congress held at Chulalongkorn University, Bangkok, Thailand, vol 20, pp74–81.

Church, M. (2002) 'Geomorphic thresholds in riverine landscapes', *Freshwater Biology*, vol 47, pp541–557.

Courtenay, C.I. and Lookingbill, T.R. (2014) 'Designing a regional trail network of high conservation value using principles of green infrastructure', *Southeastern Geographer*, vol 54, pp270–290.

Curtis, W.J., Gebhard, A.E. and Perkin, J.S. (2018) 'The river continuum concept predicts prey assemblage structure for an insectivorous fish along a temperate riverscape', *Freshwater Science* vol 37, no 3, pp618–630.

Deitrich, J.T. (2016) 'Riverscape mapping with helicopter-based structure-from-motion photogrammetry', *Geomorphology*, vol 252, pp144–157.

Demarchi, L., van de Bund, W. and Pistocchi, A. (2020) 'Object-based ensemble learning for pan-European riverscape units mapping based on Copernicus VHR and EU-DEM data fusion', *Remote Sensing*, vol 12, art 1222.

Doyle, M.W. and Bernhardt, E.S. (2011) 'What is a stream?', *Environmental Science & Technology*, vol 45, pp354–359.

Dugdale, S.J., Bergeron, N.E. and St-Hilaire, A. (2015) 'Spatial distribution of thermal refuges analysed in relation to riverscape hydromorphology using airborne thermal infrared imagery', *Remote Sensing of Environment*, vol 160, pp43–55.

Dunham, J.B., Angermeier, P.L., Crausby, S.D., Cravens, A.E., Gosnell, H., McEvoy, J., Moritz, M.A., Raheem, N. and Sanford, T. (2018) 'Rivers are social-ecological systems: time to integrate human dimensions into riverscape ecology and management', *WIREs Water*, vol 5, no 4, e1291.

Dyer, J.J. and Brewer, S.K. (2020) 'Seasonal movements and tributary-specific fidelity of blue sucker *Cycleptus elongatus* in a southern plains riverscape', *Journal of Fish Biology*, vol 97, no 1, pp 279–292.

Ebrahimian, A., Gulliver, J.S. and Wilson, B.N. (2016) 'Effective impervious area for runoff in urban watersheds', *Hydrological Processes*, vol 30, no 20, pp3717–3729.

Elmore, A.J. and Kaushal, S.S. (2008) 'Disappearing headwaters: patterns of stream burial due to urbanization', *Frontiers in Ecology and the Environment*, vol 6, pp308–312.

Elosegi, A., Diez, J. and Mutz, M. (2010) Effects of hydromorphological integrity on biodiversity and functioning of river ecosystems. *Hydrobiologia*, vol 65, no 1, pp199–215.

Erős, T. and Grant, E.H.C. (2015) 'Unifying research on the fragmentation of terrestrial and aquatic habitats: patches, connectivity, and the matrix in riverscapes', *Freshwater Biology*, vol 60, pp1487–1501.

Erős, T. and Lowe, W.H. (2019) 'The landscape ecology of rivers: from patch-based to spatial network analyses', *Current Landscape Ecology Reports*, vol 4 no 4, pp103–112.

Erős, T., Olden, J.D., Schick, R.S., Schmera, D. and Fortin, M.J. (2012) 'Characterizing connectivity relationships in freshwaters using patch-based graphs', *Landscape Ecology*, vol 2, no 2, pp303–317.

Erős, T., O'Hanley, J.R. and Czegledi, I. (2017) 'A unified model for optimizing riverscape conservation', *Journal of Applied Ecology*, vol 55, no 4, pp 1871–1883.

Fausch, K.D., Torgersen, C.E., Baxter, C.V. and Li, H.W. (2002) 'Landscapes to riverscapes: bridging the gap between research and conservation of stream fishes', *BioScience*, vol 52, no 6, pp483–498.

Ferrari, J.R., Lookingbill, T.R., McCormick, B., Townsend, P.A. and Eshleman, K.N. (2009) Surface mining and reclamation effects on flood response of watersheds in the central Appalachian Plateau region', *Water Resources Research*, vol 45, art W04407.

Freeman, M.C., Pringle, C.M. and Jackson, C.R. (2007) 'Hydrologic connectivity and the contribution of stream headwaters to ecological integrity at regional scales', *Journal of the American Water Resources Association*, vol 43, no 1, pp5–14.

Frissell, G.A., Liss, W.J., Warren, G.E. and Hurley, M. (1986) 'A hierarchical framework for stream habitat classification: viewing streams in a watershed context', *Environmental Management*, vol 10, pp199–214.

Gao, J.B., Wu, Q., Li, Q.L., Ma, J., Xu, Q.F., Groffman, P.M. and Yu, S. (2013) 'Preliminary results from monitoring of stream nitrogen concentrations, denitrification, and nitrification potentials in an urbanizing watershed in Xiamen, southeast China', *International Journal of Sustainable Development and World Ecology*, vol 20, pp223–230.

Gardner, J., Yang, X., Topp, S., Ross, M.R.V. and Pavelsky, T. (2021). 'The color of rivers', *Geophysical Research Letters*, vol. 48, e2020GL08894.

Gergel, S.E., Turner, M.G., Miller, J.R., Melack, J.M. and Stanley, E.H. (2002) 'Landscape indicators of human impacts to riverine systems', *Aquatic Sciences*, vol 64, no 2, pp118–128.

Grabowski, Z.J., Denton, A., Rozance, M.A., Matsler, M. and Kidd, S. (2017) 'Removing dams, constructing science: coproduction of undammed riverscapes by politics, finance, environment, society and technology', *Water Alternatives*, vol 10, no 3, pp769–795.

Graf, W.L. (2006) 'Downstream hydrologic and geomorphic effects of large dams on American rivers', *Geomorphology*, vol 79, no 3–4, pp336–360.

Guida-Johnson, B., Faggi, A.M. and Zuleta, G.A. (2017) 'Effects of urban sprawl on riparian vegetation: is compact or dispersed urbanization better for biodiversity?', *River Research and Applications*, vol 33, no 6, pp959–969.

Hand, B.K., Flint, C.G., Frissel, C.A., Muhlfeld, C.C., Devlin, S.P., Kennedy, B.P., Crabtree, R.L., McKee, W.A., Luikart, G. and Stanford, J. (2018) 'A social-ecological perspective for riverscape management in the Columbia River Basin', *Frontiers in Ecology*, vol 16, no S1, ppS23–S33.

Hawkes, H.A. (1975) 'River zonation and classification', in B.A. Whitton (ed) *River Ecology*. Blackwells, Oxford.

Hope, A.J., McDowell, W.H. and Wollheim, W.M. (2014) 'Ecosystem metabolism and nutrient uptake in an urban, piped headwater stream', *Biogeochemistry*, vol 121, no 1, pp167–187.

Jantz, C.A., Goetz, S.J., Donato, D. and Claggett, P. (2010) 'Designing and implementing a regional urban modeling system using the SLEUTH cellular urban model', *Computers Environment and Urban Systems*, vol 34, no 1, pp1–16.

Jardine, T.J., Bond, N.R., Burford, M.A., Kennard, M.J., Ward, D.P., Bayliss, P., Danies, P.M., Douglas, M.M., Hamilton, S.K., Melack, J.M., Naiman, R.J., Pettit, N.S., Pusey, B., Warfe, D.M. and Bunn, S.E. (2015) 'Does flood rhythm drive ecosystem responses in tropical riverscapes?', *Ecology*, vol 96, no 3, pp684–692.

Julian, J.P., Stanley, E.H. and Doyle, M.W. (2008) 'Basin-scale consequences of agricultural land use on benthic light availability and primary production along a sixth-order temperate river', *Ecosystems*, vol 11, no 7, pp1091–1105.

Julian, J.P., Seegert, S.Z., Powers, S.M., Stanley, E.H. and Doyle, M.W. (2011) 'Light as a first-order control on ecosystem structure in a temperate stream', *Ecohydrology*, vol 4, pp422–432.

Julian, J.P., Wilgruber N.A., de Beurs K.M., Mayer, P.M. and Jawarneh, R.N. (2015) 'Long-term impacts of land cover changes on stream channel loss', *Science of the Total Environment*, vol 537, pp399–410.

Julian, J.P., Podolak, C.J.P., Meitzen, K.M., Doyle, M.W., Manners, R.B., Hester, E.T., Ensign, S. and Wilgruber, N.A. (2016) 'Shaping the physical template: biological, hydrological, and geomorphic connections in stream channels', in J.B. Jones and E.H. Stanley (eds) *Stream Ecosystems in a Changing Environment*. Elsevier, London.

Julian J.P., Daly, G. and Weaver, R. (2018) 'University students' social demand of a blue space and the influence of life experiences', *Sustainability*, vol 10, art 3178.

Junk, W.J., Bayley, P.B. and Sparks, R.E. (1989) 'The flood pulse concept in river-floodplain systems', in Proceedings of the International Large River Symposium, vol 106, pp110–127.

King, R.S., Baker, M.E., Kazyak, P.F. and Weller, P.F. (2011) 'How novel is too novel? Stream community threshold at exceptionally low levels of catchment urbanization', *Ecological Applications*, vol 21, no 5, pp1659–1678.

Kupfer, J.A., Meitzen, K.M. and Gao, P. (2015) 'Flooding and surface connectivity of Taxodium-Nyssa stands in a southern floodplain forest ecosystem', *River Research and Applications*, vol 31, pp1299–1310.

Lapointe, K.M. (2011) 'Regional variability in Atlantic salmon (*Salmo salar*) riverscapes: a simple landscape ecology model explaining the large variability in size of salmon runs across Gaspe watersheds, Canada', *Ecology of Freshwater Fish*, vol 20, pp144–156.

Leopold, L.B. and Maddock, T., Jr. (1953) 'The hydraulic geometry of stream channels and some physiographic implications', *USGS Professional* Paper, vol 252, pp1–57.

Levin-Keitel, M. (2014) 'Managing urban riverscapes: towards a cultural perspective of land and water governance', *Water International*, vol 39, pp842–857.

Lininger, K.B. and Wohl, E. (2019) 'Floodplain dynamics in North American permafrost regions under a warming climate and implications for organics carbon stocks: a review and synthesis', *Earth-Science Reviews*, vol 193, pp24–44.

Lookingbill, T.R., Kaushal, S.S., Elmore, A.J. Gardner, R., Eshleman, K.N., Hilderbrand, R.H., Morgan, R.P., Boynton, W.R., Palmer, M.A. and Dennison, W.C. (2009) 'Altered ecological flows blur boundaries in urbanizing watersheds', *Ecology and Society*, vol 14, no 2, art 10.

Lowe, W.H. and Likens, G.E. (2005) 'Moving headwater streams to the head of the class', *Bioscience*, vol 55, no 3, pp196–197.

Machlis, G.E., Force, J.E. and Burch, W.R. (1997) 'The human ecosystem part I: the human ecosystem as an organizing concept in ecosystem management', *Society and Natural Resources*, vol 10, pp347–367.

Manning, A., Julian, J.P and Doyle, M.W. (2020) 'Riparian vegetation as an indicator of stream channel presence and connectivity in arid environments', *Journal of Arid Environments*, vol 178, art 104167.

McCluney, K.E., Poff, N.L., Palmer, M.A., Thorp, J.H., Poole, G.C., Williams, B.S., Williams, M.R. and Baron, J.S. (2014) 'Riverine macrosystems ecology: sensitivity, resistance, and resilience of whole river basins with human alterations', *Frontiers in Ecology and the Environment*, vol 12, no 1, pp48–58.

McManamay, R.A., Troia, M.J., DeRolph, C.R., Sheldon, A.O., Barnett, A.R., Kao, S.C. and Anderson, M.G. (2018) 'A stream classification system to explore the physical habitat diversity and anthropogenic impacts in riverscapes of the eastern United States', *PLoS One*, vol 13, no 6, art e0198439.

Meitzen, K.M., Doyle, M.W., Thoms, M.C. and Burns, C.E. (2013) 'Geomorphology in the interdisciplinary context of environmental flows', *Geomorphology*, vol 200, pp143–154.

Meitzen, K.M., Hardy, T. and Jensen, J. (2018a) 'Floodplain inundation analysis of the lower Guadalupe River: linking hydrology and floodplain-dependent resources', *Final Report for Texas Water Development Board Contract No. 1448311791*, Prepared for Texas Parks and Wildlife Department Contract No. 474120.

Meitzen, K.M., Kupfer, J.A. and Gao, P. (2018b) 'Modeling hydrologic connectivity and virtual fish movement across a large southeastern floodplain', *Aquatic Sciences*, vol 80, art 5.

Melles, S.J., Jones, N.E. and Schmidt, B. (2012) 'Review of theoretical developments in stream ecology and their influence on stream classification and conservation planning', *Freshwater Biology*, vol 57, pp415–434.

Merriam, E.R., Petty, J.T., Strager, M.P., Maxwell, A.E. and Ziemkiewicz, P.F. (2015) 'Complex contaminant mixtures in multistressor Appalachian riverscapes', *Environmental Toxicology and Chemistry*, vol 34, no 11, pp2603–2610.

Meyer, J.L., Strayer, D.L., Wallace, J.B., Eggert, S.L., Helfman, G.S. and Leonard, N.E. (2007) 'The contribution of headwater streams to biodiversity in river networks', *Journal of the American Water Resources Association*, vol 43, pp86–103.

Milinkovic, M. Corovic, D. and Vuksanovic-Macura, Z. (2019) 'Historical enquiry as a critical method in urban riverscape revisions: the case of Belgrade's confluence', *Sustainability*, vol 11, art 1177.

Moss, M.R. (1975) 'Biophysical land classification schemes: their relevance and applicability to agricultural development in the humid tropics', *Journal of Environmental Management*, vol 3, pp287–307.

Napieralski, J.A. and Carvalhaes, T. (2016) 'Urban stream deserts: mapping a legacy of urbanization in the United States', *Applied Geography*, vol 67, pp129–139.

Naveh, Z. and Lieberman, A.S. (1994) *Landscape Ecology: Theory and Application*. Springer, New York.

Palmer, M.A. (2009) 'Reforming watershed restoration: science in need of application and applications in need of science', *Estuaries and Coasts*, vol 32, no 1, pp1–17.

Palmer, M.A., Menninger, H.L. and Bernhardt, E. (2010) 'River restoration, habitat heterogeneity and biodiversity: a failure of theory or practice?', *Freshwater Biology*, vol 55, pp205–222.

Paul, M.J. and Meyer, J.L. (2001) 'Streams in the urban landscape', *Annual Review of Ecology and Systematics*, vol 32, pp333–365.

Perkin, J.S., Gibbs, W.K., Ridgway, J.L. and Cook, S.B. (2019a) 'Riverscape correlates for distribution of threatened spotfin chub *Esox monachus* in the Tennessee River Basin', *Endangered Species Research*, vol 40, pp91–105.

Perkin, J.S., Murphy, S.P., Murray, C.M., Gibbs, W.K. and Gebhard, A.E. (2019b) 'If you build it, they will go: a case study of stream fish diversity loss in an urbanizing riverscape', *Aquatic Conservation: Marine Freshwater Ecosystems*, vol 29, pp623–638.

Peters, D.P.C., Bestelmeyer, B.T. and Turner, M.G. (2007) 'Cross-scale interactions and changing pattern-process relationships: consequences for system dynamics', *Ecosystems*, vol 10, pp790–796.

Pettit, R.J., Naiman, R.J., Warfe, D.M., Jardine, T.D., Douglas, M.M., Bunn, S.E. and Davies, P.M. (2017) 'Productivity and connectivity in tropical riverscapes of Northern Australia: ecological insights for management', *Ecosystems*, vol 20, pp492–514.

Pomeroy, J.W., Green, M.B. and Fitzgibbon, J.E. (1983) 'Evaluation of urban riverscape aesthetics in the Canadian prairies', *Journal of Environmental Monitoring*, vol 17, pp263–276.

Poole, G.C. (2002) 'Fluvial landscape ecology: addressing uniqueness within the river discontinuum', *Freshwater Biology*, vol 47, pp641–660.

Rajesh, T., Martin, T. and Parsons, M. (2016) 'An adaptive cycle hypothesis of semi-arid floodplain vegetation productivity in dry and wet resource states', *Ecohydrology*, vol 9, pp39–51.

Ramírez, A., Rosas, K.G., Lugo, A.E. and Ramos-González, O.M. (2014) 'Spatio-temporal variation in stream water chemistry in a tropical urban watershed', *Ecology and Society*, vol 19, art 11.

Redman, C.L., Grove, J.M. and Kuby, L.H. (2004) 'Integrating social science into the Long-Term Ecological Research (LTER) network: social dimensions of ecological change and ecological dimensions of social change', *Ecosystems*, vol 7, pp161–171.

Richardson, M.L., Wilson, B.A., Aiuto, D.A.S., Crosby, J.E., Alonso, A., Dallmeier, F. and Golinski, G.K. (2017) 'A review of the impact of pipelines and power lines on biodiversity and strategies for mitigation', *Biodiversity and Conservation*, vol 26, pp1801–1815.

Rinaldi, M., Gurnell, A.M., del Tanago, M.G., Bussettini, M. and Hendriks, D. (2016) 'Classification of river morphology and hydrology to support management and restoration', *Aquatic Sciences*, vol 78, no 1, pp17–33.

Rios, S.L. and Bailey, R.C. (2006) 'Relationship between riparian vegetation and stream benthic communities at three spatial scales', *Hydrobiologia*, vol 553, pp153–160.

Robertson, C.R., Aziz, K., Buckmeier, D.L., Smith, N.G. and Raphelt, N. (2018) 'Development of a flow-specific floodplain inundation model to assess Alligator gar recruitment success', *Transaction of the American Fisheries Society*, vol 147, no 4, pp 674–686.

Roebuck, J.A., Seidel, M., Dittmar, T. and Jaffe, R. (2020) 'Controls of land use and the river continuum concept on dissolved organic matter composition in an anthropogenically disturbed subtropical watershed', *Environmental Science & Technology*, vol 54, no 1, pp195–206.

Roy, A.H. and Shuster, W.D. (2009) 'Assessing impervious surface connectivity and applications for watershed management', *Journal of the American Water Resources Association*, vol 45, pp198–209.

Roy, A.H., Dybas, A.L., Fritz, K.M. and Lubbers, H.R. (2009) 'Urbanization affects the extent and hydrologic permanence of headwater streams in a midwestern US metropolitan area', *Journal of the North American Benthological Society*, vol 28, pp911–928.

RVAH2O (2017) *RVA Clean Water Plan*. Prepared for the City of Richmond Department of Public Utilities. Richmond, VA.

Saunders, D.L., Meeuwig, J.J. and Vincent, A.C.J. (2002) 'Freshwater protected areas: strategies for conservation', *Conservation Biology*, vol 16, pp30–41.

Schlosser, I.J. (1991) 'Stream fish ecology: a landscape perspective', *BioScience*, vol 41, pp704–712.

Schoonover, J.E. and Lockaby, B.G. (2006) 'Land cover impacts on stream nutrients and fecal coliform in the lower Piedmont of West Georgia', *Journal of Hydrology*, vol 331, pp371–382.

Schueler, T.R. (1994) 'The importance of imperviousness', *Watershed Protection Techniques*, vol 1, no 3, pp100–111.

Schumm, S.A. (1977) *The Fluvial System*. Wiley, New York.

Shields, C. and Tague, C. (2015) 'Ecohydrology in semiarid urban ecosystems: modeling the relationship between connected impervious area and ecosystem productivity', *Water Resources Research*, vol 51, no 1, pp302–319.

Soranno, P.A., Cheruvelil, K.S., Bissell, E.G., Bremigan, M.T., Downing, J.A., Fergus, C.E., Webster, K.E. 2014. Cross-scale interactions: quantifying multi-scaled cause-effect relationships in macrosystems. *Frontiers in Ecology and the Environment*, 12(1): 65–73.

Stammler, K.L., Yates, A.G. and Bailey, R.C. (2013) 'Buried streams: uncovering a potential threat to aquatic ecosystems', *Landscape and Urban Planning*, vol 114, pp37–41.

Strahler, A.N. (1957) 'Quantitative analysis of watershed geomorphology', *Transactions of the American Geophysical Union*, vol 38, pp913–920.

Thorp, J.H., Thoms, M.C. and Delong, M.D. (2006) 'The riverine ecosystem synthesis: biocomplexity in river networks across space and time', *River Research and Applications*, vol 22, pp123–147.

Tracy-Smith, E., Galat, D.L. and Jacobson, R.B. (2012) 'Effects of flow dynamics on the aquatic-terrestrial transition zone (ATTZ) of the lower Missouri River sandbars with implications for selected biota', *River Research and Applications*, vol 28, no 7, pp793–813.

Uphoff, J.H., McGinty, M., Lukacovic, R., Mowrer, J. and Pyle, B. (2011) 'Impervious surface, summer dissolved oxygen, and fish distribution in Chesapeake Bay subestuaries: linking watershed development, habitat conditions, and fisheries management', *North American Journal of Fisheries Management*, vol 31, no 3, pp554–566.

Utz, R.M., Hilderbrand, R.H. and Boward, D.M. (2009) 'Identifying regional differences in threshold responses of aquatic invertebrates to land cover gradients', *Ecological Indicators*, vol 9, pp556–567.

Vannote, R.L., Minshall, G.W., Cummins, K.W., Sedell, J.R. and Cushing, C.E. (1980). 'The river continuum concept', *Canadian Journal of Fisheries and Aquatic Sciences*, vol 37, pp130–137.

Vietz, G.V., Sammonds, M.J., Walsh, C.J., Fletcher, T.D., Rutherfurd, I.D. and Stewardson, M.J. (2014) 'Ecologically relevant geomorphic attributes of streams are impaired by even low levels of watershed effective imperviousness', *Geomorphology*, vol 206, pp67–78.

Walsh, C.J. and Kunap, J. (2009) 'The importance of upland flow paths in determining urban effects on stream ecosystems', *Journal of North American Benthological Society*, vol 28, pp977–990.

Walsh, C.J., Roy, A.H., Feminella, J.W., Cottingham, P.D., Groffman, P.M. and Morgan, R.P. (2005) 'The urban stream syndrome: current knowledge and the search for a cure', *Journal of the North American Benthological Society*, vol 24, pp706–723.

Walsh, C.J., Waller, K.A., Gehlin, J. and MacNally, R. (2007) 'Riverine invertebrate assemblages are degraded more by catchment urbanisation than by riparian deforestation', *Freshwater Biology*, vol 52, pp574–587.

Ward, J.V. (1989) 'The 4-dimensional nature of lotic ecosystems', *Journal of the North American Benthological Society*, vol 8, no 1, pp2–8.

Ward, J.V. (1998) 'Riverine landscapes: biodiversity patterns, disturbance regimes, and aquatic conservation', *Biological Conservation*, vol 83, pp269–278.

Ward, J.V. and Stanford, J.A. (1995) 'The serial discontinuity concept: Extending the model to floodplain rivers', *River Research and Applications*, vol 10, pp159–168.

Weitzell, R.E., Kaushal, S.S., Lynch, L.M., Guinn, S.M. and Elmore, A.J. (2016) 'Extent of stream burial and relationships to watershed area, topography, and impervious surface area', *Water*, vol 8, art 22.

Wellemeyer, J.C., Perkin, J.S., Jameson, M.L., Costigan, K.H., and Waters, R. (2019) 'Hierarchy theory reveals multiscale predictors of Arkansas darter (*Etheostoma cragini*) abundance in a Great Plains riverscape'. *Freshwater Biology*, vol 64, pp659–670.

Wiens, J.A. (2002) 'Riverine landscapes: taking landscape ecology into the water', *Freshwater Biology*, vol 47, pp501–515.

Wohl, E. (2013) 'Floodplains and wood', *Earth-Science Reviews*, vol 123, pp194–212.

Wohl, E. (2014) *Rivers in the Landscape: Science and Management*. Wiley-Blackwell, Oxford.

Wohl, E. (2019) 'Forgotten legacies: understanding and mitigating historical human alterations of river corridors', *Water Resources Research*, vol 55, no 7, pp5181–5201.

Yang, G., Bowling, L.C., Cherkauer, K.A., Pijanowski, B.C. and Niyogi, D. (2010) 'Hydroclimatic response of watersheds to urban intensity: an observational and modeling-based analysis for the White River Basin, Indiana', *Journal of Hydrometeorology*, vol 11, pp122–138.

22

Landscape restoration

Aveliina Helm

Increasing importance of restoration

Intensive anthropogenic land-use, resulting in loss and degradation of habitats and homogeniza-tion of landscapes, is the major driver of biodiversity decline and an important contributor to climate change (Elbakidze et al., 2018; Newbold et al., 2015; Sala et al., 2000). Humans have altered 75% of the land surface (Díaz et al., 2019), and different models estimate the proportion of land degraded by human use to be between 7% and 40% globally (Gibbs and Salmon, 2015). Land-use contributes ca. 23% of total anthropogenic greenhouse gas emissions (IPCC, 2019), but when accounting for the lost potential of current agricultural areas to store carbon as forests or other native ecosystems, the role of land-use is considerably larger (Searchinger et al., 2018).

Loss of biodiversity and habitats at the landscape scale reduces the ability of agricultural or forestry systems to withstand fluctuations in environmental conditions, increases the prob-ability of pest and disease outbreaks, and increases the vulnerability to climate change (Isbell et al., 2015; Mori et al., 2017). Landscape homogenization can also have a substantial impact on local and indigenous people, as it can be related to deterioration of cultural diversity and diminish different opportunities for land-use and obtaining livelihoods (Olden et al., 2005). The Intergovernmental Science-Policy Platform on Biodiversity and Ecosystem Services (IPBES) assessment report of land degradation and restoration concluded that current degradation of land and marine ecosystems undermines the well-being of 3.2 billion people and costs about 10% of the annual global gross product in loss of species and ecosystems services (Montanarella et al., 2018).

Ecological restoration is defined as the process of assisting the recovery of an ecosystem that has been degraded, damaged, or destroyed to reflect values regarded as inherent in the ecosystem and to provide goods and services that people value (Martin, 2017; Gann et al., 2019). In degraded areas, there are now a multitude of examples confirming the positive impacts of ecological restoration on biodiversity and local people through improved ecosystem service provision and enhancement of human livelihoods (Shimamoto et al., 2018; Jones et al., 2018). At the same time, the priority must be on preserving and conserving extant ecosystems and biodiverse landscapes. Restoration, even if meticulously executed, cannot fully replace lost ecosystems, and it can take hundreds or even thousands of years to restore functions and services comparable to those of ancient ecosystems (Jones et al., 2018). For example, restored grasslands have been shown to require at least a century, and more often millennia, to recover their former richness (Nerlekar and Veldman, 2020).

The need to raise global attention to conserving and restoring ecosystems is highlighted by multiple recent scientific overviews and reports (Leclère et al., 2020; World Wildlife Fund, 2020; Díaz et al., 2019; Montanarella et al., 2018). The upcoming UN Decade on Ecosystem Restoration 2020–2030 has a main focus on landscape restoration, aiming to massively scale up the restoration of degraded and destroyed ecosystems as a proven measure to fight climate change and enhance food security, water supply, and biodiversity.

Why restoration needs to focus on landscapes

Healthy and resilient ecosystems can only thrive in ecologically well-functioning and well-connected landscapes. Landscape scale is the scale 'where ecological, social and economic priorities can be balanced' (*The Global Partnership on Forest and Landscape Restoration,* https:// infoflr.org), and it is also the scale where restoration has the potential to deliver the largest benefits for ecological functioning as well as for people (Metzger and Brancalion, 2016). The importance of landscape-scale processes in determining restoration success has long been acknowledged (Crouzeilles and Curran, 2016; Helm, 2015; Prach et al., 2015; Shackelford et al., 2013), indicating that a focus on the landscape rather than individual habitat patches is needed during restoration (Aavik and Helm, 2017; Besseau et al., 2018; Brudvig, 2011). However, restoration is still often largely confined to the site scale (Watson et al., 2017), and scaling up from local habitat patches to the landscape level is considered a key challenge in restoration and conservation (Besseau et al., 2018). Scale is also a fuzzy concept in landscape restoration, as the perception of landscape can be quite different for different species depending on their dispersal ability (Metzger and Brancalion, 2016). For example, for a beetle that is a forest habitat specialist, the relevant scale might not exceed 50 meters; for soil fauna, the relevant scale can be considered in a few meters; whereas for a mammal in the same forest, landscape structure and composition can be relevant in more than a 10 km radius. Knowledge of ecosystem functioning and species requirements will help to determine the particular scales at which the restorations are relevant. Thus, for meeting the needs of a variety of species in ecosystems, appropriate scales in landscape restoration can range from a few hundred meters to 10 kilometers or more (Metzger and Brancalion, 2016; Crouzeilles and Curran, 2016).

Focusing on the landscape scale during restoration has the potential to ensure the recovery of species- and gene-level biodiversity of impoverished and degraded habitats, enhance connectivity between protected areas, protect and regenerate water and soil resources, reinforce cultural values, improve conditions for future sustainable land-use, enhance the livelihoods and well-being of local human populations, and build up climate change resilience (Chazdon et al., 2017; Cohen-Shacham et al., 2017; Erbaugh and Oldekop, 2018). Ecological (both biotic and abiotic) processes that operate at relatively large spatial scales and therefore, have to be addressed at a landscape scale include nutrient flows, movements of fertilizers and pesticides, water flows (including floods and erosion), disturbances (including fire regimes), dispersal and reproduction (including movement of seed dispersers and genetic material), and ecological interactions (including pollination, herbivory, predation) (Holl et al., 2003; Ekroos et al., 2016). Thus, landscape restoration requires developing and balancing a mosaic of interacting land uses and management practices, ranging from intact ecosystem conservation to regenerating degraded lands and habitats. The large-scale approach is particularly topical and acute in homogeneous or heavily degraded agricultural regions but also in intensively managed forest/plantation systems and in urban areas (Ekroos et al., 2016).

Aveliina Helm

Landscape ecology can guide successful restoration

Ecological restoration, both at a patch level and at a landscape scale, is a 'local' action—meaning that it needs to be designed and applied in full recognition and appreciation of natural and socio-economic conditions in the given landscape. This requires answers to many questions: Which biological and socio-economic drivers have led to the degradation of the landscape? What is needed to mitigate the impact of drivers? Which habitats have been lost or degraded? Which ecosystem functions and services are in need of restoration? What are the expectations of local communities, and how can they be engaged? Depending on the specific drivers of degradation, actions needed for landscape-scale restoration can include efficient and inclusive landscape or urban planning that would solve possible land-use and land-ownership conflicts, facilitation of practices towards sustainable food production, support for capacity-building among local people, restoration of specific habitats or conditions for particular species, removal of pollution and polluting sources, re-introduction of natural or semi-natural disturbance regimes (fires, grazing), establishment of ecological corridors, or setting aside areas for natural regeneration.

For the successful recovery of biodiversity, it is necessary to take into account the knowledge regarding biological functioning of landscapes and ecosystems that has been mounting up over the past decades (Brudvig, 2011). Principles developed in landscape ecology can help to identify which habitats are in need of restoration, which drivers have led to the degradation of habitats and landscapes, and what are the suitable actions for restoring environmental conditions and ecosystem functions (Holl et al., 2003; Metzger and Brancalion, 2016).

Landscape composition (i.e. which land-use types and habitats occur in the landscape) and structure (how habitats are spatially located in the landscape, and what kind of matrix is between them) have strong influences on the functioning of ecosystems, on the ecological processes occurring in the landscape (e.g. dispersal, reproduction, population dynamics), and on a provision of ecosystem services. Thus, core questions that have to be answered during landscape restoration from the landscape ecology perspective include the following: (1) Does restoration result in sufficiently large good-quality habitat patches that support viable populations? (2) Are all desired species and genetic material already present in the landscape, or are active introduction measures necessary? (3) Are restored habitat patches sufficiently connected to other similar habitats in the landscape, warranting the dispersal of individuals, propagules, and pollen? and (4) Does the restoration lead to the recovery or preservation of other relevant ecosystem-level functions and processes (biotic interactions, nutrient cycling, soil formation, suitable disturbance regime)?

Knowledge of ecological processes at a landscape scale helps to understand whether it is possible to rely on spontaneous succession or whether active species introduction measures are needed. It allows one to assess which land-use practices support long-term sustainability following restoration and what should be the desired post-restoration landscape structure, habitat area, and connectivity for maintaining well-functioning ecosystems. For example, when there are seed sources nearby in the landscape, then projects that use a natural regeneration approach are more likely to be successful. In the following sections, I will introduce some ecological processes and concepts that should be considered when planning and executing restoration at the landscape scale.

Restoring landscape-scale processes

Natural or assisted restoration—are relevant species and genes already present in the landscape?

All habitats have their own habitat-specific species pool, which can be defined as a group of native species that are able to inhabit and coexist under particular environmental conditions in a given region (Pärtel and Zobel, 1998; Zobel et al., 2011). For example, the habitat-specific species pool of a floodplain meadow in Yorkshire, United Kingdom, contains all the species that have been found to inhabit similar good-quality meadows in Northern England. Thus, if the aim is to restore a floodplain in the region, this is the set of species that needs to colonize the restored area.

Biodiversity encompasses genetic diversity, and while this is often not specifically targeted during restoration planning or monitoring, the long-term viability of established populations depends on their genetic makeup (Brudvig, 2011; Helsen et al., 2013; Mijangos et al., 2015). Similarly to the species pool approach, the concept of gene pools (i.e. the set of genotypes and alleles that are characteristic of naturally occurring populations in a particular region) can be applied in landscape restoration. If this is not specifically tackled, restored populations can suffer genetic bottlenecks or founder effects: the loss of genetic variation when populations go through a severe decline or when a new population is established by a very small number of individuals, respectively. Low genetic diversity makes established populations vulnerable to effects of inbreeding and stochastic environmental conditions, threatening the long-term persistence of species. Genetic diversity provides the adaptive potential necessary for responding to environmental change and helps to buffer populations against ongoing climate and land-use change (Hoffmann and Sgró, 2011).

In recent years, discussions regarding the effectiveness and applicability of 'passive' versus 'active' restoration activities have emerged (Crouzeilles et al., 2016; Prach et al., 2020). Passive restoration, recently termed 'natural restoration' (sensu Atkinson and Bonser, 2020), removes drivers of degradation and assumes that species can recolonize naturally. Active restoration involves significant alterations to abiotic conditions and targeted re-introduction of species.

The presence of species from habitat-specific species and gene pools in the landscape makes it possible to apply passive restoration and ensure spontaneous colonization of desired species and sufficient genetic diversity of populations in restored habitats. For passive recolonization to succeed, it is relevant to know the following information: 1) whether the desired species are actually present in the landscape, able to reach restored habitat, and able to re-establish landscape-scale dispersal, which is needed for long-term persistence (Török and Helm, 2017), and 2) whether the genetic diversity of extant and restored populations is sufficient to ensure long-term viability or whether active measures are needed to overcome potential bottlenecks in population genetic structure in impoverished conditions (Aavik and Helm, 2017).

Passive restoration can be cost-effective and impactful. For example, Prach et al. (2015) studied the recovery of grassland vegetation following the restoration of 82 formerly arable fields in the Carpathian Mountains. They found that despite the use of different species introduction techniques (sowing of different seed mixtures), target species presence in the landscape around the restoration site was the most important determinant of restoration success. As another example, a large meta-analysis of studies observing naturally regenerating and actively restored (planted) forests worldwide found that tree planting does not result in consistently faster or better results, and just simply ending the degrading land use can be sufficient for forests to recover in many cases (Meli et al., 2017).

How to ensure sufficient connectivity for landscape-scale dispersal?

For restoration to succeed, species (or genes) that are expected to inhabit restored habitat also need to be able to move between habitat patches. Movement of individuals, propagules, and genetic material across the landscape is the key process linked to recovery and persistence of biodiversity. Isolated habitats, even when initial introductions of species have been made during restoration, fail to build up high species- and gene-level diversity due to low connectivity of habitats across the landscape. In short, both short-term and long-term restoration success often depends on the success of restoring and maintaining landscape-scale dispersal.

A number of studies have indicated that most species groups require a relatively 'compact' habitat network for effective movement between habitat patches and landscape-scale functional connectivity (Auffret et al., 2017; Winsa et al., 2015). In a global meta-analysis, Crouzeilles and Curran (2016) showed that the restoration success of forests, estimated via changes in diversity of multiple species groups (mammals, birds, invertebrates, herpetofauna, plants), was directly dependent on increasing undisturbed forest cover in a 10-km radius. They emphasized the necessity to incorporate landscape context in the restoration and in the case of forests, ensure that the cover of contiguous forest habitat up to 10 km around restoration site is at least 50%. For dispersal of many plant species, a surrounding area of about 9 km^2 appeared sufficient to represent the colonization potential of the landscape (Prach et al., 2015). In agricultural landscapes with fragmented forest patches in Belgium, colonization rates were only a few meters per year for 85% of the forest plant species (Honnay et al., 2002), and the probability of occurrence of isolation-sensitive species in forest fragments was almost zero when the nearest source patch was farther than 200 m (Butaye et al., 2001).

Thus, restoration planning should pay considerable attention to establishing migration between extant and restored habitats at the scales relevant to species in the system. Where possible, restored areas should be in the direct neighborhood of potential colonization sources. In highly degraded areas that are distant from potential colonization sources, active species introductions (planting, seed sowing, re-introduction of characteristic animal species and soil biota) are necessary. Introduction activities must also focus on the landscape scale, creating a connected network of habitats and integrating all restored ecosystems into a larger ecological matrix. Complementing site-level restoration with the restoration and maintenance of similar habitat types in the surrounding landscape and the creation of wildlife-friendly corridors between them can not only substantially benefit the functional connectivity of species but also increase the recovery and maintenance of important ecological interactions and viable ecosystem services.

Restoring landscape-scale ecological interactions

Ecosystems and landscapes form complex systems of interacting species, populations, and individuals. Loss and decline of interactions, from pollination and predation to deterioration of more hidden relationships belowground, severely alters the functioning of ecosystems and landscapes. The recovery of interactions and trophic networks is a crucial part of restoration, but our understanding of cascading impacts in ecosystems, as well as practical experience of how to facilitate the recovery of interactions, is yet to be fully realized (Perring et al., 2015). Pollination, predation, herbivory, seed dispersal, cascading impacts of so-called ecosystem engineers through trophic levels—these are interactions that 'act' at a landscape scale of a few hundred meters to a few kilometers.

Humans have had a selective negative impact on animals with larger body sizes and larger home ranges, which has triggered cascading extinctions among other species groups and declines in important ecosystem processes and functions (Dirzo et al., 2005). Trophic rewilding is a core concept emphasizing the necessity to ensure recoveryof such species groups that act as ecosystem engineers (i.e. species that regulate community structure and interactions among trophic levels, for example large herbivores) (Svenning et al., 2016; Pereira and Navarro, 2015). There are already a number of successful rewilding examples, and the approach is currently hotly debated regarding its applicability in different regions and landscapes around the world. In 2021, based on rewilding concept, Denmark started planning of 13 new national parks that would form contiguous areas that provide space for nature to develop on its own terms together with large grazing animals and their predators.

In addition to large mammals, keystone species can include a myriad of species from different trophic levels (Hale and Koprowski, 2018); species initiating cascading effects to other species in an ecosystem can be relatively subtle and overlooked. For example, grasslands restored on intensive agricultural land developed high biodiversity of beetles of various trophic levels only if the food plants of beetles were seeded into the restoration site (Woodcock et al., 2012).

Belowground diversity and mutualistic relationships (e.g. mycorrhiza, nitrogen fixation) play a substantial role in shaping the aboveground communities—an aspect that we are only starting to understand in the context of restoration (Kardol and Wardle, 2010; Neuenkamp et al., 2019) and for which knowledge of landscape-scale impacts is still scarce. For example, it has been found only recently that arbuscular mycorrhizal (AM) fungi likely disperse quite rapidly to early-successional or restored ecosystems (e.g. García de León et al., 2016), but for restoration of completely degraded landscapes such as post-mining areas, it might still be necessary to (re-)introduce plants together with their symbiotic fungi for establishing species-rich vegetation (Vahter et al., 2020).

Pollination is one ecosystem process that is likely to strongly depend on landscape structure and the success of landscape-scale restoration (Betts et al., 2019). Insect pollinators depend on availability of food plants and nesting places within their foraging distance. Loss, degradation, and fragmentation of habitats, homogenization of landscapes, and impacts of agrochemicals are among the main drivers of the pollinator crisis, all of which are manifested at a landscape scale (Willmer, 2011). From recovery of specialized interactions between specific plants and pollinators, to effective functioning of pollinator networks at an ecosystem level, a focus on restoring habitats at the landscape scale and ensuring the high-quality 'matrix' between such habitats is vital (Betts et al., 2019; Tonietto and Larkin, 2018). Recently, quantitative models of pollination services across landscapes have started to emerge, enabling the identification of landscapes and regions where pollination service is hindered and where specific restoration activities are needed (e.g. Häussler et al., 2017).

Seed dispersal is also an ecosystem process that can depend on efficient functioning of ecological interactions. For example, many plant species require specific animal vectors for successful dispersal. Grassland species often disperse with the help of mammals; thus, traditional shepherding and restoration of landscape-scale movement of grazing animals could help to disperse the seeds over longer distances, even in greatly fragmented landscapes (Auffret et al., 2012). Seed dispersal by fruit-eating birds is known to be crucial for recovery of rainforest plant communities. In tropical and subtropical Asia, elephants, gibbons, civets, fruit bats, macaques, rodents, bears, and many other species play a key role in seed dispersal (Corlett, 2017). Thus, although it might be one of the most challenging tasks in restoration, facilitation of a diverse community of vertebrates that form mutualistic relationships with plants is essential (Galatowitsch, 2012). Until now, however, the knowledge and experience regarding re-establishment of specialized dispersal networks at a landscape scale have been scarce, although the recovery of robust and functionally diverse seed dispersal networks might be among the most important indicators of restoration success (Howe, 2016; Ribeiro da Silva et al., 2015).

The importance of small landscape elements

While it is essential to keep the focus of conservation on well-preserved and large natural eco-systems and prioritize them in restoration, it is also necessary to acknowledge the role of small landscape elements. Their importance in supporting biodiversity and the provision of ecosystem services in a landscape is often underestimated (Hunter et al., 2017). Road verges, ditches, ponds, uncropped areas in agricultural lands, city parks, cemeteries, power line corridors, flowerbeds, individual large trees, hedges, and front- or backyard gardens can all act as corridors or stepping stones for dispersal of seeds and pollen, insects, birds, and even small mammals. They provide some additional square meters or hectares of habitat for many species suffering habitat loss and fragmentation in simplified landscapes, and they can even be the last refugia for some species. Although such small natural features do not substitute for larger and intact ecosystems, they sig-nificantly increase the range of environmental conditions needed by different groups of species (Manning et al., 2006).

The role of small landscape elements is particularly large in urban and agricultural land-scapes (Hall et al., 2017; Poschlod and Braun-Reichert, 2017). Sometimes, landscape ele-ments supporting biodiversity can be quite surprising. For example, in agricultural areas in Poland, old farmsteads and farmhouses were shown to be important habitats for many farmland birds, and there was a marked negative association between the richness and abun-dance of bird species and the increasing proportion of new and renovated homesteads (Rosin et al., 2016). Even individual trees can serve as refugia for biodiversity, deliver a variety of ecosystem services (Manning et al., 2006), and support livelihoods (Ndayambaje et al., 2013). Recently, Poschlod and Braun-Reichert (2017) termed such landscape features 'keystone structures' (derived from the term 'keystone species'), emphasizing their disproportionate significance.

In restoration, small landscape elements can provide valuable source populations that sup-port the recovery of landscape-scale biodiversity; thus, they need to be taken into account while planning and implementing restoration activities. The creation and preservation of such small features are in the hands of all land-users, from people tending small individual gardens to farmers, enterprises, and municipalities making decisions for larger areas. In Estonia, we have termed such small-scale activities by local people, municipalities, or companies 'Everyone's Nature Conservation', with a key message that everyone's decisions and actions matter, and each square meter can be designed to benefit biodiversity. Since 2019, the potential actions related to maintaining, recognizing, and creating small habitat patches and landscape elements have been promoted by the Ministry of Environment as an important complementary mecha-nism in addition to traditional conservation measures to safeguard biodiversity in Estonia (see Box 22.1).

Box 22.1 Tools that focus on local conditions can guide landscape-scale restoration. Example of 'Greenmeter' from Estonia

In order to help local communities and municipalities understand the ecological condition of their surroundings and plan landscape-scale restoration, we developed a web-based tool called Greenmeter at the University of Tartu, Estonia. The tool is publicly accessible to everyone at the web address www.greenmeter.eu. Greenmeter compiles information from more than 60 detailed

---

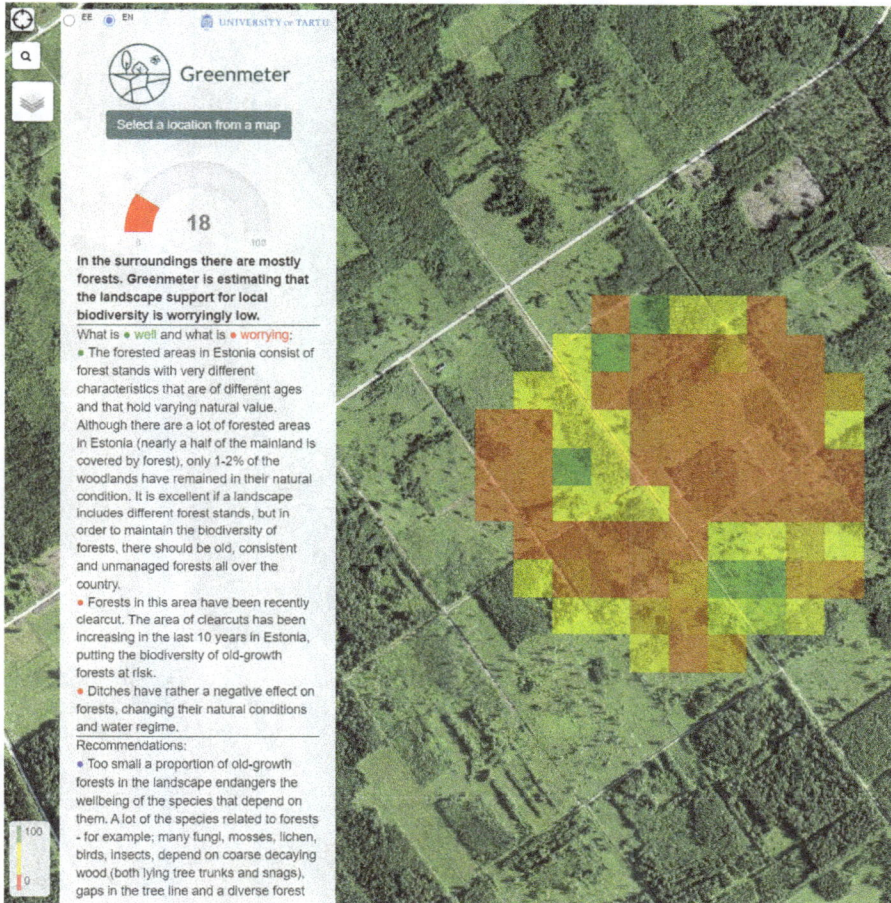

Figure 22.1 Forest landscape (500-m radius) in Estonia with low support for forest biodiversity, mostly due to fragmentation of forest areas, recent logging activities, and lack of high-quality habitats (among other indicators).

geographic information system (GIS) layers expressing land-use parameters and biodiversity indicators that are relevant in determining the condition of habitats and landscapes. If users click on their home or any other location, the tool provides an index of landscape-level support for biodiversity in a 500-meter radius. These kinds of tools have the potential to help with the planning and execution of landscape-scale restoration if they are built on good-quality data and on a good understanding of factors that influence biodiversity in a particular region. Local-language information regarding ecological shortcomings in particular landscapes, as well as suggestions for actions and activities that are applicable by most land-users, are likely to engage and empower local people (see 'Everyone's Nature Conservation' in the section titled 'The importance of small landscape elements').

Figure 22.2 Forest landscape (500-m radius) in Estonia with high support for forest biodiversity, as indicated (among many other factors) by good connectivity of forest habitats, lack of recent disturbance, and high incidence of protected species in the observed landscape.

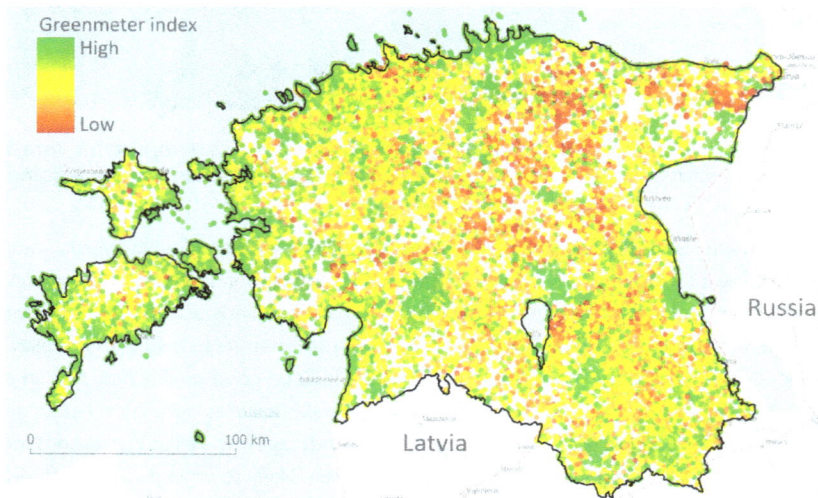

Figure 22.3 User requests from Greenmeter between April and September 2020 in Estonia. Green regions indicate locations where landscape-scale biodiversity support is measured as high and red regions those where it was low and where landscape restoration is required.

What to be aware of during landscape restoration?

Setting goals for restoration

Motivations for ecological restoration can be very diverse, depending on local environmental and socio-economic requirements and conditions. Along with a shift in conservation paradigms and the increasing intensity of the global environmental and climate crisis, the motivations for restoration also change over time. Restoration activities range from reinstating populations of threatened species and improving quality of habitats to enhancing ecosystem services (e.g. planting to increase carbon sequestration, reduce erosion, or tackle pollution) (Perring et al., 2015). The changing climate might make it increasingly necessary to re-evaluate current knowledge and set dynamic restoration goals that take near-future climate and environmental conditions into account. For example, it might be necessary to consider climate-adjusted provenancing by obtaining genetic material from different locations to capture adaptive variation within species (Prober et al., 2015). So far, however, there is no consensus among restoration ecologists and conservationists regarding the potential threats and benefits of rather radical approaches such as assisted migration or using non-native taxon substitutions to maintain key functions (Corlett, 2016; Higgs et al., 2014).

Due to a changing climate, it is probably unrealistic and perhaps even unreasonable to set a target of 'turning back time' to fully reinstate some previous, pre-degradation state of ecosystems. At the same time, knowledge of historical land-use dynamics and biodiversity is crucial for successful landscape-scale restoration (Higgs et al., 2014). An increasing number of studies show strong relationships between species diversity and historical land-use patterns, indicating the long legacy of favorable land-use practices for biodiversity, on the one hand, and the very slow response of species dynamics to landscape changes, on the other hand (Essl et al., 2015; Kuussaari et al., 2009). Historical habitat structure and landscape composition, for example the period with the longest-lasting stability prior to recent massive land-use changes, can guide spatial planning for landscape restoration. Each landscape can have remnant landscape elements, seed banks, or soil microbiota that have remained even if most habitat is destroyed. Due to these legacies and the very slow response of some species groups to landscape-scale changes, planning restoration according to historical landscape structure can substantially improve the restoration outcome (Török and Helm, 2017).

Recovery takes time

Everything takes time. 'Building up' characteristic diversity and achieving the full potential for ecosystem recovery following restoration can take decades or even centuries, so patience is needed for achieving the expected outcome (Redhead et al., 2014; Török and Helm, 2017). For example, Kolk and Naaf (2015) studied the colonization of forest species in younger and older forest patches and found that recovery can take up to 230 years. Time-delayed impacts are rarely considered when evaluating restoration success. Most projects only monitor restoration recovery for 3–5 years at most, occasionally up to 15 years after restoration (Wortley et al., 2013). This hinders our ability to understand and assess actual dynamics of restored habitats and landscapes. It is important to take a long-term approach to assessing restoration success, take possible time-delays into account, and ensure that monitoring relies on relevant indicators, including indicators of the landscape structure, species and gene pool composition, and socio-economic impacts (Aavik and Helm, 2017; de Bello et al., 2010; Wortley et al., 2013).

Aveliina Helm

What's next for landscape restoration?

Given the urgency of biodiversity and ecosystem loss, landscape-scale restoration principles have to be applied everywhere habitat degradation has occurred, most acutely in intensively used agricultural areas, cities and suburbs, and heavily degraded industrial and mining regions. In the recent decade, major advances regarding landscape-scale restoration have occurred. For example, the Bonn Challenge is a global effort to restore 350 million hectares of forest landscapes by 2030. By 2020, more than 60 countries had pledged to restore 210 million hectares of degraded and deforested lands. However, there are also a multitude of concerns linked to such large-scale commitments, ranging from the questionable feasibility of the pledges (Fagan et al., 2020) to potential shortcomings in applying necessary ecological principles when setting the targets and carrying out restoration at the landscape scale (Temperton et al., 2020).

In coming years, with mounting climate and biodiversity crises, more and more global initiatives such as the Bonn Challenge are likely to emerge, and more studies and global maps with restoration guidelines will be developed. For example, gaining worldwide attention in 2019, Bastin et al. proposed that globally there are ca 0.9 billion hectares of land available for potential tree restoration. However, while global action is needed, it is necessary to keep in mind that efficient solutions must be locally planned and executed and must be designed for particular ecological and socio-economic conditions in each region and landscape (Figures 22.1 and 22.2). Global-scale "top-down" approaches or guidelines that ignore local conditions can be virtually worthless or even harmful. Temperton et al. (2020) emphasize that despite the urgency and enthusiasm linked to the Bonn Challenge and other forest restoration initiatives, it might be necessary to 'step back and assess *where* to restore *what*' and recognize that in many landscapes, multiple habitat types should be restored, not only forests (Figure 22.3).

Decisions regarding landscape restoration need to be focused on local communities, and landscape restoration has to be implemented with care and concern regarding sustainable land governance and land-use (Sayer et al., 2013). It has been convincingly shown that lands governed by indigenous and local peoples have fared better than other lands, and restoring indigenous land rights is critical to biodiversity conservation (Díaz et al., 2019). Dismissal of the co-operation with local land-users can lead to failure. In turn, engagement of all stakeholders, most importantly local people, enterprises, and municipalities, can be a key to successful planning and implementation of landscape restoration as well as for long-term sustainability of restored landscapes.

Landscape restoration is a multidisciplinary and complex activity covering not only the ecological features that were the topic of this chapter but also socio-economic, historical, political, and governance-related aspects. With good and inclusive planning on strong ecological grounds, it is possible to tick all the boxes and deliver all the interconnected benefits: biodiversity, a safe climate, food security, and a fair and just world for all people.

Acknowledgments

The author is grateful to Emily Minor for inspiration and support and to Karen Holl for valuable comments on the manuscript. This work was funded by the Estonian Research Council (PRG874) and by the European Regional Development Fund through the Centre of Excellence EcolChange.

Bibliography

Aavik, T. and Helm, A. (2017) 'Restoration of plant species and genetic diversity depends on landscape-scale dispersal', *Restoration Ecology*, vol 26, no 52, ppS92–S102.

Auffret, A.G., Schmucki, R., Reimark, J. and Cousins, S.A.O. (2012) 'Grazing networks provide useful functional connectivity for plants in fragmented systems', *Journal of Vegetation Science*, vol 23, pp970–977.

Auffret, A.G., Rico, Y., Bullock, J.M., Hooftman, D.A.P., Pakeman, R.J., Soons, M.B., Suárez-Esteban, A., Traveset, A., Wagner, H.H. and Cousins, S.A.O. (2017) 'Plant functional connectivity: Integrating landscape structure and effective dispersal', *Journal of Ecology*, vol 105, pp1648–1656.

Atkinson, J. and Bonser, S.P. (2020) '"Active" and "passive" ecological restoration strategies in meta-analysis', *Restoration Ecology*, vol 28, no. 5, pp1032–1035.

Bastin, J.-F., Finegold, Y., Garcia, C., Mollicone, D., Rezende, M., Routh, D., Zohner, C.M. and Crowther, T.W. (2019) 'The global tree restoration potential', *Science*, vol 365, no 6448, pp76–79.

Besseau, P., Graham, S. and Christophersen, T. (2018) *Restoring Forests and Landscapes: The Key to a Sustainable Future.* Global Partnership on Forest and Landscape Restoration, Vienna, Austria, www.iufro.org/publications/other-publications/article/2018/08/28/restoring-forests-and-landscapes-the-key-to-a-sustainable-future-1/, accessed 22 Dec 2020.

Betts, M.G., Hadley, A.S. and Kormann, U. (2019) 'The landscape ecology of pollination', *Landscape Ecology*, vol 34, pp961–966.

Brudvig, L.A. (2011) 'The restoration of biodiversity: Where has research been and where does it need to go?', *American Journal of Botany*, vol 98, pp549–558.

Butaye, J., Jacquemyn, H. and Hermy, M. (2001) 'Differential colonization causing non-random forest plant community structure in a fragmented agricultural landscape', *Ecography*, vol 24, pp369–380.

Cardinale, B.J., Duffy, J.E., Gonzalez, A., Hooper, D.U., Perrings, C., Venail, P., Narwani, A., Mace, G.M., Tilman, D., Wardle, D.A., Kinzig, A.P., Daily, G.C., Loreau, M., Grace, J.B., Larigauderie, A., Srivastava, D. and Naeem, S. (2012) 'Biodiversity loss and its impact on humanity', *Nature*, vol 486, pp59–67.

Chazdon, R.L., Brancalion, P.H.S., Lamb, D., Laestadius, L., Calmon, M. and Kumar, C. (2017) 'A policy-driven knowledge agenda for global forest and landscape restoration', *Conservation Letters*, vol 10, pp125–132.

Cohen-Shacham, E., Walters, G., Janzen, C. and Maginnis, S. (eds) (2017) *Nature-based Solutions to Address Global Societal Challenges.* IUCN, Gland.

Corlett, R.T. (2016) 'Restoration, reintroduction, and rewilding in a changing world', *Trends in Ecology and Evolution*, vol 31, pp453–462.

Corlett, R.T. (2017) 'Frugivory and seed dispersal by vertebrates in tropical and subtropical Asia: An update', *Global Ecology and Conservation*, vol 11, pp1–22.

Crouzeilles, R. and Curran, M. (2016) 'Which landscape size best predicts the influence of forest cover on restoration success? A global meta-analysis on the scale of effect', *Journal of Applied Ecology*, vol 53, pp440–448.

Crouzeilles, R., Curran, M., Ferreira, M.S., Lindenmayer, D.B., Grelle, C.E.V. and Benayas, J.M.R. (2016) 'A global meta-analysis on the ecological drivers of forest restoration success', *Nature Communications*, vol 7, pp1–8.

Dass, P., Houlton, B., Wang, Y. and Warlind, D. (2018) 'Grasslands may be more reliable carbon sinks than forests in California', *Environmental Research Letters*, vol 13, art 74027.

de Bello, F., Lavorel, S., Díaz, S., Harrington, R., Cornelissen, J., Bardgett, R., Berg, M., Cipriotti, P., Feld, C., Hering, D., Martins da Silva, P., Potts, S., Sandin, L., Sousa, J., Storkey, J., Wardle, D. and Harrison, P. (2010) 'Towards an assessment of multiple ecosystem processes and services via functional traits', *Biodiversity and Conservation*, vol 19, pp2873–2893.

Díaz, S., Settele, J., Brondízio, E., Ngo, H., Guèze, M., Agard, J., Arneth, A., Balvanera, P., Brauman, K., Butchart, S., Chan, K., Garibaldi, L., Ichii, K., Liu, J., Submranian, S., Midgley, G., Miloslavich, P., Molnár, Z., Obura, D., Pfaff, A. and Polasky, C. (2019) *Summary for Policymakers of the Global Assessment Report on Biodiversity and Ecosystem Services of the Intergovernmental Science-Policy Platform on Biodiversity and Ecosystem Services.* IPBES Secretariat, Bonn.

Dirzo, R., Young, H.S., Galetti, M., Ceballos, G., Isaac, N.J.B. and Collen, B. (2005) 'Defaunation in the anthropocene', *Science*, vol 345, no 6195, pp401–406.

Elbakidze, M., Hahn, T., Dawson, L., Zimmermann, N.E., Cudlín, P., Friberg, N., Genovesi, P., Guarino, R., Helm, A., Jonsson, B. and Lengyel, S., Leroy, B., Luzzati, T., Milbau, A., Pérez-Ruzafa, A., Roche, P., Roy, H., Sabyrbekov, R., Vanbergen, A. and Vand, V. (2018) 'Direct and indirect drivers of change in biodiversity and nature's contributions to people', in M. Rounsevell, M. Fischer, A. Torre-Marin Rando and A. Mader (eds) *The IPBES Regional Assessment Report on Biodiversity and Ecosystem Services for Europe and Central Asia*. Secretariat of the Intergovernmental Science-Policy Platform on Biodiversity and Ecosystem Services, Bonn.

Ekroos, J., Ödman, A.M., Andersson, G.K., Birkhofer, K., Herbertsson, L., Klatt, B.K., Olsson, O., Olsson, P.A., Persson, A.S., Prentice, H.C. and Rundlöf, M. (2016) 'Sparing land for biodiversity at multiple spatial scales', *Frontiers in Ecology and Evolution*, vol 3, p145.

Erbaugh, J.T. and Oldekop, J.A. (2018) 'Forest landscape restoration for livelihoods and well-being', *Current Opinion in Environment Sustainability*, vol 32, pp76–83.

Essl, F., Dullinger, S., Rabitsch, W., Hulme, P.E., Pyšek, P., Wilson, J.R.U. and Richardson, D.M. (2015) 'Historical legacies accumulate to shape future biodiversity in an era of rapid global change', *Diversity and Distributions*, vol 21, pp534–547.

Fagan, M.E., Zahawi, R.A. and Reid, J.L. (2020) 'RE: No Bonn Challenge restoration commitment exceeds a country's potential area'.

Galatowitsch, S.M. (2012) *Ecological Restoration*. Sinauer Associates, Sunderland, MA.

Gann, G.D., McDonald, T., Walder, B., Aronson, J., Nelson, C.R., Jonson, J., Hallett, J.G., Eisenberg, C., Guariguata, M.R., Liu, J., Hua, F., Echeverría, C., Gonzales, E., Shaw, N., Decleer, K. and Dixon, K.W. (2019) 'International principles and standards for the practice of ecological restoration. Second edition', *Restoration Ecology*, vol 27, ppS1–S46.

García de León, D., Moora, M., Öpik, M., Neuenkamp, L., Gerz, M., Jairus, T., Vasar, M., Bueno, C.G., Davison, J. and Zobel, M. (2016) 'Symbiont dynamics during ecosystem succession: Co-occurring plant and arbuscular mycorrhizal fungal communities', *FEMS Microbiology Ecology*, vol 92, no 7, art fiw097.

Gibbs, H.K. and Salmon, J.M. (2015) 'Mapping the world's degraded lands', *Applied Geography*, vol 57, pp12–21.

Haines-Young, R. and Potschin, M. (2010) 'The links between biodiversity, ecosystem services and human well-being', in D. Raffaelli and C. Frid (eds) *Ecosystem Ecology: A New Synthesis*. BES Ecological Reviews Series, Cambridge University Press, Cambridge.

Hale, S.L. and Koprowski, J.L. (2018) 'Ecosystem-level effects of keystone species reintroduction: A literature review', *Restoration Ecology*, vol 26, pp439–445.

Hall, D.M., Camilo, G.R., Tonietto, R.K., Ollerton, J., Ahrné, K., Arduser, M., Ascher, J.S., Baldock, K.C.R., Fowler, R., Frankie, G., Goulson, D., Gunnarsson, B., Hanley, M.E., Jackson, J.I., Langellotto, G., Lowenstein, D., Minor, E.S., Philpott, S.M., Potts, S.G., Sirohi, M.H., Spevak, E.M., Stone, G.N. and Threlfall, C.G. (2017) 'The city as a refuge for insect pollinators', *Conservation Biology*, vol 31, pp24–29.

Häussler, J., Sahlin, U., Baey, C., Smith, H.G. and Clough, Y. (2017) 'Pollinator population size and pollination ecosystem service responses to enhancing floral and nesting resources', *Ecology and Evolution*, vol 7, no 6, pp1898–1908.

Helm, A. (2015) 'Habitat restoration requires landscape-scale planning', *Applied Vegetation Science*, vol 18, pp177–178.

Helsen, K., Jacquemyn, H., Hermy, M., Vandepitte, K. and Honnay, O. (2013) 'Rapid buildup of genetic diversity in founder populations of the gynodioecious plant species *Origanum vulgare* after semi-natural grassland restoration', *PLoS One*, vol 8, pp15–18.

Higgs, E., Falk, D.A., Guerrini, A., Hall, M., Harris, J., Hobbs, R.J., Jackson, S.T., Rhemtulla, J.M. and Throop, W. (2014) 'The changing role of history in restoration ecology', *Frontiers in Ecology and the Environment*, vol 12, pp499–506.

Hoffmann, A.A. and Sgró, C.M. (2011) 'Climate change and evolutionary adaptation', *Nature*, vol 470, pp479–485.

Holl, K.D., Crone, E.E. and Schultz, C.B. (2003) 'Landscape restoration: Moving from generalities to methodologies', *BioScience*, vol 53, pp491–502.

Honnay, O., Verheyen, K., Butaye, J., Jacquemyn, H., Bossuyt, B. and Hermy, M. (2002) 'Possible effects of habitat fragmentation and climate change on the range of forest plant species', *Ecology Letters*, vol 5, pp525–530.

Howe, H.F. (2016) 'Making dispersal syndromes and networks useful in tropical conservation and restoration', *Global Ecology and Conservation*, vol 6, pp152–178.

Hunter, M.L., Acuña, V., Bauer, D.M., Bell, K.P., Calhoun, A.J.K., Felipe-Lucia, M.R., Fitzsimons, J.A., González, E., Kinnison, M., Lindenmayer, D., Lundquist, C.J., Medellin, R.A., Nelson, E.J.

and Poschlod, P. (2017) 'Conserving small natural features with large ecological roles: A synthetic overview', *Biological Conservation*, vol 211, pp88–95.

IPCC (2019) *IPCC Special Report on Climate Change, Desertification, Land Degradation, Sustainable Land Management, Food Security, and Greenhouse gas fluxes in Terrestrial Ecosystems*. IPCC WGIII, London.

Isbell, F., Craven, D., Connolly, J., Loreau, M., Schmid, B., Beierkuhnlein, C., Bezemer, T M., Bonin, C., Bruelheide, H., de Luca, E., Ebeling, A., Griffin, J.N., Guo, Q., Hautier, Y., Hector, A., Jentsch, A., Kreyling, J., Lanta, V., Manning, P., Meyer, S.T., Mori, A.S., Naeem, S., Niklaus, P.A , Polley, H.W., Reich, P.B., Roscher, C., Seabloom, E.W., Smith, M.D., Thakur, M.P., Tilman, D., Tracy, B.F., van der Putten, W.H., van Ruijven, J., Weigelt, A., Weisser, W.W., Wilsey, B. and Eisenhauer, N. (2015) 'Biodiversity increases the resistance of ecosystem productivity to climate extremes', *Nature*, vol 526, pp574–577.

Isbell, F., Gonzalez, A., Loreau, M., Cowles, J., Díaz, S., Hector, A., Mace, G.M., Wardle, D.A., Connor, M.O., Duffy, J.E., Turnbull, L.A., Thompson, P.L. and Larigauderie, A. (2017) 'Linking the influence and dependence of people on biodiversity across scales', *Nature*, vol 546, pp65–80.

IUCN and WRI (2014) *A guide to the Restoration Opportunities Assessment Methodology (ROAM): Assessing Forest Landscape Restoration Opportunities at the National or Sub-national Level*. IUCN, Gland, Switzerland.

Jones, H.P., Jones, P.C., Barbier, E.B., Blackburn, R.C., Rey Benayas, J.M., Holl, K.D., … and Mateos, D.M. (2018) 'Restoration and repair of Earth's damaged ecosystems', *Proceedings of the Royal Society B: Biological Sciences*, vol 285, no 1873, art 20172577.

Kardol, P. and Wardle, D.A. (2010) 'How understanding aboveground-belowground linkages can assist restoration ecology', *Trends in Ecology and Evolution*, vol 25, pp670–679.

Kolk, J. and Naaf, T. (2015), 'Herb layer extinction debt in highly fragmented temperate forests–Completely paid after 160 years?' *Biological Conservation*, vol 182, pp164–172.

Kuussaari, M., Bommarco, R., Heikkinen, R.K., Helm, A., Krauss, J., Lindborg, R., Öckinger, E., Pärtel, M., Pino, J., Rodà, F., Stefanescu, C., Teder, T., Zobel, M. and Steffan-Dewenter, I. (2009) 'Extinction debt: A challenge for biodiversity conservation', *Trends Ecology and Evolution*, vol 24, no 10, pp564–571.

Kolk, J. and Naaf, T. (2015) 'Herb layer extinction debt in highly fragmented temperate forests: Completely paid after 160 years?', *Biological Conservation*, vol 182, pp164–172.

Leclère, D., Obersteiner, M., Barrett, M., Butchart, S.H., Chaudhary, A., De Palma, A., DeClerck, F.A., Di Marco, M., Doelman, J.C., Dürauer, M. and Freeman, R. (2020) 'Bending the curve of terrestrial biodiversity needs an integrated strategy', *Nature*, vol 585, pp 551–556.

Manning, A.D., Fischer, J. and Lindenmayer, D.B. (2006) 'Scattered trees are keystone structures - Implications for conservation', *Biological Conservation*, vol 132, pp311–321.

Martin, D.M. (2017) 'Ecological restoration should be redefined for the twenty-first century', *Restoration Ecology*, vol 25, pp668–673.

Meli, P., Holl, K.D., Rey Benayas, J.M., Jones, H.P., Jones, P.C., Montoya, D. and Moreno Mateos, D. (2017) 'A global review of past land use, climate, and active vs. passive restoration effects on forest recovery', *PLoS ONE*, vol 12, no 2, art e0171368.

Meli, P., Schweizer, D., Brancalion, P.H.S., Murcia, C. and Guariguata, M.R. (2019) 'Multidimensional training among Latin America's restoration professionals', *Restoration Ecology*, vol 27, pp477–484.

Metzger, J.P. and Brancalion, P.H. (2016) 'Landscape ecology and restoration processes', in M.A. Palmer, J.B. Zedler and D.A. Falk (eds) *Foundations of Restoration Ecology*. Island Press, Washington, DC.

Mijangos, J.L., Pacioni, C., Spencer, P.B. and Craig, M.D. (2015) 'Contribution of genetics to ecological restoration', *Molecular Ecology*, vol 24, no (1), pp22–37.

Montanarella, L., Scholes, R. and Brainich, A. (2018) *The IPBES Assessment Report on Land Degradation and Restoration*. Secretariat of the Intergovernmental Science-Policy Platform on Biodiversity and Ecosystem Services, Bonn, Germany.

Mori, A.S., Lertzman, K.P. and Gustafsson, L. (2017) 'Biodiversity and ecosystem services in forest ecosystems: A research agenda for applied forest ecology', *Journal of Applied Ecology*, vol 54, pp12–27.

Ndayambaje, J.D., Heijman, W.J.M. and Mohren, G.M.J. (2013) 'Farm woodlots in rural Rwanda: Purposes and determinants', *Agroforestry Systems*, vol 87, pp797–814.

Nerlekar, A.N. and Veldman, J.W. (2020) 'High plant diversity and slow assembly of old-growth grasslands', *PNAS*, vol 117, pp18550–18556.

Neuenkamp, L., Zobel, M., Lind, E., Gerz, M. and Moora, M. (2019) 'Arbuscular mycorrhizal fungal community composition determines the competitive response of two grassland forbs', *PLoS ONE*, vol 14, no 7, art e0219527.

Newbold, T., Hudson, L.N., Hill, S.L.L., Contu, S., Lysenko, I., Senior, R.A., Börger, L., Bennett, D.J., Choimes, A., Collen, B., Day, J., De Palma, A., Díaz, S., Echeverria-Londoño, S., Edgar, M.J., Feldman, A., Garon, M., Harrison, M.L.K., Alhusseini, T., Ingram, D.J., Itescu, Y., Kattge, J., Kemp, V., Kirkpatrick, L., Kleyer, M., Correia, D.L.P., Martin, C.D., Meiri, S., Novosolov, M., Pan, Y., Phillips, H.R.P., Purves, D.W., Robinson, A., Simpson, J., Tuck, S.L., Weiher, E., White, H.J., Ewers, R.M., MacE, G.M., Scharlemann, J.P.W. and Purvis, A. (2015) 'Global effects of land use on local terrestrial biodiversity', *Nature*, vol 520, pp45–50.

Olden, J.D., Douglas, M. and Douglas, M.R. (2005) 'The human dimensions of biotic homogenization', *Conservation Biology*, vol 19, pp2036–2038.

Pärtel, M. and Zobel, M. (1998) 'Formation of actual species pools in calcareous grasslands: Historical and geographical aspects (Abstracts)', *Studies in Plant Ecology*, vol 20, pp101.

Pereira, H. and Navarro, L. (eds) (2015) *Rewilding European Landscapes*. Springer Cham, Heidelberg.

Perring, M.P., Standish, R.J., Price, J.N., Craig, M.D., Erickson, T.E., Ruthrof, K.X., Whiteley, A.S., Valentine, L.E. and Hobbs, R.J. (2015) 'Advances in restoration ecology: Rising to the challenges of the coming decades', *Ecosphere*, vol 6, no 8, pp1–25.

Pollock, M.M., Beechie, T.J., Wheaton, J.M., Jordan, C.E., Bouwes, N., Weber, N. and Volk, C. (2014) 'Using beaver dams to restore incised stream ecosystems', *BioScience*, vol 64, pp279–290.

Poschlod, P. and Braun-Reichert, R. (2017) 'Small natural features with large ecological roles in ancient agricultural landscapes of Central Europe: History, value, status, and conservation', *Biological Conservation*, vol 211, pp60–68.

Prach, K., Fajmon, K., Jongepierová, I. and Řehounková, K. (2015) 'Landscape context in colonization of restored dry grasslands by target species', *Applied Vegetation Science*, vol 18, pp181–189.

Prach, K., Šebelíková, L., Řehounková, K. and del Moral, R. (2019) 'Possibilities and limitations of passive restoration of heavily disturbed sites', *Landscape Research*.

Prober, S.M., Byrne, M., McLean, E.H., Steane, D.A., Potts, B.M., Vaillancourt, R.E. and Stock, W.D. (2015) 'Climate-adjusted provenancing: A strategy for climate-resilient ecological restoration', *Frontiers in Ecology and Evolution*, vol 3, pp1–5.

Redhead, J.W., Sheail, J., Bullock, J.M., Ferreruela, A., Walker, K.J. and Pywell, R.F. (2014) 'The natural regeneration of calcareous grassland at a landscape scale: 150 years of plant community re-assembly on Salisbury Plain, UK', *Applied Vegetation Science*, vol 17, pp408–418.

Ribeiro da Silva, F., Montoya, D., Furtado, R., Memmott, J., Pizo, M.A. and Rodrigues, R.R. (2015) 'The restoration of tropical seed dispersal networks', *Restoration Ecology*, vol 23, pp852–860.

Rosin, Z.M., Skórka, P., Pärt, T., Żmihorski, M., Ekner-Grzyb, A., Kwieciński, Z. and Tryjanowski, P. (2016) 'Villages and their old farmsteads are hot spots of bird diversity in agricultural landscapes', *Journal of Applied Ecology*, vol 53, pp1363–1372.

Sala, O.E., Chapin III, F.S., Armesto, J.J., Berlow, E., Bloomfield, J., Dirzo, R., Huber-Sanwald, E., Huenneke, L.F., Jackson, R.B., Kinzig, A., Leemans, R., Lodge, D.M., Mooney, H.A., Oesterheld, M., Poff, N.L., Sykes, M.T., Walker, B.H., Walker, M. and Wall, D.H. (2000) 'Global biodiversity scenarios for the Year 2100', *Science*, vol 287, no 5459, pp1770–1774.

Sayer, J., Sunderland, T., Ghazoul, J., Pfund, J.-L., Sheil, D., Meijaard, E., Venter, M., Boedhihartono, A.K., Day, M., Garcia, C., van Oosten, C. and Buck, L.E. (2013) 'Ten principles for a landscape approach to reconciling agriculture, conservation, and other competing land uses', *PNAS*, vol 110, pp8349–8356.

Searchinger, T.D., Wirsenius, S., Beringer, T. and Dumas, P. (2018) 'Assessing the efficiency of changes in land use for mitigating climate change', *Nature*, vol 564, pp249–253.

Şekercioğlu, Ç.H., Anderson, S., Akçay, E., Bilgin, R., Can, Ö.E., Semiz, G., Tavşanoğlu, Ç., Yokeş, M.B., Soyumert, A., İpekdal, K., Sağlam, İ.K., Yücel, M. and Nüzhet Dalfes, H. (2011) 'Turkey's globally important biodiversity in crisis', *Biological Conservation*, vol 144, pp2752–2769.

Shackelford, N., Hobbs, R.J., Burgar, J.M., Erickson, T.E., Fontaine, J.B., Laliberté, E., Ramalho, C.E., Perring, M.P. and Standish, R.J. (2013) 'Primed for change: Developing ecological restoration for the 21st century', *Restoration Ecology*, vol 21, pp297–304.

Shimamoto, C.Y., Padial, A.A., da Rosa, C.M. and Marques, M.C.M. (2018) 'Restoration of ecosystem services in tropical forests: A global meta-analysis', *PLoS ONE*, vol 13, no 12, art e0208523.

Svenning, J.-C., Pedersen, P.B.M., Donlan, C.J., Ejrnæs, R., Faurby, S., Galetti, M., Hansen, D.M., Sandel, B., Sandom, C.J., Terborgh, J.W. and Vera, F.W.M. (2016) 'Science for a wilder Anthropocene: Synthesis and future directions for trophic rewilding research', *PNAS*, vol 113, pp898–906.

Temperton, V.M., Buchmann, N., Buisson, E., Durigan, G., Kazmierczak, Ł., Perring, M.P., de Sá Dechoum, M., Veldman, J.W. and Overbeck, G.E. (2019) 'Step back from the forest and step up to the Bonn

Challenge: how a broad ecological perspective can promote successful landscape restoration', *Restoration Ecology*, vol 27, no. 4, pp705–719.

Tonietto, R.K. and Larkin, D.J. (2018) 'Habitat restoration benefits wild bees: A meta-analysis', *Journal of Applied Ecology*, vol 55, no 2, pp582–590.

Török, P. and Helm, A. (2017) 'Ecological theory provides strong support for habitat restoration', *Biological Conservation*, vol 206, pp85–91.

Vahter, T., Bueno, C.G., Davison, J., Herodes, K., Hiiesalu, I., Kasari-Toussaint, L., Oja, J., Olsson, P.A., Sepp, S.K., Zobel, M. and Vasar, M. (2020) 'Co-introduction of native mycorrhizal fungi and plant seeds accelerates restoration of post-mining landscapes', *Journal of Applied Ecology*, vol 57, no 9, pp1741–1751.

Watson, D.M., Doerr, V.A., Banks, S.C., Driscoll, D.A., van der Ree, R., Doerr, E.D. and Sunnucks, P. (2017) 'Monitoring ecological consequences of efforts to restore landscape-scale connectivity', *Biological Conservation*, vol 206, pp201–209.

Willmer, P. (2011) *Pollination and Floral Ecology*. Princeton University Press, Princeton.

Winsa, M., Bommarco, R., Lindborg, R., Marini, L. and Öckinger, E. (2015) 'Recovery of plant diversity in restored semi-natural pastures depends on adjacent land use', *Applied Vegetation Science*, vol 18, no 3, pp413–422.

Woodcock, B.A., Bullock, J.M., Mortimer, S.R. and Pywell, R.F. (2012) 'Limiting factors in the restoration of UK grassland beetle assemblages', *Biological Conservation*, vol 146, pp136–143.

Wortley, L., Hero, J.M. and Howes, M. (2013) 'Evaluating ecological restoration success: A review of the literature', *Restoration Ecology*, vol 21, pp537–543.

World Wildlife Fund (2020) *Living Planet Report 2020: Bending the Curve of Biodiversity Loss*. R.E.A. Almond, M. Grooten and T. Petersen (eds). WWF, Gland, Switzerland.

Zobel, M., Szava-Kovats, R.C. and Pärtel, M. (2011) 'Regional influences on local species composition', in R. Baxter (ed) *Encyclopedia of Life Sciences*. Wiley Online Library.

23

Landscapes and climate change – case studies from Europe

B.C. Meyer and G. Mezosi

Introduction

Both landscape and climate change are complex concepts. Landscapes are a complex system of natural (e.g., geomorphic, climatic, hydrographic, and vegetation features) and social (e.g., land use) elements. Climate change, at least in the Anthropocene, typically involves rising temperatures, changes in precipitation volume and timing, and increased frequency and severity of extreme weather events (e.g., intense rainfall events, strong winds). Landscapes can be evaluated in many ways, for example from aesthetic, ecological, or cultural points of view (including concepts such as landscape function and ecosystem services). As discussed in other chapters of this volume, there are many possible methods for evaluating these landscape factors, including fieldwork (Chapter 11), landscape metrics (Chapter 15), and modelling (Chapter 16). Climate change can also be characterized in multiple ways, often using quantitative parameters that describe key variables (e.g., number and amount of rainy days, number of summer days (>25°) derived from the outputs of broad-scale, general circulation models (Meyer et al. 2017, Mezösi et al. 2015, 2016).

The impacts of fast-rising temperatures – and related changes in precipitation and wind – on land surfaces are envisaged in the Intergovernmental Panel on Climate Change (IPCC) (2019) report on 'Climate Change and Land'. The report explores the impacts of climate change on the 'land' (including the problems of land use change) by focusing on the problems of (1) desertification, (2) land degradation, (3) sustainable land management, (4) food security, and (5) greenhouse gas fluxes in terrestrial ecosystems. The report considers the broader concept of 'land' with very limited reference to landscapes or landscape systems and no reference to seascapes. For these five problems, the report indicates model-based pathways of change with chapters on (1) people, land, and climate in a warming world; (2) adaptation and mitigation response options; (3) enabling response options; and (4) action in the near term. The earlier IPCC report on *Managing the Risks of Extreme Events and Disasters to Advance Climate Change Adaptation* (Field et al. 2012) focused attention on climate-driven problems related to increased risks of extreme events and disasters affecting multiple land uses and landscapes globally. Together, these reports give a clear orientation to the need for landscape science and landscape ecological research to improve understanding of the impacts of climate change (in terms of both expected averages and extremes) on land use, landscape function, and ecosystem services. An important goal of landscape science should be the scaling down of the broad IPCC (2019) scenario pathways to the landscape scale by linking landscape-level socioeconomic development, mitigation responses, and land use change.

Landscapes are defined as spatial entities of land shaped by geology, relief, climate, water, soil, and ecological communities superimposed human modifications by land use (Bastian and Steinhardt 2002). Of these, climate change most directly affects the climate factor, with changing temperature, precipitation, and wind affecting many processes at the landscape scale. Landscape systems are commonly analyzed by characterizing or quantifying landscape functions, ecosystem services, or risks due to environmental hazards (Bastian and Steinhardt 2002). System changes of water functions modified by climate change generally occur over shorter temporal scales than system changes in (for example) vegetation or soil structure (Farina 2006). Thresholds are commonly used to specify the rate of internal regulation and to clarify levels of a system change, and 'tipping points' are also used as descriptors of the point at which a system changes from one behavioral status (or state) to another (Bowman et al. 2015). However, it is often difficult to specify the integrative numerical thresholds (or tipping points) applicable to a landscape as a total system. Consequently, landscape states are often estimated on the basis of individual thresholds of inherent landscape parameters or indicator variables including, for example, groundwater level change, soil erosion rates, and vegetation degradation (Bridges and Oldeman 1999, Kairis et al. 2014, Salvati and Forino 2014). By synthesizing data on several of these indicators, assessments of the status of challenges such as 'land degradation' can be evaluated (Kertész 2009).

Climate change is a large-scale and encompassing issue that is likely to have many impacts on many different aspects of landscapes. Consequently, many challenges exist when aiming to improve understanding and ensure sustainable management of landscapes and land use change under the focus of climate change. This chapter first considers how data from global climate change scenarios and models can be used at regional and landscape scales. It then presents case studies of the impacts of climate change on aridification and drought, wind erosion, and land use. The case studies are located in central European landscapes, but many of the issues discussed are relevant to other landscapes in other regions. We conclude by considering how climate changes may influence related landscape risks and hazards.

Downscaling global climate change to landscapes: a European example

Global physical models used to simulate climate change scenarios (e.g., General Circulation Models) are based on large-scale data that poorly reflect the complexity of landscapes at the local scale (Stocker et al. 2013). To downscale global physical models, two further modelling steps are needed: (i) from global physical models to continental-scale regional climate change models, and (ii) the linkage and combination of landscape information with these regional models. A range of competing regional models for Europe (and for all regions in the world) have been developed in the last two decades. These regional models are 'forced' by (i.e., their boundary conditions are provided by) general circulation models to provide a set of regional data at a data scale of 25—50 km². These regional data are intended to enable improved climate change understanding at regional and landscape scales (e.g. as in ensemble model work for Europe; Van der Linden and Mitchell 2009) as well as providing data for meteorological assessments and policy advice in spatial and landscape planning (e.g. EEA 2015).

The following example gives insights into the methodological steps developed in an assessment of regional climate change impacts on Hungarian landscapes (Mezösi et al. 2013) and is based on the methodology first developed by Rannow et al. (2010) for German landscapes. Using multiple indicators, and accounting for both exposure and sensitivity to climate change, Rannow et al. (2010) developed a framework to assess climate impacts that are important for spatial planning. The framework uses input data for multiple physical processes to classify the exposure and sensitivity of spatial regions (specifically for Germany at the Nomenclature of Territorial Units for Statistics regional level) to different indicators before combining this into a final potential climate impact assessment.

Building on the framework approach developed by Rannow et al. (2010) and following the discussion in Reed et al. (2011), Mezösi et al. (2013) aimed to develop a countrywide assessment of climate change impacts on Hungarian landscapes. The study combined a sensitivity analysis of important landscape functions based on simple predictive modelling of hazardous landscape processes (e.g. flash floods) with the aim of developing a specific method for the investigation of effects of changing weather extremes to determine the regional-scale landscape vulnerability (Mezösi et al. 2015). Climatic exposure parameter analysis was analyzed in Mezösi et al. (2015) on the basis of the original data of the meso-scale IPCC A1B climate scenario from the REMO and ALADIN regional models. The ALADIN and REMO models were deemed most suitable in this example, but a large number of competing regional models have been developed in the last decade, all with the aim of downscaling climate signals from different global physically based models (Dosio 2016). Climate exposure parameters from the REMO and ALADIN models for extreme events important at the landscape scale (e.g. days with precipitation greater than 30 mm, heat waves) and for daily temperature and precipitation were analyzed. Landscape functions influenced by climate change are proxies of the primary challenges affecting Hungary. Soil erosion caused by water and wind, drought, mass movements, and flash floods were investigated for 1961–1990, 2021–2050, and 2071–2100 based on methods described in Mezösi et al. (2013). Sensitivity thresholds taken from scientific literature were used to make impact assessments. The landscape sensitivity indicators were interpreted, and an integrative summary of the five selected indicators was made by mapping the regions facing only a few of multiple problematic sensitivities. Figure 23.1 shows an example of landscape-based regional types of climate change exposure as a result of cluster analysis based on all factors regionalized.

The cluster analysis results shown in Figure 23.1 for modelled climate change exposure reveals four main climate-based regional types across Hungary:

- Regional type 1 is expected to have moderate average temperature increases but with some distinct changes in temperature events. Future annual precipitation is expected to increase in total but with moderate changes in extreme rainfall events. This regional type creates a western-central corridor from the north to the south of Hungary;
- Regional type 2 lies in the north-east of Hungary, along the Slovakian border. It has the lowest annual mean temperatures, the highest intraregional temperature variation, and moderate precipitation totals. These factors result in greater humidity when compared with the other regions. A slight increase in the precipitation total and the (few) extreme event days predicted may have positive effects on agriculture;
- Regional type 3 covers a large area of Hungary in the center and the south-east. This region type has a flat topography with the highest temperatures and the highest modelled temperature increase. Significant changes in extreme temperature events, including a decline in the number of frost days, are expected. This regional type also receives the lowest annual precipitation totals and expects the greatest precipitation decline. An increase in heavy-rainfall days indicates a growing concentration of rainfall and correspondingly, longer drought periods;
- Regional type 4 covers two meso-regions in the western, hilly area of Hungary. This regional type is characterized by lower temperatures, less temperature increase, and less expected change in temperature extremes. This type is expected to be more humid in the scenario, with higher precipitation totals but also smaller precipitation-change ratios and smaller change rates regarding heavy rain events.

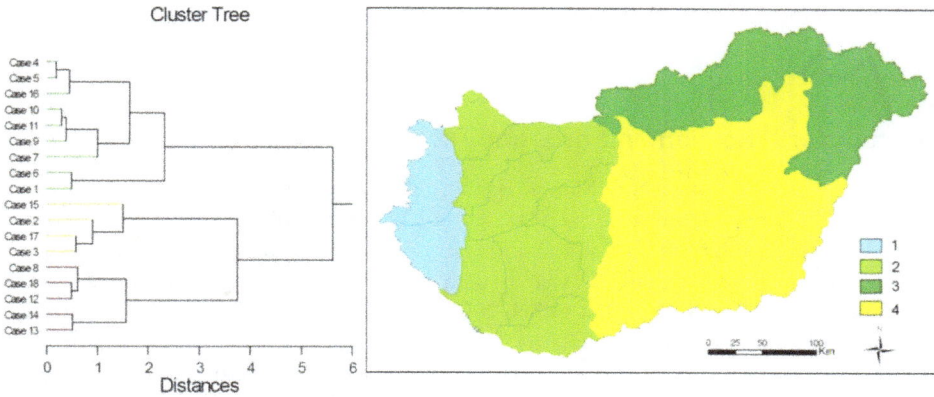

Figure 23.1 The four main Hungarian regional types of climate change exposure as a result of cluster analysis using all climatic factors based on regional climate models (Mezösi et al. 2013). Regions 1–17 are the meso-scale regions of Hungary. The main regional type clusters are differentiated into region types.

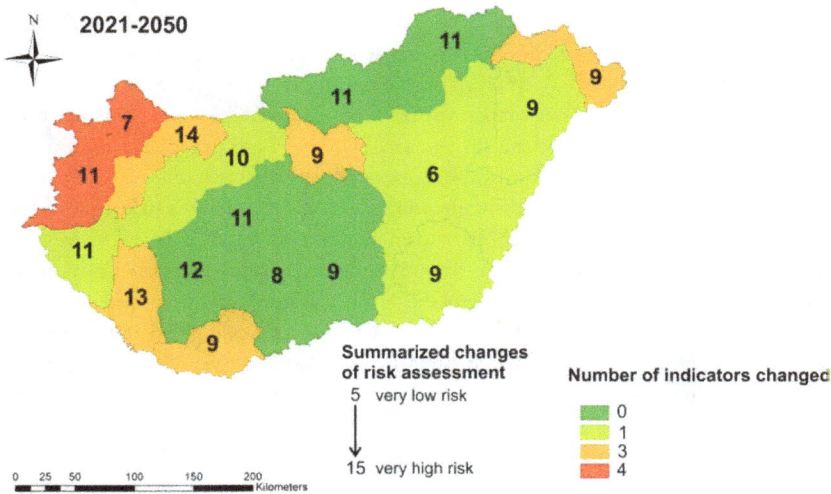

Figure 23.2 Number of hazard class changes of indicators and additive sensitivity assessment of meso-regional hazard for 2021–2050 scenarios compared with the period 1961–1990 for Hungary. (From Mezösi et al. 2013.)

Figure 23.2 shows the additive sensitivity assessment of the meso-regional hazards based on the five investigated indicators: (i) soil erosion caused by water; (ii) soil erosion caused by wind; (iii) drought; (iv) mass movement; and (v) flash floods. The figure is based on an additive summary of changes in the classes for 2021–2050 (Mezösi et al. 2013). Such qualitative assessment methods are suitable for usage in adaptive planning for regions. From a research and scientific modelling viewpoint, further detailed investigations based on integrative landscape and land use models are required to deepen knowledge and understanding of processes at the landscape scale. Further research is also needed on how landscape maintenance is linked to interactions between the investigated functions (e.g. between wind erosion and drought or between water erosion and flash floods).

Case study: aridification and drought

The challenges of aridification and drought as landscape degradation processes are highlighted in the 2019 IPCC report on Climate Change and Land (Shukla et al. 2019). Landscapes represent a complex entity with material and energy flow across their borders, and as such, a landscape is an open and dynamic system with parameters whose interrelationships lead to high complexity (Brunsden 2001, Usher 2001). As a system, a landscape has an internal regulation (feedback) of its own, which can shift it partly towards a stable state, i.e. balance, or in the case of positive feedback, towards a state of imbalance. Degradation negatively impacts a landscape's ability to return to a stable and balanced state. We understand and define 'landscape degradation' as an irreversible or non-resilient system change within a landscape that affects the landscape system components (i.e. their abiotic and biotic elements, land use, and interlinkages) and the natural and cultural capacities of the landscape in terms of structure, processes, and landscape functions (productive, ecological, and social). Aridification refers generally to a regionally drier environment caused by climate change and human impacts (Meyer et al. 2017).

A study on landscape degradation at different spatial scales caused by aridification (Meyer et al. 2017) provides an example of the risk of climate change pushing the landscape system into a new equilibrium state. Such system transformation may cause irreversible changes. The study focuses on the sand-covered landscapes of the Hungarian Danube-Tisza Interfluve region and the influence of potential aridification on landscape degradation processes at the regional, landscape, and local site scales. Variables included in the analysis included changes in groundwater level (well data), lake surface area (Modified Normalized Difference Water Index), and vegetation cover (Enhanced Vegetation Index) based on data measurements on a timescale between 12 and 60 years. Significant regional variations in decreasing groundwater levels were observed. The 'lake surface area' parameter is based on remote sensing data and shows a good representative capability, since several local lakes that had dried out in the landscapes were identified. Based on the results of changes in groundwater, vegetation, land cover, and land use indicators, land degradation was generally identified by the methods and data explained in Meyer et al. (2017). However, the height of water in a single local lake did not fall during the critical period because it has been regulated artificially by the local water management activities; at the landscape scale, changes in the water balance were interpreted as phases of system stability and system transformation. Improvements in understanding thresholds are needed to better identify the aridification problem and to better support regional policy and management towards a neutrality of landscape degradation as demanded by the sustainability development goals (Goal 15.3). Furthermore, a scenario analysis for climate change in this region has not yet been applied because forecasts are not yet available for future groundwater levels. The Meyer et al. (2017) aridification study, nevertheless, clarifies the need for running further detailed landscape-scale investigations into drought with the aim of spatializing drought sensitivities and vulnerability at the landscape scale for climate change scenarios.

A second study on projected changes in the drought hazard in Hungary due to climate change (Blanka et al. 2013) shows that drought is a severe natural hazard likely to cause challenges in the Carpathian Basin. Over the next century, drought is likely to remain one of the most serious natural hazards in the region. Motivated by this problem, the analysis presented by

Blanka et al. (2013) outlines spatial and temporal changes in drought hazards until the end of this century based on the REMO and ALADIN regional climate model simulations. The aim of the study was to assess the state of drought hazards in Hungary and to identify potentially vulnerable spatial areas due to climate change in the periods 2021–2050 and 2071–2100 for the IPCC's A1B emission scenario. In the study of Blanka et al. (2013), the magnitude of drought hazard was calculated using the aridity Index of De Martonne (IDM), the Pálfai Drought Index (PaDI$_0$), and the standardized anomaly index (SAI). Technically, an aridity index characterizes the climate of a region, but it is still useful for drought hazard assessment. Both PaDI$_0$ and SAI are relative indicators and characterize drought in agricultural years.

The results from Blanka et al. (2013) show how the identification of critical drought hazard regions can be used in spatial planning to develop optimal land uses and to adapt water management in relation to climate changes. For example, Figure 23.3 (after Blanka et al 2013) shows the expected changes across Hungarian landscapes based on values of the PaDI$_0$ index in the periods of 1961–1990, 2021–2050, and 2071–2100. The PaDI$_0$ index uses monthly temperature and precipitation data, well suited to clarifying changes based on climate change scenarios.

During the 21st century, drought hazard is expected to increase in a spatially heterogeneous manner across Hungary due to climate change (Blanka et al. 2013). Based on temperature and precipitation data, the largest increase in projected drought hazard by the end of the 21st century is expected for the Great Hungarian Plain. Moreover, the changes in extreme indices (e.g. in the reduction of the number of days with precipitation greater than 30 mm) suggest that the trend of frequency and duration of drought periods is increasing. The drought hazard is projected to be lowest in the westernmost part of Hungary. The result of this study is based on qualitative and quantitative analyses that showed changes in precipitation, temperature, and extreme indices.

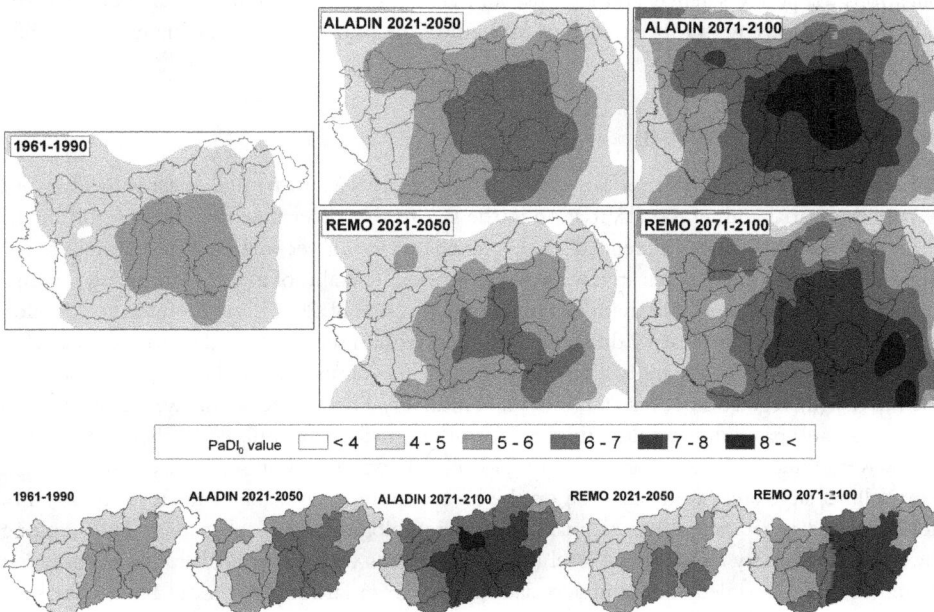

Figure 23.3 Changes in the PaDI$_0$ index for the periods 2021–2050 and 2071–2100. (After Blanka et al. 2013.)

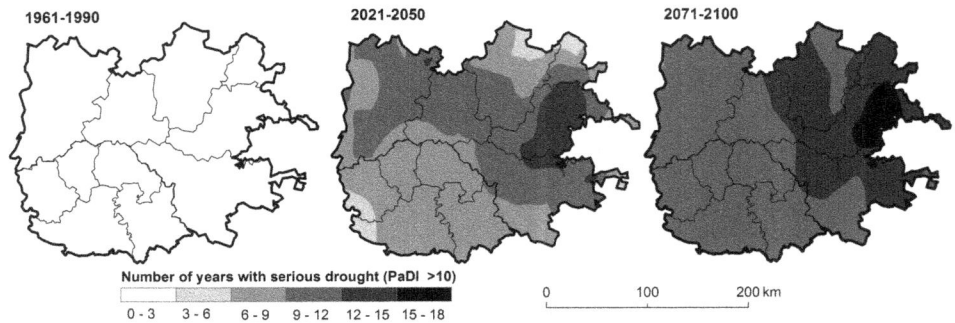

Figure 23.4 Number of years with serious drought (PaDI values >10) in the periods 1961–1990, 2021–2050, and 2071–2100, based on the averages of the REMO and ALADIN model simulations. (From Mezösi et al. 2016.)

These results demonstrate how drought may cause problems in land use more frequently in the future. In the Carpathian Basin, aridification may result in groundwater level and water surface area changes, potentially modifying the local site or the landscape-scale levels through a complete ecological transformation, creating land use changes by management activities. In some situations, aridification may result in a system transformation to another equilibrium state (Mezősi et al. 2015). A more detailed investigation in the South Hungarian–Serbian–Romanian border region was performed by Mezösi et al. (2016) on the expected mid- and long-term changes in drought hazard for the south-eastern Carpathian Basin. The results clearly show the dramatic increase in serious droughts expected in the A1B scenario (Figure 23.4). However, the implications for policy, science, and land use and landscapes are not fully understood, as is the case in many regions, with further analysis of the possible impacts of climate change on aridification and drought needed.

Case study: wind erosion

Wind erosion is one of the major landscape problems and serious natural hazards in Hungary. However, the spatial and temporal variation in the factors that determine the location and intensity of wind erosion damage influenced by climate change is not yet well known, and nor are the changes in regional and local sensitivities to wind erosion. The aim of a study estimating regional differences in wind erosion sensitivity in Hungary (Mezösi et al. 2015) was, therefore, to develop a method to estimate the regional differences in wind erosion sensitivity and exposure. The wind erosion sensitivity analysis and modelling considered the most important and complex factors of (i) soil sensitivity, (ii) vegetation cover, and (iii) wind erodibility. These three factors were first analyzed based on relevant spatial data, resulting in three factor sensitivity maps. Subsequently, these sensitivity maps were linked by fuzzy logic into a regional-scale wind erosion sensitivity map. Through this modelling, large areas were analyzed on the basis of publicly available datasets in the form of remotely sensed vegetation data, soil maps, and meteorological wind speed data. The resulting wind erosion sensitivity at a regional level (Figure 23.5) was verified by field studies in three test areas and through recorded economic loss data (mostly from farming) from the Hungarian state insurance company. The spatial resolution of the sensitivity map is appropriate for regional applications. Sensitive areas are identified using the map, providing a good overview for development control and protection measures, policy programs, and adapting management activities.

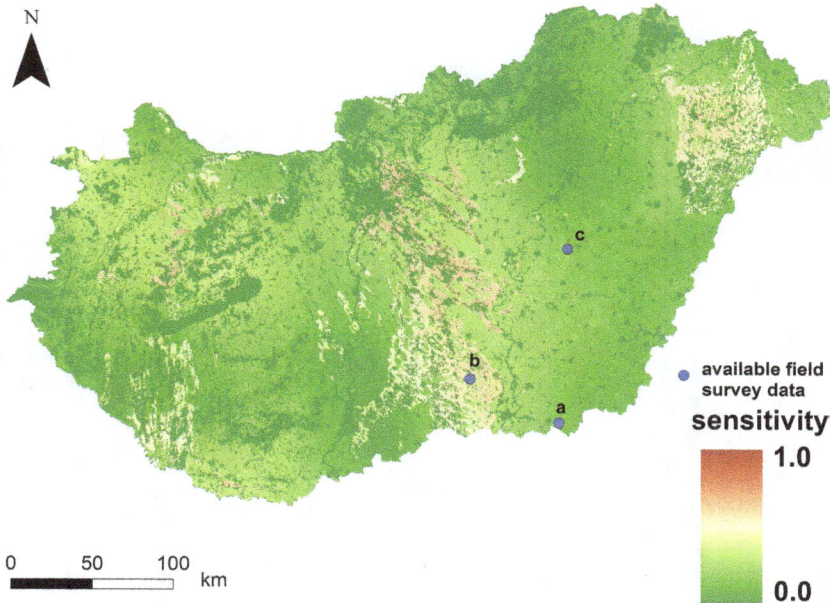

Figure 23.5 Wind erosion sensitivity at a regional level overlaid with a land cover mask to exclude forest and urbanized areas. The study sites for validation where field survey data are available (a: Apatfalva, b: Kömpöc, c: Tisza plain). (From Mezösi et al. 2015.)

Mezösi *et al.* (2014) took a similar approach to examine possible climate change impacts on environmental hazards on the Great Hungarian Plain through to the end of the 21st century. This region is most susceptible to wind erosion in March and April. For this period, ALADIN and REMO model simulations (as explained in detail by Mezösi et al. 2014 projected a slight (2%) increase in wind erosion hazard for 2071–2100 compared with the 1961–1990 reference period. However, no increase is projected for the 2021–2050 period by ALADIN model simulations, while REMO projects a greater increase (7%) than for the later period of 2071–2100. To normalize projections for further analysis, extreme values (<−10% and >15%) were removed. After the normalization, the rate of changes varied between the maxima of −10% and 15% in the first period (2021–2050) but only between −5% and 5% in the second period (2071–2100). In the earlier period, the models exhibit quite different patterns of change (REMO model simulation data have greater variation than ALADIN model simulation data: −5% and 15% compared with −10% and 0%). According to the ALADIN model, 83% of arable land is projected to have an insignificant decrease (−5% to 0%) in wind erosion hazard. In contrast, the REMO model projects a decrease in wind erosion hazard over only 6% of the area, while half of the area (49%) will be affected by a more than 10% increase.

These results demonstrate how projections of landscape change due to climate change can be strongly dependent on the climate model chosen. However, results from Mezösi et al. (2014) indicate that the spatial distribution of wind erosion is also largely dependent on the erodibility of the soils, as demonstrated in Figure 23.6. Here, the greatest wind erosion hazard was found in the north-eastern and central areas of the study area, especially in the sandy areas in the Great Hungarian Plain, as expected from other studies (Lóki 2011). For the period 2071–2100, the

a,

b,

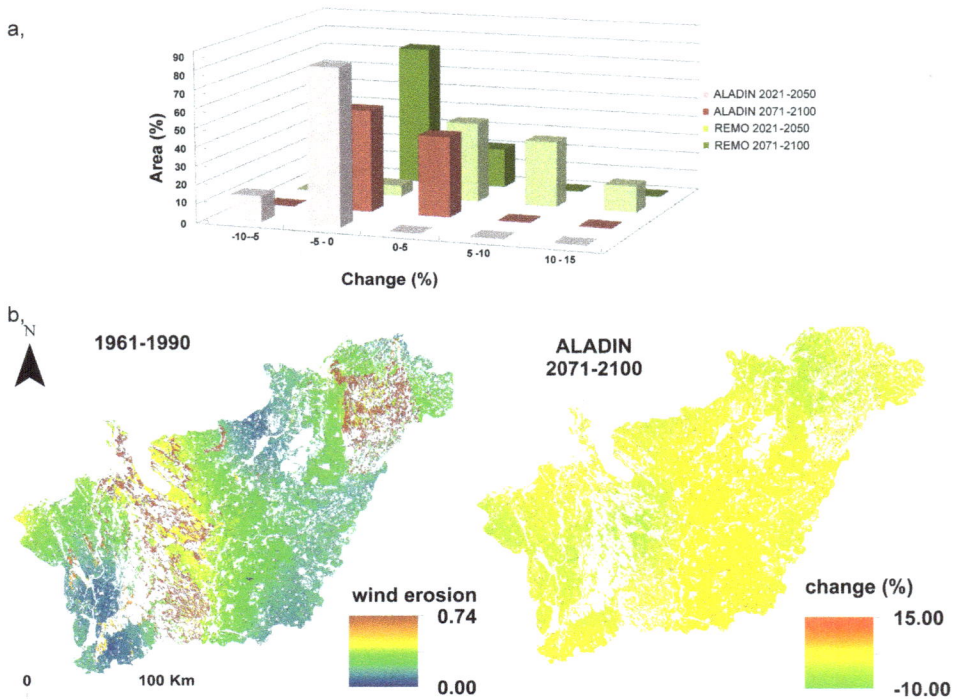

Figure 23.6 (a) The area of arable lands (% compared with the total arable lands) in 5% intervals of the changes of wind erosion hazard between 1961–1990 and the two future periods based on ALADIN and REMO models; (b).wind erosion hazard on arable lands on the Great Hungarian Plain in the reference period (1961–1990) and its changes (%) for the future period of 2071–2100 based on ALADIN model simulation. (From Mezösi et al. 2014.)

highest increase is expected in the south-eastern part due to the significant changes in climate forcing parameters (Figure 23.6).

Case study: land use

A crucial question is how climate change impacts will vary between land uses and in turn, how climate change impacts will influence further land use changes. Climate change is likely to influence changes in all land uses (e.g. forestry, agriculture, urban) as well as in protected landscapes (e.g. nature conservation, water protection, heritage areas). Multi-indicator analyses are therefore important for understanding how climate change may influence future land use. For example, Ladányi et al. (2015) conducted a multi-indicator sensitivity analysis of climate change effects on landscapes in the Kiskunság National Park, Hungary, including assessing the drought sensitivity of different landscape types and land use types.

The Kiskunság National Park contains important biotopes, including alkaline lakes, sand dunes, wetlands, dry steppes, and woods, in a matrix of intensive agricultural land and forest plantations. The indicators used in the study refer to (1) the soil water regime, (2) changes in groundwater resources, (3) biomass production of vegetation, and (4) wind erosion hazard and are examined for climate change scenarios (Figure 23.7). Ladányi et al. (2015) estimated future drought hazard as an indicator of climate change using the regionalization and downscaling of the REMO and ALADIN regional climate model scenario for two future time periods (2021–2050 and 2071–2100). As Figure 23.8 shows, the results indicate an increase in drought hazard for the entire region, with landscapes relatively more exposed to drought in the northern park territory. The most sensitive areas identified were located within a transitional zone from sand to alluvial plain in mostly wetland and sand-dune areas. The results strongly indicate that conservation management should especially focus on the northern part of the Kiskunság National Park as the area most at risk of the expected increasing drought under future climate change. The outcomes of this research highlight the applicability and suitability of the multi-indicator and multilayered landscape sensitivity approach for identifying adaptation strategies needed to mitigate the complex effects of climate change on nature conservation.

In the context of their results, Ladányi et al. (2015) discuss four adaptation strategies commonly adopted to address the adverse effects of climate change: (1) climate change–oriented water management, (2) integrated research activities, (3) enhanced cooperation, and (4) a dynamic, pro-active approach to nature conservation practices. Of these, they argued that climate change–oriented water management is the most practical approach and is well aligned with national (Hungarian) and European strategies for water management. Given the expected increases in drought severity, the availability of adapted or protected water resources is the greatest challenge facing the studied region. Therefore, Ladányi et al. (2015) argue that the retention of precipitation and effluent waters should be a priority for land management and conservation planners. Furthermore, the efficient use of both surface water and groundwater resources should be promoted, and specific adaptive environmental management of the arable lands in the matrix of the National Park will be vital.

Integrated research activities concerning the consequences of climate change and the identification of suitable adaptation measures for all biotopes and landscape types in the research area will also be important. Such integrated research requires a systems approach and should examine interdependencies between the changing factors as well as their temporal dynamics (Little et al. 2009). Integrated landscape research of this type is often unexplored despite the existence of considerable specialized research within individual fields and sectors (e.g. agricultural production, water production, nature conservation, landscape planning, recreation planning, climate modelling). Ladányi et al. (2015) also argue that enhanced cooperation between governance, research, and civil sectors is needed to improve data availability for research and to enable the findings from research to effectively support the realization of water management in practice. To support these data and information flows, improvements in metadata and spatial planning information systems (including geographic information systems [GIS]) to provide data freely online will provide the basis for more effective cooperation. Finally, Ladányi et al. (2015) suggest that a dynamic, pro-active approach to nature conservation is the most challenging management issue. A dynamic approach will be needed because protected areas are threatened and will alter in response to climate change. For example, in Hungary, a drying climate means that the extent of wetland habitats is likely to be generally smaller in the future. If these existing wetland areas become unsustainable despite the management practices undertaken, the actual goals of conservation could disappear. In this circumstance, considering multifunctional land uses as an adaptation response to the multilayered effects of climate change may be valuable. In general, future land use management will

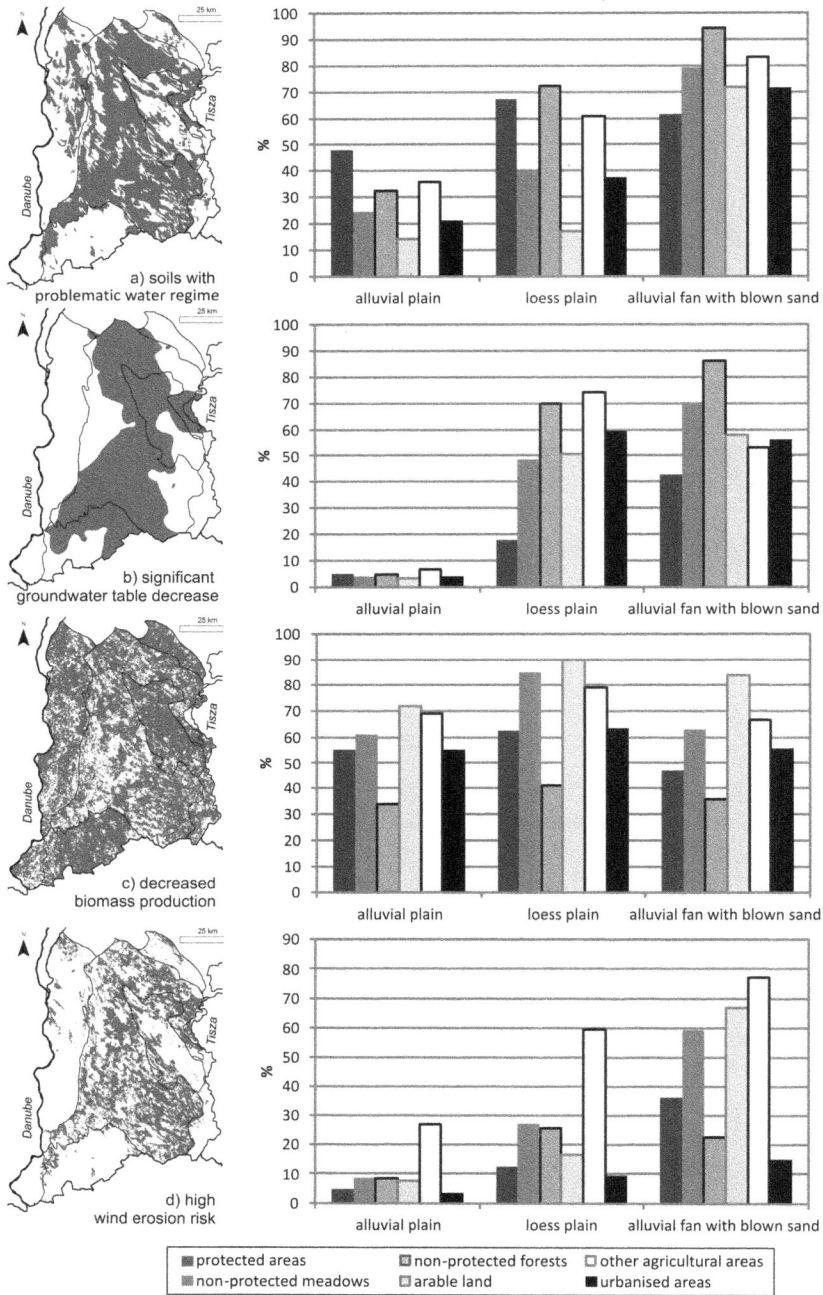

Figure 23.7 Spatial pattern of the indicators and their distribution (%) according to landscape types: (a) soils with problematic water regime, (b) the areas affected by significant groundwater decrease, (c) the vegetation showing a high decrease of biomass production in arid years, (d) areas with high wind erosion hazard. (From Ladányi et al. 2015.)

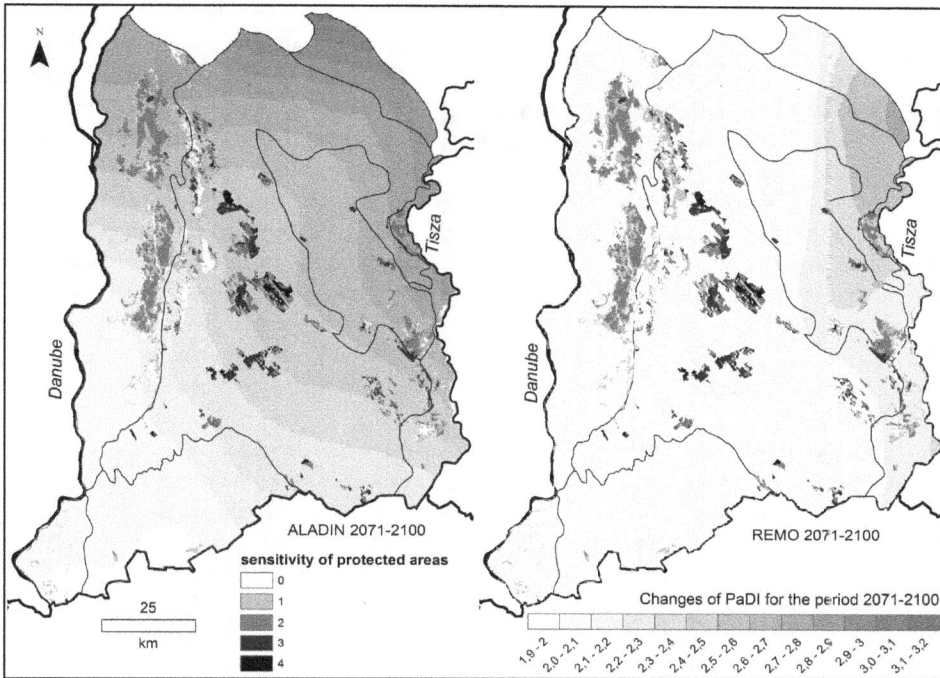

Figure 23.8 Overlay of the sensitivity of protected areas of the Kiskunság National Park areas and the predicted increase of drought hazard (PaDI$_0$) for the period 2071–2100 based on ALADIN-REMO regional climate model simulations. (From Ladányi et al. 2015.)

need a more dynamic, pro-active approach to robustly address the challenges envisaged due to climate change.

Summary: identifying climate change effects

The aim of this chapter was to examine how the interdependencies of landscapes, land uses, and the hazards associated with land use are affected by regional climate change. The studies summarized here are based on examples of the relationship and interlinkages between landscape and climate processes and demonstrate the effects of climate change including the landscape system responses. Understanding how landscapes respond to climate change is important because of the ecological and economic damage that may be caused. Over the past 30 years, the number of documented droughts and extreme temperatures has increased by 30% (Zommers and Singh 2014, Montz et al. 2017). Globally, similar results can be seen in the UN database (UNSTATS no date), indicating the impact of climate change. This is mainly due to extreme temperatures, pests, floods, and fires, which are linked to climate change (Ritchie and Roser 2020).

Based on these, when analyzing the effects of climate change on a landscape, it is first necessary to decide whether to analyze it in an integrated way, as a complex system, or as a set of individual parameters. The latter is the simpler task. In this case, we examine the effects of climate change parameters individually (or in combination) on specific elements of the landscape to estimate

457

the results of the impacts on the landscape. For example, the landscape-related consequences of climate change–induced temperature increase and (in Europe) drought as a result of a modified rainfall distribution should also be investigated. The exemplary case studies included here analyze the local and regional consequences of parameters (e.g. temperature, evaporation) due to climate change. In several of the case studies presented, regional climate models were used to calculate changes in the major climate parameters of the coming decades. There are dozens of such regional models, and although the average values of many models can be produced, the authors suggest the usage of more independent models and the analysis of their uncertainties one by one.

The other scientific entrance for the understanding of possible climate change impacts treats the landscape as an integrated spatial entity. In this case, the system changes can only be analyzed at the general level. Therefore, in this approach, neutral landscape-specific categories are confronted with climate change. Studies attempt to describe landscape changes with neutral categories such as vulnerability, changes in natural hazards and landscape sensitivity, and changes in the value and status of ecosystem services.

Another way of considering the consequences of climate change impacts is to assess ecological versus economic impacts. It can be difficult to prioritize between these. From an ecological point of view, the landscape and its components can be damaged by climate change, which is difficult to quantify in monetary terms. The societal consequences of such impacts can also be catastrophic. Such was the case, for example, with heatwaves and droughts in western Europe in 2003 and in Russia in 2010 (Munich Re 2020). The effects of these processes on a global scale can be particularly serious in human terms (WMO database). In the landscape context, the effect may primarily be a potential deterioration of the system (e.g. by reduced stability) that may cause the original system to disintegrate in the long run. In financial terms, the cost of damage is growing exponentially. For example, according to data from major international reinsurers, 7 of the 11 most damaging effects of some hazards (tropical cyclone and drought) majority caused by climate change (Munich Re 2020[1]). Of these, the number of hydrological hazards in Asia and the number of meteorological and hydrological hazards in Europe are particularly notable.

Analyzing the effects of climate change on landscape is a difficult task because both the causes and the affected unit are under similar anthropogenic influence. Separating the physical effects of changes in climate parameters from changes in landscape characteristics such as land use intensity makes it difficult to distinguish the importance of each. The examples presented here offer some insights into how this can be done, but further study across landscapes with a range of varying intensities of climate change (e.g. variation in rainfall) and anthropic change (e.g. within vs. outwith protected areas) will improve our understanding in this area and better enable mitigation and management efforts.

Summary

Main aspects shown in the chapter:

- Landscapes and climate change: identifying, on the basis of European case studies, climate change effects and differentiating the landscapes on different scales, which have affected the related risks analyzed by parameters, functions, and hazards towards changes in ecosystem services as well as the management options causing land use changes;
- Regionalization of climate change: clarifying the methodological problem of downscaling global climate change impacts to landscapes by regional climate change modelling, model

frameworks, and statistical methods and data for determining the changes in temperature, precipitation, and wind causing at the same time changes in landscape systems;

- Case study: aridification and drought: determining the factors modelled by a regional climate change scenario in the context of measurements on the landscape scale and describing the expected general change by the example of regional-scale aridification and landscape-scale drought hazards;
- Case study: wind erosion: scaling down the expected changes in wind erosion hazards based on a regional climate change scenario by modelling and assessment of the factor changes of the wind erosion loss equation;
- Case study: scaling down the expected land use changes by sensitivities and landscape in a National Park.

Note

1 https://www.munichre.com/en/solutions/for-industry-clients/natcatservice.html

References

Bastian, O. and Steinhardt, U. (eds) (2002) *Development and Perspectives of Landscape Ecology*. Springer, Dordrecht.

Blanka, V., Mezősi, G. and Meyer, B. (2013) 'Projected changes in the drought hazard in Hungary due to climate change', *Quarterly Journal of the Hungarian Meteorological Service*, vol 117, no 2, pp219–237.

Bowman, D.M., Perry, G.L.W. and Marston, J.B. (2015) 'Feedbacks and landscape-level vegetation dynamics', *Trends in Ecology & Evolution*, vol 30, no 5, pp255–260.

Bridges, E.M. and Oldeman, L.R. (1999) 'Global assessment of human-induced land degradation', *Arid Soil Research and Rehabilitation*, vol 13 pp319–325.

Brunsden, D. (2001) 'A critical assessment of the sensitivity concept in geomorphology', *Catena*, vol 42 no 2–4, pp99–123.

Dosio, A. (2016) 'Projections of climate change indices of temperature and precipitation from an ensemble of bias-adjusted high-resolution EURO-CORDEX regional climate models', *Journal of Geophysical Research: Atmospheres*, vol 121, no 10, pp5488–5511.

EEA (2015) 'Climate change impacts and adaptation', www.eea.europa.eu/soer-2015/europe/climate-change-impacts-and-adaptation, accessed 23 Dec 2020.

Farina, A. (2006) *Principles and Methods in Landscape Ecology. Towards a Science of the Landscape*. Springer, Dordrecht.

Field, C.B., Barros, V., Stocker, T.F., Qin, D., Dokken, D.J., Ebi, K.L., Mastrandrea, M.D., Mach, K.J., Plattner, G.-K., Allen, S.K., Tignor, M. and Midgley, P.M. (eds) (2012) *Managing the Risks of Extreme Events and Disasters to Advance Climate Change Adaptation. A Special Report of Working Groups I and II of the Intergovernmental Panel on Climate Change*. Cambridge University Press, Cambridge.

Kairis, O., Kosmas, C., Karavitis, C.H., Ritsema, C., Salvati, L., et al. (2014) 'Evaluation and selection of indicators for land degradation and desertification monitoring: types of degradation, causes, and implications for management', *Environmental Management*, vol 54, no 5, pp971–982.

Kertesz, Á. (2009) 'The global problem of land degradation and desertification', *Hungarian Geographical Bulletin*, vol 58, no 1, pp19–31

Ladányi, Z., Blanka, V., Meyer, B., Mezősi, G. and Rakonczai, J. (2015) 'Multi-indicator sensitivity analysis of climate change effects on landscapes in the Kiskunság National Park, Hungary,' *Ecological Indicators*, vol 58, pp8–20.

Little, J., Garland, S., Jelinek, A., Woods, G., Robertson, G., Schulz, S. (2009) *DECC Adaptation Strategy for Climate Change Impacts on Biodiversity*. Department of Environment and Climate Change NSW, Sydney.

Lóki, J. (2011) 'Research of the land forming activity of wind and protection against wind erosion in Hungary', *Riscuri Si Catastrofe*, vol 9, no 1, pp83–97.

Meyer, B.C., Mezősi, G. and Kovács, F. (2017) 'Landscape degradation at different spatial scales caused by aridification', *Moravian Geographical Reports*, vol 25, no 4, pp271–281.

Mezösi, G., Meyer, B.C., Loibl, W., Aubrecht, C., Csorba, P. and Bata, T. (2013) 'Assessment of regional climate change impacts on Hungarian landscapes', *Regional Environmental Change*, vol 13, no 4, pp797–811.

Mezösi, G., Bata, T., Meyer, B.C., Blanka, V. and Ladányi, Z. (2014) 'Climate change impacts on environmental hazards on the Great Hungarian Plain, Carpathian Basin', *International Journal of Disaster Risk Science*, vol 5, no 2, pp136–146.

Mezősi, G., Blanka, V., Bata, T., Kovács, F., Meyer, B.C. (2015) 'Estimation of regional differences in wind erosion sensitivity in Hungary', *Natural Hazards and Earth Systems Sciences*, vol 15, pp97–107.

Mezősi, G., Blanka, V., Ladanyi, Z., Bata, T., Urdea, P., Frank, A. and Meyer, B.C. (2016) 'Expected mid-and long-term changes in drought hazard for the South-Eastern Carpathian Basin', *Carpathian Journal of Earth and Environmental Sciences*, vol 11, no 2, pp355–366.

Montz, B.E., Tobin, G.A. and Hagelman III, R.R. (2017) *Natural Hazards: Explanation and Integration*. The Guilford Press, New York.

Rannow, S., Loibl, W., Greiving, S., Gruehn, D. and Meyer, B.C. (2010) 'Potential impacts of climate change in Germany: identifying regional priorities for adaptation activities in spatial planning', *Landscape and Urban Planning*, vol 98, no 3–4, pp160–171.

Reed, M.S., Buenemann, M., Atlhopheng, J., Akhtar-Schuster, M., Bachmann, F., Bastin, G. Bigas, H. *et al.* (2011) 'Cross-scale monitoring and assessment of land degradation and sustainable land management: a methodological framework for knowledge management', *Land Degradation and Development*, vol 22, pp261–271.

Ritchie, H. and Roser, M. (2020) 'Natural disasters', https://web.archive.org/web/20201211163343/https://ourworldindata.org/natural-disasters, accessed 22 Dec 2022.

Salvati, L. and Forino, G. (2014) 'A 'laboratory' of landscape degradation: social and economical implications for sustainable development in peri-urban areas', *International Journal of Innovation Sustainable Development*, vol 8, pp232–249.

Shukla, P.R., Skea, J., Calvo Buendia, E., Masson-Delmotte, V., Pörtner, H.-O., Roberts, D.C., Zhai, P., Slade, R., Connors, S., van Diemen, R., Ferrat, M., Haughey, E., Luz, S., Neogi, S., Pathak, M., Petzold, J., Portugal Pereira, J., Vyas, P., Huntley, E., Kissick, K., Belkacemi, M., Malley, J. (eds) (2019) *Climate Change and Land: an IPCC Special Report on Climate Change, Desertification, Land Degradation, Sustainable Land Management, Food Security, and Greenhouse Gas Fluxes in Terrestrial Ecosystems*. Cambridge University Press, Cambridge.

Stocker, T.F., Qin, D., Plattner, G.-K., Tignor, M., Allen, S.K., Boschung, J., Nauels, A., Xia, Y., Bex V. and Midgley P.M. (eds) (2013) *Climate Change 2013: The Physical Science Basis. Contribution of Working Group I to the Fifth Assessment Report of the Intergovernmental Panel on Climate Change*. Cambridge University Press, Cambridge.

UNSTATS - United Nations Statistics Division (n.d.) 'Statistics Division: environment statistics', https://unstats.un.org/unsd/envstats/climatechange.cshtml, accessed 22 Dec 2020.

Usher, M.B. (2001) 'Landscape sensitivity: from theory to practice', *Catena*, vol 42, no 2–4, pp375–383.

Van der Linden, P. and Mitchell, J.F.B. (eds) (2009) *ENSEMBLES: Climate Change and Its Impacts: Summary of Research and Results from the ENSEMBLES Project*. Met Office Hadley Centre, Exeter.

Zommers, Z. and Singh, A. (eds) (2014) *Reducing Disaster: Early Warning Systems for Climate Change*. Springer, Dordrecht. Information Classification: General

Index

Page numbers in *italic* represent photographs or figures, while those in **bold** mark the locations of tables.

For Product Safety Concerns and Information please contact our EU
representative GPSR@taylorandfrancis.com
Taylor & Francis Verlag GmbH, Kaufingerstraße 24, 80331 München, Germany